MOLECULAR DIAGNOSTICS

MOLECULAR DIAGNOSTICS

Edited by

GEORGE P. PATRINOS
Faculty of Medicine and Health Sciences
Department of Cell Biology and Genetics
Erasmus Medical Centre
Rotterdam, The Netherlands

WILHELM ANSORGE
European Molecular Biology Laboratory
Biochemical Instrumentation—Functional Genomics Technology Programme
Heidelberg, Germany

ELSEVIER
ACADEMIC
PRESS

Amsterdam • Boston • Heidelberg • London • New York • Oxford • Paris • San Diego
San Francisco • Singapore • Sydney • Tokyo

Elsevier Academic Press
30 Corporate Drive, Suite 400, Burlington, MA 01803, USA
525 B Street, Suite 1900, San Diego, California 92101-4495, USA
84 Theobald's Road, London WC1X 8RR, UK

This book is printed on acid-free paper.

Library of Congress Cataloging-in-Publication Data
APPLICATION SUBMITTED

British Library Cataloguing in Publication Data
A catalogue record for this book is available from the British Library

ISBN: 0-12-546661-7

For all information on all Elsevier Academic Press publications visit our Web site at
www.books.elsevier.com

Printed in the United States of America
05 06 07 08 09 10 9 8 7 6 5 4 3 2 1

Contents

List of Contributors

Numbers in parentheses indicate the chapter number and page number(s) on which the contribution begins.

Dr. AMIR ABDOLLAHI (16-201), Senior Scientist, Department of Radiation Oncology, German Cancer Research Center (DKFZ), 280 Im Neuenheimer Feld, D-69120 Heidelberg, Germany

Prof. Dr. FARID E. AHMED (19-235), Professor, Department of Radiation Oncology, Leo W. Jenkins Cancer Center, East Carolina University, The Brody School of Medicine, 600 Moye Boulevard, LSB 014, Greenville, NC 27858, USA. Telephone: +1-252-744-4636, Fax: +1-252-744-3775, E-mail: ahmedf@mail.ecu.edu

Dr. IOANNIS AMARANTOS (16-201), Scientific Researcher, European Molecular Biology Laboratory, Biochemical Instrumentation Programme, Meyerhofstrasse 1, D-69117, Heidelberg, Germany

Dr. CLAUS L. ANDERSEN (15-185), Assistant Professor, Molecular Diagnostic Laboratory, Department of Clinical Biochemistry, Aarhus University Hospital, Skejby, 8200-Aarhus N, Denmark

Dr. ALEXANDRA ANSORGE (16-201), Research Scientist, European Molecular Biology Laboratory, Biochemical Instrumentation Programme, Meyerhofstrasse 1, D-69117, Heidelberg, Germany

Prof. Dr. WILHELM ANSORGE (1-1, 16-201), Programme Coordinator, European Molecular Biology Laboratory, Biochemical Instrumentation—Functional Genomics Technology Programme, Meyerhofstrasse 1, D-69117 Heidelberg, Germany. Telephone: +49-6221-387.355, Fax: +49-6221-387.306, E-mail: ansorge@embl.de

Prof. Dr. AGLAIA ATHANASSIADOU (3-23), Professor of Molecular Genetics, Head of Dept. of General Biology, University of Patras, Faculty of Medicine, Asclepiou street, Panepistimioupolis, GR- 26110, Patras Greece. Telephone: +30-2610-996-169, Fax: +30-2610-991-769, E-mail: athanass@med.upatras.gr

Prof. Dr. FRANCISCO E. BARALLE (14-171), Head, Molecular Pathology Laboratory, Director, International Centre for Genetic Engineering and Biotechnology (ICGEB), Padriciano, 99, 34012 Trieste, Italy. Telephone: +39-040-375-7337, Fax: +39-040-375-7361, E-mail: baralle@icgeb.org

Dr. WALTER BELL (33-421), Assistant Professor of Pathology, Division of Anatomic Pathology, University of Alabama at Birmingham, Kracke Building—Room 532A, University Station, Birmingham, AL 35294-6823, USA

Dr. FERNANDO BENAVIDES (24-311), Assistant Professor, Department of Carcinogenesis, University of Texas, M.D. Anderson Cancer Center, Science Park, 1808 Park Road 1C—PO Box 389, Smithville, TX 78957, USA

Dr. CHRISTOPHE BEROUD (25-319), Assistant Professor, Laboratoire de Génétique Moléculaire et Chromosomique, CHU de Montpellier, Institut Universitaire de Recherche Clinique, 641, Avenue du Doyen Gaston Giraud, 34093 Montpellier Cedex 5, France. Telephone: +33-4-67-41-53-60, Fax: +33-4-67-41-53-65, E-mail: Christophe.beroud@igh.cnrs.fr

Prof. Dr. WILLIAM L. BIGBEE (17-211), Professor and Leader, Molecular Biomarkers Group, University of Pittsburgh Cancer Institute—The Hillman Cancer Center, UPCI Research Pavilion Suite 2.26d, 5117 Centre Avenue, Pittsburgh, PA 15213-1863, USA

Dr. JONATHON BLAKE (16-201), Computational Biologist, European Molecular Biology Laboratory, Biochemical Instrumentation Programme, Meyerhofstrasse 1, D-69117, Heidelberg, Germany

LAURYN BLAKESLEY (5-41), Scientific Technician, Basic Science Division, Fox Chase Cancer Center, 700 Lower State Road, North Wales, PA 19454, USA

Prof. Dr. ANJA KATRIN BOSSERHOFF (7-67), Professor of Molecular Pathology, University of Regensburg, Institute of Pathology, Franz-Josef-Strauß-Allee 11, D-93042 Regensburg, Germany

Dr. BRUCE BUDOWLE (21-267), Senior Scientist, Federal Bureau of Investigation, Laboratory Division, 2501 Investigation Parkway, Quantico, VA, 22135, USA. Telephone: +1-703-632-8386, Fax: +1-703-632-7817, E-mail: bruce.budowle@ic.fbi.gov

Dr. CHINH THIEN BUI (5-41), Group Leader, Genomic Disorders Research Centre, 7th Floor, Daly Wing, St. Vincent's Hospital, 41 Victoria Parade, Fitzroy 3065, Melbourne, Victoria, Australia. Telephone: +61-3-9288-297, Fax: +61-3-9288-2989, E-mail: chinhbui@medstv.unimelb.edu.au

Dr. DAVID BURNETT (34-429), Consultant in Quality and Accreditation Systems, Lindens Lodge, Bradford Place, Penarth, CF64 1LA, United Kingdom. Telephone: +44-29-2070-0521, Fax: +44-29-2070-0521, E-mail: davidburnett4@btinternet.com

Dr. ROWAN S. CAMPBELL (21-267), Research Associate, University of North Texas, School of Medicine, P.O. Box 311277, Denton, TX 76203, USA

Dr. TIMOTHY CAULFIELD (30-391), Canada Research Chair in Health Law and Policy, Research Director, Health Law Institute,

Associate Professor, Faculty of Law, Faculty of Medicine and Dentistry, 461 Law Centre, University of Alberta, Edmonton, Alberta, T6G 2H5, Canada. Telephone: +1-780-492-8358, Fax: +1-780-492-9575, E-mail: tcaulfld@law.ualberta.ca

MARIA CHIOTIS (5-41), *Research Assistant, Genomic Disorders Research Centre, 7th Floor, Daly Wing, St. Vincent's Hospital, 41 Victoria Parade, Fitzroy 3065, Melbourne, Victoria, Australia*

Prof. Dr. RICHARD G. H. COTTON (5-41), *Director, Genomic Disorders Research Centre, 7th Floor, Daly Wing, St. Vincent's Hospital, 41 Victoria Parade, Fitzroy 3065, Melbourne, Victoria, Australia*

HEIKO DRZONEK (16-201), *Technical Assistant, European Molecular Biology Laboratory, Biochemical Instrumentation Programme, Meyerhofstrasse 1, D-69117, Heidelberg, Germany*

Dr. LARS DYRSKJØT (15-185), *Assistant Professor, Molecular Diagnostic Laboratory, Department of Clinical Biochemistry, Aarhus University Hospital, Skejby, 8200-Aarhus N, Denmark*

Dr. ARTHUR J. EISENBERG (21-267), *Associate Professor and Director, DNA Identity Laboratory, Department of Pathology and Anatomy, University of North Texas Health Science Center, 3500 Camp Bowie Boulevard, Ft. Worth, TX 76107, USA*

BARRY ENG (2-15), *Technical Specialist, Genetic Services, Hamilton Regional Laboratory Medicine Program, 1200 Main Street West, Hamilton, Ontario, L8N 3Z5, Canada*

Prof. Dr. PAOLO FORTINA (28-349), *Professor of Medicine, Thomas Jefferson University, Director of Genomics and Diagnostics Center for Translational Medicine, Department of Medicine, 408 College Building, 1025 Walnut Street, Philadelphia, PA 19107, USA. Telephone: +1-215-955-0683, Fax: +1-215-955-6905, E-mail: Paolo.fortina@jefferson.edu*

Dr. JERRY FREDENBURGH (33-421), *Chief Executive Officer, Richard Allan Scientific, 4481 Campus Drive, Kalamazoo, MI 49008, USA*

Dr. STEFAN FRUEHAUF (26-327), *Staff Scientist, Department of Internal Medicine V, University of Heidelberg, Hospitalstrasse 3, D-69115 Heidelberg, Germany. Telephone: +49-6221-562781, Fax: +49-6221-565722, E-mail: Stefan.Fruehauf@med.uni-heidelberg.de*

Dr. GEORGE A. GARINIS (6-55), *Scientific Researcher, Erasmus MC, Faculty of Medicine and Health Sciences, MGC-Department of Cell Biology and Genetics, PO Box 1738, 3000 DR, Rotterdam, The Netherlands*

BERNHARD GENTNER (26-327), *Research Assistant, Innovative Cancer Diagnostic and Therapy, German Cancer Research Center, Im Neuenheimer Feld 280, D-69120 Heidelberg, Germany*

Dr. ANNA PAULA GIULIETTI (10-109), *Scientific Researcher, Laboratory for Experimental Medicine and Endocrinology*

(LEGENDO), Catholic University of Leuven, U.Z. Gasthuisberg, Herestraat 49, B-3000 Leuven, Belgium

Dr. ANDREW K. GODWIN (5-41), *Member, Fox Chase Cancer Center, 333 Cottman Avenue, Philadelphia, PA 19111-2497, USA*

Dr. BERT GOLD (27-337), *Staff Scientist, Human Genetics Section, Laboratory of Genomic Diversity, National Cancer Institute at Frederick, Building 560, Room 21-21, Frederick, MD, 21702, USA. Telephone: +1-301-846-5098, Fax: +1-301-846-1909, E-mail: goldb@ncifcrf.gov*

Prof. Dr. WILLIAM E. GRIZZLE (17-211, 33-421), *Professor of Pathology, Department of Pathology, University of Alabama at Birmingham, Zeigler Research Building—Room 408, 703 South 19th Street, Birmingham, AL 35294-0007, USA. Telephone: +1-205-934-4214, Fax: +1-205-975-7128, E-mail: grizzle@path.uab.edu*

Dr. PIOTR GRODZINSKI (28-349), *Technical Staff Member, Bioscience Division, MS J586, Michelson Resource, B-4, Los Alamos National Laboratory, Los Alamos, NM 87545, USA*

Dr. JEAN-LOUIS GUÉNET (24-311), *Director, Unit of Mammalian Genetics, Pasteur Institute, 25 Rue du Docteur Roux, 75724 Paris Cedex 15, France. Telephone: +33-1-45-68-85-55, Fax: +33-1-45-68-86-34, E-mail: guenet@pasteur.fr*

Prof. Dr. SAMIR HANASH (18-223), *Professor, Department of Pediatrics, University of Michigan, 1150 West Medical Center Drive, Ann Arbor, MI 48109, USA. Telephone: +1-734-763-9311, Fax: +1-734-647-148, E-mail: shanash@med.umich.edu*

Dr. CLAUS HELLERBRAND (7-67), *Lecturer, University of Regensburg, Department of Internal Medicine I, Franz-Josef-Strauß-Allee 11, D-93042 Regensburg, Germany. Telephone: +49-941-944-7104, Fax: +49-941-944-7002, E-mail: Claus.hellerbrand@klinik.uni-regensburg.de*

Prof. Dr. ANTHONY D. HO (16-201, 26-327), *Director, Department of Internal Medicine V, University of Heidelberg, Hospitalstrasse 3, D-69115 Heidelberg, Germany*

Dr. PETER HUBER (16-201), *Senior Scientist, Department of Radiation Oncology, German Cancer Research Center (DKFZ), 280 Im Neuenheimer Feld, D-69120 Heidelberg, Germany*

Prof. Dr. JENS L. JENSEN (15-185), *Professor, Molecular Diagnostic Laboratory, Department of Clinical Biochemistry, Aarhus University Hospital, Skejby, 8200-Aarhus N, Denmark*

Dr. ANJA KELLERMANN (16-201), *Engineer for Biotechnology and Economics, German Genome Resource Center (RZPD), Heubnerweg 6, Berlin, D-14059, Germany*

Dr. KAREN KOED (15-185), *Assistant Professor, Molecular Diagnostic Laboratory, Department of Clinical Biochemistry, Aarhus University Hospital, Skejby, 8200-Aarhus N, Denmark*

Dr. BERNHARD KORN (16-201), *Chief Research Officer, German Genome Resource Center (RZPD), Neuenheimer Feld 580, Heidelberg, D-69120, Germany*

Dr. MOGENS KRUHØFFER (15-185), *Associate Professor, Molecular Diagnostic Laboratory, Department of Clinical Biochemistry, Aarhus University Hospital, Skejby, 8200-Aarhus N, Denmark*

ANDREANA LAMBRINAKOS (5-41), *Ph.D. Scholar, Genomic Disorders Research Centre, 7th Floor, Daly Wing, St. Vincent's Hospital, 41 Victoria Parade, Fitzroy 3065, Melbourne, Victoria, Australia*

Prof. Dr. INGE LIEBAERS (23-297), *Professor and Director, Centre for Medical Genetics and Research Centre Reproduction and Genetics, Medical School and University Hospital of the Dutch-speaking Brussels Free University (Vrije Universiteit Brussel, VUB), Laarbeeklaan 101, 1090 Brussels, Belgium*

Prof. Dr. KLAUS LINDPAINTNER (20-249), *Professor of Medicine, Harvard Medical School, Boston, MA, Vice-President Research and Head, Roche Genetics and Roche Center for Medical Genomics, F. Hoffmann-La Roche Ltd., Bldg 93/804A, CH-4070 Basel, Switzerland. Telephone: +41-61-688-0254, Fax: +41-61-688-1929, E-mail: Klaus.lindpaintner@roche.com*

Dr. ROBIN LIU (28-349), *Manager, Applied Nanobioscience Center, Arizona State University, P.O. Box 874004, Tempe, AZ 85287-4004, USA*

Dr. CHRISTIAN MAERCKER (16-201), *Research Scientist, German Genome Resource Center (RZPD), Neuenheimer Feld 580, Heidelberg, D-69120, Germany*

Dr. STEPHANIE MAIER-LAUFS (26-327), *Scientific researcher, Innovative Cancer Diagnostic and Therapy Unit, German Cancer Research Center, Im Neuenheimer Feld 280, D-69120 Heidelberg, Germany*

Dr. GUNJAN MALIK (17-211), *Scientific Researcher, Department of Microbiology and Molecular Cell Biology, Eastern Virginia Medical School, 700 West Olney Road, Norfolk, VA 23507, USA*

BARKHA MANNE (17-211), *Research Assistant, Department of Pathology, University of Alabama at Birmingham, Zeigler Research Building—Room 210, 703 South 19th Street, Birmingham, AL 35294-0007, USA*

Dr. UPENDER MANNE (17-211), *Assistant Professor, Department of Pathology, University of Alabama at Birmingham, Kracke Building—Room 505, 1922 7th Avenue South, Birmingham, AL 35294-6823, USA*

Dr. CHANTAL MATHIEU (10-109), *Associate Professor of Medicine, Laboratory for Experimental Medicine and Endocrinology (LEGENDO), Catholic University of Leuven, U.Z. Gasthuisberg, Herestraat 49, B-3000 Leuven, Belgium. Telephone: +32-16-345-970, Fax: +32-16-345-934, E-mail: Chantal.Mathieu@uz.kuleuven.ac.be*

Dr. JOHANNES MAURER (16-201), *Scientific Managing Director, German Genome Resource Center (RZPD), Heubnerweg 6, Berlin, D-14059, Germany*

Dr. RONALD C. Mc GLENNEN (29-365), *Associate Professor, Molecular Diagnostics Laboratory, Department of Laboratory Medicine and Pathology, University of Minnesota Medical School, President, Access Genetics, LLC, 7159 Shady Oak Road, Eden Prairie, MN 55344, USA. Telephone: +1-952-942-0671, Fax: +1-952-942-0703, Email: rcmcglennen@access-genetics.com*

Dr. PANAYIOTIS G. MENOUNOS (6-55), *Director, Laboratory of Research, Nursing Military Academy, SAKETA "A" barrack, Vironas GR-16201, Athens, Greece*

Dr. ELIZABETH MILLER (17-211), *Fellow, Department of Pathology, University of Alabama at Birmingham, Zeigler Research Building—Room 428, 703 South 19th Street, Birmingham, AL 35294-0007, USA*

K. ZSUZSANNA NAGY (26-327), *Research Assistant, Innovative Cancer Diagnostic and Therapy Unit, German Cancer Research Center, Im Neuenheimer Feld 280, D-69120 Heidelberg, Germany*

Dr. EMMANUELLE NICOLAS (5-41), *Research Associate, Fox Chase Cancer Center, 333 Cottman Avenue, Philadelphia, PA 19111-2497, USA*

DENISE K. OELSCHLAGER (17-211), *Research Assistant, Department of Pathology, University of Alabama at Birmingham, Lyons-Harrison Research Building—Room 534, 700 South 19th Street, Birmingham, AL 35294-0007, USA*

Prof. Dr. TORBEN F. ØRNTOFT (15-185), *Professor and Head of Department, Molecular Diagnostic Laboratory, Department of Clinical Biochemistry, Aarhus University Hospital, Skejby, 8200-Aarhus N, Denmark. Telephone: +45-89-495-100, Fax. +45-89-496-018, E-mail: orntoft@kba.sks.au.dk*

Dr. LUT OVERBERGH (10-109), *Scientific Researcher, Laboratory for Experimental Medicine and Endocrinology (LEGENDO), Catholic University of Leuven, U.Z. Gasthuisberg, Herestraat 49, B-3000 Leuven, Belgium*

Dr. FRANCO PAGANI (14-171), *Staff Scientist, International Centre for Genetic Engineering and Biotechnology (ICGEB), Padriciano, 99, 34012 Trieste, Italy*

Dr. ANDREA FARKAS PATENAUDE (32-409), *Director of Psycho-Oncology Research and Assistant Professor of Psychology, Division of Pediatric Oncology, Dana-Farber Cancer Institute; Department of Psychiatry, Harvard Medical School, 44 Binney Street, Boston MA 02445, USA. Telephone: +1-617-632-3314, Fax: +1-617-713-4466, E-mail: Andrea_patenaude@dfci.harvard.edu*

Dr. GEORGE P. PATRINOS (1-1, 6-55), *Senior Research Scientist, MGC-Department of Cell Biology and Genetics, Erasmus MC, Faculty of Medicine and Health Sciences, PO Box 1738, 3000 DR, Rotterdam, The Netherlands. Telephone: +31-10-408-7454, Fax: +31-10-408-9468, E-mail: g.patrinos@erasmusmc.nl, g.patrinos@goldenhelix.org*

Dr. HARTMUT PETERS (8-83), *Research Scientist, Institute of Medical Genetics, Charité University Hospital, Humboldt University Berlin, Augustenburger Platz 1, D-13353, Berlin, Germany*

Dr. MARY PETROU (31-399), *Head, Haemoglobinopthy Genetics Center (Perinatal Center), University College Hospital NHS*

Trust and University College London Medical School, Perinatal Centre, Department of Obstetrics and Gynaecology, 86–96 Chenies Mews, London WC1E 6HX, United Kingdom. Telephone: +44-207-388-9246, Fax: +44-207-380-9864, E-mail: M.PETROU @UCL.AC.UK

Dr. JOHN V. PLANZ (21-267), *Assistant Professor and Associate Director, DNA Identity Laboratory, Department of Pathology and Anatomy, University of North Texas Health Science Center, 3500 Camp Bowie Boulevard, Ft. Worth, TX 76107, USA*

Dr. UWE RADELOF (16-201), *Chief Production Officer, German Genome Resource Center (RZPD), Heubnerweg 6, Berlin, D-14059, Germany*

KYLEE REES (5-41), *Ph.D. Scholar, Genomic Disorders Research Centre, 7th Floor, Daly Wing, St. Vincent's Hospital, 41 Victoria Parade, Fitzroy 3065, Melbourne, Victoria, Australia*

Dr. PETER N. ROBINSON (8-83), *Staff Scientist, Institute of Medical Genetics, Charité University Hospital, Humboldt University Berlin, Augustenburger Platz 1, D-13353, Berlin, Germany. Telephone: +49-30-450-569-124, Fax: +49-30-450-569-915, E-mail: Peter.robinson@charite.de*

Dr. EEVA-LIISA ROMPPANEN (4-31), *Clinical Biochemist, Department of Clinical Chemistry, Kuopio University Hospital, P.O. Box 1777, FIN-70211 Kuopio, Finland. Telephone: +358-44-717-4318, Fax: +358-17-173-200, E-mail: eeva-liisa.romppanen @phks.fi*

GEORGINA SALLMANN (5-41), *Research Assistant, Genomic Disorders Research Centre, 7th Floor, Daly Wing, St. Vincent's Hospital, 41 Victoria Parade, Fitzroy 3065, Melbourne, Victoria, Australia*

Dr. CHRISTIAN SCHWAGER (16-201), *Computational Biologist, European Molecular Biology Laboratory, Biochemical Instrumentation Programme, Meyerhofstrasse 1, D-69117, Heidelberg, Germany*

KATHRYN E. SCOTT (28-349), *Molecular Research Specialist, Thomas Jefferson University, Department of Medicine Center for Translational Medicine Genomics Center, 415 College Building, 1025 Walnut Street, Philadelphia, PA 19107, USA*

JAN SELIG (16-201), *Engineer, European Molecular Biology Laboratory, Biochemical Instrumentation Programme, Meyerhofstrasse 1, D-69117, Heidelberg, Germany*

Dr. LORYN SELLNER (13-159), *Medical Scientist, Princess Margaret Hospital, Roberts Road, SUBIACO WA 6008, Australia*

Dr. O. JOHN SEMMES (17-211), *Associate Professor, Department of Microbiology and Molecular Cell Biology, Eastern Virginia Medical School, Norfolk, VA 23507, USA*

Dr. KAREN SERMON (23-297), *Research Scientist, Centre for Medical Genetics and Research Centre Reproduction and Genetics, Medical School and University Hospital of the Dutch-speaking Brussels Free University (Vrije Universiteit Brussel,*

VUB), *Laarbeeklaan 101, 1090 Brussels, Belgium. Telephone: +32-2-477-6073, Fax: +32-2-477-6068, E-mail: Karen.sermon@ az.vub.ac.be*

Dr. YOUSIN SUH (9-97), *Assistant Professor of Physiology, University of Texas Health Science Center, STCBM Building, Suite 2.200, San Antonio, TX 78245, USA*

Prof. Dr. SAUL SURREY (28-349), *Professor of Medicine and Associate Director of Research, Cardeza Foundation for Hematologic Research and Division of Hematology, Thomas Jefferson University, 703 Curtis Building, 1015 Walnut Street, Philadelphia, PA 19107, USA*

Dr. GRAHAM R. TAYLOR (13-159), *Scientific Director and Senior Lecturer, Regional Genetics Laboratory and Cancer Research-UK Mutation Detection Facility, St James' University Hospital, DNA Laboratory, Leeds Yorkshire LS9 7TF, United Kingdom. Telephone: +44-113-206-5217, E-mail: gtaylor@hgmp.mrc.ac.uk*

Dr. HOLGER TÖNNIES (12-139), *Research Scientist, Humboldt-University, Berlin, Charite, Campus Virchow-Klinikum, Institute of Human Genetics, Chromosome Diagnostics and Molecular Cytogenetics, Augustenburger Platz 1, D-13353 Berlin, Germany. Telephone: +49-30-450566807, Fax: +49-30-450566933, E-mail: Holger.toennies@charite.de*

Dr. NIELS TØRRING (15-185), *Senior Chemist, Molecular Diagnostic Laboratory, Department of Clinical Biochemistry, Aarhus University Hospital, Skejby, 8200-Aarhus N, Denmark*

Dr. IAN TROUNCE (5-41), *Associate Director, Genomic Disorders Research Centre, 7th Floor, Daly Wing, St. Vincent's Hospital, 41 Victoria Parade, Fitzroy 3065, Melbourne, Victoria, Australia*

DIRK VALCKX (10-109), *Research Associate, Laboratory for Experimental Medicine and Endocrinology (LEGENDO), Catholic University of Leuven, U.Z. Gasthuisberg, Herestraat 49, B-3000 Leuven, Belgium*

Prof. Dr. ALEX VAN BELKUM (22-281), *Professor of Molecular Microbiology, Erasmus MC, Department of Medical Microbiology and Infectious Diseases, Unit Research and Development, Dr. Molewaterplein 40, 3015 GD Rotterdam, The Netherlands. Telephone: +31-10-463-58-13, Fax: +31-10-463-38-75, E-mail: a.vanbelkum@erasmusmc.nl*

Prof. Dr. ANDRÉ VAN STEIRTEGHEM (23-297), *Professor and Scientific Director, Centre for Reproductive Medicine and Research Centre Reproduction and Genetics, Medical School and University Hospital of the Dutch-speaking Brussels Free University (Vrije Universiteit Brussel, VUB), Laarbeeklaan 101, 1090 Brussels, Belgium*

Prof. Dr. JAN VIJG (9-97), *Professor of Physiology, University of Texas Health Science Center and Geriatric Research Education and Clinical Center, South Texas Veterans Health Care System, STCBM Building, Suite 2.200, San Antonio, TX 78245, USA. Telephone: +1-210-562-5027, Fax: +1-210-562-5028, E-mail: vijg@uthscsa.edu*

Dr. MECHTHILD WAGNER (16-201), *Research Scientist, European Molecular Biology Laboratory, Biochemical Instrumentation Programme, Meyerhofstrasse 1, D-69117, Heidelberg, Germany*

Dr. WOLFGANG WAGNER (16-201), *Scientific Researcher, Department of Internal Medicine V, University of Heidelberg, Hospitalstrasse 3, D-69115 Heidelberg, Germany*

Dr. MICHAEL WARD (28-349), *Graduate Research Assistant, Bioscience Division, MS J586, Michelson Resource, B-4, Los Alamos National Laboratory, Los Alamos, NM 87545, USA*

Prof. Dr. JOHN S. WAYE (2-15), *Professor, Department of Pathology and Molecular Medicine, McMaster University, Head of Service, Molecular Diagnostic Genetics, Hamilton Regional Laboratory Medicine Program, McMaster University Medical Centre, Room 3N17, 1200 Main Street West, Hamilton, Ontario, Canada L8N 3Z5. Telephone: +1-905-521-2100 ext. 76273, Fax: +1-905-521-2651, E-mail: wayej@mcmaster.ca*

Dr. FRIEDRIK WIKMAN (15-185), *Senior Chemist, Molecular Diagnostic Laboratory, Department of Clinical Biochemistry, Aarhus University Hospital, Skejby, 8200-Aarhus N, Denmark*

Dr. UTE WIRKNER (16-201), *Senior Research Scientist, European Molecular Biology Laboratory, Biochemical Instru-mentation Programme, Meyerhofstrasse 1, D-69117, Heidelberg, Germany*

Dr. JOHANNA K. WOLFORD (11-127), *Associate Investigator, Head, Diabetes Research Unit, Translational Genomics Research Institute, 400 North 5th Street, Phoenix, AZ 85004, USA. Telephone: +1-602-343-8812, Fax: +1-602-343-8840, E-mail: jwolford @tgen.org*

KIMBERLY A. YEATTS (11-127), *Senior Laboratory Technician, Genetic Basis of Human Disease Division, Translational Genomics Research Institute, 400 North Fifth Street, Suite 1600, Phoenix, AZ 85004, USA*

Dr. ANTHONY T. YEUNG (5-41), *Director, Fannie E. Rippel Biochemistry and Biotechnology Facility, Member, Fox Chase Cancer Center, 333 Cottman Avenue, Philadelphia, PA 19111-2497, USA*

Prof. Dr. W. JENS ZELLER (26-327), *Group leader, Innovative Cancer Diagnostic and Therapy Unit, German Cancer Research Center, Im Neuenheimer Feld 280, D-69120 Heidelberg, Germany*

LIU ZHU (17-211), *Research Assistant, Department of Pathology, University of Alabama at Birmingham, Zeigler Research Building—Room 212, 703 South 19th Street, Birmingham, AL 35294-0007, USA*

Foreword

The recent pace of discoveries in human genetics and molecular biology is stunning. Completion of the human genome sequence analysis, successive development of diverse high-throughput analytical approaches (genomics, proteomics, transcriptomics), and most recently, developments in nano-technology have permitted innovative enhancement of genetic testing. Elucidation of the molecular bases of numerous human diseases, combined with rapidly evolving techniques in molecular biology, have brought molecular diagnostics to a high stage of sophistication that is still evolving.

This textbook on molecular diagnostics is ambitious in its scope and intended usage. It addresses not only scientists already in the field, but also those who intend to enter it. It provides a much-needed comprehensive reference publication on current and developing methods. Importantly, it also includes an innovative treatment of diverse molecular diagnostic techniques as they apply to the actual practice of clinical microbiology, forensic analysis, development of pharmaceuticals, and other fields.

The first section of the book covers the principles and applications of numerous genetic testing methodologies, in approximately the chronological order of discovery. This section includes chapters on both targeted and high-throughput diagnosis using genomic, transcriptomic, and proteomic approaches.

The second section is devoted to the application of molecular diagnostics in a variety of fields. It illustrates the vast potential as well as the versatility of molecular diagnostics, which are applicable not only to modern medical practice but also to the identification of genetically modified organisms, forensics, quality assessment of laboratory animals, pharmacogenomics, and other fields. Of special interest in this section are the discussions of recently emerged issues, such as gene patenting, ethics, safety, and the psychological context of genetic testing.

The editors succeeded in attracting more than 100 well-qualified authors to compile the 34 chapters of this book. The result is an up-to-date comprehensive treatment of molecular diagnostics, which aims to be a standard reference resource for biomedical scientists from a wide range of disciplines. It will be invaluable for clinicians and others who come into close contact with genetic testing for diagnostic purposes. It will also promote a wider understanding of molecular diagnostics among physicians, postgraduate students, researchers in the academic and corporate world, and even policymakers.

The self-explanatory nature of the book, assisted by a glossary and numerous illustrations, also will appeal to undergraduate students in the medical and life sciences. The editors can be congratulated for putting together such a valuable resource, for an area of the biomedical sciences that is of major and ever-increasing importance.

Prof. Fotis C. Kafatos
Director-General
European Molecular Biology Laboratory
Heidelberg, Germany

August 2004

Preface

This book is the result of the initiative to edit a reference publication dealing exclusively with all the aspects of modern molecular diagnostics. The need for such a book derives from the intellectual revolution in biomedical science and the realization that human diseases, both congenital and inherited, are rapidly becoming susceptible to molecular quantitative analysis. Our effort has been assisted by many world-leading scientists, experts in their field, who have kindly accepted our invitation to compile the 34 chapters of this book and share with us and our readers their expertise, experience, and results, sometimes even unpublished ones.

In the post-genomic era at the dawn of the twenty-first century, the deciphering of the information encrypted into the human genome sequence, together with the expansion of our knowledge on the molecular basis of human inherited disorders, brought up the need to provide diagnosis at the molecular level for inherited diseases easily, accurately, and fast. As a consequence, an increasing number of diagnostic and research laboratories, both academic and private in the United States and in Europe, are preparing to deal with the demands and challenges this new era sets forth for medicine and health sciences. They move in that direction not only by establishing innovative approaches for mutation detection, or improving existing ones, but also by implementing novel technologies for the benefit of the patient.

Despite the fact that molecular diagnostics has already entered its golden era, the scientific literature lacks a specialized reference book, dealing exclusively with the technology and techniques available for the detection of the sequence variations leading to inherited diseases. We consider this reference book, together with the underlying framework, to be fundamental for the fruitful synthesis of all aspects of modern molecular diagnostics to result in a well-orchestrated entity. There are only a few books available in the literature that allude to these issues in the topics they deal with, and none is dedicated exclusively to diagnosis at the molecular level. On the other hand, a great number of Internet sites are scattered throughout the World Wide Web, which serve as tools, for example, for implementing mutation screening methodologies. Unfortunately, in their vast majority, these sites possess the inherent dangers of being outdated, unclear, and hence confusing, and are sometimes difficult and time-consuming to find. Also, they often lack the important aspect of being compiled

and based on many years of relevant experience of well-known expert researchers, leaders in their fields.

The innovative aspect of this book is its structure. It contains an expert introduction to each subject, next to only few protocols and technical details, which can be found in comprehensive reference lists at the end of each chapter. The contents of this book are divided in two parts. The first part is dedicated to the battery of the most widely used molecular biology techniques. Their arrangement is in a more or less chronologic order of their development. Starting from the first ones with relatively low throughput, like single-strand conformation polymorphism analysis, allele-specific amplification, temperature and denaturing gradient gel electrophoresis, and others, to more modern techniques, such as fluorescence *in situ* hybridization, denaturing high performance liquid chromatography, two-dimensional gene scanning, microarrays, mass spectrometry, proteomic analysis, and others. These techniques are still evolving fast, becoming more cost-effective, and are characterized by high throughput. Each of these chapters includes the principle and a brief description of the technique, followed by examples from the area of expertise from the selected contributor. The selected contributors are well-known experts in their fields and come from a variety of disciplines, so that the book covers efficiently not only a great number of techniques applied to molecular diagnosis but also their applications to a variety of inherited disorders.

The second part attempts to integrate the previously mentioned techniques to the different aspects of molecular diagnosis, such as identification of genetically modified organisms, stem cells, pharmacogenetics, modern forensic science, genetic quality of laboratory animals, molecular microbiology, and preimplantation genetic diagnosis. We believe that issues such as personalized medicine, pharmacogenetics, and integration of diagnostics and therapeutics will be subjects for debate in the following years. In addition, various everyday issues in a diagnostic laboratory are discussed, such as the establishment of mutation databases to store and organize the continuously growing information on gene variation, genetic counseling, patenting of genes and of genetic tests, safety and quality management, and various ethical considerations and psychological issues pertaining to diagnostics. We feel that the inclusion of the latter issues in this reference book have great relevance to our society and the aim is to assist in finding answers to some of the problematic questions that undoubtedly will arise, since application of molecular diagnostics may precipitate an important ethical crisis that physicians and the communities they serve will be confronted with.

The intended audience of this book is university postgraduate students from various life sciences disciplines, physicians, scientists in human molecular genetics and medicine, professionals working in diagnostic laboratories in academia or industry, academic institutions, hospital libraries, biotechnology and pharmaceutical companies. We believe that this book will be of help to decision-making advisors in Medical Insurance companies. In addition, undergraduate medical and life science students will find very useful the description and explanation of recent modern techniques in life sciences. A major concern was to formulate the book contents in such a way that the notions described therein are clear and explained in a simple language and terminology. The numerous illustrations of the book are comprehensive and self-descriptive, and the glossary at the end of this book provides a brief explanation of the most commonly found terms.

We expect that some points in this book can be further improved. We would welcome comments and criticism from the attentive readers, which will contribute to the improvement of the content of this book in its future editions as well as helping to establish it as a reference publication in the field of molecular diagnostics.

Without the support and contributions of many people the completion of this book would not be possible. We wish to thank the anonymous referees who supported our proposal and subsequently the people at Elsevier, in particular Dr. Tessa Picknett and her department, whose interest in this title has encouraged us to proceed.

We are also grateful to the editors, Drs. Claire Minto and Tari Paschall, and the senior production manager, Sarah Hajduk, at Elsevier, who helped us in a close collaboration to solve and overcome encountered difficulties. Their efficiency, pleasant manner, and patience added immensely to the smooth completion of this project from the very start.

We also express our gratitude to all contributors for delivering outstanding compilations that summarize their experience and many years of hard work in their field of research. We are indebted to Julio Esperas who was responsible for the design and the cover of this book and to the copy editor, Adrienne Rebello, who has refined the final manuscript prior to going into production. We also thank our university colleagues, in particular Prof. Miel Ribbe from the Medical School of the University of Amsterdam, for discussions of the content of this book and ensuring its greatest possible relevance. We owe our thanks to the academic reviewers for their constructive criticisms on the chapters.

We wish also to thank our families, from whom we have taken considerable amounts of time to dedicate to this work and whose patience and support have been conducive to the successful completion of this project.

George P. Patrinos
Erasmus Medical Centre
Rotterdam, The Netherlands

Wilhelm Ansorge
European Molecular
Biology Laboratory
Heidelberg, Germany

June 2005

CHAPTER 1

Molecular Diagnostics: Past, Present, and Future

GEORGE P. PATRINOS[1] AND WILHELM ANSORGE[2]

[1] *Erasmus University Medical Center, Faculty of Medicine and Health Sciences, MGC Department of Cell Biology and Genetics, Rotterdam, The Netherlands;*

[2] *European Molecular Biology Laboratory, Biological Structures and Biocomputing Programme, Heidelberg, Germany*

TABLE OF CONTENTS

1.1 INTRODUCTION

Molecular or nucleic acid-based diagnosis of human disorders is referred to as the detection of the various pathogenic mutations in DNA and/or RNA samples in order to facilitate detection, diagnosis, subclassification, prognosis, and monitoring response to therapy. Molecular diagnostics combines laboratory medicine with the knowledge and technology of molecular genetics and has been enormously revolutionized over the last decades, benefiting from the discoveries in the field of molecular biology (see Fig. 1.1). The identification and fine characterization of the genetic basis of the disease in question is vital for accurate provision of diagnosis. Gene discovery provides invaluable insights into the mechanisms of disease, and gene-based markers allow physicians not only to assess disease predisposition but also to design and implement improved diagnostic methods. The latter is of great importance, as the plethora and variety of molecular defects demands the use of multiple rather than a single mutation detection platform. Molecular diagnostics is currently a clinical reality with its roots deep into basic study of gene expression and function.

1.2 HISTORY OF MOLECULAR DIAGNOSTICS: INVENTING THE WHEEL

In 1949, Pauling and his colleagues introduced the term *molecular disease* in the medical vocabulary, based on their discovery that a single amino acid change at the β-globin chain leads to sickle cell anemia, characterized mainly by recurrent episodes of acute pain due to vessel occlusion. In principle, their findings have set the foundations of molecular diagnostics, although the big revolution occurred many years later. At that time, when molecular biology was only hectically expanding, the provision of molecular diagnostic services was inconceivable and technically not feasible.

The first seeds of molecular diagnostics were provided in the early days of recombinant DNA technology, with many scientists from various disciplines working in concert. cDNA cloning and sequencing were at that time invaluable tools for providing the basic knowledge on the primary sequence of various genes. The latter provided a number of DNA probes, allowing the analysis via southern blotting of genomic regions, leading to the concept and application of restriction fragment length polymorphism (RFLP) to track a mutant allele from heterozygous parents to a high-risk pregnancy. In 1976, Kan and colleagues carried out, for the first time, prenatal diagnosis of α-thalassemia, using hybridization on DNA isolated from fetal fibroblasts. Also, Kan and Dozy, in 1978, implemented RFLP analysis to pinpoint sickle cell alleles of African descent. This breakthrough provided the means of establishing similar diagnostic approaches for the characterization of other genetic diseases, such as phenylketonurea (Woo *et al.*, 1983), cystic fibrosis (Farrall *et al.*, 1986), and so on.

At that time, however, a significant technical bottleneck had to be overcome. The identification of the disease causing mutation was possible only through the construction of a genomic DNA library from the affected individual, in order to first clone the mutated allele and then determine its

DATE	DISCOVERY
1949	Characterization of sickle cell anemia as a molecular disease
1953	Discovery of the DNA double helix
1958	Isolation of DNA polymerases
1960	First hybridization techniques
1969	In situ hybridization
1970	Discovery of restriction enzymes and reverse transcriptase
1975	Southern blotting
1977	DNA sequencing
1983	First synthesis of oligonucleotides
1985	Restriction fragment length polymorphism analysis
1985	Invention of PCR
1986	Development of fluorescent in situ hybridization (FISH)
1988	Discovery of the thermostable DNA polymerase – Optimization of PCR
1992	Conception of real time PCR
1993	Discovery of structure-specific endonucleases for cleavage assays
1996	First application of DNA microarrays
2001	First draft versions of the human genome sequence
2001	Application of protein profiling in human diseases

FIGURE 1.1   The timeline of the principal discoveries in the field of molecular biology, which influenced the development of molecular diagnostics.

nucleotide sequence. Again, many human globin gene mutations were among the first to be identified through such approaches (Busslinger *et al.*, 1981; Treisman *et al.*, 1983). In 1982, Orkin and his colleagues showed that a number of sequence variations were linked to specific β-globin gene mutations. These groups of RFLPs, termed *haplotypes* (both intergenic and intragenic), have provided a first-screening approach in order to detect a disease-causing mutation. Although this approach enabled researchers to predict which β-globin gene would contain a mutation, significantly facilitating mutation screening, no one was in the position to determine the exact nature of the disease-causing mutation, as many different β-globin gene mutations were linked to a specific haplotype in different populations (further information is available at http://globin.cse.psu.edu/hbvar; Hardison *et al.*, 2002; Patrinos *et al.*, 2004).

At the same time, in order to provide a shortcut to DNA sequencing, a number of exploratory methods for pinpointing mutations in patients' DNA were developed. The first methods involved mismatch detection in DNA/DNA or RNA/DNA heteroduplexes (Myers *et al.*, 1985a; Myers *et al.*, 1985b) or differentiation of mismatched DNA heteroduplexes using gel electrophoresis, according to their melting profile (Myers *et al.*, 1987). Using this laborious and time consuming approach, a number of mutations or polymorphic sequence variations have been identified, which made possible the design of short synthetic oligonucleotides that were used as allele-specific probes onto genomic

Southern blots. This experimental design was quickly implemented for the detection of β-thalassemia mutations (Orkin *et al.*, 1983; Pirastu *et al.*, 1983).

Despite the intense efforts from different laboratories worldwide, diagnosis of inherited diseases on the DNA level was still underdeveloped and therefore still not ready to be implemented in clinical laboratories for routine analysis of patients due to the complexities, costs, and time requirements of the technology available. It was only after a few years that molecular diagnosis entered its golden era with the discovery of the most powerful molecular biology tool since cloning and sequencing, the Polymerase Chain Reaction (PCR).

1.3 THE PCR REVOLUTION: GETTING MORE OUT OF LESS

The discovery of PCR (Saiki *et al.*, 1985; Mullis and Faloona, 1987) and its quick optimization, using a thermostable *Taq* DNA polymerase from *Thermus aquaticus* (Saiki *et al.*, 1988) has greatly facilitated and in principle revolutionized molecular diagnostics. The most powerful feature of PCR is the large amount of copies of the target sequence generated by its exponential amplification (see Fig. 1.2), which allows the identification of a known mutation within a single day, rather than months. Also, PCR has markedly decreased or even diminished the need for radioactivity for routine molecular diagnosis. This has allowed molecular diagnostics to enter the clinical laboratory for the provision of genetic services, such as carrier or population screening for known mutations, prenatal diagnosis of inherited diseases, or in recent years, identification of unknown mutations, in close collaboration with research laboratories. Therefore, being moved to their proper environment, the clinical laboratory, molecular diagnostics could provide the services for which they have been initially conceived.

The discovery of PCR also has provided the foundations for the design and development of many mutation detection schemes, based on amplified DNA. In general, PCR either is used for the generation of the DNA fragments to be analyzed, or is part of the detection method. The first attempt was the use of restriction enzymes (Saiki *et al.*, 1985) or oligonucleotide probes, immobilized onto membranes or in solution (Saiki *et al.*, 1986) in order to detect the existing genetic variation, in particular the sickle cell disease-causing mutation. In the following years, an even larger number of mutation detection approaches have been developed and implemented. These techniques can be divided roughly into three categories, depending on the basis for discriminating the allelic variants:

1. *Enzymatic-based methods*. RFLP analysis was historically the first widely used approach, exploiting the alter-

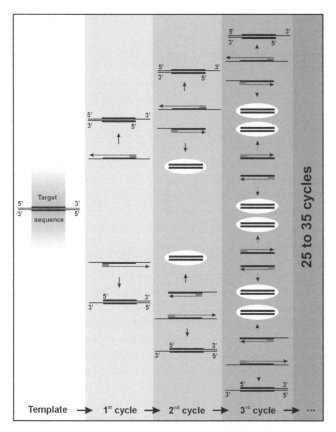

FIGURE 1.2 The PCR principle. Thick and thin black lines correspond to the target sequence and genomic DNA, respectively; gray boxes correspond to the oligonucleotide primers, and the correct size PCR products are included in the white ellipses. Dashed lined arrows depict the elongation of the template strand.

ations in restriction enzyme sites, leading to the gain or loss of restriction events (Saiki *et al.*, 1985). Subsequently, a number of enzymatic approaches for mutation detection have been conceived, based on the dependence of a secondary structure on the primary DNA sequence. These methods exploit the activity of resolvase enzymes T4 endonuclease VII, and more recently, T7 endonuclease I to digest heteroduplex DNA formed by annealing wild type and mutant DNA (Mashal *et al.*, 1995). Digestion fragments indicate the presence and the position of any mutations. A variation of the theme involves the use of chemical agents for the same purpose (Saleeba *et al.*, 1992; see also Chapter 5). Another enzymatic approach for mutation detection is the oligonucleotide ligation assay (Landegren *et al.*, 1988, Chapter 4). In this technique, two oligonucleotides are hybridized to complementary DNA stretches at sites of possible mutations. The oligonucleotide primers are designed such that the 3′ end of the first primer is immediately adjacent to the 5′ end of the second primer. Therefore, if the first primer matches completely with the target DNA, then the primers can be ligated by DNA ligase. On the other hand, if a mismatch occurs at the 3′ end of the first primer, then no ligation products will be obtained.

2. *Electrophoretic-based techniques.* This category is characterized by a plethora of different approaches designed for screening of known or unknown mutations, based on the different electrophoretic mobility of the mutant alleles, under denaturing or nondenaturing conditions. Single strand conformation polymorphism (SSCP) and heteroduplex (HDA) analyses (Orita *et al.*, 1989; see Chapter 6) were among the first methods designed to detect molecular defects in genomic loci. In combination with capillary electrophoresis (see Chapter 7), SSCP and HDA analysis now provide an excellent, simple, and rapid mutation detection platform with low operation costs and, most interestingly, the potential of easily being automated, thus allowing for high-throughput analysis of patients' DNA. Similarly, Denaturing and Temperature Gradient Gel Electrophoresis (DGGE and TGGE, respectively) can be used equally well for mutation detection (see Chapter 8). In this case, electrophoretic mobility differences between a wild type and mutant allele can be "visualized" in a gradient of denaturing agents, such as urea and formamide, or of increasing temperature. Finally, an increasingly used mutation detection technique is the two-dimensional gene scanning (see Chapter 9), based on two-dimensional electrophoretic separation of amplified DNA fragments, according to their size and base pair sequence. The latter involves DGGE, following the size separation step.

3. *Solid phase-based techniques.* This set of techniques consists of the basis for most of the present-day mutation detection technologies, since they have the extra advantage of being easily automated and hence are highly recommended for high throughput mutation detection or screening. A fast, accurate, and convenient method for the detection of known mutations is reverse dot-blot, initially developed by Saiki and colleagues (1989) and implemented for the detection of β-thalassemia mutations. The essence of this method is the utilization of oligonucleotides, bound to a membrane, as hybridization targets for amplified DNA. Some of this technique's advantages is that one membrane strip can be used to detect many different known mutations in a single individual (a one strip-one patient type of assay), the potential of automation, and the ease of interpretation of the results, using a classical avidin-biotin system. However, this technique cannot be used for the detection of unknown mutations. Continuous development has given rise to allele-specific hybridization of amplified DNA (PCR-ASO, Chapter 2) on filters and recently extended on DNA oligonucleotide microarrays (see Chapter 15) for high throughput mutation analysis (Gemignani *et al.*, 2002; Chan *et al.*, 2004). In particular, oligonucleotides of known sequence are immobilized onto appropriate

surfaces and hybridization of the targets to the microarray is detected, mostly using fluorescent dyes.

The choice of the mutation detection method is dependent upon a number of variables, including the mutation spectrum of a given inherited disorder, the available infrastructure, and the number of tests performed in the diagnostic laboratory, and recently with issues of intellectual properties (see also Section 1.5.1 and Chapter 30). Most of the clinical diagnostic laboratories have not invested to expensive high technology infrastructure, since the test volumes, that is, the number of tests expected to be performed, have not been large enough to justify the capital outlay. Therefore, simple screening tests such as SSCP and HDA were and still are the methods of choice for many clinical laboratories, as they allow for rapid and simultaneous detection of different sequence variations at a detection rate of close to 100%.

Although PCR has significantly facilitated the expansion of molecular diagnostics, it nonetheless has a number of limitations. First of all, amplification of CG repeat-rich regions can be problematic for *Taq* Polymerase, which sometimes leads to the classic alternative of Southern blot analysis. Also, *Taq* Polymerase is error-prone at a range of 10^{-4} to 10^{-5} per nucleotide, which is strongly influenced by the conditions of the amplification reaction, such as magnesium or deoxyribonucleotide concentration, pH, temperature, and so on. Polymerase errors can contribute to unspecific background, depending on the detection method, resulting in limiting the detection level. To overcome these technical problems, positive results should be confirmed by alternative methods or by using high fidelity thermostable polymerases.

Finally, it needs to be stressed that despite the wealth of mutation detection methodologies, DNA sequencing is still considered the golden standard and the definitive experimental procedure for mutation detection. However, the costs for the initial investment and the difficulties for standardization and interpretation of ambiguous results has restricted its use only to basic research laboratories.

1.4 MOLECULAR DIAGNOSTICS IN THE POST-GENOMIC ERA

In February 2001, with the announcement of the first draft sequence of the human genome (International Human Genome Sequencing Consortium, 2001; Venter *et al.*, 2001) and subsequently with the genomic sequence of other organisms, molecular biology has entered into a new era with unprecedented opportunities and challenges. These tremendous developments put pressure on a variety of disciplines to intensify their research efforts to improve by orders of magnitude the existing methods for mutation detection, to make available data sets with genomic variation and analyze these sets using specialized software, to standardize and

commercialize genetic tests for routine diagnosis, and to improve the existing technology in order to provide state-of-the-art automated devices for high throughput genetic analysis.

The biggest challenge, following the publication of the human genome draft sequence, was to improve the existing mutation detection technologies to achieve robust cost-effective, rapid, and high-throughput analysis of genomic variation. In the last couple of years, technology has improved rapidly and new mutation-detection techniques have become available, whereas old methodologies have evolved to fit into the increasing demand for automated and high throughput screening. The chromatographic detection of polymorphic changes of disease-causing mutations using denaturing high performance liquid chromatography (DHPLC; for review, see Xiao and Oefner, 2001) is one of the new technologies that emerged. DHPLC reveals the presence of a genetic variation by the differential retention of homo- and heteroduplex DNA on reversed-phase chromatography under partial denaturation. Single-base substitutions, deletions, and insertions can be detected successfully by UV or fluorescence monitoring within two to three minutes in unpurified PCR products as large as 1.5-kilo bases. These features, together with its low cost, make DHPLC one of the most powerful tools for mutational analysis. Also, pyrosequencing, a non–gel-based genotyping technology, provides a very reliable method and an attractive alternative to DHPLC (see Chapter 11). Pyrosequencing detects *de novo* incorporation of nucleotides based on the specific template. The incorporation process releases a pyrophosphate, which is converted to ATP and followed by luciferase stimulation. The light produced, detected by a charge couple device camera, is "translated" to a pyrogram, from which the nucleotide sequence can be deduced (Ronaghi *et al.*, 1996).

The use of the PCR in molecular diagnostics is considered the gold standard for detecting nucleic acids and it has become an essential tool in the research laboratory. Real-time PCR (Holland *et al.*, 1991) has engendered wider acceptance of the PCR due to its improved rapidity, sensitivity, and reproducibility (see Chapter 10). The method allows for the direct detection of the PCR product during the exponential phase of the reaction, thereby combining amplification and detection in one single step. The increased speed of real-time PCR is due largely to reduced cycles, removal of post-PCR detection procedures, and the use of fluorogenic labels and sensitive methods of detecting their emissions. Therefore, real-time PCR is a very accurate and sensitive methodology with a variety of applications in molecular diagnostics, allows a high throughput, and can easily be automated and performed on very small volumes, which makes it the method of choice for many modern diagnostic laboratories.

Above all, the DNA microarray-based genotyping approach offers simultaneous analysis of many polymorphisms and sequence alterations (see Chapter 15). Based more or less on the reverse dot-blot principle, microarrays consist of hundreds of thousands of oligonucleotides attached on a solid surface in an ordered array. The DNA sample of interest is PCR amplified and then hybridized to the microarray. Each oligonucleotide in the high-density array acts as an allele-specific probe and therefore perfectly matched sequences hybridize more efficiently to their corresponding oligonucleotides on the array. The hybridization signals, obtained from allele-specific arrayed primer extension (AS-APEX) (Pastinen *et al.*, 2000), are quantified by high-resolution fluorescent scanning and analyzed by computer software, resulting in the identification of DNA sequence alterations. Therefore, using a high-density microarray makes possible the simultaneous detection of a great number of DNA alterations, hence facilitating genome-wide screening. Several arrays have been generated to detect variants in the HIV genome (Kozal *et al.*, 1996; Wen *et al.*, 2000), human mitochondria mutations (Erdogan *et al.*, 2001), β-thalassemia (Chan *et al.*, 2004), and glycose-6-phosphate dehydrogenase (G-6-PD) deficiency mutations (Gemignani *et al.*, 2002), and so on.

In recent years, there has been a significant development of proteomics, which has the potential to become an indispensable tool for molecular diagnostics. A useful repertoire of proteomic technologies is available, with the potential to undergo significant technological improvements, which would be beneficial for increased sensitivity and throughput while reducing sample requirement (see Chapter 18). The improvement of these technologies is a significant advance toward the need for better disease diagnostics. The detection of disease-specific protein profiles goes back to the use of two-dimensional protein gels (Hanash, 2000), when it was demonstrated that leukemias could be classified into different subtypes based on the different protein profile (Hanash *et al.*, 2002). Nowadays, mass spectrometers are able to resolve many protein and peptide species in body fluids, being virtually set to revolutionize protein-based disease diagnostics (see Chapter 17). The robust and high-throughput nature of the mass spectrometric instrumentation is unparalleled and imminently suited for future clinical applications, as elegantly demonstrated by many retrospective studies in cancer patients (reviewed in Petricoin *et al.*, 2002). Also, high-throughput protein microarrays, constructed from recombinant, purified, and yet functional proteins, allow the miniaturized and parallel analysis of large numbers of diagnostic markers in complex samples. The first pilot studies on disease tissues are already starting to emerge, such as assessing protein expression profiles in tissue derived from squamous cell carcinomas of the oral cavity (Knezevic *et al.*, 2001), or the identification of proteins that induce an acute antibody response in autoimmune disorders, using auto-antigen arrays (Robinson *et al.*, 2002). These findings indicate that proteomic pattern analysis ultimately might be applied as a screening tool for cancer in high-risk and general populations.

The development of state-of-the-art mutation detection techniques has not only a positive impact on molecular genetic testing of inherited disorders, but also provides the technical means to other disciplines. Mutation detection schemes are applicable for the identification of genetically modified (GM) products, which may contaminate non-GM seeds, or food ingredients containing additives and flavorings that have been genetically modified or have been produced from GM organisms (see Chapter 19). The same techniques can ascertain the genotype of an animal strain (see Chapter 24). Another research area that benefits from the continuous development of mutation detection strategies is pharmacogenetics (see Chapter 20), referred to as the effort to define the interindividual variations that are expected to become integral for treatment planning, in terms of efficacy and adverse effects of drugs. This approach uses the technological expertise from high-throughput mutation detection techniques, genomics, and functional genomics to define and predict the nature of the response of an individual to a drug treatment, and to rationally design newer drugs or improve existing ones. Ultimately, the identified genomic sequence variation is organized and stored into specialized mutation databases, enabling a physician or researcher to query upon and retrieve information relevant to diagnostic issues (see Chapter 25).

Finally, and for the last 20 years, DNA analysis and testing has also significantly revolutionized the forensic sciences. The technical advances in molecular biology and the increasing knowledge of the human genome has had a major impact on forensic medicine (see Chapter 21). Genetic characterization of individuals at the DNA level enables identity testing from a minimal amount of biological specimen, such as hair, blood, semen, bone, and so forth, in cases of sexual assault, homicide, and unknown human remains, and paternity testing is also changing from the level of gene products to the genomic level. DNA testing is by far more advantageous over the conventional forensic serology, and over the years has contributed to the acquittal of falsely accused people (saving most of them even from death row) and the identification of the individual who had committed criminal acts (Cohen, 1995), and even helped to specify identities of unknown human remains, such as those from the victims at Ground Zero in New York, or from the skeletons of the Romanov family members (Gill *et al.*, 1994).

1.5 FUTURE PERSPECTIVES: WHAT LIES BEYOND

As an intrinsic part of DNA technology, molecular diagnostics are rooted in the April 1953 discovery of the DNA double helix. Today it is clear that they embody a set of notable technological advances allowing for thousands of diagnostic reactions to be performed at once and for a range of mutations to be simultaneously detected. The reasons for this dramatic increase are two-fold. First of all, the elucidation of the human genomic sequence, as well as that of other species such as bacterial or viral pathogens, has led to an increased number of diagnostically relevant targets. Second, the molecular diagnostic testing volume is rapidly increasing. This is the consequence of a better understanding of the basis of inherited diseases, therefore allowing molecular diagnostics to play a key role in patient or disease management.

Presently, a great number of blood, hair, semen, and tissue samples are analyzed annually worldwide in both public and private laboratories, and the number of genetic tests available is steadily increased year by year (see Fig. 1.3). Taking these premises into account, we can presume that it is only a matter of time before molecular diagnostic laboratories become indispensable in laboratory medicine. In the post-genomic era, genetic information will have to be examined in multiple health care situations throughout people's lives. Currently, newborns can be screened for phenylketonurea and other treatable genetic diseases (Yang *et al.*, 2001). It is also possible that in the not-so-distant future, children at high risk for coronary artery disease will be identified and treated to prevent changes in their vascular walls during adulthood. Similarly, parents will have the option of being informed about their carrier status for many recessive diseases before they decide to start a family. Although not widely accepted, this initiative has already started to be implemented in Cyprus, where a couple at risk for thalassemia syndrome is advised to undergo a genetic

test for thalassemia mutations before the marriage (see also Chapter 31). Also, for middle-aged and older populations, scientists will be able to determine risk profiles for various late-onset diseases, preferably before the appearance of symptoms, which at least could be partly prevented through dietary or pharmaceutical interventions. In the near future, the monitoring of individual drug response profiles throughout life, using genetic testing for the identification of their individual DNA signature, will be part of the standard medical practice. Soon, genetic testing will comprise a wide spectrum of different analyses with a host of consequences for individuals and their families, which is worth emphasizing when explaining molecular diagnostics to the public (see also Chapter 32). All these issues are discussed in detail next. However, and in order to be more realistic, many of these expectations still are based on promises, though quite optimistic ones. Thus, some of the new perspectives of the field could be a decade away, and several challenges remain to be realized.

1.5.1 Commercializing Molecular Diagnostics

Currently, clinical molecular genetics is part of the mainstream healthcare worldwide. Almost all clinical laboratories have a molecular diagnostic unit or department. Although in recent years the notion of molecular diagnostics has increasingly gained interest, genetic tests are still not generally used for population screening, but rather for diagnosis, carrier screening, and prenatal diagnosis, and only on a limited basis. Therefore, and in order to make molecular diagnostics widely available, several obstacles and issues need to be taken into consideration and resolved in the coming years.

The first important issue is the choice of the mutation detection platform. Despite the fact that there are over 50 different mutation detection and screening methods, there is no single platform or methodology that prevails for genetic testing. Genotyping can be done using different approaches, such as filters, gels, microarrays, microtiter plates; different amplification-based technologies; different separation techniques, such as blotting, capillary electrophoresis, microarrays, mass spectroscopy; and finally different means for labeling, such as radioactive, fluorescent, chemiluminescent, or enzymatic substances. The variety of detection approaches makes it not only difficult but also challenging to determine which one is better suited for a laboratory setting. The initial investment costs and the expected test volume are some of the factors that need to be taken into consideration prior to choosing the detection technique. Related issues are also the costs of the hardware and software, testing reagents, and kits. The latter is of great importance, since the fact that most of the diagnostic laboratories today are running "home-brew" assays—for example, not using well-standardized genetic testing kits due to cost bar-

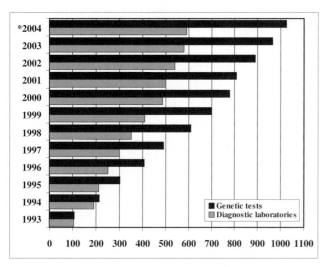

FIGURE 1.3 Number of genetic tests provided by diagnostic laboratories. Data are from the NIH-funded genetic testing information databases, GeneTests (www.genetests.org) and GeneClinics (www.geneclinics.org). (Note: Information available until April 2004.)

riers, which brings to surface the issue of quality control of the reagents (see Chapter 34) and of safety (see Chapter 33). Currently, there are several clinical and technical recommendations for genetic testing for monogenic disorders that have been issued by several organizations (see Table 1.1).

Another very important issue is training the personnel of a molecular diagnostic laboratory, reflecting in the quality and the correct interpretation of the results. Continuous education of the personnel of the diagnostic laboratory is crucial for the accuracy of the results provided (see also Chapter 34). Many times, such as in the case of prenatal or preimplantation diagnosis, irrevocable decisions need to be made, most of the time based on a simple test result. In the past five years, there has been a significant reduction of the number of incorrect genotypes diagnosed (from 30% to only 5% by the year 2000), as a result of continuous training and proficiency testing schemes (Dequeker and Cassiman, 2000). In the United States, there is a voluntary biannual proficiency testing for molecular diagnostic laboratories, and recently the European Molecular genetics Quality Network (http://www.emqn.org) has been founded to promote quality in molecular genetic testing through the provision of external quality assessment (proficiency testing schemes) and the organization of best practice meetings and publication of guidelines. It is generally true that many geneticists and nongeneticist physicians would benefit from continuous education regarding the appropriate use of molecular diagnostic tests, which is necessary to evaluate the method preanalytically and to interpret results. Also, replacing manual with automated testing would help decrease some of the analytical errors, but such investment would be justified only by a large test volume. Therefore, it is not surprising that manual methods still exist in the vast majority of diagnostic laboratories.

The legal considerations and the ethical concerns are also hurdles that need to be overcome in the coming years. One issue is reimbursing of the diagnosis costs. At present, there are no insurance companies that reimburse the costs for molecular testing to the people insured; the necessary regulatory and legal framework remains to be established. "Legalizing" molecular testing, by the adoption of the relevant regulations, would probably result in an increase of the test volume and at the same time it can pose an immense barrier to uncontrolled genetic testing. Similarly, the need to obtain an informed consent from the patient to be analyzed is also of great importance and should be encouraged and facilitated by the diagnostic laboratory.

On the other hand, the issue of intellectual properties hampers the wide commercialization of molecular diagnostics. Almost all the clinically relevant genes have been now patented and the terms that the patent holders offer vary considerably (see Chapter 30). Among the difficulties that this issue imposes is the limiting choice of mutation detection platforms, the large loyalties for reagent use, and the exclusive sublicenses that many companies grant to clinical laboratories, leading eventually to monopolies. Since one of the biggest challenges that the clinical laboratory is facing is patent and regulatory compliance, partnerships and collaborations may be envisaged in order to take the technology licenses to the diagnostic laboratory that will subsequently develop, standardize, and distribute the assays. These will partly alleviate some of the intellectual properties issues.

Finally, the issue of the medical genetics specialty is more urgent than ever. In the United States, medical genetics has been formally recognized as a medical specialty only within the past 10 years, and in Europe, medical genetics only recently has been formally recognized as a specialty (http://www.eshg.org). The implementation of this decision is still facing substantial difficulties (http://www.eshg.org/geneticseurope.htm), which will probably take years to bypass. With the completion of the Human Genome Project, genetics has become the driving force in medical research and is now poised for integration into medical practice. An increase in the medical genetics workforce, including geneticists and genetic counselors, will be necessary in the coming years. After all, the Human Genome Project has made information of inestimable diagnostic and therapeutic importance available and therefore the medical profession now has the obligation to rise to both the opportunities and challenges that this wealth of genetic information presents.

TABLE 1.1 Clinical and technical recommendations for genetic testing for monogenic disorders. ACMG: American College of Medical Genetics; ASHG: American Society of Human Genetics.

Disease/ Syndrome	Gene	References
Alzheimer	ApoE	ACMG, 1995
Canavan	ASPA	ACMG, 1998
Cystic fibrosis	CFTR	Dequeker E. *et al.*, 2000 Grody *et al.*, 2001a
Thrombophilia	Factor V Leiden	Grody *et al.*, 2001b
Fragile X	FMR1	Maddalena *et al.*, 2001
Prader-Willi/ Angelman	15q11-q13	ASHG/ACMG, 1996
Multiple endocrine neoplasia	MEN 1/2	Brandi *et al.*, 2001
Tuberous sclerosis	TSC 1/2	Roach *et al.*, 1999
Breast cancer	BRCA1	Sorscher and Levonian, 1997

1.5.2 Personalized Medicine

The term "personalized medicine" refers to the practice of medicine where patients receive the most appropriate medical treatment, fitting dosage, and combination of drugs based on their genetic background. Some of the reasons for many types of adverse drug reactions are already known and often related to polymorphic gene alleles of drug metabolizing enzymes (Nebert and Menon, 2001; Risch *et al.*, 2002). The application of high throughput genotyping tools for the identification and screening of single nucleotide polymorphisms (SNPs) eventually can lead to the determination of the unique molecular signature of an individual in a relatively short period of time. This way, individual drug responses can be predicted from predetermined genetic variances correlated with a drug effect. In other words, this will allow the physician to provide the patient with a selective drug prescription. A handful of pharmaceutical companies are developing a precise haplotyping scheme to identify individuals/patients who will derive optimal benefit from drugs currently under development. Clinicians will facilitate this effort by importing clinical data into this haplotyping system for a complete patient analysis and drug evaluation. In addition to these efforts, there is a growing need to incorporate this increasingly complex body of knowledge to standard medical practice. Incorporating pharmacogenomics-related courses in the standard curriculum of medical schools potentially can ensure that the forthcoming generation of clinicians and researchers will be familiar with the latest developments in that field and will be capable of providing patients with the expected benefits of personalized medicine.

However, there are growing concerns on the ethical aspects of personalized medicine. First of all, equality in medical care needs to be ensured, when genetics foretell clinicians which patients would be less likely to benefit from a particular drug treatment. Second, it will become increasingly vital to devise operational tools for the prevention of stigmatization and discrimination of different populations, in particular on ethnic grounds (van Ommen, 2002), and therefore every precaution should be taken to eliminate all lingering prejudice and bias associated with the study of human genetic variation. Other dilemmas include the right to deny an available treatment from specific patient populations according to genetic-derived indications, as currently is the case with prenatal diagnosis (see also Chapter 31). Appropriate guidelines will be crucially needed for the successful implementation of pharmacogenomics into clinical practice.

1.5.3 Theranostics: Integrating Diagnostics and Therapeutics

The ultimate goal in healthcare over the next decades will be the efficient integration of molecular diagnostics with therapeutics. Experts believe that in less than a decade, people will be able to have their own genomes sequenced for under $1,000. This is going to involve sequencing technology that is a lot cheaper and faster than today's machines.

When that point is reached, this can ultimately be translated in a patient being able to carry a microarray, like an ordinary credit card, with all his or her genetic information encoded on it. Such microarrays can be constructed by blood sampling of the individual, sequencing of the functional DNA, and identifying the genetic variations in the genes.

In the future, a person may appear at the clinic for treatment, carrying a microarray with its entire genome. Provided that this person is diagnosed accurately for specific mutations, alleles, or even polymorphic changes pertaining to a specific disease, then his or her response to treatment can be vastly improved. Therefore, gene-based disease management and treatment, incorporating molecular analysis, will be able to predict the efficacy, and at the same time the safety, of a specific therapeutic product. Relevant efforts have been already in progress for Alzheimer's disease, where ApoE genotype tests are extremely useful for predicting the response to certain drugs, and therefore can be used as diagnostic markers for clinical trials. In particular, the apoE4 allele has been significantly associated with a decreased response to tacrine (Poirier *et al.*, 1995) and is one of the most important risk factors for Alzheimer's disease (reviewed in Myers and Goate, 2001). In principle, such efforts will emerge as an integrated healthcare system, in which genetic screening and therapeutics will enable prevention and molecular diagnostics-based therapy.

1.6 CONCLUSIONS

In the coming years, molecular diagnostics will continue to be of critical importance to public health worldwide. Molecular genetic testing will facilitate the detection and characterization of disease, as well as monitoring of the drug response, and will assist in the identification of genetic modifiers and disease susceptibility. A wide range of molecular-based tests is available to assess DNA variation and changes in gene expression. However, there are major hurdles to overcome before the implementation of these tests in clinical laboratories, such as which test to employ, the choice of technology and equipment, and issues such as cost-effectiveness, accuracy, reproducibility, personnel training, reimbursement by third-party payers, and intellectual property. At present, PCR-based testing predominates; however, alternative technologies aimed at exploring genome complexity without PCR are anticipated to gain momentum in the coming years. Furthermore, development of integrated chip devices ("lab-on-a-chip") should facilitate genetic readouts from single cells and molecules. Together with proteomic-based testing, these advances will improve molecular diagnostics and will present additional challenges for implementing such technology in public or private research units, hospitals, clinics, and pharmaceutical industries.

References

American College of Medical Genetics. (1998). Position statement on carrier testing for canavan disease. www.faseb.org/genetics/acmg/pol-31.htm.

American College of Medical Genetics/American Society of Human Genetics Working Group on ApoE and Alzheimer disease. (1995). Statement on use of apolipoprotein E testing for Alzheimer disease. *JAMA* 274, 1627–1629.

American College of Medical Genetics/American Society of Human Genetics. (1996). Diagnostic testing for Prader-Willi and Angleman syndromes: Report of the ASHG/ACMG test and technology transfer committee. *Am. J. Hum. Genet.* 58, 1085–1088.

Brandi, M. L., Gagel, R. F., Angeli, A., Bilezikian, J. P., Beck-Peccoz, P., Bordi, C., Conte-Devolx, B., Falchetti, A., Gheri, R. G., Libroia, A., Lips, C. J., Lombardi, G., Mannelli, M., Pacini, F., Ponder, B. A., Raue, F., Skogseid, B., Tamburrano, G., Thakker, R. V., Thompson, N. W., Tomassetti, P., Tonelli, F., Wells, S. A. Jr., and Marx, S. J. (2001). Guidelines for diagnosis and therapy of MEN type 1 and type 2. *J. Clin. Endocrinol. Metab.* 86, 5658–5671.

Busslinger, M., Moschonas, N., and Flavell, R. A. (1981). Beta + thalassemia: aberrant splicing results from a single point mutation in an intron. *Cell* 27, 289–298.

Chan, K., Wong, M. S., Chan, T. K., and Chan, V. (2004). A thalassaemia array for Southeast Asia. *Br. J. Haematol.* 124, 232–239.

Cohen, J. (1995). Genes and behavior make an appearance in the O. J. trial. *Science* 268, 22–23.

Dequeker, E., and Cassiman, J. J. (2000). Genetic testing and quality control in diagnostic laboratories. *Nat. Genet.* 25, 259–260.

Dequeker, E., Cuppens, H., Dodge, J., Estivill, X., Goossens, M., Pignatti, P. F., Scheffer, H., Schwartz, M., Schwarz, M., Tummler, B., and Cassiman, J. J. (2000). Recommendations for quality improvement in genetic testing for cystic fibrosis. European Concerted Action on Cystic Fibrosis. *Eur. J. Hum. Genet.* 8(Suppl. 2), S2–24.

Erdogan, F., Kirchner, R., Mann, W., Ropers, H. H., and Nuber, U. A. (2001). Detection of mitochondrial single nucleotide polymorphisms using a primer elongation reaction on oligonucleotide microarrays. *Nucleic Acids Res.* 29, E36.

Farrall, M., Rodeck, C. H., Stanier, P., Lissens, W., Watson, E., Law, H. Y., Warren, R., Super, M., Scambler, P., Wainwright, B., and Williamson R. (1986). First-trimester prenatal diagnosis of cystic fibrosis with linked DNA probes. *Lancet* 327, 1402–1405.

Gemignani, F., Perra, C., Landi, S., Canzian, F., Kurg, A., Tonisson, N., Galanello, R., Cao, A., Metspalu, A., and Romeo, G. (2002). Reliable detection of beta-thalassemia and G6PD mutations by a DNA microarray. *Clin. Chem. 2002*, 48, 2051–2054.

Gill, P., Ivanov, P. L., Kimpton, C., Piercy, R., Benson, N., Tully, G., Evett, I., Hagelberg, E., and Sullivan, K. (1994). Identification of the remains of the Romanov family by DNA analysis. *Nat. Genet.* 6, 130–135.

Grody, W. W., Cutting, G. R., Klinger, K. W., Richards, C. S., Watson, M. S., and Desnick, R. J.; Subcommittee on Cystic Fibrosis Screening, Accreditation of Genetic Services Committee, ACMG. American College of Medical Genetics.

(2001a). Laboratory standards and guidelines for population-based cystic fibrosis carrier screening. *Genet. Med.* 3, 149–154.

Grody, W. W., Griffin, J. H., Taylor, A. K., Korf, B. R., and Heit, J. A.; ACMG Factor V. Leiden Working Group. (2001b). American College of Medical Genetics consensus statement on factor V Leiden mutation testing. *Genet. Med.* 3, 139–148.

Hanash, S. M. (2000). Biomedical applications of two-dimensional electrophoresis using immobilized pH gradients: current status. *Electrophoresis* 21, 1202–1209.

Hanash, S. M., Madoz-Gurpide, J., and Misek, D. E. (2002). Identification of novel targets for cancer therapy using expression proteomics. *Leukemia* 16, 478–485.

Hardison, R. C., Chui, D. H., Giardine, B., Riemer, C., Patrinos, G. P., Anagnou, N., Miller, W., and Wajcman, H. (2002). HbVar: A relational database of human hemoglobin variants and thalassemia mutations at the globin gene server. *Hum. Mutat.* 19, 225–233.

Holland, P. M., Abramson, R. D., Watson, R., and Gelfand, D. H. (1991). Detection of specific polymerase chain reaction product by utilizing the $5'$–$3'$ exonuclease activity of Thermus aquaticus. *Proc. Natl. Acad. Sci. USA* 88, 7276–7280.

International Human Genome Sequencing Consortium. (2001). Initial sequencing and analysis of the human genome. *Nature* 409, 860–921.

Kan, Y. W., Golbus, M. S., and Dozy, A. M. (1976). Prenatal diagnosis of alpha-thalassemia. Clinical application of molecular hybridization. *N. Engl. J. Med.* 295, 1165–1167.

Kan, Y. W., and Dozy, A. M. (1978). Polymorphism of DNA sequence adjacent to human beta-globin structural gene: relationship to sickle mutation. *Proc. Natl. Acad. Sci. USA* 75, 5631–5635.

Knezevic, V., Leethanakul, C., Bichsel, V. E., Worth, J. M., Prabhu, V. V., Gutkind, J. S., Liotta, L. A., Munson, P. J., Petricoin, E. F. 3rd, and Krizman, D. B. (2001). Proteomic profiling of the cancer microenvironment by antibody arrays. *Proteomics* 1, 1271–1278.

Kozal, M. J., Shah, N., Shen, N., Yang, R., Fucini, R., Merigan, T. C., Richman, D. D., Morris, D., Hubbell, E., Chee, M., and Gingeras, T. R. (1996). Extensive polymorphisms observed in HIV-1 clade B protease gene using high-density oligonucleotide arrays. *Nat. Med.* 2, 753–759.

Landegren, U., Kaiser, R., Sanders, J., and Hood, L. (1988). A ligase-mediated gene detection technique. *Science* 241, 1077–1080.

Maddalena, A., Richards, C. S., McGinniss, M. J., Brothman, A., Desnick, R. J., Grier, R. E., Hirsch, B., Jacky, P., McDowell, G. A., Popovich, B., Watson, M., and Wolff, D. J. (2001). Technical standards and guidelines for fragile X: the first of a series of disease-specific supplements to the Standards and Guidelines for Clinical Genetics Laboratories of the American College of Medical Genetics. Quality Assurance Subcommittee of the Laboratory Practice Committee. *Genet. Med.* 3, 200–205.

Mashal, R. D., Koontz, J., and Sklar, J. (1995). Detection of mutations by cleavage of DNA heteroduplexes with bacteriophage resolvases. *Nat. Genet.* 9, 177–183.

Mullis, K. B., and Faloona, F. A. (1987). Specific synthesis of DNA in vitro via a polymerase-catalyzed chain reaction. *Methods Enzymol.* 155, 335–350.

Myers, A. J., and Goate, A. M. (2001). The genetics of late-onset Alzheimer's disease. *Curr. Opin. Neurol.* 14, 433–440.

Myers, R. M., Larin, Z., and Maniatis, T. (1985a). Detection of single base substitutions by ribonuclease cleavage at mismatches in RNA:DNA duplexes. *Science* 230, 1242–1246.

Myers, R. M., Lumelsky, N., Lerman, L. S., and Maniatis, T. (1985b). Detection of single base substitutions in total genomic DNA. *Nature* 1985, 313, 495–498.

Myers, R. M., Maniatis, T., Lerman, L. S. (1987). Detection and localization of single base changes by denaturing gradient gel electrophoresis. *Methods Enzymol.* 155, 501–527.

Nebert, D. W., and Menon, A. G. (2001). Pharmacogenomics, ethnicity, and susceptibility genes. *Pharmacogenomics J.* 1, 19–22.

Orita, M., Iwahana, H., Kanazawa, H., Hayashi, K., and Sekiya, T. (1989). Detection of polymorphisms of human DNA by gel electrophoresis as single-strand conformation polymorphisms. *Proc. Natl. Acad. Sci. USA* 86, 2766–2770.

Orkin, S. H., Kazazian, H. H. Jr., Antonarakis, S. E., Goff, S. C., Boehm, C. D., Sexton, J. P., Waber, P. G., and Giardina, P. J. (1982). Linkage of beta-thalassaemia mutations and beta-globin gene polymorphisms with DNA polymorphisms in human beta-globin gene cluster. *Nature* 296, 627–631.

Orkin, S. H., Markham, A. F., and Kazazian, H. H. Jr. (1983). Direct detection of the common Mediterranean beta-thalassemia gene with synthetic DNA probes. An alternative approach for prenatal diagnosis. *J. Clin. Invest.* 71, 775–779.

Pastinen, T., Raitio, M., Lindroos, K., Tainola, P., Peltonen, L., and Syvanen, A. C. (2000). A system for specific, high-throughput genotyping by allele-specific primer extension on microarrays. *Genome Res.* 10, 1031–1042.

Patrinos, G. P., Giardine, B., Riemer, C., Miller, W., Chui, D. H., Anagnou, N. P., Wajcman, H., and Hardison, R. C. (2004). Improvements in the HbVar database of human hemoglobin variants and thalassemia mutations for population and sequence variation studies. *Nucleic Acids Res.* 32, D537–D541.

Pauling, L., Itano, H. A., Singer, S. J., and Wells, I. C. (1949). Sickle cell anemia, a molecular disease. *Science* 110, 543–548.

Petricoin, E. F., Zoon, K. C., Kohn, E. C., Barrett, J. C., and Liotta, L. A. (2002). Clinical proteomics: translating benchside promise into bedside reality. *Nat. Rev. Drug. Discov.* 1, 683–695.

Pirastu, M., Kan, Y. W., Cao, A., Conner, B. J., Teplitz, R. L., and Wallace, R. B. (1983). Prenatal diagnosis of beta-thalassemia. Detection of a single nucleotide mutation in DNA. *N. Engl. J. Med.* 309, 284–287.

Poirier, J., Delisle, M. C., Quirion, R., Aubert, I., Farlow, M., Lahiri, D., Hui, S., Bertrand, P., Nalbantoglu, J., Gilfix, B. M., and Gauthier, S. (1995). Apolipoprotein E4 allele as a predictor of cholinergic deficits and treatment outcome in Alzheimer disease. *Proc. Natl. Acad. Sci. USA* 92, 12260–12264.

Risch, N., Burchard, E., Ziv, E., and Tang, H. (2002). Categorization of humans in biomedical research: genes, race and disease. *Genome Biol.* 3, 1–12.

Roach, E. S., DiMario, F. J., Kandt, R. S., Northrup, H. (1999). Tuberous Sclerosis Consensus Conference: recommendations for diagnostic evaluation. National Tuberous Sclerosis Association. *J. Child Neurol.* 14, 401–407.

Robinson, W. H., DiGennaro, C., Hueber, W., Haab, B. B., Kamachi, M., Dean, E. J., Fournel, S., Fong, D., Genovese, M. C., de Vegvar, H. E., Skriner, K., Hirschberg, D. L., Morris, R. I., Muller, S., Pruijn, G. J., van Venrooij, W. J., Smolen, J. S., Brown, P. O., Steinman, L., and Utz, P. J. (2002). Autoantigen microarrays for multiplex characterization of autoantibody responses. *Nat. Med.* 8, 295–301.

Ronaghi, M., Uhlen, M., and Nyren, P. (1998). A sequencing method based on real-time pyrophosphate. *Science* 281, 363–365.

Saiki, R. K., Scharf, S., Faloona, F., Mullis, K. B., Horn, G. T., Erlich, H. A., and Arnheim, N. (1985). Enzymatic amplification of beta-globin genomic sequences and restriction site analysis for diagnosis of sickle cell anemia. *Science* 230, 1350–1354.

Saiki, R. K., Bugawan, T. L., Horn, G. T., Mullis, K. B., and Erlich, H. A. (1986). Analysis of enzymatically amplified beta-globin and HLA-DQ alpha DNA with allele-specific oligonucleotide probes. *Nature* 324, 163–166.

Saiki, R. K., Gelfand, D. H., Stoffel, S., Scharf, S. J., Higuchi, R., Horn, G. T., Mullis, K. B., and Erlich, H. A. (1988). Primer-directed enzymatic amplification of DNA with a thermostable DNA polymerase. *Science* 239, 487–491.

Saiki, R. K., Walsh, P. S., Levenson, C. H., and Erlich, H. A. (1989). Genetic analysis of amplified DNA with immobilized sequence-specific oligonucleotide probes. *Proc. Natl. Acad. Sci. USA* 86, 6230–6234.

Saleeba, J. A., Ramus, S. J., and Cotton, R. G. (1992). Complete mutation detection using unlabeled chemical cleavage. *Hum. Mutat.* 1, 63–69.

Sorscher, S., and Levonian, P. (1997). BCRA 1 testing guidelines for high-risk patients. *J. Clin. Oncol.* 15, 1711.

Treisman, R., Orkin, S. H., and Maniatis, T. (1983). Specific transcription and RNA splicing defects in five cloned beta-thalassaemia genes. *Nature* 302, 591–596.

van Ommen, G. J. (2002). The Human Genome Project and the future of diagnostics, treatment and prevention. *J. Inherit. Metab. Dis.* 25, 183–188.

Venter, J. C., Adams, M. D., Myers, E. W., Li, P. W., Mural, R. J., Sutton, G. G., Smith, H. O., Yandell, M., Evans, C. A., Holt, R. A., Gocayne, J. D., Amanatides, P., Ballew, R. M., Huson, D. H., Wortman, J. R., Zhang, Q., Kodira, C. D., Zheng, X. H., Chen, L., Skupski, M., Subramanian, G., Thomas, P. D., Zhang, J., Gabor Miklos, G. L., Nelson, C., Broder, S., Clark, A. G., Nadeau, J., McKusick, V. A., Zinder, N., Levine, A. J., Roberts, R. J., Simon, M., Slayman, C., Hunkapiller, M., Bolanos, R., Delcher, A., Dew, I., Fasulo, D., Flanigan, M., Florea, L., Halpern, A., Hannenhalli, S., Kravitz, S., Levy, S., Mobarry, C., Reinert, K., Remington, K., Abu-Threideh, J., Beasley, E., Biddick, K., Bonazzi, V., Brandon, R., Cargill, M., Chandramouliswaran, I., Charlab, R., Chaturvedi, K., Deng, Z., Di Francesco, V., Dunn, P., Eilbeck, K., Evangelista, C., Gabrielian, A. E., Gan, W., Ge, W., Gong, F., Gu, Z., Guan, P., Heiman, T. J., Higgins, M. E., Ji, R. R., Ke, Z., Ketchum, K. A., Lai, Z., Lei, Y., Li, Z., Li, J., Liang, Y., Lin, X., Lu, F., Merkulov, G. V., Milshina, N., Moore, H. M., Naik, A. K., Narayan, V. A., Neelam, B., Nusskern, D., Rusch, D. B., Salzberg, S., Shao, W., Shue, B., Sun, J., Wang, Z., Wang, A., Wang, X., Wang, J., Wei, M., Wides, R., Xiao, C., Yan, C., Yao, A., Ye, J., Zhan, M., Zhang, W., Zhang, H., Zhao, Q., Zheng, L., Zhong, F., Zhong, W., Zhu, S., Zhao, S., Gilbert, D., Baumhueter, S., Spier, G., Carter, C., Cravchik, A., Woodage, T., Ali, F., An, H., Awe, A., Baldwin, D., Baden, H., Barnstead, M., Barrow, I., Beeson, K., Busam, D., Carver, A., Center, A., Cheng, M. L., Curry, L., Danaher, S., Davenport, L., Desilets, R., Dietz, S., Dodson, K., Doup, L., Ferriera, S., Garg, N.,

Gluecksmann, A., Hart, B., Haynes, J., Haynes, C., Heiner, C., Hladun, S., Hostin, D., Houck, J., Howland, T., Ibegwam, C., Johnson, J., Kalush, F., Kline, L., Koduru, S., Love, A., Mann, F., May, D., McCawley, S., McIntosh, T., McMullen, I., Moy, M., Moy, L., Murphy, B., Nelson, K., Pfannkoch, C., Pratts, E., Puri, V., Qureshi, H., Reardon, M., Rodriguez, R., Rogers, Y. H., Romblad, D., Ruhfel, B., Scott, R., Sitter, C., Smallwood, M., Stewart, E., Strong, R., Suh, E., Thomas, R., Tint, N. N., Tse, S., Vech, C., Wang, G., Wetter, J., Williams, S., Williams, M., Windsor, S., Winn-Deen, E., Wolfe, K., Zaveri, J., Zaveri, K., Abril, J. F., Guigo, R., Campbell, M. J., Sjolander, K. V., Karlak, B., Kejariwal, A., Mi, H., Lazareva, B., Hatton, T., Narechania, A., Diemer, K., Muruganujan, A., Guo, N., Sato, S., Bafna, V., Istrail, S., Lippert, R., Schwartz, R., Walenz, B., Yooseph, S., Allen, D., Basu, A., Baxendale, J., Blick, L., Caminha, M., Carnes-Stine, J., Caulk, P., Chiang, Y. H., Coyne, M., Dahlke, C., Mays, A., Dombroski, M., Donnelly, M., Ely, D., Esparham, S., Fosler, C., Gire, H., Glanowski, S., Glasser, K., Glodek, A., Gorokhov, M., Graham, K., Gropman, B., Harris, M., Heil, J., Henderson, S., Hoover, J., Jennings, D., Jordan, C., Jordan, J., Kasha, J., Kagan, L., Kraft, C., Levitsky, A., Lewis, M., Liu, X., Lopez, J., Ma, D., Majoros, W., McDaniel, J., Murphy, S., Newman, M., Nguyen, T., Nguyen, N., Nodell, M., Pan, S., Peck, J., Peterson, M., Rowe, W., Sanders, R., Scott, J., Simpson, M., Smith, T., Sprague, A., Stockwell, T., Turner, R., Venter, E., Wang, M., Wen, M., Wu, D., Wu, M., Xia, A., Zandieh, A., and Zhu, X. (2001). The sequence of the human genome. *Science* 291, 1304–1351.

Wen, W. H., Bernstein, L., Lescallett, J., Beazer-Barclay, Y., Sullivan-Halley, J., White, M., and Press, M. F. (2000). Comparison of TP53 mutations identified by oligonucleotide microarray and conventional DNA sequence analysis. *Cancer Res.* 60, 2716–2722.

Woo, S. L., Lidsky, A. S., Guttler, F., Chandra, T., and Robson, K. J. (1983). Cloned human phenylalanine hydroxylase gene allows prenatal diagnosis and carrier detection of classical phenylketonuria. *Nature* 306, 151–155.

Xiao, W., and Oefner, P. J. (2001). Denaturing high-performance liquid chromatography: A review. *Hum. Mutat.* 17, 439–474.

Yang, Y., Drummond-Borg, M., and Garcia-Heras, J. (2001). Molecular analysis of phenylketonuria (PKU) in newborns from Texas. *Hum. Mutat.* 17, 523.

Molecular Diagnostic Technology

CHAPTER 2

Allele-Specific Mutation Detection by PCR-ARMS and PCR-ASO

JOHN S. WAYE[1,2] AND BARRY ENG[2,3]

[1]*Department of Pathology and Molecular Medicine, McMaster University;*
[2]*Molecular Diagnostic Genetics, Hamilton Regional Laboratory Medicine Program;*
[3]*Genetic Services, Hamilton Regional Laboratory Medicine Program, Hamilton, Ontario, Canada*

TABLE OF CONTENTS

2.1 INTRODUCTION

A primary function of molecular diagnostics is the detection of mutations and single nucleotide polymorphisms (SNPs) that are associated with particular phenotypes. This chapter provides a description of two relatively simple PCR-based techniques that can be applied to detect known point mutations or SNPs in DNA. The first approach, known as the amplification refractory mutation system, or PCR-ARMS, is based on the principle that a mismatch between the 3′ nucleotide of a PCR primer and the template reduces or prevents primer extension by *Taq* polymerase. A variety of strategies have been developed using primers that are complementary to, and allow for the specific amplification of, individual alleles.

The second approach described in this chapter is based on hybridization of PCR products to allele-specific oligonucleotide probes, or PCR-ASO. This method can be applied in two formats: the forward ASO approach where PCR products are immobilized on membrane and hybridized to labeled ASO probes; and the reverse ASO approach, where ASO probes are immobilized on the membrane and hybridized to labeled PCR products.

PCR-ARMS and PCR-ASO have been widely used in research and molecular diagnostics since their initial development in the late 1980s. The attraction of these methods lies in their simplicity and applicability to the analysis of virtually any known point mutation or SNP. Moreover, these methods do not require expensive and sophisticated instrumentation.

2.2 PCR-ARMS

2.2.1 Origin of Method

In 1989, several independent groups described a PCR-based approach for analyzing known point mutations in DNA and distinguishing between normal, heterozygous, and homozygous mutant genotypes (Newton *et al.*, 1989; Nichols *et al.*, 1989; Okayama *et al.*, 1989; Sommer *et al.*, 1989; Wu *et al.*, 1989). The method is most commonly referred to as PCR-ARMS or ARMS (Amplification Refractory Mutation System) (Newton *et al.*, 1989). It has also been referred to as PASA (PCR Amplification of Specific Alleles) by Sommer *et al.* (1989) and as ASPCR (Allele-Specific PCR) by Wu *et al.* (1989). The initial reports demonstrated the applicability of this approach to analyze known disease mutations associated with α1-antitrypsin deficiency, amyloidotic polyneuropathy, phenylketonuria, and sickle cell anemia. Subsequent validation studies quickly established that this strategy is applicable to the analysis of any known point mutation or SNP (Sommer *et al.*, 1992).

2.2.2 Basic Principles

PCR-ARMS is based on the observation that PCR amplification is inefficient or completely refractory if there is a mismatch between the 3′ terminal nucleotide of a PCR primer and the corresponding template (Newton *et al.*, 1989). *Taq* DNA polymerase lacks a 3′ to 5′ exonuclease activity, and therefore cannot correct mismatches at the 3′ terminus of the primer. As such, complementary base-pairing at the 3′ end of the primer is required for efficient amplification by *Taq* DNA polymerase and is a strong determining factor of template specificity. Amplification of the normal allele, and not that of the mutant, is accomplished using a primer that is complementary to the normal allele and has a mismatch between the 3′ residue and the mutant allele. Conversely, only the mutant will be amplified if the 3′ residue of the primer is complementary to the mutant allele and not the normal allele. The specificity or discriminating power of the 3′ terminal nucleotide can be enhanced further by incorporating an additional mismatch positioned near the 3′ nucleotide (Newton *et al.*, 1989). The basic concept is illustrated in Fig. 2.1.

Various studies have attempted to quantify the inhibitory effect of different 3′ mismatches on PCR amplification (Kwok *et al.*, 1990; Sarkar *et al.*, 1990; Huang *et al.*, 1992; Ayyadevara *et al.*, 2000). Although some trends have emerged, the results were remarkably discordant. Sarkar and colleagues (1990) concluded that PCR is inhibited by mismatches between the template and the 3′ or 3′ penultimate nucleotide of the primer. Under relatively relaxed stringency conditions for primer:template annealing, Kwok and colleagues (1990) demonstrated a 20-fold reduction in amplification efficiency with A:A (primer:template) mismatches, 100-fold reductions with A:G, G:A, and C:C mismatches, and little or no reduction with any other mismatches. Under higher stringency conditions, Huang and colleagues (1992) showed some degree of inhibition with every combination of 3′ mismatch. The weakest inhibition, about 100-fold reduction, was associated with C:T mismatches. There was

approximately 10^3-fold reduction with A:C, C:A, G:T, and T:G mismatches; 10^3 to 10^4-fold reduction with T:C and T:T mismatches; and at least 10^6-fold reduction with A:A, G:A, A:G, G:G, and C:C mismatches. Ayyadevara and colleagues (2000) systematically varied both the 3′ terminal and penultimate nucleotides of primers, all under relatively high stringency conditions. Their study indicated that primers ending with 3′ A are moderately inferior to those ending in other nucleotides. Allele-specific amplification had 40 to 100-fold reduction when the mismatched primer had T, G, or C at the 3′ terminal position. They also concluded that the penultimate 3′ nucleotide plays a minor role in mismatch discrimination, and that amplification efficiency is reduced when A (and to a lesser extent T) occupies the penultimate 3′ position.

2.2.3 PCR-ARMS Design Concepts

The design and optimization of PCR-ARMS protocols is primarily a function of the target sequence and the nucleotide differences that define the alleles. In addition to mismatches between the 3′ terminal base of the primer and the target, single mismatches should be incorporated at several positions from the 3′ terminus. Apart from the theoretical considerations relating to the 3′ terminal position of the allele-specific primers, the design and optimization of PCR-ARMS primers follows the same considerations used for any other type of PCR. Primers are chosen to have comparable theoretical melting temperatures (T_m). Primer lengths are generally 20 nucleotides or greater, although the length is less important than the T_m. Primers should not have self-complementary sequences of 4 nucleotides or more, nor should they have more than 4 nucleotide complementarity between their 3′ ends.

As with any PCR-based strategy, false negative results could result from the presence of polymorphisms that negatively impact on the primer annealing and/or amplification. This potential problem can be overcome by targeting the opposite strand for amplification, or by incorporating a degenerate nucleotide into the primer.

For single mutation ARMS, the PCR conditions can be established by titrating the $MgCl_2$ concentration and/or primer concentrations, at constant annealing temperature. For multiplex ARMS, the first step is to optimize the PCR conditions for sensitive and specific detection of each allele. The objective is to define the PCR cycling parameters and $MgCl_2$ concentration under which all of the alleles will be amplified in an efficient and specific manner. It may be necessary to redesign one or more primer pair to achieve allele-specific amplification under one set of PCR conditions. Once the PCR parameters have been established, the primer pairs can be combined to evaluate the performance of the multiplex ARMS assay. Primer concentrations should be adjusted such that each of the alleles is amplified to a comparable degree. The specificity of the ARMS assay should

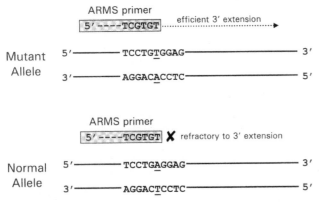

FIGURE 2.1 Schematic of PCR-ARMS. Schematic representation of a PCR-ARMS assay for the detection of a single base mutation (underlined). The 3′ terminal nucleotide of the ARMS primer is complementary to the mutant allele. The ARMS primer has an additional mismatch positioned three bases from the 3′ terminal nucleotide (not shown).

be evaluated using samples from normal controls and known carriers of the mutations. Specificity can also be tested using serial dilutions of mutant DNA mixed with normal DNA (e.g., mutant:normal = 1:1, 1:2, 1:4, 1:8, etc.). Uniform and specific amplification of each allele may require further manipulation of the cycling parameters or the concentration of one or more reagents (primers, *Taq* polymerase, MgCl$_2$).

2.2.4 Single Mutation PCR-ARMS

A common application of PCR-ARMS is the detection of individual point mutations in DNA. Primers are designed that will preferentially amplify the mutant allele, while being refractory to amplification of the normal allele. Included in the reaction mix is a second set of primers that are specific for a heterologous locus that serves as a positive control for PCR amplification. Conventional agarose or polyacrylamide gel electrophoresis systems are used to resolve the control amplicon from the mutant amplicon. Since the efficiency of amplification is inversely proportional to the length of the amplicon, the control amplicon should be larger or close in size to the mutant amplicon. Figure 2.2 illustrates the use of this approach to detect the most common *DHCR7* gene mutation associated with the Smith-Lemli-Opitz syndrome (Nowaczyk *et al.*, 2001).

This assay distinguishes between samples that are positive for the IVS 8-1 G→C mutation (heterozygous or homozygous) and those that are negative for the mutation. This is sufficient for applications where it is not necessary to distinguish between heterozygotes and homozygotes (e.g., mutation analysis of unaffected carriers of a recessive disorder). For other applications, such as mutation analysis for individuals affected with a recessive disorder, it may be desirable to genotype individuals who are positive for the mutation. This can be accomplished by screening all positive samples with a second ARMS assay that is specific for the normal allele and refractory for the mutant allele. Heterozygotes will be positive for both ARMS assays, whereas homozygotes will be positive for only the mutant ARMS assay.

The specificity of PCR-ARMS is such that pools of samples can be screened to identify rare carriers of specific mutations. This approach is capable of detecting a single positive sample in pools of 30 or more samples. This eliminates the need to test large numbers of samples individually to establish population frequencies for individual mutant alleles (Nowaczyk *et al.*, 2001; Waye *et al.*, 2002).

2.2.5 Multiplex PCR-ARMS

Many genetic disorders are characterized by a small number of common mutations that account for a significant proportion, and in some instances, the majority of mutant alleles represented in a given population. Once the spectrum of mutations and the frequencies of individual alleles have

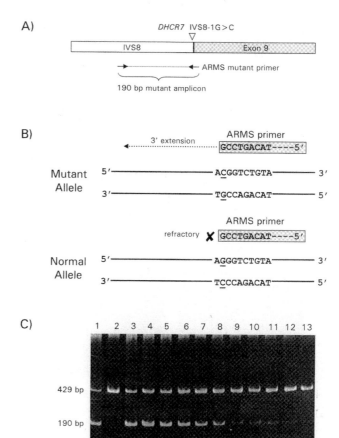

FIGURE 2.2 PCR-ARMS detection of a single *DHCR7* mutation. A. Schematic representation of the boundary between intervening sequence 8 and exon 8 of the *DHCR7* gene showing the position of a common mutation causing Smith-Lemli-Opitz syndrome. The mutation alters the canonical splice acceptor sequence (AG→AC) and can be detected using an ARMS primer specific for the mutant allele. B. PCR-ARMS detection of the IVS 8-1 G→C mutation (underlined). The 3′ terminal nucleotide of the ARMS primer is complementary to the mutant allele, and an additional mismatch is incorporated three bases from the 3′ nucleotide (not shown). C. Analysis of PCR-ARMS products by nondenaturing polyacrylamide gel electrophoresis, visualized by ethidium bromide staining and UV fluorescence. The 190 bp fragment is specific for the IVS 8-1 G→C mutant allele. The 429 bp fragment corresponds to a region of the *HFE* gene that serves as a positive internal control for PCR amplification. Lane 1: IVS 8-1 G→C heterozygote, lane 2: normal control, lane 3: 1:1 mixture of carrier: normal DNA, lanes 4–13: 1:2^n mixture (N = 1 – 10) of carrier: normal DNA, respectively (1:2 to 1:1024).

been established, PCR-ARMS can be used to simultaneously screen for the most common mutations. In practice, multiplex assays can easily be developed for single-tube detection of four to six different mutations (Old *et al.*, 1990). Figure 2.3 shows two multiplex PCR-ARMS panels that detect eleven common *DHCR7* gene mutations. Collectively, these mutations account for more than 85% of the

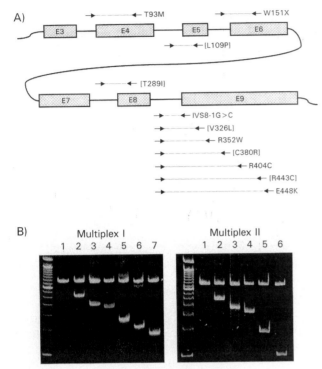

FIGURE 2.3 Multiplex PCR-ARMS detection of 11 *DHCR7* mutations. A. Schematic representation of the *DHCR7* gene (introns not drawn to scale) showing the seven coding exons and PCR-ARMS strategies for detecting 11 point mutations associated with Smith-Lemli-Opitz syndrome. The mutations are detected in two multiplex PCR-ARMS assays (multiplex II mutations indicated in brackets). B. Analysis of PCR-ARMS multiplex products. Each multiplex includes an internal control amplicon from the *HFE* gene. Multiplex I: lane 1: normal control; lane 2: E448K carrier; lane 3: R404C; lane 4: W151X; lane 5: R352W; lane 6: T93M; lane 7: IVS 8-1 G→C. Multiplex II: lane 1: normal control; lane 2: R443C carrier; lane 3: T289I; lane 4: C380R; lane 5: V326L; lane 6: L109P.

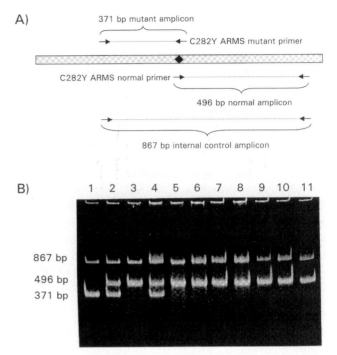

FIGURE 2.4 Genotyping of *HFE* C282Y mutation. A. Schematic representation of a portion of the *HFE* gene showing the most common mutation associated with hereditary hemochromatosis (C282Y). The multiplex analysis contains primers for two different PCR-ARMS assays, run off opposite strands. One PCR-ARMS assay is specific for the mutant allele, whereas the other assay is specific for the normal allele. The PCR products for the mutant and normal alleles have different lengths and can easily be distinguished by gel electrophoresis. B. Analysis of the PCR-ARMS products. Lane 1, C282Y homozygote; lane 2, C282Y heterozygote; lane 3, normal control; lanes 4–11, samples.

mutant alleles detected in North American Smith-Lemli-Opitz syndrome patients.

2.2.6 Genotyping with PCR-ARMS

PCR-ARMS can be used to determine genotypes for individual mutations or SNPs; that is, distinguishing heterozygotes from homozygotes. This can be accomplished using two separate ARMS assays: one specific for the mutant allele and the other specific for the normal allele. Alternatively, PCR-ARMS systems have been developed for single-tube genotyping of mutations or SNPs. The simplest system involves bidirectional amplification, with the normal allele amplified using one strand as the template and the mutant allele amplified off the complementary strand (Ye *et al.*, 2001; Waterfall and Cobb, 2001). The primers are designed such that the lengths of the amplicons can easily be resolved by conventional gel electrophoresis. This strategy has a built-in positive control resulting from amplification

between the outermost primers of the normal and mutant amplicons. Figure 2.4 shows how PCR-ARMS can be used to genotype samples for the most common mutation associated with hereditary hemochromatosis (*HFE* C282Y).

2.2.7 Haplotyping with PCR-ARMS

PCR-ARMS can be used to establish haplotypes of individuals in the absence of samples from relatives. This is particularly useful for haplotyping SNPs that are located within distances that are amenable to PCR amplification. Consider the case of adjacent bi-allelic SNPs, where the alleles are designated Aa and Bb. PCR-ARMS using the four possible combinations of ARMS primers specific for the SNP A and SNP B alleles (AB, Ab, aB, ab) can be used to establish the haplotype (Eitan and Kashi, 2002).

2.2.8 Advantages and Limitations

PCR-ARMS is ideally suited for many molecular diagnostic applications, particularly those requiring detection of relatively small numbers of point mutations and having low-to-moderate throughputs. The primary advantage of PCR-ARMS is the ease with which multiplex assays can be developed, validated, and implemented. Moreover, the tests are nonradioactive and do not require expensive and sophisticated detection systems.

The most significant limitation of PCR-ARMS is that it can be used to detect only known mutations and polymorphisms. As such, it is usually necessary to combine PCR-ARMS with other molecular diagnostic strategies (e.g., sequencing) to provide comprehensive mutation detection. In its simplest formats, such as those described in this chapter, PCR-ARMS may be impractical for applications involving large numbers of mutations or high throughput. For such applications, consider using ARMS assays with allele detection strategies that are more amenable to automation. One such approach is to use ARMS primers labeled with fluorescent dyes.

2.3 PCR-ASO

2.3.1 Origin of Method

The analysis of point mutations in DNA using hybridization with Allele-Specific Oligonucleotide (ASO) probes is based on the principle that even single nucleotide mismatches between a probe and its target can destabilize the hybrid. ASO probes can be designed to be complementary and specific for the various alleles, thus providing a simple methodology to detect any known mutation or SNP.

The use of ASO probes actually predates PCR, and was a commonly used approach to analyze cloned DNA. Radioactively labeled ASO probes have even been used to diagnose genetic disease using nonamplified genomic DNA that has been immobilized on a membrane after restriction endonuclease digestion and electrophoretic separation (Conner *et al.*, 1983; Orkin *et al.*, 1983; Pirastu *et al.*, 1983). With the advent of PCR amplification (Saiki *et al.*, 1985; Saiki *et al.*, 1988a; Saiki *et al.*, 1988b), PCR-ASO became one the first approaches used to analyze known point mutations within amplified DNA fragments (Saiki *et al.*, 1986).

2.3.2 PCR-ASO Design Concepts

The design of ASO probes is largely dependent on the sequence of the region being targeted for analysis. ASO probes are generally short oligonucleotides (15- to 17-mers) with 30–50% G+C content, designed with the discriminating nucleotide located near the middle of the probe. Longer probes can be used to compensate for regions that have low G+C content. G:T and G:A mismatches are slightly destabilizing, whereas the effect is significantly greater for A:A, T:T, C:T, and C:A mismatches (Ikuta *et al.*, 1987). Therefore, the choice of the sense or anti-sense strand may affect

specificity (e.g., a C-A mismatch is often easier to discriminate than a G-T mismatch).

The choice of ASO primer sequences can involve a considerable amount of trial-and-error testing of candidate probes. Candidate probes should be evaluated individually using known positive and negative control samples, all under fixed hybridization and wash stringency conditions. For some mutations, it may be necessary to synthesize several versions of a given probe to attain allele-specificity under the assay conditions.

Care should be taken to avoid sequences that are associated with polymorphisms. False negative results could occur if a polymorphism impairs annealing and/or extension of a primer used for amplification, or if a polymorphism lies near the mutation being tested and destabilizes the ASO/target hybrid.

2.3.3 Forward ASO Format

The forward ASO format involves immobilizing PCR-amplified DNA fragments on a nylon membrane, and hybridizing the membrane to a labeled oligonucleotide probe that is complementary and specific for a given sequence. The membrane is then washed at the appropriate stringency to dissociate any probe molecules that are not perfectly matched to the target. The first-generation PCR-ASO protocols utilized oligonucleotide probes that were phosphorylated at their 5′ termini with [γ-^{32}P], and exposure to X-ray film to detect the membrane bound probe-target

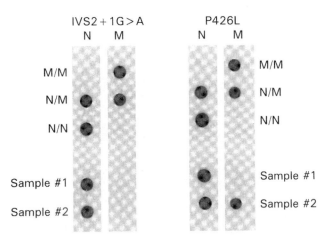

FIGURE 2.5 Forward PCR-ASO of *ARSA* mutations. PCR-ASO analysis of the two most common *ARSA* mutations (IVS2+1 G→A and P426L) causing metachromatic leukodystrophy. Corresponding regions of the *ARSA* were amplified by PCR and immobilized to nylon membranes. The membranes were hybridized to nonradioactively labeled ASO probes that are complementary to the normal or mutant alleles. Probe/target hybrids were detected by chemiluminescence. Each test strip contains amplified DNA from known controls [homozygous mutant (M/M), heterozygous mutant (N/M), normal (N/N)] and two samples.

hybrids (Saiki *et al.*, 1986). Subsequent protocols employed ASO probes that have biotin conjugated to their 5′ termini (Saiki *et al.*, 1988b). Following stringency washes, the probe-target hybrid is detected using streptavidin conjugated with horseradish peroxidase (HRP). The HRP activity can then be detected using a colorimetric detection with tetramethylbenzidine and hydrogen peroxide. Alternatively the HRP activity can be detected using chemiluminescent substrates.

The forward ASO format is most useful when large numbers of samples are being screened for a small number of mutant alleles. Figure 2.5 shows examples of forward dot-blots for the two most common *ARSA* gene mutations in patients with metachromatic leukodystrophy (Polten *et al.*, 1991).

2.3.4 Reverse ASO Format

The forward ASO format requires separate labeled probes and hybridization cycles for each allele being tested, making this technique overly cumbersome for applications involving multiple mutations. The solution to this problem was simple; immobilize an array of ASO probes to a membrane strip and hybridize the strip to PCR-amplified DNA that is labeled. The original reverse ASO format, or reverse dot-blot, employed probes that had poly(dT) tails added to their 3′ termini and were immobilized on the nylon membranes by UV cross-linking (Saiki *et al.*, 1989). The method subsequently was improved by covalent binding of the ASO probes to membranes via 5′ amino linkers (Zhang *et al.*, 1991; Chehab and Wall, 1992).

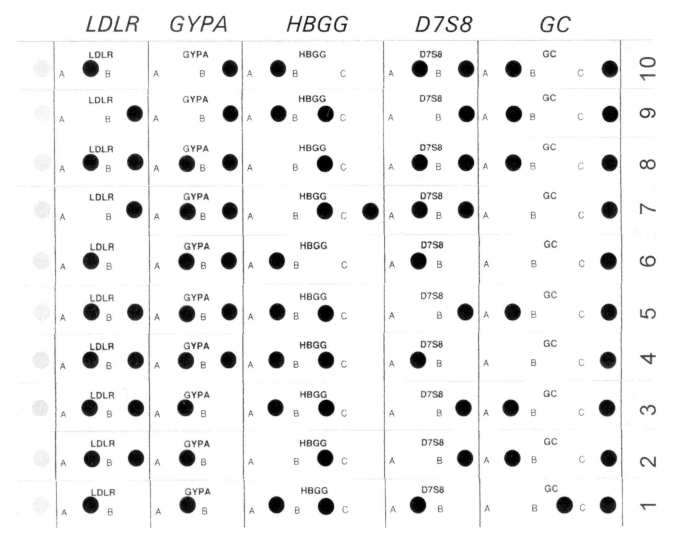

FIGURE 2.6 Reverse PCR-ASO of SNPs. PCR-ASO analysis of SNPs at five independent loci using the AmpliType PolyMarker™ reverse dot-blot system (Perkin-Elmer). The corresponding regions of the genome are amplified as a multiplex PCR using biotinylated primer sets. The PCR products are then hybridized to membrane strips containing ASO probes that are complementary to the various alleles. Biotinylated PCR products that hybridize to the membrane-bound ASO probes are detected using a nonradioactive colorimetric reaction. SNP profiles are shown for 10 unrelated individuals. A, B, C on the strips correspond to the different alleles at each locus.

Over the past decade, the reverse ASO format has become a widely used tool for routine screening of genes that have numerous mutant alleles (reviewed in Gold, 2003; see also Chapter 27). Reverse PCR-ASO test kits have evolved for mutation screening for several genetic diseases, including α-thalassemia (Chan *et al.*, 1999; Foglietta *et al.*, 2003), β-thalassemia (Chehab, 1993; Cai *et al.*, 1994) and cystic fibrosis (Chehab and Wall, 1992; Makowski *et al.*, 2003). Test kits also have been developed for genotyping SNPs used for forensic identity testing (Budowle *et al.*, 1995; see also Chapter 21). Figure 2.6 shows commercial ASO reverse dot-blots for SNPs associated with five independent loci.

2.3.5 Advantages and Limitations

The PCR-ASO method, and particularly the reverse ASO format, provides a simple approach for simultaneous genotyping of large numbers of mutations and polymorphisms. The method can be applied to any known point mutation or SNP, is nonradioactive, and does not require specialized instrumentation to detect the alleles.

A potential drawback to the PCR-ASO strategy is the amount of developmental work needed to identify a panel of oligonucleotide probes that are allele-specific under the same hybridization and wash stringency conditions. For small laboratories with limited resources, this initial investment may preclude the development of in-house PCR-ASO assays. For the same reasons, it is unlikely that commercial PCR-ASO will become available for rare diseases having limited market potential.

References

Ayyadevara, S., Thaden, J. J., and Shmookler Reis, R. J. (2000). Discrimination of primer 3′-nucleotide by Taq DNA polymerase during polymerase chain reaction. *Anal. Biochem.* 284, 11–18.

Budowle, B., Lindsey, J. A., DeCou, J. A., Koons, B. W., Giusti, A. M., and Comey, C. T. (1995). Validation and population studies of the loci LDLR, GYPA, HBGG, D7S8, and GC (PM loci), and HLA-DQα using a multiplex amplification and typing procedure. *J. Forensic Sci.* 40, 45–54.

Cai, S. P., Wall, J., Kan, Y. W., and Chehab, F. F. (1994). Reverse dot blot probes for the screening of β-thalassemia mutations in Asians and American Blacks. *Hum. Mutat.* 3, 59–63.

Chan, V., Yam, I., Chen, F. E., and Chan, T. K. (1999). A reverse dot-blot for rapid detection of non-deletion α thalassaemia. *Br. J. Haematol.* 104, 513–515.

Chehab, F. F., and Wall, J. (1992). Detection of multiple cystic fibrosis mutations by reverse dot blot hybridization: a technology for carrier screening. *Hum. Genet.* 89, 163–168.

Chehab, F. F. (1993). Molecular diagnostics: Past, present, and future. *Hum. Mutat.* 2, 331–337.

Conner, B. J., Reyes, A. A., Morin, C., Itakurs, K., Teplitz, R. L., and Wallace, R. B. (1983). Detection of sickle cell S-globin allele by hybridization with synthetic oligonucleotides. *Proc. Natl. Acad. Sci. USA* 80, 278–282.

Eitan, Y., and Kashi, Y. (2002). Direct micro-haplotyping by multiple double PCR amplifications of specific alleles (MD-PASA). *Nucleic Acids Res.* 30, e62.

Foglietta, E., Bianco, I., Maggio, A., and Giambona, A. (2003). Rapid detection of six common Mediterranean and three non-Mediterranean α-thalassemia point mutations by reverse dot-blot analysis. *Am. J. Hematol.* 74, 191–195.

Gold, B. (2003). Origin and utility of the reverse dot-blot. *Expert Rev. Mol. Diagn.* 3, 143–152.

Huang, M.-M., Arnheim, N., and Goodman, M. F. (1992). Extension of base mispairs by Taq DNA polymerase: implications for single nucleotide discrimination in PCR. *Nucleic Acids Res.* 20, 4567–4573.

Ikuta, S., Tagaki, K., Wallace, R. B., and Itakura, K. (1987). Dissociation kinetics of 19 base paired oligonucleotide-DNA duplexes containing different single mismatched base pairs. *Nucleic Acids Res.* 15, 797–811.

Kwok, S., Kellogg, D. E., McKinney, N., Soasic, D., Goda, L., Levenson, C., and Sninsky, J. J. (1990). Effects of primer-template mismatches on the polymerase chain reaction: human immunodeficiency virus type 1 model studies. *Nucleic Acids Res.* 18, 999–1005.

Makowski, G. S., Nadeau, F. L., and Hopfer, S. M. (2003). Single tube multiplex PCR detection of 27 cystic fibrosis mutations and 4 polymorphisms using neonatal blood samples collected on Guthrie cards. *Ann. Clin. Lab. Sci.* 33, 243–250.

Newton, C. R., Graham, A., Heptinstall, L. E., Powell, S. J., Summers, C., Kalsheker, N., Smith, J. C., and Markham, A. F. (1989). Analysis of any point mutation in DNA. The amplification refractory mutation system (ARMS). *Nucleic Acids Res.* 17, 2503–2516.

Nichols, W. C., Liepieks, J. J., McKusick, V. A., and Benson, M. D. (1989). Direct sequencing of the gene for Maryland/German familial amyloidotic polyneuropathy type II and genotyping by allele-specific enzymatic amplification. *Genomics* 5, 535–540.

Nowaczyk, M. J., Nakamura, L. M., Eng, B., Porter, F. D., and Waye, J. S. (2001). Frequency and ethnic distribution of the common DHCR7 mutation in Smith-Lemli-Opitz syndrome. *Am. J. Med. Genet.* 102, 383–386.

Okayama, H., Curiel, D. T., Brantly, M. L., Holmes, M. D., and Crystal, R. D. (1989). Rapid nonradioactive detection of mutations in the human genome by allele-specific amplification. *J. Lab. Clin. Med.* 114, 105–113.

Old, J. M., Varawalla, N. Y., and Weatherall, D. J. (1990). Rapid detection and prenatal diagnosis of β-thalassaemia: studies in Indian and Cypriot populations in the UK. *Lancet* 336, 834–837.

Orkin, S. H., Markham, A. F., and Kazazian, H. H., Jr. (1983). Direct detection of the common Mediterranean beta-thalassemia gene with synthetic DNA probes: an alternate approach for prenatal diagnosis. *J. Clin. Invest.* 71, 775–779.

Pirastu, M., Kan, Y. W., Cao, A., Conner, B. J., Teplitz, R. L., and Wallace, R. B. (1983). Prenatal diagnosis of β-thalassemia: detection of a single nucleotide mutation in DNA. *N. Engl. J. Med.* 309, 284–287.

Polten, A., Fluharty, A. L., Fluharty, C. B., Kappler, J., von Figura, K., and Gieselmann, V. (1991). Molecular basis of different forms of metachromatic leukodystrophy. *N. Engl. J. Med.* 324, 18–24.

Saiki, R. K., Scharf, S., Faloona, F., Mullis, K. B., Horn, G. T., Erlich, H. A., and Arnheim, N. (1985). Enzymatic amplification of β globin sequences and restriction site analysis for diagnosis of sickle cell anemia. *Science* 230, 1350–1354.

Saiki, R. K., Bugawan, T. L., Horn, G. T., Mullis, K. B., and Erlich, H. A. (1986). Analysis of enzymatically amplified β-globin and HLA-DQα DNA with allele-specific oligonucleotide probes. *Nature* 324, 163–166.

Saiki, R. K., Gelfand, D. H., Stoffel, S., Scharf, S. J., Higuchi, R., Horn, G. T., Mullis, K. B., and Erlich, H. A. (1988a). Primer-directed enzymatic amplification of DNA with a thermostable DNA polymerase. *Science* 239, 487–491.

Saiki, R. K., Chang, C.-A., Levenson, C. H., Warren, T. C., Boehm, C. D., Kazazian, H. H., Jr., and Erlich, H. A. (1988b). Diagnosis of sickle cell anemia and β-thalassemia with enzymatically amplified DNA and nonradioactive allele-specific oligonucleotide probes. *N. Engl. J. Med.* 319, 537–541.

Saiki, R. K., Walsh, P. S., Levenson, C. H., and Erlich, H. A. (1989). Genetic analysis of amplified DNA with immobilized sequence-specific oligonucleotide probes. *Proc. Natl. Acad. Sci. USA* 86, 6230–6234.

Sarkar, G., Cassady, J., Bottema, C. D. K., and Sommer, S. S. (1990). Characterization of polymerase chain reaction amplification of specific alleles. *Anal. Biochem.* 186, 64–68.

Sommer, S. S., Cassady, J. D., Sobell, J. L., and Bottema, C. D. (1989). A novel method for detecting point mutations or polymorphisms and its application to population screening for carriers of phenylketonuria. *Mayo Clin. Proc.* 64, 1361–1372.

Sommer, S. S., Groszbach, A. R., and Bottema, C. D. (1992). PCR amplification of specific alleles (PASA) is a general method for rapidly detecting known single-base changes. *Biotechniques* 12, 82–87.

Waterfall, C. M., and Cobb, B. D. (2001). Single tube genotyping of sickle cell anaemia using PCR-based SNP analysis. *Nucleic Acids Res.* 29, e119.

Waye, J. S., Nakamura, L. M., Eng, B., Hunnisett, L., Chitayat, D., Costa, T., and Nowaczyk, M. J. (2002). Smith-Lemli-Opitz syndrome: carrier frequency and spectrum of DHCR7 mutations in Canada. *J. Med. Genet.* 39, e31.

Wu, D. Y., Ugozzoli, L., Pal, B. K., and Wallace, R. B. (1989). Allele-specific enzymatic amplification of beta-globin genomic DNA for diagnosis of sickle cell anemia. *Proc. Natl. Acad. Sci. USA* 86, 2757–2760.

Ye, S., Dhillon, S., Ke, X., Collins, A. R., and Day, I. N. M. (2001). An efficient procedure for genotyping single nucleotide polymorphisms. *Nucleic Acids Res.* 29, e88.

Zhang, Y., Coyne, M. Y., Will, S. G., Levenson, C. H., and Kawasaki, E. S. (1991). Single-base mutational analysis of cancer and genetic diseases using membrane-bound modified oligonucleotides. *Nucleic Acids Res.* 19, 3929–3933.

CHAPTER **3**

Competitive Oligopriming

AGLAIA ATHANASSIADOU

Department of General Biology, Faculty of Medicine, School of Health Sciences, University of Patras, Patras, Greece

3.1 INTRODUCTION

Competitive oligonucleotide priming of DNA synthesis has been described for the first time by Gibbs and colleagues in 1989. It is a strategy for the detection of known point mutations or single base polymorphisms, based on allele-specific amplification, by the use of allele-specific oligonucleotides (ASO, see also Chapter 2). However, there is a fundamental difference between the usual methods involving allele-specific amplification (Nollau and Wagener, 1997) and competitive oligonucleotide priming: in the former, mismatching between primer and template DNA prevents extension of DNA synthesis, whereas in the competitive oligopriming system mismatching prevents primer annealing. This is true because in the allele-specific amplification methods, the mismatching is formed at the 3′ end of the primer, but in the competitive oligopriming, mismatching occurs within the primer, usually in the middle. The oligonucleotides that detect DNA sequence alterations in competitive oligopriming are a pair of synthetic short DNA sequences (competitive oligoprimers or COP primers), carrying the mutant or the normal configuration at the mutation site in the middle of their sequence and capable of discriminating between mutated and normal template DNA. Thus, competitive oligopriming is a system of allele-specific amplification, through differential primer annealing. Differential primer annealing in competitive oligopriming is achieved by the use of three primers of DNA synthesis instead of two, in one polymerase chain reaction. The pair of the competitive oligonucleotides are used as forward primers and the third primer is serving as a common reverse one, for the respective PCR. Thus, allele-specific amplification occurs for both alleles in the same reaction, which is the important feature of the method.

Once the competitive amplification of DNA has been completed, detection of the mutant versus the normal amplified allele among the products of a competitive PCR is rendered crucial. To this end a number of approaches have been applied, involving differential labeling of the competitive primers prior to their use in competitive PCR.

The system has been applied for the detection of mutations in the α- and β-globin genes and the CMV genome (Chehab and Kan, 1989), and in the diseases β-thalassemia (Athanassiadou *et al.*, 1995) and post-succinylcholine apnea (Yen *et al.*, 2003).

In later years, competitive allele-specific PCR also has been made possible with the use of competitive primers forming the mismatch at their 3′ end that prevents DNA synthesis (Germer *et al.*, 1999; Koch *et al.*, 2000; Giffard *et al.*, 2001; Myakishev *et al.*, 2001; McClay *et al.*, 2002). (See Table 3.1.) This approach is a development that combines features from the competitive oligopriming and the amplification refractory mutation system (ARMS, see also Chapter 2).

3.2 THE COMPETITIVE OLIGOPRIMING ASSAY

The competitive oligopriming assay (COP assay) is carried out in two consecutive stages. The focus of the first stage is on the DNA amplification by competitive oligopriming; the second stage involves the detection of the genetic identity of the amplified material, with reference to the sequence alteration under investigation. Detailed presentation of the two stages is given in the following paragraphs.

3.2.1 DNA Amplification *in vitro* by Competitive Oligopriming

3.2.1.1 PRINCIPLE

The classic ASO hybridizations approach for the detection of single base changes in amplified DNA fixed on solid support makes use of a pair of synthetic oligonucleotides, consisting of one oligonucleotide for the mutated sequence and one for the corresponding normal sequence. It is based on two separate, single-oligonucleotide hybridizations in stringent conditions, followed by comparative estimation of the results.

If each of the two allele-specific oligonucleotides is used in a DNA synthesis reaction with template DNA containing the respective sequence, it can also serve as an efficient primer for DNA synthesis. Furthermore, in ASO hybridizations there is usually a residual hybridization with mismatching oligonucleotides; that is, when the mutated oligonucleotide is hybridized to the normal template DNA and vice versa. This is also the case with priming DNA synthesis. Usually both allele-specific oligonucleotides can drive DNA synthesis of the normal and the mutated template, albeit not with the same efficiency.

If the conditions are formulated, in which both allele-specific oligonucleotides—the mutated and the normal one—are used in the same DNA synthesis reaction, then a competition arises between them for binding on the target template sequence and in that course, the binding of the perfectly matched primer to the template DNA is strongly favored relative to the primer differing by a single base. This primer competition for annealing to the template DNA results in the "correct"—the fully matched—oligonucleotide driving DNA synthesis.

In the ASO setting, the DNA around the mutation site refers to both strands of the double stranded DNA and for each strand a normal and a mutated sequence can be specified. So, on the whole, four kinds of allele-specific oligonucleotides are possible—two normal and two mutated ones—and any pair of them containing a normal and a mutated sequence may be used in allele-specific hybridizations, whether they derive from the same strand of DNA or one oligonucleotide from one strand of DNA and the other oligonucleotide from the other strand. However, when such oligonucleotides are used to drive DNA synthesis (e.g., as forward primers in a PCR), they must have the appropriate orientation to facilitate DNA amplification in concert with the reverse primer used in the reaction. So far, only one (common) reverse primer has been used in a competitive oligopriming PCR and in this case, both allele-specific primers must derive from the same (opposite) strand of DNA sequence. One nevertheless can formulate a competitive PCR, in which the competitive oligoprimers are

of opposite direction, but this necessitates the use of two reverse primers, too, and most probably, elaborate optimization procedures.

3.2.1.2 COMPETITIVE OLIGOPRIMING SYSTEM

Conventionally, two complementary reactions are required to determine the zygosity of DNA in a two-allele system, with allele-specific amplification, each reaction containing one of the two-allele specific primers and a common reverse primer. In competitive oligopriming all three primers are used in one reaction (see Fig. 3.1) with a given template DNA, namely the common reverse primer and the two

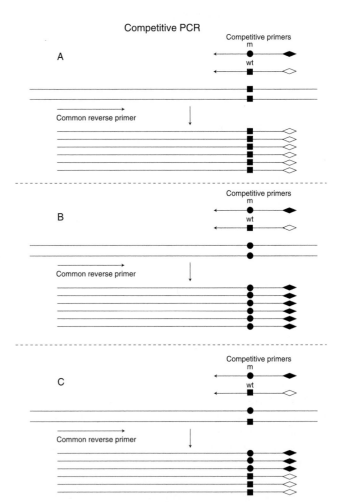

FIGURE 3.1 The three possible outcomes of a competitive PCR using a pair of allele-specific competitive primers: dark dots (m): mutant, and dark squares (wt): normal; as well as a common reverse primer, with template DNA A: wild type (dark squares only), B: homozygous mutant (dark dots only), and C: heterozygous (dark squares and dark dots). Competitive PCR products in each case show correct priming as a result of efficient competition between the two primers for correct annealing on the template DNA. Dark and empty romvoids at the end of primers and PCR products represent differential labeling for the discrimination between normal and mutated amplified DNA.

primers for the mutation site, one carrying the mutated sequence (m) and the other carrying the normal counterpart (wt), both complementary to the same template DNA strand and functioning as forward primers.

In this setting competitive primers (m) and (wt) anneal only or mainly with the mutated and normal DNA, respectively, promoting correct priming and not vice versa, which would result in "mismatch" priming. The results of competitive oligopriming in PCR fall into three categories, depending on the genotype of the template DNA and judged by the kind of competitive primer that has incorporated into the amplified DNA.

- Amplified DNA contains only (or mainly) the normal primer (wt). Then the template DNA used is wild-type for the mutation under investigation.
- Amplified DNA contains only (or mainly) the mutated primer (m). Then the template DNA used is homozygous for the mutation under investigation.
- Amplified DNA contains both m and wt primers in equal (or almost equal) amounts. Then the template DNA used is heterozygous for the mutation site under investigation.

It is clear that by determining the identity of the incorporated competitive primer (mutant versus normal) in the respective PCR product, one obtains conclusive evidence for the genetic constitution of the template DNA, at the mutation site under investigation.

3.2.1.3 PARAMETERS AFFECTING THE EFFICIENCY OF COMPETITIVE OLIGOPRIMING

Gibbs and colleagues (1989) have shown that successful competitive oligopriming is dependent on *primers' length*. Competition experiments between oligoprimers of different length have shown that 12-mers bear a greater potential for correct priming than 20-mers. In the recent work of McClay and colleagues (2002), the use of 13-, 15-, 18-, and 19-mers with mismatch at the 3′ end, showed that 18-mers were most efficient in competition. However, the precise relationship between the length of competitive primers and their discriminatory potential depends on their overall sequence, as well as on the individual base mismatch.

Another important parameter in competitive oligopriming is the *annealing temperature*. In the work of Gibbs and colleagues (1989), effective competition occurs at low stringency for shorter types of primers (12-mers). In the author's experience (Athanassiadou *et al.*, 1995; and unpublished data) with 19-mers in the detection of β-globin gene mutations, effective competition can be achieved only in higher temperatures, close to high stringency conditions employed when the respective oligonucleotides are used in ASO hybridization experiments. Recently (Yen *et al.*, 2003), this system was applied for the detection of mutations at the butylcholinesterase (BCHE) gene with 16-mers used at annealing temperatures equal to their Tm. Finally, in the pre-

viously mentioned work of McClay and colleagues (2002), fairly stringent annealing temperatures were the optimum to use for the 18-mers' competition, among a range of the different length oligoprimers.

Primer concentration is another parameter to be considered. With an excess primer to template ratio, the correct primer will compete 100-fold better than the mispriming one, even in the presence of 100-fold excess of the latter (Gibbs *et al.*, 1989). In the author's laboratory, increasing primer concentration from 25 to 100 pmoles in low stringency reaction had practically no effect, whereas it was important, and in one instance crucial, in high stringency reactions (Athanassiadou *et al.*, 1995).

The *mismatch position* within the competitive oligoprimers, whether in the middle or in any other position along the oligonucleotide sequence, may be significant but no data are available on this issue. Considering the *nature of the mismatch* in the initial description of the technique by Gibbs and colleagues (1989), successful competition was shown for base changes that generate the mismatched A/A, G/G, G/A, T/C, C/A, and T/G between primer and template DNA.

3.2.2 Detection Systems

A detection system of the genetic identity of the competitive PCR products is necessary and it relies on the possibility for differential detection of the identity (mutant versus normal) of the incorporated competitive primer. This can be achieved by differential labeling of the two competitive primers, as the size of the PCR products is the same in all three cases A, B, and C (see Fig. 3.1). The two competitive primers must be labeled in a way such that not only can the mutant be discriminated against the normal in the respective homozygous states, but also the heterozygote wt/m, represented in a mixture of equal amounts of the two primers, can be recognized as a separate state from the other two.

Methods that have been employed so far include various approaches: A color complementation assay was developed by Chehab and colleagues (1989), which allows discrimination between fluorescent oligonucleotide primers. Differential end labeling of the competitive oligoprimers with compounds that are recognized by different antibodies also has been applied (Athanassiadou *et al.*, 1995). The addition of a 5′ GC tail to one of the primers so that it can be distinguished by Tm shift was employed by Germer and colleagues (1999), and allele-specific PCR with universal energy transfer labeled primers was developed by Myakishev and colleagues (2001).

3.3 APPLICATION OF COP ASSAY FOR β-THALASSEMIA MUTATIONS

The COP assay has been applied for the detection of known mutations in the human β-globin gene that causes β-thalassemia, using, for every mutation, 19-mers as competitive oligoprimers, as well as a reverse common primer (Athanassiadou *et al.*, 1995). The choice of 19-mers was based on the possibility of using the same diagnostic oligonucleotides in COP assay, as well as in ASO hybridizations, maximizing their potential. Development of the COP assay was carried out in the following steps. First, single COP oligoprimer PCRs were performed in order to test the respective level of mismatch priming of DNA synthesis, followed by COP PCR, in which both competitive primers as well as the common reverse primer were used in the same reaction. Finally, detection of the COP PCR products was carried out.

3.3.1 COP PCRs

In single COP primer PCRs, a series of six PCRs was carried out for each mutation under study, so that each of the two COP primers was used with all three genetic constitutions of template DNA, namely the wild-type (n/n), the heterozygous (m/n), and the homozygous mutant (m/m).

Mismatch priming is observed in all cases of single COP primer reactions. Specifically, there is extensive mispriming at annealing temperatures of 10°C lower than the Tm of the competitive oligoprimers (low stringency) and low degree of mispriming at annealing temperatures close to the primers' Tm. The lowest degree of mispriming is observed in high stringency PCRs, that is one or two degrees C below the primers' T_m. There is also a variation of mispriming even within the same pair of competitive primers, and this is evident in Fig. 3.2 for mutation IVSI-110 G→A, where the mutant COP-primer shows a low degree of mispriming with the wild-type template DNA, and the normal COP-primer

shows a mismatch priming equal to the correct priming of the mutant COP-primer with the homozygous mutant template DNA. Competitive oligoprimers for mutation IVSI-6 T→C show a much lower, but by no means absent, mismatch priming.

In COP PCRs, a series of three reactions are carried out for each mutation, using template DNA of the three different genetic constitutions just given (see Fig. 3.3a). COP PCRs carried out at low annealing temperatures (low stringency) show very poor competition of the 19-mer COP oligoprimers for correct priming. As annealing temperature increases, competition for correct priming becomes more efficient and attains discriminatory potential at high stringency.

3.3.2 Detection of the COP PCR Products

Detection of the identity (normal versus mutant) of the COP primers incorporated in each case is carried out by means of differential 5′ labeling of the COP primers (mutant primer with dansylchloride and normal primer with FITC (Athanassiadou *et al.*, 1995)) that were subsequently recognized by specific antibodies on a solid support. In this system the common reverse primer is biotinylated so as to facilitate the formation of a conjugate on the amplified DNA.

Three sets of results are shown (see Fig. 3.3b), two sets for mutation IVSI-110 G→A (I and II) and one for mutation IVSI-6 T→C (III). The two sets of mutation IVSI-110 G→A differ from each other only by 1°C in the annealing temperature; that is, case I is carried out at 1°C and case II at

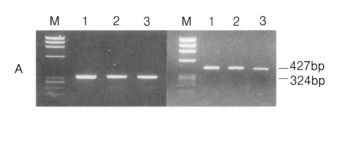

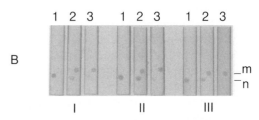

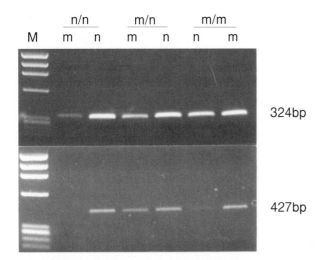

FIGURE 3.2 Results of single COP primer for two β-thalassemia mutations, IVSI-110 G→A upper panel and IVSI-6 T→C lower panel, both at high stringency; m: mutant, n: normal. Intense correct priming is evident but also a variable degree of mismatch priming.

FIGURE 3.3 A. Competitive PCRs with (1) wild type, (2) heterozygous, and (3) homozygous mutant template DNA, for mutations IVSI-110 G→A (left) and IVSI-6 T→C (right). B. Detection of the mutant (m) and normal (n) products of the competitive PCRs for mutation IVSI-110 G→A (I and II) and IVSI-6 T→C (III), all at primer annealing temperatures at or close to T_m (high stringency).

TABLE 3.1   Genetic loci for which the competitive oligopriming approach for mutation detection was applied. The position of the mismatch within COP primers, and the potential of multiplexing and/or high throughput screening are recorded. HPRT: Hypoxanthine Phosphoribosyltransferase; BCHE: Butyrylcholinesterase; ALDH2: Aldehyde dehydrogenase; PON: Paraoxonase/arylesterase.

Gene	Mismatch position	Multiplexing	High throughput	References
HPRT	Middle	No	No	Gibbs *et al.*, 1989
α-, β-globin	Middle	Yes	No	Chehab *et al.*, 1989
β-globin	Middle	Yes	No	Athanassiadou *et al.*, 1995
PON-Apolipo-protein B	3′ end	Yes	Yes	Germer *et al.*, 1999
ALDH2	3′ end	No	No	Koch *et al.*, 2000
	3′ end	No	Yes	McClay *et al.*, 2002
MCAD Factor V	3′ end	No	No	Giffard *et al.*, 2001
Various SNPs	3′ end	No	Yes	Myakishev *et al.*, 2001
BCHE	Middle	No	No	Yen *et al.*, 2003

2°C below high stringency. It is noticeable that a very low degree of mispriming persists for the n primer for mutation IVSI-110 G→A, like the practically negligible level of cross-hybridization of these primers when used in ASO hybridization, but this does not interfere with interpretation of results.

Therefore, it can be concluded that detection of β-thalassemia mutations is efficient and reliable with the COP assay applied. The COP assay, as exemplified here, is a robust and reliable method of mutation detection.

3.4 LIMITATIONS OF THE COP ASSAY

The main difficulty of the method, as it has been implemented in the case of β-thalassemia mutations and a few other cases so far, arises in the detection system, which is easy to use and reliable but not readily available. In general, the detection system has been a problem, as there is a need to differentiate between equal size DNA molecules, but of different genetic identity, produced in one PCR. This difficulty has been interfering with the widespread use of competitive oligopriming.

Second, COP assay is applicable exclusively with DNA changes whose precise nature is known in advance, so that the COP oligoprimers can be designed. This is an important obstacle, when the method is compared with methods in which prior knowledge of the nature of the mutation is not an absolute requirement, like Denaturing Gradient Gel Electrophoresis (DGGE, see Chapter 8), or Single Strand Conformation Polymorphism (SSCP) analysis (see Chapter 6), which can be applied for known and unknown DNA sequence changes.

Finally, an aspect that makes COP assay difficult to apply is the rather long and laborious optimization procedure that is required, in order for every mutation, to specify the exact

conditions for a COP PCR. This includes the elimination of mispriming and of improper differential detection of PCR products, which may interfere with interpretation of the results.

3.5 MULTIPLEX COP ASSAY

True multiplexing refers to the simultaneous DNA amplification of more than one genetic site, for the purpose of performing multiple analysis on one DNA sample. Such need arises in genetic diseases caused by multiple mutations, for example β-thalassemia. In the original COP assay (Gibbs *et al.*, 1989) it was stated that multiple oligonucleotides could be used to simultaneously test different DNA regions, but the maximum number of oligonucleotides that could be employed in a single reaction has not yet been established. Multiplex COP assay is dependant on the system by which the genetic identity of the COP PCR products is determined. Thus, when the length of the COP PCR products is different for the different pairs of COP primers used, it is possible to determine the identity of the different products by performing gel electrophoresis. However, a less complex detection method than gel electrophoresis is usually desirable.

A color complementation assay as a detection system was developed by Chehab and colleagues (1989) for the simultaneous amplification of DNA loci for two β-thalassemia mutations and an α-globin gene deletion (five primers in total) that were simultaneously analyzed in one-tube reaction. Each of the five primers was conjugated to a different dye and the fluorescence of each dye was scored in a fluorometer.

When a discrimination system for various PCR products, generated in a one-tube multiplex PCR, is not available, multiplex COP assay is possible only as a system of a number of COP PCR tubes, designed to operate in one PCR machine with the same PCR program. In the case of β-thalassemia mutation detection (Athanassiadou *et al.*, 1995), screening for three different mutations was possible in three COP assays performed concurrently in the same machine with the same PCR program. The labeling for the mutant COP primer was the same in all three cases, as was the labeling for the normal one; this was subsequently recognized by the respective antibodies, fixed on a predetermined position on a solid support.

The main feature that should be pointed out concerning both these multiplex assays is their requirement for the same annealing temperature for efficient competition of all COP primer pairs used. This is perhaps due to the fact that in both cases the COP primers were 19-mers. Shorter primers are expected to compete efficiently in not just one annealing temperature value, but rather in a spectrum of temperatures (albeit in lower stringency). It may be also possible that several COP primer pairs could be used in one PCR, provided there is an appreciable overlap among their respective annealing temperature for efficient competition. The multiplex system in this case would be more tolerant with respect to the annealing temperatures of the individual COP PCRs, than with the longer COP primers.

Evidently, every disease entity caused by multiple mutations needs its own analysis, which will determine the conditions for simultaneous detection of as many mutations as possible.

3.6 DEVELOPMENT OF COMPETITIVE OLIGOPRIMING WITH A 3′ END MISMATCH

The COP assay described thus far makes use of competitive oligoprimers that carry the mutation site in the middle of their length, and the mismatch prevents primer annealing in a competitive setting.

Recently, competitive oligopriming was applied in systems in which the two forward primers (mutant and normal) are designed to form the mismatch at their 3′ end, thus promoting differential DNA synthesis rather than differential primer annealing.

Such a system was used by Koch and colleagues (2000) for the purpose of allele association studies. The detection of the genetic identity of the competitive PCR products was by individual fluorescent labeling of the competitive primers at the 5′ end, which was then detected on an automated fluorescent DNA sequencer.

Another system, developed by Germer and colleagues (1999) and termed T_m-shift genotyping, is a single-tube assay that uses two forward primers each containing a 3′-terminal base that corresponds to one of the two single nucleotide polymorphism (SNP) allelic variants, a common reverse primer, and a fluorescent dye that detects double-stranded DNA. In this system, detection of primer identity within competitive PCR products is based on the addition of a GC-tail in one of the allele-specific primers. This results in a shift in the Tm of the PCR product that has incorporated the primer with the GC tail. Interestingly similar systems have been reported for high-throughput SNP genotyping; they are presented later.

The question of the efficiency, in terms of accuracy and specificity, of the competitively primed versus the conventional single allele-specific DNA amplification was addressed by Giffard and colleagues (2001). In a simulation study, it has been argued that although conventional allele-specific amplification has somewhat higher inherent specificity than competitive oligopriming reaction, it actually may be easier to optimize the latter ones to offer greater reproducibility and tolerance to alterations in target amounts without any significant loss of specificity. This was supported by the observation that specificity may be increased by adopting a conventional COP strategy.

3.7 HIGH-THROUGHPUT APPROACHES

Since it has become apparent that SNPs are the most common source of human genetic variation, there is a need for high-throughput SNP genotyping by reliable and cost-effective methods.

Two high-throughput approaches involving competitive oligopriming have been developed lately, both using competitive oligoprimers with the mismatch at their 3′ end. One of them (Myakishev *et al.*, 2001) involved PCR amplification of genomic DNA with two-tailed allele-specific primers, which introduce priming sites for universal energy-transfer labeled primers. The accuracy of the method is greater in conditions of competitive oligopriming than when single allele-specific primer is used, thus supporting the work of Giffard and colleagues (2001), with little if any non-specific signal recorded. In addition to that, the method shows high sensitivity with small amounts (0.4 ng) of genomic DNA per reaction required and is cost effective to the extent that one set of energy-transfer primers can be used for all SNPs analyzed. The method as presented is tailored to meet the requirements of testing many samples for one SNP rather than testing many SNPs at one go, as amplification reaction for various SNPs needs to run in multiple thermal cyclers and the fluorescence image of each SNP assay needs to be recorded individually. A second approach on high throughput SNP genotyping (McClay *et al.*, 2002) is using competitively a pair of allele-specific primers also with a 3′ end mismatch and a fluorescent label at their 5′ end, so that PCR products can be detected by conventional automatic sequencer. The SniPTag used in this method is capable of discriminating among all combinations of nucleotide substitutions. A very interesting feature of this approach is that it is possible to increase throughput by loading the PCR products for several markers-loci in a single lane on the automated sequencer, provided that these products have been designed to be of different length.

Neither of the two previously mentioned high-throughput methods includes a setting for true multiplexing—the simultaneous detection of more than one SNP—due to the narrow optimum annealing temperature for each marker.

3.8 CONCLUSIONS

Competitive oligopriming is a method with a comparably high degree of accuracy and specificity, which has not been fully exploited yet, presumably because the detection systems are somewhat elaborate and need special equipment. The development of high-throughput systems brings this method forward for the purpose of large-scale SNP genotyping that concern genome scans and DNA diagnostics of genetic, acquired, and infectious diseases.

The formulation of multiplex PCRs, in which two or more mutations in the same or different genetic loci can be analyzed simultaneously, is desirable. This implies that a number of primers have to be optimized to function in the same PCR conditions. Thus, optimization of primers to this direction prior to their use in large-scale experiments is absolutely necessary, with most important parameters being the length of the primer and the annealing temperature.

Competitively primed PCR seems to show higher tolerance in terms of the range of annealing temperature when the original method is used, in which the mismatch is designed to be in the middle of the COP primers' sequence, than with the later methods where the mismatch occurs at the 3′ end of the competitive primers. This is shown by the fact that the only multiplex attempts so far are presented within the context of differential primer annealing in COP PCR. In the original work of Gibbs and colleagues (1989), efficient competition was obtained with 12-mer to 16-mer oligoprimers in low stringency. It is, therefore, worthwhile to combine this framework with current, easier, and more available detection systems of the genetic identity of COP PCR products, for the purposes of formulating a true multiplex, competitive oligopriming system.

References

Athanassiadou, A., Papachatzopoulou, A., and Gibbs R. A. (1995). Detection of genetic analysis of β-thalassemia mutations by competitive oligopriming. *Hum. Mutat.* 6, 30–35.

Chehab, F. F., and Kan, Y. W. (1989). Detection of specific DNA sequences by fluorescence amplification: a color complementation assay. *Proc. Natl. Acad. Sci. USA* 86, 9178–9182.

Germer, S., and Higuchi, R. (1999). Single-tube genotyping without oligonucleotide probes. *Genome Res.* 9, 72–78.

Gibbs, R. A., Nguyen, Phi-Nga, and Caskey, T. C. (1989). Detection of single DNA base differences by competitive oligonucleotide priming. *Nucleic Acids Res.* 17, 2437–2448.

Giffard, P. M., McMahon, J. A., Gustafson, H. M., Barnard, R. T., and Voisey J. (2001). Comparison of competitively primed and conventional allele-specific nucleic acid amplification. *Anal. Biochem.* 292, 207–215.

Koch, H. G., McClay, J., Loh, E.-W., Higuchi, S., Zhao, J.-H., Sham, P., Ball, D., and Craig, I. W. (2000). Allele association studies with SSR and SNP markers at known physical distances within a 1Mb region embracing the ALDH2 locus in Japanese, demonstrates linkage desequilibrium extending up to 400 kb. *Hum. Mol. Genet.* 9, 2993–2999.

McClay, J. L., Sugden, K., Koch, H. G., Higuchi, S., and Craig, I. W. (2002). High-throughput single-nucleotide polymorphism genotyping by fluorescent competitive allele-specific polymerase chain reaction (SniPTag). *Anal. Biochem.* 301, 200–206.

Myakishev, M. V., Khripin, Y., Hu, S., and Hamer, D. H. (2001). High-throughput SNP genotyping by allele-specific PCR with universal energy-transfer-labeled primers. *Genome Res.* 11, 163–169.

Nollau, P., and Wagener, C. (1997). Methods for detection of point mutations: performance and quality assessment. *Clin. Chem.* 43, 1114–1128.

Soukup, G. A., Ellington, A. D., and Maher, L. J. (1996). Selection of RNAs that bind to duplex DNA at neutral pH. *J. Mol. Biol.* 259, 216–228.

Yen, T., Hightingale, N., Burns, J. C., Sullivan, D. R., and Stewart, P. M. (2003). Butyrylcholinesterase (BCHE) genotyping for post-succinylcholine apnea in an Australian population. *Clin. Chem.* 49, 1297–1308.

CHAPTER 4

Oligonucleotide Ligation Assays for the Diagnosis of Inherited Diseases

EEVA-LIISA ROMPPANEN
Department of Clinical Chemistry, Kuopio University Hospital, Kuopio, Finland

TABLE OF CONTENTS

4.1 INTRODUCTION

The ability of DNA ligases to join short oligonucleotides covalently to each other has been utilized in a number of assays where the ligation of oligonucleotides in some way reflects the genotype of the target DNA. The joining of two oligonucleotide probes by DNA ligase is dependent on three events: the probes have to be hybridized to the target DNA, the probes must lie directly adjacent to one another in a 5′ to 3′ orientation, and the probes must have perfect base-pair complementarity with the target DNA at the site of their join. The oligonucleotide ligation assay (OLA) takes advantage of these three events resulting in reliable, accurate, and reproducible discrimination of sequence variants, which is essential in clinical molecular genetic testing. The reaction can be monitored by introducing a detectable function in either of the probes in the ligation reaction. Several different analytical techniques are in use, differing from each other in their detection methods and instrumentation needs.

Several applications of OLA for the detection of molecular defects causing human diseases are in use in molecular diagnostic laboratories on a global scale. Due to the fact that numerous mutations can contribute to inherited diseases, there is often a need to scan many loci simultaneously on a patient sample. The clinical application of polymerase chain reaction-OLA (PCR-OLA) in greatest use is the multiplex analysis for the most common mutations causing cystic fibrosis. The best throughput of PCR-OLA has been achieved using biosensor chips enabling the detection of hundreds of sequence variants within one assay. However, before this and the other new high-throughput PCR-OLA methodologies can be used for molecular diagnostics in clinical laboratories, more research and thorough assay validation is needed.

4.2 HISTORY OF OLIGONUCLEOTIDE LIGATION ASSAYS

DNA ligases are found in all organisms, and are important components of the natural repair mechanism for DNA. These enzymes have several roles *in vivo*, and lately they have become an important enzymatic tool in molecular biology laboratories. They catalyze the synthesis of a phosphodiester bond between directly adjacent 3′-hydroxyl and 5′-phosphoryl groups of DNA segments. The enzymes act only when the DNA segments are perfectly hybridized to a complementary DNA sequence. Even a single base pair mismatch between two strands decreases significantly the efficiency of this enzyme, thus preventing the ligation (Wu and Wallace, 1989).

In 1988, this enzymatic reaction was applied to ligate short single-stranded DNA probes as a means to detect sequence variants (Alves and Carr, 1988; Landegren *et al*,

1988). This was the beginning of the development of several analytical methods utilizing DNA ligases, for example, OLA and ligase chain reaction (LCR). Thermostable DNA ligase became available commercially in 1990 and enabled the temperature cycling of the ligation reaction. In contrast to PCR or LCR, in OLA only the target DNA serves as a template throughout the rounds of amplification since the oligonucleotide probes used in the reaction are complementary solely to one strand of the target DNA. Due to that, linear amplification of the ligation product is achieved in the ligation reaction.

The breakthrough in OLA began in 1990 when PCR was coupled to the assay prior to ligation reaction (Nickerson *et al.*, 1990). PCR amplification of the target DNA enhanced the sensitivity of the assay, enabling nonradioactive detection of OLA results. Further enhancement was achieved by thermostable DNA ligase-mediated cycling of the ligation reaction. These improvements made this assay more useful for DNA-based diagnostics of inherited diseases in the clinical laboratories. Nowadays the method is used in numerous clinical and research laboratories worldwide for the detection of a wide variety of medically relevant mutations. There are also commercially available genotyping kits based on PCR-OLA for the detection of mutations of medical interest.

4.3 PRINCIPLE OF THE OLIGONUCLEOTIDE LIGATION ASSAY

The specificity and fidelity of both mesophilic and thermostable DNA ligases have been utilized in several genetic assays developed for the detection of single base substitutions, small insertions, and deletions. In these assays, the ligation of the probes provides a highly specific detection of sequence variants permitting a convenient way for distinction between gene alleles.

In most applications, the relevant gene fragment is first amplified in a PCR, although in some applications genomic DNA has been directly subjected to the ligation reaction. The template DNA for OLA is denatured to form a single-stranded target DNA. Two adjacent oligonucleotide probes (approximately 20-mers) and DNA ligase enzyme are needed for each ligation reaction. The oligonucleotides for OLA are designed so that the juxtaposition of the adjacent ends occurs at a previously identified mutation site on the template. The probes are hybridized to the target DNA in a head-to-tail (5′ to 3′) orientation. If the probes and the target DNA are perfectly complementary to each other, the DNA ligase enzyme forms a phosphodiester bond between the probes. If there is even a single base pair mismatch on either side or close to the nick between the probes, the efficiency of the enzyme is significantly decreased and the ligation prevented; therefore it can be distinguished from a perfect match (see Fig. 4.1; Wu and Wallace, 1989; Pritchard and

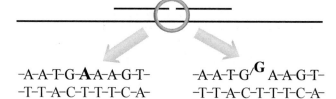

-A-A-T-G-**A**-A-A-G-T-   -A-A-T-G-/ᴳ-A-A-G-T-
-T-T-A-C-T-T-T-C-A-   -T-T-A-C-T-T-T-C-A-

FIGURE 4.1 Principle of oligonucleotide ligation. Template DNA corresponding to the wild-type (WT) allele is denatured to form a single-stranded target DNA. Allele-specific and common probes are hybridized to the target DNA. The WT allele-specific and the common probes are ligated to each other by DNA ligase enzyme since the template DNA is complementary to the WT allele-specific probe (left). No ligation occurs between the mutant allele-specific and the common probe, since there is a mismatch between the template DNA and the mutant allele-specific probe (right).

Southern, 1997). It is essential for DNA ligase that the probe-to-target hybrids are sufficiently long to accommodate the footprint of the enzyme used. In general, 20 base pairs around a nick are satisfactory for most DNA ligases. Furthermore, ligation reactions should be performed at a temperature where all probes hybridize stringently to their complementary sequence (Jarvius *et al.*, 2003).

Three synthetic oligonucleotide probes are used for the detection of one bi-allelic sequence variant. One allele-specific probe is needed for each allele to be detected. In the case of a point mutation, these allele-specific probes differ from each other only by a single base complementary to either the wild-type (WT) allele or the mutant allele, where, as in cases of a deletion or an insertion, the sequence of allele-specific probes may differ more from each other. The third, common probe then hybridizes to the target DNA sequence immediately adjacent to the allele-specific probes and is completely complementary to either allelic target. Although the most commonly used enzymes, T4 DNA ligase and Tth DNA ligase, are more discriminating toward mismatches at the 3′-end compared to the 5′-end of the junction, the site of the mutation can be on either site of the nick (Jarvius *et al.*, 2003). In Fig. 4.1, the site of the mutation is situated at the 3′-end of the allele-specific probe and the ligation occurs between the 3′-end of the allele-specific probe and the 5′-end of the common probe.

If a thermostable DNA ligase is used in the reaction, the denaturation and the ligation steps can be cycled to achieve a linear amplification of the ligated probes, thus enhancing the sensitivity of the assay. The use of thermostable enzyme also enables one to carry out the ligation reactions at high temperatures, which further increases the specificity of the ligation. Furthermore, if the oligonucleotides used for the PCR amplification of the target DNA differ by their melting temperature from the oligonucleotides used for the ligation,

PCR and ligation can be combined into a single reaction (Eggerding, 1995; Chen *et al.*, 1998). In such a case, the temperature profile of the reaction cycles initially favors the amplification of the target DNA and then subsequently the oligonucleotide ligation due to the higher annealing temperature of PCR primers than OLA probes. Combined PCR and ligation diminish the workload and decrease the risk of sample-to-sample contamination and therefore may facilitate the use of OLA in clinical laboratories.

4.4 DETECTION METHODS FOR LIGATED OLIGONUCLEOTIDES

Either an allele-specific or a common probe in a ligation reaction, or both of them, have to be constructed to enable the resolution and detection of ligated probes from non-ligated probes. In the first report describing OLA, a radioactive label (^{32}P) was used to label the common oligonucleotide and the detection was based on autoradiography of immobilized probes (Landegren *et al.*, 1988). However, several different nonisotopic detection systems based on colorimetry, fluorometry, time-resolved fluorometry, and chemiluminometry have been described and these are more convenient for routine use in clinical laboratories.

The assay formats underlying these detection methods can be divided into three groups: assays based on solid-phase capture of the probes, assays based on electrophoretical separation of the probes, and assays based on homogenous detection requiring no separation step (see Table 4.1).

A separate ligation reaction can be set up and run for both the WT and the mutant allele to confirm the absence or presence of both of the alleles. In that case, a label or some modifications can be introduced, either at the allele-specific or at the common probe. Oligonucleotide ligation assays that enable the detection of several variants in one ligation reaction have been introduced for the diagnosis of inherited diseases. In those applications it is necessary to distinguish between allele-specific probes.

A choice between different detection methods in each laboratory is dependent on many factors such as available instrumentation in the laboratory, number of samples to be analyzed, and number of genetic variations to be analyzed per sample, as well as costs and the need for speed, simplicity, and robustness.

4.4.1 Assays Based on Solid-Phase Capture of Ligated Probes

One of the oligonucleotide probes used for ligation reaction can be attached directly to a solid-phase support, for example, paramagnetic particle or biosensor ship (Martinelli *et al.*, 1996; Zhong *et al.*, 2003). However, a more common approach is to capture the probe on solid-phase by hybridi-

TABLE 4.1 Assay formats and detection methods used in OLA.

Assay format	Detection method	Alleles detected per reaction[a]
Solid-phase capture on		
Membranes	Autoradiography	1
Microplate wells, pins, or particles	Colorimetry, chemiluminescence, or time-resolved fluorescence measurement in microplate reader	1–3
Microspheres	Fluorescence in flow cytometer	>10
Biosensor chips	Colorimetric detection by digital camera	>300
Electrophoresis		
Polyacrylamide gel or capillary	Fluorescence in DNA sequencer	>30
Homogenous assays		
Solution-phase homogenous	Fluorescence resonance energy transfer in real-time PCR apparatus	1–2

[a] Number of alleles or genetic variants that can be detected in a single ligation reaction by using differently modified probes.

zation or streptavidin-biotin interaction (Nickerson *et al.*, 1990; Iannone *et al.*, 2000). Streptavidin and avidin are proteins that bind biotin with very high affinity, and the binding is virtually irreversible.

Several detection methods utilizing 96- or 386-well microplates have been published. The signal detection on a microplate can be performed by colorimetry, time-resolved fluorometry, or chemiluminometry. Most of these methods utilize indirect detection; that is, the labeled probes are recognized by antibodies, and the enzymes that are conjugated to the antibodies catalyze the formation of a colored or chemiluminogenic product.

4.4.1.1 COLORIMETRIC DETECTION ON MICROPLATE

The principle of colorimetric enzyme-linked immunosorbent assay (ELISA) based detection (Nickerson *et al.*, 1990) for the detection of WT and mutant alleles in two different ligation reactions is described in Fig. 4.2. A biotin molecule at the 5′-end of one oligonucleotide acts as a chemical hook capturing this probe on a streptavidin-coated microplate well. The PCR product and the unligated probes are washed away by a denaturing wash solution. In the case of a successful ligation, the digoxigenin- or fluorescein-labeled

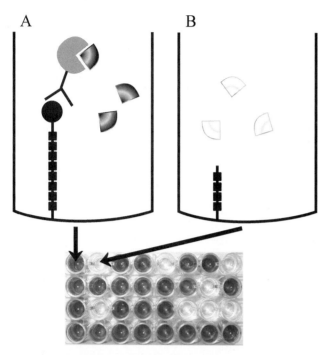

FIGURE 4.2 Colorimetric ELISA in the detection of ligated probes. Biotinylated allele-specific probes are captured on streptavidin-coated microplate wells, the enzyme-coupled antibody (⤚ᐸ) is attached to the label (•) in the common probe, and finally substrate (▱) for the enzyme is added to the well. In reaction A, a colored signal is detected due to the ligation of the WT allele-specific probe and the common probe, whereas in reaction B no colour is detected.

reporter probe is covalently attached to the biotinylated probe and therefore remains in the well. Antidigoxigenin or antifluorescein antibodies coupled with an enzyme (e.g., alkaline phosphatase or horseradish peroxidase) are added into the well. After the incubation and washes, an enzyme-specific substrate is added into the well. The presence of the labeled reporter is seen as a color formation within the well, whereas in the case of unsuccessful ligation, no color is detected (see Fig. 4.2). If a sample is from a normal individual (homozygous for the WT allele), the color is formed only in the reaction where a WT allele-specific probe was used. If a sample is from a homozygous case for the mutant allele, color is formed in the mutant allele-specific reaction. A heterozygous sample is distinguished by the colored end product in both of the ligation reactions, respectively.

If two different haptens—for example, digoxigenin and fluorescein—are used to label allele-specific probes, both alleles can be detected in a single ligation reaction. In this kind of a procedure, the biotinylated common probe is captured into a streptavidin-coated well and the presence of the covalently ligated allele-specific oligonucleotides is detected sequentially using enzyme-coupled antibodies and appropriate enzyme substrates (Tobe *et al.*, 1996).

4.4.1.2 CHEMILUMINOMETRIC DETECTION ON MICROPLATE

Chemiluminescence is the light emission that occurs when chemically excited molecules return to the electronic ground state. ELISA-based detection can utilize enzyme-specific substrates that are converted to molecules producing chemiluminescence (Tannous *et al.*, 2003). Alternatively, the detection can be obtained by another interaction between molecules; for example, binding of avidin with a chemiluminogenic molecule to biotin (Tannous *et al.*, 2003) or by labeling one of the probes directly with a molecule producing a chemiluminescence signal; for example, acridinium esters (Martinelli *et al.*, 1996). If the allele-specific probes are labeled with molecules that differ from one another in their light-emission kinetics, the analysis of both WT and mutant alleles in a single microplate well is possible.

4.4.1.3 TIME-RESOLVED FLUOROMETRY ON MICROPLATE

The third possibility for the detection of ligation products captured to streptavidin-coated microplate wells is time-resolved fluorometry produced by lanthanide-labeled probes. Samarium (Sm), europium (Eu), and terbium (Tb) ions emit narrow-peaked fluorescence of specific wavelength when they coordinate to specific ligands. Lanthanide complexes have long fluorescent lifetimes (several hundreds microseconds) compared to traditional fluorophores. Due to this property, the complexes can be excited by a flash of light (flash excitation) and the fluorescence can be measured after a delay time. This procedure reduces various backgrounds caused by fluorescent molecules of short lifetimes and enables highly specific and sensitive detection. In PCR-OLA, WT and mutant allele-specific probes are differently labeled with europium and samarium. The biotinylated probes are captured to the streptavidin-coated wells and after denaturing washes, the ligation products remaining in the wells are quantified by time-resolved fluorescence measurements (Hansen *et al.*, 1995). Biotinylated probes can also be captured on a streptavidin or avidin-coated 96-pin capture manifold in which the avidin is captured on a polystyrene support designed to fit into individual microplate wells (Samiotaki *et al.*, 1994). Pins can be washed extensively and transferred to microplate wells in which lanthanide ions are released at low pH. After the removal of the pin manifold, the fluorescence of lanthanide chelates is quantitated.

4.4.1.4 FLUOROMETRIC DETECTION ON MICROSPHERES

Sample throughput in PCR-OLA with detection on microplates is limited; therefore this methodology is suitable only in laboratories where the sample-load is reasonable. One way to increase the throughput of PCR-OLA is flow cytometry, a methodology designed primarily for cellular analysis. Flow cytometers can detect particles in the lower micron range, including inert microspheres of different sizes dyed with various fluorochromes. For the detection of ligated probes by flow cytometry, the common probes are labeled with fluorescein at the 3′-end and every allele-specific probe has a tail of a unique 25-base sequence at the 5′-end. After a ligation reaction, capture probes are hybridized to the tails of the allele-specific probe. Capture probes are attached to fluorescent latex microspheres that can be identified individually by their different ratios of red and orange fluorescence. After the hybridization, the microspheres are washed, suspended, and subjected to flow cytometric analysis in which the fluorescence of each individual microsphere is determined. If the ligation reaction has occurred then both green fluorescence signal of fluorescein and red or orange signal of the fluorescence of microsphere are detected, but in a case of no ligation, only the fluorescence of the microsphere will be detected (Iannone *et al.*, 2000). Flow cytometer can differentiate dozens of microspheres from each other and therefore can be used for the detection in the multiplex oligonucleotide ligation assay.

4.4.1.5 COLORIMETRIC DETECTION OF BIOSENSOR CHIPS

Recently, a novel methodology utilizing optical thin-film biosensor chips and colorimetric detection has been described, which raises the assay's throughput to an even higher level (Zhong *et al.*, 2003). In this methodology, the ligation reactions are performed on a biosensor chip surface into which the allele-specific probes are covalently attached. The common probes are biotinylated and the colored signal for successful ligation is developed by antibiotin antibodies labeled with horseradish peroxidase and a suitable substrate for the enzyme. The ligation results on the biosensor chips can be visualized via digital images. Several hundreds of sequence variants can be detected on a single chip. The chip dimension allows for the placement of chips into 96-well microplates. This enables the use of robotic pipetting devices in the process, increasing overall throughput of the assay. Few reports describing DNA microarray methods for multiplex fluorescent detection of genomic variations also have been described although no reports of their use in clinical diagnostics or carrier screening have been published so far (Gunderson *et al.*, 1998; Gerry *et al.*, 1999).

4.4.2 Electrophoretic Assays

One alternative for the detection of ligated probes is the observation of change in the probe size upon ligation. The ligation products are resolved from each other and from the nonligated probes under denaturing conditions on high-resolution polyacrylamide gels or capillary electrophoresis (see Chapter 7) by their unique electrophoretic mobility. The detection method can be either autoradiography or fluorescence, depending on the label used in the common probes. Electrophoretic mobility of the ligation product is modified by using allele-specific probes of different lengths (Day *et*

al., 1995) or by attaching a non-nucleotide tail to the allele-specific probe (Grossman *et al.*, 1994). Only a few sequence variants per ligation reaction can be detected when a single label is used in the common probes. However, dozens of genetic variations have been scanned in a single ligation reaction by tagging the common probes by one of three fluorescent dyes (Brinson *et al.*, 1997). The combination of a particular electrophoretic mobility with the specific fluorescent dye provides a unique address for each allelic ligation product. When detected on an automated four-color DNA sequencer, the fourth fluorescent tag is used for the molecular size standard. This allows precise scaling of electrophoretic mobility correcting interrun and interlane variation. The genotype assignment can be automated via computer analysis but validation of the genotype data must be assigned by a qualified diagnostician.

4.4.3 Homogenous Assays

In homogenous assays, there is no necessity to remove or resolve nonligated probes from the covalently ligated probes. A detectable function of one probe can be linked to a retrievable function of some other probe through ligation. Fluorescence resonance energy transfer (FRET) is observed when two florescent dyes are in close proximity and emission spectrum of the first dye (the donor) overlaps with the excitation spectrum of the second dye (the acceptor). In such cases, the emission energy of the donor is transferred to the acceptor, resulting in quenching of the fluorescence of the donor and concomitantly the increase of the fluorescence emission of the acceptor (see Fig. 4.3). In PCR-OLA, the ligation of two probes, one of which is labeled with the donor and the other with the acceptor, brings the probes so close to each other that FRET is detected. Three dye-labeled

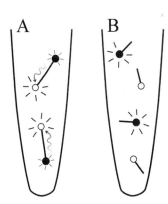

FIGURE 4.3 Fluorescence energy resonance transfer (FRET) in the detection of ligated probes. Ligation of the probe is detected real-time in PCR apparatus. In reaction A, fluorescence of the WT allele-specific probe (acceptor probe) is detected since the WT allele-specific probe and the common probe are in close proximity to each other due to the successful ligation. In reaction B, only the fluorescence of the common probe (donor probe) is detected.

ligation probes are needed for each mutation site if one wishes to detect both alleles of a biallelic sequence variant in the same reaction. 5'-Carboxyfluorescein (FAM) can be used as donor dye in the common probe, whereas different rhodamine dyes (ROX and TAMRA) can be used as acceptor dyes in the allele-specific probes. The PCR and ligation reactions are coupled together in the same reaction tube using a two-stage thermal cycling protocol. During the first stage, the PCR is preferred due to the high annealing temperature, but during the second stage the ligation reaction is favored due to the lower annealing temperature. Fluorescence intensities are monitored in real-time at the annealing temperature by a PCR apparatus with a fluorescence spectrophotometer (Chen *et al.*, 1998).

FRET also has been utilized in a recently described method combining OLA to rolling circle amplification. In this method, only one probe containing an allele-specific sequence at one end and the common sequence at the other end is used for the detection of each allele. No label is attached to this probe. The sequence of the probe is designed so that its hybridization to the target sequence brings the 5'- and 3'-ends next to each other. Ligation of these ends produces a circular probe. Successfully circularized probes are detected by amplification in the presence of two amplification primers. The first primer hybridizes to its complementary region on the circularized probe. The primer is amplified by strand-displacing DNA polymerase, resulting in the generation of a single-stranded concatamer of the ligated probe. The second amplification primer hybridizes to each tandem repeat of the original probe. This primer contains a 5'-hairpin loop that is labeled with a donor and an acceptor. During elongation of the primer, new recognition site for the first primer is exposed, resulting in the opening of the hairpin and detection of the fluorescence of the donor (see Fig. 4.4; Pickering *et al.*, 2002).

4.5 ADVANTAGES AND LIMITATIONS OF OLIGONUCLEOTIDE LIGATION ASSAYS

PCR-OLA reliably distinguishes heterozygous and homozygous cases from normal individuals. The specificity of the sequence discrimination is based on DNA ligase enzyme that accurately detects the mismatches between the target DNA and the oligonucleotides. The use of thermostable DNA ligase improves the specificity of the reaction since the ligation can be performed at higher temperatures, thus increasing the annealing stringency (Barany, 1991). The assay can be performed over a broad range of ligation temperatures, salt concentrations, and DNA ligase concentrations without any loss of sensitivity (Hansen *et al.*, 1995). This eliminates the risk of false-negative or false-positive reactions due to variations in these factors. Amplification of the signal is obtained by using thermostable DNA ligase to cycle the ligation reaction, and also in the ELISA format by

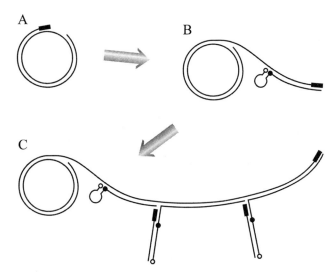

FIGURE 4.4 Detection of the ligated probes by rolling circle amplification and FRET. A circular probe formed in the ligation serves as a target in the amplification. The first primer (━) hybridizes to the circularized probe and initiates the amplification (A). While the primer is elongated by strand-displacing DNA polymerase, a single-stranded concatamer of the ligated probe is formed. The second primer (♉) containing a complementary sequence for the amplification product and a 5'-hairpin structure, in which there is a fluorescent dye in both ends, hybridizes to each tandem repeat of the original probe (B). When the hairpin structure is present FRET is detected due to the close proximity of two fluorescent dyes. However, during the elongation of the second primer, new recognition site for the first primer is exposed. While this primer is elongated, the hairpin loop is opened and only the fluorescence of the donor is detected in the reaction (C).

amplifying the formation of a colored end product, which also further enhances the sensitivity of PCR-OLA. Although radioactive labels are more sensitive, the use of nonradioactive labels is preferable, offering the advantages of safe handling, stable reagents, and the possibility for multiplex analysis (Landegren *et al.*, 1988).

Colorimetry, chemiluminometry, and time-resolved fluorometry based detection of PCR-OLA can be performed in the microplate format by immobilizing one of the probes directly or by biotin-streptavidin interaction into a well, a particle, or a pin. Since the covalent bond joining the probes is stable, extensive washes of the wells or particles can be performed, thus reducing the nonspecific signal. One major disadvantage of these applications is that they require multiple phases. Fortunately, the microplate format enables automation of the pipetting and washing steps. The automation diminishes human intervention, thus reducing the error rate and permitting faster and more efficient processing of samples. The automated interpretation of the results is enabled by the numeric values measured for each reaction. The numeric readings can be translated automatically to genotypes of the samples by calculating a numeric ratio of the WT and the mutant reaction for each sample. The translation of these ratios is based on the method-dependent threshold settings achieved during method validation in each laboratory. For example, in PCR-OLA with colorimetric detection, a ratio of >5 is interpreted as a normal individual, 0.5–2 as a heterozygote for the mutation, and a ratio of <0.2 as a mutant homozygote (threshold settings in Kuopio University Hospital). Samples with ratios outside these limits normally are subjected to reanalysis. When the interpretation of the assay results is based on specific software, it is independent from interference by users.

PCR-OLA can be multiplexed for the detection of several alleles or loci in the same reaction. ELISA-based detection of fluorescein- and digoxigenin-labeled oligonucleotides has been used to detect two alleles in the same reaction (Tobe *et al.*, 1996). With most commonly used lanthanide labels (europium, samarium, and terbium), three alleles or three loci can be detected simultaneously (Samiotaki *et al.*, 1994). Further multiplexing in the range of tens of sequence variants per reaction can be achieved by electrophoretic separation of ligation products containing fluorescent tags. The electrophoretic separation can be enhanced by using nucleotide or non-nucleotide tails as mobility modifiers. However, polyadenosine tails may interact with the amplified target causing a mismatch ligation (Day *et al.*, 1995); non-nucleotide tails are essentially inert in their interaction with target DNA (Grossman *et al.*, 1994). Although the rate-limiting step of the electrophoresis is the capacity of the analyzer, that is, an automated DNA sequencer, this methodology is a good choice in cases where several genetic variations are to be analyzed in each sample. Multiple fluorescently labeled probes required for the assay raise the analytical costs. Yet, when compared to other PCR-based methods (minisequencing and 5′-nuclease assay), PCR-OLA is the least expensive (Aydin *et al.*, 2001).

DNA microarrays with fluorometric or colorimetric detection of the ligated probes have been developed (Gunderson *et al.*, 1998; Gerry *et al.*, 1999; Zhong *et al.*, 2003). These assay formats enable detection of dozens or hundreds of sequence variants in one assay. However, before these methods are used for clinical diagnostics or carrier screening of inherited diseases more research and better validation of the assays are needed.

OLA applications using PCR amplicons as the target DNA suffer similar limitations as other diagnostic methods incorporating PCR amplification. The PCR amplification step necessary to achieve sensitive genotyping is the principal factor that limits the throughput of PCR-OLA. Multiplex PCR amplification is difficult to carry out in a reproducible manner owing to the generation of spurious amplification products. However, due to the specificity of DNA ligase, PCR-OLA enables correct genotyping even when nonspecific products or primer-dimers have been formed during PCR amplification (Delahunty *et al.*, 1995; Zotz *et al.*, 1997). In multiplex PCR reactions, all amplicons usually are not amplified with the same efficacy. The sensitivity of PCR-OLA enables genotyping of poorly amplified samples, but despite that the possibility of allele drop-out must be taken into account when interpreting PCR-OLA results from multiplex assays. Furthermore, the risk of contamination must be minimized by taking adequate precautions. The performance of the PCR amplification, the ligation reaction, even detection of the ligated probes in the same tube (Chen *et al.*, 1998) can reduce the manual work and sample handling, thus minimizing the risk of PCR carryover contamination.

4.6 PCR-OLA APPLICATIONS FOR THE DIAGNOSIS OF INHERITED DISORDERS

The first report describing OLA referred to its use in the diagnosis of sickle cell anaemia (MIM 603903), a haemoglobinopathy caused by a point mutation in the human β-globin gene (Landegren *et al.*, 1988). Since then OLA has been used for prenatal screening of this disease (Day *et al.*, 2002). In addition to this application, OLA has been used to detect a wide variety of medically relevant mutations causing inherited diseases (see Table 4.2) or mutations contributing to a common disease. Some of these mutations are of interest in large populations; for example, mutations causing sickle cell anaemia and cystic fibrosis. PCR-OLA applications also have been developed for the detection of mutations occurring in restricted populations; for example, mutations causing inherited diseases enriched into the Finnish population.

Most of the OLA applications are developed and validated in-house, but several commercial genotyping kits are also available for the diagnosis of inherited disorders. Furthermore, there are several kits for the assessing the risk of an individual for a particular disease—kits for detecting the common genetic variants of clotting factors II (prothrombin) and V, as well as methylene tetrahydrofolate reductase, which predisposes an individual to thrombophilia. Even these commercial kits have to be validated for clinical use in each laboratory.

4.6.1 Cystic Fibrosis

Cystic fibrosis (CF, MIM 602421) is the most common autosomal recessive disorder in the Caucasian population, with a reported frequency of 1 in 2,500 to 3,300 live births (Richards *et al.*, 2002). It is characterized by viscous mucus in the lungs with involvement of digestive and reproductive systems and sweat glands. The cystic fibrosis transmembrane conductance regulator gene (*CFTR*) is located on chromosome 7 and encodes for a chloride ion channel regulating chloride ion conductance. More than 900 mutations have been reported in the *CFTR* gene and the severity of CF depends on the type and location of the mutation.

TABLE 4.2   PCR-OLA applications for the detection of molecular defects causing inherited diseases.

Inherited disease	Cause of a disease	References
α1-antitrypsin deficiency	Lack of α1-antitrypsin activity	Samiotaki *et al.*, 1994
Aspartylglycosaminuria (AGU)	Lack of glycosylasparaginase activity	Delahunty *et al.*, 1995
Beta thalassemia	Decreased production of β-globin chain	Chen and Kwok, 1999
Sickle cell anaemia	Deformation of red cells	Landegren *et al.*, 1988; Day *et al.*, 2002
Congenital nephrosis of Finnish type (CNF)	Defect in membrane protein	Romppanen and Mononen, 2000
Cystic fibrosis	Defective chloride ion transport	Brinson *et al.*, 1997
Familial hypercholesteremia	Impaired function of low-density-lipoprotein (LDL) receptor or defect in apolipoprotein B-100	Baron *et al.*, 1996
Hereditary haemochromatosis	Defective iron metabolism	Feder *et al.*, 1996
Infantile neuronal ceroid lipofuscinosis (INCL)	Lack of palmitoyl-protein thioesterase activity	Romppanen *et al.*, 1998b
MCAD deficiency	Dysfunction in fatty acid oxidation	Romppanen *et al.*, 1998a
Steroid 21-hydroxylase deficiency	Defect in cortisol and aldosterone biosynthesis	Day *et al.*, 1995

The major mutation in *CFTR* gene, deletion of three base pairs resulting in the loss of phenylalanine (delta F508 mutation), accounts for 30–88% of CF chromosomes worldwide. However, no other single mutation accounts for more than 5% of CF chromosomes in most populations, and most mutations are found at frequencies of less than 1% (Richards *et al.*, 2002). There are ethnic discrepancies in the mutation spectrum and laboratories providing genetic testing for CF should be aware of mutation frequencies in their testing population. Due to the diversity of mutations, multiplex analysis, which can detect large numbers of potential mutations in the one run, is appropriate for molecular diagnosis, carrier screening, and neonatal screening for CF. One approach to achieve high throughput is multiplex PCR followed by multiplex allele-specific oligonucleotide ligation and fluorescent electrophoresis. This method has been used to simultaneously detect 29 normal and 31 mutant alleles that together account for the most frequent CF mutant chromosomes (Brinson *et al.*, 1997). PCR-OLA has been used successfully for neonatal screening together with a biochemical test with no false negative results reported in the screening (sensitivity of the combined assay 100%; Padoan *et al.*, 2002). There are also commercially available kits based on these multiplex assays for PCR-OLA to perform the assay. Standards and guidelines including technique-specific guidelines have been published for *CFTR* mutation testing (Richards *et al.*, 2002).

4.6.2 Medium-Chain Acyl-CoA Dehydrogenase Deficiency

Medium-chain acyl-CoA dehydrogenase (MCAD) deficiency (MIM 201450) is an autosomal recessive disorder of mitochondrial fatty acid oxidation. It is characterized by fasting intolerance, recurrent episodes of hypoglycemic coma, and hepatic dysfunction. The acute episodes can be fatal, although after the diagnosis, the prognosis is generally good if preventive treatment is carefully managed. The estimated prevalence of MCAD deficiency in Caucasian populations varies from 1 in 6,400 to 18,800 (Roe and Coates, 1995). Although several mutations have been described in the *MCAD* gene in chromosome 1, a single point mutation causing a lysine to glutamic acid (K304E) accounts for about 90% of the disease-causing alleles in Caucasian patients substitution (Matsubara *et al.*, 1990). About 80% of the patients are homozygous for this mutation, and most of the remaining 20% are compound heterozygotes. The high prevalence of only one mutation means that molecular techniques are particularly useful in the diagnosis of MCAD deficiency and in the detection of asymptomatic carriers of the disease (Romppanen *et al.*, 1998a).

4.6.3 Hereditary Haemochromatosis

Hereditary haemochromatosis (HH, MIM 235200) is an autosomal recessive disease of iron metabolism that results in excessive iron deposition in a variety of organs and leads to dysfunction of many organs. The disorder is common in individuals of Northern European descent with an incidence of 1 : 200 to 400. A point mutation in the *HFE* gene in chromosome 6 resulting in a cysteine to tyrosine substitution (C282Y) is found in the majority of homozygous HH patients. Another point mutation causing a histidine to aspartic acid substitution (H63D) is found in some patients, although the contribution to HH is less clear (Feder *et al.*, 1996). Since many complications of haemochromatosis cannot be reversed, early diagnosis and treatment with scheduled phlebotomy are essential. The diagnosis of HH is based on clinical, biochemical, and histological evidence of iron overload and molecular testing of C282Y and H63D mutations. About 90% of the patients are either homozygotes for the C282Y mutation or compound heterozygotes for the mutations C282Y and H63D. PCR-OLA methods have been described for the detection of both mutations (Feder *et al.*, 1996) and PCR-OLA based on colorimetric ELISA is also commercially available. The identification of *HFE* mutations allows a better screening of HH and provides the opportunity to characterize more precisely other iron overload syndromes where the *HFE* gene implication has been controversial.

4.6.4 Finnish Disease Heritage

A group of over 30 mainly recessively inherited diseases that have a higher frequency in Finland than in any other population are collectively referred to as the Finnish disease heritage. These disorders have become enriched into the Finnish population mainly due to the genetic drift maintained by national and local isolation of the population. Furthermore, although a spectrum of different mutations is identified in non-Finnish patients, one or only a few founder mutations are found in nearly all disease alleles in Finland (Peltonen, 1997). In Kuopio University Hospital, PCR-OLA with colorimetric detection has been used for diagnosis and carrier detection of aspartylgaucosaminuria (AGU, MIM 208400), infantile neuronal ceroid lipofuscinosis (INCL, MIM 256730), and congenital nephrotic syndrome of the Finnish type (CNF, MIM 256300) (Delahunty *et al.*, 1995; Romppanen *et al.*, 1998b; Romppanen and Mononen, 2000). The diagnostic tests have been performed for confirmation of the diagnosis of symptomatic patients and for prenatal diagnostic testing in cases with positive family history. Furthermore, PCR-OLA has been used for detection of the heterozygous carriers of individuals with positive family history. For large scale carrier screening, the throughput of PCR-OLA in microplate may not be sufficient. In such cases, novel array-based technologies could open up new prospects.

References

Alves, A. M., and Carr, F. J. (1988). Dot blot detection of point mutations with adjacently hybridising synthetic oligonucleotide probes. *Nucleic Acids Res.* 16, 8723.

Aydin, A., Baron, H., Bahring, S., Schuster, H., and Luft, F. C. (2001). Efficient and cost-effective single nucleotide polymorphism detection with different fluorescent applications. *Biotechniques* 31, 920–928.

Barany, F. (1991). Genetic disease detection and DNA amplification using cloned thermostable ligase. *Proc. Natl. Acad. Sci. USA* 88, 189–193.

Baron, H., Fung, S., Aydin, A., Bahring, S., Luft, F. C., and Schuster, H. (1996). Oligonucleotide ligation assay (OLA) for the diagnosis of familial hypercholesterolemia. *Nat. Biotechnol.* 14, 1279–1282.

Brinson, E. C., Adriano, T., Bloch, W., Brown, C. L., Chang, C. C., Chen, J., Eggerding, F. A., Grossman, P. D., Iovannisci, D.

M., Madonik, A. M., Sherman, D. G., Tam, R. W., Winn-Deen, E. S., Woo, S. L., Fung, S., and Iovannisci, D. A. (1997). Introduction to PCR/OLA/SCS, a multiplex DNA test, and its application to cystic fibrosis. *Genet. Test.* 1, 61–68.

Chen, X., and Kwok, P. Y. (1999). Homogeneous genotyping assays for single nucleotide polymorphisms with fluorescence resonance energy transfer detection. *Genet. Anal.* 14, 157–163.

Chen, X., Livak, K. J., and Kwok, P. Y. (1998). A homogeneous, ligase-mediated DNA diagnostic test. *Genome Res.* 8, 549–556.

Day, D. J., Speiser, P. W., White, P. C., and Barany, F. (1995). Detection of steroid 21-hydroxylase alleles using gene-specific PCR and a multiplexed ligation detection reaction. *Genomics* 29, 152–162.

Day, N. S., Tadin, M., Christiano, A. M., Lanzano, P., Piomelli, S., and Brown, S. (2002). Rapid prenatal diagnosis of sickle cell diseases using oligonucleotide ligation assay coupled with laser-induced capillary fluorescence detection. *Prenat. Diagn.* 22, 686–691.

Delahunty, C. M., Ankener, W., Brainerd, S., Nickerson, D. A., and Mononen, I. T. (1995). Finnish-type aspartylglucosaminuria detected by oligonucleotide ligation assay. *Clin. Chem.* 41, 59–61.

Eggerding, F. A. (1995). A one-step coupled amplification and oligonucleotide ligation procedure for multiplex genetic typing. *PCR Methods Appl.* 4, 337–345.

Feder, J. N., Gnirke, A., Thomas, W., Tsuchihashi, Z., Ruddy, D. A., Basava, A., Dormishian, F., Domingo, R., Jr., Ellis, M. C., Fullan, A., Hinton, L. M., Jones, N. L., Kimmel, B. E., Kronmal, G. S., Lauer, P., Lee, V. K., Loeb, D. B., Mapa, F. A., McClelland, E., Meyer, N. C., Mintier, G. A., Moeller, N., Moore, T., Morikang, E., Prass, C. E., Quintana, L., Starnes, S. M., Schatzman, R. C., Brunke, K. J., Drayna, D. T., Risch, N. J., Bacon, B. R., and Wolff, K. (1996). A novel MHC class I-like gene is mutated in patients with hereditary haemochromatosis. *Nat. Genet.* 13, 399–408.

Gerry, N. P., Witowski, N. E., Day, J., Hammer, R. P., Barany, G., and Barany, F. (1999). Universal DNA microarray method for multiplex detection of low abundance point mutations. *J. Mol. Biol.* 292, 251–262.

Grossman, P. D., Bloch, W., Brinson, E., Chang, C. C., Eggerding, F. A., Fung, S., Iovannisci, D. M., Woo, S., Winn-Deen, E. S., and Iovannisci, D. A. (1994). High-density multiplex detection of nucleic acid sequences: oligonucleotide ligation assay and sequence-coded separation. *Nucleic Acids Res.* 22, 4527–4534.

Gunderson, K. L., Huang, X. C., Morris, M. S., Lipshutz, R. J., Lockhart, D. J., and Chee, M. S. (1998). Mutation detection by ligation to complete n-mer DNA arrays. *Genome Res.* 8, 1142–1153.

Hansen, T. S., Petersen, N. E., Iitia, A., Blaabjerg, O., Hyltoft-Petersen, P., and Horder, M. (1995). Robust nonradioactive oligonucleotide ligation assay to detect a common point mutation in the *CYP2D6* gene causing abnormal drug metabolism. *Clin. Chem.* 41, 413–418.

Iannone, M. A., Taylor, J. D., Chen, J., Li, M. S., Rivers, P., Slentz-Kesler, K. A., and Weiner, M. P. (2000). Multiplexed single nucleotide polymorphism genotyping by oligonucleotide ligation and flow cytometry. *Cytometry* 39, 131–140.

Jarvius, J., Nilsson, M., and Landegren, U. (2003). Oligonucleotide ligation assay. *Methods Mol. Biol.* 212, 215–228.

Landegren, U., Kaiser, R., Sanders, J., and Hood, L. (1988). A ligase-mediated gene detection technique. *Science* 241, 1077–1080.

Martinelli, R. A., Arruda, J. C., and Dwivedi, P. (1996). Chemiluminescent hybridization-ligation assays for delta F508 and delta I507 cystic fibrosis mutations. *Clin. Chem.* 42, 14–18.

Matsubara, Y., Narisawa, K., Miyabayashi, S., Tada, K., Coates, P. M., Bachmann, C., Elsas, L. J., 2nd, Pollitt, R. J., Rhead, W. J., and Roe, C. R. (1990). Identification of a common mutation in patients with medium-chain acyl-CoA dehydrogenase deficiency. *Biochem. Biophys. Res. Commun.* 171, 498–505.

Nickerson, D. A., Kaiser, R., Lappin, S., Stewart, J., Hood, L., and Landegren, U. (1990). Automated DNA diagnostics using an ELISA-based oligonucleotide ligation assay. *Proc. Natl. Acad. Sci. US.A* 87, 8923–8927.

Padoan, R., Genoni, S., Moretti, E., Seia M., Giunta, A., and Corbetta, C. (2002). Genetic and clinical features of false-negative infants in a neonatal screening programme for cystic fibrosis. *Acta Paediatr.* 91, 82–87.

Peltonen, L. (1997) Molecular background of the Finnish disease heritage. *Ann. Med.* 29, 553–556.

Pickering, J., Bamford, A., Godbole, V., Briggs, J., Scozzafava, G., Roe, P., Wheeler, C., Ghouze, F., and Cuss, S. (2002). Integration of DNA ligation and rolling circle amplification for the homogeneous, end-point detection of single nucleotide polymorphisms. *Nucleic Acids Res.* 30, e60.

Pritchard, C. E., and Southern, E. M. (1997). Effects of base mismatches on joining of short oligodeoxynucleotides by DNA ligases. *Nucleic Acids Res.* 25, 3403–3407.

Richards, C. S., Bradley, L. A., Amos, J., Allitto, B., Grody, W. W., Maddalena, A., McGinnis, M. J., Prior, T. W., Popovich, B. W., Watson, M. S., and Palomaki, G. E. (2002). Standards and guidelines for *CFTR* mutation testing. *Genet. Med.* 4, 379–391.

Roe, C. R., and Coates, P. M. (1995). In *The metabolic and molecular bases of inherited disease*, C. R. Scriver, A. L. Beaudet, W. S. Sly, D. Valle, eds. Mitochondrial fatty acid oxidation disorders, pp. 1501–1533. McGraw-Hill, Inc., New York.

Romppanen, E. L., Mononen, T., and Mononen, I. (1998a). Molecular diagnosis of medium-chain acyl-CoA dehydrogenase deficiency by oligonucleotide ligation assay. *Clin. Chem.* 44, 68–71.

Romppanen, E. L., Valtonen, P., Mononen, T., and Mononen, I. (1998b). Molecular diagnosis of Finnish type infantile neuronal ceroid lipofuscinosis by restriction fragment length polymorphism and oligonucleotide ligation assay. *Clin. Chem.* 44, 2373–2376.

Romppanen, E. L., and Mononen, I. (2000). Detection of the Finnish-type congenital nephrotic syndrome by restriction fragment length polymorphism and dual-color oligonucleotide ligation assays. *Clin. Chem.* 46, 811–816.

Samiotaki, M., Kwiatkowski, M., Parik, J., and Landegren, U. (1994). Dual-color detection of DNA sequence variants by ligase-mediated analysis. *Genomics* 20, 238–242.

Tannous, B. A., Verhaegen, M., Christopoulos, T. K., and Kourakli, A. (2003). Combined flash- and glow-type chemiluminescent reactions for high-throughput genotyping of biallelic polymorphisms. *Anal. Biochem.* 320, 266–272.

Tobe, V. O., Taylor, S. L., and Nickerson, D. A. (1996). Single-well genotyping of diallelic sequence variations by a two-color ELISA-based oligonucleotide ligation assay. *Nucleic Acids Res.* 24, 3728–3732.

Wu, D. Y., and Wallace, R. B. (1989). The ligation amplification reaction (LAR)–amplification of specific DNA sequences using sequential rounds of template-dependent ligation. *Genomics* 4, 560–569.

Zhong, X. B., Reynolds, R., Kidd, J. R., Kidd, K. K., Jenison, R., Marlar, R. A., and Ward, D. C. (2003). Single-nucleotide polymorphism genotyping on optical thin-film biosensor chips. *Proc. Natl. Acad. Sci. USA* 100, 11559–11564.

Zotz, R. B., Giers, G., Maruhn-Debowski, B., and Scharf, R. E. (1997). Genetic typing of human platelet antigen 1 (HPA-1) by oligonucleotide ligation assay in a specific and reliable semi-automated system. *Br. J. Haematol.* 96, 198–203.

CHAPTER **5**

Enzymatic and Chemical Cleavage Methods to Identify Genetic Variation

CHINH T. BUI,[1] EMMANUELLE NICOLAS,[2] GEORGINA SALLMANN,[1] MARIA CHIOTIS,[1] ANDREANA LAMBRINAKOS,[1] KYLEE REES,[1] IAN TROUNCE,[1] RICHARD G. H. COTTON,[1] LAURYN BLAKESLEY,[2] ANDREW K. GODWIN,[2] AND ANTHONY T. YEUNG[2]

[1] Genomic Disorders Research Centre and The University of Melbourne, Fitzroy, Melbourne, Australia;
[2] Fox Chase Cancer Center, Philadelphia, PA, USA

TABLE OF CONTENTS

5.1 INTRODUCTION

Although there are several methods for the high sensitivity detection of known mutations, unknown mutations are more difficult to uncover. Recently, the latter methods have been improved significantly and utilized in practical applications. These include the establishment of a single nucleotide polymorphism (SNP) map for a specific part of the genome for a specific animal population or the screening for unknown mutations in important genes, such as cancer susceptibility genes (Neuhausen and Ostrander, 1997; Warner et al., 1999; Oleykowski et al., 1998; Hartge et al., 1999). These genes are large, with many exons, and thus hundreds of possible mutations that affect the functions of those proteins can be found. Also, these methods can serve as tools for reverse genetics to screen for chemically induced point mutations in specific regions of specific genes (Colbert et al., 2001) and for screening for specific mutations in the genome of emerging pathogenic microorganisms (Sokurenko et al., 2001). In this chapter, two of these mutation detection methods will be discussed in detail.

A mutation is by definition a change in the DNA sequence as compared to the population at large. Historically, mutations are detected by phenotypic changes. At the DNA level, it is often, but not always, revealed by direct DNA sequencing. Indeed, most researchers involved in mutation screening have encountered mutations that are not detectable by Sanger dideoxy-DNA sequencing, but obvious to complementary approaches that detect changes in the physical properties of the DNA helix when a mismatch is present. In mutation screening methods not based on enzymatic DNA polymerization, the mutated DNA helix often is first converted to mismatch heteroduplexes when the polymerase chain reaction (PCR; Mullis and Faloona, 1987) products of two alleles of the gene of interest are amplified by PCR, mixed, denatured, and rehybridized. Next the chemical or enzymatic properties of a mismatched base are exploited to lead to a break in the DNA strand near the mismatch. Finally, a suitable fragment analysis method is used to visualize the shortened DNA fragments, in either single-stranded form or the double-stranded form, that are produced by the DNA break near the mismatch residue. The suitability of a method for each laboratory depends on the existing skills in that laboratory and the available analysis platform.

The mixing of two alleles of a gene of interest in PCR results in heteroduplexes at the end of a denaturation and renaturation step. However, homozygous mutations and hemizygous mutations will not lead to the formation of mismatch heteroduplexes. Thus, it is important that a mutation detection method be sufficiently sensitive to detect a mutation in a pool of multiple normal alleles so that statistically a mutation will not be missed.

5.2 CHEMICAL PROPERTIES OF MISMATCHES

Mismatches can be either insertions/deletions or base substitutions. The latter include A/A, G/G, C/C, T/T, A/G, A/C, G/T, C/A, and C/T. It is important to realize that no two mismatches have the same chemistry and structure, therefore most mutation detection methods may recognize either one or more of the lesions with high specificity but not all of them, or conversely, recognize all the lesions at only moderate specificity.

A mismatched base pair presents minimal information to the mutation detection system as compared to the recognition of multiple base pair palindrome sequences by restriction enzymes. In a mismatch, often only two opposing bases in the two DNA strands are involved with modification in the hydrogen bonding, leading to enhanced chemical reactivity. Alternatively, a mismatched base may flip out of the helix at some frequency for enzyme recognition. The mismatched base may also destabilize local regions of the DNA helix to create single-strandedness. The single-strandedness of insertion/deletions can also be exploited (Burdon and Lees, 1985). However, the natural tendency of DNA helix to undergo breathing at AT-rich sequences and to exhibit secondary structures like hairpins and cloverleaf structures at some sequences can create false positive signals for mismatch detection systems. Other sequences have structures that can hide a mismatch and cause some mismatch detection methods to give false negative results.

5.3 CHEMICAL CLEAVAGE OF MISMATCH METHOD FOR MUTATION DETECTION

Detection of unknown mutations is a complex and expensive task, particularly for screening kilobase lengths of DNA

sequence for a single base change and/or small insertions and deletions. Chemical Cleavage of Mismatch (CCM) technology, developed by Cotton and colleagues (1988), theoretically establishes simple and cost-effective chemical means to detect all types of mismatches, and thus, mutations at this point in time (Cotton *et al.*, 1988). The protocol employs two commercially available chemicals, hydroxylamine (NH_2OH) and potassium permanganate ($KMnO_4$) (or osmium tetroxide (OsO_4) in an earlier version) to react with mismatched cytosine and thymine residues, respectively (Ramus and Cotton, 1996; Cotton, 1999). The modified mismatched DNA becomes highly susceptible to cleavage by piperidine. The resulting DNA fragments simply are analyzed by denaturing polyacrylamide gel or capillary electrophoresis (Ren, 2001) to identify the mutation sites. Due to its simplicity in manipulation and high sensitivity, the method has been continuously improved and documented in very simple protocols by many users (see Table 5.1). For example, $KMnO_4$ was first introduced in 1990 to replace the hazardous chemical OsO_4. Radioactive-labeled (^{32}P and ^{35}S) DNA probes were replaced with fluorescence-labeled primers. DNA was also incorporated onto streptavidin-coated magnetic beads to bypass multiple washing steps. Both chemical reactions (with $KMnO_4$ and NH_2OH) were simplified by being carried out in a single tube. Recently, the DNA samples were adsorbed onto silica beads under high salt concentrations (3 M TEAC solution) before undergoing the chemical modification steps (Bui *et al.*, 2003a). The post-cleavage washing steps also were eliminated by incubation of piperidine together with the loading dye buffer. In addition, various types of amine-bases also were reported to improve cleavage reactions used in CCM methods (Block, 1999). All relevant protocols are suitable for analyzing mismatches located on long stretches of DNA (up to 2.0 kb). For the purpose of practical guidance, this chapter describes two established protocols, the solid phase

TABLE 5.1   Protocols used in chemical cleavage of mismatches.

Milestones	Protocols	References
1988	The first described CCM technique (hydroxylamine and osmium tetroxide)	Cotton *et al.*, 1988
1990	$KMnO_4$/TMAC (tetramethyl ammonium chloride) was introduced to replace the hazardous chemical OsO_4	Gogos *et al.*, 1990
1991	^{32}P and ^{35}S labeled probes for PCR products	Saleeba, 1991; Cotton *et al.*, 1988
1995	Use of fluorescent primers	Verpy, 1994; Haris, 1994
1996	Single tube CCM method	Ramus and Cotton, 1996
1998	Development of solid phase CCM	Rowley *et al.*, 1995; Roberts *et al.*, 1997
1999	$KMnO_4$/TEAC (tetraethyl ammonium chloride) condition was applied	Roberts *et al.*, 1997; Lambrinakos *et al.*, 1999
1999	Cleavage reactions with amine-bases	Block, 1999
2001	Solid phase CCM method (silica bead)	Bui *et al.*, 2003a
2001	Piperidine in cleavage loading dye solution	Cotton and Bray, 2001
2001	Capillary electrophoresis with laser-induced fluorescence detection	Ren, 2001

and liquid phase protocols, which most commonly are used in our laboratories and others (Gogos *et al.*, 1990; Rowley *et al.*, 1995; Roberts *et al.*, 1997; Lambrinakos *et al.*, 1999).

The formation of heteroduplexes can be performed by mixing and reannealing equimolar amounts of wild type and mutant DNA, thereby resulting in mismatched base pairs. In principle, the imperfect duplexes are different from their corresponding perfect duplexes in terms of their local conformational changes, physical and chemical properties (Bui *et al.*, 2002). These changes are reflected at lower melting temperatures in the imperfect duplexes (Patel *et al.*, 1982), less stability as indicated by thermodynamic constants (Patel *et al.*, 1982) and extra-helical or "flip-out" phenomena of mismatched bases (Kao *et al.*, 1993; Roberts and Cheng, 1998). However, such discrepancies induced by mismatches are so small that the mismatched sites can be recognized and cleaved only by enzymatic means and not by chemical reagents (Kennard, 1988). For this reason, the mismatched sites require further modification by chemicals before the chemical cleavage reaction (piperidine) can take place. Two common chemicals, KMnO₄ and NH₂OH, are the most effective for modifying thymine and cytosine mismatches, respectively, in this regard. The former reaction leads to the formation of a mixture of a thymine glycol and a ketone analog (some evidence indicates that KMnO₄ also reacts with cytosine but to a lesser extent; Bui and Cotton, 2002) and the latter gives rise to a modified cytosine containing a hydroxylamine moiety (see Fig. 5.1). Physical data established on the model for short mismatched oligonucleotides (38 bp) indicate that the differences in melting temperatures and gel mobility of the chemically modified heteroduplex samples are significant compared to the unmodified one (Bui *et al.*, 2003b). The results suggest that the destabilized mismatched site favors the site-selective cleavage reaction of

piperidine and the mismatch can be pinpointed on a denaturing gel as cleaved bands.

CCM is considered as the method of choice as it can detect all key types of mismatch (T/G, T/C, C/C, A/C, and T/T), which represent all eight possible mispairs that can be generated from the heteroduplex formation (see Fig. 5.2). The other mismatches (A/G, G/G, and A/A) can be detected via the complementary heteroduplexes (i.e., A/A will be detected by T/T). It is also emphasized that some neighboring matched bases also respond to the reactions due to instability of the whole region near the mismatched site. In addition, when both mutant and wild type DNA are labeled, the chance of detecting mutations will be doubled (see Fig. 5.2).

5.3.1 Liquid Phase Protocol

The standard liquid phase CCM protocol consists of six steps: Chemical modification with KMnO₄ and hydroxylamine, termination, separation, washing, cleavage, and gel-electrophoresis (see Table 5.2). The wild type and mutant DNA samples are amplified (by PCR) using the fluorescence-labeled primers (6-FAM and HEX at 5′ and 3′ ends, respectively). Subsequently, the amplified DNA samples are purified, using either a commercially available purification kit (Stratagene™ PCR Purification Kit, CA, USA) or agarose gel electrophoresis. The resulting wild type and mutant DNA samples are mixed in equal amounts to form the heteroduplexes prior to the assay. In the first step of the described protocol, the reactions of heteroduplex DNA with KMnO₄ and hydroxylamine usually are carried out in separate tubes. It is also noted that both reactions can be optimized and carried out in a single tube protocol (but this will not be discussed in this chapter). When the reactions are completed, the DNA samples are separated by ethanol precipitation and washed carefully before the next cleavage step. To simplify the protocol, the gel loading dye (blue

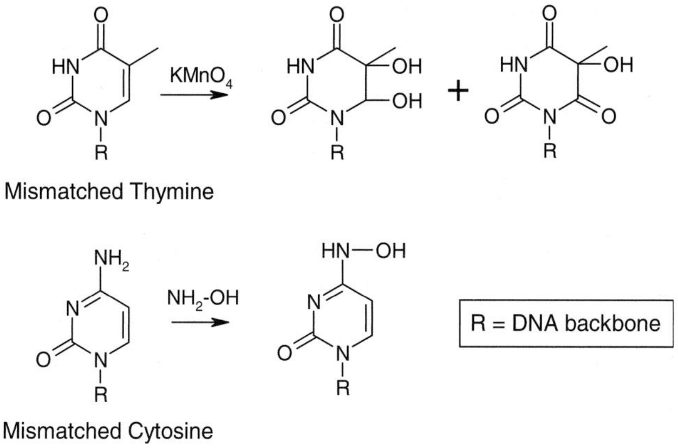

FIGURE 5.1 Chemical reactions involved in the CCM method. KMnO₄ selectively reacts with mismatched thymine to afford a mixture of thymine glycol and a ketone analog, and hydroxylamine reacts with mismatched cytosine to afford a single monosubstituted product under the described conditions.

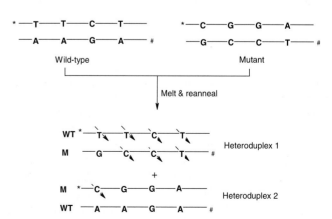

FIGURE 5.2 Formation of heteroduplexes from the wild type (WT) and mutant (M) homoduplexes. * indicates 5′-FAM and # indicates 3′-HEX fluorescence-labeled DNA strands.

TABLE 5.2   The liquid phase protocol.

Steps	Procedure[a]
KMnO₄ reaction	0.2 μL of 100 mM KMnO₄ + 19.8 μl of 3 M TEAC. Incubate at 25°C for 10 min.
Hydroxylamine reaction	20 μL of 4.2 M hydroxylamine solution (pH = 6.0 with triethylamine). Incubate at 37°C for 40 min.
Termination	Add 200 μl of STOP buffer (25 μg/ml tRNA, 0.1 mM EDTA, 0.3 M sodium acetate, pH 5.2).
Separation	Add 750 μl of ice-cold 100% ethanol (at −20°C for at least 30 min). Centrifuge for 20 min at 14,000 rpm to collect the precipitates.
Washing	Rewash the pellet with 200 μl of 70% ethanol. Centrifuge it again for 10 min (14,000 rpm). Air-dry the pellet for 10 min.
Cleavage	Add 10 μl of cleavage loading dye solution [20 μl piperidine, 64 μl formamide and 16 μl dye (50 mg blue dextran in 1 ml distilled water)] to the DNA pellets and incubate at 90°C for 30 min.
Gel separation[b]	Load samples onto a denaturing gel and analyzed on an ABI 377 DNA sequencer (2 μl of sample is needed for each well).

[a] 6 μl of homoduplex and heteroduplex DNA (0.6 μg DNA) in TE buffer (10 mM Tris-HCl, 1 mM EDTA, ethylene-diamine-tetra-acetic acid, pH 8.0) is used for each of the reactions.
[b] The denaturing polyacrylamide gel 4.25% (acrylamide: *bis*-acrylamide, 19 : 1), 6 M urea gel in the TBE buffer (16.2 g Tris-base, 8.1 g boric acid and 1.12 g EDTA in 1500 ml distilled water, pH = 8.0).

TABLE 5.3   The solid phase protocol.

Steps	Procedure
Loading DNA onto solid supports	Mix 3 μl of Ultra-bind bead suspension with 1 μl of DNA samples (0.1 to 0.2 μg of homoduplex or heteroduplex DNA) on shaker for 1 hour at 25°C. Centrifuge at 14,000 rpm, the pellets are collected. Wash the beads by resuspending in the Ultra-wash solutions (2 × 500 μL). Centrifuge the mixture, discard the supernatant, and air-dry the beads at 25°C for 15 min.
KMnO₄ reaction	Mix the beads with 0.3 μl of 100 mM KMnO₄ in 29.7 μl of 3 M TEAC solution. Allow to stand at 25°C for 5 min.
Hydroxylamine reaction	Mix the beads with 15 μl of 4.2 M hydroxylamine solution in 15 μl of 3 M TEAC solution. Allow to stand at 37°C for 40 min.
Washing	The beads are separated by centrifugation and washed twice with the Ultra-wash solution (200 μL per wash) and the pellets are air-dried at 25°C for 15 min.
Cleavage	Add 10 μL of the cleavage dye solution to each reaction tube and vortex well. Incubate the tubes at 90°C for 30 min. Mix tubes by flicking occasionally. Cool the tubes on ice and the supernatant is separated by centrifugation.
Gel separation	Load samples onto a denaturing gel and analyze on an ABI 377 DNA sequencer (2 μl of sample is needed for each well).

dextran) is added to the cleavage solution and the reaction mixtures can be directly loaded onto a denaturing gel immediately after the cleavage step. The DNA fragments are analyzed by an ABI 377–DNA sequencer without further purification (only 2 μl of sample is needed for loading). Two types of size standards (Tamra 500 and Tamra 2500) are added to the gel. Therefore, analysis with the ABI sequencer allows identification of the positions of the mismatches without the use of sequencing.

5.3.2 Solid Phase Protocol

In order to bypass the separation and washing steps, the solid phase protocols have been developed by immobilizing DNA on silica solid supports (see Table 5.3). Attachment of DNA on silica in a high salt solution has been established and well practiced as an effective purification technique (Bui *et al.*, 2003a). In the described protocol, the commercially available silica beads are used (MO BIO Laboratories Inc. CA, USA) to bind to the DNA, and the DNA-attached beads are then sequentially treated with chemicals, washing solutions, and piperidine for cleavage. In the last cleavage step, the DNA fragments are cleaved by piperidine and released simultaneously from the solid supports. The resulting super-

natant is isolated and loaded directly onto a denaturing polyacrylamide gel.

Mismatch detection is based on the comparison of the homoduplex and heteroduplex traces. A mutation is identified by cleavage peaks present in the trace of heteroduplex sample but not in the control homoduplex sample.

In the authors' laboratories, both liquid phase and solid phase protocols are routinely used for mismatch detection, and some typical examples are described next.

The liquid and solid phase protocols were successfully carried out on 547 bp DNA fragments derived from the cloned mouse β-globin promoter DNA to detect T/C and T/G mismatches respectively (see Figs. 5.3 and 5.4).

Detection of single base insertion and deletion (C base) was carried out with DNA fragments (893 bp in Fig. 5.5 and 660 bp in Fig. 5.6) derived from human mitochondrial DNA by using the liquid phase protocol with hydroxylamine reaction.

5.4 ADVANTAGES AND LIMITATIONS

The major advantage of the solid phase protocol is that it is fast and simple in manipulation compared to the liquid phase

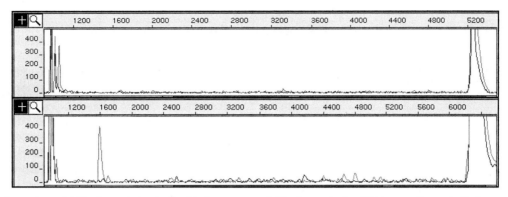

FIGURE 5.3   Detection of T/C mismatch by the liquid phase protocol. The heteroduplex DNA trace (bottom) displays a strong cleavage peak at the mismatched C base of the 3′-HEX sequence, induced by hydroxylamine/piperidine. The control trace (top) shows no cleavage peak.

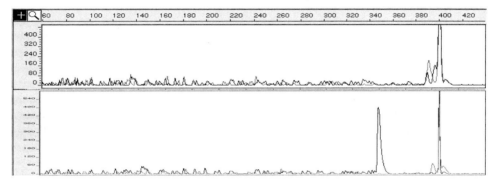

FIGURE 5.4   Detection of T/G mismatch by the solid phase protocol. The heteroduplex DNA trace (bottom) displays a strong cleavage peak at the mismatched T base of the 5′-FAM sequence, induced by KMnO$_4$/piperidine. The control trace (top) shows no cleavage peak.

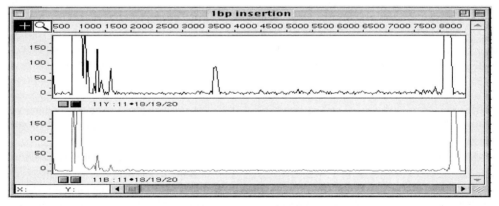

FIGURE 5.5   Detection of a single base insertion (C-base) by the liquid phase protocol. The DNA trace (top) displays a strong cleavage peak at the insertion position (C base) of the 5′-FAM sequence, induced by hydroxylamine/piperidine. The control trace (bottom) shows no cleavage peak.

protocol. The protocol bypasses the washing steps and DNA does not require separation by ethanol precipitation techniques. However, the solid phase protocol is most suitable for short DNA sequences (up to around 500 bp). Adsorption of long DNA on beads is not sufficient due to limit of surface area of the beads. Expensive cost of the silica beads and the washing liquid are also taken into account.

In addition, no false positive and negative results have been reported in the CCM method so far. Usage of 5′-FAM and 3′-HEX fluorescence-labeled primers will offer two chances of a mutation being detected. A typical example is the detection of a C/C mismatch of 547 bp DNA fragment by using hydroxylamine. The mismatched DNA trace (heteroduplex) displays two strong cleavage peaks of the 5′FAM

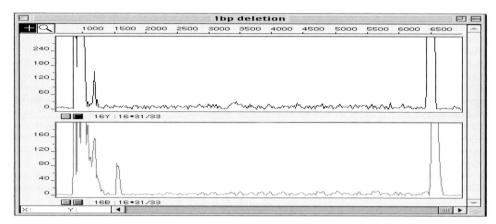

FIGURE 5.6   Detection of a single base deletion (C-base) by the liquid phase protocol. The DNA trace (bottom) displays a strong cleavage peak at the deletion position (C base) of the 3′-HEX sequence, induced by hydroxylamine/piperidine. The control trace (top) shows no cleavage peak.

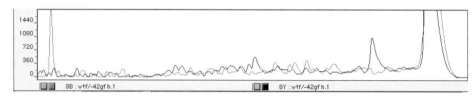

FIGURE 5.7   Detection of C/C mismatch by the solid-phase protocol. The heteroduplex DNA trace displays two strong cleavage peaks of the 3-HEX and 5′-FAM sequences induced by hydroxylamine/piperidine reactions. The control trace is not shown.

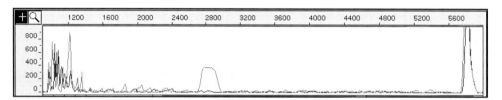

FIGURE 5.8   Multiple cleavage reactions at the AT-rich sequence by KMnO₄/piperidine. The homoduplex DNA trace displays a plateau peak due to cleavage reactions at the AT-rich sequence.

sequence and 3′HEX sequence at the mismatched C base (see Fig. 5.7).

Optimization is recommended when starting to use the method. Prolonged incubation leads to multiple cleavage peaks on the background due to overreactions. No cleavage peaks may be the result of short incubations. Controls are needed to assure that the chemicals are active. Adjacent matched bases appear to react with KMnO₄/piperidine and the trace may display multiple peaks. These unusual cases have been observed in a recent study, suggesting the strong destacking phenomenon (bubble formation) within the mismatched site enhanced the reactivity (Lambrinakos *et al.*, 1999).

AT-rich sequences are also susceptible to multiple modification/cleavage reactions and a peak with a plateau shape may be observed. This was observed on the homoduplex 500 bp DNA sample containing 18 consecutive A/T matched pairs (see Fig. 5.8). KMnO₄–susceptibility to AT-

rich sequences was also suggested due to the local destacking nature (curvature) of the repeated A-track region (De Santis *et al.*, 1990).

Finally, to improve the quality of the result, PCR products of mutant and wild type DNA usually are purified prior to any subsequent treatment, and the KMnO₄ solution should be freshly made before use because an aging solution turns brown-yellow with the precipitation of MnO₂ after few days.

5.5 ENZYMATIC CLEAVAGE OF MISMATCH METHODS

5.5.1 The mutHLS System of Mismatch Detection

The post-replicative mismatch repair enzyme systems were explored as potential tools for mismatch detection (Lu and

Hsu, 1992; Lishanski *et al.*, 1994; Wagner *et al.*, 1995). Because the MutHLS system is tightly coupled to the DNA replication machinery *in vivo*, it is not meant to screen the genome as a whole for the presence of mismatches. The crystal structure further illuminated that MutS bends the DNA only slightly, allowing it to recognize almost all of the mismatches, but not with very high specificity or affinity (Natrajan *et al.*, 2003). As a result, mutation screening based on MutS protein has thus far been unsatisfactory.

5.5.2 DNA N-glycosylase Approaches of Mismatch Detection

The DNA N-glycosylases have evolved nucleotide-binding pockets with very tight fit and nucleotide specificity, including the ability to insert a peptide loop into the base that has been displaced. Unfortunately, nature needs only two mismatches to be recognized by DNA glycosylases. The thymine DNA N-glycosylase recognizes the T/G mismatch that results from the deamination of 5-methy-cytosine (Hardeland *et al.*, 2001), and the MutY and its homologs recognize the A/G mismatch that occurs from the misincorporation of 8-oxo guanine across from the A residue (Sanchez *et al.*, 2003). Thus, methods of mutation detection that use DNA glycosylases do not utilize all the mismatch heteroduplexes that are present during the PCR amplification of two alleles. Moreover, the DNA N-glycosylases do not detect insertion/deletion mutations, or mutations that do not form T/G and A/G mismatches. Because a DNA N-glycosylase only creates an apurinic site but will not break the DNA chain, the detection of DNA truncation at the mismatch site requires either the addition of an apurinic endonuclease or a base treatment step such as piperidine at elevated temperatures, and DNA cleanup steps. Double-stranded DNA breaks are not produced, thus requiring the use of denaturing conditions for the shortened single-stranded products to be analyzed. Successful application of the DNA glycosylase methods have been documented in numerous publications (Zhang *et al.*, 2002).

5.5.3 The Resolvase Approach of Mismatch Detection

Another enzymatic approach involves the use of DNA resolvases that recognize the DNA distortion created by mismatches mimicking the DNA recombination intermediate structures. Successful applications of this approach use the T4 endonuclease VII system (Mashal *et al.*, 1995; Youil *et al.*, 1995; 1996) and the related T7 endonuclease I system (Babon *et al.*, 2003). These enzymes cut the mismatch duplexes within a few nucleotides of the mismatch sites, and usually lead to double-stranded breaks in the heteroduplex. The resolvases also lead to nonspecific DNA cutting at some unknown DNA sequences and imperfect products of PCR reactions (Norberg *et al.*, 2001).

5.5.4 The Endonuclease V Plus Ligation Method

Although endonuclease V itself is not very mismatch-specific, its incision within one or two nucleotides of a mismatch base allows a DNA ligase to distinguish between the mismatch product nicks that cannot be ligated and the nicks made at nonmismatch sites that can be repaired by ligation (Huang *et al.*, 2002). Some sequences may contain mismatches that are not detectable by Endo V or ligatable due to mismatch slippage. In common with other approaches, there is nonspecific nicking at AT-rich regions, and the ligation repair reaction is lengthy.

5.5.5 The Plant Mismatch Endonuclease Method

The enzymatic mutation detection approach that currently shows much potential for exploitation in a number of applications is the plant mismatch endonucleases exemplified by the CEL I endonuclease of celery (Oleykowski *et al.*, 1998; 1999; Yang *et al.*, 2000; Kulinski *et al.*, 2000). These plant nucleases apparently belong to a subgroup in the S1 nuclease family and are induced during plant senescence and remodeling. The in mutation detection is extremely simple, applicable on various fragment analysis platforms, whereas the detection of the products of DNA truncation at the mismatch site in either the single-stranded form or the double-stranded forms gives the potential of multiple detection formats. Moreover, the high precision of the CEL I nuclease to cut at the 3′ phosphodiester bond immediately next to the mismatched base allows DNA ligation repair to be used more effectively if desired. The latter was used in the early experiments during the development of the CEL I platform, but is not necessary for most CEL I applications. The high mismatch-specificity of CEL I nuclease is believed to come from the enzyme binding to both bases of a base-substitution mismatch at the same time.

Current CEL I mutation detection assay is exemplified in Fig. 5.9, in which a DNA heteroduplex contains a mismatch. The ability of CEL I to form single-stranded DNA nick in short incubations but convert to double-strand DNA truncation mode under conditions of longer incubations or enzyme excess has allowed two powerful assay approaches to be developed.

5.5.5.1 SINGLE-STRANDED DNA TRUNCATION ASSAY
This assay mode is widely used in fragment analysis platforms like ABI-377 slab gel system, the LiCor infrared slab gel system, the ABI-3100/3730 capillary DNA sequencers, and the Beckman CEQ8000 infrared DNA sequencer. In this assay, the PCR primers for a given target region, often under 600 bp long, are labeled with one color for the forward primer and a second color for the reverse primer. Color combinations used for the 377 system is 6-FAM/TET, for the

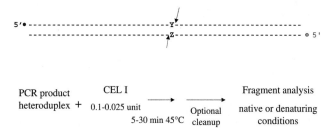

PCR product   CEL I   Fragment analysis
heteroduplex + 0.1–0.025 unit   ———→   ———→ native or denaturing
                  5–30 min 45°C   Optional   conditions
                                           cleanup

FIGURE 5.9 Schematic of the CEL I mismatch nuclease mutation detection procedure. The top panel illustrates that in a DNA heteroduplex containing a mismatch Y/Z, CEL I makes either one or two incisions in the same DNA molecule denoted by the two arrows. The bottom panel illustrates the simplicity of the CEL I mutation detection method. A denaturing fragment analysis method may use 5′ termini radioactive labeling or fluorescence two-color labeling followed by resolution of an automated DNA sequencer. The nondenaturing fragment analysis methods may use SYBR-green fluorescence staining and resolution on an agarose gel or PAGE or an Agilent BioAnalyzer lab on a chip.

ABI 3100/3730 is 6-FAM/HEX, and for the Beckman CEQ 8000 is Cy5/Cy 5.5 or D3/D4. For the latter, the Cy5/Cy5.5 combination is easier in DNA synthesis and purification. Purification of the infrared dye primers is not necessary because the presence of reporterless primers enhance the PCR efficiency without diminishing the mutation detection sensitivity. As illustrated in Fig. 5.9, the two strands will be differentially labeled after PCR and heteroduplexes will be formed, either during PCR or thereafter by denaturation and renaturation. The heteroduplex is treated with CEL I, without further purification, for about 5–30 min. The cut DNA can be loaded onto a DNA sequencer/fragment analyzing system without further purification. A purification step enhances the performance of some fragment analysis platforms like the CEQ 8000 and the LiCor sequencer. The CEL I truncation bands of two colors are measured on the fragment analysis system. The sum of the lengths of the bands of the two colors correlated to the same SNP is equal to the full-length PCR primer plus one nucleotide, or more if the insertion involves more than one base. CEL I cuts an insertion at the phosphodiester bond at the 3′ end of the loop, and then shortens the single-stranded region slowly thereafter. This assay is simple, sensitive, and easily automated for high throughput. The two color cuts each originate from a different DNA molecule in the case of the single-strand cut assay, and thus their presence represents independent confirmation of the presence of the mismatch. This assay is used routinely in *BRCA1* and *BRCA2* genetic screening in Fox Chase Cancer Center, and in the Tilling reverse genetics procedure (Colbert *et al.*, 2001) that offers another method of targeted gene knockout in plants, zebra fish, mice, ES cells, and other organisms.

An important point in the single-strand mismatch nicking assay is that the nicked DNA may be a minor population in

the case of a weak mismatch substrate like the base substitutions T/T and G/T mismatches. Although the insertion/deletions produce restriction-enzyme-like strong signals with CEL I single-strand nicking assay, many base substitutions require rescaling of the fluorescence intensity axis (see Figs. 5.10 and 5.11). Experiments in Figs. 5.11 and 5.12 use a 500 bp PCR product from the exon 11.4 of the BRCA1 gene (Oleykowski *et al.*, 1998). This human genomic fragment, containing two T→C base substitutions and one G→A base substitution, is a very demanding standard routinely used in the authors' laboratory for the comparison of the mutation detection performance of different assay systems and conditions. The CEL I mismatch-specific signals may be small when the PCR product and the primer peaks are expressed in full scale (see Fig. 5.11, insert), but they are still many times higher in signal than the background peaks. Figure 5.11 illustrates that even in the case of three mismatches being present in the same PCR fragment, they are decisively identified in the CEL I mismatch detection assay on the CEQ 8000. The same performance was previously reported for the ABI 377 platform (Kulinski *et al.*, 2000), and also routinely obtained for the ABI 3100 capillary sequencing system (see Fig. 5.12).

A PCR product from two alleles form heteroduplexes of two combinations for any given base substitution (e.g., a C→T transition produces both C/A and T/G mismatches and four recognition sites for CEL I). Because CEL I mismatch nuclease can recognize all mismatches and can cut at either strand or both strands of a mismatch, all mismatches have multiple chances to be detected with this system. When used carefully, this system does not produce false positives and false negatives.

Table 5.4 shows the assay conditions used for the CEQ 8000 system (Yeung laboratory) and the conditions routinely used for BRCA1/BRCA2 mutation detection on the ABI 3100 system (Godwin laboratory), respectively. The CEQ 8000 system apparently produces slightly sharper peaks than the ABI 3100, but is less tolerant to salt and primer contaminations. Therefore, an ethanol precipitation step is needed prior to loading of the CEL I reaction products onto the CEQ 8000. On the contrary, the ABI 3100 apparently can load the CEL I reaction product with no sample purification. Gel filtration desalting is not suitable for the CEQ 8000 system in which the primers and PCR products contain very hydrophobic infrared cyanine dyes that lead to binding to the Sephadex used in the gel filtration columns. Addition of acetonitrile to the buffers diminishes the sample loss, but also decreases the gel filtration performance.

5.5.5.2 DOUBLE-STRANDED DNA TRUNCATION ASSAY

Above single-stranded end-labeled primer CEL I assay uses partial mismatch digestion to minimize the fluorescence label loss to the 5′ exonuclease activity of CEL I. This activity appears to cut within a few nucleotides of the 5′ termini

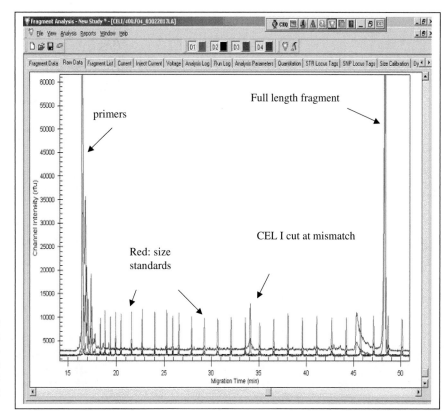

FIGURE 5.10 CEL I mutation detection of a C→T base substitution mutation in a 500 bp PCR product of human BRCA1 gene. A screen shot of the unprocessed data of the CEL I mutation detection product resolved in the infrared capillary sequencer Beckman CEQ 8000 is shown.

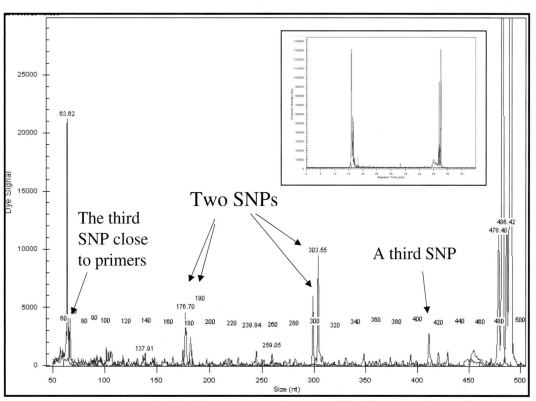

FIGURE 5.11 CEL I mutation detection of three SNPs in the same fragment of BRCA1 exon 11.4. CEQ 8000 analysis close-up of CEL I cut at three SNPs, two T to C, and one G to A base substitutions in the same fragment. The sizes of the standard ladder are shown by numbers in the figure. Incubation was 0.025 units of CEL I for 30 min at 45°C. The insert represents the unprocessed chromatogram in full scale display, illustrating that the CEL I cuts at multiple SNPs are small signals between tall peaks of full length PCR product and the PCR primer peaks.

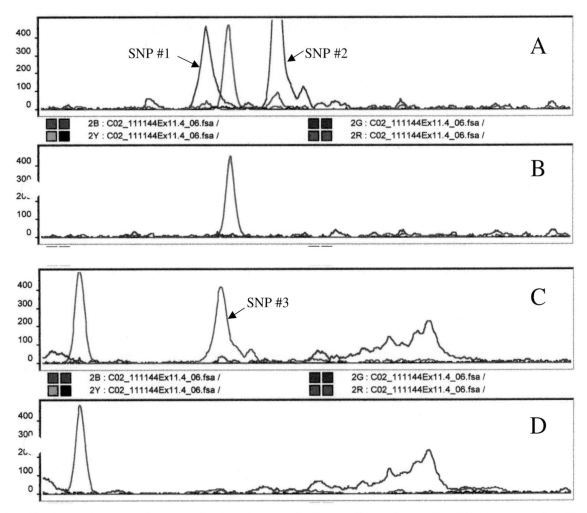

FIGURE 5.12 CEL I mismatch nuclease mutation detection using the Genescan® procedure in the ABI 3100 automated capillary DNA sequencer. The same PCR product fragment containing three SNPs used in the Fig. 5.10 experiment is shown. The forward and reverse primers are labeled at the 5′ termini with 6-FAM and HEX, respectively. The assay conditions are specified in Table 5.2 (under Godwin laboratory procedures). A. The two SNPs detected by the forward primer. B. The size marker and the low background noise. C. The third SNP detected by the reverse primer. D. Size marker and background noise corresponding to panel C.

slowly, and limits the accumulation of the mismatch incision products with the fluorescent reporter still attached. This 5′ endonuclease activity is minimized in the presence of 0.5 units of Taq DNA polymerase in the CEL I incubation, or the presence of a cyanine dye reporter at the 5′ end of the PCR product. This partial digestion, on the positive side, allows multiple mutations to be detected in the same PCR fragment, even when they are three nucleotides apart. The use of initial kinetics in the mismatch digestion also leads to high signal-to-noise ratio, and a broad window of enzyme quantity or digestion time of at least 20 fold and still producing useful mutation detection data. This 5′ exonuclease effect can be avoided when a DNA staining procedure is used instead of the 5′ termini label. For example, using SYBRgreen staining of the double-stranded DNA fragments, the 5′ exonuclease-processed mismatch incision

fragments can still be visualized to produce much stronger signal-to-noise ratio in the mutation detection assay than the single-strand assay method (see Fig. 5.13). This is one of the principles developed in the Giraff method of CEL I genomic scanning for bacterial mutations (Sokurenko *et al.*, 2001). In Fig. 5.13, the CEL I incised double-stranded DNA is analyzed in a native PAGE visualized with SYBRgreen fluorescence staining. A second benefit of this native gel approach is that any single-stranded nonspecific nicking by CEL I in the double-stranded DNA does not contribute to background unless it leads to double-stranded truncation. Double-stranded truncation at a mismatch by CEL I can be obtained by using the same amount of CEL I as in the single-strand nicking assay but extending the incubation for 16 hours. The use of a higher concentration of CEL I for a shorter incubation produces much higher signal-to-noise

TABLE 5.4 CEL I mismatch nuclease single-strand mismatch nicking protocols for capillary DNA sequencers.

Capillary sequencer CEL protocols	Yeung laboratory, Beckman CEQ 8000	Godwin laboratory, ABI 3100
PCR reaction	20 μL	20 μL
Human genomic DNA	60–75 ng	75–150 ng
Reaction buffer, 10×	MgCl$_2$, 2 mM final	MgCl$_2$
DMSO	5% final	6.7% final
dNTP	200 μM	80 μM
Primers, each	1.5 pmoles	4 pmoles
DNA polymerase	1 U AmpliTaq Gold	1 U AmpliTaq
PCR cycles	94°C for 4 min	94°C for 4 min
	97°C for 1 min	20 cycles of:
Exons that are prone to deletion are checked with agarose gels		94°C for 5 sec
		65°C for 1 min, touchdown at −0.5°C/cycle
		72°C for 1 min
	30 cycles of:	32 cycles of:
	94°C for 10 sec	94°C for 5 sec
	55°C for 20 sec	55°C for 1 min
	72°C for 45 sec	72°C for 1 min
	72°C for 5 min, then 4°C	72°C for 4 min, then 4°C
Denaturation/renaturation step	94°C 1 min then cooled to 4°C over 30 min	none
CEL I reaction	20 μL rxn	10 μL rxn
PCR reaction product	5 μL	5 μL
Water	12 μL	3.1 μL
10× CEL I buffer	2 μL	0.9 μL
CEL I enzyme	0.1 units	0.025 units
Incubation	45°C for 7 min	45°C for 1 hr
Stopping CEL I reaction	5 μL of (0.1 mM EDTA, 1.2 M Na-acetate, 15 v/v glycogen, 1.7% v/v of NF co-precipitant) to the 20 μL CEL I rxn 60 μL of 95% ethanol, to a final volume of 85 μL. Centrifuge for 15 min 4°C at 21,000xg Wash pellet 2× with 70% ethanol, dry in speedvac	Add 3 μL 5 mM o-phenantholine or just add formamide tracking dye. No ethanol ppt or centrifugation.
Preparation for capillary	Resuspend pellet in 40 μL of sample loading solution (SLS)	5 μL of stopped enzyme rxn is added to 15 μL of marker mix
Sequencer loading	Add 1.5 μL of suspension to 28.5 μL of SLS. Add 10 uL marker mix Load all 40 μL to the CEQ8000 Sample well ~0.25 uL of the original PCR reaction	Load all 20 μL to the ABI 3100 Sample well approximately 1.92 μL of the original PCR reaction

ratio. The new CEL nuclease SURVEYOR™ agarose gel kit supplied by Transgenomic Inc. has been further optimized for this assay and can detect base substitutions in long PCR products (i.e., 4 kb in length). For shorter fragments, a pool depth of 1 mutant in 40 normal alleles can be obtained. For biological DNA, the Giraff approach has demonstrated mismatch detection in heteroduplexes longer than 10 kb (Sokurenko *et al.*, 2001).

The full exploitation of the agarose/acrylamide gel assay format is still in progress. For example, the use of the Agilent BioAnalyzer lab-on-a-chip system for this assay would produce automation and very sensitive detection of PCR products up to 12 kb, provides options for background subtraction, and takes less than an hour. Because the proof-reading DNA polymerase Optimase is used in the SURVEYOR™ kit, the PCR products can be longer, while it allows trans-intronic PCR reactions to be performed for inbred strains. Besides leading to higher screening efficiency, longer PCR products also allow the positioning of the query region near the center of the PCR product, thereby making this assay easy and efficient even when an ordinary inexpensive agarose gel apparatus is used. This mutation detection format is truly fast, convenient, and amenable to automation because no harsh chemicals are used, all enzyme

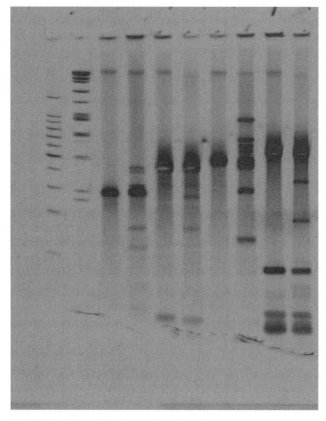

1 2 3 4 5 6 7 8 9 10

FIGURE 5.13 CEL I mutation detection double-strand DNA truncation assay of SNPs in the BRCA1 gene. Each PCR product was incubated with 0.025 units of CEL I for 17 hours at 37°C and resolved on 8% native PAGE in 1X TBE buffer. DNA bands were visualized with SYBRgreen fluorescence staining. Lanes 1. 100 base marker; 2. 1 kb marker; 3. Exon 2 control; 4. Exon 2 AG deletion; 5. Exon 20 control; 6. Exon 20 C insertion; 7. Exon 11.9 control; 8. Exon 11.9 4 bp deletion; 9. Exon 11.4 control; 10. Exon 1 1.4 G→A base substitution.

steps can be done in the same tube, it takes little over an hour, and processing of the CEL nuclease incised DNA is unnecessary prior to loading onto the native condition fragment analysis systems.

5.6 CONCLUSIONS

There has been dramatic improvement in the methods for screening for unknown mutations. These improvements already have translated into exciting applications in numerous research centers. For both the chemical and the enzymatic methods of mutation detection, PCR artifacts and DNA polymerization errors often limit the sensitivity these assays can deliver. One approach that potentially can remove these errors is the use of hairpin primers in PCR that

effectively link each pair of semiconservative DNA replication strands as one duplex (Kaur and Makrigiorgos, 2003). Any duplex that contains a PCR error can be removed with some mismatch-specific method to be determined so that error-free PCR products can be collected and used in high-sensitivity mutation detection. It is expected that more improvements will be forthcoming so that effective mutational screening will become a routine research and diagnostic tool for modern genetics.

References

Babon, J. J., McKenzie, M., and Cotton, R. G. (2003). The use of resolvases T4 endonuclease VII and T7 endonuclease I in mutation detection. *Mol. Biotechnol.* 23, 73–81.

Block, W. (1999). Improved detection of mutations in nucleic acids by chemical cleavage. Patent application number PCT/US98/16385.

Bui, C. T., and Cotton, R. G. H. (2002). Comparative study of permanganate oxidation reactions of nucleotide bases by spectroscopy. *Bioorg. Chem.* 30, 133–137.

Bui, C. T., Rees, K., Lambrinakos, A., Bedir, A., and Cotton, R. G. H. (2002). Site selective reactions of imperfectly matched DNA with small chemical molecules: applications in mutation detection. *Bioorg. Chem.* 30, 216–232.

Bui, C. T., Lambrinakos, A., Babon, J. J., and Cotton, R. G. H. (2003a). Chemical cleavage reactions of DNA on solid support: application in mutation detection. *BMC Chem. Biol.* 3, 1.

Bui, C. T., Lambrinakos, A., and Cotton, R. G. H. (2003b). Spectroscopic study in permanganate oxidation reactions of oligonucleotides containing single base mismatches. *Biopolymers* 70, 628–636.

Burdon, M. G., and Lees, J. H. (1985). Double-strand cleavage at a 2-base deletion mismatch in a DNA heteroduplex by Nuclease-S1. *Bioscience Reports* 5, 627–632.

Colbert, T. G., Till, B., Tompa, R., Reynolds, S. H., Steine, M., Yeung, A. T., McCallum, C. M., Comai, L., and Henikoff, S. (2001). High-throughput screening for induced point mutations. *Plant Physiol.* 126, 480–484.

Cotton, R. G. H., Rodrigues, H. R., and Campbell, R. D. (1988). Reactivity of cytosine and thymine in single base-pair mismatches with hydroxylamine and osmium tetroxide and its application to the study of mutations. *Proc. Natl. Acad. Sci. USA.* 85, 4397–4401.

Cotton, R. G. H. (1999). In *The Nucleic Acid Protocols Handbook*, R. Rapley, ed. Methods in Molecular Biology Series. Detection of mutations in DNA and RNA by chemical cleavage. pp. 685–693, John Wiley & Sons, Inc.

Cotton, R. G. H., and Bray, P. J. (2001). Using CCM and DHPLC to detect mutations in the glucocorticoid receptor in atherosclerosis: a comparision. *J. Biochem. Biophys. Methods* 47, 91–100.

De Santis, P., Palleschi, A., Savino, M., and Scipioni, A. (1990). Validity of the nearest-neighbor approximation in the evaluation of the electrophoretic manifestations of DNA curvature. *Biochemistry* 29, 9269–9273.

Gogos, J. A., Karayiorgou, M., Aburatani, H., and Kafatos, F. C. (1990). Detection of single base mismatches of thymine and cytosine residues by potassium permanganate and hydroxylamine in the presence of tetralkylammonium salts. *Nucleic Acids Res.* 18, 6807–6812.

Hardeland, U., Bentele, M., Lettieri, T., Steinacher, R., Jiricny, J., and Schar, P. (2001). Thymine DNA glycosylase. *Prog. Nucleic Acid Res. Mol. Biol.* 68, 235–253.

Haris, I. I., Green, P. M., Bentley, D. R., and Giannelli, F. (1994). Mutation detection by fluorescent chemical cleavage: application to hemophillia B. *PCR Meth. Appl.* 3, 268–271.

Hartge, P., Struewing, J. P., Wacholder, S., Brody, L. C., and Tucher, M. A. (1999). The prevalence of common BRCA1 and BRCA2 mutations among Ashkenazi Jews. *Am. J. Hum. Genet.* 64, 963–970.

Huang, J., Kirk, B., Favis, R., Soussi, T., Paty, P., Cao, W., and Barany, F. (2002). An endonuclease/ligase based mutation scanning method especially suited for analysis of neoplastic tissue. *Oncogene* 21, 1909–1921.

Kao, J. Y., Goljer, I., Phan, T. A., and Bolton, P. H. (1993). Characterization of the effects of a thymine glycol residue on the structure, dynamics and stability of duplex DNA by NMR. *J. Biol. Chem.* 268, 17787–17793.

Kaur, M., and Makrigiorgos, G. M. (2003). Novel amplication of DNA in a hairpin structure: towards a radical elimination of PCR errors from amplified DNA. *Nucleic Acids Res.* 31, e26.

Kennard, O. (1988). R. H. Sarma, and M. H. Sarma, eds. Structural studies of base pair mismatches and their relevance to theories of mismatches formation and repair. *Structure & Expression*, Vol. 2 in DNA and its drugs complexes. pp. 1–25, Adenine Press.

Kulinski, J., Besack, D., Oleykowski, C. A., Godwin, A. K., and Yeung, A. T. (2000). The CEL I Enzymatic Mutation Detection Assay. *Biotechniques* 29, 44–48.

Lambrinakos, A., Humphrey, K. E., Babon, J. J., Ellis, T. P., and Cotton, R. G. H. (1999). Reactivity of potassium permanganate and tetraethylammonium chloride with mismatched bases and a simple mutation detection protocol. *Nucleic Acids Res.* 27, 1866–1874.

Lishanski, A., Ostrander, E. A., and Rine, J. (1994). Mutation detection by mismatch binding-protein, MutS, in amplified DNA–Application to the Cystic-Fibrosis Gene. *Proc. Natl. Acad. Sci. USA.* 91, 2674–2678.

Lu, A. L., and Hsu, I. C. (1992). Detection of single DNA-base mutations with mismatch repair enzymes. *Genomics* 14, 249–255.

Mashal, R. D., Koontz, J., and Sklar, J. (1995). Detection of Mutations by Cleavage of DNA Heteroduplexes with Bacteriophage Resolvases. *Nat. Genet.* 9, 177–183.

Mullis, K. B., and Faloona, F. A. (1987). Specific synthesis of DNA in vitro via a polymerase-catalyzed chain reaction. *Methods Enzymol.* 155, 335–350.

Natrajan, G., Lamers, M. H., Enzlin, J. H., Winterwerp, H. H., Perrakis, A., and Sixma, T. K. (2003). Structures of Escherichia coli DNA mismatch repair enzyme MutS in complex with different mismatches: a common recognition mode for diverse substrates. *Nucleic Acids Res.* 31, 4814–4821.

Neuhausen, S. L., and Ostrander, E. A. (1997). Mutation testing of early-onset breast cancer genes BRCA1 and BRCA2. *Genet. Test.* 1, 75–83.

Norberg, T., Klaar, S., Lindqvist, L., Lindahl, T., Ahlgren, J., and Bergh, J. (2001). Enzymatic mutation detection method evaluated for detection of p53 mutations in cDNA from breast cancers. *Clin. Chem.* 47, 821–828.

Oleykowski, C. A., Bronson Mullins, C. R., Godwin, A. K., and Yeung, A. T. (1998). Mutation detection using a novel plant endonuclease. *Nucleic Acids Res.* 26, 4597–4602.

Oleykowski, C. A., Bronson Mullins, C. R., Chang, D. W., and Yeung, A. T. (1999). Incision at nucleotide insertions/deletions and basepair mismatches by the SP Nuclease of Spinach. *Biochemistry* 38, 2200–2205.

Patel, D. J., Kozlowski, S. A., Marky, L. A., Rice, J. A., Broka, C., Dallas, J., Itakura, K., and Breslauer, K. J. (1982). Structure, dynamics, and energetics of deoxyguanosine. Thymidine wobble base pair formation in the self-complementary d(CGTGAATTCGCG) duplex in solution. *Biochemistry* 21, 437–444.

Ramus, S. J., and Cotton, R. G. H. (1996). Single-tube chemical cleavage of mismatch: Successive treatment with hydroxylamine and osmium tetroxide. *Biotechniques* 21, 216–220.

Ren, J. (2001). In *Methods in Molecular Biology*, Vol. 163, K. R. Mitchelson and J. Cheng, eds. Chemical mismatch cleavage analysis by capillary electrophoresis with laser-induced fluorescence detection. Humana Press Inc. pp. 231–239. Totowa, NJ.

Roberts, R. J., and Cheng, X. (1998). Base flipping. *Annu. Rev. Biochem.* 67, 181–198.

Roberts, E., Deeble, V. J., Woods, C. G., and Taylor, G. R. (1997). Potassium permanganate and tetraethylammonium chloride are a safe and effective substitute for osmium tetroxide in solid-phase fluorescent chemical cleavage of mismatch. *Nucleic Acids Res.* 25, 3377–3378.

Rowley, G., Saad, S., Giannelli, F., and Green, P. M. (1995). Ultra rapid mutation detection by multiplex, solid-phase chemical cleavage. *Genomics* 30, 574–582.

Saleeba, J. A., and Cotton, R. G. H. (1991). ^{35}S-labelled probes improve detection of mismatched base pairs by chemical cleavage. *Nucleic Acids Res.* 19, 127–128.

Sanchez, A. M., Volk, D. E., Gorenstein, D. G., and Lloyd, R. S. (2003). Initiation of repair of A/G mismatches is modulated by sequence context. *DNA Repair* 2, 863–878.

Sokurenko, E. V., Tchesnokova, V., Yeung, A. T., Oleykowshi, C. A., Trintchina, E., Hughes, K. T., Rashid, R. A., Brint, J. M., Moseley, S. L., and Lory, S. (2001). Detection of simple mutations and polymorphisms in large genomic regions. *Nucleic Acids Res.* 29, e111.

Verpy, E., Biasotto, M., Meo, T., and Tosi, M. (1994). Efficient detection of point mutations on color-coded strands of target DNA. *Proc. Natl. Acad. Sci. USA.* 91, 1873–1877.

Wagner, R., Debbie, P., and Radman, M. (1995). Mutation Detection using Immobilized Mismatch Binding-Protein (MutS). *Nucleic Acids Res.* 23, 3944–2948.

Warner, E., Foulkes, W., Goodwin, P., Meschino, W., Blondal, J., Paterson, C., Ozcelik, H., Gross, P., Allingham-Hawkins, D., Hamel, N., Di Prospero, L., Contiga, V., Serruya, C., Klein, M., Moslehi, R., Honeyford, J., Liede, A., Glendon, G., Brunet, J. S., and Narod, S. (1999). Prevalence and penetrance of BRCA1 and BRCA2 gene mutations in unselected Ashkenazi Jewish women with breast cancer. *J. Natl. Cancer Inst.* 91, 1241–1247.

Yang, B., Wen, X., Oleykowski, C. A., Kodali, N. A., Miller, C. G., Kulinski, J., Besack, D., Yeung, J. A., Kowalski, D., and Yeung, A. T. (2000). Purification, cloning and characterization of the CEL I nuclease. *Biochemistry* 39, 3533–3541.

Youil, R., Kemper, B. W., and Cotton, R. H. (1995). Screening For Mutations by Enzyme Mismatch Cleavage With T4 Endonuclease-7. *Proc. Natl. Acad. Sci. USA.* 92, 87–91.

Youil, R., Kemper, B., and Cotton, R. (1996). Detection of 81 of 81 known mouse β-globin promoter mutations with T4 endonuclease VII-the EMC method. *Genomics* 32, 431–435.

Zhang, Y., Kaur, M., Price, B. D., Tetradis, S., and Makrigiorgos, G. M. (2002). An amplification and ligation-based method to scan for unknown mutations in DNA. *Hum. Mutat.* 20, 139–147.

CHAPTER **6**

Mutation Detection by Single Strand Conformation Polymorphism and Heteroduplex Analysis

GEORGE A. GARINIS,[1] PANAYIOTIS G. MENOUNOS,[2] GEORGE P. PATRINOS[1]
[1] Erasmus University Medical Center, Faculty of Medicine, and Health Sciences, MGC-Department of Cell Biology and Genetics, Rotterdam, The Netherlands;
[2] Nursing Military Academy, Laboratory of Research, Athens, Greece

TABLE OF CONTENTS

6.1 INTRODUCTION

Single strand conformation polymorphism (SSCP) and heteroduplex analysis (HDA) are two of the most popular electrophoresis-based mutation detection methods. Coupled to DNA amplification of the sequence to be analyzed, these techniques have become the methods of choice for a number of molecular diagnostic laboratories. This can be explained mainly by the numerous advantages, namely their technical simplicity and relatively high specificity for the detection of sequence variations, the low operation costs, and the potential for automation for high-throughput mutation analysis. However, there are several factors that influence sensitivity, and therefore need to be taken into account in order to obtain reproducible results as well as to maximize the sensitivity of mutation detection. In the following pages, the theory and practice of both SSCP and HDA will be discussed. In particular, emphasis will be given to the principle, the parameters influencing the sensitivity and reproducibility of the results, the available detection schemes, and the limitations of both techniques. Finally, a number of applications for screening genomic loci in order to investigate the underlying molecular heterogeneity will be discussed, together with the automation potential of SSCP, allowing for high throughput analysis of the genomic variation.

6.2 PRINCIPLES OF SINGLE STRAND CONFORMATION POLYMORPHISM ANALYSIS

The analysis of single strand conformation polymorphism (SSCP) has been established by Orita and colleagues (1989) as a simple, efficient, and reliable method for the detection of sequence alterations in genomic loci. Based on polymerase chain reaction (PCR), SSCP was developed soon after the introduction of PCR technology, and relied on the fact that relatively short single-stranded DNA fragments can migrate in a nondenaturing gel not only as a function of their size but also of their sequence. In other words, following amplification of any given DNA sequence, the amplified DNA fragments are subjected to denaturation with either heat or chemical agents, such as formamide. Subsequently, the denatured DNA fragments are electrophoresed through a native (nondenaturing) polyacrylamide gel. During electrophoresis, single-stranded DNA fragments adopt a specific three-dimensional shape according to their nucleotide sequence, and hence exhibit a unique conformation. Therefore, their electrophoretic mobility is dependent upon the previously mentioned three-dimensional shape (see Fig. 6.1). Based on these principles, it is well understood that even a single base difference between a DNA fragment being tested and its wild type counterpart is sufficient to adopt a different

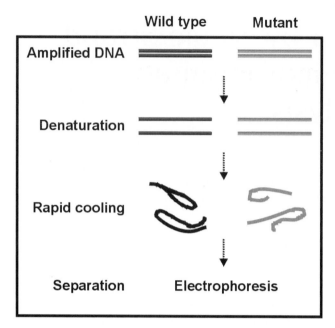

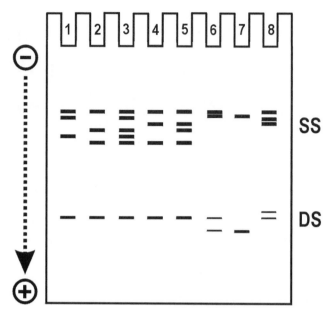

FIGURE 6.1 Schematic representation of the SSCP principle. Amplified DNA for both the wild type and mutant alleles is subjected to denaturation and then to immediate cooling, where the denatured single stranded alleles adopt a specific conformation. As the single stranded wild type and mutant alleles have different conformations, they can be distinguished when electrophoresed through a native polyacrylamide gel.

FIGURE 6.2 Schematic drawing of a typical result from a SSCP analysis. After staining, the gel can be divided into two parts: the part with the bands corresponding to the single-stranded alleles (SS) and the other part with the double-stranded bands (DS), as complete denaturation is sometimes not feasible. In this example, lane 1 corresponds to the wild type pattern for any given locus analyzed. If lane 2 corresponds to a homozygous case for mutation A, then lane 3 corresponds to a heterozygous case for the same mutation, and lanes 4 and 5 correspond to a homozygous case for mutation B and a compound heterozygous case for mutations A and B, respectively. Lanes 6 to 8 are characteristic examples of the electrophoretic pattern of small deletions or insertions, which can also be distinguished at the DS part of the gel (see also Fig. 6.4b): Lanes 6 and 7, heterozygous and homozygous cases for a small deletion, respectively; Lane 8, heterozygous case for a small insertion.

conformation and thus to migrate at a different position during electrophoresis (see Fig. 6.2).

SSCP analysis is extremely advantageous for fast mutation screening of known loci because it is easy to use, precludes the use of radioactive substances for detection (see also Section 6.3), and requires no special equipment. On the other hand, SSCP is not considered the method of choice for the analysis of unknown sequences, as there is no theoretical background established so far that could enable one to predict the exact electrophoretic mobility of a given DNA fragment according to its sequence (as is the case with Denaturing Gradient Gel Electrophoresis (DGGE; see also Chapter 8). It is noteworthy though, that despite the fact that SSCP is still under constant development, this assay has enjoyed already an enormous success in identifying sequence variations within the various PCR products being tested.

There are several parameters of SSCP analysis influencing the pattern of single strand conformation that will eventually affect electrophoretic migration. These are DNA amplification (such as the size of PCR products), denaturation, and the electrophoretic conditions (the length and duration of gel run, the ionic strength in the buffer being used, the temperature as well as the gel matrix composition). Another important aspect of SSCP analysis is the visualization of the single-stranded DNA fragments, using a number

of detection methods. All these aspects will be discussed in detail in the following paragraphs.

6.2.1 DNA Amplification

Certain key aspects during amplification of the DNA fragment in question are of great importance in the performance of the SSCP analysis. Preferably, the size of the DNA fragment to be amplified should be between 150 and 350bp. With certain exceptions, where mutation detection results have been reported for fragments of approximately 550bp, this range is relatively safe and has reached a consensus between different laboratories (Hayashi, 1991; Hayashi and Yandell, 1993). Typically, one should start to screen DNA fragments that will lie within the range of 100 to 350bp, although this is often dependent on the sequence of the DNA fragment, the GC content, as well as the various electrophoretic conditions being used. However, if longer PCR fragments are to be screened, it is advisable to digest the fragment with restriction enzymes. The choice of restriction enzymes may affect considerably the resulted conformation and hence the performance of the SSCP assay. In general, the longer the fragments, the harder it is for any given single nucleotide change to have an effect in the conformation of the fragment, although under certain circumstances detection efficiency is not uniformly decreased with increasing DNA fragment length. It is noteworthy that sensitivity of SSCP can be greatly improved even for fragments as big as 800 bp, by running the electrophoresis in low pH buffer systems and at a fixed temperature (Kukita *et al.*, 1997).

After amplification, all PCR products should be screened on an agarose gel for the desired product length. If, despite considerable efforts, undesired side products are difficult to eliminate, a PCR product purification protocol should be applied. In addition, a negative and where possible, a positive control of each PCR product should be included during each amplification step, since this will be the only indirect evidence that the screening assay is performed on the desired PCR products. PCR products can be stored on DNase-free tubes at 4°C. However, bear in mind that prolonged storage should be avoided and the subsequent steps of SSCP protocol should be performed as soon as possible.

6.2.2 Denaturation

In SSCP analysis, it is important to achieve complete and as much irreversible denaturation of the DNA strands as possible. Incomplete denaturation, partial folding, and reannealing to double-stranded DNA will greatly reduce the amount of single-stranded DNA in the assay and will subsequently affect detection of the single-stranded molecules. Usually, denaturation of the PCR products is carried out by incubation to high temperature, that is, 95°C for 5 to 7 min, and immediately after are chilled on ice for approximately 10 min. Alternatively, there are numerous denaturing agents, such as formamide, methyl-mercuric hydroxide, sodium hydroxide, and urea, which seem to perform well (Humphries *et al.*, 1997). Formamide is the most commonly used denaturing agent. In certain conditions, it is advisable to use 5–10% glycerol prior to loading the samples on the gel. This strategy was previously shown to produce sharper DNA bands, which greatly facilitate subsequent interpretation of results.

6.2.3 Electrophoretic Parameters

Prior to reviewing the various electrophoretic parameters that may influence the SSCP analysis, it is worthwhile mentioning that currently there is neither adequate theory nor any physicochemical model available that could allow one to predict the three-dimensional structure of any given single-stranded DNA fragment, and as a result, its electrophoretic mobility. Apart from the size of the DNA fragment and its GC content, the following parameters have been empirically found to affect the sensitivity of SSCP analysis: the gel matrix composition, the buffer composition (ionic strength, the pH and buffer supplements, such as glycerol), the duration of gel run, the gel length, the DNA concentration, and the electrophoresis temperature. The effect of these factors on SSCP resolution and sensitivity is outlined next.

1. *Gel matrix composition.* The most common and widely accepted matrix is a cross-linked acrylamide polymer (8–12%). The small pore size of acrylamide-derived matrices makes it ideal for enhanced resolution and discrimination even at the nucleotide level. Higher resolution can be achieved upon addition of 10–15% of sucrose or glycerol. It has been previously shown that the mutation detection enhancement (MDE) gel (FMC Bioproducts) had a mutation detection rate of approximately 95% (Ravnik-Glavac *et al.*, 1994). Although there is a considerable variability in the percentage of mutations detected with this commercially available gel matrix, it should be noted that in a considerable and rather growing number of studies for which SSCP has been implemented, MDE gels were used instead of the standard acrylamide.

2. *Buffer composition.* So far, the Tris-borate buffer is the buffer of choice for most investigators in SSCP analysis. However, in certain instances HEPES buffer has been demonstrated to offer an alternative solution that may increase the sensitivity of the SSCP. The addition of 5% glycerol has been shown to lower the pH and to decrease the electrostatic repulsion between the negatively charged phosphates in the nucleic acid backbone, resulting in a higher resolution between the mutant and wild type DNA fragments. Conformational structures can also be more compacted by increasing the salt concentration. Finally, buffer systems with low pH have been shown to increase the sensitivity for mutation detection in larger DNA fragments (Kukita *et al.*, 1997).

3. *Gel length and duration of gel run.* There is a considerable variation in the duration of the gel run that has been adopted by the various diagnostic laboratories. It is inevitable that the time of electrophoresis is dependent on both the length of the gel as well as the applied voltage. It is preferable to start the electrophoresis with a relatively moderate voltage and increase it as soon as the PCR fragments have been migrated into the gel. The length could vary between 10 and 40 cm. In general the bigger the gel length, the better the resolution, since at several occasions the conformational changes of the wild type and mutant single-stranded alleles are so minor that they may migrate at very close proximity to each other.

4. *Temperature.* Several laboratories have demonstrated in the past the importance of temperature on the conformational changes of the DNA fragments. It is conceivable that the known temperature effect on the stabilization of

the secondary structure of single-stranded DNA fragments may affect to a varying degree (depending on the primary sequence) the SSCP results. In addition to a lower pH, decreasing the temperature to 4°C has been shown to enhance the stability of the conformation for any given single-stranded DNA fragment. It is advisable that one should try a gradual temperature decrease, starting from 15°C and descending with an increment of 2–4°C at a time.

5. *DNA concentration.* In the past, several investigators have used high DNA concentrations in order to enhance detection. Unfortunately, this has often led to a decrease in the specific concentration of the single-stranded DNA. Even after the addition of formamide, it has been empirically shown that at high concentrations, the two single-stranded molecules tend to reanneal and form a double-stranded DNA. It is preferable, therefore, to keep the DNA concentration relatively low in the loading buffer. Also, gel overloading can sometimes result in abnormal migration of the bands, leading to decreased resolution.

6.2.4 Detection

Most molecular diagnostic laboratories are well adapted in the use of radioactivity and have the equipment and expertise required for this type of protocol. Typically, autoradiography will require immobilization of the gel, a drying step, and finally film exposure. However, if radioactivity has not been used before, it will be rather difficult to establish the protocol. Moreover, it is more time-consuming than some of the more recently described protocols that utilize silver staining or detection with fluorescent dyes, such as the SYBR Green. Alternatively, SSCP analysis can be done on automated sequencers with fluorescent dye-labeled DNA fragments. In relative terms, and although ethidium bromide staining can be considered an attractive alternative to the use of fluorescent substances (Yap and McGee, 1992a), the silver staining approach comprises one of the most straightforward, fast, as well as sensitive methods for the visualization of bands on SSCP gels. In brief, after electrophoresis, the polyacrylamide gels are first fixed with 10% acetic acid for approximately 30 min at room temperature and subsequently washed with water. Depending on the concentration of silver nitrate, incubation with the silver nitrate solution can last for approximately 60 min (on a 0.001% $AgNO_3$, 0.036% formaldehyde solution). This incubation step is performed in the dark, while avoiding any contamination with protein-containing solution (proteins are stained extensively with silver nitrate leading to immense background). Subsequently, the polyacrylamide gels are washed with water and color development is performed by incubating the gel for 5 to 10 min with a color development solution (containing 2.5% Na_2CO_3, 0.036% formaldehyde and 0.002% sodium thiosulfate). Color development can be stopped with a solution containing a chelating agent (such as 1.5% EDTA). Gels can be subsequently fixed with 30% ethanol and 4% glycerol. The stained gels are transferred to a vacuum dryer and are immobilized to a porous paper. Results are interpreted visually or can be analyzed by means of an image analysis system.

6.3 HETERODUPLEX ANALYSIS FOR MUTATION DETECTION

The principle of heteroduplex analysis (HDA) is simple and closely related to that of SSCP. In brief, heteroduplexes are formed between different DNA alleles; for example, by mixing wild type and mutant amplified DNA fragments, followed by denaturation at 95°C and slow reannealing to room temperature (White *et al.*, 1992). If the target DNA consists of different alleles already—for example, a heterozygous case—then heteroduplexes are formed automatically during the amplification step. The result is the formation of two homoduplexes and two heteroduplexes, which are retarded during electrophoresis in native polyacrylamide gels (see Fig. 6.3). There are two types of heteroduplex molecules, depending on the type of the mutation, which in turn, reflects on their stability. In other words, small deletions or insertions create stable heteroduplexes, termed *bulge type* heteroduplexes, which have been verified by electron microscopy (Wang *et al.*, 1992). On the other hand, single base substitutions form the so-called *bubble type* heteroduplexes, which are much more difficult to visualize, and optimization of the experimental conditions is required to achieve optimal resolution of this type of heteroduplex.

A number of molecular diagnostic techniques, based on heteroduplex formation, are reported in the literature, such as enzymatic or chemical cleavage (see Chapter 5) and Denaturing Gradient Gel Electrophoresis (DGGE; see Chapter 8), but HDA appears to be the most attractive one, as it can be performed rapidly on short gels without the need of specialized equipment and the use of radioactivity. The most typical example of the use of HDA for mutation screening is the rapid detection of the 3-bp $\triangle$F508 deletion in the CFTR gene, leading to cystic fibrosis (Wang *et al.*, 1992).

6.4 SENSITIVITY AND LIMITATIONS

Several factors can affect the sensitivity of both SSCP and HDA analysis and their optimization is highly empirical, as there is no adequate theoretical basis or type of algorithm (as in DGGE; Myers *et al.*, 1987) that would enable researchers to predict the three-dimensional conformation of the single-stranded DNA fragments under specific experimental procedures. Those elements that frequently affect the sensitivity of both methods will be discussed in this section.

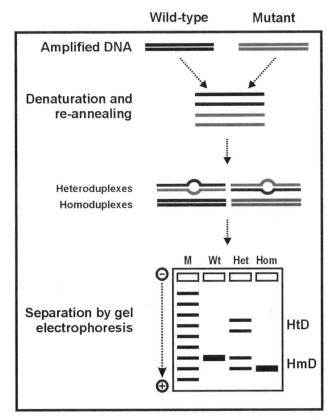

FIGURE 6.3 The HDA principle. Unlike SSCP analysis, amplified DNA is now subjected to denaturation followed by slow reannealing of the denatured alleles, leading to both homoduplexes (HmD) and heteroduplexes (HtD). The latter migrate slower during gel electrophoresis, due to their sequence mismatch(es) and therefore, heterozygous (Het) and homozygous (Hom) cases can be easily distinguished from the wild type (Wt) based on their electrophoretic pattern. M: size marker.

6.4.1 Optimizing SSCP Sensitivity

In general, most of the single-stranded DNA fragments will have a more compacted conformation at lower temperatures or in the presence of higher salt concentrations. It follows, however, that by increasing the salt concentration, the conductivity is also increased, thus having a rather significant effect in temperature. Depending on these parameters, which largely have been determined empirically, it is suggested that electrophoresis be performed at room temperature with 5–10% glycerol or at 4°C without addition of glycerol or salt; this is considered a good starting point. In addition, it is highly suggestive that an empirical estimate of SSCP sensitivity will be based on the likelihood of detecting known sequence variations under controlled conditions. This is expected to provide a valuable means by which the effect on electrophoretic mobility of known sequence variations will be determined. Under no circumstances should a perfect overall detection efficiency be anticipated. Hayashi and Yandell (1993) have conducted a valuable study to deter-

mine which are the most effective sets of parameters influencing the detection of a single mutation. Given that the latter is largely dependent on DNA fragment's size, it is expected that for most fragments shorter than 200 nucleotides, more than 90% of the sequence variations will be detected. Gradually, as the size of the PCR fragment increases to 300–350 nucleotides, a safe prediction is that more than 80% of mutations will be detected. It should be noted however, that within this range, Sheffield and colleagues (1993) have shown that the overall sensitivity of the SSCP is not affected by the position of a given mutation.

Alternatively, a more quantitative method for estimating the sensitivity of the SSCP detection method based on statistical arguments has been developed. This method is based on the fact that the chance of any given strand to exhibit a mobility shift is independent from the other strand. On this assumption, the probability of observing shifts in both strands (P_2), in one strand (P_1), or in none of the strands (P_0), is equal to x^2, $2x(1 - x)$ and $(1 - x)^2$, respectively, where x is the sensitivity of the technique when only one of the strands is labeled. The ratio of P_2 to P_1 is equal to the observed number of the mutations on both strands over the number of the observed mutations on the single strand [$r = (x/2(1 - x)$]. The useful sensitivity of the technique can be calculated as being equal to $1 - P_0$. Based on this probabilistic theory, the estimates from previous studies are that the sensitivity for 100–200 bp fragments is approximately 96%, regardless of the presence or absence of glycerol (Hayashi and Yandell, 1993). However, by adding glycerol, the sensitivity is still high for fragments ranging from 200 to 300 bp, but is decreasing when glycerol is not used. The latter may indicate the inability of the former calculation to depict the actual electrophoretic mobility, which to some extent is expected. In addition, it may also explain the fact that substitutions, which induce significant conformational changes, are most likely to have an effect, to a variable extent, on the other strand. Despite this discrepancy, it can be safely concluded that decreasing the fragment size will greatly enhance the sensitivity. It seems rational that the overall effect of a given mutation is displayed more efficiently when the total number of nucleotides surrounding that particular mutation is less. It is also profound that glycerol greatly increases the overall sensitivity when electrophoresis is performed at room temperature. Inevitably, if a fragment is larger than the recommended size, it is advisable to use restriction enzymes (where restriction sites are readily available), thus further facilitating mutation detection.

Practical observations suggest that any fragment that exhibits a differential mobility often will migrate very close to the reference fragment. However, the overall fragment number is not always predictable in advance. Any given number of conformations may be supported to a variable extent by the applied electrophoretic parameters. Furthermore, the band intensity is irrelevant to allelic differences, as it is strictly dependent on the different conformations. Therefore, SSCP is not a safe method to predict gene-dosage effects. In addition, the simultaneous detection of more than one mutation in a single DNA sample is not easily predictable, as previous data have shown that the electrophoretic pattern may vary considerably within different experiments.

Attempts to further improve the sensitivity of the SSCP have led to the development of the RNA-SSCP approach (Sarkar *et al.*, 1992). Here, although the method is essentially the same, the double-stranded DNA is converted to the corresponding single-stranded RNA by means of one of the two primers that has phage promoter sequences on its 5′ end. The amplified product is served as a product for *in vitro* transcription. This method has shown a higher sensitivity compared to the previously described SSCP methodology. Its sensitivity may rely on the fact that the *in vitro* transcribed strand has no complementary strand to reanneal with. Therefore, sufficient amounts of the *in vitro* transcribed product can be electrophoresed and easily analyzed even by ethidium bromide staining (Sarkar *et al.*, 1992). So far, there has been much discussion concerning the false negative results of this assay and possible ways to minimize them. However, bear in mind that false positive results also affect the net outcome of the SSCP analysis. In order to minimize the frequency of reporting false positive results, it is advisable to perform repeated SSCP electrophoretic runs (particularly when SSCP data are of clinical interest). An additional way is to determine the minimum mobility variation, which is detectable within the context of laboratory SSCP conditions. In practical terms, mobility differences of 3 mm are generally clear, but a detection difference of only 2 mm requires excellent gel running conditions and is often subjective. Therefore, any difference smaller or equal to 2 mm can be considered only with reservation.

Reproducibility is the last and perhaps most essential parameter applicable to most techniques in molecular diagnostics. In general, if conditions are kept constant then the resulting reproducibility is usually high. Nevertheless, Hayashi (1991) has previously suggested that DNA sequences may have different stable conformations. The latter is thus interpreted as a variable that may compromise SSCP reproducibility. It relies on the possibility that when the free energy difference between different conformations is small, an oscillation between different structures of comparable energies may be observed.

6.4.2 Sensitivity of HDA

So far, the sensitivity of the HDA methodology has not been determined to the extent of the SSCP analysis. However, Rossetti and colleagues (1995) compared directly both SSCP and HDA assays for the detection of known mutations in a panel of four genes. Despite the fact that none of the assays was performed with 100% efficiency, HDA detected slightly more mutations than SSCP in the same samples. Interestingly, these authors suggested that both techniques could be used in concert to detect all mutations.

Another aspect to improve the performance of HDA is the gel matrix. As in SSCP, the use of MDE (derived from HydroLink D5000™; Keen *et al.*, 1991) has basically made HDA a valuable mutation detection technique. Today, the majority of the diagnostic laboratories, in which HDA is the method of choice to detect genomic variation, are employing MDE gels.

Finally, as already mentioned in Section 6.3, the *bubble type* heteroduplexes, which are formed due to the presence of single base substitutions, are much more difficult to be visualized compared to the *bulge type* heteroduplexes. In order to overcome this bottleneck, and based on the observation that heteroduplexes are much easier to visualize when a deletion or insertion mutation is involved, the Universal Heteroduplex Generator (UHG) was conceived (Wood *et al.*, 1993a; Wood *et al.*, 1993b). In brief, the UHG consists of a synthetic DNA fragment, which bears a small (that is, 2 to 5-bp) deletion. This synthetic fragment is amplified by the use of the same oligonucleotide primers as the DNA under study. After amplification, the test amplicon is mixed with the amplified UHG, denatured, and then slowly reannealed, followed by electrophoresis. If no mutation is present in the test DNA, then only a bulge type heteroduplex will be present, slightly retarded compared to the homoduplex. If, however, a single base substitution is also present, then the resulting heteroduplex will have two mismatches: a bulge and a bubble type, which will result in the heteroduplex migrating significantly lower, compared to the simple bulge type heteroduplex. The use of UHG has been reported for the detection of known mutations within a number of loci, such as von Willebrand disease (Wood *et al.*, 1995), phenylketonurea (Wood *et al.*, 1993a), and for prenatal determination of blood group alleles (Stoerker *et al.*, 1996).

6.5 DETECTION OF THE UNDERLYING GENOMIC VARIATION USING SSCP AND HDA

The wealth of genomic sequence information, made available from the completion of the Human Genome Project, has propelled the implementation of a number of mutation detection techniques for the identification of the underlying genomic variation. Due to their numerous advantages, SSCP and HDA analysis are nowadays the methods of choice in a growing number of private or public molecular diagnostic laboratories to either interrogate known mutations or scan for known or unknown mutations in short stretches of DNA and in relatively short time. Similarly, SSCP and HDA also have been proven to be invaluable tools for basic science, enabling both the identification of causative genes for human hereditary diseases and mapping of genomic loci.

A short summary of the existing applications of SSCP and HDA follows in the next paragraphs, which is only indicative for the applicability of these techniques in almost every genomic locus.

6.5.1 Applications in Basic Science

The utilization of SSCP in basic science as a tool for genomic DNA analysis is well established. Mutations in several key candidate genes implicated in various cell processes have now been identified, and the extent of those mutations as well as their frequency has given an insight into the role of these molecules in the relevant processes.

Damage to DNA is considered to be the main initiating event by which genotoxins cause hereditary effects and cancer. An accumulation of mutations throughout the genome will eventually result in cell death or in a cascade of events, which in turn may initiate malignant transformation. Therefore, it is not surprising that SSCP analysis was first used in screening candidate genes in tumourigenesis (Suzuki *et al.*, 1990; Yap and McGee, 1992b). Since then, several tissue specimens have been examined and nearly all possible tumour types, isolated from a variety of tissue resources, have been considered for mutations in several key suspect genes. A difficulty often was the fact that many genes such as the pRB or BRCA genes have an enormous size, comprising several exons, and their analysis was often cumbersome if not impossible. Nevertheless, a number of research groups, including ours, have identified a multitude of p53 and nm-23 genomic alterations in almost every tumor type, such as in breast, colorectal (see Fig. 6.4), prostate, ovarian, and so on. Where SSCP failed to identify mutations, several investigators were able to point out abnormalities at the transcriptional or protein levels. In effect, this methodology proved to be particularly useful in revealing, in a stepwise approach, that the altered expression of several cell cycle regulatory molecules either at the genomic or transcriptional and protein levels may exert a synergetic effect on tumor growth and chromosomal instability on breast cancer and non-small–cell lung and colorectal carcinomas.

In addition to the utilization of SSCP methodology as a screening tool, several investigators previously have employed this technique for gene mapping in mouse genes (Beier, 1993 and references therein). The methodology is based on the fact that a given polymorphism readily can be found in noncoding regions of genes such as the 3′ untranslated regions or introns, between alleles of mouse species, and in several occasions between inbred strains as well. The segregations of these polymorphisms can be analyzed with recombinant inbred or interspecific crosses and the strain distribution pattern obtained can be compared with that for other markers and analyzed by standard linkage analysis algorithms. For instance, SSCP has previously been employed to localize 39 mouse-specific sequence-tagged sites (STSs), generated from mouse-hamster somatic cell

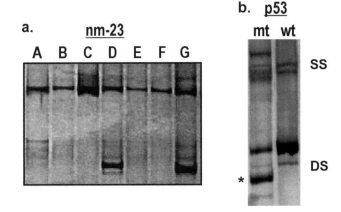

FIGURE 6.4 Typical examples of silver stained SSCP gels, in which different genomic loci, responsible for tumourigenesis, are analyzed. **a.** Analysis of the nm-23 gene. Lanes A, D, and G correspond to three different heterozygous cases; lanes B, C, E, and F are normal individuals (Garinis G., unpublished). **b.** Analysis of the human p53 locus. Different electrophoretic mobility of both the single-stranded (SS) and double-stranded (DS) alleles, due to a small deletion (depicted by an asterisk) in a heterozygous breast cancer affected individual (mt), compared to the wild type (wt) electrophoretic pattern (adapted from Patrinos *et al.*, 1999, with permission).

hybrids. These were subsequently integrated with other markers to generate a high-density map of mouse chromosome 1 containing over 100 markers typed on a single interspecific backcross (Watson *et al.*, 1992). SSCP analysis also has been applied in relatively fewer cases for the linkage analysis of human genes (Nishimura *et al.*, 1993; Avramopoulos *et al.*, 1993).

6.5.2 Molecular Diagnostic Applications

Both SSCP and HDA can be successfully used for the detection of known mutations in any genomic locus. A brief summary of the numerous applications of SSCP analysis for various human genes is given in Table 6.1. When SSCP analysis is coupled to nonradioactive detection schemes, then it most certainly becomes the method of choice for routine molecular diagnostic analysis (see Fig. 6.5 for representative examples from G6PD mutation screening; Menounos *et al.*, 2000). Additionally, HDA strategies have been designed for the detection on genetic defects in a number of human genes, such as CFTR (Rommens *et al.*, 1990), neurofibromatosis type 1 (Shen *et al.*, 1993) and type 2 (Sainz *et al.*, 1994), APC (Paul *et al.*, 1993), PKU (Wood *et al.*, 1993a), and so on.

Furthermore, SSCP and HDA can be used equally well for molecular typing in clinical microbiology (see also Chapter 22). Recently, Nair and colleagues (2002) have investigated the possibility of using SSCP analysis for the detection of nucleotide variation at the groEL gene, in order

TABLE 6.1 Summary of the majority of genomic loci, for which a SSCP mutation analysis strategy is designed and implemented.

Genomic loci	Disease/syndrome	References
Ras	Various types of cancer	Suzuki *et al.*, 1990
TP53		Sheffield *et al.*, 1993; Kutach *et al.*, 1999
BRCA1	Susceptibility to breast cancer	Castilla *et al.*, 1994
BRCA2		Phelan *et al.*, 1996
RB1	Retinoblastoma	Hogg *et al.*, 1992; Shimizu *et al.*, 1994
APC	Adenomatous polyposis coli	Groden *et al.*, 1993; Varesco *et al.*, 1993
CYP21	Congenital adrenal hyperplasia	Bobba *et al.*, 1997
PAH	Phenylketonurea	Dockhorn-Dworniczak *et al.*, 1991; Labrune *et al.*, 1991
G6PD	Glycose-6-dehydrogenase deficiency	Calabro *et al.*, 1993
HFE	Hereditary hemochromatosis	Hertzberg *et al.*, 1998
Factor VII	Hemophilia A	Economou *et al.*, 1992
Factor IX	Hemophilia B	David *et al.*, 1993
α-spectrin	Hereditary elliptocytosis	Maillet *et al.*, 1996
β-spectrin	and spherocytosis	
Ankyrin-1	Hereditary spherocytosis	Eber *et al.*, 1996
Band-3		Jarolim *et al.*, 1996
α-globin	α-thalassemia	Harteveld *et al.*, 1996
β-globin	β-thalassemia	Takahashi-Fujii *et al.*, 1994
FAA	Fanconi anemia	Levran *et al.*, 1997
CFTR	Cystic fibrosis	Claustres *et al.*, 1993
PKD1	Polycystic kidney disease 1	Afzal *et al.*, 1999
PKD2	Polycystic kidney disease 2	Veldhuisen *et al.*, 1997
LDL receptor	Familial hypercholesterolemia	Day *et al.*, 1995
NF1	Neurofibromatosis type 1	Upadhyaya *et al.*, 1995
NF2	Neurofibromatosis type 2	MacCollin *et al.*, 1996
MECP2	Rett syndrome	Zappella *et al.*, 2003
Glucocerebrosidase	Gaucher disease	Kawame *et al.*, 1992
Dystrophin	Duchenne/Becker muscular dystrophy	Tuffery *et al.*, 1993
WFS1	Wolfram syndrome	Strom *et al.*, 1998
Cx26	Autosomal recessive nonsyndromic hearing loss	Scott *et al.*, 1998
GJA1 (Cx43)		Liu *et al.*, 2001

to differentiate salmonella strains both at the interserovar and the intraserovar levels. In this study, SSCP analysis has exhibited the potential to complement classic typing methods such as serotyping and phage typing for the identification of salmonella serovars, due to its rapidity and simplicity. SSCP analysis of the groEL genes of carious Salmonella serovars produced various SSCP profiles, which indicates the potential of this technique to differentiate various salmonella serovars.

6.6 CONCLUSIONS AND FUTURE ASPECTS

As previously mentioned, SSCP and HDA are simple, reliable, and sensitive methods for the detection of nucleotide sequence changes in genomic loci. The methods can detect single nucleotide substitutions, insertions, or deletions of a short nucleotide sequence accurately and in a relatively short time, and therefore they can be used for DNA analysis of human cancers and other genetic disorders. Almost all the polymorphic base substitutions, thought to be present every

few hundred base pairs in genomic DNA, also can be detected and used as genetic markers. In addition, the identification of known mutations and polymorphisms can readily be standardized, and a role of SSCP and/or HDA analysis in routine laboratory screening procedures has been already demonstrated.

Finally, the generation of the draft human genome sequence has imposed an increasing need for methods and technology that can be used for high-throughput mutation screening in a large number of DNA samples. Initially, improvements in SSCP included sample handling, allowing mutation analysis in an extremely rapid manner (Whittal *et al.*, 1995). In particular, a 96-well format was employed for storing DNA and subsequently for amplification and mutation screening, requiring no tube labeling and reducing the probability of contamination by using a multipipette. In addition, higher throughput also was achieved by modification of the gel system. Ultimately, capillary electrophoresis coupled with SSCP analysis offers unprecedented opportunities for large-scale mutation screening. Moreover, performance of capillary electrophoresis/SSCP in a microarray

G6PD

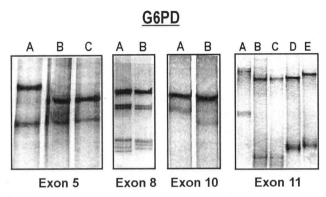

Exon 5   Exon 8   Exon 10   Exon 11

FIGURE 6.5 Analysis of the coding region of the human G6PD locus (adapted from Menounos *et al.*, 2000, with permission). Exon 5: Lanes B and C correspond to the wild type pattern and lane A corresponds to a heterozygous case for a point mutation. Exon 8: Lane A: normal individual, Lane B: heterozygous case. Exon 10: Lane A: heterozygous case for a point mutation, Lane B: normal individual. Exon 11: Lanes B and C correspond to the wild type pattern, Lanes A and D correspond to heterozygous cases for different mutations, and Lane E corresponds to a heterozygous case with two point mutations *in cis*, one of which is the same as in Lane D (note the minor mobility difference of the bands for the single-stranded alleles between Lanes D and E).

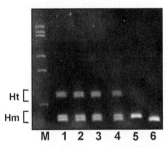

FIGURE 6.6 Mutation screening in the Exon 10 of the CFTR gene for the △F508 mutation, leading to cystic fibrosis, using HDA. Electrophoresis is performed in a nondenaturant 8% acrylamide gel. The amplification product of the wild type allele is 97 bp and that of the mutant allele is 94 bp. Lanes 1 to 4 correspond to △F508 heterozygotes, Lane 6 corresponds to a △F508 homozygote, and Lane 5 to a wild type individual. The electrophoretic mobility of the heteroduplexes (Ht) is retarded compared to the homoduplexes (Hm). M: φX174/HaeIII size marker (photo courtesy of Dr. Angeliki Balassopoulou, Athens, Greece).

format also has been shown to decrease the overall analysis time, without any significant loss in resolution (Medintz *et al.*, 2000; Medintz *et al.*, 2001). These advances will be discussed extensively in the following chapter. One way or another, one can safely predict that SSCP and HDA will continue to be extremely popular mutation detection methodologies and they will substantially contribute to our growing knowledge of the human genetic variation as well as of other organisms.

References

Afzal, A. R., Hand, M., Ternes-Pereira, E., Saggar-Malik, A., Taylor, R., and Jeffery, S. (1999). Novel mutations in the 3 region of the polycystic kidney disease 1 (PKD1) gene. *Hum. Genet.* 105, 648–653.

Avramopoulos, D., Cox, T., Kraus, J. P., Chakravarti, A., and Antonarakis, S. E. (1993). Linkage mapping of the cystathionine beta-synthase (CBS) gene on human chromosome 21 using a DNA polymorphism in the 3′ untranslated region. *Hum. Genet.* 90, 566–568.

Beier, D. R. (1993). Single-strand conformation polymorphism (SSCP) analysis as a tool for genetic mapping. *Mamm. Genome.* 4, 627–631.

Bobba, A., Iolascon, A., Giannattasio, S., Albrizio, M., Sinisi, A., Prisco, F., Schettini, F., and Marra, E. (1997). Characterisation of CAH alleles with non-radioactive DNA single strand conformation polymorphism analysis of the CYP21 gene. *J. Med. Genet.* 34, 223–228.

Calabro, V., Mason, P. J., Filosa, S., Civitelli, D., Cittadella, R., Tagarelli, A., Martini, G., Brancati, C., and Luzzatto, L. (1993). Genetic heterogeneity of glucose-6-phosphate dehydrogenase deficiency revealed by single-strand conformation and sequence analysis. *Am. J. Hum. Genet.* 52, 527–536.

Castilla, L. H., Couch, F. J., Erdos, M. R., Hoskins, K. F., Calzone, K., Garber, J. E., Boyd, J., Lubin, M. B., Deshano, M. L., Brody, L. C., Collins F. S., and Weber, B. L. (1994). Mutations in the BRCA1 gene in families with early-onset breast and ovarian cancer. *Nat. Genet.* 8, 387–391.

Claustres, M., Laussel, M., Desgeorges, M., Giansily, M., Culard, J. F., Razakatsara, G., and Demaille, J. (1993). Analysis of the 27 exons and flanking regions of the cystic fibrosis gene: 40 different mutations account for 91.2% of the mutant alleles in southern France. *Hum. Mol. Genet.* 2, 1209–1213.

David, D., Rosa, H. A., Pemberton, S., Diniz, M. J., Campos, M., and Lavinha, J. (1993). Single-strand conformation polymorphism (SSCP) analysis of the molecular pathology of hemophilia B. *Hum. Mutat.* 2, 355–361.

Day, I. N., Whittall, R., Gudnason, V., and Humphries, S. E. (1995). Dried template DNA, dried PCR oligonucleotides and mailing in 96-well plates: LDL receptor gene mutation screening. *Biotechniques* 18, 981–984.

Dockhorn-Dworniczak, B., Dworniczak, B., Brommelkamp, L., Bulles, J., Horst, J., and Bocker, W. W. (1991). Non-isotopic detection of single strand conformation polymorphism (PCR-SSCP): a rapid and sensitive technique in diagnosis of phenylketonuria. *Nucleic Acids Res.* 19, 2500.

Eber, S. W., Gonzalez, J. M., Lux, M. L., Scarpa, A. L., Tse, W. T., Dornwell, M., Herbers, J., Kugler, W., Ozcan, R., Pekrun, A., Gallagher, P. G., Schroter, W., Forget, B. G., and Lux, S. E. (1996). Ankyrin-1 mutations are a major cause of dominant and recessive hereditary spherocytosis. *Nat. Genet.* 13, 214–218.

Economou, E. P., Kazazian, H. H. Jr., and Antonarakis, S. E. (1992). Detection of mutations in the factor VIII gene using single-stranded conformational polymorphism (SSCP). *Genomics* 13, 909–911.

Groden, J., Gelbert, L., Thliveris, A., Nelson, L., Robertson, M., Joslyn, G., Samowitz, W., Spirio, L., Carlson, M., Burt, R., Leppert, M., and White, R. (1993). Mutational analysis of patients with adenomatous polyposis: identical inactivating mutations in unrelated individuals. *Am. J. Hum. Genet.* 52, 263–272.

Harteveld, K. L., Heister, A. J., Giordano, P. C., Losekoot, M., and Bernini, L. F. (1996). Rapid detection of point mutations and polymorphisms of the alpha-globin genes by DGGE and SSCA. *Hum. Mutat.* 7, 114–122.

Hayashi, K., and Yandell, D. W. (1993). How sensitive is PCR-SSCP? *Hum. Mutat.* 2, 338–346.

Hayashi, K. (1991). PCR-SSCP: a simple and sensitive method for detection of mutations in the genomic DNA. *PCR Methods Appl.* 1, 34–38.

Hertzberg, M. S., McDonald, D., and Mirochnik, O. (1998). Rapid diagnosis of hemochromatosis gene Cys282Tyr mutation by SSCP analysis. *Am. J. Hematol.* 57, 260–261.

Hogg, A., Onadim, Z., Baird, P. N., and Cowell, J. K. (1992). Detection of heterozygous mutations in the RB1 gene in retinoblastoma patients using single-strand conformation polymorphism analysis and polymerase chain reaction sequencing. *Oncogene* 7, 1445–1451.

Humphries, S. E., Gudnason, V., Whittall, R., and Day, I. N. (1997). Single-strand conformation polymorphism analysis with high throughput modifications, and its use in mutation detection in familial hypercholesterolemia. International Federation of Clinical Chemistry Scientific Division: Committee on Molecular Biology Techniques. *Clin. Chem.* 43, 427–435.

Jarolim, P., Murray, J. L., Rubin, H. L., Taylor, W. M., Prchal, J. T., Ballas, S. K., Snyder, L. M., Chrobak, L., Melrose, W. D., Brabec, V., and Palek, J. (1996). Characterization of 13 novel band 3 gene defects in hereditary spherocytosis with band 3 deficiency. *Blood* 88, 4366–4374.

Kawame, H., Hasegawa, Y., Eto, Y., and Maekawa, K. (1992). Rapid identification of mutations in the glucocerebrosidase gene of Gaucher disease patients by analysis of single-strand conformation polymorphisms. *Hum. Genet.* 90, 294–296.

Keen, J., Lester, D., Inglehearn, C., Curtis, A., and Bhattacharya, S. (1991). Rapid detection of single base mismatches as heteroduplexes on Hydrolink gels. *Trends Genet.* 7, 5.

Kukita, Y., Tahira, T., Sommer, S. S., and Hayashi, K. (1997). SSCP analysis of long DNA fragments in low pH gel. *Hum. Mutat.* 10, 400–407.

Kutach, L. S., Bolshakov, S., and Ananthaswamy, H. N. (1999). Detection of mutations and polymorphisms in the p53 tumor suppressor gene by single-strand conformation polymorphism analysis. *Electrophoresis* 20, 1204–1210.

Labrune, P., Melle, D., Rey, F., Berthelon, M., Caillaud, C., Rey, J., Munnich, A., and Lyonnet, S. (1991). Single-strand conformation polymorphism for detection of mutations and base substitutions in phenylketonuria. *Am. J. Hum. Genet.* 48, 1115–1120.

Levran, O., Erlich, T., Magdalena, N., Gregory, J. J., Batish, S. D., Verlander, P. C., and Auerbach, A. D. (1997). Sequence variation in the Fanconi anemia gene FAA. *Proc. Natl. Acad. Sci. USA* 94, 13051–13056.

Liu, X. Z., Xia, X. J., Adams, J., Chen, Z. Y., Welch, K. O., Tekin, M., Ouyang, X. M., Kristiansen, A., Pandya, A., Balkany, T., Arnos, K. S., and Nance, W. E. (2001). Mutations in GJA1 (connexin 43) are associated with non-syndromic autosomal recessive deafness. *Hum. Mol. Genet.* 10, 2945–2951.

MacCollin, M., Ramesh, V., Jacoby, L. B., Louis, D. N., Rubio, M. P., Pulaski, K., Trofatter, J. A., Short, M. P., Bove, C., Eldridge,

R., Parry D. M., and Gusella J. F. (1994). Mutational analysis of patients with neurofibromatosis 2. *Am. J. Hum. Genet.* 55, 314–320.

Maillet, P., Alloisio, N., Morle, L., and Delaunay, J. (1996). Spectrin mutations in hereditary elliptocytosis and hereditary spherocytosis. *Hum. Mutat.* 8, 97–107.

Medintz, I., Wong, W. W., Sensabaugh, G., and Mathies, R. A. (2000). High speed single nucleotide polymorphism typing of a hereditary haemochromatosis mutation with capillary array electrophoresis microplates. *Electrophoresis* 21, 2352–2358.

Medintz, I. L., Paegel, B. M., Blazej, R. G., Emrich, C. A., Berti, L., Scherer, J. R., and Mathies, R. A. (2001). High-performance genetic analysis using microfabricated capillary array electrophoresis microplates. *Electrophoresis* 22, 3845–3856.

Menounos, P., Zervas, C., Garinis, G., Doukas, C., Kolokithopoulos, D., Tegos, C., and Patrinos, G. P. (2000). Molecular heterogeneity of the glucose-6-phosphate dehydrogenase deficiency in the Hellenic population. *Hum. Hered.* 50, 237–241.

Myers, R. M., Maniatis, T., and Lerman, L. S. (1987). Detection and localization of single base changes by denaturing gradient gel electrophoresis. *Methods Enzymol.* 155, 501–527.

Nair, S., Lin, T. K., Pang, T., and Altwegg, M. (2002). Characterization of Salmonella serovars by PCR-single-strand conformation polymorphism analysis. *J. Clin. Microbiol.* 40, 2346–2351.

Nishimura, D. Y., Purchio, A. F., and Murray, J. C. (1993). Linkage localization of TGFB2 and the human homeobox gene HLX1 to chromosome 1q. *Genomics* 15, 357–364.

Orita, M., Iwahana, H., Kanazawa, H., Hayashi, K., and Sekiya, T. (1989). Detection of polymorphisms of human DNA by gel electrophoresis as single-strand conformation polymorphisms. *Proc. Natl. Acad. Sci. USA* 86, 2766–2770.

Patrinos, G. P., Garinis, G., Kounelis, S., Kouri, E., and Menounos, P. (1999). A novel 23-bp deletion in exon 5 of the p53 tumor suppressor gene. *J. Mol. Med.* 77, 686–689.

Paul, P., Letteboer, T., Gelbert, L., Groden, J., White, R., and Coppes, M. J. (1993). Identical APC exon 15 mutations result in a variable phenotype in familial adenomatous polyposis. *Hum. Mol. Genet.* 2, 925–931.

Phelan, C. M., Lancaster, J. M., Tonin, P., Gumbs, C., Cochran, C., Carter, R., Ghadirian, P., Perret, C., Moslehi, R., Dion, F., Faucher, M. C., Dole, K., Karimi, S., Foulkes, W., Lounis, H., Warner, E., Goss, P., Anderson, D., Larsson, C., Narod, S. A., and Futreal, P. A. (1996). Mutation analysis of the BRCA2 gene in 49 site-specific breast cancer families. *Nat. Genet.* 13, 120–122.

Ravnik-Glavac, M., Glavac, D., and Dean, M. (1994). Sensitivity of single-strand conformation polymorphism and heteroduplex method for mutation detection in the cystic fibrosis gene. *Hum. Mol. Genet.* 3, 801–807.

Rommens, J., Kerem, B. S., Greer, W., Chang, P., Tsui, L. C., and Ray, P. (1990). Rapid nonradioactive detection of the major cystic fibrosis mutation. *Am. J. Hum. Genet.* 46, 395–396.

Rossetti, S., Corra, S., Biasi, M. O., Turco, A. E., and Pignatti, P. F. (1995). Comparison of heteroduplex and single-strand conformation analyses, followed by ethidium fluorescence visualization, for the detection of mutations in four human genes. *Mol. Cell. Probes* 9, 195–200.

Sainz, J., Huynh, D. P., Figueroa, K., Ragge, N. K., Baser, M. E., and Pulst, S. M. (1994). Mutations of the neurofibromatosis type 2 gene and lack of the gene product in vestibular schwannomas. *Hum. Mol. Genet.* 3, 885–891.

Sarkar, G., Yoon, H. S., and Sommer, S. S. (1992). Screening for mutations by RNA single-strand conformation polymorphism (rSSCP): comparison with DNA-SSCP. *Nucleic Acids Res.* 20, 871–878.

Scott, D. A., Kraft, M. L., Carmi, R., Ramesh, A., Elbedour, K., Yairi, Y., Srisailapathy, C. R., Rosengren, S. S., Markham, A. F., Mueller, R. F., Lench, N. J., Van Camp, G., Smith, R. J., and Sheffield, V. C. (1998). Identification of mutations in the connexin 26 gene that cause autosomal recessive nonsyndromic hearing loss. *Hum. Mutat.* 11, 387–394.

Sheffield, V. C., Beck, J. S., Kwitek, A. E., Sandstrom, D. W., and Stone, E. M. (1993). The sensitivity of single-strand conformation polymorphism analysis for the detection of single base substitutions. *Genomics* 16, 325–332.

Shen, M. H., Harper, P. S., and Upadhyaya, M. (1993). Neurofibromatosis type 1 (NF1): the search for mutations by PCR-heteroduplex analysis on Hydrolink gels. *Hum. Mol. Genet.* 2, 1861–1864.

Shimizu, T., Toguchida, J., Kato, M. V., Kaneko, A., Ishizaki, K., and Sasaki, M. S. (1994). Detection of mutations of the RB1 gene in retinoblastoma patients by using exon-by-exon PCR-SSCP analysis. *Am. J. Hum. Genet.* 54, 793–800.

Stoerker, J., Hurwitz, C., Rose, N. C., Silberstein, L. E., and Highsmith, W. E. (1996). Heteroduplex generator in analysis of Rh blood group alleles. *Clin. Chem.* 42, 356–360.

Strom, T. M., Hortnagel, K., Hofmann, S., Gekeler, F., Scharfe, C., Rabl, W., Gerbitz, K. D., and Meitinger, T. (1998). Diabetes insipidus, diabetes mellitus, optic atrophy and deafness (DIDMOAD) caused by mutations in a novel gene (wolframin) coding for a predicted transmembrane protein. *Hum. Mol. Genet.* 7, 2021–2028.

Suzuki, Y., Orita, M., Shiraishi, M., Hayashi, K., and Sekiya, T. (1990). Detection of ras gene mutations in human lung cancers by single-strand conformation polymorphism analysis of polymerase chain reaction products. *Oncogene* 5, 1037–1043.

Takahashi-Fujii, A., Ishino, Y., Kato, I., and Fukumaki, Y. (1994). Rapid and practical detection of beta-globin mutations causing beta-thalassemia by fluorescence-based PCR-single-stranded conformation polymorphism analysis. *Mol. Cell. Probes* 8, 385–393.

Tuffery, S., Moine, P., Demaille, J., and Claustres, M. (1993). Base substitutions in the human dystrophin gene: detection by using the single-strand conformation polymorphism (SSCP) technique. *Hum. Mutat.* 2, 368–374.

Upadhyaya, M., Maynard, J., Osborn, M., Huson, S. M., Ponder, M., Ponder, B. A., and Harper, P. S. (1995). Characterisation of germline mutations in the neurofibromatosis type 1 (NF1) gene. *J. Med. Genet.* 32, 706–710.

Varesco, L., Gismondi, V., James, R., Robertson, M., Grammatico, P., Groden, J., Casarino, L., De Benedetti, L., Bafico, A.,

Bertario, L., Sala, P., Sassatelli, R., Ponz de Leon, M., Biasco, G., Allergetti, A., Aste, H., De Sanctis, S., Rossetti, C., Illeni, M. T., Sciarra, A., Del Porto, G., White, R., and Ferrara, G. B. (1993). Identification of APC gene mutations in Italian adenomatous polyposis coli patients by PCR-SSCP analysis. *Am. J. Hum. Genet.* 52, 280–285.

Veldhuisen, B., Saris, J. J., de Haij, S., Hayashi, T., Reynolds, D. M., Mochizuki, T., Elles, R., Fossdal, R., Bogdanova, N., van Dijk, M. A., Coto, E., Ravine, D., Norby, S., Verellen-Dumoulin, C., Breuning, M. H., Somlo, S., and Peters, D. J. (1997). A spectrum of mutations in the second gene for autosomal dominant polycystic kidney disease (PKD2). *Am. J. Hum. Genet.* 61, 547–555.

Wang, Y. H., Barker, P., and Griffith, J. (1992). Visualization of diagnostic heteroduplex DNAs from cystic fibrosis deletion heterozygotes provides an estimate of the kinking of DNA by bulged bases. *J. Biol. Chem.* 267, 4911–4915.

Watson, M. L., D'Eustachio, P., Mock, B. A., Steinberg, A. D., Morse, H. C. 3rd, Oakey, R. J., Howard, T. A., Rochelle, J. M., and Seldin, M. F. (1992). A linkage map of mouse chromosome 1 using an interspecific cross segregating for the gld autoimmunity mutation. *Mamm. Genome* 2, 158–171.

White, M. B., Carvalho, M., Derse, D., O'Brien, S. J., and Dean, M. (1992). Detecting single base substitutions as heteroduplex polymorphisms. *Genomics* 12, 301–306.

Whittall, R., Gudnason, V., Weavind, G. P., Day, L. B., Humphries, S. E., and Day, I. N. (1995). Utilities for high throughput use of the single strand conformational polymorphism method: screening of 791 patients with familial hypercholesterolaemia for mutations in exon 3 of the low density lipoprotein receptor gene. *J. Med. Genet.* 32, 509–515.

Wood, N., Tyfield, L., and Bidwell, J. (1993a). Rapid classification of phenylketonuria genotypes by analysis of heteroduplexes generated by PCR-amplifiable synthetic DNA. *Hum. Mutat.* 2, 131–137.

Wood, N., Standen, G., Hows, J., Bradley, B., and Bidwell, J. (1993b). Diagnosis of sickle-cell disease with a universal heteroduplex generator. *Lancet* 342, 1519–1520.

Wood, N., Standen, G. R., Murray, E. W., Lillicrap, D., Holmberg, L., Peake, I. R., and Bidwell, J. (1995). Rapid genotype analysis in type 2B von Willebrand's disease using a universal heteroduplex generator. *Br. J. Haematol.* 89, 152–156.

Yap, E. P., and McGee, J. O. (1992a). Nonisotopic SSCP detection in PCR products by ethidium bromide staining. *Trends Genet.* 8, 49.

Yap, E. P., and McGee, J. O. (1992b). Nonisotopic SSCP and competitive PCR for DNA quantification: p53 in breast cancer cells. *Nucleic Acids Res.* 20, 145.

Zappella, M., Meloni, I., Longo, I., Canitano, R., Hayek, G., Rosaia, L., Mari, F., and Renieri, A. (2003). Study of MECP2 gene in Rett syndrome variants and autistic girls. *Am. J. Med. Genet.* 119B, 102–107.

CHAPTER **7**

Capillary Electrophoresis

ANJA BOSSERHOFF[1] AND CLAUS HELLERBRAND[2]

[1] *Institute of Pathology and* [2] *Department of Internal Medicine I, University of Regensburg, Regensburg, Germany*

TABLE OF CONTENTS

7.1 INTRODUCTION

Diagnosis of inherited diseases or cancer predisposition frequently involves analysis of specific mutations or polymorphisms. The number of characterized monogenetic and polygenetic diseases is rising significantly every year. In consequence, molecular diagnostics is faced with an increasing number of patient samples and with a rising complexity of genetic diagnostics. In order to apply genetic analyses for large groups of patients or population screening, automation

of a sensitive and precise method is highly desirable. With capillary electrophoresis (CE), an analytical technique was developed that can process a large number of patient samples rapidly in an automated fashion.

7.2 HISTORY, PRINCIPLE, AND POTENTIAL APPLICATIONS OF CAPILLARY ELECTROPHORESIS

7.2.1 History

Electrophoretic separation of molecules in a glass tube and subsequent detection of the separated compounds by ultraviolet absorption was first described by Hjerten in 1967. In his communication, separation of serum proteins, inorganic and organic ions, peptides, nucleic acids, viruses, and bacteria are described. However, electrophoresis in a tube did not become popular until 1981, when Jorgenson and Lucaks demonstrated the high-resolution power of capillary zone electrophoresis (CZE; Jorgenson *et al.*, 1981). In the late 1980s, the first commercial CE instrument was offered on the market. Since then, many advances and applications have taken place with tremendous impact on the progress of science. The most recent example is the progress of the Human Genome Project (HGP), leading to the generation of the draft human genomic sequence in an astonishingly short period of time (Lander *et al.*, 2001; Venter *et al.*, 2001). The acceleration and rapid success of this project was possible due only to the introduction of CE-based sequencers, allowing the production of 14.9 billion base pairs of sequence in just nine months.

7.2.2 Principle

In principle, all CE techniques are carried out using the same equipment:

1. A high voltage power supply.
2. The anode and cathode buffer reservoirs with corresponding electrodes.
3. The separation chamber; that is, the capillary tube.
4. The injection system.
5. The detector (see Fig. 7.1).

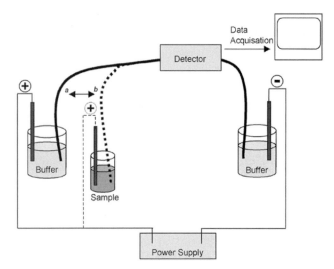

FIGURE 7.1 Principle of capillary electrophoresis. A capillary tube is placed between the two buffer reservoirs, and an electric field is applied by means of a high power supply (a). Then, a defined volume of the analysate is introduced into the capillary by replacing one buffer reservoir with the sample vial (b). The result of the electrophoretic separation is measured by a detector, and the sample data are collected and stored by a computer.

TABLE 7.1   Subtypes of the CE methodology.

Subtype	Abbreviation
Capillary Zone Electrophoresis	CZE
Capillary Gel Electrophoresis	CGE
Micellar Electrokinetic Chromatography	MEKC
Capillary Isoelectric Focusing	CIEF
Capillary Isotachophoresis	CITP
Constant Denaturant Capillary Electrophoresis	CDCE

A capillary tube is filled with buffer, placed between the two buffer reservoirs, and an electric field is applied by means of a high power supply. Then, defined volume is introduced into the capillary by replacing one buffer reservoir with the sample vial. A detector is located at the opposite side of the capillary as the injection site. The sample data are collected and stored by computer, and analyzed using dedicated software.

CE is based on the movement of molecules in an electric field. Initially simple buffer solutions were employed to separate ionic water-soluble solutes. However, different than classical electrophoresis, CE is not restricted to the separation of large molecules based on size and charge. Several modifications of CE have been developed to separate molecules with low molecular weight or neutral charge (some of them are summarized in Table 7.1).

Capillary zone electrophoresis (CZE) is the simplest form of CE. A capillary is filled with a relatively low viscosity buffer and the analysates migrate from one end of the capillary to the other with velocities determined by the charge-to-mass ratio to form discrete peaks (zones). In contrast, *capillary gel electrophoresis* (CGE) separates analytes, such as DNA or proteins, based on molecular weight. A variety of compounds, such as bis-polyacrylamide, agarose, or methylcellulose, have been used to act as a molecular sieve to cause separation. In addition, *micellar electrokinetic chromatography* (MEKC) was introduced for the separation of neutral compounds by the addition of a micelle sodium dodecyl sulfate (SDS) to the buffer solution. Similar to classic isoelectric focusing, in *capillary isoelectric focusing* (CIEF) a pH gradient is created inside a capillary to separate peptides and proteins on the basis of their isoelectric point. In *capillary isotachophoresis* (CITP), a combination of two buffers with different mobilities is used causing the analytes of interest to be concentrated in zones between the leading and terminating constituents. Constant denaturant capillary electrophoresis (CDCE) permits high-resolution separation of single-base variations occurring in an approximately 100 bp isomelting DNA sequence based on their differential melting temperatures (Li-Sucholeiki *et al.*, 1999). It requires different degrees of partial denaturation at a given temperature in the presence of denaturing agents (Khrapko *et al.*, 1994).

7.2.3 Applications

CE has been used successfully in a wide range of application areas. There was a steady, though not spectacular, growth in CE during the 1990s. Routine methods were established in a number of application areas including the analysis of pharmaceuticals, DNA, chiral compounds, proteins, peptides, clinical and forensic samples, metal ions, and inorganic anions.

For clinical diagnostics, CE is widely used for analyzing proteins in physiological matrices, such as serum, urine, and cerebrospinal fluid. Furthermore, CE is used in clinical and molecular biology settings for DNA analysis and also for forensic science and therapeutic drug screening. In addition to clinical applications, CE finds wide use in the analysis of pesticides, food content and composition, and pollutants in water and soil samples.

More and more, CE is challenging the use of traditional gel electrophoresis methods. CE especially benefits from high resolution, simplicity, versatility, low operating costs, and the possibility of direct sample injection without complex sample pretreatment.

CE is characterized by high sensitivity, allowing analysis of small amounts of samples with a minimal amount of reagents. The rugged nature of the CE capillary allows it to be rinsed and cleaned between injections with relatively harsh solutions, such as NaOH. This can allow direct injection of biofluids with possible reductions in sample pretreatment requirements, which are attractive in clinical chemistry where sample throughput is high. High throughput screening involves testing of a large number of samples in a rapid and often automated fashion. Due to the possibility for automation and parallelization, CE allows fast, large scale and high-throughput analysis. Sophisticated CE autosamplers are available that allow quantitative injection and analysis of a large number of samples in an unattended sequence.

Furthermore, in molecular diagnostics, CE has several advantages over classical techniques like slab polyacrylamide gel electrophoresis (PAGE), southern blotting, sequencing, or conventional gel electrophoresis. CE has not only the ability to analyze nonideal tissue samples such as archival paraffin embedded formalin fixed tissue, but also automated digital imaging capabilities that can use either peak height or peak area in a semiquantitative or quantitative manner.

7.3 CAPILLARY ELECTROPHORESIS IN MOLECULAR DIAGNOSTICS

Nucleotides have been quantified in different matrices, including tissue and cell extracts and several DNA and RNA sources. Therapeutic antisense oligonucleotides are of interest for many applications and CE can be used to characterize and quantify these materials (Righetti *et al.*, 1998).

DNA separations generally are conducted by CGE, in which gel- or polymer-filled capillaries act as sieving media to resolve the different lengths of DNA. DNA of different lengths has similar electrophoretic mobility as each increase in size is accompanied by a corresponding increase in the number of negative charges. Therefore, the separation by mobility differences is not the preferred approach and the majority of separations are achieved using a sieve mechanism. The capillary is filled with a matrix of synthetic or natural polymer. The various DNA fragments migrate through this matrix and become entangled with or trapped in the matrix. The migration of the larger DNA fragment is retarded to a greater extent, which results in a size-based separation mechanism.

A variety of compounds, such as bis-polyacrylamide, agarose, or methylcellulose, have been used to act as a molecular sieve to cause separation (Ren, 2000). One of the most active research areas for compound separation has been the investigation of alternative sieving matrices to replace gels (Cottet *et al.*, 1998; Magnusdottir *et al.*, 1998).

Generally, it is difficult to prepare gel-filled capillaries manually due to bubble formation. Further, gels have a limited lifetime, since they are easily destroyed by high current and joule heating. CE requires a liquid medium, and the best choice appears to be a semidilute solution of polymers, which is very much similar to gel, since the molecules are entangled with one another in a way that they create a fine mesh (Sunada *et al.*, 1997). Entangled polymer solutions (ESCE, Entangled polymer solutions CE) can be

replaced after each run and are therefore much easier to operate. The recent development of *capillary nongel sieving electrophoresis* (CNGSE) has facilitated the application of CE for detection of mutations and polymorphisms in human molecular diagnostics.

7.4 MODES OF APPLICATION

Most protocols adapted to automated CE represent analyses of DNA fragment length or DNA restriction patterns (RFLP), analyses of single-strand conformation polymorphism (SSCP), and microsatellite analyses.

7.4.1 DNA Sequencing

DNA sequence analysis is the golden standard for molecular testing. It yields the greatest amount of information since it identifies the order of each deoxynucleotide base of a particular target, usually amplified DNA or cDNA. Although PCR techniques made sequencing routine, detection and analysis of the resulting DNA fragments became the major bottleneck for large-scale DNA sequencing (Dovichi, 1997; Schmalzing *et al.*, 1999). Although the identification of the human genome sequence is now almost completely achieved, the importance of DNA sequencing has not been diminished. Sequencing of other organisms' genomes, comparative genomics, or screening for human genetic defects are still challenging fields of application for the future.

DNA sequencing using dye-labeled didesoxynucleotides and the method developed by Sanger and colleagues (1992) was adapted to CE. The single-base resolving capability of CE permits all four fluorochrome products to be separated simultaneously compared to classical sequencing with PAGE, where each chain terminator has to be loaded in separate lanes. The result is a multicolored ladder where each color represents a different base. The order of the bases is then analyzed using secondary software that characterizes the sequence in terms of identity, and relatedness to prototypical sequences in a database. Today CE serves as the routine high throughput technique for sequencing. CE systems with multi-array capillaries (Tan *et al.*, 1998; Zhang *et al.*, 1999) will further enhance high throughput.

7.4.2 Analyses of DNA Fragment Length or Restriction Patterns and Microsatellites

Cleavage of DNA products and analysis of the DNA cleavage pattern by CE currently is widely used to detect gene mutations (Andersen *et al.*, 1998; Sell *et al.*, 1999a). The method is applicable in cases where, due to mutations or polymorphisms, restriction sites are lost or gained.

In contrast to DNA sequencing, fragment analysis incorporates one fluorochrome-labeled PCR primer in a standard PCR reaction, after which the product(s) are separated and visualized. Fragment analysis is more rapid and less expensive because the fluorochrome is incorporated in the initial PCR reaction, thus eliminating the need for sequencing reactions that incorporate the dye terminating bases. Interpretation of the fragment analysis data is also easier than sequence analysis because the fragments are readily identified by the imaging software, and the need for secondary sequence software analysis is not required.

Using this technique, some research groups established protocols for high-speed separation that allow distinguishing between wild type and mutant PCR products extremely rapidly (Chan *et al.*, 1996) with a short effective length capillary and high field strength.

Although less expensive and more rapid, fragment analysis does not provide exact, detailed sequence data. In addition, PCR fragments can result from erroneous amplification, and thus verification using sequencing techniques or independent probes sometimes is recommended for quality assurance purposes. In these cases, sequencing is often used to confirm the identity of the PCR product. Because of problems due to incomplete cleavage the method is being replaced by SSCP where possible.

Furthermore, microsatellite analysis originally was performed using gel electrophoresis and radioactively labeled PCR products of defined areas of the genome. Application on CE made this method faster and more reliable.

7.4.3 Analyses of Single-strand Conformation Polymorphism

Genetic diagnosis of an inherited disease or cancer predisposition often involves a search for unknown point mutations in several genes. SSCP, a method exploiting visualization of different secondary structures under native conditions (see also previous chapter), combined with CE was shown by many groups to be a rapid and automated technique for processing large numbers of samples. Larsen and colleagues (1999) used the method to detect point mutations associated with the inherited cardiac disorders long QT syndrome (LQTS) and hypertrophic cardiomyopathy (HCM). Sensitivity has been reported to be almost 100% when 34 different point mutations were analyzed, and 10 previously unknown variants were found. These results clearly demonstrate that the method has high resolution, good reproducibility, and is very robust. Based on the possibility of automation and short time of analysis, the method should be suitable for high-throughput applications such as genetic screening of large populations. Several other groups have already established this method for specific screening of mutations in other diseases (see next).

7.4.4 Other Applications

CDCE, combined with high fidelity DNA amplification, can be used to detect mutants of a fraction of 10^{-6} (Muniappan *et al.*, 1999). Future applications of this method include studies of somatic mitochondrial mutations with respect to aging, measurement of mutational spectra of nuclear genes in healthy human tissues, and population screening for disease-associated single nucleotide polymorphisms (SNPs) in large pooled samples. The technique of CDCE also was shown to be applicable for detection of mutations in the K-ras and N-ras genes (Bjorheim *et al.*, 1998; Kumar *et al.*, 1995).

Kuypers and colleagues (1996) developed a method for online melting of double-stranded DNA for SSCP-CE, while they further improved the integration of PCR and CE by composing a contamination-free, automated PCR in the CE apparatus (Kuypers *et al.*, 1998).

Furthermore, a CE system combined with a computer-controlled temperature gradient was recently published (Schell *et al.*, 1999). Using this method this group performed mutation analysis in a DNA fragment of the prion protein gene, which was not resolved by previously described techniques.

Further, RT-PCR followed by CE had been used to quantify the expression levels of specific gene products (Borson *et al.*, 1998; Butler, 1998; Odin *et al.*, 1999). Bor and colleagues (2000) described a protocol for simultaneous quantification of several mRNA species, after calibrating RT-PCR with CE. In a study by Schummer and colleagues (1999), the clinical relevance of multidrug resistance-associated protein (MRP) gene expression was correlated with chemoresistance in prostate carcinoma. De Cremoux and colleagues (1999) compared the *c-erbB-2* gene amplification in breast cancer with the expression of c-erbB-2 protein evaluated by immunohistochemistry and found a concordance of 91% between the two techniques. They concluded that *c-erbB-2* gene amplification can be quantitated accurately by competitive PCR followed by CE, and that this method is also suitable for small, fixed tissue specimens. Stanta and Bonin (1998) also used CE to quantify specific RT-PCR products and were able to show that the quantitation by CE (mean standard deviation ±3%) is more reliable than by dot blot (mean standard deviation ±10%).

Finally, Kuypers and colleagues (1994) compared the use of CE with that of slab gel electrophoresis for quantification of chromosomal translocations in lymphoma, showing that the results of both methods are comparable.

7.5 SPECIFIC DIAGNOSTIC APPLICATIONS

7.5.1 Diagnosis of Neoplastic Disorders

In the diagnosis of neoplastic disorders, CE has been used for the detection of chromosomal aberrations, microsatellite instability, clonality assays, and the detection of several other cancer-relevant genetic mutations, like SNPs.

Colorectal cancer (CRC) is the leading cause of death related to cancer in western countries (Weir *et al.*, 2003). Molecular biology studies have led to the identification of two broad categories of molecular alterations in colorectal cancer. Loss of heterozygosity (LOH) represents 80% of colorectal cancers and is characterized by aneuploidy or allelic losses. The second group displays phenotypic microsatellite instability (MSI-positive tumors), has a near-diploid karyotype, and a relatively low frequency of allelic losses. It accounts for 15% of all colorectal cancers. As a consequence of these two phenomena, other specific genetic events occur at high frequency. These include inactivation of tumor suppressor genes by genetic (deletion or mutation) or epigenetic events (reviewed in Garinis *et al.*, 2002), activation of proto-oncogenes by mutation, and dysregulated expression of several other molecules known to be involved in the development of colorectal cancer.

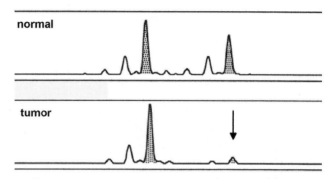

FIGURE 7.2 Analysis for LOH by CE. Genomic DNA is isolated from normal or tumor tissue, amplified with specific primers for defined chromosomal loci and the PCR products are subsequently separated by CE. The two alleles are represented by two peaks in DNA isolated from normal tissue, automatically marked with grey color by the analysis software. In contrast, in DNA from the tumor tissue, one peak is missing (arrow).

7.5.1.1 LOSS OF HETEROZYGOSITY

Aneuploidy indicates gross losses or gains in chromosomal DNA and often is seen in many human primary tumors and premalignant conditions. Loss of one allele at a chromosomal locus may imply the presence of a tumor suppressor gene at that site. Loss of both alleles at a given locus (homozygous deletion) is an even stronger indicator of the existence of a tumor suppressor gene. Many of these loci already are associated with one or more known candidate tumor suppressor genes, including 17p (p53 gene), 5q21 (APC gene), 3p21 (β-catenin gene), 9p (p16 and p15 genes), and 13q (retinoblastoma gene).

LOH analysis originally was performed using gel electrophoresis and radioactively labeled PCR products of defined areas to the genome. The use of CE made this analysis more reliable and faster (Canzian *et al.*, 1996). CE was used for the detection of the loss of several tumor suppressor genes in micro-dissected cancerous tissue (Marsh *et al.*, 2003). Gene fragments were amplified using PCR with flanking oligonucleotides bearing fluorescent labels and subsequently separated and analyzed by CE.

The p53 gene locus is the most common site demonstrating loss of heterozygosity. p53 is a DNA binding protein, which is a transcriptional activator and can cause cell cycle arrest in response to DNA damage.

A second tumor suppressor gene adenomatous polyposis coli (APC) is inactivated in more than 80% of early colorectal cancers. An important function of the APC gene is to prevent the accumulation of molecules associated with cancer, such as catenins. Familial adenomatous polyposis (FAP) is an autosomal dominant disease, caused by germline mutations in the APC gene. FAP is a rare condition in which hundreds or thousands of polyps develop along the length of the colon, and, if left untreated, lead to colon cancer. Figure 7.2 depicts LOH of the APC locus in tumor compared to normal tissue.

An additional application of LOH-assays is the identification of chromosomal differences between normal and tumor tissue. This is useful in distinguishing between tumor recurrence vs. *de novo* cancer formation (Rolston *et al.*, 2001; Sasatomi *et al.*, 2002).

7.5.1.2 MICROSATELLITE INSTABILITY

Microsatellite Instability (MSI), or replication error, comprises length alterations of oligonucleotide repeat sequences that occur somatically in human tumors. They are the manifestation of genomic instability where tumor cells have a decreased overall ability to faithfully replicate DNA.

MSI is a frequent, if not obligatory, surrogate marker of underlying functional inactivation of one of the human DNA mismatch repair genes (MLH1, MSH2, MSH6, PMS1, PMS2, reviewed in Jacob *et al.*, 2002). DNA mismatch repair enzymes normally remove misincorporated single or multiple nucleotide bases as a result of random errors during recombination or replications. Functional loss of mismatch repair occurs due to biallelic inactivation via combination of gene mutation, LOH, and/or promoter methylation (Herman *et al.*, 1998; Jacob *et al.*, 2002).

Germline mutation of a mismatch repair gene has been shown to be the autosomal dominant genetic defect in most hereditary nonpolyposis colorectal cancer (HNPCC) patients (Marra *et al.*, 1995; Saletti *et al.*, 2001). A second hit incurred in tumor cells in HNPCC individuals results in biallelic inactivation of the specific MMR gene. This results in loss of faithful replication of microsatellite DNA in tumor (Saletti *et al.*, 2001). Bethesda criteria (Henson *et al.*, 1995) have been outlined to guide the identification of this syndrome. However, finding MSI positive tumors was shown to be the best predictor of germline mutation (Liu *et al.*, 1999). Although implicating a germline defect in HNPCC patients, MSI also is found in 15–20% of sporadic colon

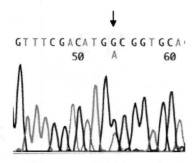

FIGURE 7.3 Mutation analysis by sequencing using CE. Genomic DNA is amplified, followed by a second round of amplification performed with deoxynucleotides along with limiting amounts of chain terminating dideoxynucleotides, labeled with different flourophores on each base. This second PCR reaction results in a series of different colored, truncated DNA products due to a random incorporation of the dideoxynucleotides in competition with the normal deoxynucleotides. Subsequently, these products are separated by CE.

The result of the sequence analysis of 19bp of the coding region of the MSH2 gene is shown. Clearly, at one position (arrow) two peaks appear, indicating the presence of a mutation (G to A in one allele) in heterozygous state. (Courtesy of W. Dietmaier, University of Regensburg, Germany.)

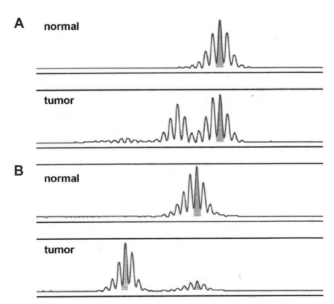

FIGURE 7.4 Analysis of microsatellite instability by CE. Detection of microsatellite instability is performed using genomic DNA isolated from normal or tumor tissue. Defined chromosomal loci are amplified by PCR and the PCR products are separated by CE. Panels A and B depict the results of the analysis of two different genetic loci. In both cases, abnormalities in the displayed CE pattern of the tumor are detectable, in comparison to the normal tissue. (Courtesy of W. Dietmaier, University of Regensburg, Germany.)

cancer (Goel *et al.*, 2003), where the finding reflects an overall increase in genomic instability.

CE has been used for microsatellite analysis as well as for sequencing of DNA mismatch repair genes. Figure 7.3 shows an example for mutation detection in the human MSH-2 gene, using CE-based DNA sequencing; Fig. 7.4 depicts the result of the analysis of 2 marker regions of MSI, comparing normal and tumorous tissue.

Detection of LOH and replication errors in HNPCC has been enhanced by the availability of two complete kits (RER/LOH Assay: Roche Diagnostics; Perkin Elmer Applied Biosystems).

Many different microsatellite markers and loci have been used to identify MSI in tumors. In 1997, the national cancer institute (NCI) recommended a panel of microsatellite markers for the use in colorectal cancer MSI testing (Boland *et al.*, 1998). Berg and colleagues (2000) described a fluorescent multiplex PCR-capillary electrophoresis (FM-CE) assay that permits the simultaneous detection of all five loci proposed by the NCI.

A protocol for SSCP-CE screening for alterations in the exon of the MSH2 gene leading to the possibility of high throughput screening was recently published (Merkelbach-Bruse *et al.*, 2000). This protocol significantly reduces time and expenses, compared to conventional sequence determination.

7.5.1.3 MONOCLONALITY ASSAYS

Rearrangements of antigen receptor genes in B- and T-cells generate products of unique length and sequence. To iden-

tify clonal lymphocyte populations in the majority of clonal B and T cell malignancies, B-cell and T-cell clonality assays have been developed. Monoclonal populations of B-cells are detected through analysis of changes in the Ig heavy chain or kappa chain. T-cell monoclonality can be detected by amplifying T-cell gamma or beta-receptors.

Multiple primer sets are needed to detect clonal rearrangements or chromosomal translocations within these antigen receptor loci. Using standard methods, products of the individual reactions must be analyzed separately, and small clonal populations remain difficult to identify. Miller and colleagues (Miller *et al.*, 1999) developed an integrated fluorescence-based approach using CE to increase amplicon resolution, analytical sensitivity, and overall assay throughout. The newly developed method showed 94% agreement to individual B- and T-cell PCR assays, and had an overall monoclonal detection rate of almost 100%. Munro and colleagues (Munro *et al.*, 1999) also reported a microarray-based CE assay for T- and B-cell proliferative disorders, which again dramatically reduced time without loss of diagnostic accuracy.

7.5.1.4 ANALYSIS OF TUMOR-RELATED MUTATIONS

In addition to chromosomal and allelic alteration, CE also can be used to detect other genetic alterations with implications on cancer development and prognosis. Wenz and colleagues (1998) have proposed to use CE for detection of p53 mutations and have shown in a small study that all 10 samples could be identified correctly as mutated or not. Furthermore, different kinds of mutations could be separated. Also, Atha and colleagues (1998) and Kuypers and colleagues (1993) identified known mutations in the p53 gene by SSCP-CE, and recently, Ekstrom and colleagues (1999) used the system of CDCE to identify mutations in exon 8 of p53 by heteroduplex analysis.

7.5.1.5 SINGLE NUCLEOTIDE POLYMORPHISMS

Detection of SNPs in cancer relevant genes can have important clinical implications. For example, TP53, which is associated with a "normal" genotype, has been shown to predict prolonged survival of patients with certain tumors (Bandoh *et al.*, 2002; Kandioler *et al.*, 2002).

Furthermore, determination of SNPs can help to identify the most appropriate treatment of certain types of cancers (see also Chapter 20), as shown for the treatment with a specific tyrosine kinase inhibitor in gastrointestinal stromal tumors expressing mutant c-kit (Miettinen *et al.*, 2002). In addition, analysis of SNPs is used for assessing the risk in developing cancer. BRCA 1 and 2 variants can be used to help identify women with enhanced risk for developing breast or ovarian cancer (Nicoletto *et al.*, 2001).

Matrix metalloproteinases (MMPs) are a family of closely related enzymes that degrade the extracellular matrix. MMPs are implicated in connective tissue destruction during cancer invasion and metastasis of tumor cells. A guanosine insertion/deletion polymorphism within the promoter region of MMP-1 influences the transcription of the gene (the insertion-type (2G) promoter possesses higher transcriptional activity than the deletion-type (G) promoter). The 2G-genotype has been found to be a genetic risk factor for the development of several cancers, including colon cancer (Ghilardi *et al.*, 2001), ovarian carcinoma (Kanamori *et al.*, 1999), and lung cancer (Zhu *et al.*, 2001), and was found to be associated with a bad prognosis in several tumors. Different MMP-1 genotypes can be identified by CE-based DNA sequencing, as shown in Fig. 7.5.

7.5.2 Diagnosis of Hereditary Diseases and Prenatal Testing

Nowadays, mutation detection using CE-based approaches is widespread in diagnostic laboratories. For obvious reasons, only a fraction of the available mutation detection approaches will be briefly outlined here and in Table 7.2.

Fragile X syndrome is a common form of inherited mental retardation with an incidence of 1 in 4,000 to 5,000 males. Almost all cases are caused by expansion of a $(CGG)_n$ trinucleotide repeat within the 5′ untranslated region of the FMR1 (fragile X mental retardation) gene transcript. Until today the disease reliably was diagnosed by Southern blot-

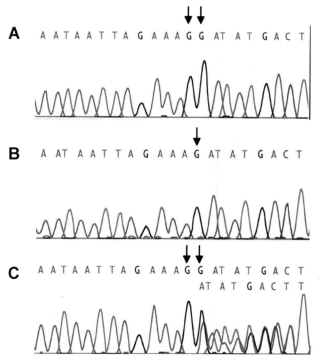

FIGURE 7.5 Analysis of polymorphisms using CE. The results of three different CE analyses for a known SNP (insertion of a guanine (G) nucleotide at −1607 bp in the MMP-1 promoter sequence) is shown. In panel A, homozygosity for the insertion of a G is shown. In panel B, the wild type sequence with only one G is depicted. Finally, in panel C, the analysis reveals a heterozygous genotype. Note the frameshift in the heterozygous sample, as depicted by the appearance of double peaks immediately adjacent to the insertion site (indicated by the arrows).

ting, requiring large samples and high input of time. Recently, automated CE was employed by Larsen and colleagues (1997) for accurate and high-throughput analysis of the FRAXA $(CGG)_n$ region in the normal and permutation range. Their method is based on PCR amplification of extracted genomic DNA followed by automated CE and detection of multicolor fluorescence. The method proved to be useful in both research and clinical mutation screening when a large number of samples, predominantly in the normal range of amplification, are to be analyzed.

Also, *Huntington disease* (HD) belongs to the group of neurodegenerative disorders characterized by unstable expanded trinucleotide repeats. In the case of HD, the expansion of a CAG repeat occurs in the IT15 gene. Williams and colleagues (1999) have established CE analysis for sizing CAG repeats and showed that it enables confident use in sizing HD alleles. Further applications of CE involve multiallelic specific amplification in the analysis of the 21-hydroxylase gene for patients with *congenital adrenal hyperplasia* (Barta *et al.*, 1998) and prenatal diagnosis of thalassemia where many different mutations need to be detected (Trent *et al.*, 1998). Geisel and colleagues

TABLE 7.2 Indicative list of a number of molecular diagnostic applications of CE.

Diagnostic Applications	References
1. Neoplastic disorders	
LOH detection	Canzian *et al.*, 1996; Miller *et al.*, 1999
Microsatellite instability	Berg *et al.*, 2000; Merkelbach-Bruse *et al.*, 2000
Analysis of monoclonality	Miller *et al.*, 1999; Munro *et al.*, 1999
Detection of tumor-related mutations (SNPs)	Nicoletto *et al.*, 2001; Bandoh *et al.*, 2002; Kandioler *et al.*, 2002
2. Diagnosis of hereditary diseases and prenatal testing	
Hereditary hemochromatosis	Lupski *et al.*, 1991; Bosserhoff *et al.*, 1999
Fragile X syndrome	Barta *et al.*, 1998
Thalassemia	Geisel *et al.*, 1999
Congenital adrenal hyperplasia	Trent *et al.*, 1998
Charcot Marie Tooth 1A	Tsui *et al.*, 1992a; Thormann *et al.*, 1999
Cystic fibrosis	Tsui *et al.*, 1992b
Huntington disease	Williams *et al.*, 1999
Risk for coronary heart diseases or thrombosis	Baba *et al.*, 1995; Benson *et al.*, 1999; Sell *et al.*, 1999
3. Diagnosis of infectious diseases	
Bacterial infections	
Mycobacterial species	Hernandez *et al.*, 1999
Pseudomonas sp./gram-negative non-fermenting bacteria	Ghozzi *et al.*, 1999
Listeria sp.	Sciacchitano *et al.*, 1998
Viral diseases	
Herpes simplex	Pancholi *et al.*, 1997
Hepatitis C virus	Pancholi *et al.*, 1997; Doglio *et al.*, 1998
HIV-1	Kolesar *et al.*, 1997
4. Identity testing	
Forensic applications	LaFountain *et al.*, 1998; Pouchkarev *et al.*, 1998
Paternity testing	
Identifying of suitable recipients of organ transplantation	
Bone marrow engraft analysis	

(1999) described a SSCP-CE method to screen for unknown mutations in the low density lipoprotein (LDL) receptor gene. They PCR-amplified the promoter region as well as all 18 exons and tested the accuracy of the developed technique by reproducing 61 known genetic variations by a distinct abnormal SSCP pattern.

The C677T mutation of the methylenetetrahydrofolate reductase (MTHFR) gene is a nutrient-oriented mutation that is associated with elevated levels of homocysteine and an increased risk for *coronary heart disease*. An optimized assay for automated PCR-RFLP genotyping of the MTHFR gene was established by Sell and Lugemwa (Sell *et al.*, 1999b). Following amplification, the resulting PCR product was digested by the Hinfl restriction endonuclease, and the resulting fragments were analyzed by CE. The method was shown to be suitable for high-throughput screening and will support the screening of large sample sizes. Benson and colleagues (1999) performed multiplex analysis of mutations in factor V Leiden (G1691A), prothrombin (G20210A), 5,10-methylenetetrahydrofolate reductase (C677T), and cystathionine beta-synthetase (844ins68), which have been associated with an enhanced risk of thrombosis, using a CE setting. They proved increasing throughput without compromising precision.

Hereditary hemochromatosis (HH) represents an autosomal recessive disease in which increased iron absorption causes iron overload and irreversible tissue damage. Recently, two point mutations in the HFE gene on chromosome 6p have been found to be associated to HH and led to the possibility of patient screening before the onset of irreversible tissue damage. Jackson and colleagues (Jackson *et al.*, 1997) have developed a heteroduplex analysis using capillary electrophoresis for the detection of the Cys282Tyr mutation. An SSCP-CE approach has been recently adapted for the detection of point mutations in codon 63 or 282 of HH patients (Bosserhoff *et al.*, 1999), indicating that SSCP-CE is a reliable, cost-effective, sensitive, and rapid method for genotyping HFE mutations (see Fig. 7.6). Nevertheless, CE can be performed equally well for RFLP analysis in order to provide diagnosis of HFE gene mutations (see Fig. 7.7).

The hereditary *Charcot Marie Tooth 1A neuropathy* (CMT1A) has been shown to be to the result of a 1.5 Mb duplication in chromosome 17p11.2–p12 (Lupski *et al.*, 1991). Suitable dinucleotide markers in the region have been published and can be used to detect the three distinct alleles after PCR and CE (Thormann *et al.*, 1999).

Finally, a commercially available kit based on multiplex oligonucleotide ligation assay (see also Chapter 4) and sequence-coded separation was developed by Perkin Elmer Applied Biosystems for screening of 31 different mutations in the cystic fibrosis gene, corresponding to approximately 95% of the mutated cystic fibrosis alleles (Tsui *et al.*, 1992a; 1992b).

7.5.3 Diagnosis of Infectious Diseases

Diagnosis of infectious diseases is a fast-growing field for CE applications. In general, molecular methods do not require the presence of viable organisms, permitting the identification of bacteria, viruses, and fungi that are difficult, if not impossible, in culture. Molecular identification of infectious agents can be used for both diagnostic and therapeutic purposes and CE has the main advantages to include

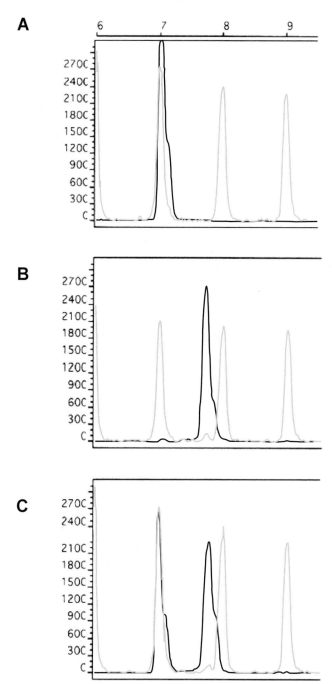

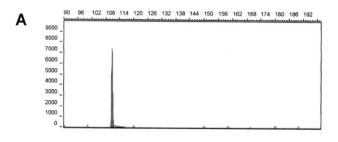

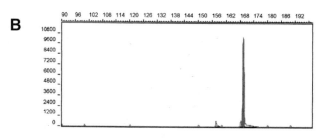

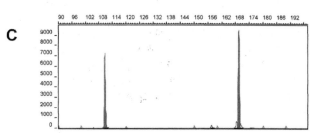

FIGURE 7.7 RFLP analysis using CE. RFLP analysis of genomic DNA by classical gel electrophoresis can be replaced by CE as mentioned in the text. PCR reaction and digestion of the PCR product is performed following the standard protocol but using one fluorescent-labeled primer. Subsequent analysis by CE reveals characteristic patterns for the individual genotypes. As an example, analysis of the HFE gene is shown. The HFE mutation at codon 282 generates a new restriction site. Digestion with the according restriction enzyme results in two shorter DNA fragments (only the one with the aligning fluorescent-labeled primer at one end can be detected by CE). In panels A and B, only one peak at different positions is seen, corresponding to the wild type (digested) and mutated sequence of the HFE gene (undigested), respectively. In panel C, the heterozygous genotype is displayed (two peaks). (Courtesy of A. Hartmann, University of Regensburg, Germany.)

FIGURE 7.6 Analysis of codon 282 of the HFE gene in HH patients with a combined SSCP-CE analysis. Three typical SSCP profiles of codon 282 PCR fragments are shown (black). Panel A shows the analysis of a normal individual. Panel B displays the result of a patient with the mutation at codon 282 of the HFE gene in the homogyzous state (one peak at a different position than in the wild-type sequence). In panel C, a heterozygous genotype is seen (two peaks). Together with the PCR-products analyzed, the HD-400-ROX standard was applied (displayed in gray), clearly allowing the identification of the characteristic peaks of the individual genotypes.

higher throughput and sensitivity than conventional methods.

Since it has been shown that different HCV genotypes are associated with distinct profiles of pathogenicity and responses to antiviral treatment, demand for genotyping hepatitis C virus (HCV) has increased. Thus, Doglio and colleagues (1998) developed a CE-based detection mode, which in combination with direct cycle sequencing provides a simple, rapid, and convenient method for routine HCV genotyping analysis. Also, Pancholi and colleagues (1997) reported the diagnostic detection of herpes simplex and hepatitis C viral amplicons by CE.

The technique of enterobacterial repetitive intergenic consensus (ERIC)-PCR produces genomic DNA fingerprints that allow discrimination between bacterial species and strains. Sciacchitano (1998) applied this technique coupled to CE to differentiate Listeria monocytogenes, an important food-borne pathogen implicated in numerous cases of listeriosis.

Kolesar and colleagues (1997) used CE coupled to laser-induced fluorescence technique (CE-LIF) to directly quantify HIV-1 RNA. They developed a fluorescent-labeled DNA probe with optimal stability and sensitivity for RNA hybridization of HIV-1 RNA isolated from plasma and showed that as little as 19fg of HIV RNA could be reliably and quantitatively detected.

Hernandez and colleagues (1999) developed a scheme for rapid identification of mycobacterium species based on a combined restriction enzyme-CE analysis of PCR-amplified DNA and showed this detection method to be comparable to conventional methods for identification of mycobacteria.

Finally, SSCP-CE was used by Ghozzi and colleagues (1999) to rapidly identify Pseudomonas aeruginosa and other gram-negative nonfermenting bacillii from patients with cystic fibrosis. They have shown that this approach is suited for rapid identification of the main gram-negative nonfermenting bacteria.

7.5.4 Identity Testing and Forensic Science Applications

In 1985, multilocus and single-locus DNA probes for the detection of RFLPs first were applied to identify specific individuals. Later, the introduction of PCR for specific amplification of short tandem repeat (STR) polymorphisms (microsatellites) represented a major break-through. Multiplex PCR for several loci, combined with automated CE, is used efficiently for this purpose, reducing time and costs of analysis (Klintschar et al., 1999; Lazaruk et al., 1998). Currently six to 16 fluorescently labeled STR loci are analyzed simultaneously with a single PCR amplification. As part of multiplex PCR kits, sex determination of forensic samples can be obtained using CE-based analysis of the X-Y homologous gene amelogenin (LaFountain et al., 1998; Pouchkarev et al., 1998). The analysis of six to 10 STRs provides a random match probability of approximately one in five billion. Applications range from identifying suitable recipients for organ transplantation and bone marrow engraft analysis to paternity testing and forensic testing.

Also, in the past years, sequencing of mitochondrial DNA (mtDNA) has become a routine forensic casework application (see also Chapter 21). This method is used when only limited quantities or poor quality of DNA is available for testing. It can be applied to hair shafts, bones, teeth, and other samples that are not suitable for routine STR analysis (Pouchkarev et al., 1998). The two hypervariable regions HV1 and HV2 of the polymorphic control region of mito-

chondrial DNA often are used in forensic applications to differentiate among individuals within a population. The analysis of mitochondrial DNA sequences currently provides a power of discrimination of approximately 1 in 3,000 due to the limited size of mitochondrial sequence databases. Butler and colleagues (1998) successfully attempted to replace expensive sequencing of the amplified PCR products by the use of restriction endonuclease digestion followed by CE to separate and size the PCR-RFLP fragments. This offers a rapid alternative method for screening of polymorphisms.

DNA analysis has become a key element in forensic applications and the use of CE-based DNA-typing as scientific evidence has been accepted in American courts of law (Marchi et al., 1997) as well as in many other countries. CE as a rapid, cheap, and high-throughput, high-resolution method for the analysis of amplified DNA fragments currently is being used in many forensic laboratories for casework applications and paternity testing.

7.6 FUTURE IMPROVEMENTS

Undoubtedly the future direction of CE will lie in improved instrumentation and enhanced method development possibilities.

One of the most promising processes has been made in the area of micro-fabricated capillary array electrophoresis (CAE). CAE represents a new means to perform CE even faster (<160 sec) and to analyze more different samples in parallel (Behr et al., 1999; Gao et al., 1999; Woolley et al., 1997). Genetic mapping and DNA sequencing can be completed more rapidly by using this means with up to 48 to 96 capillaries running simultaneously. As described by Mansfield and colleagues (1997) the CAE system has the capacity to generate up to 5.5 million genotypes per year. For detecting HH mutations, CAE already was shown to be feasible (Simpson et al., 1998).

To provide even smaller micro-machined analytic systems, nonelectronic chips will be developed in the future. Generally, entire CE systems are constructed on glass or plastic chips. Gel-filled plastic channels in the microchips can be used for DNA analysis. These devices contain buffer and sample channels and have a connection to a detection system. The first systems already established are microchip-based, which analyze blood parameters in clinical diagnostics. Integration of enzymatic DNA digestion and CE on-chip was first performed by Jacobson and colleagues (1996). The DNA sample and the restriction enzyme reacted on the device in a 0.7 nl reaction chamber for 10 s followed by resolution on a 1% cellulose gel within three minutes. The fragments were detected by laser-induced fluorescence. Also outstanding progress toward the lab-on-chip concept was made by the functional integration of a PCR heating chamber coupled to a CE separation channel on a chip (Woolley et al., 1996).

A further technological advance would be the introduction of a fixed loop injection device, which would improve the injection precision in CE, which is currently poorer than HPLC. Development of a fixed volume injector prototype was first reported in 1987 and hopefully a commercial device may evolve—possibly from the microchip CE research efforts.

Further improvements in the kind of sieving matrix will lead to an even longer lifetime, easier operation, lower costs, and higher resolution and speed (Liang *et al.*, 1999; Shen *et al.*, 1999).

In essence, CE and its variations offer great possibilities in the area of molecular diagnostics, and their further improvement will eventually lead to fast and inexpensive test equipment for clinical analysis.

References

Andersen, P. S., Larsen, L. A., Kanters, J., Havndrup, O., Bundgaard, H., Brandt, N. J., Vuust, J., and Christiansen, M. (1998). Mutation Detection by Cleavase in Combination with Capillary Electrophoresis Analysis: Application to Mutations Causing Hypertrophic Cardiomyopathy and Long-QT Syndrome. *Mol. Diagn.* 3, 105–111.

Atha, D. H., Wenz, H. M., Morehead, H., Tian, J., and O'Connell, C. D. (1998). Detection of p53 point mutations by single strand conformation polymorphism: analysis by capillary electrophoresis. *Electrophoresis* 19, 172–179.

Baba, Y., Tomisaki, R., Sumita, C., Morimoto, I., Sugita, S., Tsuhako, M., Miki, T., Ogihara, T. (1995). Rapid typing of variable number of tandem repeat locus in the human apolipoprotein B gene for DNA diagnosis of heart disease by polymerase chain reaction and capillary electrophoresis. *Electrophoresis* 16, 1437–1440.

Bandoh, N., Hayashi, T., Kishibe, K., Takahara, M., Imada, M., Nonaka, S., and Harabuchi, Y. (2002). Prognostic value of p53 mutations, bax, and spontaneous apoptosis in maxillary sinus squamous cell carcinoma. *Cancer* 94, 1968–1980.

Barta, C., Sasvari-Szekely, M., and Guttman, A. (1998). Simultaneous analysis of various mutations on the 21-hydroxylase gene by multi-allele specific amplification and capillary gel electrophoresis. *J. Chromatogr. A.* 817, 281–286.

Behr, S., Matzig, M., Levin, A., Eickhoff, H., and Heller, C. (1999). A fully automated multicapillary electrophoresis device for DNA analysis. *Electrophoresis* 20, 1492–1507.

Benson, J. M., Ellingsen, D., Renshaw, M. A., Resler, A. G., Evatt, B. L., and Hooper, W. C. (1999). Multiplex analysis of mutations in four genes using fluorescence scanning technology. *Thromb. Res.* 96, 57–64.

Berg, K. D., Glaser, C. L., Thompson, R. E., Hamilton, S. R., Griffin, C. A., and Eshleman, J. R. (2000). Detection of microsatellite instability by fluorescence multiplex polymerase chain reaction. *J. Mol. Diagn.* 2, 20–28.

Bjorheim, J., Lystad, S., Lindblom, A., Kressner, U., Westring, S., Wahlberg, S., Lindmark, G., Gaudernack, G., Ekstrom, P., Roe, J., Thilly, W. G., and Borresen-Dale, A. L. (1998). Mutation analyses of KRAS exon 1 comparing three different techniques: temporal temperature gradient electrophoresis, constant denaturant capillary electrophoresis and allele specific polymerase chain reaction. *Mutat. Res.* 403, 103–112.

Boland, C. R., Thibodeau, S. N., Hamilton, S. R., Sidransky, D., Eshleman, J. R., Burt, R. W., Meltzer, S. J., Rodriguez-Bigas, M. A., Fodde, R., Ranzani, G. N., and Srivastava, S. (1998). A National Cancer Institute Workshop on Microsatellite Instability for cancer detection and familial predisposition: development of international criteria for the determination of microsatellite instability in colorectal cancer. *Cancer Res.* 58, 5248–5257.

Bor, M. V., Sorensen, B. S., and Nexo, E. (2000). Simultaneous quantitation of several mRNA species by calibrated reverse transcription polymerase chain reaction and capillary electrophoresis: analysis of the epidermal growth factor receptor and its activating ligands EGF, TGF-alpha, and HB-EGF in rat liver. *Lab. Invest.* 80, 983–986.

Borson, N. D., Strausbauch, M. A., Wettstein, P. J., Oda, R. P., Johnston, S. L., and Landers, J. P. (1998). Direct quantitation of RNA transcripts by competitive single-tube RT-PCR and capillary electrophoresis. *Biotechniques* 25, 130–137.

Bosserhoff, A. K., Seegers, S., Hellerbrand, C., Scholmerich, J., and Buttner, R. (1999). Rapid genetic screening for hemochromatosis using automated SSCP-based capillary electrophoresis (SSCP-CE). *Biotechniques* 26, 1106–1110.

Butler, J. M. (1998). Rapid assessment of PCR product quality and quantity by capillary electrophoresis. *Methods Mol. Biol.* 98, 39–47.

Butler, J. M., Wilson, M. R., and Reeder, D. J. (1998). Rapid mitochondrial DNA typing using restriction enzyme digestion of polymerase chain reaction amplicons followed by capillary electrophoresis separation with laser-induced fluorescence detection. *Electrophoresis* 19, 119–124.

Canzian, F., Salovaara, R., Hemminki, A., Kristo, P., Chadwick, R. B., Aaltonen, L. A., and de la Chapelle, A. (1996). Semiautomated assessment of loss of heterozygosity and replication error in tumors. *Cancer Res.* 56, 3331–3337.

Chan, K. C., Muschik, G. M., Issaq, H. J., Garvey, K. J., and Generlette, P. L. (1996). High-speed screening of polymerase chain reaction products by capillary electrophoresis. *Anal. Biochem.* 243, 133–139.

Cottet, H., Gareil, P., and Viovy, J. L. (1998). The effect of blob size and network dynamics on the size-based separation of polystyrenesulfonates by capillary electrophoresis in the presence of entangled polymer solutions. *Electrophoresis* 19, 2151–2162.

de Cremoux, P., Martin, E. C., Vincent-Salomon, A., Dieras, V., Barbaroux, C., Liva, S., Pouillart, P., Sastre-Garau, X., and Magdelenat, H. (1999). Quantitative PCR analysis of c-erb B-2 (HER2/neu) gene amplification and comparison with p185(HER2/neu) protein expression in breast cancer drill biopsies. *Int. J. Cancer* 83, 157–161.

Doglio, A., Laffont, C., Thyss, S., and Lefebvre, J. C. (1998). Rapid genotyping of hepatitis C virus by direct cycle sequencing of PCR-amplified cDNAs and capillary electrophoresis analysis. *Res. Virol.* 149, 219–227.

Dovichi, N. J. (1997). DNA sequencing by capillary electrophoresis. *Electrophoresis* 18, 2393–2399.

Ekstrom, P. O., Borresen-Dale, A. L., Qvist, H., Giercksky, K. E., and Thilly, W. G. (1999). Detection of low-frequency mutations in exon 8 of the TP53 gene by constant denaturant capillary electrophoresis (CDCE). *Biotechniques* 27, 128–134.

Gao, Q., Pang, H. M., and Yeung, E. S. (1999). Simultaneous genetic typing from multiple short tandem repeat loci using a 96-capillary array electrophoresis system. *Electrophoresis* 20, 1518–1526.

Garinis, G. A., Patrinos, G. P., Spanakis, N. E., and Menounos, P. G. (2002). DNA hypermethylation: when tumour suppressor genes go silent. *Hum. Genet.* 111, 115–127.

Geisel, J., Walz, T., Bodis, M., Nauck, M., Oette, K., and Herrmann, W. (1999). Fluorescence-based single-strand conformation polymorphism analysis of the low density lipoprotein receptor gene by capillary electrophoresis. *J. Chromatogr. B. Biomed. Sci. Appl.* 724, 239–247.

Ghilardi, G., Biondi, M. L., Mangoni, J., Leviti, S., DeMonti, M., Guagnellini, E., and Scorza, R. (2001). Matrix metallo-proteinase-1 promoter polymorphism 1G/2G is correlated with colorectal cancer invasiveness. *Clin. Cancer Res.* 7, 2344–2346.

Ghozzi, R., Morand, P., Ferroni, A., Beretti, J. L., Bingen, E., Segonds, C., Husson, M. O., Izard, D., Berche, P., and Gaillard, J. L. (1999). Capillary electrophoresis-single-strand conformation polymorphism analysis for rapid identification of Pseudomonas aeruginosa and other gram-negative nonfermenting bacilli recovered from patients with cystic fibrosis. *J. Clin. Microbiol.* 37, 3374–3379.

Goel, A., Arnold, C. N., Niedzwiecki, D., Chang, D. K., Ricciardiello, L., Carethers, J. M., Dowell, J. M., Wasserman, L., Compton, C., Mayer, R. J., Bertagnolli, M. M., and Boland, C. R. (2003). Characterization of sporadic colon cancer by patterns of genomic instability. *Cancer Res.* 63, 1608–1614.

Henson, D. E., Fielding, L. P., Grignon, D. J., Page, D. L., Hammond, M. E., Nash, G., Pettigrew, N. M., Gorstein, F., and Hutter, R. V. (1995). College of American Pathologists Conference XXVI on clinical relevance of prognostic markers in solid tumors. Summary. Members of the Cancer Committee. *Arch. Pathol. Lab. Med.* 119, 1109–1112.

Herman, J. G., Umar, A., Polyak, K., Graff, J. R., Ahuja, N., Issa, J. P., Markowitz, S., Willson, J. K., Hamilton, S. R., Kinzler, K. W., Kane, M. F., Kolodner, R. D., Vogelstein, B., Kunkel, T. A., and Baylin, S. B. (1998). Incidence and functional consequences of hMLH1 promoter hypermethylation in colorectal carcinoma. *Proc. Natl. Acad. Sci. USA* 95, 6870–6875.

Hernandez, S. M., Morlock, G. P., Butler, W. R., Crawford, J. T., and Cooksey, R. C. (1999). Identification of Mycobacterium species by PCR-restriction fragment length polymorphism analyses using fluorescence capillary electrophoresis. *J. Clin. Microbiol.* 37, 3688–3692.

Hjerten, S. (1967). Free zone electrophoresis. *Chromatogr. Rev.* 9, 122–219.

Jackson, H. A., Bowen, D. J., and Worwood, M. (1997). Rapid genetic screening for haemochromatosis using heteroduplex technology. *Br. J. Haematol.* 98, 856–859.

Jacob, S., and Praz, F. (2002). DNA mismatch repair defects: role in colorectal carcinogenesis. *Biochimie* 84, 27–47.

Jacobson, S. C., and Ramsey, J. M. (1996). Integrated microdevice for DNA restriction fragment analysis. *Anal. Chem.* 68, 720–723.

Jorgenson, J. W., and Lukacs, K. D. (1981). Free-zone electrophoresis in glass capillaries. *Clin. Chem.* 27, 1551–1553.

Kanamori, Y., Matsushima, M., Minaguchi, T., Kobayashi, K., Sagae, S., Kudo, R., Terakawa, N., and Nakamura, Y. (1999). Correlation between expression of the matrix metalloproteinase-1 gene in ovarian cancers and an insertion/deletion polymorphism in its promoter region. *Cancer Res.* 59, 4225–4227.

Kandioler, D., Zwrtek, R., Ludwig, C., Janschek, E., Ploner, M., Hofbauer, F., Kuhrer, I., Kappel, S., Wrba, F., Horvath, M., Karner, J., Renner, K., Bergmann, M., Karner-Hanusch, J., Potter, R., Jakesz, R., Teleky, B., and Herbst, F. (2002). TP53 genotype but not p53 immunohistochemical result predicts response to preoperative short-term radiotherapy in rectal cancer. *Ann. Surg.* 235, 493–498.

Khrapko, K., Hanekamp, J. S., Thilly, W. G., Belenkii, A., Foret, F., and Karger, B. L. (1994). Constant denaturant capillary electrophoresis (CDCE): a high resolution approach to mutational analysis. *Nucleic Acids Res.* 22, 364–369.

Klintschar, M., Ebner, A., and Reichenpfader, B. (1999). Population genetic studies on nine tetrameric short tandem repeat loci using fluorescence dye-labeled primers and capillary electrophoresis in the Austrian population. *Electrophoresis* 20, 1740–1742.

Kolesar, J. M., Allen, P. G., and Doran, C. M. (1997). Direct quantification of HIV-1 RNA by capillary electrophoresis with laser-induced fluorescence. *J. Chromatogr. B. Biomed. Sci. Appl.* 697, 189–194.

Kumar, R., Hanekamp, J. S., Louhelainen, J., Burvall, K., Onfelt, A., Hemminki, K., and Thilly, W. G. (1995). Separation of transforming amino acid-substituting mutations in codons 12, 13 and 61 the N-ras gene by constant denaturant capillary electrophoresis (CDCE). *Carcinogenesis* 16, 2667–2673.

Kuypers, A. W., Willems, P. M., van der Schans, M. J., Linssen, P. C., Wessels, H. M., de Bruijn, C. H., Everaerts, F. M., and Mensink, E. J. (1993). Detection of point mutations in DNA using capillary electrophoresis in a polymer network. *J. Chromatogr.* 621, 149–156.

Kuypers, A. W., Meijerink, J. P., Smetsers, T. F., Linssen, P. C., and Mensink, E. J. (1994). Quantitative analysis of DNA aberrations amplified by competitive polymerase chain reaction using capillary electrophoresis. *J. Chromatogr. B. Biomed. Appl.* 660, 271–277.

Kuypers, A. W., Linssen, P. C., Willems, P. M., and Mensink, E. J. (1996). On-line melting of double-stranded DNA for analysis of single-stranded DNA using capillary electrophoresis. *J. Chromatogr. B. Biomed. Appl.* 675, 205–211.

Kuypers, A. W., Linssen, P. C., Lauer, H. H., and Mensink, E. J. (1998). Contamination-free and automated composition of a reaction mixture for nucleic acid amplification using a capillary electrophoresis apparatus. *J. Chromatogr. A.* 806, 141–147.

LaFountain, M., Schwartz, M., Cormier, J., and Buel, E. (1998). Validation of capillary electrophoresis for analysis of the X-Y homologous amelogenin gene. *J. Forensic Sci.* 43, 1188–1194.

Lander, E. S., Linton, L. M., Birren, B., Nusbaum, C., Zody, M. C., Baldwin, J., Devon, K., Dewar, K., Doyle, M., FitzHugh, W., Funke, R., Gage, D., Harris, K., Heaford, A., Howland, J., Kann, L., Lehoczky, J., LeVine, R., McEwan, P., McKernan, K., Meldrim, J., Mesirov, J. P., Miranda, C., Morris, W., Naylor, J., Raymond, C., Rosetti, M., Santos, R., Sheridan, A., Sougnez, C., Stange-Thomann, N., Stojanovic, N., Subramanian, A., Wyman, D., Rogers, J., Sulston, J., Ainscough, R., Beck, S.,

Bentley, D., Burton, J., Clee, C., Carter, N., Coulson, A., Deadman, R., Deloukas, P., Dunham, A., Dunham, I., Durbin, R., French, L., Grafham, D., Gregory, S., Hubbard, T., Humphray, S., Hunt, A., Jones, M., Lloyd, C., McMurray, A., Matthews, L., Mercer, S., Milne, S., Mullikin, J. C., Mungall, A., Plumb, R., Ross, M., Shownkeen, R., Sims, S., Waterston, R. H., Wilson, R. K., Hillier, L. W., McPherson, J. D., Marra, M. A., Mardis, E. R., Fulton, L. A., Chinwalla, A. T., Pepin, K. H., Gish, W. R., Chissoe, S. L., Wendl, M. C., Delehaunty, K. D., Miner, T. L., Delehaunty, A., Kramer, J. B., Cook, L. L., Fulton, R. S., Johnson, D. L., Minx, P. J., Clifton, S. W., Hawkins, T., Branscomb, E., Predki, P., Richardson, P., Wenning, S., Slezak, T., Doggett, N., Cheng, J. F., Olsen, A., Lucas, S., Elkin, C., Uberbacher, E., Frazier, M., Gibbs, R. A., Muzny, D. M., Scherer, S. E., Bouck, J. B., Sodergren, E. J., Worley, K. C., Rives, C. M., Gorrell, J. H., Metzker, M. L., Naylor, S. L., Kucherlapati, R. S., Nelson, D. L., Weinstock, G. M., Sakaki, Y., Fujiyama, A., Hattori, M., Yada, T., Toyoda, A., Itoh, T., Kawagoe, C., Watanabe, H., Totoki, Y., Taylor, T., Weissenbach, J., Heilig, R., Saurin, W., Artiguenave, F., Brottier, P., Bruls, T., Pelletier, E., Robert, C., Wincker, P., Smith, D. R., Doucette-Stamm, L., Rubenfield, M., Weinstock, K., Lee, H. M., Dubois, J., Rosenthal, A., Platzer, M., Nyakatura, G., Taudien, S., Rump, A., Yang, H., Yu, J., Wang, J., Huang, G., Gu, J., Hood, L., Rowen, L., Madan, A., Qin, S., Davis, R. W., Federspiel, N. A., Abola, A. P., Proctor, M. J., Myers, R. M., Schmutz, J., Dickson, M., Grimwood, J., Cox, D. R., Olson, M. V., Kaul, R., Raymond, C., Shimizu, N., Kawasaki, K., Minoshima, S., Evans, G. A., Athanasiou, M., Schultz, R., Roe, B. A., Chen, F., Pan, H., Ramser, J., Lehrach, H., Reinhardt, R., McCombie, W. R., de la, B. M., Dedhia, N., Blocker, H., Hornischer, K., Nordsiek, G., Agarwala, R., Aravind, L., Bailey, J. A., Bateman, A., Batzoglou, S., Birney, E., Bork, P., Brown, D. G., Burge, C. B., Cerutti, L., Chen, H. C., Church, D., Clamp, M., Copley, R. R., Doerks, T., Eddy, S. R., Eichler, E. E., Furey, T. S., Galagan, J., Gilbert, J. G., Harmon, C., Hayashizaki, Y., Haussler, D., Hermjakob, H., Hokamp, K., Jang, W., Johnson, L. S., Jones, T. A., Kasif, S., Kaspryzk, A., Kennedy, S., Kent, W. J., Kitts, P., Koonin, E. V., Korf, I., Kulp, D., Lancet, D., Lowe, T. M., McLysaght, A., Mikkelsen, T., Moran, J. V., Mulder, N., Pollara, V. J., Ponting, C. P., Schuler, G., Schultz, J., Slater, G., Smit, A. F., Stupka, E., Szustakowski, J., Thierry-Mieg, D., Thierry-Mieg, J., Wagner, L., Wallis, J., Wheeler, R., Williams, A., Wolf, Y. I., Wolfe, K. H., Yang, S. P., Yeh, R. F., Collins, F., Guyer, M. S., Peterson, J., Felsenfeld, A., Wetterstrand, K. A., Patrinos, A., Morgan, M. J., Szustakowki, J., de Jong, P., Catanese, J. J., Osoegawa, K., Shizuya, H., and Choi, S. (2001). Initial sequencing and analysis of the human genome. *Nature* 409, 860–921.

Larsen, L. A., Gronskov, K., Norgaard-Pedersen, B., Brondum-Nielsen, K., Hasholt, L., and Vuust, J. (1997). High-throughput analysis of fragile X (CGG)n alleles in the normal and premutation range by PCR amplification and automated capillary electrophoresis. *Hum. Genet.* 100, 564–568.

Larsen, L. A., Christiansen, M., Vuust, J., and Andersen, P. S. (1999). High-throughput single-strand conformation polymorphism analysis by automated capillary electrophoresis: robust multiplex analysis and pattern-based identification of allelic variants. *Hum. Mutat.* 13, 318–327.

Lazaruk, K., Walsh, P. S., Oaks, F., Gilbert, D., Rosenblum, B. B., Menchen, S., Scheibler, D., Wenz, H. M., Holt, C., and Wallin, J. (1998). Genotyping of forensic short tandem repeat (STR) systems based on sizing precision in a capillary electrophoresis instrument. *Electrophoresis* 19, 86–93.

Li-Sucholeiki, X. C., Khrapko, K., Andre, P. C., Marcelino, L. A., Karger, B. L., and Thilly, W. G. (1999). Applications of constant denaturant capillary electrophoresis/high-fidelity polymerase chain reaction to human genetic analysis. *Electrophoresis* 20, 1224–1232.

Liang, D., Song, L., Zhou, S., Zaitsev, V. S., and Chu, B. (1999). Poly(N-isopropylacrylamide)-g-poly(ethyleneoxide) for high resolution and high speed separation of DNA by capillary electrophoresis. *Electrophoresis* 20, 2856–2863.

Liu, T., Tannergard, P., Hackman, P., Rubio, C., Kressner, U., Lindmark, G., Hellgren, D., Lambert, B., and Lindblom, A. (1999). Missense mutations in hMLH1 associated with colorectal cancer. *Hum. Genet.* 105, 437–441.

Liu, T., Wahlberg, S., Burek, E., Lindblom, P., Rubio, C., and Lindblom, A. (2000). Microsatellite instability as a predictor of a mutation in a DNA mismatch repair gene in familial colorectal cancer. *Genes Chromosomes Cancer* 27, 17–25.

Lupski, J. R., de Oca-Luna, R. M., Slaugenhaupt, S., Pentao, L., Guzzetta, V., Trask, B. J., Saucedo-Cardenas, O., Barker, D. F., Killian, J. M., and Garcia, C. A. (1991). DNA duplication associated with Charcot-Marie-Tooth disease type 1A. *Cell* 66, 219–232.

Magnusdottir, S., Viovy, J. L., and Francois, J. (1998). High resolution capillary electrophoretic separation of oligonucleotides in low-viscosity, hydrophobically end-capped polyethylene oxide with cubic order. *Electrophoresis* 19, 1699–1703.

Mansfield, E. S., Vainer, M., Harris, D. W., Gasparini, P., Estivill, X., Surrey, S., and Fortina, P. (1997). Rapid sizing of polymorphic microsatellite markers by capillary array electrophoresis. *J. Chromatogr. A.* 781, 295–305.

Marchi, E., and Pasacreta, R. J. (1997). Capillary electrophoresis in court: the landmark decision of the People of Tennessee versus Ware. *J. Capillary Electrophor.* 4, 145–156.

Marra, G., and Boland, C. R. (1995). Hereditary nonpolyposis colorectal cancer: the syndrome, the genes, and historical perspectives. *J. Natl. Cancer Inst.* 87, 1114–1125.

Marsh, J. W., Finkelstein, S. D., Demetris, A. J., Swalsky, P. A., Sasatomi, E., Bandos, A., Subotin, M., and Dvorchik, I. (2003). Genotyping of hepatocellular carcinoma in liver transplant recipients adds predictive power for determining recurrence-free survival. *Liver Transpl.* 9, 664–671.

Merkelbach-Bruse, S., and Buettner, R. (2000). Detection of HNPCC patients by multiple use of capillary electrophoresis techniques. *Combin. Chem. High Throughput Screen.* 3, 519–524.

Miettinen, M., Majidi, M., and Lasota, J. (2002). Pathology and diagnostic criteria of gastrointestinal stromal tumors (GISTs): a review. *Eur. J. Cancer* 38 Suppl. 5, S39–51.

Miller, J. E., Wilson, S. S., Jaye, D. L., and Kronenberg, M. (1999). An automated semiquantitative B and T cell clonality assay. *Mol. Diagn.* 4, 101–117.

Muniappan, B. P., and Thilly, W. G. (1999). Application of constant denaturant capillary electrophoresis (CDCE) to mutation detection in humans. *Genet. Anal.* 14, 221–227.

Munro, N. J., Snow, K., Kant, J. A., and Landers, J. P. (1999). Molecular diagnostics on microfabricated electrophoretic devices: from slab gel- to capillary- to microchip-based assays for T- and B-cell lymphoproliferative disorders. *Clin. Chem.* 45, 1906–1917.

Nicoletto, M. O., Donach, M., De Nicolo, A., Artioli, G., Banna, G., and Monfardini, S. (2001). BRCA-1 and BRCA-2 mutations as prognostic factors in clinical practice and genetic counselling. *Cancer Treat. Rev.* 27, 295–304.

Odin, E., Wettergren, Y., Larsson, L., Larsson, P. A., and Gustavsson, B. (1999). Rapid method for relative gene expression determination in human tissues using automated capillary gel electrophoresis and multicolor detection. *J. Chromatogr. B. Biomed. Sci. Appl.* 734, 47–53.

Pancholi, P., Oda, R. P., Mitchell, P. S., Landers, J. P., and Persing, D. H. (1997). Diagnostic Detection of Herpes Simplex and Hepatitis C Viral Amplicons by Capillary Electrophoresis: Comparison With Southern Blot Detection. *Mol. Diagn.* 2, 27–37.

Pouchkarev, V. P., Shved, E. F., and Novikov, P. I. (1998). Sex determination of forensic samples by polymerase chain reaction of the amelogenin gene and analysis by capillary electrophoresis with polymer matrix. *Electrophoresis* 19, 76–79.

Ren, J. (2000). High-throughput single-strand conformation polymorphism analysis by capillary electrophoresis. *J. Chromatogr. B. Biomed. Sci. Appl.* 741, 115–128.

Righetti, P. G., and Gelfi, C. (1998). Recent advances in capillary zone electrophoresis of DNA. *Forensic Sci. Int.* 92, 239–250.

Rolston, R., Sasatomi, E., Hunt, J., Swalsky, P. A., and Finkelstein, S. D. (2001). Distinguishing de novo second cancer formation from tumor recurrence: mutational fingerprinting by microdissection genotyping. *J. Mol. Diagn.* 3, 129–132.

Salas-Solano, O., Carrilho, E., Kotler, L., Miller, A. W., Goetzinger, W., Sosic, Z., and Karger, B. L. (1998). Routine DNA sequencing of 1000 bases in less than one hour by capillary electrophoresis with replaceable linear polyacrylamide solutions. *Anal. Chem.* 70, 3996–4003.

Saletti, P., Edwin, I. D., Pack, K., Cavalli, F., and Atkin, W. S. (2001). Microsatellite instability: application in hereditary nonpolyposis colorectal cancer. *Ann. Oncol.* 12, 151–160.

Sanger, F., Nicklen, S., and Coulson, A. R. (1992). DNA sequencing with chain-terminating inhibitors [classical article: 1977]. *Biotechnology* 24, 104–108.

Sasatomi, E., Finkelstein, S. D., Woods, J. D., Bakker, A., Swalsky, P. A., Luketich, J. D., Fernando, H. C., and Yousem, S. A. (2002). Comparison of accumulated allele loss between primary tumor and lymph node metastasis in stage II non-small cell lung carcinoma: implications for the timing of lymph node metastasis and prognostic value *Cancer Res.* 62, 2681–2689.

Schell, J., Wulfert, M., and Riesner, D. (1999). Detection of point mutations by capillary electrophoresis with temporal temperature gradients. *Electrophoresis* 20, 2864–2869.

Schmalzing, D., Koutny, L., Salas-Solano, O., Adourian, A., Matsudaira, P., and Ehrlich, D. (1999). Recent developments in DNA sequencing by capillary and microdevice electrophoresis. *Electrophoresis* 20, 3066–3077.

Schummer, B., Siegsmund, M., Steidler, A., Toktomambetova, L., Kohrmann, K. U., and Alken, P. (1999). Expression of the gene for the multidrug resistance-associated protein in human prostate tissue. *Urol. Res.* 27, 164–168.

Sciacchitano, C. J. (1998). DNA fingerprinting of Listeria monocytogenes using enterobacterial repetitive intergenic consensus (ERIC) motifs-polymerase chain reaction/capillary electrophoresis. *Electrophoresis* 19, 66–70.

Sell, S. M., and Lugemwa, P. R. (1999). Development of a highly accurate, rapid PCR-RFLP genotyping assay for the methylenetetrahydrofolate reductase gene. *Genet. Test.* 3, 287–289.

Shen, Y., Xu, Q., Han, F., Ding, K., Song, F., Fan, Y., Zhu, N., Wu, G., and Lin, B. (1999). Application of capillary nongel sieving electrophoresis for gene analysis. *Electrophoresis* 20, 1822–1828.

Simpson, P. C., Roach, D., Woolley, A. T., Thorsen, T., Johnston, R., Sensabaugh, G. F., and Mathies, R. A. (1998). High-throughput genetic analysis using microfabricated 96-sample capillary array electrophoresis microplates. *Proc. Natl. Acad. Sci. USA* 95, 2256–2261.

Stanta, G., and Bonin, S. (1998). RNA quantitative analysis from fixed and paraffin-embedded tissues: membrane hybridization and capillary electrophoresis. *Biotechniques* 24, 271–276.

Sunada, W. M., and Blanch, H. W. (1997). Polymeric separation media for capillary electrophoresis of nucleic acids. *Electrophoresis* 18, 2243–2254.

Tan, H., and Yeung, E. S. (1998). Automation and integration of multiplexed on-line sample preparation with capillary electrophoresis for high-throughput DNA sequencing. *Anal. Chem.* 70, 4044–4053.

Thormann, W., Wey, A. B., Lurie, I. S., Gerber, H., Byland, C., Malik, N., Hochmeister, M., and Gehrig, C. (1999). Capillary electrophoresis in clinical and forensic analysis: recent advances and breakthrough to routine applications. *Electrophoresis* 20, 3203–3236.

Trent, R. J., Le, H., and Yu, B. (1998). Prenatal diagnosis for thalassaemia in a multicultural society. *Prenat. Diagn.* 18, 591–598.

Tsui, L. C. (1992a). Mutations and sequence variations detected in the cystic fibrosis transmembrane conductance regulator (CFTR) gene: a report from the Cystic Fibrosis Genetic Analysis Consortium. *Hum. Mutat.* 1, 197–203.

Tsui, L. C. (1992b). The spectrum of cystic fibrosis mutations. *Trends Genet.* 8, 392–398.

Venter, J. C., Adams, M. D., Myers, E. W., Li, P. W., Mural, R. J., Sutton, G. G., Smith, H. O., Yandell, M., Evans, C. A., Holt, R. A., Gocayne, J. D., Amanatides, P., Ballew, R. M., Huson, D. H., Wortman, J. ., Zhang, Q., Kodira, C. D., Zheng, X. H., Chen, L., Skupski, M., Subramanian, G., Thomas, P. D., Zhang, J., Gabor Miklos, G. L., Nelson, C., Broder, S., Clark, A. G., Nadeau, J., McKusick, V. A., Zinder, N., Levine, A. J., Roberts, R. J., Simon, M., Slayman, C., Hunkapiller, M., Bolanos, R., Delcher, A., Dew, I., Fasulo, D., Flanigan, M., Florea, L., Halpern, A., Hannenhalli, S., Kravitz, S., Levy, S., Mobarry, C., Reinert, K., Remington, K., Abu-Threideh, J., Beasley, E., Biddick, K., Bonazzi, V., Brandon, R., Cargill, M., Chandramouliswaran, I., Charlab, R., Chaturvedi, K., Deng, Z., Di, F., Dunn, P., Eilbeck, K., Evangelista, C., Gabrielian, A. E., Gan, W., Ge, W., Gong, F., Gu, Z., Guan, P., Heiman, T. J., Higgins, M. E., Ji, R. R., Ke, Z., Ketchum, K. A., Lai, Z., Lei, Y., Li, Z., Li, J., Liang, Y., Lin, X., Lu, F., Merkulov, G. V., Milshina, N., Moore, H. M., Naik, A. K., Narayan, V. A., Neelam, B., Nusskern, D., Rusch, D. B., Salzberg, S., Shao, W., Shue, B., Sun, J., Wang, Z., Wang, A., Wang, X., Wang, J., Wei,

M., Wides, R., Xiao, C., Yan, C., Yao, A., Ye, J., Zhan, M., Zhang, W., Zhang, H., Zhao, Q., Zheng, L., Zhong, F., Zhong, W., Zhu, S., Zhao, S., Gilbert, D., Baumhueter, S., Spier, G., Carter, C., Cravchik, A., Woodage, T., Ali, F., An, H., Awe, A., Baldwin, D., Baden, H., Barnstead, M., Barrow, I., Beeson, K., Busam, D., Carver, A., Center, A., Cheng, M. L., Curry, L., Danaher, S., Davenport, L., Desilets, R., Dietz, S., Dodson, K., Doup, L., Ferriera, S., Garg, N., Gluecksmann, A., Hart, B., Haynes, J., Haynes, C., Heiner, C., Hladun, S., Hostin, D., Houck, J., Howland, T., Ibegwam, C., Johnson, J., Kalush, F., Kline, L., Koduru, S., Love, A., Mann, F., May, D., McCawley, S., McIntosh, T., McMullen, I., Moy, M., Moy, L., Murphy, B., Nelson, K., Pfannkoch, C., Pratts, E., Puri, V., Qureshi, H., Reardon, M., Rodriguez, R., Rogers, Y. H., Romblad, D., Ruhfel, B., Scott, R., Sitter, C., Smallwood, M., Stewart, E., Strong, R., Suh, E., Thomas, R., Tint, N. N., Tse, S., Vech, C., Wang, G., Wetter, J., Williams, S., Williams, M., Windsor, S., Winn-Deen, E., Wolfe, K., Zaveri, J., Zaveri, K., Abril, J. F., Guigo, R., Campbell, M. J., Sjolander, K. V., Karlak, B., Kejariwal, A., Mi, H., Lazareva, B., Hatton, T., Narechania, A., Diemer, K., Muruganujan, A., Guo, N., Sato, S., Bafna, V., Istrail, S., Lippert, R., Schwartz, R., Walenz, B., Yooseph, S., Allen, D., Basu, A., Baxendale, J., Blick, L., Caminha, M., Carnes-Stine, J., Caulk, P., Chiang, Y. H., Coyne, M., Dahlke, C., Mays, A., Dombroski, M., Donnelly, M., Ely, D., Esparham, S., Fosler, C., Gire, H., Glanowski, S., Glasser, K., Glodek, A., Gorokhov, M., Graham, K., Gropman, B., Harris, M., Heil, J., Henderson, S., Hoover, J., Jennings, D., Jordan, C., Jordan, J., Kasha, J., Kagan, L., Kraft, C., Levitsky, A., Lewis, M., Liu, X., Lopez, J., Ma, D., Majoros, W., McDaniel, J., Murphy, S., Newman, M., Nguyen, T., Nguyen, N., and Nodell, M. (2001). The sequence of the human genome. *Science* 291, 1304–1351.

Weir, H. K., Thun, M. J., Hankey, B. F., Ries, L. A., Howe, H. L., Wingo, P. A., Jemal, A., Ward, E., Anderson, R. N., and Edwards, B. K. (2003). Annual report to the nation on the status of cancer, 1975–2000, featuring the uses of surveillance data for cancer prevention and control. *J. Natl. Cancer Inst.* 95, 1276–1299.

Wenz, H. M., Ramachandra, S., O'Connell, C. D., and Atha, D. H. (1998). Identification of known p53 point mutations by capillary electrophoresis using unique mobility profiles in a blinded study. *Mutat. Res.* 382, 121–132.

Williams, L. C., Hedge, M. R., Herrera, G., Stapleton, P. M., and Love, D. R. (1999). Comparative semi-automated analysis of (CAG) repeats in the Huntington disease gene: use of internal standards. *Mol. Cell Probes.* 13, 283–289.

Woolley, A. T., Hadley, D., Landre, P., deMello, A. J., Mathies, R. A., and Northrup, M. A. (1996). Functional integration of PCR amplification and capillary electrophoresis in a microfabricated DNA analysis device. *Anal. Chem.* 68, 4081–4086.

Woolley, A. T., Sensabaugh, G. F., and Mathies, R. A. (1997). High-speed DNA genotyping using microfabricated capillary array electrophoresis chips. *Anal. Chem.* 69, 2181–2186.

Zhang, J., Voss, K. O., Shaw, D. F., Roos, K. P., Lewis, D. F., Yan, J., Jiang, R., Ren, H., Hou, J. Y., Fang, Y., Puyang, X., Ahmadzadeh, H., and Dovichi, N. J. (1999). A multiple-capillary electrophoresis system for small-scale DNA sequencing and analysis. *Nucleic Acids Res.* 27, e36.

Zhu, Y., Spitz, M. R., Lei, L., Mills, G. B., and Wu, X. (2001). A single nucleotide polymorphism in the matrix metalloproteinase-1 promoter enhances lung cancer susceptibility. *Cancer Res.* 61, 7825–7829.

CHAPTER 8

Temperature and Denaturing Gradient Gel Electrophoresis

HARTMUT PETERS AND PETER N. ROBINSON
Institute of Medical Genetics, Charité University Hospital, Berlin, Germany

TABLE OF CONTENTS

8.1 INTRODUCTION

Temperature gradient gel electrophoresis (TGGE) and the related method denaturing-gradient gel electrophoresis (DGGE) are both based on the principle that the electrophoretic mobility of double-stranded DNA fragments is significantly reduced by their partial denaturation. Owing to the sequence dependence of the melting properties of DNA fragments, sequence variations can be detected. Although the sensitivity of TGGE and DGGE in detecting point mutations in genetic disorders and other settings has been reported to be close to 100%, these methods have never become as popular as other mutation detection methods such as SSCP (see Chapter 6), which may be related to the perception that it is difficult to design adequate PCR primers and set up the assays.

In this chapter, the basic principles of TGGE/DGGE will be discussed and procedures for setting up assays will be described, including how to design and test PCR primers suitable for TGGE/DGGE analysis. Furthermore, studies on the sensitivity of TGGE/DGGE for mutation analysis of genetic disorders will be reviewed and an overview of variations on the basic TGGE/DGGE method will be provided. TGGE and DGGE are robust and highly sensitive methods for mutation screening of genetic disorders that have many advantages that counterbalance the extra effort required in establishing the method.

8.2 THE THEORY OF TEMPERATURE-GRADIENT GEL ELECTROPHORESIS

8.2.1 Melting Behavior of Short Double-stranded DNA Fragments

Myers and colleagues (1985b) originally developed a method of separating DNA fragments differing by single nucleotide substitutions in denaturing gradient gels. The method was based on the notion that the denaturation (melting) of DNA fragments can be regarded as an equilibrium for each base pair (bp) between two distinct states: 1) double helical, and 2) a more random state in which bases are neither paired nor stacked on adjacent bases in any orderly way (Myers *et al.*, 1987). The change from the first to the second state is caused by increasing temperature or increasing concentration of denaturing agents.

In the case of single-nucleotide substitutions, the replacement of an A:T bp (two hydrogen bonds) by a G:C pair (three hydrogen bonds) generally will be expected to increase the temperature at which the corresponding DNA sequence melts. The context of the nucleotide substitution also plays a role, and substitutions of A:T by T:A pairs, or

G : C by C : G pairs, also can affect the temperature at which a DNA sequence dissociates.

Furthermore, a DNA fragment dissociates in a stepwise fashion as the temperature is gradually increased. Dissociation occurs nearly simultaneously in distinct, approximately 50 to 300 nucleotide long regions, termed "melting domains." All nucleotides in a given melting domain dissociate in an all-or-nothing manner within a narrow temperature interval.

The melting temperature (T_m) indicates the temperature at which 50% of the individual molecules are dissociated in the given melting domain, and 50% are double helical. As indicated earlier, the T_m is strongly dependent on the individual DNA sequence and can be altered significantly by small changes in the DNA sequence including single nucleotide substitutions.

8.2.2 Electrophoretic Mobility and the Melting State of DNA Fragments

TGGE is based on detecting differences in the electrophoretic mobility between molecules that may differ only at a single position. DNA fragments produced by the polymerase chain reaction (PCR) are subjected to electrophoresis through a linearly increasing gradient of temperature (or concentration gradient of denaturing agents such as urea and formamide for DGGE). Nucleotide substitutions and other small changes in the DNA sequence are associated with additional bands following TGGE.

The electrophoretic mobility of DNA fragments differs according to whether the fragment is completely double helical, if one or more melting domains has dissociated, or if complete dissociation to two single-stranded molecules has occurred. Each of these states can be visualized using a perpendicular TGGE experiment, as will be discussed further in Section 8.3.2.

The electrophoretic mobility of a double helical (nondenatured) DNA fragment is not significantly altered by single nucleotide substitutions within it, but is primarily dependent on the length and perhaps the curvature of the fragment (Haran *et al.*, 1994). Therefore, assuming that PCR products contain a mixture of two DNA fragments that differ at a single position, as would be the case for a heterozygous point mutation, both fragments initially will progress through the gel at the same speed.

When the molecules reach that point in the gel where the temperature equals their T_m, the molecules will experience a decrease in mobility owing to a transition from a completely duplex (double helical) conformation to a partially denatured one. Dissociation of the first or first few melting domains generally results in a dramatic reduction in the mobility of the DNA fragment, because the fragment takes on a complex, branched conformation.

Due to the strong sequence dependence of the melting temperature, branching (dissociation) and consequent retar-

dation of electrophoretic mobility occurs at different levels of the temperature gradient associated with bands at different positions in the gel (Myers *et al.*, 1987). In addition to the two homoduplex molecules (wt/wt and mt/mt), two different heteroduplex molecules (wt/mt and mt/wt) can be formed by dissociating and reannealing DNA fragments containing a heterozygous mutation prior to performing TGGE (see Fig. 8.1). In practice, it is also possible to perform 40 cycles of PCR; the activity of the *Taq* polymerase is exhausted in the final cycles of PCR, such that heteroduplices are formed as efficiently as is one performed denaturation and reannealing following PCR. Heteroduplex fragments then contain unpaired bases or "bulges" in the otherwise double helical DNA, resulting in a significant reduction in the T_m of the affected melting domain (Ke and Wartell, 1995). The melting temperatures of the two heteroduplex molecules are generally different from one another, so that each heteroduplex is separately visible in the gel. A heterozygous point mutation will thus be visualized by the appearance of four bands: a band representing the normal allele (homoduplex), a band representing the mutant homoduplex that will lie above or underneath the wild-type homoduplex band, depending on the effect of the mutation on the T_m, and two heteroduplex bands that are always above the homoduplex bands (see Fig. 8.2; Myers *et al.* 1987). Mutant and wild-type homoduplex bands generally are separated by 2–10 mm, and the heteroduplex bands are often three or more cm above the homoduplex bands.

8.2.3 Mutations Are Detectable Only in the Lowest Melting Domain(s)

In the preceding discussion, a significant issue is that mutations are detectable only in the melting domain(s) with the lowest melting temperature. If, however, a DNA molecule contains several melting domains with different melting

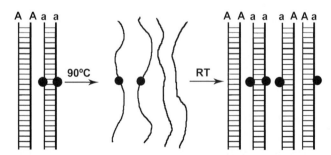

FIGURE 8.1 Mechanism of heteroduplex formation. In the case of heterozygous point mutations in genetic disorders, PCR produces two alleles differing only at the position of the point mutation. A wild-type (AA) and a mutant (aa) molecule are present at an approximately 1 : 1 ratio. Denaturation followed by reannealing of these molecules produces both wild-type (AA) and mutant (aa) homoduplex molecules, as well as two heteroduplex molecules, consisting of a wild-type and a mutant strand (Aa and aA).

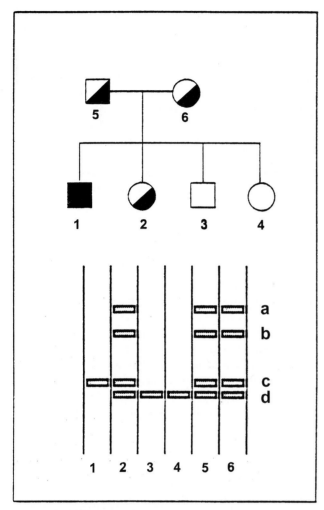

FIGURE 8.2 Parallel TGGE/DGGE. Mutation screening generally is performed with the temperature or denaturing gradient parallel to the direction of electrophoresis. In this example, results of electrophoresis from top to bottom for a hypothetical family segregating an autosomal recessive disorder are shown. Cases 3 and 4 are normal, carrying only the wild-type allele (d). Cases 2, 5, and 6 are heterozygous for a point mutation resulting in the appearance of an additional homoduplex band (c), as well as two additional heteroduplex bands (a and b). Case 1, who is homozygous for the mutation, shows just the mutant homoduplex band (c).

temperatures, it is generally not possible to visualize mutations located elsewhere than in the melting domain with the lowest T_m. Once the DNA fragments reach the temperature at which the first melting domain dissociates, the mobility of the fragment is greatly reduced so that it may not reach temperatures relevant for the higher T_m domains under the conditions of the experiment. Also, dissociation of the highest T_m domain results in complete dissociation of the DNA fragment into two single-stranded DNA molecules. Single-stranded DNA, like completely double helical DNA, does not demonstrate differences in electrophoretic mobility, owing to small sequence changes, and hence there is no

possibility of distinguishing two sequences once complete dissociation has occurred.

The consequence of these observations is that only mutations in the lowest T_m domain can be detected reliably by TGGE or DGGE (Myers *et al.*, 1987).

8.2.4 GC- and Psoralen Clamps Extend the Usefulness of TGGE

Myers and colleagues (1985a) presented an extension of the original DGGE protocol that allowed mutations in every region of the DNA fragment under analysis to be detected. These researchers attached a 135 bp, GC-rich sequence, known as GC-clamp, to the β-globin promoter region in which mutations were being sought. The β-globin promoter region was found to contain two melting domains; without the GC-clamp, only mutations in the domain with the lower T_m could be visualized in the gel. Owing to its high GC content, the GC-clamp has a significantly higher melting temperature than most naturally occurring sequences. The attachment of the GC-clamp was found to significantly alter the melting properties of the β-globin sequence and mutations in the entire β-globin sequence could be experimentally detected (Myers *et al.*, 1985a). By adding a 40 nt G+C rich sequence to one of the two PCR primers, a GC-clamp can conveniently be added to any DNA fragment produced by PCR (Sheffield *et al.*, 1989). It is also possible to use a universal GC-clamp that is incorporated into amplified DNA fragments during PCR, thereby avoiding the expense of synthesizing long primers (Top, 1992).

Psoralen-modified PCR primers are an alternative to GC-clamps. One of the two PCR primers is 5′ modified by 5-(ω-hexyloxy)-psoralen. The 5′ terminus of the primer should have two adenosine residues; if the natural sequence does not have AA, this sequence should be appended to the specific DNA sequence of the primer. Psoralens are bifunctional photoreagents that can form covalent bonds with pyrimidine bases (especially thymidine). If intercalated at 5′-TpT in double helical DNA (this will be the complementary sequence of the 3′ terminus of the other strand following PCR), psoralen forms a covalent bond with thymidine after photoinduction (Costes *et al.*, 1993b). Photoinduction can be performed by exposing to the PCR products to a source of UV light (365) for 5 to 15 minutes, which can be done conveniently in the original PCR tubes or 96-well plates.

In general, psoralen clamping provides comparable results to GC clamping, except that cross-linking of the PCR fragments is only approximately 85% efficient, so that one observes single-stranded, denatured DNA fragments running below the main bands in the TGGE. Psoralen clamping sometimes is preferred over GC-clamping because the PCR is often easier to optimize, and bipolar clamping is possible if necessary (see section 8.3.4). Psoralen modification of primers is available from many commercial oligonucleotide sources.

8.3 THE PRACTICE OF TEMPERATURE-GRADIENT GEL ELECTROPHORESIS

Detailed protocols for TGGE and DGGE are available elsewhere (Kang *et al.*, 1995; Murdaugh and Lerman, 1996). In the following sections, the most important issues concerning how to set up TGGE or DGGE assays successfully are discussed, including especially the issues related to primer design and optimization procedures. Several points that apply only to DGGE are discussed in Section 8.4.

8.3.1 Primer Design for TGGE/DGGE

One of the first and most widely used computer programs to design primers for TGGE was the Melt87 package by Lerman and Silverstein (1987). An updated version of this program (Melt94) is available at http://web.mit.edu/osp/www/melt.html. The Melt87 program calculates the T_m for each bp in the DNA fragment; that is, the temperature at which 50% of the individual molecules are double helical and 50% of the molecules are in a fully disordered, melted state. The results of such a calculation are termed "melting map" (see Fig. 8.3). DNA fragments typically are divided

into distinct melting domains of about 50 to 300 bp in length, in which all base pairs have nearly an identical T_m. The melting map demonstrates the lowest melting domain in the DNA fragment; as mentioned earlier, only mutations in this region will be visible by TGGE analysis.

A further useful program in the Melt87 package is SQHTX. This program calculates the expected displacement in the gradient for a single-nucleotide mismatch (as would be the case for a heteroduplex molecule with a single-nucleotide substitution) at every position in the fragment. This analysis provides the clearest indication of the position in the fragment, where mutations will be detectable by TGGE analysis (Lerman and Silverstein, 1987). Figure 8.4 provides an example of a displacement map calculated with SQHTX.

The Melt87 programs are DOS-based and difficult to use for those with little experience with DOS and menu-based programs. Melt87 has no graphic capabilities of its own, and users need to process its output with a graphics program of their choice. For this reason, several freely available and proprietary programs have become available, which are significantly easier to use (see Table 8.1).

Users should load a DNA sequence encompassing the DNA fragment to be analyzed (e.g., an exon with flanking intron sequences) together with about 100 nucleotides' "extra" sequence to either side of the fragment of interest.

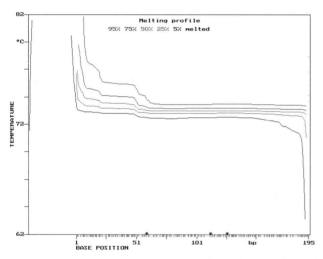

FIGURE 8.3 Melting map. This graphic represents a fragment from exon 14 of the *NF1* gene and was produced using TGGE-Star. Each tick on the x-axis represents a base pair. The base pairs are numbered from 1 to 195. The y-axis shows the temperature where the probability for a bp to be melted has the value 0.95, 0.75, 0.5, 0.25, and 0.05, respectively. The 5′-terminus of the fragment corresponds to a GC-clamp. Additionally, one can distinguish two further melting domains: from the 5′-terminus to the 50th bp and from the 50th bp to the 3′-terminus. The difference between these two melting domains is small and the sensitivity of TGGE is not disturbed. If the difference between these two plateaus in the curve were higher, both regions would need to be tested in two different PCR-TGGE steps. Mutations were detected in both regions of this fragment: three asterisks above the x-axis mark positions of mutations detected with this assay (See color plate).

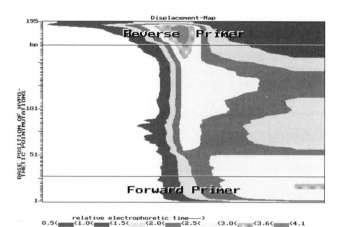

FIGURE 8.4 Displacement maps calculated using the program SQHTX, and graphic created with TGGE-Star. In the case of a heterozygous mutation, two heteroduplex bands occur. Heteroduplices do not migrate as far as the wild type fragments because they melt at lower temperatures. The distance of heteroduplex bands and wild type bands depends on the electrophoretic duration (x-axis) and the base position (y-axis). A mutation can be detected only when the displacement is higher than the resolution of the gel. The color codes indicate different electrophoretic times, and the width of each band of color indicates the expected displacement (in arbitrary units) in the gel for a point mutation at the corresponding position in the sequence (See color plate).

TABLE 8.1　Programs for the design of PCR primers for use in TGGE/DGGE

Name	Comment	URL
Melt94	DOS-based	http://web.mit.edu/osp/www/melt.html
TGGE-Star	DOS-based, freely available user-friendly wrapper for Melt87 (Gille and Gille, 2002)	http://www.charite.de/bioinf/tgge/
Poland	Server-based implementation of Poland's algorithm (Steger, 1994)	http://www.biophys.uni-duesseldorf.de/local/POLAND/poland.html
MELTingeny	A commercial, Java-based GUI program with flexible routines for designing DGGE/TGGE primers	http://www.ingeny.com
WinMelt, MacMelt	Commercial GUI programs for melting profile analysis	http://www.medprobe.com/uk/melt.html

The previously mentioned programs can be used to find primers that result in a DNA fragment with melting properties adequate for TGGE or DGGE. In general, some amount of trial and error is needed to find optimal primers for any given sequence. Users need to decide both the position of the forward and reverse primers as well as whether the GC-clamp is to be placed on the 5′ or 3′ PCR primer or both (see later for a discussion of bipolar clamping). Programs such as TGGE-Star and MELTingeny facilitate this process by allowing users to easily shift primer positions and recalculate the melting maps. It should be mentioned that a 40-nucleotide GC-clamp can be substituted for a psoralen clamp in the computer analysis.

8.3.2　Perpendicular TGGE for the Determination of the T_m

In most cases in which TGGE is used for mutation analysis, parallel electrophoresis with simultaneous analysis of multiple samples will be performed. For each such assay, the optimal temperature gradient and run time must be determined experimentally. The procedures used for this purpose are described in this and the following section.

The optimization process begins with a perpendicular TGGE experiment, in which electrophoresis is performed perpendicularly to the temperature gradient (see Fig. 8.5). Perpendicular TGGE is used to verify the reversible melting behavior of the DNA fragment and to determine its T_m under the experimental conditions. Perpendicular TGGE is run with a gradient of 20°C–60°C, which will be adequate for the vast majority of PCR fragments. Electrophoresis is initially performed at room temperature for 10–15 minutes to run the sample into the gel. Then, electrophoresis is stopped while a temperature gradient of 20°C–60°C is established, after which electrophoresis should be continued for 90–120 minutes. Figure 8.5 demonstrates the use of this analysis to determine the T_m of the DNA fragment being analyzed.

8.3.3　Travel Schedule Experiments

Up to three novel bands are observed upon TGGE/DGGE analysis of a heterozygous mutation or polymorphism. The separation will begin to become apparent when the heteroduplex molecules have reached their T_m, as their mobility will be retarded by partial denaturation. Separation of the homoduplex molecules will occur in a region of the gradient surrounding the T_m of the lowest melting domain of the DNA fragment. Therefore, TGGE assays are set up to avoid a long running time before the samples reach the effective range of separation. One should choose the temperature gradient such that the effective range of separation is approximately in the middle or somewhat above the middle of the gel, and that the upper and lower temperature ranges are separated by about 15°C from the T_m of the DNA fragment.

Once an appropriate temperature gradient has been chosen, the optimal running time can be determined by a travel-schedule experiment—a parallel TGGE experiment in which samples are applied every 30 minutes for three hours (or longer), such that the last sample to be loaded has run 30 minutes, and the first sample, three hours. Usually, one will see a reduction on electrophoretic mobility of samples after a certain period of time (generally 60 to 90 minutes if the temperature gradient was chosen correctly). Samples often do not continue to wander in the gel with any significant velocity once their melting temperature has been reached. These gels generally are run for about 30 minutes longer than the time determined in this manner (see Fig. 8.6). Different choices of the range and starting point of the temperature gradient affect both the range in the gel at which mutations will be visible as well as the optimal running time (see Fig. 8.7).

8.3.4　Bipolar Clamping

Occasionally, TGGE analysis will result in fuzzy bands that are difficult to evaluate, despite apparently adequate melting behavior, as predicted by Melt94 or other programs. Bipolar clamping of PCR products, by means of attaching a psoralen clamp to each of the two PCR primers rather than just one, is an efficacious method to improve melting characteristics of PCR fragments that are otherwise not amenable to TGGE/DGGE analysis (Gille *et al.*, 1998). Bipolar clamping is a simple procedure that can significantly improve

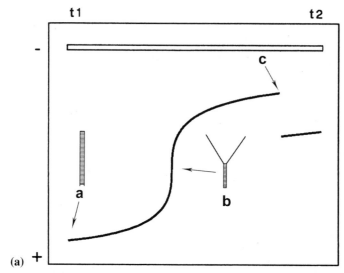

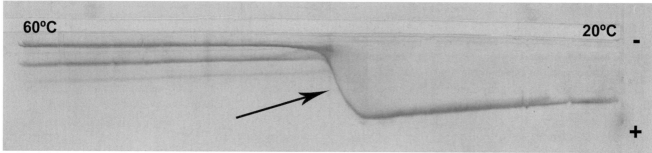

(b)

FIGURE 8.5 **a)** Schematic drawing of a perpendicular TGGE/DGGE gel. A temperature gradient from t1 (e.g., 20°C) to t2 (e.g., 60°C) is established perpendicularly to the direction of electrophoresis (indicated by − and +). Fragments at lower temperatures remain completely double helical and have a relatively high electrophoretic mobility (a). Once the melting temperature of the lowest-temperature melting domain is reached, partial denaturation of the DNA fragment (b) causes a significant reduction of electrophoretic mobility. The temperature at which 50% of individual molecules are melted is denoted as the melting temperature (T_m), and is indicated by the arrow in the figure (b). A reversible denaturation step is observed as a continuous transition (curve). Once the temperature of the highest-melting domain is reached, irreversible melting occurs, causing a discontinuous transition in the melting curve (c). **b)** Perpendicular TGGE gel. In this example, a PCR fragment corresponding to *NF1* gene exon 14 was analyzed. PCR product was applied and run into the gel at 10°C for 15 minutes. Then, a temperature gradient from 20°C to 60°C was established perpendicularly to the direction of electrophoresis, which was then performed for an additional 60 minutes. One observes a high electrophoretic mobility in portions of the gel with temperatures below the T_m of the fragment. The gradual decrease in mobility around the middle of the gel indicates reversible melting of the lowest-temperature melting domain. In portions of the gel with temperatures above the T_m of the fragment, partial denaturation of the fragment leads to a significantly reduced electrophoretic mobility. The arrow at the midway point of the curve indicates the T_m of the fragment under the experimental conditions (approximately 39°C).

results of TGGE analysis in cases where analysis with only one clamp has yielded suboptimal results. Programs such as TGGE-Star (Gille and Gille, 2002) offer the possibility of computer analysis with two clamps, and may suggest the use of bipolar clamping for amplicons whose predicted melting properties are otherwise not satisfactory.

8.4 DENATURING GRADIENT GEL ELECTROPHORESIS (DGGE)

The theory of DGGE/TGGE is described in detail in the first part of this chapter. Parallel DGGE is a form of polyacry-

lamide gel electrophoresis in which a double-stranded DNA fragment migrates into a gradient of linearly increasing denaturing conditions. The denaturing gradient is functionally equivalent to the temperature gradient of TGGE. The denaturants used are heat (a constant temperature of generally 60°C) and a fixed ratio of formamide (ranging from 0–40%) and urea (ranging from 0–7 M). The temperature of 60°C was empirically chosen to exceed the melting temperature of an AT-rich DNA fragment in the absence of a denaturant. For extremely GC-rich DNA sequences, higher temperatures (e.g., 75°C) can be used. To achieve a uniform temperature distribution the electrophoresis unit is attached to a circulating water bath.

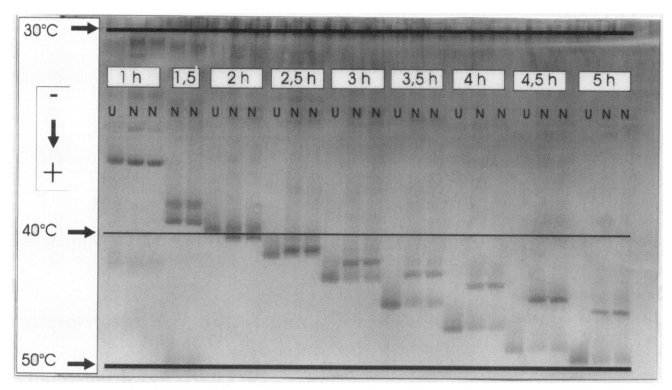

FIGURE 8.6 Travel schedule experiment. This experiment is used to determine the optimal running time of a TGGE experiment. Fragments, corresponding to exon 19a of the *NF1* gene, in which one of the primers was modified with psoralen (see Section 8.2.4), were applied at intervals of 30 minutes, such that the first fragments had a total running time of 5 hours, and the last fragments to be applied had a running time of 60 minutes. Lanes labeled U contain PCR fragments that were not UV-irradiated to effect psoralen-mediated crosslinking, and lanes labeled N (two lanes were loaded for each timepoint) contain irradiated PCR fragments. One sees that the fragments initially are completely double-helical (1 h), such that irradiated and nonirradiated fragments display the same band pattern. Starting at the T_m of this fragment (40°C), the nonirradiated fragments (U) undergo complete dissociation so that only a single-strand band running well below the main band of the irradiated (cross-linked) fragments is visible (compare the time points at 2.5 and 3 hours. Additionally, the irradiated fragments (N) show a large reduction in electrophoretic mobility following partial denaturation at about 40°C. Under these conditions, an optimal running time would be 3 hours, although the running time could be reduced by adjusting the temperature gradient (see also Fig. 8.7). (See color plate)

8.4.1 Optimization of Gel Running Conditions

The computer programs (e.g., Melt94) described earlier reduce the number of preliminary experiments required for optimization of the gel running conditions. However, it is still necessary to run some preliminary gels to determine the optimal electrophoresis conditions and running times and to confirm that the optimal denaturing gradient has been chosen. The aim of these travel schedule gels is to have well-separated bands (normal and mutation positive control are simultaneously loaded on the gels) that are focused by the gradient. PCR products with two low-melting domains require different gel conditions for the analysis of each domain.

The choice of the denaturant concentration range can be determined as follows. The differences in gradient depth (the displacement) between a fragment and the same fragment with a change at a specified bp are calculated by the program

SQHTX (Lerman and Silverstein, 1987) as described in Section 8.3.1. SQHTX calculates the displacement as the difference in temperature at which the wild-type homoduplex and the heteroduplex molecules partially melt (see Fig. 8.4). To convert between the temperature values and the denaturant concentration, a difference of 1°C is converted to a difference of 3% denaturant concentration (approximately equivalent to 1 cm distance within a 20% urea gradient gel). An experimental determination of gradient behavior can be achieved by perpendicular gel electrophoresis. Data from the perpendicular gels help to estimate the denaturant concentration range to use in parallel gel electrophoresis. For parallel gels, the gradient initially should be chosen with a 25% to 30% difference in denaturant concentration centered around the melting temperature of the domain (Myers *et al.*, 1987). Once optimized gel running conditions have been established, the method can be used for mutation screening.

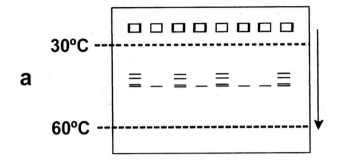

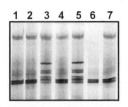

FIGURE 8.8 A heterozygous point mutation in exon 31 of the *NF1* gene detected by TGGE. The third and fifth lanes display a classical four-band appearance due to the presence of the heterozygous mutation 5839C→T.

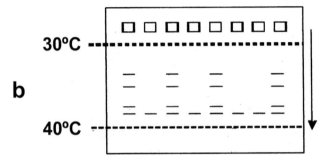

FIGURE 8.7 The effect of different temperature gradients on separation of samples. A relatively wide gradient (a) leads to a relatively small separation of homoduplex and heteroduplex bands in the case of a heterozygous mutation, while a shallower gradient (b) increases the separation. The lower temperature of the gradient can also be adjusted to reduce the amount of time needed before the T_m of the lowest melting domain is reached.

8.5 THE USE OF TGGE/DGGE FOR MUTATION DETECTION

TGGE and DGGE have been used to investigate a large number of disease genes, some of which are listed in the following sections. Due to the relative ease of detecting heterozygous mutations owing to the occurrence of up to three novel bands, TGGE and DGGE have been particularly useful for disorders characterized by heterozygous mutations or frequent *de novo* mutations (reviewed in Fodde and Losekoot, 1994). In light of the effort involved in designing primers and optimizing conditions, TGGE or DGGE generally is reserved for situations when large numbers of samples are to be screened for mutations.

Most mutation screening protocols involve the simultaneous analysis of 24 or more samples on one parallel TGGE/DGGE. In general, altered band patterns are easy to spot. The classic appearance of heterozygous mutations (see Figs. 8.2 and 8.8) is due to the appearance of three additional bands. With some mutations, only one or two additional bands are seen. In the authors' experience, the

specificity of TGGE/DGGE is exquisitely high. In other words, a false-positive four-band pattern occurs rarely if at all.

8.6 DETECTION RATE AND SENSITIVITY

By using DGGE Myers and colleagues (1985b) detected an estimated 40% of the sequence variants in a DNA fragment up to 500 bp in their initial study. The use of GC-clamps, psoralen clamps, or bipolar clamping, which aid the formation of uniform low melting domains, significantly improved the detection rate of TGGE/DGGE, which in many cases approaches nearly 100%.

The sensitivity of TGGE/DGGE for detecting known mutations is generally reported to be nearly 100%, generally performing as well as or better than other mutation detection methods (Abrams *et al.*, 1990; Ferec *et al.*, 1992; Gelfi *et al.*, 1997; Gejman *et al.*, 1998; Tchernitchko *et al.*, 1999; Zschocke *et al.*, 2000; Breton *et al.*, 2003). In one study with a panel of known mutations, DGGE detected 201 of 201 known mutations in the CFTR gene (Macek *et al.*, 1997). The reasons for lower reported detection rates of unknown mutations in some studies has been speculated to be due to genetic heterogeneity (Ferec *et al.*, 1999), clinical overdiagnosis (Katzke *et al.*, 2002), or location of mutations in intronic or promoter regions that were not included in the screening program. Optimization of the TGGE/DGGE assay conditions and primers, perhaps including the use of bipolar clamping (Gille and Gille, 2002), may increase sensitivity. In summary, the sensitivity of TGGE/DGGE, when properly used, is close to 100%.

TGGE/DGGE also has been shown to be very sensitive in the detection of mutations in situations where the mutation sequence is present in proportions less than 50% (as is generally the case when heterozygous mutations are sought in genomic DNA). This has proved useful in detection of heteroplasmy in mitochondrial disorders with heteroplasmic proportions as low as 1% (Tully *et al.*, 2000), as well as in testing for residual disease in cancer (Ahnhudt *et al.*, 2001; Alkan *et al.*, 2001).

8.7 RELATED TECHNIQUES AND VARIANTS

A wide range of improvements and further developments of the principles underlying DGGE and TGGE have appeared in the last decade, the most important of which are briefly summarized here.

Broad range DGGE. A single gel and a single set of conditions is used to screen all the exons of one gene (Guldberg and Guttler, 1994; Hayes *et al.*, 1999).

Multiplex DGGE. Several exons are simultaneously analyzed in one DGGE gel (Costes *et al.*, 1993a).

Genomic DGGE (gDGGE). Genomic DNA is digested with a restriction enzyme, electrophoresed by DGGE, transferred to nylon membrane, and hybridized to a unique DNA probe (Borresen *et al.*, 1988).

Constant DGGE (cDGGE). Gels contain constant concentrations of denaturants. This allows an increased resolution of mutant fragments since they will constantly migrate with a different electrophoretic mobility through the whole length of the gel (Hovig *et al.*, 1991).

Constant denaturant capillary electrophoresis (CDCE). DNA migrates through a 30 cm quartz capillary of 75 μm inner diameter, filled with a viscous polyacrylamide solution. A 10 cm part of the capillary, prior to the detector, is heated to a temperature permitting partial melting (see also previous chapter). Usually the DNA is fluorescein-labeled and detected by laser-induced fluorescence (Khrapko *et al.*, 1994). Separation of DNA fragments is achieved by the differential velocity of partly melted DNA in a medium with uniform denaturant concentration.

Temporal temperature gradient gel electrophoresis (TTGE). A constant concentration of urea or formamide is used as in cDGGE, but the temperature during the run is increased gradually (Yoshino *et al.*, 1991; Wiese *et al.*, 1995). The denaturant concentration (usually 6–8% urea) used in TTGE can be determined either from the theoretical melting curve or experimentally from a perpendicular DGGE.

Microtemperature-gradient gel electrophoresis (μTGGE). A minimized gel (20 × 20 × 0.5 mm) leads to the reduction of the amount of DNA required and to shorter running times (approximately 12 min at 100 V, 10 mA). The method was used in microbial ecology and epidemiology (Biyani and Nishigaki, 2001).

Double-gradient, denaturing gradient gel electrophoresis (DG-DGGE). In addition to the chemical denaturing gradient (formamide and urea) a second sieving gradient (e.g., 6%–12% polyacrylamide gradient) is used (Cremonesi *et al.*, 1997).

Two-dimensional DNA fingerprinting/two-dimensional gene scanning (TDGS, 2D-DNA typing). Combines size fractionation of DNA fragments in the first dimension with their sequence-specific separation through DGGE in the second dimension (see also next chapter).

Denaturing HPLC (dHPLC). Uses an ion-pair chromatography separation principle, combined with a precise control of the column temperature and optimized mobile phase gradient for separation of mutant DNA molecules (reviewed in Xiao and Oefner, 2001).

Cycling gradient capillary electrophoresis (CGCE). DNA sequence variants are detected based on their differential migration in a polymer-filled capillary system. A cycling (oscillating) temporal temperature gradient

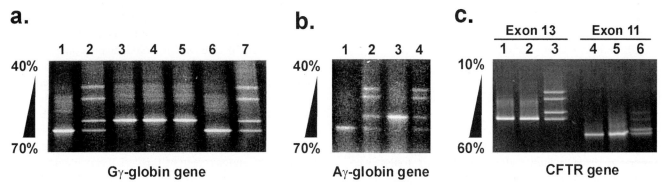

FIGURE 8.9 Mutation detection using DGGE analysis. **a)** Screening for the Gγ −158 C→T polymorphism in the promoter region of the human Gγ-globin gene. Lanes 3, 4, and 5 correspond to homozygous samples for that polymorphism; lanes 2 and 7 correspond to heterozygous samples; and lanes 1 and 6 correspond to samples that do not carry this polymorphism to either of the two alleles (photo courtesy of Dr. George P. Patrinos). **b)** DGGE analysis of the promoter region of the human Aγ-globin gene. Lanes 2 and 4 correspond to heterozygous cases for the Aγ −117 G→A mutation, leading to the Greek type of nondeletional Hereditary Persistence of Fetal Hemoglobin (nd-HPFH); lane 3 corresponds to a homozygous case for the same mutation; lane 1 corresponds to a wild-type control (photo courtesy of Dr. George P. Patrinos). **c)** Mutation analysis of exons 11 and 13 of the CFTR gene. Lanes 1, 2, 4, and 5 correspond to wild-type cases; lane 3 corresponds to a heterozygous case for the E822X nonsense mutation; lane 6 corresponds to a heterozygous case for the G542X nonsense mutation, leading to cystic fibrosis (photo courtesy of Dr. Angeliki Balassopoulou, Athens, Greece). The gradient of denaturing agents is depicted at the left side of each gel.

is applied. This improvement enables utilization of a multiple injection technique, in which multiple samples are injected into the same capillary (or set of capillaries) separated by predefined time intervals of partial electrophoresis. A 96-capillary system is able to screen over 15,000 samples in 24 h (Minarik *et al.*, 2003).

8.8 TECHNICAL EQUIPMENT FOR TGGE/DGGE

In general, for DGGE, preexisting vertical electrophoresis equipment with buffer-tank and combined heater/stirrer thermostat can be adapted. For TGGE, special equipment to achieve a constant temperature gradient is necessary.

The **Biometra TGGE** (Goettingen, Germany; www.biometra.de) system uses a temperature block powered by Peltier technology, which enables a strictly linear gradient that may allow more reproducible conditions than with conventional chemical gradients or temperature gradients using water baths. The Biometra TGGE system is available in two formats: A TGGE "mini" system operates small gels and is therefore suitable for fast, serial experiments. A TGGE maxi system provides a large separation distance and allows high parallel sample throughput.

The **DCode mutation detection system** (Bio-Rad Laboratories, Hercules USA) can be used to screen mutations by DGGE, TGGE, CDGE, TTGE, and other techniques. The system performs TTGE by controlling the buffer temperature during the electrophoresis run. A temperature control module regulates the rate of temperature increase in a uniform and linear fashion.

Sooner Scientific (Garvin, USA, www.soonersci.com) offers five different-sized DGGE Systems variants (for 2, 4, or 8 smaller gels or one large gel).

The **INGENYphorU system** (Ingeny International, GP Goes, The Netherlands; www.ingeny.com) is suitable for DGGE, TGGE, CDGE, and other techniques.

8.9 APPLICATIONS OF TGGE/DGGE AND RELATED METHODS

TGGE/DGGE has been applied in an increasing number of studies. A recent search in PubMed database found over 1,100 citations. The following applications have been described:

- Screening for polymorphisms in human genes; for example, *COL1A2* gene (Borresen *et al.*, 1988), alpha-1-antitrypsin (Hayes, 2003), human γ-globin genes (Patrinos *et al.*, 1998; 2001; Fig. 8.9a,b)
- Mutation detection in human genes; for example, p53 (Pignon *et al.*, 1994), *FBN1* (Tiecke *et al.*, 2001; Katzke *et al.*, 2002; Robinson *et al.*, 2002), *NF1* (Peters *et al.*,

1999; Fahsold *et al.*, 2000), dystrophin gene (Hofstra *et al.*, 2004), δ-globin (Papadakis *et al.*, 1997) and β-globin genes (Losekoot *et al.*, 1990), *CFTR* gene (Fig. 8.9c), and so on
- Mutation and polymorphism detection in mitochondrial DNA (Hanekamp *et al.*, 1996; Chen *et al.*, 1999)
- Analysis in microbial ecology, determination of biodiversity of bacterial populations in soil, fresh, or salt water (Muyzer and Smalla, 1998; van Elsas *et al.*, 2002)
- Genome profiling and provisional microbial species identification on the basis of random PCR and TGGE (Watanabe *et al.*, 2002)
- Determination of biodiversity in fecal or intestinal microflora (Tannock, 2002)
- HLA typing (Uhrberg *et al.*, 1994)
- Analysis of proteins and antibody binding (Riesner *et al.*, 1991; Arakawa *et al.*, 1993)
- Clonality analysis of T-cell or T-cell receptors (Plonquet *et al.*, 2002; Lukowsky, 2003)
- Mutation detection and detection of variation between genomes of viral strains (Lu *et al.*, 2002; Motta *et al.*, 2002)
- Analysis of biodiversity and polymorphisms in plants (Gomes *et al.*, 2003; Nikolcheva *et al.*, 2003)
- Examination of the fidelity of DNA polymerases (Keohavong and Thilly, 1989)

8.10 CONCLUSIONS

TGGE/DGGE and related methods provide a very high sensitivity and are relatively easy and cheap to perform once the assays have been designed and optimized. The main advantages are in the high detection rate and specificity and improved heterozygote detection. The methodology is simple, nonradioactive, and relatively nontoxic. The disadvantages of TGGE and DGGE include mainly the limitation of PCR fragment length to about 500 nucleotides, the difficulties of analyzing GC-rich fragments, and the need for computer analysis of potential PCR fragments (which on the other hand can save time and money by eliminating the use of inadequate primers). However, once primers and conditions have been chosen, TGGE/DGGE is a robust and easy-to-perform mutation screening method. It is particularly well suited for the detection of known and unknown mutations in large genes, where high sensitivity is required and when large numbers of samples are to be tested.

Acknowledgments

The authors would like to thank Christoph Gille, Anja Klose, Angelika Pletschacher, and Horst Schlechte.

References

Abrams, E. S., Murdaugh, S. E., and Lerman, L. S. (1990). Comprehensive detection of single base changes in human genomic DNA using denaturing gradient gel electrophoresis and a GC clamp. *Genomics* 7, 463–475.

Ahnhudt, C., Muche, J. M., Dijkstal, K., Sterry, W., and Lukowsky, A. (2001). An approach to the sensitivity of temperature-gradient gel electrophoresis in the detection of clonally expanded T-cells in cutaneous T-cell lymphoma. *Electrophoresis* 22, 33–38.

Alkan, S., Cosar, E., Ergin, M., and Hsi, E. (2001). Detection of T-cell receptor-gamma gene rearrangement in lymphoproliferative disorders by temperature gradient gel electrophoresis. *Arch. Pathol. Lab. Med.* 125, 202–207.

Arakawa, T., Hung, L., Pan, V., Horan, T. P., Kolvenbach, C. G., and Narhi, L. O. (1993). Analysis of the heat-induced denaturation of proteins using temperature gradient gel electrophoresis. *Anal. Biochem.* 208, 255–259.

Biyani, M., and Nishigaki, K. (2001). Hundredfold productivity of genome analysis by introduction of microtemperature-gradient gel electrophoresis. *Electrophoresis* 22, 23–28.

Borresen, A. L., Hovig, E., and Brogger, A. (1988). Detection of base mutations in genomic DNA using denaturing gradient gel electrophoresis (DGGE) followed by transfer and hybridization with gene-specific probes. *Mutat. Res.* 202, 77–83.

Breton, J., Sichel, F., Abbas, A., Marnay, J., Arsene, D., and Lechevrel, M. (2003). Simultaneous use of DGGE and DHPLC to screen TP53 mutations in cancers of the esophagus and cardia from a European high incidence area (Lower Normandy, France). *Mutagenesis* 18, 299–306.

Chen, T. J., Boles, R. G., and Wong, L. J. (1999). Detection of mitochondrial DNA mutations by temporal temperature gradient gel electrophoresis. *Clin. Chem.* 45, 1162–1167.

Costes, B., Fanen, P., Goossens, M., and Ghanem, N. (1993a). A rapid, efficient, and sensitive assay for simultaneous detection of multiple cystic fibrosis mutations. *Hum. Mutat.* 2, 185–191.

Costes, B., Girodon, E., Ghanem, N., Chassignol, M., Thuong, N. T., Dupret, D., and Goossens, M. (1993b). Psoralen-modified oligonucleotide primers improve detection of mutations by denaturing gradient gel electrophoresis and provide an alternative to GC-clamping. *Hum. Mol. Genet.* 2, 393–397.

Cremonesi, L., Firpo, S., Ferrari, M., Righetti, P. G., and Gelfi, C. (1997). Double-gradient DGGE for optimized detection of DNA point mutations. *Biotechniques* 22, 326–330.

Fahsold, R., Hoffmeyer, S., Mischung, C., Gille, C., Ehlers, C., Kucukceylan, N., Abdel-Nour, M., Gewies, A., Peters, H., Kaufmann, D., Buske, A., Tinschert, S., and Nurnberg, P. (2000). Minor lesion mutational spectrum of the entire NF1 gene does not explain its high mutability but points to a functional domain upstream of the GAP-related domain. *Am. J. Hum. Genet.* 66, 790–818.

Ferec, C., Audrezet, M. P., Mercier, B., Guillermit, H., Moullier, P., Quere, I., and Verlingue, C. (1992). Detection of over 98% cystic fibrosis mutations in a Celtic population. *Nat. Genet.* 1, 188–191.

Ferec, C., Raguenes, O., Salomon, R., Roche, C., Bernard, J. P., Guillot, M., Quere, I., Faure, C., Mercier, B., Audrezet, M. P., Guillausseau, P. J., Dupont, C., Munnich, A., Bignon, J. D., and Le Bodic, L. (1999). Mutations in the cationic trypsinogen gene and evidence for genetic heterogeneity in hereditary pancreatitis. *J. Med. Genet.* 36, 228–232.

Fodde, R., and Losekoot, M. (1994). Mutation detection by denaturing gradient gel electrophoresis (DGGE). *Hum. Mutat.* 3, 83–94.

Gejman P. V., Cao Q., Guedj F., and Sommer S. (1998). The sensitivity of denaturing gradient gel electrophoresis: a blinded analysis. *Mutat. Res.* 382, 109–114.

Gelfi, C., Righetti, S. C., Zunino, F., Della Torre, G., Pierotti, M. A., and Righetti, P. G. (1997). Detection of p53 point mutations by double-gradient, denaturing gradient gel electrophoresis. *Electrophoresis* 18, 2921–2927.

Gille, C., Gille, A., Booms, P., Robinson, P. N., and Nurnberg, P. (1998). Bipolar clamping improves the sensitivity of mutation detection by temperature gradient gel electrophoresis. *Electrophoresis* 19, 1347–1350.

Gille, C., and Gille, A. (2002). TGGE-STAR: primer design for melting analysis using PCR gradient gel electrophoresis. *Biotechniques* 32, 264, 266, 268.

Gomes, N. C., Fagbola, O., Costa, R., Rumjanek, N. G., Buchner, A., Mendona-Hagler, L., and Smalla, K. (2003). Dynamics of fungal communities in bulk and maize rhizosphere soil in the tropics. *Appl. Environ. Microbiol.* 69, 3758–3766.

Guldberg, P., and Guttler, F. (1994). "Broad-range" DGGE for single-step mutation scanning of entire genes: application to human phenylalanine hydroxylase gene. *Nucleic Acids Res.* 22, 880–881.

Hanekamp, J. S., Thilly, W. G., and Chaudhry, M. A. (1996). Screening for human mitochondrial DNA polymorphisms with denaturing gradient gel electrophoresis. *Hum. Genet.* 98, 243–245.

Haran, T. E., Kahn, J. D., and Crothers, D. M. (1994). Sequence elements responsible for DNA curvature. *J. Mol. Biol.* 244, 135–143.

Hayes, V. M. (2003). Genetic diversity of the alpha-1-antitrypsin gene in Africans identified using a novel genotyping assay. *Hum. Mutat.* 22, 59–66.

Hayes, V. M., Wu, Y., Osinga, J., Mulder, I. M., van der Vlies, P., Elfferich, P., Buys, C. H., and Hofstra, R. M. (1999). Improvements in gel composition and electrophoretic conditions for broad-range mutation analysis by denaturing gradient gel electrophoresis. *Nucleic Acids Res.* 27, e29.

Hofstra, R. M., Mulder, I. M., Vossen, R., de Koning-Gans, P. A., Kraak, M., Ginjaar, I. B., van der Hout, A. H., Bakker, E., Buys, C. H., van Ommen, G. J., van Essen, A. J., and den Dunnen, J. T. (2004). DGGE-based whole-gene mutation scanning of the dystrophin gene in Duchenne and Becker muscular dystrophy patients. *Hum. Mutat.* 23, 57–66.

Hovig, E., Smith-Sorensen, B., Brogger, A., and Borresen, A. L. (1991). Constant denaturant gel electrophoresis, a modification of denaturing gradient gel electrophoresis, in mutation detection. *Mutat. Res.* 262, 63–71.

Kang, J., Kühn, J. E., Schäfer, P., Immelmann, A., and Henco, K. (1995) Quantification of DNA and RNA by PCR. In *A Practical Approach*, M. J. McPherson, B. D. Hames, G. R. Taylor, eds. PCR 2, pp. 119–133. IRL Press, Oxford.

Katzke, S., Booms, P., Tiecke, F., Palz, M., Pletschacher, A., Turkmen, S., Neumann, L. M., Pregla, R., Leitner, C., Schramm, C., Lorenz, P., Hagemeier, C., Fuchs, J., Skovby, F., Rosenberg, T., and Robinson, P. N. (2002). TGGE screening of the entire FBN1 coding sequence in 126 individuals with marfan syndrome and related fibrillinopathies. *Hum. Mutat.* 20, 197–208.

Ke, S. H., and Wartell, R. M. (1995). Influence of neighboring base pairs on the stability of single base bulges and base pairs in a DNA fragment. *Biochemistry* 34, 4593–4600.

Keohavong, P., and Thilly, W. G. (1989). Fidelity of DNA polymerases in DNA amplification. *Proc. Natl. Acad. Sci. USA* 86, 9253–9257.

Khrapko, K., Hanekamp, J. S., Thilly, W. G., Belenkii, A., Foret, F., and Karger, B. L. (1994). Constant denaturant capillary electrophoresis (CDCE): a high resolution approach to mutational analysis. *Nucleic Acids Res.* 22, 364–369.

Lerman, L. S., and Silverstein, K. (1987). Computational simulation of DNA melting and its application to denaturing gradient gel electrophoresis. *Methods Enzymol.* 155, 482–501.

Losekoot, M., Fodde, R., Harteveld, C. L., van Heeren, H., Giordano, P. C., and Bernini, L. F. (1990). Denaturing gradient gel electrophoresis and direct sequencing of PCR amplified genomic DNA: a rapid and reliable diagnostic approach to beta thalassaemia. *Br. J. Haematol.* 76, 269–274.

Lu, Q., Hwang, Y. T., and Hwang, C. B. (2002). Mutation spectra of herpes simplex virus type 1 thymidine kinase mutants. *J. Virol.* 76, 5822–5828.

Lukowsky, A. (2003). Clonality analysis by T-cell receptor gamma PCR and high-resolution electrophoresis in the diagnosis of cutaneous T-cell lymphoma (CTCL). *Methods Mol. Biol.* 218, 303–320.

Macek, M., Jr., Mercier, B., Mackova, A., Miller, P. W., Hamosh, A., Ferec, C., and Cutting, G. R. (1997). Sensitivity of the denaturing gradient gel electrophoresis technique in detection of known mutations and novel Asian mutations in the CFTR gene. *Hum. Mutat.* 9, 136–147.

Minarik, M., Minarikova, L., Bjorheim, J., and Ekstrom, P. O. (2003). Cycling gradient capillary electrophoresis: a low-cost tool for high-throughput analysis of genetic variations. *Electrophoresis* 24, 1716–1722.

Motta, F. C., Rosado, A. S., and Couceiro, J. N. (2002). Standardization of denaturing gradient gel electrophoresis for mutant screening of influenza A (H3N2) virus samples. *J. Virol. Methods* 101, 105–115.

Murdaugh, S. E., and Lerman, L. S. (1996) DGGE. Denaturing Gradient Gel Electrophoresis and Related Techniques. In *Laboratory Protocols for Mutation Detection*, U. Landegren, ed., pp. 33–37, Oxford University Press, Oxford.

Muyzer, G., and Smalla, K. (1998). Application of denaturing gradient gel electrophoresis (DGGE) and temperature gradient gel electrophoresis (TGGE) in microbial ecology. *Antonie Van Leeuwenhoek* 73, 127–141.

Myers, R. M., Fischer, S. G., Lerman, L. S., and Maniatis, T. (1985a). Nearly all single base substitutions in DNA fragments joined to a GC-clamp can be detected by denaturing gradient gel electrophoresis. *Nucleic Acids Res.* 13, 3131–3145.

Myers, R. M., Lumelsky, N., Lerman, L. S., and Maniatis, T. (1985b). Detection of single base substitutions in total genomic DNA. *Nature* 313, 495–498.

Myers, R. M., Maniatis, T., and Lerman, L. S. (1987). Detection and localization of single base changes by denaturing gradient gel electrophoresis. *Methods Enzymol.* 155, 501–527.

Nikolcheva, L. G., Cockshutt, A. M., and Barlocher, F. (2003). Determining diversity of freshwater fungi on decaying leaves: comparison of traditional and molecular approaches. *Appl. Environ. Microbiol.* 69, 2548–2554.

Papadakis, M., Papapanagiotou, E., and Loutradi-Anagnostou, A. (1997). Scanning method to identify the molecular heterogeneity of delta-globin gene especially in delta-thalassemias: detection of three novel substitutions in the promoter region of the gene. *Hum. Mutat.* 9, 465–472.

Patrinos, G. P., Kollia, P., Loutradi-Anagnostou, A., Loukopoulos, D., and Papadakis, M. N. (1998). The Cretan type of non-deletional hereditary persistence of fetal hemoglobin [Agamma-158C→T] results from two independent gene conversion events. *Hum. Genet.* 102, 629–634.

Patrinos, G. P., Kollia, P., Papapanagiotou, E., Loutradi-Anagnostou, A., Loukopoulos, D., and Papadakis, M. N. (2001). Agamma-haplotypes: a new group of genetic markers for thalassemic mutations inside the 5′ regulatory region of the human Agamma-globin gene. *Am. J. Hematol.* 66, 99–104.

Peters, H., Hess, D., Fahsold, R., and Schulke, M. (1999). A novel mutation L1425P in the GAP-region of the NF1 gene detected by temperature gradient gel electrophoresis (TGGE). Mutation in brief no. 230. Online. *Hum. Mutat.* 13, 337.

Pignon, J. M., Vinatier, I., Fanen, P., Jonveaux, P., Tournilhac, O., Imbert, M., Rochant, H., and Goossens, M. (1994). Exhaustive analysis of the P53 gene coding sequence by denaturing gradient gel electrophoresis: application to the detection of point mutations in acute leukemias. *Hum. Mutat.* 3, 126–132.

Plonquet, A., Gherardi, R. K., Creange, A., Antoine, J. C., Benyahia, B., Grisold, W., Drlicek, M., Dreyfus, P., Honnorat, J., Khouatra, C., Rouard, H., Authier, F. J., Farcet, J. P., Delattre, J. Y., and Delfau-Larue, M. H. (2002). Oligoclonal T-cells in blood and target tissues of patients with anti-Hu syndrome. *J. Neuroimmunol.* 122, 100–105.

Riesner, D., Henco, K., and Steger, G. (1991) Temperature-gradient gel electrophoresis: A method for the analysis of conformational transitions and mutations in nucleic acids and proteins. In *Advances in electrophoresis*, A. Chrambach, M. J. Dunn, B. J. Radola, eds., Vol. 4. VCH, pp. 169–250. Verlagsgesellschaft, Weinhein.

Robinson, P. N., Booms, P., Katzke, S., Ladewig, M., Neumann, L., Palz, M., Pregla, R., Tiecke, F., and Rosenberg, T. (2002). Mutations of FBN1 and genotype-phenotype correlations in Marfan syndrome and related fibrillinopathies. *Hum. Mutat.* 20, 153–161.

Sheffield, V. C., Cox, D. R., Lerman, L. S., and Myers, R. M. (1989). Attachment of a 40-base-pair G + C-rich sequence (GC-clamp) to genomic DNA fragments by the polymerase chain reaction results in improved detection of single-base changes. *Proc. Natl. Acad. Sci. USA* 86, 232–236.

Steger, G. (1994). Thermal denaturation of double-stranded nucleic acids: prediction of temperatures critical for gradient gel electrophoresis and polymerase chain reaction. *Nucleic Acids Res.* 22, 2760–2768.

Tannock, G. W. (2002). Analysis of the intestinal microflora using molecular methods. *Eur. J. Clin. Nutr.* 56(Suppl. 4), S44–49.

Tchernitchko, D., Lamoril, J., Puy, H., Robreau, A. M., Bogard, C., Rosipal, R., Gouya, L., Deybach, J. C., and Nordmann, Y. (1999). Evaluation of mutation screening by heteroduplex analysis in acute intermittent porphyria: comparison with denaturing gradient gel electrophoresis. *Clin. Chim. Acta.* 279, 133–143.

Tiecke, F., Katzke, S., Booms, P., Robinson, P. N., Neumann, L., Godfrey, M., Mathews, K. R., Scheuner, M., Hinkel, G. K., Brenner, R. E., Hovels-Gurich, H. H., Hagemeier, C., Fuchs, J., Skovby, F., and Rosenberg, T. (2001). Classic, atypically severe and neonatal Marfan syndrome: twelve mutations and genotype-phenotype correlations in FBN1 exons 24–40. *Eur. J. Hum. Genet.* 9, 13–21.

Top, B. (1992). A simple method to attach a universal 50-bp GC-clamp to PCR fragments used for mutation analysis by DGGE. *PCR Methods Appl.* 2, 83–85.

Tully, L. A., Parsons, T. J., Steighner, R. J., Holland, M. M., Marino, M. A., and Prenger, V. L. (2000). A sensitive denaturing gradient-Gel electrophoresis assay reveals a high frequency of heteroplasmy in hypervariable region 1 of the human mtDNA control region. *Am. J. Hum. Genet.* 67, 432–443.

Uhrberg, M., Hinney, A., Enczmann, J., and Wernet, P. (1994). Analysis of the HLA-DR gene locus by temperature gradient gel electrophoresis and its application for the rapid selection of unrelated bone marrow donors. *Electrophoresis* 15, 1044–1050.

van Elsas, J. D., Garbeva, P., and Salles, J. (2002). Effects of agronomical measures on the microbial diversity of soils as related to the suppression of soil-borne plant pathogens. *Biodegradation* 13, 29–40.

Watanabe, T., Saito, A., Takeuchi, Y., Naimuddin, M., and Nishigaki, K. (2002). A database for the provisional identification of species using only genotypes: web-based genome profiling. *Genome Biol.* 3, RESEARCH0010.

Wiese, U., Wulfert, M., Prusiner, S. B., and Riesner, D. (1995). Scanning for mutations in the human prion protein open reading frame by temporal temperature gradient gel electrophoresis. *Electrophoresis* 16, 1851–1860.

Xiao, W., and Oefner, P. J. (2001). Denaturing high-performance liquid chromatography: A review. *Hum. Mutat.* 17, 439–474.

Yoshino, K., Nishigaki, K., and Husimi, Y. (1991). Temperature sweep gel electrophoresis: a simple method to detect point mutations. *Nucleic Acids Res.* 19, 3153.

Zschocke, J., Quak, E., Guldberg, P., and Hoffmann, G. F. (2000). Mutation analysis in glutaric aciduria type I. *J. Med. Genet.* 37, 177–181.

CHAPTER **9**

Two-Dimensional Gene Scanning

JAN VIJG[1,2] AND YOUSIN SUH[1]

[1] *Department of Physiology, University of Texas Health Science Center, 15355 Lambda Drive, STCBM 2.200, San Antonio, TX 78245, USA;*
[2] *Geriatric Research Education and Clinical Center, South Texas Veterans Health Care System, San Antonio, TX 78229-7762, USA*

TABLE OF CONTENTS

9.1 INTRODUCTION

With the completion of the human genome project, consensus sequences of all human genes may soon be available. In theory, this would open up the possibility to identify all possible gene variants in different human populations, determine the functional impact of such variation, if any, and associate their presence with individual phenotypes, including disease. In practice, many hurdles remain before such a straightforward approach has a chance of being successful. For example, not all human genes have been identified; only part of the known genes have an assigned biochemical function and only a fraction of those are associated with a phenotype (Freimer and Sabatti, 2003). Even if phenotypes have been assigned, a full understanding of the role of a gene in human physiology is constrained by the multiple phenotypes that may be associated with its different allelic variants. Moreover, genes rarely act alone and interactions between different gene variants in a pathway or across different pathways need to be elucidated. Hence, the rapid emergence of gene sequence information has surpassed our capabilities for assessing genotype-phenotype relationships.

9.2 STATISTICAL STRATEGIES TO IDENTIFY HUMAN DISEASE GENES

Although the lack of phenotypic insight is a challenge currently addressed by efforts to develop efficient tools for assessing phenotypes (Bochner, 2003), the association of human gene variants with disease phenotypes suffers from a lack of suitable strategies. Traditionally, the main tool for discovering gene-disease relationships involved linkage analysis and positional cloning. In this approach, the location of an unknown disease gene is determined on the basis of its cosegregation with a polymorphic marker in affected families. This approach has been successful for those rare illnesses that are caused by mutations in single genes. Linkage analysis is inherently unsuitable to find disease genotypes involving many loci; that is, with individual genes exerting only small effects. Although some successes have been achieved for uncommon subsets of generally more common disorders, such as maturity-onset diabetes of youth, other approaches are now generally more favorably considered (Risch, 2000; Tabor *et al.*, 2002). Especially whole genome association studies, based on linkage disequilibrium, have been discussed as a general approach to determine the gene loci involved in human chronic diseases with a genetic component (Kruglyak, 1999). Even though in the past, linkage disequilibrium studies could be carried out only for a very limited number of sequence variants of one or few genes, the International Haplotype Map or HapMap project, should soon have organized the 10 million single nucleotide polymorphisms (SNPs) in the human genome into a few hundred thousand haplotype blocks such that a small fraction of SNPs (tag SNPs) represents most of common haplotypes (Daly *et al.*, 2001; Patil, 2001; Gabriel *et al.*, 2002; Phillips *et al.*, 2003). These tag SNPs may be able to capture most of the common genetic variants contributing to complex human disease.

Although this approach is unbiased and allows a comprehensive genome-wide survey, some significant uncertainties do not make it the method of choice for population-based studies. The main uncertainty involves the lack of detailed information regarding the extent of linkage disequilibrium in the human genome. Since complete linkage disequilibrium between markers and disease gene is unlikely, sample sizes, which should be large to begin with in order to account for multiple testing when so many markers are used, need to be increased even further (McCarthy and Hilfiker, 2000; Risch, 2000). This problem is exacerbated when the disease gene variant occurs at low frequency (Cardon and Abecasis, 2003). The need of large group sizes make linkage disequilibrium mapping inherently unsuitable for studies with relatively small numbers of individuals (several hundred). Furthermore, current genotyping assays still have not reached the stage where a few hundred thousand SNPs can be scored in thousands of individuals at a reasonable cost.

9.3 CANDIDATE PATHWAY APPROACH

Linkage-based positional cloning and linkage disequilibrium-based whole genome association studies are statistical methods that make no prior assumptions about the identity of the genes to be found. This offers obvious advantages, but also poses, probably insurmountable, obstacles to their application to the unraveling of genotype-phenotype interactions involved in human complex disease. In contrast, candidate gene approaches are based on the study of one or more genes selected on the basis of hypotheses about the disease phenotype. The candidate gene approaches were initially criticized because of the small numbers of genes and only one or a few SNPs within or close to a gene of interest were assessed for association with a phenotype (Goldstein, 2003a). However, with the emergence of consensus sequence information and databases containing SNP and SNP haplotype (allele) information on a great many genes, obtained on the basis of the resequencing of such genes in a panel of unrelated individuals (e.g., http://www.niehs.nih.gov/envgenom, http://pga.gs.washington.edu), it has become feasible to do much larger studies based on entire pathways of genes. As the most promising approach to statistically powerful candidate gene association studies (Johnson *et al.*, 2001; Goldstein, 2001; 2003b), it was suggested to:

1. Select candidate genes in pathways appropriate for the condition of interest on the basis of different types of information, including available linkage results and results obtained with mouse models of the phenotype.
2. Determine their haplotype structures in control individuals from target populations for these genes.
3. Select sets of tag SNPs in an attempt to represent all the common haplotypes in the gene.
4. Test for association between phenotype and haplotype status.

5. Scrutinize positively associated haplotypes to identify the actual location of true causal variants followed by functional assays using a variety of short-term gene functionalization methods, including *in silico* methods.

All these steps greatly increase the rational basis of such selected haplotypes in candidate pathways being involved in the disease phenotype of interest.

There are several drawbacks of such a "candidate pathway" approach using tag SNPs for unraveling the genetic etiology of complex disease. First, to increase the likelihood that the gene(s) of interest is among the ones acting in the selected pathway(s), the number of such genes need to be large, probably at least 100. With the increased annotation efforts currently ongoing it is likely that such informed guesses will become more and more reliable, which may reduce the number of genes to be screened. Second, if susceptibilities for common complex traits are due to large numbers of rare variants at many loci, this strategy would fail, as no single haplotype would be strongly associated with the complex traits (Pritchard, 2001) and the contribution of most individual variants would be too small (Slager *et al.*, 2000). Third, currently only common SNPs and SNP haplotypes for a small fraction of all human genes can be found in databases; that is, with frequencies above 10% (Schneider *et al.*, 2003). This, of course, is due to the high cost of resequencing entire genes of multiple individuals. In general, the resequencing panels are not larger than about 100, although greater depths have been obtained in a number of cases. Gene SNP haplotypes that are relatively rare will not be found in these databases. Finally, it is possible that ethnic, geographical, and other characteristics of the study population greatly influence the frequencies of the different SNP haplotypes and tag SNPs identified in one population may not necessarily perform well in another. This suggests that exclusive use of SNP databases for associating genotype with phenotype will prove to be too limited. Direct detection of all possible DNA variations in candidate genes in all individuals of the case control study would address these concerns and guarantee that no important gene variants will be missed.

Current methods for the discovery of new gene sequence variants are not optimized for the generation of highly accurate data on multiple genes in hundreds of individuals in population-based studies or in the clinical setting. The most reliable system for resequencing genes in multiple individuals is still nucleotide sequencing itself. Unfortunately, the elegant principle of the most commonly used method (Sanger *et al.*, 1977), which, in semi-automated form, remains the basis of all current sequencing protocols, is still not compatible with cost-effective large-scale population-based genetic screening. Hence, the future of human DNA variation studies, in which hundreds of thousands of individuals need to be screened for, in principle, all possible sequence variants in essentially all genes, critically depends on the availability of more cost-effective, yet highly accurate, methods for gene resequencing. Of the many alternatives currently available (Vijg and Suh, 2003), a high-throughput method has been designed, based on denaturing gradient gel electrophoresis (DGGE, see previous chapter). By applying DGGE in a two-dimensional format in combination with multiplex PCR, a system was obtained that provides high accuracy at low cost. In this chapter, the different components of this system, Two-Dimensional Gene Scanning (TDGS), will be reviewed, as well as its future prospects as a high-throughput system for screening for the discovery of novel SNPs in population-based studies.

9.4 DENATURING GRADIENT GEL ELECTROPHORESIS (DGGE)

DGGE, first described by Fischer and Lerman (1979) and discussed in detail in Chapter 8, generally is considered to be the most accurate method for detecting DNA sequence variation. Indeed, to some extent it is even more accurate than sequencing (Gejman *et al.*, 1998), which requires special software to accurately detect heterozygous mutations (Phelps *et al.*, 1995). Unlike sequencing, DGGE detects mutations, including base pair substitutions, on the basis of differences in the melting temperature of the target fragments. Since this is described in detail in the previous chapter, it will be discussed only briefly here. A given DNA fragment comprises one or more domains, each representing a stretch of between 50 and 300 base pairs with equal melting temperature (the temperature at which each base pair has a 50% probability of being in either the helical or the denatured state). Therefore, when a double-stranded DNA fragment migrates through a gel, parallel to a gradient of increased denaturants (urea/formamide or temperature), the lowest melting domain will denature first and, because of branching of the DNA strands, the fragment will be retarded in the gel. The melting temperature of a domain is dependent on the base composition of the fragment (G-C base pairs have a higher melting temperature than A-T base pairs) and the stacking interactions of the bases. Therefore, DNA sequence variation in a fragment will be reflected by its position in the gel.

The sensitivity of DGGE in detecting mutations can be increased dramatically by attaching so-called GC-clamps— a stretch of 30 to 50 G and C bases—to the target DNA fragments (Myers *et al.*, 1985) by making it part of one of the primers in PCR amplification (Sheffield *et al.*, 1989). Without GC-clamping, a DNA fragment consisting of one melting domain will become completely single stranded upon denaturation and run off the gel. By adding a GC-clamp, a single high melting domain is artificially created at one end of the target fragment. As the GC-clamped target fragment migrates through the gradient of denaturants,

melting of the target domain causes partial branching and halting of the fragment in the gel. Thus, one function of the GC-clamp is to ensure branch formation after melting of the target fragment (see also Section 8.2.4). However, when the target DNA fragment consists of multiple melting domains, only mutations in the lowest melting domain are readily detected. To facilitate detection of all possible mutations, it is imperative that the target fragment represent only one melting domain. Fortunately, since the addition of a GC-clamp allows for stacking interactions with neighboring bases, the entire fragment often will behave as one melting domain. However, this is not always the case, and in practice, the target fragment needs to be "designed," for example, through the strategic positioning of PCR primers, to achieve the ideal single melting domain (see later).

The sensitivity of DGGE for detecting variants is enhanced further by the introduction of a heteroduplexing step using one round of denaturation/renaturation, usually at the end of PCR amplification of the target fragment. In this manner, a heterozygous mutation is revealed as four different double-stranded fragments: two homoduplex molecules (one homoduplex wild type and one homoduplex mutant) and two heteroduplex molecules (each comprising one wild type and one mutant strand). The less stable mismatched heteroduplex molecules will melt earlier than the two homoduplex molecules (see Fig. 8.2).

Although DGGE has the crucial advantage of virtually 100% sensitivity in detecting mutations, typically it has been applied in a serial fashion, for example, on a fragment-by-fragment basis. For analyzing large genes or multiple genes this is not practical, especially since it is not uncommon to use different denaturing gradient or other electrophoresis conditions for different fragments. An additional complicating factor is the need for optimization of the PCR primer positions, not only to create optimal melting behavior, but also optimal amplification conditions. For mutation detection in large human disease genes, the DGGE principle has been applied in the format as it was originally described, as a two-dimensional system of separation by size followed by DGGE (Fischer and Lerman, 1979). However, to successfully implement DGGE as a high-throughput mutation detection system, a number of technical issues needed to be addressed. First, a practical 2D format suitable for the analysis of multiple PCR fragments in parallel was developed. Second, an efficient multiplex PCR protocol was created to obtain the necessary target sequences in parallel. Third, the design of DGGE-optimized PCR fragments was automated and expanded to rapidly create a test for multiple fragments under the same set of conditions.

9.5 TWO-DIMENSIONAL GENE SCANNING

TDGS is based on DGGE as the mutation detection principle, in combination with multiplex PCR amplification to prepare the target sequences. The procedure is schematically depicted in Fig. 9.1.

The major advantage of two-dimensional electrophoresis is that it provides a high-resolution system to screen multiple fragments under the same conditions. It has been demonstrated that DGGE provides virtually 100% mutation detection sensitivity even when applied with a broad-range gradient of denaturants (Guldberg *et al.*, 1993; Sheffield *et al.*, 1993; Moyret *et al.*, 1994; Guldberg and Guttler, 1994). This opens up the possibility to analyze multiple fragments for all possible mutations under the same set of experimental conditions. The total number of target fragments that can be analyzed simultaneously depends on the resolution of the gel system used. Although high resolution can be obtained by using one-dimensional denaturing gradient gels, two-dimensional separation allows characterization of each fragment on the basis of two independent criteria, size and melting temperature. In practice, by using the 2D system, it is possible to completely visualize all fragments corresponding to an entire gene for a particular DNA sample and

immediately recognize each exon and variants therein. This has been demonstrated for several large human disease genes, including CFTR (Wu *et al.*, 1996; Lee, 2003), RB1 (Van Orsouw *et al.*, 1996), MLH1 (Nystrom-Lahti *et al.*, 1996; Smith, 1998), TP53 (Rines *et al.*, 1998), BRCA1 (Van Orsouw, 1999), and BRCA2 in combination with BRCA1 (Bounpheng *et al.*, 2003), as well as for a part of the mitochondrial genome (van Orsouw *et al.*, 1998a).

The simplest procedure for 2D DNA electrophoresis is to use separate gels for the first dimension separation, according to size, and the subsequent separation of the same fragments by DGGE, on the basis of their melting temperature. Originally, the first dimension separation was carried out on slab gels, which required post-staining of the gel to excise the lane to be transferred to the second dimension denaturing gradient gel. However, routine application of TDGS requires standardization and automation, which is incompatible with the labor-intensive step of manual interference between the first and second dimension separation. A simple automated 2D instrument has been developed in the authors'

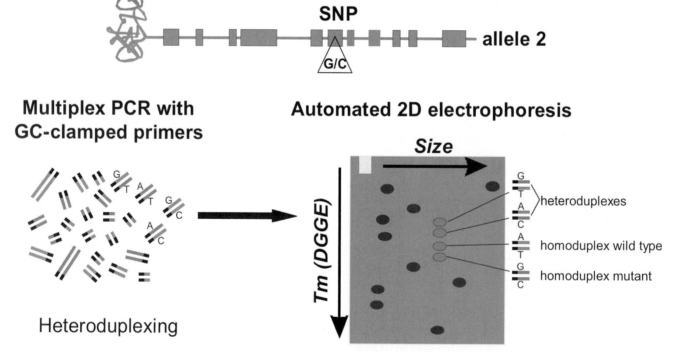

FIGURE 9.1 Schematic representation of the different steps involved in Two-Dimensional Gene Scanning (TDGS). The target regions, such as gene coding regions or gene regulatory regions, are amplified by PCR in two steps. First, a long distance PCR generates overrepresentation of the target regions as part of large PCR amplicons. Then, using the large amplicons as a template, the target regions are amplified in an extensive multiplex PCR, with the primers chosen in a way to create optimal conditions for detecting all possible sequence variants using the denaturing gradient principle. For this purpose it is also essential that all primers are attached to an approximately 40-bases-long GC-rich fragment (GC-clamp). At the end of the PCR process, all amplicons are heteroduplexed by a round of complete denaturation and renaturation, and then separated based on size and base pair sequence in a denaturing gradient gel. The presence of sequence variation is revealed by four different double-stranded fragments: two homoduplex molecules and two heteroduplex molecules. (See color plate)

laboratories in collaboration with Dr. Charles Scott (C.B.S. Scientific, Del Mar, CA), which is based on an existing vertical electrophoresis system with an isolated horizontal unit on top (see Fig. 9.2a and Fig. 9.2b; Dhanda *et al.*, 1998). The necessary contacts between the outer buffer chambers of the top unit and the gel are provided by two strategically located openings in the inner glass plate (B in Fig. 9.2a). The current version of this instrument is capable of running eight 2D gels in parallel (see Fig. 9.2c; McGrath *et al.*, 2001).

The automated system described has decreased the labor associated with TDGS and greatly facilitates standardization. This is significant both in clinical and financial terms. Without manual interference between the first and second dimensions, the possibility of errors is eliminated and the labor costs can be reduced.

Another technical improvement that has been implemented is the substitution of fluorescently labeled PCR primers for the labor-intensive step of staining the gels. To stain the gels they first have to be removed from between the glass plates and incubated with a solution containing ethidium bromide or the more sensitive dye Sybr Green. Dye primer technology for the in-gel detection of 2D spot patterns is an obvious strategy for large-scale application of TDGS. Introduction of fluorescent detection offers two advantages over gel staining. First, the reduction in labor is considerable and loss of gels due to breakage is prevented. Second, since there is no need to release the gel from between the glass plates it has become possible to use thinner gels, which will allow shorter electrophoresis times. Indeed, TDGS instruments with a high-voltage power

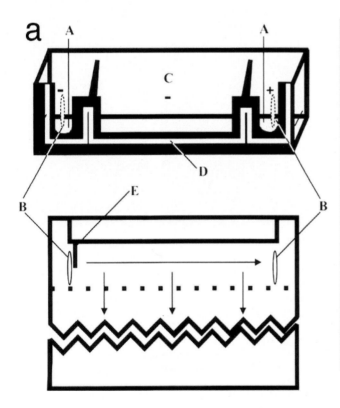

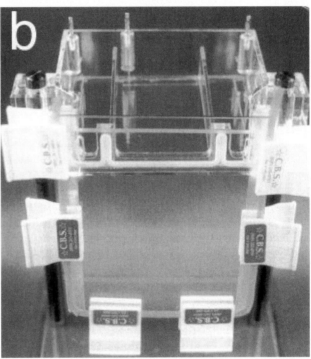

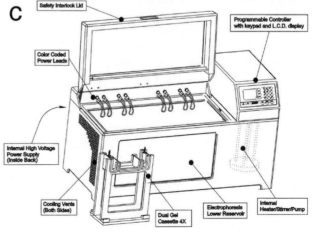

FIGURE 9.2 Automated 2D electrophoresis. **a)** Principle of automated 2D elecrophoresis using a 1D electrophoresis unit on top of an existing vertical PAGE instrument. Buffer chambers for the first dimension separation (A) are connected with the gel through openings in the (inner) glass plate (B). During the first dimension electrophoresis, the middle buffer chamber (C) is isolated from the outer chambers. For the second dimension run, buffer chamber C is flooded with buffer and the upper electrode is turned on in conjunction with a positive electrode in the lower reservoir (not shown in this figure). The gel cassette is sealed to the top unit with a serpentine silicone gasket (D), and sample is loaded in the single slot (E). The dotted line indicates the beginning of the gradient of urea/formamide. **b)** First hand-made automated TDGS system (Dhanda *et al.*, 1998). **c)** Current, high-power TDGS system with 8-gel capacity (McGrath *et al.*, 2001).

supply are now available (see Fig. 9.2c), reducing electrophoresis time from a typical 10 hours to less than half (McGrath *et al.*, 2001).

To increase the efficiency even further, different fluorophores can be used in the same gel. An example is a recently designed TDGS test for the breast cancer genes BRCA1 and BRCA2 (Bounpheng *et al.*, 2003). PCR amplification of these very large genes yielded a total of 115 gene-coding fragments, labeled in three groups using dye-primers: HEX (green spots, representing 40 BRCA1 fragments; three more than in the original single-gene test), FAM (blue spots, representing 25 BRCA2 fragments), and Texas Red (red spots, representing 50 BRCA2 fragments). All 115 gene-coding fragments together can be subjected to automatic 2-D DGGE. After fluorescent scanning of the gel, the complete three-color pattern first is visualized on the computer screen (see Fig. 9.3a). The pattern of each color group can then be interpreted separately by eye directly from the screen and sequence variants (indicated by circles) scored electronically (see Fig. 9.3b–d). Current fluorescence imagers have the option to analyze at least four different fluorophores in the same gel.

9.6 MULTIPLEX PCR AMPLIFICATION

An obvious strategy to increase the efficiency of PCR amplification of the coding regions of large human disease genes from genomic DNA is to coamplify multiple target sequences in one single reaction; that is, multiplex PCR. However, the design of PCR primers for an extensive multiplex reaction is not trivial. With each primer set added to the multiplex reaction, the permissive reaction conditions that allow each fragment to reach its melting temperature while evading spurious amplification products become increasingly less flexible. Ultimately, this lack of flexibility, due to the complexity of the genomic sequence context, would allow ample opportunity for nonspecific priming (Edwards and Gibbs, 1994). This explains why most multiplex PCR systems described thus far are small, generally not more than four fragments per reaction.

By using the two-step protocol first described by Li and Vijg (1996), extensive multiplex PCR is possible. For the first step, preamplification of target sequences by long-distance PCR (Foord and Rose, 1994), primers are designed in the intronic regions to amplify groups of exons in fragments of up to 20 kb. For large genes, this can be done in multiplex groups so that all exons of the gene are encompassed. For example, all coding regions of the BRCA1 gene have been successfully coamplified in a multiplex of seven long-distance PCR fragments (see later). Provided sufficient sequence information is available, primer design for the long-distance PCR step generally is not a problem. The over-represented (relative to the genomic environment) exon-containing genomic fragments form a highly specific

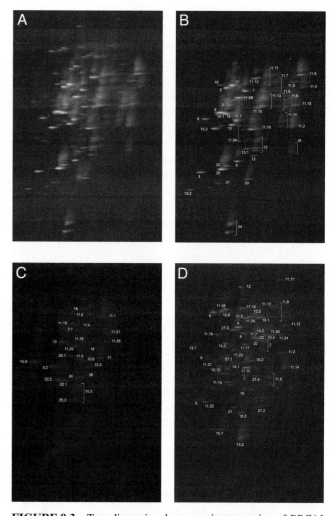

FIGURE 9.3 Two-dimensional gene variant scanning of *BRCA1* and *BRCA2* in a single assay (Bounpheng *et al.*, 2003). A. Complete separation of 115 fragments corresponding to the entire coding regions of *BRCA1* and *BRCA2*. Primers for each TDGS fragment were designed according to optimal melting behavior, with a GC-clamp attached to one primer of each pair. Primers were fluorescently labeled with HEX (green spots, representing 40 *BRCA1* fragments), FAM (blue spots, representing 25 *BRCA2* fragments), and Texas Red (red spots, representing 50 *BRCA2* fragments). B. Separate TDGS pattern for 40 HEX labeled *BRCA1* fragments. Sequence variants were detected in exons 11.6, 11.7, 11.8, 11.9A, 11.11, 11.13, 11.14, 13, 16, and 24. C. Separate visualization of the TDGS pattern for 25 FAM labeled *BRCA2* fragments. A sequence variant was detected in exon 10.3. D. Separate TDGS pattern for 50 Texas Red labeled *BRCA2* fragments (See color plate).

template for the second step, the generation of DGGE-optimized exonic target fragments. Extensive multiplex PCR groups allow the preparation of large numbers of target fragments simultaneously, thereby greatly reducing the costs associated with reagents and labor. An example is the RB1 gene consisting of 27 exons. Using multiplex PCR as described earlier, a total of 26 fragments could be amplified in one single reaction (Van Orsouw *et al.*, 1996). Multiplex groups of on average 10 fragments are now routinely obtained. In combination with 2D DNA electrophoresis, extensive multiplex PCR provides a parallel system for mutation detection in large human disease genes.

9.7 COMPUTER-AUTOMATED DESIGN OF TDGS TESTS

A potential hindrance to the widespread application of TDGS to multiple, novel genes involves the difficulties in the design of PCR primers generating single-domain fragments, which can be resolved under one set of electrophoretic conditions. To design complete gene tests for mutational analysis by TDGS, an automated, generally applicable, computer program has been developed based on a commercially available primer design program (Primer Designer 3; Scientific and Educational Software, State Line, PA), the melting routine MELT87 (Lerman and Silverstein, 1987), and a newly generated spot distribution routine (van Orsouw *et al.*, 1998b). After entering a gene's coding sequence as exons with their flanking intronic sequences, a rank of suitable PCR primers for each exon is designed by the PCR design subroutine. The optimal primer pair is then used in the melt subroutine to check for a one-domain target fragment. The program uses different GC-clamps at either the 5′ or 3′ end of the target fragment and, if necessary, additional small GC- or AT-clamps at either side of the target fragment. If it is impossible to design a one-domain fragment, the next optimal primer pair is tested and so on. If a primer pair suitable to create a one-domain fragment cannot be found, the exon is split. An example is BRCA1 exon 11, which has been split into 16 overlapping fragments (see later). As soon as primers fulfill PCR and melting criteria, the fragment is positioned according to its size (x) and melting (y) coordinates. The spot distribution routine then checks for possible overlap.

The program has recently been modified to account for additionally relevant sequences, such as regulatory regions as determined on the basis of human-mouse sequence homology. The new version first determines all possible optimal one-domain fragments in the target sequence. These fragments subsequently are tested for PCR design, which is substantially less critical due to the greatly increased specificity provided by the two-step multiplex PCR format (see earlier).

9.8 IDENTIFICATION OF SEQUENCE VARIANTS AND HAPLOTYPING: BRCA1 AS AN EXAMPLE

A critical issue in the application of TDGS in screening various populations for gene sequence variation is the identification of unique sequence variants and haplotyping. In

this respect it should be realized that TDGS is first and foremost a prescreening method. Each new sequence variation detected requires nucleotide sequencing for its final identification. This situation will remain for newly discovered sequence variants. Recurrent variants, however, including common polymorphisms, can be recognized on the basis of their often unique pattern in TDGS gels. In addition, multiple SNPs can be organized into haplotypes on the basis of the spot position identification of each allelic variant. This will be illustrated later on the hand of the BRCA1 gene.

A TDGS test for BRCA1 alone (without BRCA2) has been described previously (Van Orsouw *et al.*, 1999) and has been used extensively—over 2,000 women have been screened with this test for mutations and/or SNPs. The accuracy of the test in detecting unknown mutations recently has been evaluated in comparison to other methods and appeared to vary from 80% to over 90%, with some mutations missed due to interpretational error or because they involved large deletions removing the entire exon (Eng *et al.*, 2001; Andrulis *et al.*, 2002). Although DGGE, on which TDGS is based, is supposed to be 100% accurate, we found that misinterpretation, a flawed design (of PCR primers), or administrative errors could significantly lower accuracy rate.

PCR amplification of the BRCA1 gene required a multiplex of seven long fragments serving as template for a total of 37 short PCR fragments. These fragments covered all coding exons of the gene (the noncoding exons 1a, 1b, and the noncoding part of exon 24 were excluded). Sixteen overlapping fragments were needed to cover the large exon 11. Two-dimensional separation of these HEX-labeled fragments in DGGE gels followed by image analysis provides for the visualization of the entire BRCA1 coding region in the form of a pattern of 37 spots. This is demonstrated in Fig. 9.4a, showing the BRCA1-TDGS pattern of a control sample, containing various heterozygous polymorphisms in exons 11, 13, and 16. Heterozygous mutations or polymorphisms are the easiest to detect, that is, as four rather than one fragment (see also Fig. 9.1), representing the two homoduplex and the two heteroduplex variants. Sometimes the homoduplex or heteroduplex variants are not well separated, resulting in the appearance of three or two instead of four spots.

Figure 9.4b shows how haplotypes can be deduced once a sufficient number of individuals have been analyzed. In the upper panel, the two homoduplex fragments corresponding to the heterozygous SNP in exon fragment 11.13 represent the two alleles. Using other fragments—fragments 11.12 and 11.10—as reference markers it is easy to recognize the two different alleles, termed H and L for 11.13 also in case of homozygosities. In the lower panel, this is further illustrated by the heterozygosity in fragment 11.11. In this case, three alleles, termed L, M, and H can clearly be distinguished using the homozygous fragments 11.3 and 11.5

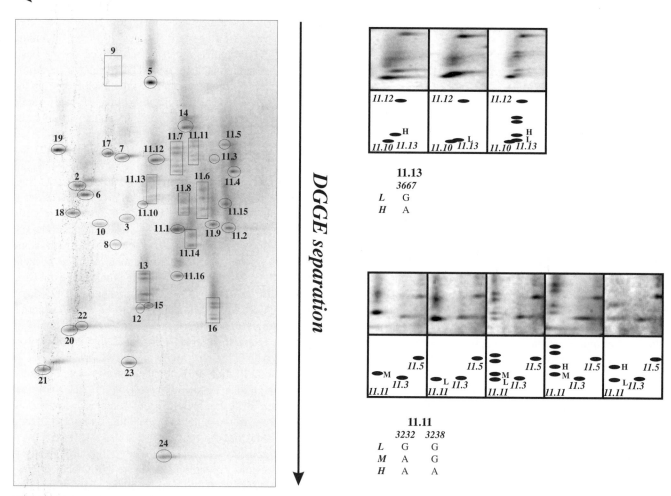

FIGURE 9.4 BRCA1 mutation scanning by TDGS. A. TDGS pattern of 37 fragments covering the entire BRCA1 coding region including splice sites. Each amplicon is indicated by an oval, whereas heterozygous variants are indicated by squares. DGGE = denaturing gradient gel electrophoresis. B. Details of individual fragments representing parts of exon 11 with schematic representations. The top panel shows a heterozygous SNP in fragment 11.13 indicating alleles H and L (right), which can be observed in homozygous state on the left (H) and in the middle section (L) using fragments 11.12 and 11.10 as markers. The lower panel shows a SNP heteryzygosity in fragment 11.11 with three alleles: H, M, and L. Alleles M and L can be detected in homozygous state in the first (M) and second (L) section. In this way, allele assignments and haplotype inferences can be made quickly, based on the relative position and configuration of fragments throughout the BRCA1 gene.

as markers. In this case, three different polymorphisms can be distinguished in the same fragment, one with alleles M and L, one with alleles M and H, and one with alleles L and H.

By recognizing the different alleles of different SNPs in the gene in their homozygous state haplotypes can be deduced for the entire gene. A total of seven frequently occurring SNPs, giving rise to amino acid changes, were arranged into 11 different haplotypes, a–k (see Fig. 9.5). Haplotypes were assigned directly from the 2D gel patterns as described previously (Van Orsouw, 1999).

9.9 PROSPECTS FOR ROUTINE APPLICATION OF TDGS

In view of its low cost and simplicity, TDGS has the potential of becoming a method of choice for large-scale genetic epidemiological research until more advanced sequencing methods currently under development become available. Based on the empirically determined cost of about $50 for screening the 115 fragments of the large BRCA1 and BRCA2 genes on one gel (Bounpheng *et al.*, 2003), the costs of analyzing a gene with an average coding region of about

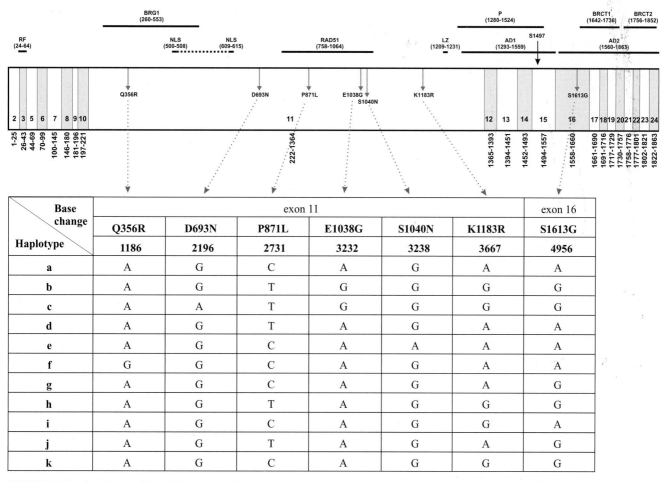

FIGURE 9.5 Combining SNP alleles as identified from TDGS gels into haplotypes. The seven SNPs used to generate the 11 haplotypes are indicated in a schematic depiction of the BRCA1 gene. Relevant domains of the protein are indicated on top. Amino acid numbers are indicated below the exon numbers.

2,000 bp (corresponding with 20 or less TDGS fragments) would be only about $10. This would permit population studies of significant sample size and number of candidate genes. For example, a study of 100 genes and a sample size of 10,000 individuals would cost less than $10 million.

Similarly, for future application in clinical practice the theoretical cost of screening an individual during his/her lifetime has been calculated for a total of 5,000 genes of average size. In this respect the assumption was made that only a fraction of all human genes will prove to be diagnostically relevant and that 5,000 genes would be sufficient to account for such phenotypes as predisposition, drug efficacy, and toxicity associated with the most common diseases. Based on the same 20 TDGS fragments per 2,000 bp average gene coding region, the maximum cost of a lifetime of diagnostic genotyping would be about $40,000. Against an average lifetime, healthcare cost of over $350,000 in the US (based on a life expectancy of about 80 years and the figures for the year 2000 from the US Health Care Financing Administration; www.hcfa.gov), this seems a modest amount, provided it

would lead to an increased quality of life and extensive cost-savings due to increased opportunities to prevent costly interventions and/or a more effective use of drugs.

The assay on which these costs are based utilizes robotics to set up PCR amplifications, controlled, automated 2D electrophoresis at high voltage and three-color fluorescence imaging. To further decrease costs and increase reliability a logical next step would be integration and miniaturization of the entire process on a lab-on-a-chip. Indeed, a two-dimensional electrophoresis chip with a temperature gradient feature has been designed for SNP screening (Liu *et al.*, 2002). In such a design also the spot pattern interpretation would occur automatically using a customized image processing system currently under development.

Acknowledgments

Part of this research, examples of which are discussed in this chapter, was supported by the NIH, the Korean Ministry of

Health and Welfare (01-PJ11-PG9-01BT05-0003), and research funds from Accelerated Genomics Inc. The authors have a commercial interest in TDGS. They also wish to thank all present and former coworkers and collaborators for sharing their results with them.

References

Andrulis, I. L., Anton-Culver, H., Beck, J., Bove, B., Boyd, J., Buys, S., Godwin, A. K., Hopper, J. L., Li, F., Neuhausen, S. L., Ozcelik, H., Peel, D., Santella, R. M., Southey, M. C., van Orsouw, N. J., Venter, D. J., Vijg, J., and Whittemore, A. S.; Cooperative Family Registry for Breast Cancer studies. (2002). Comparison of DNA- and RNA-based methods for detection of truncating BRCA1 mutations. *Hum. Mutat.* 20, 65–73.

Bochner, B. R. (2003). New technologies to assess genotype-phenotype relationships. *Nat. Rev. Genet.* 4, 309–314.

Bounpheng, M., McGrath, S., Macias, D., van Orsouw, N., Suh, Y., Rines, D., and Vijg, J. (2003). Rapid, inexpensive scanning for all possible BRCA1 and BRCA2 gene sequence variants in a single assay: implications for genetic testing. *J. Med. Genet.* 40, e33.

Cardon, L. R., and Abecasis, G. R. (2003). Using haplotype blocks to map human complex trait loci. *Trends Genet.* 19, 135–140.

Daly, M. J., Rioux, J. D., Schaffner, S. F., Hudson, T. J., and Lander, E. S. (2001). High-resolution haplotype structure in the human genome. *Nat. Genet.* 29, 229–232.

Dhanda, R., Smith, W. M., Scott, C. B., Eng, C., and Vijg, J. (1998). A simple system for automated two-dimensional electrophoresis: applications to genetic testing. *Genet. Test.* 2, 67–70.

Edwards, M. C., and Gibbs, R. A. (1994). Multiplex PCR: advantages, development, and applications. *PCR Methods Appl.* 3, S65–75.

Eng, C., Brody, L. C., Wagner, T. M., Devilee, P., Vijg, J., Szabo, C., Tavtigian, S. V., Nathanson, K. L., Ostrander, E., and Frank, T. S. (2001). Interpreting epidemiological research: blinded comparison of methods used to estimate the prevalence of inherited mutations in BRCA1. *J. Med. Genet.* 38, 824–833.

Fischer, S. G., and Lerman, L. S. (1979). Length-independent separation of DNA restriction fragments in two-dimensional gel electrophoresis. *Cell* 16, 191–200.

Foord, O. S., and Rose, E. A. (1994). Long-distance PCR. *PCR Methods Appl.* 3, S149–161.

Freimer, N., and Sabatti, C. (2003). The human phenome project. *Nat. Genet.* 34, 15–21.

Gabriel, S. B., Schaffner, S. F., Nguyen, H., Moore, J. M., Roy, J., Blumenstiel, B., Higgins, J., DeFelice, M., Lochner, A., Faggart, M., Liu-Cordero, S. N., Rotimi, C., Adeyemo, A., Cooper, R., Ward, R., Lander, E. S., Daly, M. J., and Altshuler, D. (2002). The structure of haplotype blocks in the human genome. *Science* 296, 2225–2229.

Gejman, P. V., Cao, Q., Guedj, F., and Sommer, S. (1998). The sensitivity of denaturing gradient gel electrophoresis: a blinded analysis. *Mutat. Res.* 382, 109–114.

Goldstein, D. (2001). Islands of linkage disequilibrium. *Nat. Genet.* 29, 109–111.

Goldstein, D. (2003a). Pharmacogenetics in the laboratory and the clinic. *N. Engl. J. Med.* 348, 553–556.

Goldstein, D., Ahmadi, K. R., Weale, M. E., and Wood, N. W. (2003b). Genome scan and candidate gene approaches in the study of common diseases and variable drug response. *Trends Genet.* 19, 615–622.

Guldberg, P., Henriksen, K. F., and Guttler, F. (1993). Molecular analysis of phenylketonuria in Denmark: 99% of the mutations detected by denaturing gradient gel electrophoresis. *Genomics* 17, 141–146.

Guldberg, P., and Guttler, F. (1994). "Broad-range" DGGE for single-step mutation scanning of entire genes: application to human phenylalanine hydroxylase gene. *Nucleic Acids Res.* 22, 880–881.

Johnson, G. C., Esposito, L., Barratt, B. J., Smith, A. N., Heward, J., Di Genova, G., Ueda, H., Cordell, H. J., Eaves, I. A., Dudbridge, F., Twells, R. C., Payne, F., Hughes, W., Nutland, S., Stevens, H., Carr, P., Tuomilehto-Wolf, E., Tuomilehto, J., Gough, S. C., Clayton, D. G., and Todd, J. A. (2001). Haplotype tagging for the identification of common disease genes. *Nat. Genet.* 29, 233–237.

Kruglyak, L. (1999). Prospects for whole-genome linkage disequilibrium mapping of common disease genes. *Nat. Genet.* 22, 139–144.

Lee, J. H., Choi, J. H., Namkung, W., Hanrahan, J. W., Chang, J., Song, S.-Y., Park, S. W., Kim, D. S., Yoon, J. H., Suh, Y., Cho, M.-O., Jang, I. J., Nam, J. H., Kim, S. J., Lee, J. E., Kim, K. W., and Lee, M. G. (2003). A haplotype-based approach for analyzing CFTR mutations associated with respiratory and pancreatic disease. *Hum. Mol. Genet.* 12, 2321–2332.

Lerman, L. S., and Silverstein, K. (1987). Computational simulation of DNA melting and its application to denaturing gradient gel electrophoresis. *Methods Enzymol.* 155, 482–501.

Li, D., and Vijg, J. (1996). Multiplex co-amplification of 24 retinoblastoma gene exons after pre-amplification by long-distance PCR. *Nucleic Acids Res.* 24, 538–539.

Liu, P., Xing, W.-L., Liang, D., Huang, G.-L., and Cheng, J., eds. (2002). Fast screening of single-nucleotide polymorphisms using chip-based temperature gradient electrophoresis. New York, Kluwer.

McCarthy, J. J., and Hilfiker, R. (2000). The use of single-nucleotide polymorphism maps in pharmacogenomics. *Nat. Biotechnol.* 18, 505–508.

McGrath, S. B., Bounpheng, M., Torres, L., Calavetta, M., Scott, C. B., Suh, Y., Rines, D., van Orsouw, N., and Vijg, J. (2001). High-speed, multicolor fluorescent two-dimensional gene scanning. *Genomics* 78, 83–90.

Moyret, C., Theillet, C., Puig, P. L., Moles, J. P., Thomas, G., and Hamelin, R. (1994). Relative efficiency of denaturing gradient gel electrophoresis and single strand conformation polymorphism in the detection of mutations in exons 5 to 8 of the p53 gene. *Oncogene* 9, 1739–1743.

Myers, R. M., Fischer, S. G., Lerman, L. S., and Maniatis, T. (1985). Nearly all single base substitutions in DNA fragments joined to a GC-clamp can be detected by denaturing gradient gel electrophoresis. *Nucleic Acids Res.* 13, 3131–3145.

Nystrom-Lahti, M., Wu, Y., Moisio, A. L., Hofstra, R. M., Osinga, J., Mecklin, J. P., Jarvinen, H. J., Leisti, J., Buys, C. H., de la Chapelle, A., and Peltomaki, P. (1996). DNA mismatch repair gene mutations in 55 kindreds with verified or putative hereditary non-polyposis colorectal cancer. *Hum. Mol. Genet.* 5, 763–769.

Patil, N., Berno, A. J., Hinds, D. A., Barrett, W. A., Doshi, J. M., Hacker, C. R., Kautzer, C. R., Lee, D. H., Marjoribanks, C., McDonough, D., Nguyen, B. T., Norris, M. C., Sheehan, J. B., Shen, N., Stern, D., Stokowski, R. P., Thomas, D. J., Trulson, M. O., Vyas, K. R., Frazer, K. A., Fodor, S. P., and Cox, D. R. (2001). Blocks of limited haplotype diversity revealed by high-resolution scanning of human chromosome 21. *Science* 294, 1719–1723.

Phelps, R. S., Chadwick, R. B., Conrad, M. P., Kronick, M. N., and Kamb, A. (1995). Efficient, automatic detection of heterozygous bases during large-scale DNA sequence screening. *Biotechniques* 19, 984–989.

Phillips, M. S., Lawrence, R., Sachidanandam, R., Morris, A. P., Balding, D. J., Donaldson, M. A., Studebaker, J. F., Ankener, W. M., Alfisi, S. V., Kuo, F. S., Camisa, A. L., Pazorov, V., Scott, K. E., Carey, B. J., Faith, J., Katari, G., Bhatti, H. A., Cyr, J. M., Derohannessian, V., Elosua, C., Forman, A. M., Grecco, N. M., Hock, C. R., Kuebler, J. M., Lathrop, J. A., Mockler, M. A., Nachtman, E. P., Restine, S. L., Varde, S. A., Hozza, M. J., Gelfand, C. A., Broxholme, J., Abecasis, G. R., Boyce-Jacino, M. T., and Cardon, L. R. (2003). Chromosome-wide distribution of haplotype blocks and the role of recombination hot spots. *Nat. Genet.* 33, 382–387.

Pritchard, J. K. (2001). Are rare variants responsible for susceptibility to complex diseases? *Am. J. Hum. Genet.* 69, 124–137.

Rines, R. D., van Orsouw, N. J., Sigalas, I., Li, F. P., Eng, C., and Vijg, J. (1998). Comprehensive mutational scanning of the p53 coding region by two-dimensional gene scanning. *Carcinogenesis* 19, 979–984.

Risch, N. J. (2000). Searching for genetic determinants in the new millennium. *Nature* 405, 847–856.

Sanger, F., Nicklen, S., and Coulson, A. R. (1977). DNA sequencing with chain-terminating inhibitors. *Proc. Natl. Acad. Sci. USA* 74, 5463–5467.

Schneider, J. A., Pungliya, M. S., Choi, J. Y., Jiang, R., Sun, X. J., Salisbury, B. A., and Stephens, J. C. (2003). DNA variability of human genes. *Mech. Ageing Dev.* 124, 17–25.

Sheffield, V. C., Cox, D. R., Lerman, L. S., and Myers, R. M. (1989). Attachment of a 40-base-pair G + C-rich sequence (GC-clamp) to genomic DNA fragments by the polymerase chain reaction results in improved detection of single-base changes. *Proc. Natl. Acad. Sci. USA* 86, 232–236.

Sheffield, V. C., Beck, J. S., Kwitek, A. E., Sandstrom, D. W., and Stone, E. M. (1993). The sensitivity of single-strand conformation polymorphism analysis for the detection of single base substitutions. *Genomics* 16, 325–332.

Slager, S. L., Huang, J., and Vieland, V. J. (2000). Effect of allelic heterogeneity on the power of the transmission disequilibrium test. *Genet. Epidemiol.* 18, 143–156.

Smith, W. M., Van Orsouw, N. J., Fox, E. A., Kolodner, R. D., Vijg, J., and Eng, C. (1998). Accurate, high-throughput "snapshot" detection of hMLH1 mutations by two-dimensional DNA electrophoresis. *Genet. Test.* 2, 43–53.

Tabor, H. K., Risch, N. J., and Myers, R. M. (2002). Opinion: Candidate-gene approaches for studying complex genetic traits: practical considerations. *Nat. Rev. Genet.* 3, 391–397.

Van Orsouw, N. J., Li, D., van der Vlies, P., Scheffer, H., Eng, C., Buys, C. H., Li, F. P., and Vijg, J. (1996). Mutational scanning of large genes by extensive PCR multiplexing and two-dimensional electrophoresis: application to the RB1 gene. *Hum. Mol. Genet.* 5, 755–761.

Van Orsouw, N. J., Zhang, X., Wei, J. Y., Johns, D. R., and Vijg, J. (1998a). Mutational scanning of mitochondrial DNA by two-dimensional electrophoresis. *Genomics* 52, 27–36.

Van Orsouw, N. J., Dhanda, R. K., Rines, R. D., Smith, W. M., Sigalas, I., Eng, C., and Vijg, J. (1998b). Rapid design of denaturing gradient-based two-dimensional electrophoretic gene mutational scanning tests. *Nucleic Acids Res.* 26, 2398–2406.

Van Orsouw, N. J., Dhanda, R. K., Elhaji, Y., Narod, S. A., Li, F. P., Eng, C., and Vijg, J. (1999). A highly accurate, low cost test for BRCA1 mutations. *J. Med. Genet.* 36, 747–753.

Vijg, J., and Suh, Y. (2004). Screening for mutations in cancer-predisposition genes. In *Genetic Predisposition to Cancer*, R. A. Eeles, D. F. Easton, C. Eng, B. Ponder, eds., pp. 48–60. 2004. London, Arnold.

Wu, Y., Hofstra, R. M., Scheffer, H., Uitterlinden, A. G., Mullaart, E., Buys, C. H., and Vijg, J. (1996). Comprehensive and accurate mutation scanning of the CFTR gene by two-dimensional DNA electrophoresis. *Hum. Mutat.* 8, 160–167.

CHAPTER **10**

Real-Time Polymerase Chain Reaction

LUT OVERBERGH, ANNA-PAULA GIULIETTI, DIRK VALCKX, AND CHANTAL MATHIEU
Laboratory for Experimental Medicine and Endocrinology (LEGENDO), Catholic University of Leuven, Leuven, Belgium

TABLE OF CONTENTS

10.1 HISTORY OF PCR

The polymerase chain reaction (PCR) is a technique based on the exponential amplification of DNA by the thermostable *Thermus aquaticus* (Taq) polymerase. The method uses a pair of synthetic oligonucleotide primers, each hybridizing to one strand of a double-stranded DNA (dsDNA) target, with the pair spanning a region that will be exponentially amplified (see also Fig. 1.2). The annealed primers act as a substrate for the Taq DNA polymerase, creating a complementary DNA strand via sequential addition of deoxynucleotides. The process classically consists of three steps:

1. A denaturation step at 94 or 95°C.
2. Primer annealing to the ssDNA strands at 60°C.
3. Primer extension at 72°C.

The technique was first described by Kary Mullis in the 1980s (Mullis *et al.*, 1986), for which he received the Nobel Prize in 1993. This classic end-point PCR is a yes/no reaction because it measures DNA product formation after a fixed number of cycles, that is, in the plateau phase of the reaction, thus resulting in qualitative information on the presence or absence of a certain gene or mRNA. This PCR technique has become one of the most influential tools in the biological and medical sciences (Guyer and Koshland, 1989).

Over the years, numerous adaptations and applications to this classic end-point PCR have been described, including semiquantitative PCR, quantitative competitive PCR, and its latest innovation, real-time PCR. A first adaptation measures PCR product accumulation during the exponential phase of the reaction, resulting in semiquantitative data. This method necessitates interruption of the PCR reaction after an experimentally determined number of cycles. Furthermore, samples from a single experimental setup can be analyzed only over a relatively small linear range.

Alternatively, competitive PCR has been developed, resulting in quantitative data. This method, however, needs extensive optimization, since it requires the coamplification of an internal cDNA or RNA control (competitor) with the unknown sample in the same tube. Quantification is performed by titrating an unknown amount of target template against a dilution series of known amounts of the standard. The internal control consists of target DNA or RNA that has been slightly modified. Thus, one set of primers is designed that coamplifies the target and the competitor, with the same efficiency, although they can be distinguished from each other (by, for instance, difference in length or restriction sites; Clementi *et al.*, 1994). This method provides a strategy for accurate quantification, but the construction of internal standards is technically sophisticated and labor intensive.

For the detection of PCR products using either of these methods several detection techniques can be used, which all require excessive post-PCR manipulations. The most classically used are agarose gel electrophoresis with ethidium bromide or SYBR Green staining, fluorescent labeling and analysis using polyacrylamide gels, radioactive labeling, and Southern blotting or detection by phosphorimaging. Major drawbacks using these classical detection systems are the use of hazardous chemicals and the potential risk for laboratory contamination. Moreover, all these post-PCR manipulations are very time consuming.

The development of a new procedure in the mid 1990s for the analysis and quantification of DNA or RNA, based on fluorescence-kinetic RT-PCR, enabled quantification of the PCR product in real-time (Higuchi *et al.*, 1993; Heid *et al.*, 1996; Gibson *et al.*, 1996). This sensitive and accurate technique permits quantification of PCR product during the exponential phase of the PCR reaction. This is in full contrast to the classic end-point assays, as they are designed to provide information as rapidly as the amplification process itself, thus requiring no post-PCR manipulations. The development of this real-time PCR again had a revolutionary impact on molecular research and diagnostics.

10.2 PRINCIPLE OF REAL-TIME PCR

10.2.1 Real-Time PCR Using Hydrolysis Probes: The Classic TaqMan System

Real-time PCR was first described using hydrolysis probes (Heid *et al.*, 1996; Gibson *et al.*, 1996). The technique is based on the coupling of two important processes. First, the construction of dual-labeled oligonucleotide probes, also called hydrolysis or TaqMan probes, which emit a fluorescent signal upon cleavage, based on the principle of fluorescence resonance energy transfer (FRET, Stryer, 1978; Cardullo *et al.*, 1988). Second, the discovery that the Taq DNA polymerase possesses a $5' \rightarrow 3'$ exonuclease activity, which can be exploited to degrade the fluorescent labeled probe (Holland *et al.*, 1991). The oligonucleotide probe used in this assay is nonextendable at its $3'$ end and is dual-labeled, with a reporter fluorochrome, for example, FAM (6-carboxyfluorescein), and a quencher fluorochrome, for example, TAMRA (6-carboxy-tetramethylrhodamine). It is designed to anneal to the target sequence internally of the primers, during the annealing and extension phase of the PCR reaction. In its free, intact form no fluorescent emis-

sion can be measured, because fluorescent emission of the reporter dye is absorbed by the quenching dye. However, upon annealing of the probe to one of the target strands, the probe will become degraded by the $5'$–$3'$ exonuclease activity of the Taq polymerase. Consequently, the reporter and quencher dye become separated and the reporter dye emission is no longer transferred to the quenching dye (no more FRET), resulting in an increase of reporter fluorescent emission (e.g., for FAM at 518 nm). This process occurs in every cycle and does not interfere with the exponential accumulation of PCR product. The increase in fluorescence is measured cycle by cycle and directly correlates with the amount of PCR product formed (Heid *et al.*, 1996; Gibson *et al.*, 1996; Fig. 10.1).

Apart from the classically used fluorescent reporter dye FAM, other reporter dyes are available. These include for instance TET (tetrachloro-6-carboxyfluorescein), JOE (2,7,-dimethoxy-4,5-dichloro-6-carboxyfluorescein), HEX (hexacholoro-6-carboxyfluorescein), or VIC. The choice of different reporter dyes, with a minor overlap in fluorescent emission spectra, makes it possible to perform multiplex PCR reactions, thus simultaneously amplifying different DNA targets. Similarly, there is a choice between different quencher dyes. DABCYL (4-(4′-dimethylaminophenylazol) benzoic acid) also can be used as a quencher dye, but its use is much more prevalent in the molecular beacons probes (see Section 10.2.4). An advantage of using DABCYL is its reduced auto-fluorescence compared to TAMRA. More

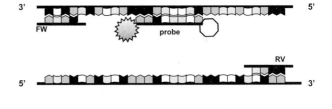

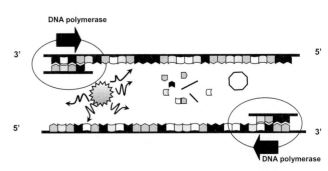

FIGURE 10.1 Hydrolysis or TaqMan probes. The dual-labeled TaqMan probe is cleaved by the $5'$ exonuclease activity of Taq DNA polymerase, during the extension step of the PCR reaction. The quencher and fluorophore are brought apart, which results in an increase in fluorescent emission. FW, forward primer; RV, reverse primer; grey 16-point star, fluorophore; white hexagon, quencher.

recently, several dark quencher fluorochromes have become available. These absorb the energy that is emitted by the reporter dye and release it as heat rather than as fluorescence. This results in a lower background signal and thus a higher sensitivity. It is therefore to be expected that dark quenching dyes will become the standard, replacing the TAMRA quenchers in the near future.

10.2.2 Real-Time PCR Using DNA Intercalating Dyes

Another widely used real-time PCR technique is based on the detection and quantification of PCR products using fluorescent DNA intercalating dyes. The principle of this technique was first described by Higuchi (1993), who monitored the increase in ethidium bromide fluorescence using a charged coupled devise camera, a method that was referred to as kinetic PCR. More recently, SYBR Green I, which is less toxic than ethidium bromide, is widely used as a dye, which incorporates into dsDNA (see Fig. 10.2). This dye incorporates into the minor groove of dsDNA, by which its fluorescence is greatly enhanced. During the PCR reaction, the amount of double-stranded target will increase exponentially, paralleled by an increase in SYBR Green I incorporation and fluorescent emission. In each cycle the fluorescent emission will increase gradually during the extension phase of the reaction, and will be low or absent during the denaturation phase.

The greatest advantage of this method as opposed to the use of fluorescent labeled probes is that it can be used with any pair of primers for any target. It is a cheaper alternative and requires less specialist knowledge than the design of fluorescent labeled probes. Consequently, specificity is diminished due to the risk of amplifying nonspecific PCR products

or primer dimers. To discriminate between specific and nonspecific PCR products, a melting curve analysis has to be performed, after termination of the PCR reaction (Ririe *et al.*, 1997). In this way, the fraction of fluorescence originating from the specific target can be distinguished from that originating from primer dimers or nonspecific amplification products. This analysis is performed by slowly increasing the temperature from 40°C to 95°C, during which fluorescent emission is monitored continuously. Fluorescent emission will be low at low temperatures—when all PCR products are double stranded—and it will increase dramatically around the melting temperature of the PCR product. The rationale behind this melting curve analysis is based on the fact that PCR products of different length will have different melting temperatures, which will result in distinct peaks when plotting the first negative derivative of the fluorescence versus the temperature.

10.2.3 Real-Time PCR Using Hybridization Probes

Real-time PCR analysis with hybridization probes uses two juxtaposed sequence-specific probes, also known as HybProbes. The development of this system was first described in the mid 1980s (Heller and Morrison, 1985). Each probe has a single fluorescent reporter, one a donor fluorophore at its 3′ end and the other an acceptor fluorophore at its 5′ end. The sequences of the two probes are designed to anneal to the target sequences in very close proximity to each other (i.e., within 1–5 nucleotides), in a head-to-tail arrangement, bringing the two dyes very close to each other (see Fig. 10.3). As long as the probe is in its free, unbound

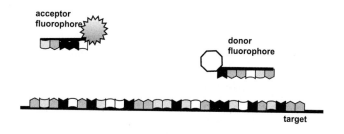

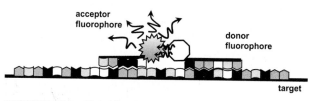

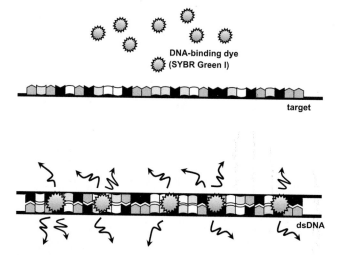

FIGURE 10.2 DNA intercalating dyes. DNA intercalating dyes incorporate into the minor groove of dsDNA. When this occurs, the dye will emit fluorescent light.

FIGURE 10.3 Hybridization probes. Two probes are used, one carrying an acceptor fluorophore and the other one carrying a donor fluorophore. When both fluorophores are brought into close proximity, i.e., when the probes anneal to the target sequence, the donor is able to excite the acceptor through FRET and fluorescence emission will occur.

form, the fluorescent signal from the reporter dye is not detected. However, during the annealing phase of the PCR reaction, the probes anneal to the target sequence and the two fluorophores come in close proximity to each other. This will result in emission of light from the donor fluorochrome, which will excite the acceptor fluorochrome, a process referred to as resonance energy transfer. The dye in one of the probes transfers energy, allowing the other one to dissipate fluorescence at a different wavelength. The amount of fluorescence emitted can be measured during the annealing phase of the PCR reaction, and is directly proportional to the amount of target DNA generated during the PCR process (Bernard and Wittwer, 2000).

A variant of these classic hybridization probes uses a fluorescent labeled primer/probe combination. In this case, a 5′-labeled hybridization probe is designed to anneal to the PCR strand in close proximity to one of the PCR primers, which has a fluorophore at its 3′ end. This method requires that the fluorescently labeled primer be positioned near the probe, usually within 5 base pairs, to allow adequate resonance energy transfer with the complementary probe (von Ahsen *et al.*, 2000).

Hybridization probes usually are constructed with FAM as the 3′ donor fluorophore, and a range of different acceptor fluorophores are commonly used (e.g., ROX, Cy5, LC Red640, LC Red705) as the 5′ acceptor fluorophore.

10.2.4 Real-Time PCR Using Molecular Beacons

Molecular beacons are probes that contain a stem-and-loop structure in their intact, unbound form. They are dual-labeled, with a fluorophore linked to one end of the molecule and a quencher linked to the other end (Tyagi *et al.*, 1996). Fluorescence is quenched when the probe is in its hairpin-like structure due to the proximity between quencher and fluorophore, resulting in complete absorption of any photons emitted by the fluorophore. When the probe sequence in the loop anneals to a complementary target sequence, a conformational change allows the formation of a linear structure whereby fluorescent energy transfer is no more occurring, resulting in an increase in fluorescence emission (see Fig. 10.4). Molecular beacons are especially suitable for identifying point mutations. They can distinguish targets that differ by only a single nucleotide and they are significantly more specific than conventional hydrolysis probes of equivalent length.

10.2.5 Real-Time PCR Using Scorpions

Real-time PCRs using the scorpions system is a relatively new system, which is carried out with two oligonucleotides: a primer and a fluorescent molecule that combines the primer and probe function (Whitcombe *et al.*, 1999). In the primer/probe combination, a primer sequence is linked to

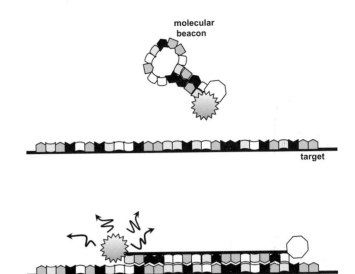

FIGURE 10.4 Molecular beacons. Molecular beacons are hairpin-shaped probes. The fluorophore and quencher are in close proximity when the probe is in its free, unbound state. When the probe anneals to the complementary target sequence, its conformation will change. Thereby the quencher and fluorophore are brought apart, causing fluorescence emission.

a specific probe sequence that is held in a hairpin-loop form. The stem-loop tail is separated from the primer by a PCR blocker to prevent the Taq DNA polymerase from amplifying the stem-loop sequence. This configuration brings the fluorophore in close proximity to the quencher and avoids fluorescence. As soon as annealing between the primer/probe and the target occurs, the hairpin is opened and the fluorophore and quencher are separated, resulting in an increase in fluorescence emission (see Fig. 10.5). Scorpions differ from molecular beacons and hydrolysis probes, in that their structure promotes a unimolecular probing mechanism. This results in a stronger fluorescent signal, especially under fast cycling conditions. Scorpions represent a relatively new chemistry, validated for mutation detection, but most likely it will be adapted to other assays (Thelwell *et al.*, 2000).

10.2.6 Real-Time PCR Using Other Detection Chemistries

More recently, other sophisticated detection chemistries, which will not be discussed in further detail here, have been developed for real-time PCR. Like hybridization probes, molecular beacons, and scorpions, these systems all rely on the FRET principle, although without the need for hydrolysis by the nuclease activity of the Taq DNA polymerase. The list of possible detection chemistries is continuously growing; some examples are Minor Groove Binding probes, ResonSense probes, Hy-beacon probes, Light-up probes, and Simple probes.

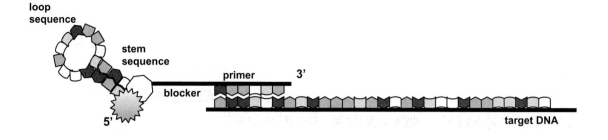

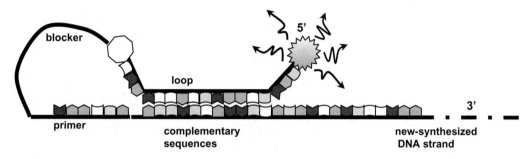

FIGURE 10.5 Scorpions. Scorpions are single-stranded dual-labeled fluorescent primer/probes with a hairpin-shaped structure. The primer/probe contains a 5′ end fluorophore and an internal quencher dye directly linked to the 5′ end of a PCR primer via a PCR blocker. In the unbound form, no fluorescent emission occurs. When binding to the target sequence, the hairpin shape unfolds and the loop-region of the probe hybridizes intramolecularly to the newly synthesized target sequence. This results in fluorescent emission, because fluorophore and quencher are separated.

10.3 REAL-TIME THERMAL CYCLERS

Since the development of real-time PCR in the mid 1990s, instrumentation systems have already undergone an extensive evolution. The first available instrument, the 7700 SDS from Applied Biosystems, took up almost an entire bench area, made use of an expensive laser as the light source, and had to be placed in an air conditioned laboratory. This instrument was unable to produce a true analysis in real-time, since data could be viewed only after termination of the PCR reaction. As the technology improved, systems became available that allowed detection of a PCR product from the moment it is formed, thus in pragmatic real-time. These newer machines are much smaller in size, and the laser has been changed for less precious tungsten-halogen or light-emitting diode lamps. Recently, instruments became available that can even be backpacked and used for on-site analysis. These instruments are able to give results within 15 min of analysis time, a rather important feature, for instance, for the on-site detection of pathogen outbreaks. Because of the wide range of instruments developed by different companies, prices have dropped significantly and instruments are becoming a standard tool for routine molecular and diagnostic laboratories.

Although all possible applications can be performed on all instruments, each one has specific advantages and disadvantages. The choice for a specific real-time thermal cycler therefore is dependent on the specific application one is focusing on. For quantification of gene expression, for instance, the length of a run is not the most important factor, although one would like to choose an instrument which can analyze a large number of samples simultaneously. Indeed, a high turnover of sample numbers will be performed more easily in 96-well or even 384-well systems. On the other hand, if the main application is pathogen identification for example, velocity is the main issue and the best choice would be a rapid thermal cycler.

Most of the thermal cyclers are able to perform multiplex PCR, are sensitive and accurate, and increasingly present user-friendly software. In Table 10.1, the main features of all presently available real-time PCR thermal cyclers are summarized.

10.4 HOW DATA ARE OBTAINED

When performing real-time PCR, the ability to monitor the amplification process of the PCR reaction in real-time revolutionizes the way in which the data are obtained. Typically, the reactions are characterized by the point in time during PCR cycling when amplification of a PCR product reaches a certain detection level, as opposed to end-point

TABLE 10.1 Available real-time thermal cyclers.

The most important features of all real-time thermal cyclers that are currently commercially available are listed. The information was retrieved from the suppliers' web sites.

Instrument	Excitation	Detection	Sample load	Sample format	Run time	Web site
ABI Prism® 7000HT SDS (Applied Biosystems)	Tungsten-halogen lamp	Four position filter wheel and CCD[a] camera	96	96 well-plate 0.2 ml tubes Strip tubes	Around 2 hours	http://home.appliedbiosystems.com/
ABI Prism® 7900HT SDS (Applied Biosystems)	Argon-ion laser source (488 nm)	Spectrograph and CCD camera (500–660 nm)	96 384	96 well-plate 384 well-plate	Around 2 hours	http://home.appliedbiosystems.com/
iCycler iQ Real-time PCR Detection System (Bio-Rad)	Tungsten-halogen lamp (400–700 nm)	CCD camera	96	96 well-plate 0.2 ml tubes Strip tubes	Around 2 hours	www.bio-rad.com
MyiQ Single-color Real-time PCR Detection System (Bio-Rad)	Tungsten-halogen lamp (400–585 nm)	CCD camera	96	96 well-plate 0.2 ml tubes Strip tubes	Around 2 hours	www.bio-rad.com
DNA Engine Opticon® Continuous Fluorescence Detection System (MJ Research)	LED light source (450–495 nm)	PMT[c] (515–545 nm)	96	96 well-plate 0.2 ml tubes Strip tubes	Around 2 hours	www.mjr.com
DNA Engine Opticon® 2 Continuous Fluorescence Detection System (MJ Research)	LED light source (470–505 nm)	PMT 2 channel detection (523–543 nm) and (540–700 nm)	96	96 well-plate 0.2 ml tubes Strip tubes	Around 2 hours	www.mjr.com
LightCycler® Instrument (Roche Applied Science)	Blue LED[b] light (470 nm)	3 channels (530 nm, 640 nm and 710 nm)	32	Glass capillaries	Less than 1 hour	www.lightcycler-online.com
LightCycler® 2.0 Instrument (Roche Applied Science)	Blue LED light (470 nm)	6 channels (530 nm, 560 nm, 610 nm, 640 nm, 670 nm)	32	Glass capillaries	Less than 1 hour	www.lightcycler-online.com
R.A.P.I.D.™ System (Idaho Technology Inc.)	LED light source (450–490 nm)	3 channels (520–540, 630–650, 690–730 nm)	32	Composite glass/ plastic vessel	Less than 30 minutes	www.idahotech.com
Rotor-Gene 3000 (Corbett Research—Pyrosequencing)	LED light (470, 530, 585, 625 nm)	510, 555, 610, 580, 610, 660 nm	36 72	0.2 ml micro tubes 0.1 ml strip-tubes	Less than 1 hour	www.corbettresearch.com
Smart Cycler® II System (Cepheid)	LED (450–495, 500–550, 565–590, 630–650 nm)	510–527, 565–590, 606–650, 670–750 nm	16 (independent wells)	0.1 ml tubes 0.025 ml tubes	Less than 1 hour	www.cepheid.com
Smart Cycler® II TD System (Cepheid)	LED (450–495, 500–550, 565–590, 630–650 nm)	510–527, 565–590, 606–650, 670–750 nm	16 (independent wells)	0.1 ml tubes 0.025 ml tubes	Less than 1 hour	www.cepheid.com
Mx3000P™ Real-Time PCR System (Stratagene)	Tungsten-halogen lamp (350–750 nm)	PMT (4 channels)	96	96 well-plate 0.2 ml tubes Strip tubes	Around 2 hours	www.mx3000p.com
Mx4000™ Multiplex Quantitative PCR System (Stratagene)	Tungsten-halogen lamp (350–750 nm)	PMT (4 channels; 350–830 nm)	96	96 well-plate 0.2 ml tubes Strip tubes	Around 2 hours	www.mx4000.com

a. CCD, charge-coupled device. b. LED, light-emitting diode. c. PMT, scanning photo multiplier tube.

detection, where the amount of product formed is measured after a fixed number of cycles. Furthermore, quantification is based on the inherent property of a PCR reaction, that the more input DNA copies one starts with, the fewer cycles of PCR amplification it takes to make a specific number of amplification product. Finally, the fact that the formation of amplification product linearly correlates with the amount of fluorescence emission is exploited in the real-time PCR assay.

In practice, using any of the developed detection chemistries on any of the available instruments, the increase in fluorescence emission can be read by a sequence detector in real-time, during the course of the reaction, and is a direct consequence of target amplification during PCR. In Fig. 10.6, a typical amplification plot is shown, in which the terms and definitions routinely used in real-time quantitative PCR are illustrated. During the initial cycles of the PCR reaction, there is little or no change in fluorescence signal. This stage is the baseline of the amplification plot. The fluorescence emission of the product at each time point is measured during PCR cycling, and is defined as Rn^+. Analogously, Rn^- is the fluorescent emission of the baseline. The increase in fluorescence is calculated by the computer software program, and is plotted on the y-axis as the ΔRn value, using the equation $\Delta Rn = Rn^+ - Rn^-$. Thus, this value directly correlates with the probe degradation (in case of

hydrolysis or TaqMan probes) during the PCR process, and consequently with the formation of specific PCR product. An arbitrary threshold is chosen, based on the variability of the baseline, usually determined as ten times the standard deviation of the baseline, set from cycle 3 to 15. This, of course, can be changed manually for each individual experiment if necessary. Threshold cycle (Ct) values are then calculated by determining the point at which the fluorescence exceeds this chosen threshold. Ct is reported as the cycle number at this point. Therefore, Ct values decrease linearly with increasing input target quantity. This is used as a quantitative measurement of the input target.

10.5 HOW DATA ARE QUANTIFIED

To quantify the results obtained by real-time PCR, two different methods are commonly used: the standard curve method and the comparative threshold method.

10.5.1 The Standard Curve Method

In the standard curve method a sample with known concentration is used to make a dilution series, which is used as a standard curve. Samples, which can be used to construct such a dilution series, are purified plasmid dsDNA, *in vitro*

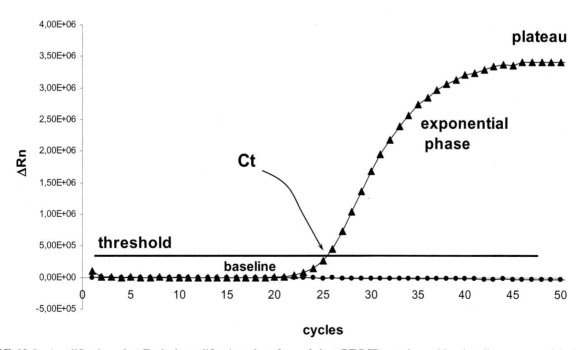

FIGURE 10.6 Amplification plot. Typical amplification plot of a real-time RT-PCR reaction, with a baseline, exponential phase, and plateau phase. Cycle number is plotted against ΔRn (the difference between the fluorescence detected at a certain point of the reaction and the initial fluorescence, or the fluorescent emission of the baseline). The threshold is chosen based on the variation of the baseline. Ct (threshold cycle) is the point where the detected fluorescence crosses this given threshold.

transcribed RNA, *in vitro* synthesized ssDNA, or any cDNA sample expressing the target gene. The concentration of these DNA or RNA samples can be measured spectrophotometrically at 260 nm and converted to the number of copies using the molecular weight of the DNA or RNA. For absolute quantification of mRNA expression absolute standards have to be used; for instance, *in vitro* transcribed RNA. Because of the labor-intensive construction of these standard clones, this method is not widely used. More often, cDNA plasmid standards are used for quantification. These are constructed by cloning a cDNA fragment into a suitable plasmid vector. For quantification of mRNA expression, however, this will result in only a relative quantification, because variations in efficiency of the reverse transcription step are not controlled.

When creating a standard curve by serial 2-fold dilutions of the standard sample, two consecutive points will have a Ct difference of 1. Similarly, Ct values from 10-fold diluted samples will differ by 3.3. This, of course, is assuming that 100% PCR efficiency is reached. Consequently, the slope of the standard curve is a measure of the efficiency of the PCR reaction. For serial 10-fold dilutions, it should ideally be −3.3. In practice, standard curves with a slope between −3.0 and −3.6 are considered acceptable. Also the sensitivity of the PCR reaction is reflected in the standard curve, by the point at which the standard curve crosses the Y-axis (Y-intercept). Indeed, the lower the Ct value at this point, the higher the sensitivity of the PCR reaction (see Fig. 10.7). By plotting the Ct value of an unknown sample on the standard curve, the amount of input target sequence in the sample can be determined. This calculation is performed automatically by the software program of the real-time PCR instrument.

10.5.2 The Comparative Ct Method

An alternative method used for relative quantification is the comparative Ct method, or the $\Delta\Delta$Ct method (Livak and Schmittgen, 2001). This method uses arithmetic formulas to calculate relative expression levels, compared to a calibrator. A nontreated control sample can be used, for instance, as a calibrator. Moreover, the value of the unknown target is normalized to an endogenous housekeeping gene. The amount of target, relative to the calibrator and normalized to an endogenous housekeeping gene, is measured by the equation $2^{-\Delta\Delta Ct}$, where $\Delta\Delta Ct = \Delta Ct_{sample} - \Delta Ct_{calibrator}$, and ΔCt is the Ct of the target gene subtracted by the Ct of the housekeeping gene. The equation thus represents the normalized expression of the target gene in the unknown sample, relative to the normalized expression of the calibrator sample. Importantly, for the $\Delta\Delta$Ct method to be applicable, the efficiency of PCR amplification for the target gene and the housekeeping gene must be approximately equal. For every real-time PCR assay that is being set up, this has to be tested, by determining how the ΔCt_{sample} and $\Delta Ct_{calibrator}$ varies with template dilution. In case the efficiencies between PCR

amplification for target and housekeeping genes are different, a new set of primer/probe combinations has to be designed. Alternatively, the standard curve method can be applied.

10.6 MULTIPLEX REAL-TIME PCR

Multiplex real-time PCR can refer either to the simultaneous amplification and detection of different target genes in one tube, or to the use of multiple fluorogenic probes for the discrimination of different alleles. The term is therefore somewhat confusing, since it is not a mere extension of classical multiplexing known in conventional PCR, which was simply the amplification of multiple templates within one reaction, using different primers.

Since diagnostic analyses are often restricted by the limited availability of bioptic material, and one of the primary goals is to decrease analysis time, multiplex PCR is considered an attractive solution. However, it is still not routinely used today.

One difficulty is associated with limitations caused by interference of multiple sets of primers, which can reduce the dynamic range of sensitivity. In practice, a problem will arise when highly differentially expressed targets have to be quantified simultaneously, since the exponential phase of amplification of the most abundant target will not overlap with that of the less abundant target. Thus, in order to be able to simultaneously amplify these different targets, an extensive optimization is required, which most often makes use of primer limiting conditions (for the most abundant target).

Other obstacles arise from the limited number of different fluorophore reporters with a good spectral resolution. Real-time PCR instrumentations contain optimized filters to minimize the overlap of the emission spectra from the different available fluorophores. Despite this, the number of fluorophores that can be combined and clearly distinguished is limited when compared with the resolution in conventional multiplex PCR. Recent improvements in the design of other probe formats as well as novel combinations of fluorophores are promising for the ability to simultaneously detect a larger number of targets. Although modern real-time PCR instruments have been significantly improved, these improvements still allow for multiplexing of only four different colors.

Another approach, which is used for the detection of human genetic diseases, is based on the discrimination between single or multiple nucleotide changes (between different alleles) by making use of the differences in melting point. In this approach, a single fluoroprobe can be used to distinguish between products, based on their distinct melting temperatures, which is reflected by the differences in thermodynamic stabilities of the perfectly complementary and the mismatched probe-target duplexes.

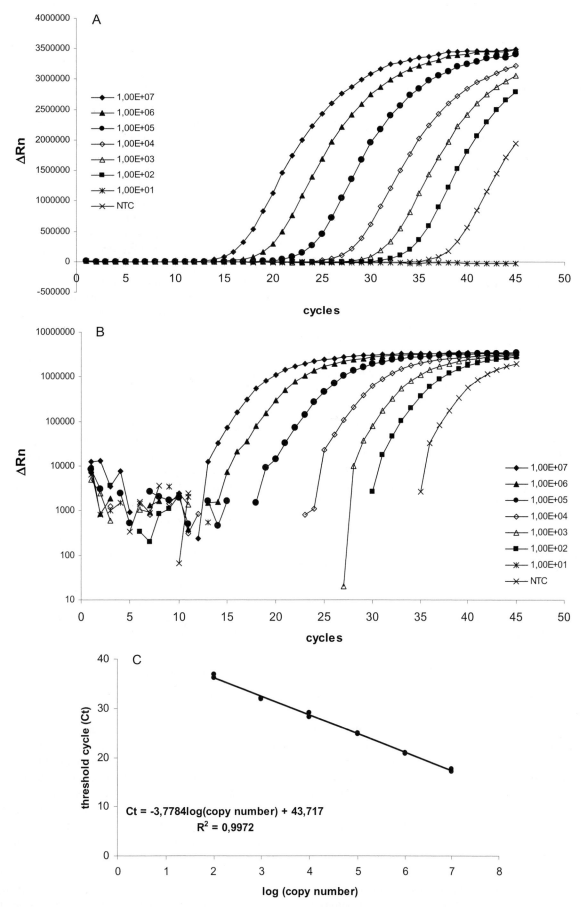

FIGURE 10.7 Standard Curve. A. Linear amplification plot. Amplification plot of β-actin plasmid cDNA. Seven points of a 10-fold dilution series of a β-actin cDNA plasmid standard, amplified using the TaqMan system. B. Logarithmic amplification plot. The same amplification curve as in A, but shown in a logarithmic scale. C. Standard curve for β-actin, constructed by plotting threshold cycle (Ct) values against input cDNA copy numbers.

Finally, the combined use of multicolor fluorimetry and fluorescence melting curve analysis can greatly increase the number of targets that can be detected simultaneously (Wittwer *et al.*, 2001). A nice example, applying this sophisticated technique, is the simultaneous detection of all 27 possible base substitutions occurring in codons 12, 13, and 61 from the wild type sequence of the *Nras* oncogene (Elenitoba-Johnson *et al.*, 2001).

10.7 APPLICATIONS IN MOLECULAR DIAGNOSTICS

10.7.1 Clinical Microbiology

Real-time PCR has been shown to be extremely useful for studies in the field of clinical virology, bacteriology, and fungal diagnosis (see also Section 22.2.3). Most of the assays developed allow an increased frequency as well as enhanced speed of pathogen detection as compared to conventional culture techniques. Moreover, quantification of pathogen load is made possible. For real-time assay development one has to take into account that most infectious agents are characterized by a high mutation rate, which can dramatically influence the pathogen load estimation. This can be overcome by designing primers in highly conserved regions. Alternatively, of course, sequence variations can provide the basis for development of subtype-specific assays.

10.7.1.1 Virology and Bacteriology

Real-time PCR is increasingly used for the diagnosis of different diseases caused by DNA viruses, such as cytomegalovirus (Tanaka and Kimura, 2000), Epstein Barr virus (Ohga and Kubo, 2001), and hepatitis B virus (Brechtuehl and Whalley, 2001). It is a very reliable technique to titer viral genome density in clinical specimens, to monitor patients with virus infections, or to monitor reactivation in patients with latent infections. Indeed, viral DNA copy numbers change proportionally with anti-viral therapy (Nitsche *et al.*, 1999). It should thus be clear that the application of real-time PCR in this field resulted in a great improvement in following the impact of the therapy.

Other applications in the field of virology for which the real-time PCR technology has been validated are the detection of Bacillus anthrax (Makino and Cheun, 2003), human papillomavirus (Hubbard, 2003), influenza virus (Ellis and Zambon, 2002), and parpovirus (Bultmann *et al.*, 2003), just to name a few.

Similarly, real-time PCR methods are used increasingly for the detection of bacterial pathogens. Validated examples are, for instance, the detection and quantification of different *Mycobacterium* species (Miller *et al.*, 2002; Rondini *et al.*, 2003), *Helicobacter pylori* (Lascols *et al.*, 2003), and *Streptococcus pneumonia* (van Haeften *et al.*, 2003).

10.7.1.2 Fungal Diagnosis

In the detection of fungal infections it has become essential to develop methods that allow for a quick, sensitive, and specific detection. Indeed, fungal infections are a major cause of morbidity and mortality in immuno-compromised patients (for instance after aggressive chemotherapy, organ transplantation, or in immuno-deficient patients). It is evident that early initiation of antifungal therapy, which critically depends on the early, fast, and accurate detection method, is essential in reducing the high mortality rates associated with fungaemia. The gold standard for detection of fungal infections has long been blood culture. But this technique is very time consuming and lacks sensitivity. Therefore the validation of real-time PCR in the detection of fungal infections essentially overcomes these limitations and is a promising tool to become the new gold standard.

Real-time PCR assays have been developed using broad-range fungal primers: for example, the fungal 18S rRNA gene. These amplify fungi-specific and highly conserved sequences of multicopy genes and are clinically useful for the detection of fungal infections. Additionally, PCR assays targeting species-specific sequences can be used to identify the pathogen. Alternatively, specific identification of the pathogen can be obtained by subsequent sequencing of the amplicon (Imhof *et al.*, 2003).

The development of these real-time PCR assays allows species determination of fungal DNA not only in human blood samples, but also in human tissue biopsies. This is a great step forward, since in this way, deep-seated fungal infections can be analyzed, which was not possible with the previous culture-based techniques (Imhof *et al.*, 2003).

Pathogenic fungi, for which detection protocols by real-time PCR have been validated, for example, are *Candida*, *Aspergillus*, *Histoplasma*, and *Conidiobolus* species (Imhof *et al.*, 2003; White *et al.*, 2003).

10.7.2 Food Microbiology

Rapid and accurate detection of bacterial pathogens from food samples is important, both for food quality assurance and to trace outbreaks of bacterial pathogens within the food supply. Real-time PCR analysis meets these criteria, since it significantly shortens analysis time compared to conventional biochemical and serological identification methods, and because the technique can be applied directly on pre-enrichment media or food products. A specific problem when analyzing food pathogens is that foods often are complex matrices, which require selective enrichment steps to overcome problems of low pathogen numbers. To address this issue, a universal enrichment broth has been developed, enabling enrichment of multiple pathogens (Bailey and Cox, 1992). An additional problem that has to be faced in the case of PCR-based detection methods for food pathogens is to discriminate between life and death pathogenic cells. One approach used is to apply a living/dead dye that covalently binds to DNA and inhibits PCR amplification from dead cells (Rudi *et al.*, 2002). Alternatively, the development of mRNA-based real-time PCR assays as opposed to DNA-based assays can be applied, which has the advantage of serving as an accurate indicator of pathogen viability (Rijpens and Herman, 2002).

Protocols have been validated for the detection of *Escherichia coli*, *Salmonella* (Bhagwat *et al.*, 2003), and *Listeria* strains (Norton, 2002) in food and environmental samples, just to name a few examples.

10.7.3 Clinical Oncology

10.7.3.1 MINIMAL RESIDUAL DISEASE
The detection of minimal residual disease (MRD)—that is, the detection of a very low number of malignant cells—significantly correlates with the clinical outcome in many hematological malignancies. Therefore, MRD monitoring is important for therapy guidance in clinical settings. Real-time PCR-based techniques for the detection of MRD have been developed, and already are used in clinical protocols. The major challenge when optimizing such assays is to reach a maximal sensitivity. Real-time PCR assays for MRD detection can be divided into three main categories, depending on the specific type of target envisioned:

1. Immunoglobulin and T-cell receptor gene rearrangements.
2. Breakpoint fusion regions of chromosome aberrations and fusion-gene transcripts.
3. Aberrant genes or aberrantly expressed genes.

The choice for a specific application is dependent on the type of disease. In hematological malignancies real-time PCR assays have been optimized; for instance for acute lymphoblastic leukemia (Pongers-Willemse *et al.*, 1998; Donovan *et al.*, 2000), acute myeloid leukemia (Jaeger and Kainz, 2003), chronic myeloid leukemia (Gonzalez *et al.*, 2003), multiple myeloma (Gerard *et al.*, 1998; Rasmussen *et al.*, 2000), and non-Hodgkin's lymphoma (Yashima *et al.*, 2003).

Importantly, in this field efforts are being made to meet standardized criteria and an international uniformity. Several European networks have been established, aiming at the development of common guidelines for data analysis and reporting MRD data. Furthermore, these networks also perform quality control rounds, which are required to monitor the performance of the participating laboratories and to further improve and standardize real-time PCR assays (van der Velden *et al.*, 2003).

10.7.3.2 SINGLE NUCLEOTIDE POLYMORPHISMS
Single nucleotide polymorphisms (SNPs), present in the human genome, are used as genetic markers to follow the inheritance patterns of chromosomal regions from generation to generation. Moreover, they consist of a powerful tool in the study of genetic factors associated with human diseases (Johnson and Todd, 2000; Risch, 2000). In a clinical setting, detection of specific SNPs can be very useful for determining the relative percentages of donor and recipient cells after allogeneic bone marrow transplantation (Oliver *et al.*, 2000).

Real-time PCR assays have been developed for SNP detection. For this application the level of discrimination between target and nontarget allele is the most important challenge when optimizing the technique. In this real-time PCR method, one set of primers and two allele-specific fluorescent labeled probes are used, making use of different reporter dyes. The two alleles can thus be distinguished by the differential fluorescent emission of the two different reporter dyes (Livak, 1999). Alternatively, SNP detection is often making use of molecular beacons (Tyagi *et al.*, 1998) or scorpion probes (Whitcombe *et al.*, 1999). For each individual SNP assay the level of discrimination has to be tested experimentally, since it is dependent on the mismatch.

10.7.3.3 CHROMOSOMAL TRANSLOCATIONS
Chromosomal translocations, which are often taking place in tumor cells, can be employed as tumor-specific PCR targets. The primers are designed so that they anneal to opposite sides of the breakpoint in the fusion gene. These fusion genes are interesting targets for the design of a PCR method. Indeed, they are directly related to the oncogenic process and therefore stable throughout the disease course. Moreover, the breakpoint fusion sites at the DNA level differ in each patient so that patient-specific real-time PCR strategies can be applied. A disadvantage of patient-specific assays is, of course, the higher analysis costs. A real-time PCR assay has been developed for the t(14,18) translocation, involving the *BCL2* and *IGH* genes (Summers *et al.*, 2001). For the design of real-time PCR assays in this setting a major difficulty is the relatively large amplicon size often necessary to span the fusion breakpoints, especially if one wants to design an assay for different patients clustering in relatively small breakpoint areas.

Alternatively, real-time PCR can be performed on the transcripts of tumor-specific fusion genes. These are disease-specific transcripts located over chromosome breakpoints, leading to an in-frame RNA product (Rowley, 1998). Interestingly, these fusion transcripts can be identical in individual patients despite distinct breakpoints, because the breakpoints often are located in introns. An advantage of this approach is that the same set of primer-probes can be used for analysis of individual patients with the same fusion-transcript, but different breakpoint translocations (van der Velden *et al.*, 2003). Examples of validated real-time PCRs using this approach are the BCR/ABL (Jones *et al.*, 2003), MLL/AF9 (Scholl *et al.*, 2003), CBFbeta/MYH11 fusion gene transcripts (Marcucci *et al.*, 2001), or the simultaneous detection of 10 different fusion gene transcripts (Osumi *et al.*, 2002).

10.7.4 Gene Therapy

The primary goal of gene therapy is to specifically deliver the therapeutic gene to the target organ in a time- and dose-dependent manner and, most importantly, to avoid delivery to nontarget organs, since this may result in toxic side-effects (see also Chapter 26). In this case, the therapeutic gene is delivered to the target cells through a vector system, most often a viral vector (Crystal, 1995). Two important parameters have to be analyzed when considering gene therapy as a drug delivery system:

1. Gene transfer estimation, which is the expression level of the therapeutic gene in regard to target tissue levels over time.
2. The bio-distribution, which is the distribution of the drug in different organs for different routes of administration.

Real-time PCR assays have been validated both in regard to gene transfer as well as bio-distribution of gene therapy vectors. The major challenge in optimizing these assays is the accuracy of quantification and the sensitivity of the assay. Real-time PCR assays using hybridization probes, for instance, have been validated for adenovirus gene transfer vectors and proven to be quantitative, reproducible, and sensitive (Senoo *et al.*, 2000; Hackett *et al.*, 2000).

10.7.5 Quantification of Gene Expression

Probably one of the most widely used applications of real-time PCR is the quantification of mRNA expression, or real-time reverse-transcriptase PCR (RT-PCR). Major challenges when optimizing real-time PCR assays for gene expression studies are the sensitivity of the assay, accurate quantification, avoiding amplification of contaminating genomic DNA, and the choice of a relevant housekeeping gene as control. The use of real-time PCR in molecular diagnostics has been reported, for instance, for the analysis of tissue-specific gene expression (Bustin, 2000) and for analyzing cytokine mRNA expression profiles (Giulietti *et al.*, 2001).

Currently, real-time RT-PCR assays are being combined with other sophisticated technologies, such as microarray analysis and laser capture micro-dissection. The availability of these high-throughput technologies enable the analysis of gene expression to measure alterations in molecular genetic events associated with the initiation and progression of a variety of diseases (Elkahloun *et al.*, 2002).

In the field of immunology, quantification of gene expression by real-time RT-PCR is widely used for the analysis of cytokines and chemokines. Cytokines are proteins, which are secreted by many different cell types, such as lymphocytes, monocytes, macrophages, dendritic cells, endothelial cells, and fibroblasts. They play a central role in modulating immune responses, including lymphocyte activation, proliferation, differentiation, survival, and apoptosis. Consequently, abnormal cytokine expression patterns will contribute to immune-mediated disorders, such as autoimmune, allergic, and infectious diseases (Borish and Steinke, 2003; Rabinovitch, 2003). In addition, they are commonly the focus of studies on immunosuppressive therapy during organ transplantation.

Chemokines, also known as chemotaxic cytokines, are secreted proteins that function as chemo-attractants, controlling the trafficking of specific subsets of leukocytes to sites of tissue damage to mediate specific immunological functions. In autoimmune diseases, the migration and accumulation of leukocytes in diseased target organs is a critical step in the development of the disease (Kunkel and Godessart, 2002).

Therefore, a reliable quantification of cytokine, cytokine receptor, chemokine, and chemokine receptor mRNA levels in diseased target organs are fundamental for our understanding of diverse pathological states and for monitoring therapeutic effects. Major challenges when optimizing real-time PCR assays for gene expression studies are the sensitivity of the assay, accurate quantification, and avoiding amplification of contaminating genomic DNA. The use of real-time RT-PCR has been validated for a large panel of cytokines and chemokines (Giulietti *et al.*, 2001; Overbergh *et al.*, 2003b). This greatly refined the role of cytokines and chemokines in numerous biological and clinical contexts; for instance, in type 1 diabetes (Gysemans *et al.*, 2003; Overbergh *et al.*, 2003a) and experimental autoimmune encephalitis (van Etten *et al.*, 2003). According to the numerous publications using real-time PCR technique for quantification of cytokine and chemokine expression, it has definitely become the gold standard in this field.

10.8 CRITERIA FOR DEVELOPING REAL-TIME PCR ASSAYS

The choice of method for performing real-time PCR reactions depends on the specific application envisioned. An important distinction has to be made, for instance, depending on whether one is dealing with measuring mRNA expression levels or with DNA copy number. Other factors, such as the number of genes, the number of targets, the importance of fast screening, allelic discrimination, accurate quantification, sensitivity, and the costs, need careful consideration.

Criteria to be taken into account when optimizing real-time PCR assays for mRNA quantification, such as primer design, template preparation, and normalization of the results, are described next. These are based mainly on our own experience in the field of quantification of mRNA expression levels of cytokines, chemokines, or other immune related targets, in the context of gaining more insight into the mechanisms of disease outcome in type 1 diabetes (Overbergh *et al.*, 2000; Gysemans *et al.*, 2000; Giulietti *et al.*, 2001).

10.8.1 Primer and Probe Design

For the design of primers and probes, different software programs are available. The most commonly used are Primer Express (Applied Biosystems) and the LightCycler Probe Design (Roche), although several useful web sites assisting in the design of primers and probes have also appeared recently. As in classical PCR reactions, the primers should be free of direct and inverted repeats, as well as of homopolymeric runs. Furthermore, primers should not be able to form inter- and intramolecular dimers. Special attention should be given to the 3′ sequence: This sequence should not form dimers or hairpins and the binding should not be too strong in order to prevent nonspecific extension. Particularly for real-time PCR detection, the amplicon length should be very short, with the common rule being "the shorter the better" (ideally 50–150 base pairs). For the design of probes (in our case TaqMan probes), a few important specifications have to be taken into account. Probes should have a melting temperature about 10°C higher than the primers in order to anneal to the target sequence during the extension phase of the PCR reaction (which is usually performed at 60°C). Furthermore, probes should not contain a guanosine at their 5′ end, and should have more cytidines than guanosines.

When designing primers/probes for the quantification of target mRNA expression, special care has to be taken to avoid coamplification of contaminating genomic DNA. An efficient way to do so is by designing primer/probe combinations on exon-intron boundaries. Sometimes this strategy is not possible, for example, in the case of an intron-less gene or in case the genomic DNA sequence is not (yet) available. In these cases one is obliged to perform a DNase treatment on the RNA samples. Different publications however have reported problems with the use of DNase-treated samples. Many researchers are indeed afraid that residual RNases will degrade their precious RNA samples or affect its long-term storage. Moreover, contradictory to what the suppliers claim, DNase treatment will not always result in complete removal of DNA (Bustin, 2002).

Since primer/probe design and experimental validation is time-consuming, a free, publicly available RTPrimerDB, a real-time PCR primer, and probe database, which provides validated real-time PCR primer and probe sequences, has been developed. At present, it contains useful user-based validated primer/probe sequences for human, mouse, rat, fruit fly, and zebrafish, using all currently available detection chemistries. These real-time PCR primer/probe records are available at http://medgen.ugent.be/rtprimerdb/ (Pattyn *et al.*, 2003).

10.8.2 Gene Expression Normalization

A reliable quantitative real-time RT-PCR method requires correction for experimental variations in individual reverse transcriptase and PCR efficiencies. Indeed, differences in efficiency of the reverse transcriptase reaction will result in an amount of cDNA that does not correspond to the starting amount of RNA. Furthermore, because of the exponential nature of the PCR reaction, minor differences in PCR amplification efficiency will result in major differences in the PCR product.

Currently the most widely applied method to correct for these variations is normalization to a housekeeping gene. In theory, an ideal housekeeping gene should be expressed at a constant level among different tissues of an organism, at all stages of development, and should not be affected by the experimental treatment itself. In practice, however, finding a gene with these characteristics is an almost impossible task. Indeed, the expression of a housekeeping gene can also be regulated by the experimental treatment or can be tissue-dependent. It is therefore important that prior to performing each specific set of experiments, a suitable housekeeping gene is chosen, with the most reliable and reproducible results. The most commonly used housekeeping genes are β-actin, glyceraldehyde-3-phosphate-dehydrogenase, rRNA, and hypoxanthine guanine phosphoribosyl transferase. Other genes, such as cyclophilin, mitochondrial ATP synthase 6, and porphobilinogen deaminase, also are used.

10.8.3 Real-Time PCR and Laser Capture Microdissection

The method of normalization is rather straightforward when the RNA samples originate from pure cell populations or cell lines. However, in case RNA extractions and real-time PCR have to be carried out on whole tissue or tissue biopsies, an additional problem arises when normalizing the results to a housekeeping gene. In the field of immunology for instance, some experimental treatments can change the size of an organ, for example, by inducing splenomegaly, rendering the tissue unsuitable for comparison to a normal control tissue. Similarly, in transplanted or auto-immune attacked organs, the diseased organ will have significantly larger immune cell infiltrates compared to a normal organ. Other treatments, such as irradiation, can change the expression level of several housekeeping genes, again making normalization very difficult. Also in cancer research the presence of normal cells in the immediate surrounding of cancer cells is a major difficulty when normalizing to a housekeeping gene. Furthermore, starting from whole tissues or tissue biopsies in these cases will result in the averaging of the expression of different cell types, and the expression profile of a specific cell type will be masked because of the bulk of surrounding cells. In addition, no information will be obtained as to which cell type is the actual producer of the measured target RNA.

An elegant approach to circumvent this problem of differences in the cell populations itself, which has received increased attention during the last couple of years, involves the combination of real-time PCR with laser capture microdissection (LCM). The technique of LCM was described for the first time in 1996 (Emmert-Buck *et al.*, 1996). It is a simple, reliable, and rapid technique, which allows the extraction of pure subpopulations of cells from heterogeneous tissue samples, with retention of cell morphology. This is performed by directing a brief laser pulse at cells within a tissue section placed on a glass slide. In this way, individual cells or groups of cells can be selected, without damaging the cells. Therefore, gene expression patterns, both at the RNA and at the protein level are maintained, and samples are suitable for molecular analysis. The combination of LCM and real-time PCR is increasingly being used, especially in the genetic analysis of cancer cells (Elkahloun *et al.*, 2002; de Preter *et al.*, 2003). This technique also has been validated in other areas, for instance, for the analysis of gene expression upon *Mycobacterium tuberculosis* infection (Zhu *et al.*, 2003) or in atherosclerotic lesions (Trogan *et al.*, 2002).

10.9 CONCLUSIONS

The introduction of the real-time PCR technology has revolutionarily simplified the quantification of DNA and RNA. This has had a great impact in the field of molecular research and diagnostics, since enormous amounts of data can be obtained within a very short research time. The decreased costs for the thermal cyclers as well as for the necessary reagents for applying the technique have aided in its rapid increase in use. Indeed, it is clear from a Medline search for the keyword "real-time PCR" that its use increased exponentially over the last couple of years (see Fig. 10.8). Therefore, it seems that this technique is becoming the gold standard for the detection and quantification of DNA and RNA in the general research or diagnostic laboratory.

Although real-time PCR assays by themselves are characterized by high precision and reproducibility, the accuracy of the data obtained are largely dependent on several other factors. In order to design and analyze experiments using real-time PCR it is not sufficient to simply extend one's knowledge of classic end-point PCR. Indeed, many other controls are needed to be certain of the accuracy of the results when using real-time PCR assays. These factors, such as sample preparation, quality of the standard, choice of a housekeeping gene, and normalization of samples, need careful consideration and optimization. Furthermore, it is important that standardized criteria and international uniformity in experimental design and data analysis is reached, in order to be able to compare data between different laboratories.

New developments, such as the combination of real-time PCR with other sophisticated techniques such as LCM, make it possible to measure gene expression or DNA copy numbers in specific cell types from *in vivo* samples, which were previously very difficult to analyze. These make the possible applications of real-time PCR even more attractive for the future.

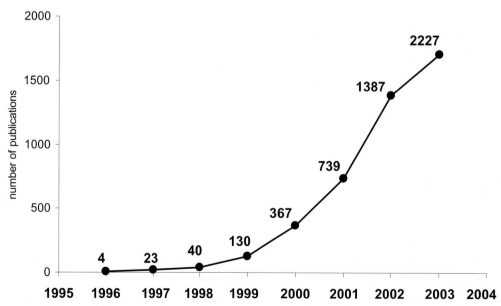

FIGURE 10.8 Real-time PCR publications. The keywords "real time" and "PCR" were typed in the PubMed search engine (http://www.ncbi.nlm.nih.gov/entrez/query.fcgi), the results were crossed with the years between 1996 (year of the first publication in real-time PCR) and 2003. The number of citations by year is shown. It is clear that the number of publications using the real-time PCR technique is increasing exponentially.

References

Bailey, J. S., and Cox, N. A. (1992). Universal pre-enrichment broth for the simultaneous detection of Salmonella and Listeria in foods. *J. Food Prot.* 55, 256–259.

Bernard, P. S., and Wittwer, C. T. (2000). Homogeneous amplification and variant detection by fluorescent hybridization probes. *Clin. Chem.* 46, 147–148.

Bhagwat, A. A. (2003). Simultaneous detection of *Escherichia coli* O157:H7, *Listeria monocytogenes* and *Salmonella* strains by real-time PCR. *Int. J. Food Microbiol.* 84, 217–224.

Borish, L. C., and Steinke, J. W. (2003). Cytokines and chemokines. *J. Allergy Clin. Immunol.* 111, S460–475.

Brechtuehl, K., and Whalley, S. (2001). A rapid real-time quantitative polymerase chain reaction for hepatitis B virus. *J. Virol. Methods* 93, 105–113.

Bultmann, B. D., Klingel, K., Sotlar, K., Bock, C. T., and Kandolf, R. (2003). Parvovirus B19: a pathogen responsible for more than heatologic disorders. *Virchows Arch.* 442, 8–17.

Bustin, S. A. (2000). Absolute quantification of mRNA using real-time reverse transcriptase polymerase chain reaction assays. *J. Mol. Endocrinol.* 25, 169–193.

Bustin, S. A. (2002). Quantification of mRNA using real-time reverse transcription PCR (RT-PCR): trends and problems. *J. Mol. Endocrinol.* 29, 23–29.

Cardullo, R. A., Agrawal, S., Flores, C., Zamecnick, P. C., and Wolf, D. E. (1988). Detection of nucleic acid hybridization by nonradioactive fluorescence resonance energy transfer. *Proc. Natl. Acad. Sci. USA* 85, 8790–8794.

Clementi, M., Bagnarelli, P., Manzin, A., and Menzo, S. (1994). Competitive polymerase chain reaction and analysis of viral activity at the molecular level. *Genet. Anal. Tech. Appl.* 11, 1–6.

Crystal, R. G. (1995). Transfer of genes to humans: early lessons and obstacles to success. *Science* 270, 404–410.

De Preter, K., Vandesompele, J., Heimann, P., Kockx, M. M., Van Gele, M., Hoebeeck, J., De Smet, E., Demarche, M., Laueys, G., Van Roy, N., De Paepe, A., and Speleman, F. (2003). Application of laser capture microdissection in genetic analysis of neuroblastoma and neuroblastoma precursor cells. *Cancer Lett.* 197, 53–61.

Donovan, J. W., Ladetto, M., Zou, G., Neuberg, D., Poor, C., Bowers, D., and Gribben, J. G. (2000). Immunoglobulin heavy-chain consensus probes for real-time PCR quantification of residual disease in acute lymphoblastic leukemia. *Blood* 15, 2651–2658.

Elenitoba-Johnson, K. S. J., Bohling, S. D., Wittwer, C. T., and King, T. C. (2001). Multiplex PCR by multicolor fluorimetry and fluorescence melting curve analysis. *Nat. Med.* 7, 249–253.

Elkahloun, A. G., Gaudet, J., Robinson, G. S., and Sgroi, D. C. (2002). In situ gene expression analysis of cancer using laser capture microdissection, microarrays and real time quantitative PCR. *Cancer Biol. Ther.* 1, 354–358.

Ellis, J. S., and Zambon, M. C. (2002). Molecular diagnosis of influenza. *Rev. Med. Virol.* 12, 375–389.

Emmert-Buck, M. R., Bonner, R. F., Smith, P. D., Chuaqui, R. F., Zhuang, Z., Goldstein, S. R., Weiss, R. A., and Liotta, L. A. (1996). Laser capture microdissection. *Science* 274, 998–1001.

Gerard, C. J., Olsson, K., Ramanathan, R., Reading, C., and Hanania, E. G. (1998). Improved quantitation of minimal resid-ual disease in multiple myeloma using real-time polymerase chain reaction and plasmid-DNA complementarity determining region III standards. *Cancer Res.* 58, 3957–3964.

Gibson, U. E. M., Heid, C. A., and Williams, P. M. (1996). A Novel Method for Real-time Quantitative RT-PCR. *Genome Res.* 6, 995–1001.

Giulietti, A., Overbergh, L., Valckx, D., Decallonne, B., Bouillon, R., and Mathieu, C. (2001). An overview of real-time quantitative PCR: applications to quantify cytokine gene expression. *Methods* 25, 386–401.

Gonzalez, D., Gonzalez, M., Alonso, M. E., Lopez-Perez, R., Balanzategui, A., Chillon, M. C., Silva, M., Garcia-Sanz, R., and San Miguel, J. F. (2003). Incomplete DJH rearrangements as a novel tumor target for minimal residual disease quantitation in multiple myeloma using real-time PCR. *Leukemia* 17, 1051–1057.

Guyer, R. L., and Koshland, Jr. (1989). The Molecule of the Year. *Science* 246, 1543–1546.

Gysemans, C. A., Waer, M., Valckx, D., Laureys, J. M., Mihkalsky, D., Bouillon, R., and Mathieu, C. (2000). Early graft failure of xenogeneic islets in NOD mice is accompanied by high levels of interleukin-1 and low levels of transforming growth factor beta mRNA in the grafts. *Diabetes* 49, 1992–1997.

Gysemans, C. A., Stoffels, K., Giulietti, A., Overbergh, L., Waer, M., Lannoo, M., Feige, U., and Mathieu, C. (2003). Prevention of primary non-function of islet xenografts in autoimmune diabetic NOD mice by anti-inflammatory agents. *Diabetologia* 46, 1115–1123.

Hackett, N. R., El Sawy, T., Lee, L. Y., Silva, I., O'Leary, J., Rosengart, T. K., and Crystal, R. G. (2000). Use of Quantitative TaqMan Real-Time PCR to track the time-dependent distribution of gene transfer vectors in vivo. *Mol. Ther.* 2, 649–656.

Heid, C. A., Stevens, J., Livak, K. J., and Williams, P. M. (1996). Real-time Quantitative PCR. *Genome Res.* 6, 986–994.

Heller, M. J., and Morrison, L. E. (1985). In D. T. Kingsbury and S. Z. Falkow, eds., *Rapid Detection and Identification of Infectious Agents.* pp. 245–256. Academic Press, New York.

Higuchi, R., Fockler, C., Dollinger, G., and Watson, R. (1993). Kinetic PCR analysis: Real-time monitoring of DNA amplification reactions. *Biotechnology* 11, 1026–1030.

Holland, P. M., Abramson, R. D., Watson, R., and Geldfand, D. H. (1991). Detection of specific polymerase chain reaction product by utilizing the 5′ to 3′ exonuclease activity of *Thermus aquaticus*. *Proc. Natl. Acad. Sci. USA* 88, 7276–7280.

Hubbard, R. A. (2003). Human papillomavirus testing methods. *Arch. Pathol. Lab. Med.* 127, 940–945.

Imhof, A., Schaer, C., Schoedon, G., Schaer, D. J., Walter, R. B., Schaffner, A., and Schneemann, M. (2003). Rapid detection of pathogenic fungi from clinical specimens using LightCycler real-time fluorescence PCR. *Eur. J. Clin. Microbiol. Infect. Dis.* 22, 558–560.

Jaeger, U., and Kainz, B. (2003). Monitoring minimal residual disease in AML: the right time for real-time. *Ann. Hematol.* 82, 139–147.

Johnson, G. C., and Todd, J. A. (2000). Strategies in complex disease mapping. *Curr. Opin. Genet. Dev.* 10, 330–334.

Jones, C. D., Yeung, C., and Zehnder, J. L. (2003). Comprehensive validation of a real-time quantitative bcr-abl assay for clinical laboratory use. *Am. J. Clin. Pathol.* 120, 42–48.

Kunkel, S. L., and Godessart, N. (2002). Chemokines in autoimmunity: from pathology to therapeutics. *Autoimmun. Rev.* 1, 313–320.

Lascols, C., Lamarque, D., Costa, J. M., Copie-Bergman, C., Le Glaunec, J. M., Deforges, L., Soussy, C. J., Petit, J. C., Delchier, J. C., and Tankovic, J. (2003). Fast and accurate quantitative detection of Helicobacter pylori and identification of Clarithromycin resistance mutations in H. pylori isolates from gastric biopsy specimens by real-time PCR. *J. Clin. Microbiol.* 41, 4573–4577.

Livak, K. J. (1999). Allelic discrimination using fluorogenic probes and the 5′ nuclease assay. *Genet. Anal.* 14, 143–149.

Livak, K. J., and Schmittgen, T. D. (2001). Analysis of relative gene expression data using real-time quantitative PCR and the $2^{-\Delta\Delta Ct}$ method. *Methods* 25, 402–408.

Makino, S., and Cheun, H. I. (2003). Application of the real-time PCR for the detection of airborne microbial pathogens in reference to the anthrax spores. *J. Microbiol. Methods* 53, 141–147.

Marcucci, G., Caligiuri, M. A., Dohner, H., Archer, K. J., Schlenk, R. F., Dohner, K., Maghraby, E. A., and Bloomfield, C. D. (2001). Quantification of CBFbeta/MYH11 fusion transcript by real-time RT-PCR in patients with INV(16) acute myeloid leukemia. *Leukemia* 15, 1072–1080.

Miller, N., Cleary, T., Kraus, G., Young, A. K., Spruill, G., and Hnatyszyn, H. J. (2002). Rapid and specific detection of Mycobacterium tuberculosis from acid-fast bacillus smear-positive respiratory specimens and BacT/ALERT MP culture bottles by using fluorogenic probes and real-time PCR. *J. Clin. Microbiol.* 40, 4143–4147.

Mullis, K., Faloona, F., Scharf, S., Saiki, R., Horn, G., and Erlich, H. (1986). Specific enzymatic amplification of DNA in vitro: the polymerase chain reaction. *Cold Spring Harb. Symp. Quant. Biol.* 51, 263–273.

Nitsche, A., Steuer, N., Schmidt, C. A., Landt, O., and Siegert, W. (1999). Different real-time PCR formats compared for the quantitative detection of human cytomegalovirus DNA. *Clin. Chem.* 45, 1932–1937.

Norton, D. M. (2002). Polymerase chain reaction-based methods for detection of Listeria monocytogenes: toward real-time screening for food and environmental samples. *J. AOAC. Int.* 85, 505–515.

Ohga, S., and Kubo, E. N. A. (2001). Quantitative monitoring of circulating Epstein-Barr virus DNA for predicting the development of post-transplantation lymphoproliferative disease. *Int. J. Hematol.* 73, 323–326.

Oliver, D. H., Thompson, R. E., Griffin, C. A., and Eshleman, J. R. (2000). Use of single nucleotide polymorphisms (SNP) and real-time polymerase chain reaction for bone marrow engraftment analysis. *J. Mol. Diagn.* 2, 202–208.

Osumi, K., Fukui, T., Kiyoi, H., Kasai, M., Kodera, Y., Kudo, K., Kato, K., Matsuyama, T., Nait, K., Tanimoto, M., Hirai, H., Saito, H., Ohno, R., and Naoe, T. (2002). Rapid screening of leukemia fusion transcripts in acute leukemia by real-time PCR. *Leuk. Lymphoma* 43, 2291–2299.

Overbergh, L., Decallonne, B., Waer, M., Rutgeerts, O., Valckx, D., Casteels, K. M., Laureys, J., Bouillon, R., and Mathieu, C. (2000). 1alpha,25-dihydroxyvitamin D3 induces an autoantigen-specific T-helper1/T-helper2 immune shift in NOD mice immunized with GAD65 (p524–543). *Diabetes* 49, 1301–1307.

Overbergh, L., Decallonne, B., Branisteanu, D. D., Valckx, D., Kasran, A., Bouillon, R., and Mathieu, C. (2003a). Acute shock induced by antigen vaccination in NOD mice. *Diabetes* 52, 335–341.

Overbergh, L., Giulietti, A., Valckx, D., Decallonne, B., and Mathieu, C. (2003b). The use of real-time reverse transcriptase PCR for the quantification of cytokine gene expression. *J. Biomol. Tech.* 14, 33–43.

Pattyn, F., Speleman, F., De Paepe, A., and Vandesompele, J. (2003). RTPrimerDB: the real-time PCR primer and probe database. *Nucleic Acids Res.* 31, 122–123.

Pongers-Willemse, M. J., Verhagen, O. J., Tibbe, G. J., Wijkhuijs, A. J., de Haas, V., Roovers, E., van der Schoot, C. E., and van Dongen, J. J. (1998). Real-time quantitative PCR for the detection of minimal residual disease in acute lymphoblastic leukemia using junctional region specific TaqMan probes. *Leukemia* 12, 2006–2014.

Rabinovitch, A. (2003). Immunoregulation by cytokines in autoimmune diabetes. *Adv. Exp. Med. Biol.* 520, 159–193.

Rasmussen, T., Poulsen, T. S., Honore, T. S., and Johnson, H. E. (2000). Quantification of minimal residual disease in multiple myeloma using an allele-specific real-time PCR assay. *Exp. Hematol.* 28, 1039–1045.

Rijpens, N. P., and Herman, L. M. (2002). Molecular methods for identification and detection of bacterial food pathogens. *J. AOAC. Int.* 85, 984–995.

Ririe, K. M., Rasmussen, R. P., and Wittwer, C. T. (1997). Product differentiation by analysis of DNA melting curves during the polymerase chain reaction. *Anal. Biochem.* 245, 154–160.

Risch, N. J. (2000). Searching for genetic determinants in the new millennium. *Nature* 405, 847–856.

Rondini, S., Mensah-Quainoo, E., Troll, H., Bodmer, T., and Pluschke, G. (2003). development and application of real-time PCR assay for quantification of *Mycobacterium ulcerans* DNA. *J. Clin. Microbiol.* 41, 4231–4237.

Rowley, J. D. (1998). The critical role of chromosome translocations in human leukemias. *Annu. Rev. Genet.* 32, 495–519.

Rudi, K., Nogva, H. K., Moen, B., Nissen, H., Bredholt, S., Moretro, T., Naterstad, K., and Holck, A. (2002). Development and application of new nucleic acid-based technologies for microbial community analyses in foods. *Int. J. Food Microbiol.* 78, 171–180.

Scholl, C., Breitinger, H., Schlenk, R. F., Döhner, H., Fröhling, S., and Döhner, K. (2003). Development of a real-time RT-PCR assay for the quantification of the most frequent MLL/AF9 fusion types resulting from translocation t(9;11)(p22;q23) in acute myeloid leukemia. *Genes, Chromosomes Cancer* 38, 274–280.

Senoo, M., Matsubara, Y., Fujii, K., Nagasaki, Y., Hiratsuka, M., Kure, S., Uehara, S., Okamura, K., Yajima, A., and Narisawa, K. (2000). Adenovirus-mediated *in utero* gene transfer in mice and guinea pigs: tissue distribution of recombinant andenovirus determined by quantitative TaqMan-polymerase chain reaction assay. *Mol. Genet. Metab.* 69, 269–276.

Stryer, L. (1978). Fluorescence energy transfer as a spectropic ruler. *Annu. Rev. Biochem.* 47, 819–846.

Summers, K. E., Goff, L. K., Wilson, A. G., Gupta, R. K., Lister, T. A., Fitzgibbon, J. (2001). Frequency of the Bcl-2/IgH rearrangement in normal individuals: implications for the monitoring of disease in patients with follicular lymphoma. *J. Clin. Oncol.* 19, 420–424.

Tanaka, N., and Kimura, H. (2000). Quantitative analysis of cytomegalovirus load using a real-time PCR assay. *J. Med. Virol.* 60, 455–462.

Thelwell, N., Millington, S., Solinas, A., Booth, J., and Brown, T. (2000). Mode of action and application of Scorpion primers to mutation detection. *Nucleic Acids Res.* 28, 3752–3761.

Trogan, E., Choudhury, R. P., Dansky, H. M., Rong, J. X., Breslow, J. L., and Fisher, E. A. (2002). Laser capture microdissection analysis of gene expression in macrophages from atherosclerotic lesions of apolipoprotein E-deficient mice. *Proc. Natl. Acad. Sci. USA* 99, 2234–2239.

Tyagi, S., and Kramer, F. R. (1996). Molecular beacons: probes that fluoresce upon hybridization. *Nat. Biotechnol.* 14, 303–308.

Tyagi, S., Bratu, D. P., and Kramer, F. R. (1998). Multicolor molecular beacons for allele discrimination. *Nat. Biotechnol.* 16, 49–53.

Van der Velden, V. H. J., Hochhaus, A., Cazzaniga, G., Szczepanski, T., Gabert, J., and van Dongen, J. J. M. (2003). Detection of minimal residual disease in hematological malignancies by real-time quantitative PCR: principles, approaches, and laboratory aspects. *Leukemia* 17, 1013–1034.

Van Etten, E., Branisteanu, D. D., Overbergh, L., Bouillon, R., Verstuyf, A., and Mathieu C. (2003). Combination of a 1,25-dihydroxyvitamin D(3) analog and a bisphosphonate prevents experimental autoimmune encephalitis and preserves bone. *Bone* 32, 397–404.

Van Haeften, R., Palladino, S., Kay, I., Keil, T., Heath, C., and Waterer, G. W. (2003). A quantitative LightCycler PCR to detect Streptococcus pneumoniae in blood and CSF. *Diagn. Microbiol. Infect. Dis.* 47, 407–414.

von Ahsen, N., Oellerich, M., and Schütz, E. (2000). Use of two reporter dyes without interference in a single-tube rapid-cycle PCR: alpha1-Antitrypsin genotyping by multiplex real-time fluorescence PCR with the LightCycler. *Clin. Chem.* 46, 156–161.

Whitcombe, D., Theaker, J., Guy, S. P., Brown, T., and Little, S. (1999). Detection of PCR products using self-probing amplicons and fluorescence. *Nat. Biotechnol.* 17, 804–807.

White, P. L., Shetty, A., and Barnes, R. A. (2003). Detection of seven *Candida* species using the Light-Cycler system. *J. Med. Microbiol.* 52, 229–238.

Wittwer, C. T., Herrmann, M. G., Cameron, N. G., and Elenitoba-Johnson, K. S. (2001). Real-time Multiplex PCR assays. *Methods* 25, 430–442.

Yashima, A., Maesawa, C., Uchiyama, M., Tarusawa, M., Satoh, T., Satoh, M., Enomoto, S., Sugawara, K., Numaoka, H., Murai, K., Utsugisawa, T., Ishida, Y., and Masuda, T. (2003). Quantitative assessment of contaminating tumor cells in autologous peripheral blood stem cells of B-cell non-Hodgkin lymphomas using immunoglobulin heavy chain gene allele-specific oligonucleotide real-time quantitative polymerase chain reaction. *Leukemia Res.* 27, 925–934.

Zhu, G., Xiao, H., Mohan, V. P., Tanaka, K., Tyagi, S., Tsen, F., Salgame, P., and Chan, J. (2003). Gene expression in the tuberculous granuloma: analysis by laser capture microdissection and real-time PCR. *Cell. Microbiol.* 5, 445–453.

CHAPTER **11**

Pyrosequencing

JOHANNA K. WOLFORD AND KIMBERLY A. YEATTS

Genetic Basis of Human Disease Division, Translational Genomics Research Institute, 445 North Fifth Street, Suite 1600, Phoenix, AZ 85004, USA

11.1 INTRODUCTION

Pyrosequencing is a real-time sequencing method that uses light release as the detection signal for nucleotide incorporation into a target DNA strand (Ronaghi *et al.*, 1998). In this method, a double-stranded DNA product is synthesized from single-stranded DNA through the iterative addition of deoxynucleotide triphosphates complementary to the DNA template. Generation of inorganic pyrophosphate during the DNA synthesis reaction forms the basis for this technique. Through a cascade of orchestrated enzymatic reactions (see Fig. 11.1), inorganic pyrophosphate is converted to adenosine triphosphate, which subsequently catalyzes the oxidation of luciferin by luciferase and results in the generation of visible light (Ronaghi *et al.*, 1998). The resulting light signal automatically is converted into a Pyrogram™ (see Fig. 11.2), allowing direct assessment of DNA sequence. The emitted light signal is proportional to the number of nucleotides incorporated into the extending primer, which is generated and visualized in real time.

11.2 TECHNOLOGY OVERVIEW

11.2.1 Template Preparation

Following polymerase chain reaction (PCR) amplification of a target DNA sequence, unincorporated nucleotides and amplification primers must be removed from the PCR product, because these components interfere with the pyrosequencing process. Two strategies are available for template preparation prior to pyrosequencing: a solid phase or an enzyme-based approach.

Solid phase template preparation requires capture of biotinylated PCR product for subsequent generation of single-stranded template. Typically, two target-specific primers, one of which is biotin-labeled, are used for DNA amplification, resulting in an amplicon with one biotinylated strand (see Fig. 11.3a). An alternative approach utilizes a universal biotinylated tag with a specific handle sequence.

In this case, two unlabeled target-specific primers are designed, one of which retains a universal handle sequence complementary to the universal biotinylated primer (Fakhrai-Rad *et al.*, 2002; Pacey-Miller and Henry, 2003). In the amplification reaction, the universal biotinylated primer and the target-specific primer without the universal handle sequence are used in equimolar concentrations, and the target-specific primer retaining the handle is present at low concentration (see Fig.11.3b). Amplification of DNA template using the three primers results in a biotinylated PCR product, which is processed according to the appropriate solid phase template procedure (discussed later). The advantage of using a universal primer is that target-specific primers can be generated quickly without the cost and delay normally associated with the synthesis of biotinylated primers.

Following DNA amplification, streptavidin-coated magnetic or sepharose beads are used to isolate the biotinylated PCR product. Both magnetic and sepharose beads operate under the same principle (i.e., capture of biotinylated PCR

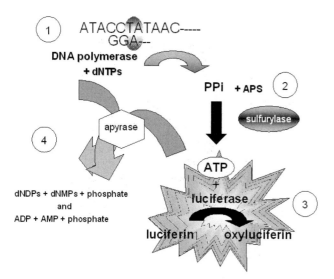

FIGURE 11.1 Principle of pyrosequencing. Step 1: in the presence of enzyme (exonuclease-deficient Klenow DNA polymerase, apyrase, luciferase, recombinant ATP sulfurylase) and substrate (magnesium acetate, bovine serum albumin, dithiothreitol, adenosine 5′-phosphosulfate or APS, polyvinylpyrrolidine, and D-luciferin) mixtures, a sequencing primer is first annealed to a PCR-generated DNA template and is extended with the addition of the complementary deoxynucleotide triphosphate (dNTP) by the action of DNA polymerase. With each nucleotide incorporated, inorganic pyrophosphate (PPi) is released in an equimolar quantity. Step 2: in the presence of APS, ATP sulfurylase converts PPi to ATP. Step 3: the ATP generated in Step 2 is used by luciferase to catalyze the conversion of luciferin to oxyluciferin, which emits light in an amount proportional to the amount of ATP utilized. Step 4: Addition of dNTPs is performed in single steps, one at a time. During the polymerization reaction, apyrase functions to degrade unincorporated dNTPs that could interfere with the next dNTP addition. Apyrase also degrades excess ATP generated from ATP sulfurylase action.

product), and the methods for template preparation using each bead type differ only in the type of tool utilized for sample transfer. The magnetic tool and the vacuum prep tool are pyrosequencing companion products for the transfer of magnetic and sepharose beads, respectively. In both approaches, biotinylated PCR product is immobilized by incubation with beads in an equal volume of 2× binding-washing buffer (10 mM Tris-HCl, 2 M NaCl, 1 mM EDTA, 0.1% Tween-20) at 65°C for 15–30 min with constant agitation. Immobilization on magnetic or sepharose beads allows transfer through subsequent denaturing, washing, and annealing steps using the appropriate tool. First, incubation with NaOH produces single-stranded DNA, which is then washed with annealing buffer (20 mM Tris-Acetate, 5 mM MgAc₂, pH 7.6). Next the beads are transferred to annealing buffer containing sequencing primer that will eventually anneal to the template and be extended by polymerization during the pyrosequencing reaction. At this point, the mixture containing single-stranded template attached to

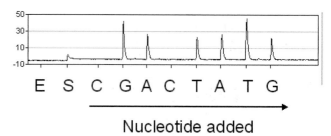

FIGURE 11.2 Sample pyrosequencing pyrogram. The light released in the luciferase-catalyzed step has a wavelength of 560 nm and can be detected by a charge-coupled device (CCD) camera, located within the PSQ instrument. As DNA synthesis progresses, the complementary strand is extended. Following the pyrosequencing reaction, the software produces peaks on a graph called a pyrogram, which is proportional to the amount of dNTP incorporated during synthesis and corresponds to the extended strand product. In this figure, the target sequence is GGATATTG (note that the heights of the first G and the second T residues are approximately double those of the second G and first T residues). In determining the nucleotide dispensation pattern, the pyrosequencing software automatically adds controls to assess template quality. In this pyrogram, the two cytosine residues represent such controls. If the DNA template had been contaminated, peaks at these residues may have been observed. E: enzyme; S: substrate.

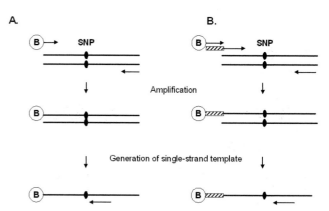

FIGURE 11.3 Amplification of DNA template for solid phase preparation. A. Using two primers: One biotinylated and one unlabeled primer are used to generate DNA template in which one of the strands carries a biotin molecule (represented by encircled B). B. Using three primers: Two unlabeled, target-specific primers, one of which contains a sequence tag (striated box) complementary to the biotinylated tag, are used in conjunction with the universal biotinylated tag to generate DNA template. In both A and B, a complementary sequencing primer anneals to the biotin-labeled strand to initiate DNA polymerization.

beads and the sequencing primer is heated briefly. After bringing the reaction components back to room temperature, the product is ready for pyrosequencing.

Although these methods generate a single-stranded DNA template for pyrosequencing, a protocol using an enzymatically purified double-stranded template has also been described (Nordstrom *et al.*, 2000a). In this method, PCR product is first incubated at room temperature with apyrase (to degrade nucleotides) and exonuclease I (to degrade oligonucleotide primers). The enzymes are subsequently heat-inactivated, sequencing primer is annealed, and the target DNA is pyrosequenced. It is also possible to genotype single nucleotide polymorphisms (SNPs) with pyrosequencing directly from PCR product without any enzymatic treatment, although sequences obtained using this method appear to be of lower quality (Nordstrom *et al.*, 2000b; Nordstrom *et al.*, 2002; Pacey-Miller and Henry, 2003). Further optimization is likely necessary before this approach can become widely utilized.

Although rapid, inexpensive, and convenient, enzymatic template preparation frequently suffers from background signal, which sometimes affects data quality. This can be ameliorated by incubating the PCR product with blocking oligonucleotides (either oligonucleotides complementary to the amplification primers or oligonucleotides with sequences identical to the amplification primers, but with modified 3′-ends), in addition to sequencing primer, apyrase, and inorganic pyrophosphatase, which degrades excess PPi (Nordstrom *et al.*, 2002). The complementary oligonucleotides prevent reannealing by hybridizing to the amplification primers and the modified same-sequence oligonucleotides prevent reannealing by competing for the same binding site on the template DNA, but because of the 3′-end modification, are unable to be extended by DNA polymerase. Inclusion of such blocking oligonucleotides therefore reduces background signal and helps to generate clean, unambiguous pyrograms.

11.2.2 Limitations of Pyrosequencing

Pyrosequencing is limited by the nonlinear light response following incorporation of more than five nucleotides (Ronaghi *et al.*, 1998). This can be problematic for accurate assessment of homopolymeric regions. Also, in SNP discovery applications, it is difficult to detect an insertion/deletion polymorphism, particularly if it shifts the sequence out of phase. This problem can be circumvented if there is prior knowledge of the existence and location of the polymorphism, in which case, a specific nucleotide dispensation can be programmed to maintain synchronization of extended products resulting from different alleles (Fakhrai-Rad *et al.*, 2002).

11.2.3 Available Pyrosequencing Systems and Software

Pyrosequencing AB currently manufactures several systems including: PSQ™96, PSQ™96MA, PSQ™HS 96A, and a customized 384-well system (McNeely, 2003). The PSQ™96 is the basic model and analyzes up to 96 samples per Pyrosequencing run. The PSQ™96MA is similar to the PSQ™96, but differs by providing multiplex analysis.

The PSQ™HS 96A is a fully automated version of PSQ™96MA.

Each instrument system utilizes a disposable inkjet cartridge containing valves that pneumatically inject enzyme, substrate, and nucleotides into a 96-well sample plate. The system dispenses enzyme and substrate first, and then distributes nucleotides into alternating wells using a pulse delay to minimize cross-talk interference between neighboring wells. When a nucleotide is incorporated into the growing strand of DNA, the light generated as a result of luciferin oxidation is detected by a high sensitivity charge-coupled device (CCD) camera, which images the plate every second. Once a run has ended, the evaluation program assigns genotypes. Measurements of signal-to-noise ratios, variance in peak heights, peak width, and agreement between expected and observed SNP flanking sequence are utilized to assess data quality.

Pyrosequencing AB offers software programs for SNP analysis (SNP software), allele frequency determination (SNP AQ software), and tag sequencing (SQA software). The SNP analysis program consists of three components including

1. SNP entry, which enables SNP sequence input for assignment of nucleotide dispensation order.
2. Instrument control, for running the pyrosequencing unit.
3. Evaluation module, which automatically converts the raw data obtained in each pyrosequencing reaction into genotypes and performs quality assessment of raw data (see Fig. 11.4).

The PSQ96MA SNP software has the additional ability to analyze insertion/deletion mutations and multiplexing reactions, and provides more standard operating options. The PSQ96MA AQ software contains a novel algorithm that automatically eliminates background noise and increases the accurate analysis of all samples.

11.2.4 Recent Developments in Technology

11.2.4.1 GENE EXPRESSION PROFILING

Signature tags are short DNA sequences (1–50 bp) that can be used for gene identification. A pyrosequencing technique has been developed that creates signature tags from isolation of 3′-cDNA ends (Agaton *et al.*, 2002). In this procedure, formation of tags is initiated by mRNA isolation from the cells of interest. Subsequently, cDNA is synthesized from mRNA template and fragmented with a frequently cutting restriction enzyme, such as *Dpn* II. Tags corresponding to the 3′-end of each cDNA are then isolated and cloned into restricted plasmids to create a cDNA library. Library clone inserts are sequenced by pyrosequencing to generate signature tags, which leads to either identification of known genes via homology searches with public sequence databases or discovery of new genes by extended sequencing of tags.

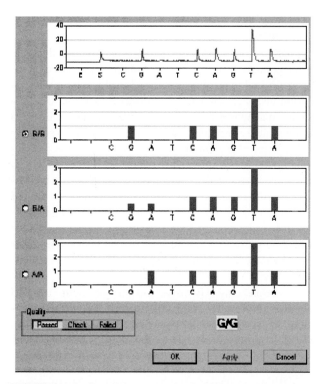

FIGURE 11.4 Sample pyrogram in the SNP evaluation module. In the SNP evaluation module, each sample SNP pyrogram can be viewed individually. At the top of the screen, the actual pyrogram is shown, below which are the patterns corresponding to the three possible genotypes. The sequence of this target DNA, with the SNP in brackets, is [G/A]CAGTTTA. The first C and T residues represent control nucleotide additions. In this example, the sample is a GG homozygote, which the SNP software has correctly called. In general, when the software calls a genotype with high confidence the result is shown in green and a passed quality designation is assigned. In some cases where the genotype may be ambiguous, the quality is flagged by a yellow-colored check or a red-colored failed and the researcher has the option of examining the raw data for the genotype in question.

The relative amount of each gene product can also be assessed using the raw data generated by pyrosequencing and a software program modeled on the Expression Profile Viewer (Larsson *et al.*, 2000), but customized for pyrosequencing (Agaton *et al.*, 2002). This method of sequencing signature tags has an advantage over other methods because it yields an average length of 25 bp, compared to 10 and 16–20 bp for serial analysis of gene expression (Velculescu *et al.*, 1995) and massively parallel signature sequencing (Brenner *et al.*, 2000), respectively. Unlike these methods, tag pyrosequencing does not rely on elaborate enzyme cleavage steps and can be evaluated in simple microtiter plate format (Agaton *et al.*, 2002).

11.2.4.2 cDNA Quantification
Pyrosequencing has also been applied to the quantification of PCR products through competitive reverse transcription

PCR (Soderback *et al.*, 2003). In this method, varying amounts of a reference oligonucleotide are added to cDNA samples and amplified using PCR. Single-stranded templates are subsequently pyrosequenced and relative peak heights plotted against the amount of reference oligonucleotide added to each cDNA sample. The point of overlap between the reference and target peaks in the pyrogram indicates the amount of target mRNA in the original cDNA sample.

11.3 APPLICATIONS OF PYROSEQUENCING

Pyrosequencing can be used in a variety of applications, including mutation detection, SNP genotyping, estimation of allele frequency, and sequencing of short sequences. In addition, pyrosequencing can be used in the analysis of gene expression, bacterial and viral typing, and mitochondrial DNA analysis. Each feature is discussed more fully in the following paragraphs.

The most common application of pyrosequencing is SNP genotyping. In this regard, pyrosequencing has distinct advantages and disadvantages relative to other SNP genotyping techniques. As a genotyping platform, pyrosequencing is accurate, allows greater flexibility in primer placement, is easily automated, and is a good choice to meet moderate throughput genotyping needs. The pyrosequencing reaction occurs in real-time and the raw data can be analyzed directly because each SNP variant yields a specific pattern. In addition, generation of sequence information, including those nucleotides flanking the SNP, is directly visualized, thus allowing accurate conclusions to be drawn about the data. Disadvantages of pyrosequencing include biotinylated primer costs, relatively labor-intensive post-PCR processing, and instrument cost.

11.3.1 SNP Genotyping

Genetic variation underlies human diversity and plays an important role in the etiology of human disease. SNPs have surpassed microsatellites as the markers of choice in gene mapping strategies due to the relative abundance of SNPs (i.e., approximately 1 SNP/kb), and the fact that SNPs frequently are located within genes and thus may directly affect gene expression or gene product (Kwok *et al.*, 1996). In addition, genotype analysis of SNPs is more amenable to automation, although at present, microsatellite markers remain the preferred choice for genome-wide linkage studies.

The technique of pyrosequencing was validated for SNP genotyping several years ago (Ahmadian *et al.*, 2000; Alderborn *et al.*, 2000). Since that time, pyrosequencing has been used extensively for SNP genotyping in both linkage disequilibrium mapping projects (e.g., Wolford *et al.*, 2001) and analyses of candidate genes. To date, SNPs have been geno-typed in biological candidate genes presumed to underlie different complex diseases including, for example, inter-leukin-6 (Vozarova *et al.*, 2003), interleukin 1b (Wieser *et al.*, 2003), phospholipase A2, group IVA (Wolford *et al.*, 2003a), and C-reactive protein (Wolford *et al.*, 2003b).

A method for a multiplex SNP genotyping assay using pyrosequencing enables simultaneous analysis of multiple SNPs (Pourmand *et al.*, 2002). This approach relies on amplification of multiple target DNA sequences using one biotinylated primer per region amplified. The biotinylated PCR product is then captured on streptavidin-coated beads and taken through the preparation process to obtain single-stranded DNA. Multiplex pyrosequencing can be performed using either one DNA template containing two (or more) SNPs or two (or more) distinct DNA templates containing one SNP each. In the first case, only one set of amplifica-tion primers is used and two (or more) sequencing primers are annealed for extension into each distinct SNP. In the second case, two or more different sets of amplification primers are used in a multiplex PCR amplification step, then sequencing primers are annealed to the complementary tem-plate sequence and extended into the appropriate SNP. In both cases, multiplex pyrosequencing is able to accurately call genotypes for all SNPs at each position.

Pyrosequencing can also be applied to the determination of haplotypes (Pettersson *et al.*, 2003). This approach com-bines allele-specific PCR with pyrosequencing analysis of multiple SNPs contained on the amplified DNA fragments. Haplotype determination using this approach was found to be highly reliable and effectively discriminated nonspecific allele amplification via incorporation of a 3′-end mismatch to improve specificity. This method is more accurate than estimation of haplotype frequencies by computational analy-sis of unphased genotypes in populations and is not as labor-intensive and costly as cloning approaches or construction of somatic cell hybrids. However, this application is relevant only for analysis of SNPs located on a single PCR template and thus is necessarily limited by the size of an amplicon that can be successfully amplified.

11.3.2 Allele Frequency Estimation

Linkage disequilibrium mapping is a potentially powerful approach for fine-scale localization of genetic variants con-ferring increased susceptibility to a complex disease. The recent identification of putative susceptibility genes for Crohn's disease (Hugot *et al.*, 2001; Ogura *et al.*, 2001) and type 2 diabetes mellitus (Horikawa *et al.*, 2000) are among the numerous examples that confirm the feasibility of this approach as a general strategy for positional cloning. The recent successes of the Human Genome Project and the SNP Consortium have resulted in the identification of vast numbers of SNPs, and development of the HapMap, which will identify key SNPs for maximum haplotype diversity in

several different ethnic groups, will undoubtedly facilitate such positional cloning efforts.

To maximize detection of linkage disequilibrium between bi-allelic markers and potential disease susceptibility alleles, a significant number of densely spaced SNPs must be geno-typed throughout the region of interest. However, genotyp-ing a large number of SNPs in sufficiently powered study groups is both costly and time-consuming. One approach to overcome the high costs associated with SNP genotyping in a large sample is to first screen strategically selected SNPs in pools of affected and unaffected individuals and deter-mine whether a significant allele frequency difference exists between the two groups. The principle of pooled DNA geno-typing is based on the assumption that the high-risk allele underlying the complex disease will be distributed differen-tially between groups of affected and unaffected individuals (see Fig. 11.5).

This strategy has been employed using a number of tech-nologic platforms, including mass spectrometry (Buetow *et al.*, 2001; Wolford *et al.*, 2001), kinetic PCR (Germer *et al.*, 2000), denaturing high performance liquid chromatography (Hoogendoorn *et al.*, 2000), the Invader assay (Ohnishi *et al.*, 2001), and the bioluminometric assay (Zhou *et al.*, 2001). Recently, several studies have investigated the per-formance of pyrosequencing in the estimation of allele fre-quency in DNA pools (Gruber *et al.*, 2002; Neve *et al.*, 2002; Wasson *et al.*, 2002). Because there is a strong correlation between peak height and allele frequency with this method, pyrosequencing lends itself well to estimation of allele fre-quency in DNA pools. As mentioned, the software program

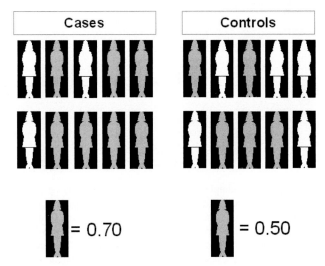

FIGURE 11.5 General principle for estimating allele frequency in DNA pools. If a variant allele is distributed differentially among affected and unaffected individuals (represented by the gray figures), then it may be possible to detect this difference by allele frequency estimation in pools of DNA comprising individuals from each group. In this example, the variant is present at a frequency of 0.70 in cases and 0.50 in controls.

for allele frequency quantification (SNPAQ) effectively determines relative allele frequencies based on raw peak height obtained from the AQ pyrogram (see Fig. 11.6).

In considering pooled DNA genotyping as an alternative to individual genotyping, the following questions needed to be addressed.

1. Is pooled DNA genotyping accurate?
2. Is pooled DNA genotyping sensitive enough to detect small differences in allele frequency between pools, which is an especially relevant consideration since large differences in allele frequency are not necessarily expected to exist in complex diseases?
3. What factors might influence the accurate estimation of allele frequency in DNA pools?

To address these questions, we selected a number of SNPs, based upon varying allele frequency and flanking base sequence, and genotyped them in three distinct pools, corresponding to two groups of unrelated individuals with early-onset type 2 diabetes mellitus and one group of unaffected controls. Allele frequency estimations derived from pooled DNA compared favorably to the actual allele frequencies measured in individual samples corresponding to each pool. In a comparison of 93 observations, allele frequencies measured in pools were found to be strongly correlated (r = 0.97; P < 0.0001) with allele frequencies obtained from individual genotypings of the DNA samples comprising each pool (see Fig. 11.7a).

Accurate measurement of allele frequency differences between groups is also an important consideration, as this is the value that will detect, or fail to detect, a potentially meaningful SNP. Thus, the difference in allele frequency was measured in affected and unaffected pools and compared to the corresponding difference as measured using individuals comprising each group. In the same set of SNPs used to estimate absolute allele frequency, we found the allele frequency difference between pools to be strongly correlated (r = 0.90; P < 0.0001) with the actual differences between case and controls as measured in individuals (see Fig. 11.7b).

To determine whether the magnitude of error in frequency estimates of pools was influenced by the frequency of the minor allele, we compared the difference in allele frequency between pools and individuals to the allele frequency derived from individual genotypes, which was assumed to be the "true" frequency (see Fig. 11.7c). We found no evidence for a systematic relationship between allele frequency and the extent of deviation of our measurement from actual allele frequency, suggesting that accuracy is not differentially affected by the frequency of the minor allele.

Deviation from a strict 1 : 1 allele peak height ratio (e.g., if comparing the signals from opposite homozygotes) might increase the error associated with accurate allele frequency estimation. In pyrosequencing, nucleotides that are both identical and adjacent cause a peak height proportional to the number of nucleotides (i.e., the height of the A peak from GAAC is approximately double that of GAC). Therefore, it is necessary to consider the context of bases directly flanking a SNP. The expected contributions of neighboring identical bases to the total height of an allele peak are shown in Fig. 11.8. SNPs were classified according to the context of flanking bases to determine whether the accurate estimation of allele frequency in pools was affected. The assays with no interference from neighboring bases (n = 7) and those in which one allele was identical to three or fewer bases of adjacent sequence (n = 5) remained correlated with actual allele frequency (r = 1.0 and 0.99, respectively; P < 0.0001 for both). For those assays in which both bases flanking the SNP were each identical to one of the SNP alleles (n = 3), allele frequency estimated from pools was still correlated with actual allele frequency, although less significantly (r = 0.88, P = 0.01). However, runs of identical bases can be avoided by designing a sequencing primer that abuts the SNP and anneals to the other bases, thereby eliminating the contribution of the flanking bases. We have found that some SNPs, particularly those contained in longer homopolymers (i.e., >3 bp), are not good candidates for pyrosequencing, in either pools or individuals, unless the sequencing primer placement can shorten or eliminate the homopolymer.

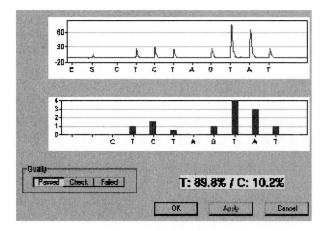

FIGURE 11.6 Sample allele quantification (AQ) pyrogram. In the evaluation module, the AQ pyrogram for each sample pool can be viewed. Nucleotide sequence is entered in the same format as used for SNP entry and control nucleotides are automatically assigned—here they correspond to cytosine and adenosine residues. The variant sequence, in this case, a C/T SNP, is shown in the pyrogram with the appropriate flanking sequence: T[C/T]GTTTTAAAT. The AQ software automatically assigns the allele distribution based on peak height (here the T allele is present at 89.8% frequency and the C allele at 10.2% frequency). In the authors' experience, a robust amplification reaction, which yields strong peak heights, is one of the best predictors of accurate allele quantification.

A.

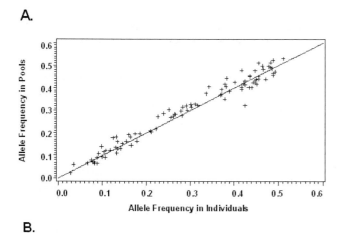

B.

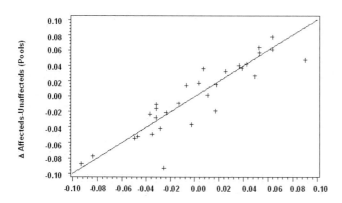

C.

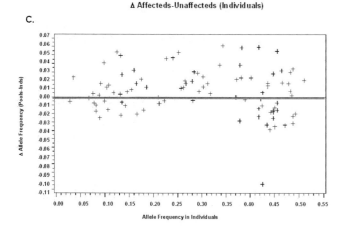

FIGURE 11.7 **A.** Accuracy of allele frequency estimation using pyrosequencing. Estimation of allele frequency in individuals versus pools. Pools were composed of 109–169 samples and allele frequency estimates represent the average of four separate amplification reactions. The allele frequency determined in those individuals contained in each pool is shown on the x-axis and the allele frequency estimated in DNA pools is shown on the y-axis. There were a total of 93 observations corresponding to 31 SNPs assayed in three pools. *Line* = line of identity; *y = x*. **B.** Estimation of allele frequency differences between pools in pools versus individuals. Allele frequency differences between affected and unaffected subjects were measured in pools and individuals. The frequency of the minor allele from the control group was subtracted from the corresponding frequency in the case group, for individuals (x-axis) and pools (y-axis). Results were derived from 31 SNP assays. *Line* = line of identity, *y = x*. **C.** Effect of minor allele frequency on disparities between pool-individual estimated allele frequencies. To determine whether pool error was related to allele frequency, the allele frequency obtained in individuals was subtracted from the frequency measured in the corresponding pool (y-axis) and this value was plotted against the allele frequency measured in individuals (x-axis), which was assumed to be the true representation of allele frequency.

# alleles matching flanking base	Genotype			# Assays	r	p
	TT	TC	CC			
0				7	1.0	<0.0001
1				5	0.99	<0.0001
2				3	0.88	0.01

FIGURE 11.8 Effect of flanking bases on the pyrosequencing assay. When a neighboring base is identical to an allele, it contributes to the relative height of the allele peak. Column 1: Number of alleles matching the flanking bases. Column 2: pyrograms representing each possible genotype corresponding to Column 1 class. The number of assays corresponds to how many of the 15 SNPs tested fit into each possible category. The correlation and p-values, based on two observations (pools) per assay, are shown in the final two columns. (Reproduced from Gruber and colleagues (2002) with permission.)

In addition to its applications in allelic association studies, allele frequency estimation in DNA pools can also be used for the rapid confirmation of potential SNPs represented in public databases (e.g., http://www.ncbi.nlm.nih.gov/SNP). These SNPs, which initially have been identified *in silico* or by resequencing efforts, can be validated quickly in DNA pools and at low cost. Moreover, allele frequencies can be estimated in sample pools representing different ethnic populations with greater

sensitivity by pyrosequencing than by conventional fluorescent dideoxy DNA sequencing (Wolford *et al.*, 2000).

11.3.3 DNA Sequencing

Pyrosequencing has also been developed for direct sequencing of PCR products (Ronaghi *et al.*, 1998). The advantage of using pyrosequencing for DNA sequencing, as opposed to other sequence detection systems, is due to the parallel processing capacity of pyrosequencing; the flexibility, accuracy, and simplicity of the assay; and the rapidity with which high quality sequence information can be obtained. Moreover, availability of a pyrosequencing DNA sequence platform is valuable for automated processing of clinical samples, rapid validation of potential polymorphisms, and rapid *de novo* sequencing of regions of interest.

The general principle of DNA sequencing using pyrosequencing is similar to other applications in that a sequencing primer is annealed to a single-stranded DNA template and through iterative addition of deoxynucleotide triphosphates, a complementary DNA strand is created. In the initial applications, the sequence read lengths using pyrosequencing were limited to approximately 50 bp, mainly due to the continuous decrease of nucleotide degrading efficiency during the sequencing process (Ronaghi *et al.*, 1998). However, a method for obtaining longer sequence reads (up to 100 bp) was recently developed, which substitutes pure 2'-deoxyadenosine-5'-O-(1-thiotriphosphate) Sp-isomer for the regular mixture of dATPαS isomers (Sp and Rp) normally used in the sequencing reaction (Gharizadeh *et al.*, 2002). High concentrations of dATPαS nucleotides are known to inhibit apyrase activity and lead to an accumulation of degradation products in the pyrosequencing reaction. In contrast, using only the Sp isomer requires much lower nucleotide concentrations for sequencing compared to the Sp and Rp mixture, leading to reduced inhibition of apyrase activity and consequently resulting in enhanced read lengths. In addition, the pure Sp solution more efficiently incorporates adenosine into homopolymeric regions and combined with the reduced apyrase inhibition, longer and high quality sequencing reads are readily obtained.

11.3.4 Other Applications

11.3.4.1 Medical Diagnostics
Screening and identification of DNA mutations in tumor tissue is an important aspect of medical diagnosis and prognosis in certain diseases, particularly cancer. Typically, potential mutations are assessed by first amplifying the region of interest using PCR, then performing mutation detection using direct sequencing, single-strand conformation polymorphism (SSCP) analysis (see also Chapter 6), or denaturing gradient gel electrophoresis (DGGE, see also Chapter 8). These methods are sensitive, accurate, and reproducible to varying degrees, but are not necessarily cost-effective, labor efficient, high throughput, or applicable for all heterogeneous situations.

The use of pyrosequencing to identify common mutations in cancer susceptibility genes, such as p53 and N-ras, has great merit in the diagnosis, prognosis, and therapeutic strategy for the management of different forms of cancer. The ability of pyrosequencing to detect mutations in the p53 gene, one of the most commonly mutated genes associated with human malignancies, was recently evaluated (Garcia *et* *al.*, 2000). Using DNA extracted from tissue biopsies of skin cancer patients, p53 mutations were detected accurately and the proportion of mutant to wild type alleles was correctly estimated (Garcia *et al.*, 2000). Similarly, a comparative analysis of SSCP and pyrosequencing was performed in a parallel assessment of the most common malignant melanoma mutations in the N-ras gene (Sivertsson *et al.*, 2002). SSCP and pyrosequencing were found to yield similar detection limits, although the ability of the latter was dependent on the ability to create a useful dispensation profile (Sivertsson *et al.*, 2002). It should also be noted that whereas pyrosequencing can assess only a few nucleotide positions in a mutation screen, SSCP can evaluate an entire amplicon. However, in specific regions that are commonly mutated (i.e., mutational hot spots), a nucleotide dispensation order can be designed to increase the sensitivity of detection in combination with identification of all relevant mutations (Sivertsson *et al.*, 2002).

The application of pyrosequencing in medical diagnostics is not limited to cancer, but also has relevance for diseases, which frequently result from germ line mutations, such as Apert syndrome. Approximately two-thirds of all cases of Apert syndrome are caused by a spontaneous C-to-G transversion at position 755 of fibroblast growth factor receptor 2 (FGFR2). In a recent report, pyrosequencing was used to quantify levels of this mutation in DNA isolated from sperm samples (Goriely *et al.*, 2003). A strong link between paternal age effect and prevalence of C755G mutation in sperm was found (intriguingly, this mutation confers a positive selection advantage in the male germ line). More importantly, however, pyrosequencing was found to provide a means for germ line screening to identify parents at high risk for passing on disease alleles to offspring.

In addition, optimized Pyrosequencing assays for SNPs in genes such as methylene tetrahydrofolate reductase (*MTHFR*), cystic fibrosis transmembrane conductance regulator (CFTR), factor V Leiden, prothrombin, apolipoprotein E, and a variety of cytochrome P450 genes (e.g., CYP2C9, CYP2C19, CYP2D6, etc.) are commercially available and provide significant value for clinical applications. Some of these assays are summarized in Table 11.1.

11.3.4.2 Pharmacogenomics
Individuals can experience significantly varied responses to pharmaceutical drugs as a result of certain genetic polymorphisms and in some cases, specific drug response based on genotypes at these loci can be predicted (Roses, 2000a; 2000b; see also Chapter 20). General genotyping methods, including pyrosequencing, have been validated for the evaluation of SNPs involved in drug response (Shi, 2001; Rose *et al.*, 2003). For example, many drugs are substrates of the multidrug-resistance protein (MDR-1) and polymorphisms within the corresponding gene, MDR1, correlate with expression and function of the protein (Hoffmeyer *et al.*, 2000). As suggested by its name, the MDR-1 protein confers

TABLE 11.1 Commercial pyrosequencing assays for molecular diagnostics. These assays are commercially available from pyrosequencing for research use only and are not intended for diagnostic purposes.

Assay	Disease Association
34 common mutations in *CFTR*	Cystic fibrosis
C677T variant in *MTHFR*	MTHFR deficiency
ARG506GLN variant in *F5* and G20210A variant in *F2* (duplex assay)	Factor V Leiden and Thrombosis
ARG506GLN variant in *F5*, G20210A variant in *F2* and 4G/5G variant in *PAI1* (triplex assay)	Factor V Leiden, Thrombosis and CAD risk
ARG158CYS and CYS112ARG variants in *APOE*	Hyperlipoproteinemia, type III

resistance against numerous drugs, including antineoplastic agents. Chemotherapy resistance based on genotype at the C3435T variant site was recently assessed in MDR1 in a group of patients with advanced breast cancer (Kafka *et al.*, 2003). From genomic DNA, a 106 bp fragment of *MDR1* exon 26 was amplified and C3435T genotype was determined by pyrosequencing. In this study, 57% of the subjects were CT heterozygotes and 22% were TT homozygotes. A significant correlation was found between clinical response to preoperative chemotherapy and TT genotype, indicating that C3435T genotype may be a marker for resistance to chemotherapy and may also provide useful information for individualized therapy.

The emergence of drug-resistant viral strains has had a severe impact on the successful application of antiretroviral drugs in fighting infection. For instance, human immunodeficiency virus type 1 (HIV-1) is commonly treated with highly active antiretroviral therapy (HAART), which includes the use of HIV protease inhibitors. Antiviral drug resistance often develops when HAART does not completely suppress viral replication and resistant strains emerge. By predetermining a patient's susceptibility to antiviral drugs, the most effective drugs can be used at the onset of treatment, thereby reducing the risk for development of antiviral drug resistance. A pyrosequencing assay was developed for the rapid characterization of resistance to HIV-1 protease inhibitors (O'Meara *et al.*, 2001). Primers were designed to analyze 33 amino acid positions likely to lead to drug resistance mutations and eight primary protease inhibitor resistance mutations, as well as several secondary mutations, were identified by pyrosequencing (O'Meara *et al.*, 2001). These results and use of this specific assay system is expected to facilitate the monitoring of HIV-1 drug resistance during therapy.

11.3.4.3 CLINICAL MICROBIOLOGY

Pyrosequencing has been applied to the rapid molecular identification and subtyping of clinical microbial isolates. The ability to characterize viral and microbial subtypes provides clinicians with the ability to make well informed decisions regarding therapeutic interventions.

Viral Typing. The hepatitis C virus (HCV) can contribute to the development of acute and chronic hepatitis, liver cirrhosis, and hepatocellular carcinoma in humans. HCV consists of six major genotypes corresponding to approximately 80 subtypes. Different subtypes can be distinguished from one another by sequencing specific sections of the HCV genome. A pyrosequencing protocol recently was developed to sequence the 140 bp segment of the HCV genome in three distinct reactions, which correctly classified viral subtypes in 98 clinical samples (Elahi *et al.*, 2003). Similarly, multiplex pyrosequencing, using one set of amplification primers and three different sequencing primers, was utilized to subtype different HCV strains in serum samples collected from 77 HCV-positive subjects (Pourmand *et al.*, 2002). Application of pyrosequencing to population-specific HCV subtyping is also possible using a well-developed multiplex assay (Elahi *et al.*, 2003). Pyrosequencing likewise has been used to classify subtypes of human papilloma virus (HPV) in clinical samples (Gharizadeh *et al.*, 2002), and assays to identify subtypes of other viruses are expected to be developed in the future. The ability to quickly and cost-effectively characterize viral subtypes holds great promise in terms of clinical diagnosis, treatment options, and prognosis.

Bacterial Typing. The rapid identification of the etiological agent of microbial infections also holds tremendous clinical benefit. Probes and assay conditions have been developed that allow provisional classification of bacterial isolates based on pyrosequencing comparisons of the 16s rRNA sequence (Jonasson *et al.*, 2002). The technique is simple and involves only four steps:

1. Picking a bacterial colony.
2. Performing PCR using a set of two universal 16s rDNA primers flanking a variable region.
3. Sequencing at least 10 nucleotides of the amplicon by pyrosequencing.
4. Matching the resultant signature sequence against a local 16s rDNA database (Jonasson *et al.*, 2002).

This approach was used to classify 500 clinical bacterial isolates, including *Staphylococcus*, *Streptococcus*, *Enterococcus*, *Enterobacteriaceae*, *Haemophilus*, and *Pseudomonas*. A similar approach was used for the rapid identification and subtyping of *Helicobacter pylori* in 23 clinical isolates obtained from gastric biopsy specimens (Monstein *et al.*, 2001). Pyrosequencing technology also has been applied to the identification of clinically relevant fungi using the 18s rRNA gene (Ronaghi and Elahi, 2002).

11.3.4.4 FORENSICS

When sources of nuclear DNA are limited, as is frequently the case in forensic investigations, mitochondrial DNA (mtDNA), which is present in cells in relative abundance, can be used instead (see also Chapter 21). The most frequently used technique for routine forensic analysis of mtDNA is direct sequencing of two hypervariable regions (HVI and HVII) within the mitochondrial D-loop; but this process, though reliable and robust, is also time-consuming and labor intensive. As an alternative to direct sequencing, a pyrosequencing assay was developed to identify polymorphisms in the mitochondrial D-loop (Andreasson *et al.*, 2002). Pyrosequencing was used to analyze mtDNA isolated from a variety of crime scene materials including clothing, wigs, mustaches, shoes, mobile phones, watches, knives, guns, fingerprints, and saliva obtained from envelopes (Andreasson *et al.*, 2002). In all cases, results obtained with pyrosequencing were comparable with direct sequencing results. In the pyrosequencing Application Note #211 (http://www.pyrosequencing.com), polymorphisms within the HVII region were interrogated using evidence material. The results were robust enough to accurately exclude two individuals, who were not suspects, and correctly identify the third individual, who was the true suspect. Thus, pyrosequencing can provide accurate and sensitive assessment of forensic materials and thus, offers an alternative platform for forensic analysis.

11.4 CONCLUSIONS

Pyrosequencing is a sequencing-by-synthesis method that has major applications in SNP genotyping, allele frequency estimation, and tag sequencing. Pyrosequencing quickly and accurately yields SNP genotype data and provides direct visualization of raw data for quality assessment. Pyrosequencing likewise provides an accurate and valid approach to allele frequency estimation in pooled DNA samples. In addition, the technique is sensitive enough to detect small differences between pools across a range of disparate allele frequencies. For studies with large numbers of SNPs, whose goal is to narrow a region of genetic linkage by linkage disequilibrium mapping, allele quantification by pyrosequencing in pools is an efficient approach to reduce the labor and expense involved in such experiments. Finally, pyrosequencing is emerging as a fast and reliable method for high quality tag sequencing. This application has strong implications for the clinical laboratory, where more conventional sequence detection systems may prove too large and cumbersome, and subsequent sequence data analysis time-consuming and complicated.

References

Agaton, C., Unneberg, P., Sievertzon, M., Holmberg, A., Ehn, M., Larsson, M., Odeberg, J., Uhlen, M., and Lundeberg, J. (2002). Gene expression analysis by signature pyrosequencing. *Gene* 289, 31–39.

Ahmadian, A., Lundeberg, J., Nyren, P., Uhlen, M., and Ronaghi, M. (2000). Analysis of the p53 tumor suppressor gene by Pyrosequencing. *Biotechniques* 28, 140–147.

Alderborn, A., Kristofferson, A., and Hammerling, U. (2000). Determination of single-nucleotide polymorphisms by real-time pyrophosphate DNA sequencing. *Genome Res.* 10, 1249–1258.

Andreasson, H., Asp, A., Alderborn, A., Gyllensten, U., and Allen, M. (2002). Mitochondrial sequence analysis for forensic identification using pyrosequencing technology. *Biotechniques* 32, 124–133.

Brenner, S., Johnson, M., Bridgham, J., Golda, G., Lloyd, D. H., Johnson, D., Luo, S., McCurdy, S., Foy, M., Ewan, M., Roth, R., George, D., Eletr, S., Albrecht, G., Vermaas, E., Williams, S. R., Moon, K., Burcham, T., Pallas, M., DuBridge, R. B., Kirchner, J., Fearon, K., Mao, J., and Corcoran, K. (2000). Gene expression analysis by massively parallel signature sequencing (mpss) on microbead arrays. *Nat. Biotechnol.* 18, 630–634.

Buetow, K. H., Edmonson, M., MacDonald, R., Clifford, R., Yip, P., Kelley, J., Little, D. P., Strausberg, R., Koester, H., Cantor, C. R., and Braun, A. (2001). High-throughput development and characterization of a genomewide collection of gene-based single nucleotide polymorphism markers by chip-based matrix-assisted laser desorption/ionization time-of-flight mass spectrometry. *Proc. Natl. Acad. Sci. USA* 98, 581–584.

Elahi, E., Pourmand, N., Chaung, R., Rofoogaran, A., Boisver, J., Samimi-Rad, K., Davis, R. W., and Ronaghi, M. (2003). Determination of hepatitis c virus genotype by pyrosequencing. *J. Virol. Methods* 109, 171–176.

Fakhrai-Rad, H., Pourmand, N., and Ronaghi, M. (2002). Pyrosequencing: An accurate detection platform for single nucleotide polymorphisms. *Hum. Mutat.* 19, 479–485.

Garcia, C. A., Ahmadian, A., Gharizadeh, B., Lundeberg, J., Ronaghi, M., and Nyren, P. (2000). Mutation detection by pyrosequencing: Sequencing of exons 5-8 of the p53 tumor suppressor gene. *Gene* 253, 249–257.

Germer, S., Holland, M. J., and Higuchi, R. (2000). High-throughput SNP allele-frequency determination in pooled DNA samples by kinetic PCR. *Genome Res.* 10, 258–266.

Gharizadeh, B., Nordstrom, T., Ahmadian, A., Ronaghi, M., and Nyren, P. (2002). Long-read Pyrosequencing using pure 2′-deoxyadenosine-5′-O′-(1-thiotriphosphate) Sp-isomer. *Anal. Biochem.* 301, 82–90.

Goriely, A., McVean, G. A., Rojmyr, M., Ingemarsson, B., and Wilkie, A. O. (2003). Evidence for selective advantage of pathogenic FGFR2 mutations in the male germ line. *Science* 301, 643–646.

Gruber, J. D., Colligan, P. B., and Wolford, J. K. (2002). Estimation of single nucleotide polymorphism allele frequency in DNA pools by using Pyrosequencing. *Hum. Genet.* 110, 395–401.

Hoffmeyer, S., Burk, O., von Richter, O., Arnold, H. P., Brockmoller, J., Johne, A., Cascorbi, I., Gerloff, T., Roots, I., Eichelbaum, M., and Brinkmann, U. (2000). Functional polymorphisms of the human multidrug-resistance gene: Multiple sequence variations and correlation of one allele with p-glycoprotein expression and activity in vivo. *Proc. Natl. Acad. Sci. USA* 97, 3473–3478.

Hoogendoorn, B., Norton, N., Kirov, G., Williams, N., Hamshere, M. L., Spurlock, G., Austin, J., Stephens, M. K., Buckland, P. R., Owen, M. J., and O'Donovan, M. C. (2000). Cheap, accurate and rapid allele frequency estimation of single nucleotide polymorphisms by primer extension and dHPLC in DNA pools. *Hum. Genet.* 107, 488–493.

Horikawa, Y., Oda, N., Cox, N. J., Li, X., Orho-Melander, M., Hara, M., Hinokio, Y., Lindner, T. H., Mashima, H., Schwarz, P. E., del Bosque-Plata, L., Oda, Y., Yoshiuchi, I., Colilla, S., Polonsky, K. S., Wei, S., Concannon, P., Iwasaki, N., Schulze, J., Baier, L. J., Bogardus, C., Groop, L., Boerwinkle, E., Hanis, C. L., and Bell, G. I. (2000). Genetic variation in the gene encoding calpain-10 is associated with type 2 diabetes mellitus. *Nat. Genet.* 26, 163–175.

Hugot, J. P., Chamaillard, M., Zouali, H., Lesage, S., Cezard, J. P., Belaiche, J., Almer, S., Tysk, C., O'Morain, C. A., Gassull, M., Binder, V., Finkel, Y., Cortot, A., Modigliani, R., Laurent-Puig, P., Gower-Rousseau, C., Macry, J., Colombel, J. F., Sahbatou, M., and Thomas, G. (2001). Association of NOD2 leucine-rich repeat variants with susceptibility to Crohn's disease. *Nature* 411, 599–603.

Jonasson, J., Olofsson, M., and Monstein, H. J. (2002). Classification, identification and subtyping of bacteria based on Pyrosequencing and signature matching of 16s rDNA fragments. *Apmis* 110, 263–272.

Kafka, A., Sauer, G., Jaeger, C., Grundmann, R., Kreienberg, R., Zeillinger, R., and Deissler, H. (2003). Polymorphism C3435T of the MDR-1 gene predicts response to preoperative chemotherapy in locally advanced breast cancer. *Int. J. Oncol.* 22, 1117–1121.

Kwok, P. Y., Deng, Q., Zakeri, H., Taylor, S. L., and Nickerson, D. A. (1996). Increasing the information content of STS-based genome maps: Identifying polymorphisms in mapped STSs. *Genomics* 31, 123–126.

Larsson, M., Stahl, S., Uhlen, M., and Wennborg, A. (2000). Expression profile viewer (exproview): A software tool for transcriptome analysis. *Genomics* 63, 341–353.

McNeely, T. (2003). Pyrosequencing AB. *Pharmacogenomics* 4, 217–221.

Monstein, H., Nikpour-Badr, S., and Jonasson, J. (2001). Rapid molecular identification and subtyping of helicobacter pylori by Pyrosequencing of the 16s rDNA variable v1 and v3 regions. *FEMS Microbiol. Lett.* 199, 103–107.

Neve, B., Froguel, P., Corset, L., Vaillant, E., Vatin, V., and Boutin, P. (2002). Rapid SNP allele frequency determination in genomic DNA pools by Pyrosequencing. *Biotechniques* 32, 1138–1142.

Nordstrom, T., Alderborn, A., and Nyren, P. (2002). Method for one-step preparation of double-stranded DNA template applicable for use with Pyrosequencing technology. *J. Biochem. Biophys. Methods* 52, 71–82.

Nordstrom, T., Nourizad, K., Ronaghi, M., and Nyren, P. (2000a). Method enabling Pyrosequencing on double-stranded DNA. *Anal. Biochem.* 282, 186–193.

Nordstrom, T., Ronaghi, M., Forsberg, L., de Faire, U., Morgenstern, R., and Nyren, P. (2000b). Direct analysis of single-nucleotide polymorphism on double-stranded DNA by Pyrosequencing. *Biotechnol. Appl. Biochem.* 31, 107–112.

Ogura, Y., Bonen, D. K., Inohara, N., Nicolae, D. L., Chen, F. F., Ramos, R., Britton, H., Moran, T., Karaliuskas, R., Duerr, R. H., Achkar, J. P., Brant, S. R., Bayless, T. M., Kirschner, B. S., Hanauer, S. B., Nunez, G., and Cho, J. H. (2001). A frameshift mutation in NOD2 associated with susceptibility to Crohn's disease. *Nature* 411, 603–606.

Ohnishi, Y., Tanaka, T., Ozaki, K., Yamada, R., Suzuki, H., and Nakamura, Y. (2001). A high-throughput SNP typing system for genome-wide association studies. *J. Hum. Genet.* 46, 471–477.

O'Meara, D., Wilbe, K., Leitner, T., Hejdeman, B., Albert, J., and Lundeberg, J. (2001). Monitoring resistance to human immunodeficiency virus type 1 protease inhibitors by Pyrosequencing. *J. Clin. Microbiol.* 39, 464–473.

Pacey-Miller, T., and Henry, R. (2003). Single-nucleotide polymorphism detection in plants using a single-stranded Pyrosequencing protocol with a universal biotinylated primer. *Anal. Biochem.* 317, 166–170.

Pettersson, M., Bylund, M., and Alderborn, A. (2003). Molecular haplotype determination using allele-specific PCR and Pyrosequencing technology. *Genomics* 82, 390–396.

Pourmand, N., Elahi, E., Davis, R. W., and Ronaghi, M. (2002). Multiplex Pyrosequencing. *Nucleic Acids Res.* 30, e31.

Ronaghi, M., Uhlen, M., and Nyren, P. (1998). A sequencing method based on real-time pyrophosphate. *Science* 281, 363–365.

Ronaghi, M., and Elahi, E. (2002). Pyrosequencing for microbial typing. *J. Chromatogr. B. Analyt. Technol. Biomed. Life Sci.* 782, 67–72.

Rose, C. M., Marsh, S., Ameyaw, M. M., and McLeod, H. L. (2003). Pharmacogenetic analysis of clinically relevant genetic polymorphisms. *Methods Mol. Med.* 85, 225–237.

Roses, A. D. (2000a). Pharmacogenetics and pharmacogenomics in the discovery and development of medicines. *Novartis Found. Symp.* 229, 63–66.

Roses, A. D. (2000b). Pharmacogenetics and the practice of medicine. *Nature* 405, 857–865.

Shi, M. M. (2001). Enabling large-scale pharmacogenetic studies by high-throughput mutation detection and genotyping technologies. *Clin. Chem.* 47, 164–172.

Sivertsson, A., Platz, A., Hansson, J., and Lundeberg, J. (2002). Pyrosequencing as an alternative to single-strand conformation polymorphism analysis for detection of n-ras mutations in human melanoma metastases. *Clin. Chem.* 48, 2164–2170.

Soderback E., Alderborn A., and Krabbe M. (2003). Gene expression analysis using Pyrosequencing technology, Pyrosequencing application note 211; http://pyrosequencing.com.

Velculescu, V. E., Zhang, L., Vogelstein, B., and Kinzler, K. W. (1995). Serial analysis of gene expression. *Science* 270, 484–487.

Vozarova, B., Fernandez-Real, J. M., Knowler, W. C., Gallart, L., Hanson, R. L., Gruber, J. D., Ricart, W., Vendrell, J., Richart, C., Tataranni, P. A., and Wolford, J. K. (2003). The interleukin-6 (-174) G/C promoter polymorphism is associated with type 2 diabetes mellitus in Native Americans and Caucasians. *Hum. Genet.* 112, 409–413.

Wasson, J., Skolnick, G., Love-Gregory, L., and Permutt, M. A. (2002). Assessing allele frequencies of single nucleotide polymorphisms in DNA pools by pyrosequencing technology. *Biotechniques* 32, 1144–1150.

Wieser, F., Hefler, L., Tempfer, C., Vlach, U., Schneeberger, C., Huber, J., and Wenzl, R. (2003). Polymorphism of the inter-leukin-1beta gene and endometriosis. *J. Soc. Gynecol. Investig.* 10, 172–175.

Wolford, J. K., Blunt, D., Ballecer, C., and Prochazka, M. (2000). High-throughput SNP detection by using DNA pooling and denaturing high performance liquid chromatography (dHPLC). *Hum. Genet.* 107, 483–487.

Wolford, J. K., Kobes, S., Hanson, R. L., Bogardus, C., and Prochazka, M. (2001). Linkage disequilibrium mapping of a putative type 2 diabetes locus (1q21-q23) using pooled DNA [abstract]. *Am. J. Hum. Genet.* 69.

Wolford, J. K., Konheim, Y. L., Colligan, P. B., and Bogardus, C. (2003a). Association of a F479L variant in the cytosolic phospho-lipase A2 gene (PLA2G4A) with decreased glucose turnover and oxidation rates in Pima Indians. *Mol. Genet. Metab.* 79, 61–66.

Wolford, J. K., Gruber, J. D., Ossowski, V. M., Vozarova, B., Antonio Tataranni, P., Bogardus, C., and Hanson, R. L. (2003b). A C-reactive protein promoter polymorphism is associated with type 2 diabetes mellitus in Pima Indians. *Mol. Genet. Metab.* 78, 136–144.

Zhou, G., Kamahori, M., Okano, K., Chuan, G., Harada, K., and Kambara, H. (2001). Quantitative detection of single nucleotide polymorphisms for a pooled sample by a bioluminometric assay coupled with modified primer extension reactions (BAMPER). *Nucleic Acids Res.* 29, E93.

CHAPTER **12**

Molecular Cytogenetics in Molecular Diagnostics

HOLGER TÖNNIES

Humboldt University Berlin, Charité School of Medicine, CVK, Institute of Human Genetics, Chromosome Diagnostics and Molecular Cytogenetics, Berlin, Germany

TABLE OF CONTENTS

12.1 INTRODUCTION

The fundamental cellular processes of DNA replication, DNA repair, mitosis, and meiosis ensure the high integrity of the human genome. However, mutations can occur in germline and somatic cells either induced or spontaneously. Depending on their physical size, these mutations manifest as altered base sequences on the DNA level (gene mutations) or gross alterations as chromosome aberrations.

Chromosome mutations can occur induced or spontaneously in germline or somatic cells. Chromosomal aberrations are the major cause for congenital anomalies, mental retardation, and infertility in humans. Furthermore, they account for approximately half of all early spontaneous pregnancy losses (Warburton, 2000). Overall, 1 in 120 liveborn children has a chromosomal abnormality, and about half of these children are phenotypically abnormal as a result of the chromosome aberration. Partial losses (monosomy) or gains (trisomy) of euchromatic chromosomal material resulting in different clinical phenotypes have been described for all human chromosomes (Schinzel, 2001). In genetic pre- and postnatal diagnostics, conventional cytogenetics using classical karyotyping of chromosomes is the most important assay to characterize numerical and structural chromosome aberrations (see Fig. 12.1).

Today, chromosome analysis, the most common genetic test in diagnostic units, is an indispensable tool for the diagnosis and prognosis of congenital (inborn) and acquired disorders (e.g., neoplasia) and disease monitoring. The precise characterization of aberrant cytogenetic findings is imperative for syndromologic assignment, phenotype-karyotype correlations, and genetic counseling.

However, limited chromosome specific banding resolution obtained by classical chromosome banding techniques makes the recognition and interpretation of masked or cryptic chromosome aberrations difficult if not impossible to ascertain in some cases. In the last two decades, molecu-

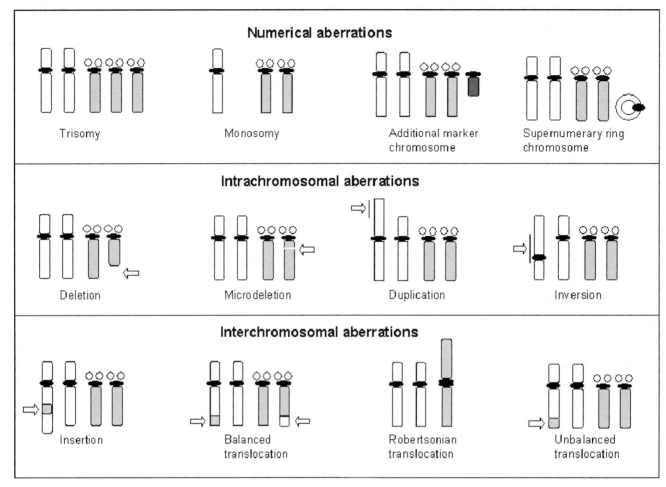

FIGURE 12.1   Schematic illustration of common numerical and structural intra- and interchromosomal aberrations in humans.

lar cytogenetic techniques based on fluorescence *in situ* hybridization (FISH) applications locating specific, fluorescence-labeled nucleic acid sequences in interphase cells or metaphase spreads, have become fast, sensitive, and important complementing tools in genetic diagnostics (reviewed in Tönnies, 2002). The use of different, specific FISH probes and multicolor assays enhances the thorough characterization of numerical and complex chromosome aberrations regardless of their complexity, filling the gap between conventional chromosome banding analysis and molecular genetic studies on the DNA level. In modern diagnostics, multiple different molecular strategies and applications are used to extend the understanding of diseases and cellular pathology and to generate routine diagnostic tests.

This chapter will introduce modern molecular-cytogenetic diagnostics and give a brief overview over the multiple molecular cytogenetic FISH techniques and multicolor assays currently used in genetic diagnostics.

12.2 FROM CONVENTIONAL TO MOLECULAR CYTOGENETICS

With the establishment of hypotonic treatment resulting in successful metaphase chromosome spreading (Hsu and Pomerat, 1953), Tijo and Levan (1956) managed to determine the correct number of chromosomes in human diploid cells to 46. Three years later, Lejeune and colleagues (1959) published their observation, that full trisomy 21 resulting in 47 chromosomes is one cause for Down syndrome. The next technical improvement in cytology was the addition of colchicine to *in vitro* cultures for the accumulation of dividing cells prior to fixation (Moorhead *et al.*, 1960). In the late 1960s, the development and establishment of conventional cytogenetic banding techniques (Casperson *et al.*, 1970) to detect numerical and structural chromosome abnormalities in humans began the era of clinical cytogenetics.

Today, the identification and characterization of numerical and structural chromosomal aberrations by chromosome karyotyping is performed routinely in cytogenetic diagnostics. Over the last decades multiple additional conventional banding techniques were introduced in routine use (for overview see Verma *et al.*, 1989; Barch *et al.*, 1997) to detect chromosomal aberrations. However, the effectiveness of conventional cytogenetics depends directly on the chromosome aberrations appearing in human cells.

Detectable chromosomal aberrations can be divided into numerical and structural aberrations (see Fig. 12.1). In humans, numerical chromosome aberrations as trisomies of whole chromosomes (e.g., trisomy 21, trisomy 18, or the chromosome constitution 45,X0) are frequent, well-characterized chromosome abnormalities and easily detectable by morphologic and numerical chromosome examination.

Structural chromosomal aberrations can be subgrouped into intrachromosomal (deletions, duplications, inversions) and interchromosomal (balanced and unbalanced translocations, insertions) aberrations affecting more than one chromosome (see Fig.12.1; ISCN 1995).

More laborious to identify by conventional cytogenetics alone are so-called marker chromosomes. Marker chromosomes are structurally abnormal chromosomes of which the origin of the euchromatic content cannot be determined by conventional cytogenetic analysis. Often supernumerary, their incidence varies from 0.3–3.7/1000 in newborns and mentally/developmentally-delayed patients (Buckton *et al.*, 1985). Ring chromosomes, some also supernumerary (numerical aberration), are mainly originated from rearranged normal chromosomes due to deletions of the telomeric ends of the chromosome with following ring formation (intrachromosomal aberration; Tönnies *et al.*, 2003). The limited chromosome-specific banding resolution and assignment obtained by chromosome banding makes the characterization and correct interpretation of complex chromosome aberrations difficult to ascertain and is therefore by nature imprecise.

Furthermore, some chromosomal aberrations detected in an affected child are *de novo*, defined by a normal constitutional karyotype of the parents. Therefore, a precise definition of the lost or duplicated chromosomal material is indispensable for any exact genotype-phenotype correlation and provides diagnostic and possibly prognostic information. By conventional cytogenetic methods, rearrangements in the size of 5–10 megabases (Mb) affecting single chromosomes are detectable if high-resolution banding is used. However, aberrations smaller than 5–10Mb (e.g. microdeletion syndromes; see Table 12.1), complex chromosome aberrations involving three or more chromosomes, and so-called marker chromosomes composed of unknown chromatic material, often gives unsatisfactory results using conventional cytogenetic techniques only. The dependency on dividing cells to prepare metaphase spreads and the occurrence of chromosome aberrations not to be characterized exactly using monochrome banding techniques enhanced the development of alternative, more sensitive approaches based on the hybridization of DNA probes, overcoming these conventional cytogenetics limitations.

12.3 FLUORESCENCE *IN SITU* HYBRIDIZATION

By *in situ* hybridization on metaphase spreads, genetic changes can be analyzed at the single cell level allowing the simultaneous assessment of different chromosomes or chromosome regions and the determination of clonal variability or mosaicism (see Fig. 12.2). Classical *in situ* hybridization is based on the binding (hybridization, annealing) of complementary, single-stranded labeled nucleic acids to the fixed and denatured target DNA of metaphase chromosomes, whole interphase nuclei, or DNA fibers. During

TABLE 12.1 Common autosomal microdeletion syndromes in clinical diagnostics.

Chromosomal localization	Name	Involved gene(s)/hybridization target	OMIM[a] entry
del(1)(p36.3)	Monosomy 1p syndrome		#607872
del(4)(p16.3)	Wolf-Hirschhorn syndrome	Wolf-Hirschhorn critical region	#194190
del(5)(p15.2p15.3)	Cri-du-chat syndrome	Cri-du-chat critical region	#123450
del(7)(q11.23q11.23)	Williams-Beuren syndrome	ELN	#194050
del(8)(q24.1q24.1)	Langer-Giedion syndrome	EXT1, TRPS I, TRPS II	#150230
del(11)(p13p13)	WAGR syndrome	WT1, PAX6	#194072
del(15)(q11q13)pat	Prader-Willi syndrome	SNRPN	#176270
del(15)(q11q13)mat	Angelman syndrome	UBE3A	#105830
del(16)(p13.3)	Rubinstein-Taybi syndrome	CBP	#180849
del(17)(p11.2p11.2)	Smith-Magenis syndrome	Smith-Magenis critical region	#182290
del(17)(p13.3)	Miller-Dieker syndrome	LIS1	#247200
del(20)(p11.23p11.23)	Alagille syndrome	JAG1	#118450
del(22)(q11.2q11.2)	VCF/DiGeorge syndrome	VCF/DiGeorge critical region	#192430
			*188400

[a] OMIM at www3.ncbi.nlm.nih.gov/Omim/searchomim.html

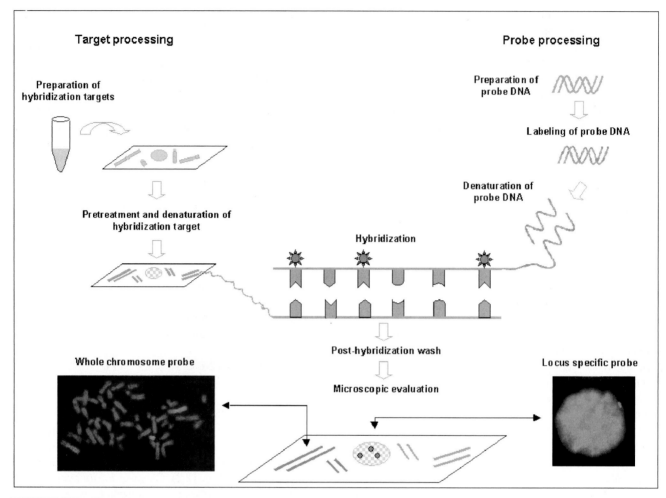

FIGURE 12.2 Flowchart summarizing the standard steps of fluorescence *in situ* hybridization. Target material (cell suspension including metaphase spreads and interphase nuclei) is dropped on glass slides, pretreated, and denatured to prepare single-stranded DNA. After hybridizing the single-stranded, labeled probe to the target material and post-hybridization wash, results can be analyzed by microscopic inspection. By using a whole chromosome probe for chromosome 3 (left), a partial trisomy is visible. Using a locus-specific probe (e.g., YAC, right), three signals are detectable in the interphase nucleus reflecting a partial trisomy 3 (See color plate).

hybridization the probe penetrates to the target material and anneals to the complementary partner resulting in DNA-duplexes of the bound probe and the former single-stranded target. After removing unbound probe material by stringent washing, results can be inspected by microscopy.

First described by Pardue and Gall (1969) using radio-labeled, repetitive DNA probes hybridizing to mice cell preparations, the detection has been carried out by autoradiography enabling the morphological (*in situ*) visualization of the presence of complementary nucleic acid sequences in the target material (John *et al.*, 1969). Hybridization and detection of the first single copy DNA sequences by autoradiography were described by Harper and Saunders (1981). Using nonradioactively labeled probes, Pinkel and colleagues (1986) introduced the basic protocol for fluorescence *in situ* hybridization (FISH), improving the spatial resolution in comparison to radio-labeled probes.

FISH is a sensitive and specific method providing precise information about the physical location of DNA sequences in cell nuclei or chromosomes.

12.4 BASIC TECHNICAL ELEMENTS AND MATERIALS

Fluorescence *in situ* hybridization, or FISH, is a technique for the visualization of labeled nucleic acid probes on target material (see Fig. 12.2). FISH is a stepwise process beginning with the selection and pretreatment of the target, followed by the production and labeling of the appropriate probe to be hybridized, the hybridization itself, and the detection and documentation of the results. For the user, a huge amount of variable starting material as different targets, probes, and labeling procedures are available nowadays. In this section, some of the different material and technical variations are discussed.

12.4.1 Targets for FISH

A wide variety of cellular materials can be used as targets for investigation by FISH. In routine diagnostics, in particular metaphase spreads and interphase nuclei of whole blood, fibroblasts, bone marrow or, in prenatal diagnostics, amniocytes and chorionic villi, cells are used after cultivation (see Fig. 12.2; for technical details see Wegner, 1999). In leukemic specimens, where the number of viable cells is sometimes low, so-called hypermetaphase spreads obtained by long-term exposure of cells to colcemide can be produced (Seong *et al.*, 1995). The yield of metaphase spread-like targets is higher in comparison to normal chromosome preparation with short-term colcemide treatment. The advantage in comparison to interphase FISH is the ability to see chromosome morphology, and to minimize false-positive results especially in tumor genetics. Furthermore without cultivation, buccal smear cells, sperm cells, and cells originating from the urinary tract can be used for interphase FISH. In preimplantation genetic diagnosis, FISH for aneuploidy screening can be performed on blastomeres (Delhanty *et al.*, 1993).

Mainly used in research rather than in clinical laboratories, FIBER-FISH using released chromatin fibers from interphase cells (Heng *et al.*, 1992; Wiegant *et al.*, 1992) is utilized to study the structure and organization of mammalian genomes with high resolution. Interphase FISH on single cells also can be performed on archived or nondividing material such as frozen and paraffin-embedded tissues and touch preparations of pathological specimen. For good hybridization results some targets need different pretreatment steps (dewaxing; proteolytic steps) for better penetration of the probe. For all target materials, a fixation step is required. In cytogenetics, conventional fixative (methanol : acetic acid = 3 : 1) is routinely used, but also ethanol-fixation (70%) provides good results. For FISH, target materials are applied to glass slides for pretreatment by RNase or protelytic digestion, if necessary. Before hybridization, the target DNA has to be denatured either by chemical treatment or by heating, melting the DNA double helices (for technical details see Schwarzacher and Heslop-Harrison, 2000).

12.4.2 DNA Probes for FISH

For FISH a variety of DNA probes established by different amplification or cloning techniques are used. Probe DNA can be prepared by locus-specific PCR amplified single genes, cloning of human DNA fragments, or by chromosome microdissection or flow-sorting of whole chromosomes. A large number of commercial FISH probes is available today (see Fig. 12.3). Region-specific chromosomal DNA can be cloned in vectors, such as cosmids, plasmids, PACs, BACs, or YACs. The choice of the vector system depends on the size of the DNA fragment to be cloned.

12.4.3 Probe Labeling

Nucleic acid FISH probes are labeled either directly using fluorochrome-conjugated nucleotides (e.g., fluorescein isothiocyanate; FITC-dUTP) or indirectly using reporter molecules (e.g., biotin-dUTP, digoxygenin (DIG)-dUTP; for details see Schwarzacher and Heslop-Harrison, 2000). Nucleic acid labeling approaches are based on the use of different enzymatic methods, and the choice of the labeling technique varies from laboratory to laboratory (for technical details see Rautenstrauß and Liehr, 2002). Most labeling

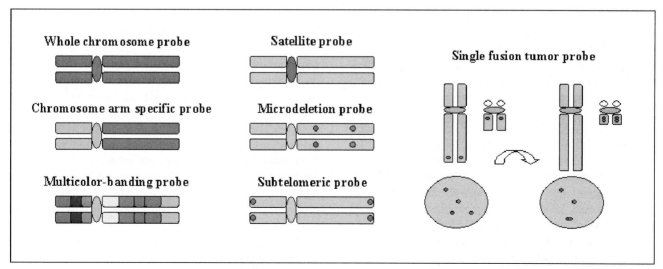

FIGURE 12.3   Schematic illustration of various FISH probes used in routine cytogenetic diagnostics (see color plate). For explanation see text. (See color plate)

strategies are based on the incorporation of label-conjugated nucleotides in a new synthesized DNA strand. Examples for these approaches are the polymerase chain reaction (PCR; amplification and labeling of the probe) using degenerated oligonucleotide primers (DOP; Telenius *et al.*, 1992), the nick-translation (no amplification but fragmentation of probe size; Rigby *et al.*, 1977), or random priming using hexamer primers (Feinberg and Vogelstein, 1983). For small oligonucleotides, terminal fluorochrome labeling reactions (Bauman *et al.*, 1980) are commonly used.

In the last years, various other labeling approaches such as chemical cross-linking have been established, and kits are commercially available to label probe DNA with high efficiency.

Until today, the number of fluorochromes with different emission wavelengths is limited. For multicolor assays (for details see Section 12.6, Multicolor-FISH screening assays), combinations of five fluorochromes can result in more than 24 colors using the ratio labeling scheme (varying ratios of different fluorochromes; Dauwerse *et al.*, 1992; Nederlof, 1992), the combinatorial labeling (combination of five fluorochromes resulting in 31 different colors) as described by Nederlof and colleagues (1990), or the COBRA (combined binary ratio labeling) strategy introduced by Tanke and colleagues (1999).

A crucial step for all labeling systems is to achieve an optimal fragment size of the labeled FISH probe, in order to assure appropriate penetration and hybridization efficiency. For FISH experiments, the fragment size after labeling should not exceed 300 bp. Therefore, some of these labeling approaches also require the enzymatic digestion of the probe DNA by DNase or ultrasonic treatment before *in situ* hybridization.

12.4.4 Hybridization, Post-hybridization Wash, Detection, and Documentation

For DNA duplex formation, probe DNA and target DNA has to be single-stranded (see Fig. 12.2). For denaturation, the probe DNA dissolved in hybridization buffer containing formamide and appropriate salt-buffer (hybridization mix) is melted in a water bath at the appropriate temperature. Many probes of higher complexity contain repetitive DNA elements that are scattered throughout the genome. In order to suppress their unspecific cross-hybridization, unlabeled competitor (Cot-1) DNA, a highly repetitive fraction of the human genome, is added to the hybridization mix. The Cot-1 DNA binds to the highly repetitive elements in the probe, suppressing their unspecific cross-hybridization to the target (chromosome *in situ* suppression hybridization; Hulten *et al.*, 1991). After appropriate prehybridization or preannealing, the liquid hybridization mix is applied onto the target material for appropriate time depending on the size and complexity of the probe. Non- and unspecific bound probe is washed off using stringent washes under conditions that

do not denature the specific DNA duplexes (wash temperature lower than melting temperature).

For direct-labeled probes, no detection steps are necessary, whereas indirect labeled probes (e.g., biotin or digoxigenin labeled probes) have to be detected using appropriate fluorochrome-conjugated antibodies. After counterstaining the target material using either DAPI (4′,6-Diamidino-2-phenylindole) or propidium iodide, and application of an antifade solution to prevent bleaching of the fluorescent signals, hybridization results can be visually inspected using an epifluorescence microscope equipped with appropriate filter sets. Electronic documentation of hybridization results is routinely done using charge-coupled device (CCD) camera, connected with a computer and appropriate analysis software.

12.5 TYPES OF FISH PROBES AND RECENT FISH APPROACHES FOR METAPHASE AND INTERPHASE FISH

The ability to detect and characterize chromosomal abnormalities in metaphase spreads and interphase cells using FISH has been greatly enhanced by the rapidly increasing availability of numerous chromosome and locus specific probes (see Fig. 12.3 and Table 12.2).

The choice of probe and the simultaneous use of multiprobe assays depend on the particular application in question.

12.5.1 Centromeric Satellite Probes

The first routinely used FISH probes were centromere-specific probes detecting highly repetitive centromeric α-satellite DNA-sequences (Cremer *et al.*, 1986). In molecular diagnostics, these probes are mainly used for chromosome enumeration and marker chromosome identification (Figs. 12.4a and b). Especially for interphase cell analysis, which can be performed on cytological preparations as well as in sections of formaldehyde-fixed and paraffin-embedded tissues, these probes are particularly attractive because of high sensitivity and excellent hybridization efficiencies. The hybridization time of these highly repetitive probes is short and provides rapid results without the need of specific cytogenetic expertise for chromosome recognition or cell culture/metaphase preparation. The so-called all-human centromeric probes, a mix of all α-satellite repeats of the human genome hybridizing to all human centromeres simultaneously, are used for the detection of dicentric chromosomes or for the definition of neocentromeres, lacking α-satellite DNA-sequences.

However, by using centromeric probes for interphase cytogenetics, centromeric polymorphisms can result in split signals and therefore in false positive findings. Additionally, for some chromosomes, sequence homologies of the satel-

TABLE 12.2   FISH probes for the characterization of specific chromosome alterations.

Probes for FISH	Hybridization target	Patient material	Aberration type detectable
Commercial probes			
Centromere specific α-satellite probes	Centromeres	Metaphase spreads Interphase cells	Numerical aberrations, identification of marker chromosome origin
Whole chromosome probes (WCP) and partial chromosome arm specific probes (PCP)	Whole chromosomes	Metaphase spreads	Numerical aberrations, interchromosomal aberrations as balanced translocations or nonhomologous insertions
Locus specific probes (microdeletion and microduplication probes)	Submicroscopic chromosomal loci	Metaphase spreads Interphase cells	Microdeletions and microduplications
Tumor probes (single fusion probes, break apart probes, double fusion probes, oncogene amplification probes)	Submicroscopic fused chromosomal loci (translocation breakpoints), homogeneously staining regions (gene amplifications)	Interphase cells (Metaphase spreads)	Chimeric gene fusions, gene amplifications, inversions, deletions
All telomeric probes (Q-FISH probes)	Telomere repeats	Metaphase spreads	Quantification of telomere size, loss of telomere
Subtelomeric probes	Subtelomeric regions	Metaphase spreads	Cryptic terminal deletions, duplications, translocations
Noncommercial probes			
PCR amplified microdissected or chromosomes or chromosomal bands (Micro-FISH)	Chromosomes or parts thereof	Patient and control metaphase spreads, forward and reverse "painting"	Marker chromosome characterization
PCR products in situ (PRINS) Cep, microdeletion, genes PCR primer for specifc chromosomal loci, genes, or alphoid sequences	Telomere repeats, microdeletions, single genes	Metaphase spreads	Telomeric aberrations, microdeletions, single gene deletions

lite sequences are resulting in cross-hybridizations (e.g., chromosomes 13/21 and 14/22), making these probes less suitable for interphase cytogenetics. Furthermore, the use of centromeric probes for supernumerary marker chromosome detection gives no information about the euchromatic chromosome contents accounting for possible phenotypic features.

12.5.2 Whole Chromosome "Painting" Probes

Whole chromosome probes (WCP), also called painting probes, are DNA libraries representing a cocktail of DNA fragments of a single human chromosome (Deaven *et al.*, 1986; Guan *et al.*, 1993; Figs. 12.4c and d). These DNA probes, obtained by chromosome flow sorting or chromosome microdissection (see Fig. 12.5; see also Section 12.5.3), allow the labeling of individual chromosomes in metaphase spreads, and subsequently the identification and characterization of both numerical and interchromosomal structural aberrations as translocations and nonhomologous insertions. Partial chromosome probes, generated by chromosome microdissection and representing the short or the long arm of chromosomes, are valuable tools for the detec-

tion of intrachromosomal pericentric inversions including the centromere.

In routine use, the application of appropriate whole chromosome probes to ascertain chromosomal aberrations needs the prior knowledge (cytogenetic expertise) of the affected chromosome(s) in question. Otherwise, for example for euchromatic marker chromosome identification, a whole repertoire of different DNA probes has to be hybridized and analyzed to narrow down or to identify the origin of unknown chromosomal material (Blennow *et al.*, 1995).

However, a locus-specific determination of the affected chromosomal region in intrachromosomal rearrangements, such as deletions and duplications, and paracentric inversions, affecting only one chromosome arm, are not detectable by whole or partial chromosome probes. The detection of small interchromosomal translocations involving only very small regions of the chromosome ends cannot be performed often with the adequate diagnostic sensitivity by using whole chromosome probes. In contrast to centromeric probes, chromosome enumeration by whole chromosome probes can be performed only on metaphase spreads, demanding the availability of proliferating material.

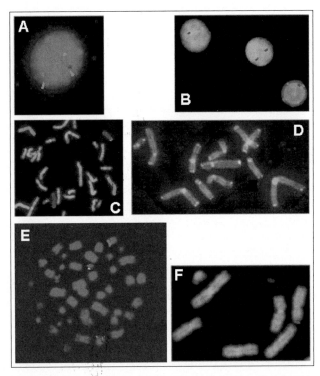

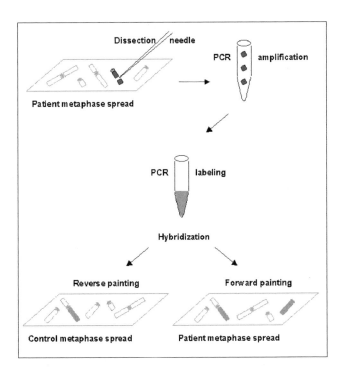

FIGURE 12.4 Example combinations of single probes used in routine diagnostics. A. Interphase nuclei of an uncultured amniocytes hybridized with three different alphoid probes (cep-X, red; cep-Y, green, cep-18, blue). A normal male signal constitution has been detected for the gonosomes X and Y. The three blue signals gave evidence for a trisomy 18 in the amniocytes (Courtesy of Markus Stumm, Berlin, Germany). B. Interphase FISH using centromeric probes for chromosomes 7 (green) and 9 (red) detecting a monosomy 7 mosaicism. C. Whole chromosome probes for chromosome 3 (green) and 11 (red) detecting a reciprocal translocation t(3;11). D. Combinatorial hybridization of whole chromosome painting for chromosome 4 (red) and an all telomere repeat probe. E. Result of a D-FISH experiment performed on a bone marrow metaphase of a chronic myelogenous leukemia (CML) patient detecting a Philadelphia-positive metaphase spread resulting in mix-color signals (fusion signals) on the derivative chromosomes der(9) and der(22) (Courtesy of Ivan Loncarevic, Institute of Human Genetics and Anthropology, Jena, Germany). F. Microdeletion of the SHOX gene (red) at the tip of one X-chromosome in a girl. As a control probe a cep-X (green) has been hybridized (See color plate).

FIGURE 12.5 Flowchart summarizing the practical steps of marker chromosome microdissection (Micro-FISH). Using a glass dissection needle that is controlled by a micromanipulator, some copies of the marker chromosome are dissected and transferred to a microcentrifuge tube. Subsequently, the chromosomal DNA is amplified and labeled by DOP-PCR. The resulting chromosome probe is hybridized to metaphases containing the marker (patient metaphase spread; forward painting) to verify the regional authenticity of the probe. Simultaneously, the chromosomal composition of the marker is determined by hybridization of the probe to normal metaphase chromosomes (reverse painting).

12.5.3 Probe Generation by Chromosome Micro-dissection (Micro-FISH)

The characterization of the composition and chromosomal origin of marker chromosomes or parts thereof can be performed straightforward by chromosome microdissection or micro-FISH (Meltzer *et al.*, 1992). This method is based on the micro-manipulated chromosome dissection of the marker chromosome followed by PCR-mediated DNA amplification. By reverse (on normal metaphase spread) or forward (on patients' metaphase spreads) hybridization, the complete euchromatic content of the marker chromosome can be explored without the use of commercial probes (see Fig. 12.5). Prerequisite for this assay is proliferating cell material of the patient. In addition to the fact that some chromosome aberrations as inversions and small duplications are not detectable by using dissected material as a probe, precise breakpoint identification is limited. Furthermore, there are practical limitations to micro-FISH for diagnostic purposes since specialized equipment, technical skills, and a profound cytogenetic knowledge for the undoubtful recognition of the marker chromosome are necessary.

12.5.4 Region and Locus-specific Probes

12.5.4.1 Locus-specific Probes

For the investigation of small, submicroscopic chromosomal loci of the human genome, which are too small to be visualized by conventional cytogenetics, a wide spectrum of so-called locus-specific probes are used routinely in diagnostic FISH laboratories. Most of these probes are vector cloned probes. For the molecular-cytogenetic verification of syndromes with submicroscopic gain or loss, such as the classical microdeletion and microduplication syndromes (see Table 12.1), single-copy probes detecting the common region of interest are hybridized to patient cells (see Fig. 12.4f). Although metaphase spreads are used routinely as targets for microdeletion syndromes, interphase nuclei are utilized for the detection of microduplications because of the better separation of closely spaced hybridization signals (Shaffer and Lupski, 2000). Today, most commercial probes include a second control probe hybridizing on the nonaffected chromosome arm for the validation of successful hybridization.

In prenatal diagnostics, different labeled single probes can be used for rapid testing for chromosomal aneuploidies on uncultivated amniocytes (Kuo *et al.*, 1991; Klinger *et al.*, 1992; Ward *et al.*, 1993; Fig. 12.4a). By this approach, a combination of locus-specific and alphoid probes of human chromosomes 21, 18, 13, X, and Y are hybridized in two batches for the detection of the most common aneuploidies in humans. However, this FISH test is not designed to detect all chromosome aneuploidies and can be utilized only as an adjunctive test to conventional cytogenetics.

12.5.4.2 TUMOR PROBES
In tumor genetics, most chromosomal rearrangements have major diagnostic and prognostic relevance. A number of single-copy locus-specific probes, detecting characteristic translocations, inversions, deletions, and oncogene amplifications on the single cell level (Wang, 2002) are commercially available. These probes allow the investigation of metaphase spreads and interphase cells in one experiment. Furthermore, the detection of minimal residual disease (see also Chapter 10) and post-transplantation follow-up can be done with high sensitivity. Some of these locus-specific probes precisely locate translocation and inversion breakpoints, reflecting chimeric gene fusions in neoplastic diseases, especially in leukemias (e.g., *BRC-ABL*). Single-fusion dual-color FISH probes (Figs. 12.3 and 12.4e), which were the first available tumor probes, detect specific translocation breakpoints in interphase nuclei by the presence of nonrandomly distributed juxtaposed signals. However, in interphase diagnostics, these probes have high rates of false positive signals. Newer types of translocation probes in cancer genetics are the break apart probes detecting simple splits of normally proximate signals and so-called double-fusion probes (D-FISH probes) detecting a fusion signal on both derivative chromosomes (Grand *et al.*, 1998). Oncogene amplification probes, commonly used in solid tumor genetics, visualize gene amplifications in metaphase and interphase cells and clarify the identity of the genes involved.

12.5.4.3 SUBTELOMERIC PROBES AND TELOMERE REPEAT PROBES
The highest gene concentrations in the human genome are in the subtelomeric regions of metaphase chromosomes

(Saccone *et al.*, 1992). These chromosomal regions are prone to rearrangements, which could give rise to cryptic aberrations probably accounting for 5% to 10% of unexplained moderate-to-severe mental retardation cases, congenital anomalies, and spontaneous abortions (Knight *et al.*, 1997; 1999; Anderlid *et al.*, 2002). Subtelomeric probes were established for all chromosome ends (excluding the short arms of acrocentric chromosomes) to screen metaphase spreads for cryptic translocations or imbalances at the terminal euchromatic parts of the chromosomes.

The noncoding telomere repeat (TTAGGG)n distal to the subtelomeric region at the end of each eukaryotic chromosome protects the chromosome against rearrangements and fusion with other chromosomes (see Fig. 12.4d). Using a repetitive all-telomere probe, ring chromosome formation can be pinpointed, especially in small supernumerary ring chromosomes.

Telomere shortening, resulting from cell divisions over time, can lead to genomic instability and neoplasia (Shay *et al.*, 1994). A quantitative FISH assay, using peptide nucleic acid (PNA) telomere oligonucleotide probes, resulting in stronger hybridization signals than standard DNA oligonucleotide probes, has been described by Lansdorp and colleagues (1996). Hybridizing these probes, the fluorescence intensity detected is directly proportional to the amount of telomere repeats.

12.5.5 Special Probe Types and Combinations

In the literature, several additional specific probe designs and combinations were published over the years, but not all these are used routinely in diagnostics. It would go beyond the scope of this chapter to list all these in detail. All the following FISH-based sensitive approaches are detecting single or few chromosome alterations at a time.

12.5.5.1 PRIMED *IN SITU* LABELING
Primed *in situ* labeling (PRINS) is a complementary FISH approach based on the *in situ* hybridization of short unlabeled primers to metaphase spreads or interphase cells and subsequent *in situ* chain elongation catalyzed by a DNA polymerase (Koch *et al.*, 1989; 1991). The product of this polymerase-chain-reaction *in situ* is visible due to incorporation of labeled nucleotides. Depending on the primers used, this time- and cost-effective method allows the detection of centromeric satellite DNA (Hindkjaer *et al.*, 1995), telomeric repeats (Krejci and Koch, 1998), microdeletions, and single copy genes (Cinti *et al.*, 1993; Kadandale *et al.*, 2000; Tharapel *et al.*, 2002). However, PRINS requires the knowledge of the high exact target sequence for the primers and high quality target material.

12.5.5.2 SINGLE-COPY FISH PROBES
Recently, Rogan and colleagues (2001) introduced the use of a new generation of small single-copy FISH (sc-FISH)

probes designed by computational sequence analysis of approximately 100-kb genomic sequences, bridging the gap between molecular genetic data and molecular cytogenetics. These labeled probes, produced by PCR, can be hybridized without preannealing or blocking, as sc-FISH probes lack repetitive DNA sequences. These short probes are produced directly from genomic DNA without recombinant DNA techniques. Applications of these probes include detection of microdeletion syndromes and submicroscopic deletions (Knoll and Rogan, 2003).

12.5.6 MACISH and FICTION: Fluorescence Immunophenotyping and Interphase FISH

MACISH (morphology, antibody, chromosomes, *in situ* hybridization) and FICTION (fluorescence immunophenotyping and interphase cytogenetics) are methods combining immunophenotyping of cells and fluorescence *in situ* hybridization (Knuutila and Teerenhovi, 1989; Weber-Mathhiesen *et al.*, 1993). These approaches allow the examination of numerical and structural chromosome abnormalities of immunologically classified cells. They especially are used to detect tumor cells carrying chromosome abnormalities in mixed cell populations contaminated with normal cells as tissue sections or cytological preparations. Recently, Martin-Subero and colleagues (2002) described the use of multicolor-FICTION, allowing the simultaneous detection of the morphological and immunophenotypic characteristics of neoplastic cells together with the most frequent chromosomal aberrations.

12.6 MULTICOLOR FISH SCREENING ASSAYS

For the simultaneous visualization of different chromosome aberrations in one experiment, a variety of multicolor FISH assays have been developed in the last decade (Liehr *et al.*, 2002; Tönnies, 2002). Some of the most important techniques for routine cytogenetics are described here (see Table 12.3). Especially in molecular diagnostics, these multicolor assays afford to perform analysis on only a few metaphases as found in bone marrow samples.

12.6.1 CenM-FISH and CM-FISH

Recently, two all-human centromere-specific multicolor-FISH approaches for the subsequent determination of the exact origin of structurally abnormal, cytogenetically unidentifiable marker chromosomes has been reported by Henegariu and colleagues (2001a; CM-FISH) and Nietzel and colleagues (2001; cenM-FISH; Fig. 12.6a). These one-step multicolor FISH assays allow the simultaneous characterization of all human centromeres using differently labeled centromeric satellite DNA as probes for all human chromo-

somes. However, by using centromeric satellite probes, only the centromeric parts of chromosomes can be explored, whereas other chromosome abnormalities affecting the euchromatin are excluded from investigation by using these assays.

Since most markers originate from the acrocentric chromosomes, Langer and colleagues (2001) developed an acroM-FISH assay using a probe mix, which consists of painting probes and centromere probes for chromosomes 13/21, 14/22, and 15, and a probe specific for rDNA, each labeled with a specific combination of fluorochromes. Using this assay the origin of approximately 80% of all markers can be identified by one hybridization assay.

12.6.2 Multicolor Assays for Subtelomeric Rearrangements

Special subtelomeric multicolor probe mixes were established to scan metaphase spreads for subtle aberrations at the gene-rich ends of the chromosomes in a scanning assay (see also Section 12.5.4.3). Henegariu and colleagues (2001b) reported the analysis of 41 chromosome ends simultaneously using a multicolor hybridization assay including DNA probes located near the end of these chromosomes (0,1-1 Mb from the telomere, TM-FISH). In the same year, Brown and colleagues (2001) presented the so-called M-TEL 12-color-FISH assay, which permits the screening of all telomeres in only two hybridizations. As an important parameter for successful hybridizations, all three techniques require well-spread metaphases (proliferating cells) with no cytoplasm.

12.7 MULTICOLOR WHOLE METAPHASE SCANNING TECHNIQUES

In addition to the various locus-specific assays described in this chapter, which often require the hybridization of a whole repertoire of probes for different chromosomes successively to narrow down the overall composition of a chromosome aberration in single FISH experiments, some unbiased whole metaphase or chromosome scanning approaches are established for the characterization of highly rearranged and unbalanced chromosome aberrations. Two major techniques, multifluor-FISH (M-FISH; Speicher *et al.*, 1996; Fig. 12.6b) and spectral karyotyping (SKY; Schröck *et al.*, 1996; Fig. 12.6c) were developed for the detection of non-homologous structural and numerical chromosomal aberrations on the single metaphase level hybridizing a 24-color whole chromosome probe mix. Each homologue chromosome is displayed in a different color, based on the use of computer-generated false-color chromosome images and karyotyping. Although the principle of hybridizing whole chromosome probes labeled by five different fluorochromes in different combinations (combinatorial labeling) is the

TABLE 12.3 Molecular cytogenetic multicolor whole metaphase/cell scanning techniques used in human molecular diagnostics.

FISH-technique[a]	Probe setup	Hybridization target (patient material)	Aberration type detectable	Specific advantages of the technique	Diagnostic limitations
CenM-FISH and CM-FISH	All centromere specific α-satellite probes	Centromeres (metaphase spreads)	Numerical aberrations, identification of marker chromosome origin	Fast characterization of supernumerary marker chromosome origin, no need of specific cytogenetic expertise	No information about euchromatic content of the marker
TM-FISH, M-Tel FISH, S-COBRA	All subtelomeric regions	Subtelomeres (metaphase spreads)	Cryptic terminal deletions, duplications and translocations	High sensitivity, simultaneous analysis of all chromosome ends involved in rearrangements	Small analysis spectrum restricted to the chromosome ends, further analysis of affected chromosome(s) necessary
SKY/M-FISH and technical modifications (CCK-FISH, COBRA-FISH, IPM-FISH)	All human wcp specific probes	Whole chromosomes (metaphase spreads)	Interchromosomal balanced and unbalanced translocations, euchromatic marker chromosome identification	Whole metaphase scanning without prior probe selection, translocation partner detection in one experiment, fast characterization of euchromatic marker chromosome content	Insensitive detection of intrachromosomal aberrations as deletions, duplications and inversions, imprecise breakpoint detection, need of nonoverlapping chromosomes
Chromosomal bar codes and Rx-FISH	YAC clones, fragment hybrids, cross species wcp-mix	Whole chromosomes (metaphase spreads)	Inter- and intrachromosomal aberrations as translocations and gross deletions, duplications, inversions	Metaphase-wide detection of gross intra- and interchromosomal chromosome alterations and marker chromosomes	Suboptimal locus-specific resolution of breakpoints, need of nonoverlapping chromosomes
Multicolor-banding, MCB	Microdissection derived partial chromosome probes	Whole chromosomes (metaphase spreads)	Inter- and intrachromosomal aberrations as translocations, deletions, and inversions	Sensitive whole karyotype detection chromosomal alterations with chromosome band specific resolution	High costs and technically demanding, need of nonoverlapping chromosomes
Comparative genomic hybridization, CGH	Patient and control DNA	Whole control chromosomes (metaphase spreads)	Chromosomal inter- and intrachromosomal imbalances, gene amplifications, marker chromosomes	Whole genome scanning technique without need of proliferating patient material, locus specific detection of gene amplification, band-specific information on imbalance size	Insensitive detection of imbalances in a subpopulation of cells, no detection of balanced aberrations, technically demanding
Matrix-CGH	Patient and control DNA	Spotted defined DNA probes	Chromosomal inter- and intrachromosomal imbalances, gene amplifications, marker chromosomes	Whole genome scanning technique without need of proliferating patient material, high-resolution target specific detection of gene amplification, submicroscopic information on imbalances	Direct dependency on the spotted targets, technically demanding, no detection of balanced translocations, missing of small mosaics

[a] Abbreviations see text.

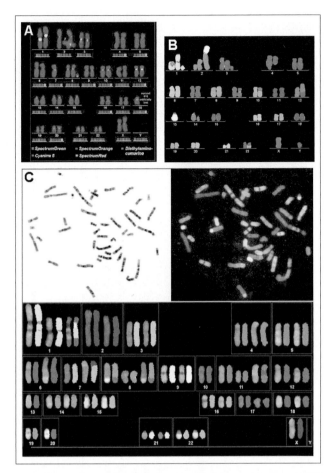

FIGURE 12.6 Different examples of multicolor FISH approaches. A. A supernumerary marker chromosome could be characterized as a derivative chromosome 8 using the cen-M-FISH approach (Courtesy of Thomas Liehr, Institute of Human Genetics and Anthropology, Jena, Germany). B. M-FISH karyotype with complex chromosomal aberrations after radiation of the lymphocyte culture (Courtesy of Christian Johannes, Essen, Germany. Hybridization and analysis carried out in collaboration with Meta-Systems, Germany). C. Spectral karyotyping (SKY)-analysis of a human multiple myeloma cell line (NCI-H929, CRL-9068, ATCC, Manassas, VA, USA). Several chromosomal aberrations were identified; e.g., translocations between chromosomes X and 7, 6 and 18, as well as 8 and 20, a dicentric chromosome 1 and a second dicentric chromosome 1 involved in a translocation or insertion with chromosome 10 (Courtesy of Isabell Grandy and Evelin Schröck, Institute of Clinical Genetics, Medical Faculty Carl Gustav Carus, Technical University, Dresden, Germany) (See color plate).

same, both approaches underlie a different technical concept: spectral karyotyping combines Fourier spectroscopy, CCD imaging, and optical microscopy to measure simultaneously at all points in the sample emission spectra in the visible and near-infrared spectral range (Schröck *et al.*, 1996). In contrast, M-FISH is based on classical epifluorescence microscopy using fluorochrome-specific optical filter sets for color discrimination (Speicher *et al.*, 1996). The fact that no prior knowledge of the affected chromo-

some(s) is required counts as an advantage for both techniques. By both approaches, the detection of chromosome rearrangements such as balanced or unbalanced translocations and euchromatic marker chromosomes in a single hybridization is possible, even if the chromosome morphology is poor. Therefore, both methods routinely are used in clinical and tumor genetics (McNeil and Ried, 2000; Schröck and Padilla-Nash, 2000; Bayani and Squire, 2002; Teixeira, 2002).

However, by both methods only the origin or constitution of an aberrant chromosome can be proved, whereas the exact locus or breakpoint assignment is not possible. Major drawbacks for routine use are the poor sensitivity to detect aberrations like intra-chromosomal deletions, duplications, and inversions. A large number of whole chromosome probes, labeled with different fluorochromes and the special technical equipment required, in particular for SKY analyses, restrict these scanning techniques to some specialized laboratories.

In 1999, Henegariu and colleagues presented a cost-effective alternative to SKY and M-FISH called color-changing karyotyping (CCK). Compared to both techniques just mentioned, CCK uses only three fluorochromes to discriminate up to 41 DNA probes. The discrimination is done through the difference in signal strength between direct and indirect labeled chromosomes. Therefore this approach can be used with a conventional 3-filter fluorescent microscope located in every molecular-cytogenetic lab. To achieve a better breakpoint resolution of translocations, Aurich-Costa and colleagues (2001) developed a multicolor karyotyping technique (IPM-FISH) that is based on the use of interspersed polymerase chain reaction generated (IRS-PCR) whole chromosome probes that display additionally to the color karyotyping a R-Band-like hybridization pattern.

12.8 MULTICOLOR CHROMOSOME BANDING TECHNIQUES

Until today, cytogenetics using black-and-white bar-coding of chromosomes is still the most important and cost-effective standard method to identify chromosome aberrations. However, due to the similar size and shade of some chromosome bands, a detailed identification and chromosomal assignment of altered chromosome regions by conventional cytogenetic banding techniques alone is not always satisfactory. Additionally, a pronounced cytogenetic expertise is needed for aberration detection and characterization, which restricts the number of persons handling these methods. Over time, the idea came up to develop a color-barcoding of chromosomes, which allows the detection of inter- and intrachromosomal aberrations as well as an automated karyotyping of the colored chromosomes. The first barcoding attempts, chromosomal bar codes (CBC; Lengauer *et al.*, 1993), were obtained by FISH with pools of Alu-PCR products from YAC clones containing human DNA inserts.

Four years later, Müller and colleagues (1997) described their chromosome bar code for human chromosomes, based on the application of a set of subregional DNA probes (human/rodent somatic cell hybrids) that distinguishes each chromosome in a single FISH assay. The result was a multicolor set of 110 distinct signals per haploid chromosome set. Using flow sorted primate chromosomes of two gibbon species, Müller and colleagues (1998) also established the so-called cross-species color segmenting or Rx-FISH. With the exception of six chromosomes, all other human chromosomes can be differentiated by this hybridization assay in at least two and up to six segments. However, allowing the detection of inter- and intrachromosomal aberrations, the resolution of classical banding patterns (approximately 400 to 600 bands/haploid genome) cannot be reached by these approaches (e.g., for breakpoint detection).

High-resolution multicolor-banding (MCB) for refined FISH analysis of human chromosomes on the band and subband level has been introduced by Chudoba and colleagues (1999) (see Figs. 12.7a and b). This technique is based on changing fluorescence intensity ratios of overlapping DNA probes, labeled by five different fluorochromes. After computer-based assignment of distinctive pseudocolors to the overlapping and nonoverlapping hybridization signals along the chromosome, this approach allows a higher resolution as the former mentioned, independent of chromosome condensation. By using the MCB technique for single chromosomes, preferentially intrachromosomal aberrations as deletions and peri- or paracentric inversions can be characterized.

Recently, Mrasek and colleagues (2001) reported the use of human MCB probes for all chromosomes, permitting the straightforward characterization not only of intrachromosomal aberrations but also of interchromosomal translocations and breakpoint mapping (see Fig. 12.7c). By using 138 microdissection derived MCB probes, a resolution of 450 bands and more can be achieved, enabling a high-resolution FISH-banding for the detection of complex chromosome rearrangements (Liehr *et al.*, 2002).

12.9 WHOLE GENOME SCANNING AND COMPARATIVE GENOMIC HYBRIDIZATION

All metaphase or chromosome scanning techniques utilize metaphases of the patient to uncover chromosomal aberrations. In tumor genetics, especially in solid tumor samples, the number of metaphases that can be prepared from these tissues is often small, if available at all. Additionally, the chromosome morphology is often poor so that karyotyping is barely possible.

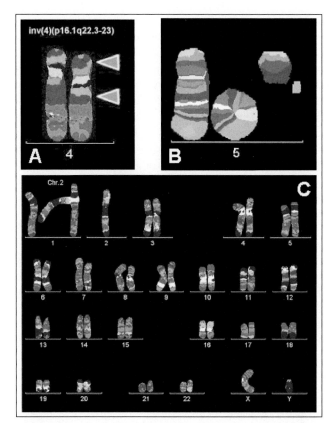

FIGURE 12.7 High-resolution multicolor-banding (MCB) for refined FISH analysis of human chromosomes on the band and subband level. A. MCB analysis showing a pericentric inversion (including the centromere) of chromosome 4 (Courtesy of Thomas Liehr). B. MCB analysis of a ring chromosome 5 and a derivative chromosome 5 induced by radiation of lymphocyte cultures (Courtesy of Christian Johannes). C. Example of the use of human MCB probes for all chromosomes simultaneously, permitting the straightforward characterization not only of intrachromosomal aberrations but also of interchromosomal translocations. By using 138 microdissection derived MCB probes a resolution of 450 bands and more can be achieved enabling a high-resolution FISH-banding for the detection of complex chromosome rearrangements (Courtesy of Thomas Liehr) (See color plate).

To circumvent these problems, Kallioniemi and colleagues (1992) established a hybridization approach, known as comparative genomic hybridization (CGH), which relies on the use of genomic tumor DNA circumventing difficult cell culture and chromosome preparations of the tumor tissue. CGH is a potent and reliable hybridization approach, linking conventional cytogenetic and molecular genetic techniques. It allows the comprehensive analysis of the entire genome from tissue samples in just one experiment providing comprehensive information about chromosomal imbalances and gene amplifications in the tumor genome. In brief, CGH is based on the cohybridization of differentially labeled whole genomic test (tumor) and normal (control)

DNA in a ratio of 1:1 to normal control metaphase spreads (see Fig. 12.8). Fluorochrome-labeled test- and control-DNA probes compete for hybridization on the target chromosomes. By visual inspection alone, the copy number changes; that is, the fluorescence ratio deviations of test DNA to control DNA cannot be evaluated sufficiently (see Fig. 12.9a). Therefore, software-assisted image capturing, karyotyping, and quantification of fluorescence intensities over the entire length of each chromosome is an integral part of this technique. The endpoint of a CGH analysis is a so-called copy number karyotype displaying all chromosomal imbalances on a conventional cytogenetic level (approximately 10 Mb; Fig. 12.9b).

To enhance sensitivity and resolution of CGH, Kirchhoff and colleagues (1998) reported on a high-resolution CGH approach based on a modified CGH software. These authors applied standard reference intervals as detection criteria and detected deletions down to 3 Mb (Kirchhoff *et al.*, 1999). CGH-based investigations provide information not only on the chromosomal assignment of a chromosomal imbalance but also on chromosomal band-specific origin. This approach allows studying DNA from any human source, even though the cells are not viable (e.g., paraffin-embedded tissue sections), or if DNA amount is limited (see also Section 12.5.3). Established in tumor genetics, a number of reports describing pre- and postnatal cytogenetic cases are available in the literature (Bryndorf *et al.*, 1995; Levy *et al.*, 1998; Tönnies *et al.*, 2001). Furthermore, even in preimplantation genetic diagnostics, first experiments have been reported, revealing chromosomal imbalances in blastomeres or polar body cells using DNA of a single cell amplified by PCR (Malmgren *et al.*, 2002; Wells *et al.*, 2002; see also Chapter 23).

Nevertheless, by CGH the detection of balanced chromosome aberrations, such as reciprocal translocations often found in routine leukemia genetics, is not possible. Furthermore, using whole genomic DNA, the information is restricted to the total amount of cells, of which the DNA is extracted. Therefore, if only a small number of cells are affected by a chromosomal imbalance (mosaicism), it will be missed due to contamination of the DNA with normal cells.

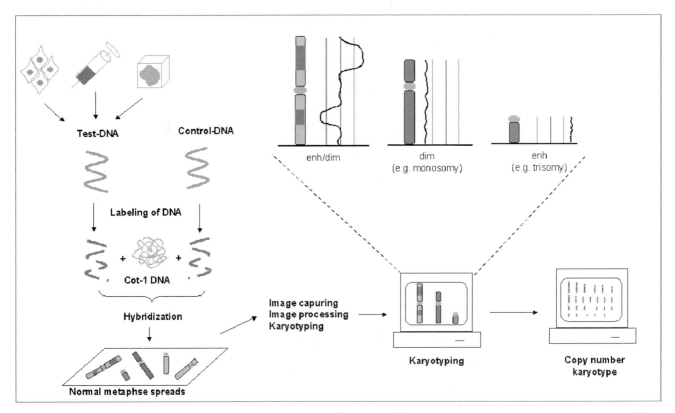

FIGURE 12.8 Simplified flowchart summarizing the steps of CGH technique. Different labeled control- and test-DNAs extracted from either tissue cultures, whole blood, or paraffin embedded material are hybridized under suppression conditions (Cot-1 DNA) on normal metaphase spreads. After image capturing and karyotyping of the metaphase chromosomes, fluorescence ratio profiles (quotient of test- to control-fluorescence intensity) for each chromosome are calculated. Mean profiles are plotted against the length of each chromosome. The center line (black) in the CGH profiles represents the balanced state of the chromosomal copy number (ratio = 1.0). An upper threshold (green) is used to define a gain of chromosomal material (ENH = enhanced), and a lower threshold (red) is used to interpret a loss of chromosomal material (DIM = diminished). CGH results are displayed in the so-called copy number karyotype (See color plate).

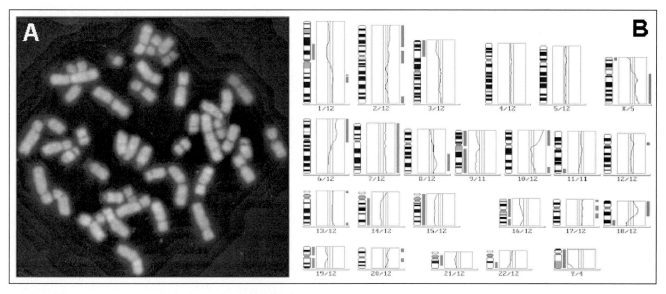

FIGURE 12.9 Comparative genomic hybridization of DNA extracted from ovarian cancer tissue. A. Metaphase spread hybridized with normal control DNA (red), tumor DNA (green), and counterstained with DAPI. A high number of fluorescence ratio differences are visible by eye inspection. B. Copy number karyotype of the same CGH analysis. Nearly all chromosomes are showing full (e.g., chromosome 7) or partial (e.g., chromosome 6) imbalances (See color plate).

12.10 DNA-ON-A-CHIP—MATRIX CGH

Molecular cytogenetic FISH technologies, as described in this chapter, are based on the combination of molecular genetic and cytogenetic techniques detecting complementary nucleic acid sequences *in situ* on the metaphase chromosome or in the cell nuclei. However, the resolution level, as described for CGH, is restricted to approximately 10 Mb. To be independent from high quality metaphases and to enhance the resolution of analyses, Solinas-Toldo and colleagues (1997) established a matrix-based CGH array (Matrix-CGH) that substitutes normal metaphase chromosomes as hybridization target by well-defined genomic DNA fragments arrayed on a solid support, allowing automated analysis of genetic imbalances as small deletions or gene amplifications (Pinkel *et al.*, 1998; Pollack *et al.*, 1999; Fig. 12.10). Recently, Snijders and colleagues (2001) established a microarray using 2,400 BAC inserts covering nearly the whole human genome for the detection of gains and losses in diploid, polyploid, and heterogeneous backgrounds. Veltman and colleagues (2002) described a DNA chip spotted with a set of 77 human chromosome specific subtelomeric probes as a target for array-based CGH. The dynamic range of the signal ratios obtained by these noncytogenetic assays is up to five times higher for matrix CGH in comparison to conventional CGH, allowing for better quantitative assessments of genetic imbalances (Wessendorf *et al.*, 2002). Amplifications and deletions of small genomic regions not detectable by chromosome analyses or CGH can be recognized with high resolution directly depending on the number of human DNA fragments spotted on the chip. By

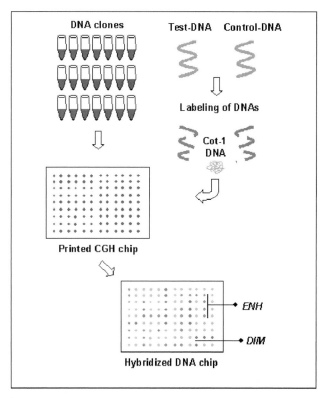

FIGURE 12.10 Schematic illustration of the matrix-comparative genomic hybridization procedure. Cloned DNAs reflecting the human genome or parts thereof are printed on a CGH microarray and serve as a target for labeled control- and test-DNAs. After hybridization, the resulting relative fluorescence intensities are measured by computer-supported scanning. DNA loss in the test-DNA results in red spots, gains in green spots (See color plate).

the use of 3,500 inserts, a resolution of 1 Mb can easily be obtained. The robustness and simplicity of these microarrays make them highly suited for routine use. If validated, these will be rapid and sensitive automated diagnostic procedures independent from cell culture.

12.11 CONCLUSIONS AND PERSPECTIVES

Recent advances, such as improved hybridization protocols and strategies, novel commercial and noncommercial probe sets, coupled with hardware development such as image analysis software and specific filter sets for newly established fluorochromes, facilitate the collection of new important data for the understanding and diagnosis of genetic diseases. Other current research-based applications, which could not be listed in this chapter, are numerous and likely to translate into a growing list of clinically useful applications in the near future. For diagnostic purposes, new time- and cost-saving approaches with the potential to investigate high numbers of genetic loci in an automated fashion and with high resolution independently of cell culture and chromosome preparation are favored.

References

Anderlid, B. M., Schoumans, J., Anneren, G., Sahlen, S., Kyllerman, M., Vujic, M., Hagberg, B., Blennow, E., and Nordenskjold, M. (2002). Subtelomeric rearrangements detected in patients with idiopathic mental retardation. *Am. J. Med. Genet.* 107, 275–284.

Aurich-Costa, J., Vannier, A., Gregoire, E., Nowak, F., and Cherif, D. (2001). IPM-FISH, a new M-FISH approach using IRS-PCR painting probes: application to the analysis of seven human prostate cell lines. *Genes Chromosomes Cancer* 30, 143–160.

Barch, M. J., Knutsen, T., and Spurbeck, J. L. (1997). *The AGT Cytogenetics Laboratory Manual.* Lippincott-Raven, Philadelphia.

Bauman, J. G., Wiegant, J., Borst, P., and van Duijn, P. (1980). A new method for fluorescence microscopical localization of specific DNA sequences by in situ hybridization of fluorochrome-labelled RNA. *Exp. Cell. Res.* 128, 485–490.

Bayani, J. M., and Squire, J. A. (2002). Applications of SKY in cancer cytogenetics. *Cancer Invest.* 20, 373–386.

Blennow, E., Nielsen, K. B., Telenius, H., Carter, N. P., Kristoffersson, U., Holmberg, E., Gillberg, C., and Nordenskjold, M. (1995). Fifty probands with extra structurally abnormal chromosomes characterized by fluorescence in situ hybridization. *Am. J. Med. Genet.* 55, 85–94.

Brown, J., Saracoglu, K., Uhrig, S., Speicher, M. R., Eils, R., and Kearney, L. (2001). Subtelomeric chromosome rearrangements are detected using an innovative 12-color FISH assay (M-TEL). *Nat. Med.* 7, 497–501.

Bryndorf, T., Kirchhoff, M., Rose, H., Maahr, J., Gerdes, T., Karhu, R., Kallioniemi, A., Christensen, B., Lundsteen, C., and Philip, J. (1995). Comparative genomic hybridization in clinical cytogenetics. *Am. J. Hum. Genet.* 57, 1211–1220.

Buckton, K. E., Spowart, G., Newton, M. S., and Evans, H. J. (1985). Forty-four probands with an additional marker chromosome. *Hum. Genet.* 69, 353–370.

Casperson, T., Zech, L., and Johhansson, C. (1970). Differential banding of alkylating fluorochromes in human chromosomes. *Exp. Cell. Res.* 60, 315–319.

Chudoba, I., Plesch, A., Lorch, T., Lemke, J., Claussen, U., and Senger, G. (1999). High resolution multicolor-banding: a new technique for refined FISH analysis of human chromosomes. *Cytogenet. Cell Genet.* 84, 156–160.

Cinti, C., Santi, S., and Maraldi, N. M. (1993). Localization of single copy gene by PRINS technique. *Nucleic Acids Res.* 21, 5799–5800.

Cremer, T., Landegent, J., Bruckner, A., Scholl, H. P., Schardin, M., Hager, H. D., Devilee, P., Pearson, P., and van der Ploeg, M. (1986). Detection of chromosome aberrations in the human interphase nucleus by visualization of specific target DNAs with radioactive and non-radioactive in situ hybridization techniques: diagnosis of trisomy 18 with probe L1.84. *Hum. Genet.* 74, 346–352.

Dauwerse, J. G., Wiegant, J., Raap, A. K., Breuning, M. H., and van Ommen, G. J. (1992). Multiple colors by fluorescence in situ hybridization using ratio-labelled DNA probes create a molecular karyotype. *Hum. Mol. Genet.* 1, 593–598.

Deaven, L. L., Van Dilla, M. A., Bartholdi, M. F., Carrano, A. V., Cram, L. S., Fuscoe, J. C., Gray, J. W., Hildebrand, C. E., Moyzis, R. K., and Perlman, J. (1986). Construction of human chromosome-specific DNA libraries from flow-sorted chromosomes. *Cold Spring Harb. Symp. Quant. Biol.* 51 Pt 1, 159–167.

Delhanty, J. D., Griffin, D. K., Handyside, A. H., Harper, J., Atkinson, G. H., Pieters, M. H., and Winston, R. M. (1993). Detection of aneuploidy and chromosomal mosaicism in human embryos during preimplantation sex determination by fluorescent in situ hybridisation, (FISH). *Hum. Mol. Genet.* 2, 1183–1185.

Feinberg, A. P., and Vogelstein, B. (1983). A technique for radio-labeling DNA restriction endonuclease fragments to high specific activity. *Anal. Biochem.* 132, 6–13.

Grand, F. H., Chase, A., Iqbal, S., Nguyen, D. X., Lewis, J. L., Marley, S. B., Davidson, R. J., Goldman, J. M., and Gordon, M. Y. (1998). A two-color BCR-ABL probe that greatly reduces the false positive and false negative rates for fluorescence in situ hybridization in chronic myeloid leukemia. *Genes Chromosomes Cancer* 23, 109–115.

Guan, X. Y., Trent, J. M., and Meltzer, P. S. (1993). Generation of band-specific painting probes from a single microdissected chromosome. *Hum. Mol. Genet.* 2, 1117–1121.

Harper, M. E., and Saunders, G. F. (1981). Localization of single copy DNA sequences of G-banded human chromosomes by in situ hybridization. *Chromosoma* 83, 431–439.

Henegariu, O., Bray-Ward, P., Artan, S., Vance, G. H., Qumsyieh, M., Ward, D. C. (2001a). Small marker chromosome identification in metaphase and interphase using centromeric multiplex fish (CM-FISH). *Lab. Invest.* 81, 475–481.

Henegariu, O., Artan, S., Greally, J. M., Chen, X. N., Korenberg, J. R., Vance, G. H., Stubbs, L., Bray-Ward, P., and Ward, D. C. (2001b). Cryptic translocation identification in human and mouse using several telomeric multiplex fish (TM-FISH) strategies. *Lab. Invest.* 81, 483–491.

Henegariu, O., Heerema, N. A., Bray-Ward, P., and Ward, D. C. (1999). Colour-changing karyotyping: an alternative to M-FISH/SKY. *Nat. Genet.* 23, 263–264.

Heng, H. H., Squire, J., and Tsui, L. C. (1992). High-resolution mapping of mammalian genes by in situ hybridization to free chromatin. *Hum. Mol. Genet.* 1, 587–591.

Hindkjaer, J., Brandt, C. A., Koch, J., Lund, T. B., Kolvraa, S., and Bolund, L. (1995). Simultaneous detection of centromere-specific probes and chromosome painting libraries by a combination of primed in situ labelling and chromosome painting (PRINS-painting). *Chromosome Res.* 3, 41–44.

Hsu, T. C., and Pomerat, C. M. (1953). Mammalian chromosomes in vitro. II. A method for spreading the chromosomes of cells in tissue culture. *J. Hered.* 44, 23–29.

Hulten, M. A., Gould, C. P., Goldman, A. S., and Waters, J. J. (1991). Chromosome in situ suppression hybridisation in clinical cytogenetics. *J. Med. Genet.* 28, 577–582.

ISCN (1995). *An International System for Human Cytogenetic Nomenclature*, F. Mitelman, ed. S. Karger, Basel.

John, H., Birnstiel, M., and Jones, K. (1969). RNA-DNA hybrids at the cytological level. *Nature* 223: 582–587.

Kadandale, J. S., Tunca, Y., and Tharapel, A. T. (2000). Chromosomal localization of single copy genes SRY and SOX3 by primed in situ labeling (PRINS*). Microb. Comp. Genomics* 5, 71–74.

Kallioniemi, A., Kallioniemi, O. P., Sudar, D., Rutovitz, D., Gray, J. W., Waldman, F. M., and Pinkel, D. (1992). Comparative genomic hybridization for molecular cytogenetic analysis of solid tumors. *Science* 258, 818–820.

Kirchhoff, M., Gerdes, T., Rose, H., Maahr, J., Ottesen, A. M., and Lundsteen, C. (1998). Detection of chromosomal gains and losses in comparative genomic hybridization analysis based on standard reference intervals. *Cytometry* 31, 163–173.

Kirchhoff, M., Gerdes, T., Maahr, J., Rose, H., Bentz, M., Dohner, H., and Lundsteen, C. (1999). Deletions below 10 megabase-pairs are detected in comparative genomic hybridization by standard reference intervals. *Genes Chromosomes Cancer* 25, 410–413.

Klinger, K., Landes, G., Shook, D., Harvey, R., Lopez, L., Locke, L., Osathanondh, R., Leverone, B., Houseal, T., Pavelka, K., and Dackowski, W. (1992). Rapid detection of chromosome aneuploidies in uncultured amniocytes by using fluorescence in situ hybridization (FISH). *Am. J. Hum. Genet.* 51, 55–65.

Knight, S. J., Horsley, S. W., Regan, R., Lawrie, N. M., Maher, E. J., Cardy, D. L., Flint, J., and Kearney, L. (1997). Development and clinical application of an innovative fluorescence in situ hybridization technique which detects submicroscopic rearrangements involving telomeres. *Eur. J. Hum. Genet.* 5, 1–8.

Knight, S. J., Regan, R., Nicod, A., Horsley, S. W., Kearney, L., Homfray, T., Winter, R. M., Bolton, P., and Flint, J. (1999). Subtle chromosomal rearrangements in children with unexplained mental retardation. *Lancet* 354, 1677–1681.

Knoll, J. H., and Rogan, P. K. (2003). Sequence-based, in situ detection of chromosomal abnormalities at high resolution. *Am. J. Med. Genet.* 121A, 245–257.

Knuutila, S., and Teerenhovi, L. (1989). Immunophenotyping of aneuploid cells. *Cancer Genet. Cytogenet.* 41, 1–17.

Koch, J. E., Kolvraa, S., Petersen, K. B., Gregersen, N., and Bolund, L. (1989). Oligonucleotide-priming methods for the chromosome-specific labelling of alpha satellite DNA in situ. *Chromosoma* 98, 259–265.

Koch, J., Hindkjaer, J., Mogensen, J., Kolvraa, S., and Bolund, L. (1991). An improved method for chromosome-specific labeling of alpha satellite DNA in situ by using denatured double-stranded DNA probes as primers in a primed in situ labeling (PRINS) procedure. *Genet. Anal. Tech. Appl.* 8, 171–178.

Krejci, K., and Koch, J. (1998). Improved detection and comparative sizing of human chromosomal telomeres in situ. *Chromosoma* 107, 198–203.

Kuo, W. L., Tenjin, H., Segraves, R., Pinkel, D., Golbus, M. S., and Gray, J. (1991). Detection of aneuploidy involving chromosomes 13, 18, or 21, by fluorescence in situ hybridization (FISH) to interphase and metaphase amniocytes. *Am. J. Hum. Genet.* 49, 112–119.

Langer, S., Fauth, C., Rocchi, M., Murken, J., and Speicher, M. R. (2001). AcroM fluorescent in situ hybridization analyses of marker chromosomes. *Hum. Genet.* 109, 152–158.

Lansdorp, P. M., Verwoerd, N. P., van de Rijke, F. M., Dragowska, V., Little, M-T., Dirks, R. W., Raap, A. K., and Tanke, H. J. (1996). Heterogeneity in telomere length of human chromosomes. *Hum. Mol. Genet.* 5, 685–691.

Lejeune, J., Turpin, R., and Gautier, M. (1959). Chromosomic diagnosis of mongolism. *Arch. Fr. Pediatr.* 16, 962–963.

Lengauer, C., Speicher, M. R., Popp, S., Jauch, A., Taniwaki, M., Nagaraja, R., Riethman, H. C., Donis-Keller, H., D'Urso, M., and Schlessinger, D. (1993). Chromosomal bar codes produced by multicolor fluorescence in situ hybridization with multiple YAC clones and whole chromosome painting probes. *Hum. Mol. Genet.* 2, 505–512.

Levy, B., Dunn, T. M., Kaffe, S., Kardon, N., and Hirschhorn, K. (1998). Clinical applications of comparative genomic hybridization. *Genet. Med.* 1, 4–12.

Liehr, T., Heller, A., Starke, H., Rubtsov, N., Trifonov, V., Mrasek, K., Weise, A., Kuechler, A., and Claussen, U. (2002). Microdissection based high resolution multicolor banding for all 24 human chromosomes. *Int. J. Mol. Med.* 9, 335–339.

Liehr, T., and Claussen, U. (2002). Multicolor-FISH Approaches for the Characterization of Human Chromosomes in Clinical Genetics and Tumor Cytogenetics. *Curr. Genomics* 3, 213–235.

Malmgren, H., Sahlen, S., Inzunza, J., Aho, M., Rosenlund, B., Fridstrom, M., Hovatta, O., Ahrlund-Richter, L., Nordenskjold, M., and Blennow, E. (2002). Single cell CGH analysis reveals a high degree of mosaicism in human embryos from patients with balanced structural chromosome aberrations. *Mol. Hum. Reprod.* 8, 502–510.

Martin-Subero, J. I., Chudoba, I., Harder, L., Gesk, S., Grote, W., Novo, F. J., Calasanz, M. J., and Siebert, R. (2002). Multicolor-FICTION: expanding the possibilities of combined morphologic, immunophenotypic, and genetic single cell analyses. *Am. J. Pathol.* 161, 413–420.

McNeil, N., and Ried, T. (2000). Novel molecular cytogenetic techniques for identifying complex chromosomal rearrangements: technology and applications in molecular medicine. *Expert Rev. Mol. Med.* 14, 1–14.

Meltzer, P. S., Guan, X. Y., Burgess, A., and Trent, J. M. (1992). Rapid generation of region specific probes by chromosome microdissection and their application. *Nat. Genet.* 1, 24–28.

Moorhead, P. S., Nowell, P. C., Mellman, W. J., Battips, D. M., and Hungerford, D. A. (1960). Chromosome preparations of leukocytes cultured from human peripheral blood. *Exp. Cell. Res.* 20, 613–616.

Mrasek, K., Heller, A., Rubtsov, N., Trifonov, V., Starke, H., Rocchi, M., Claussen, U., and Liehr, T. (2001). Reconstruction of the female gorilla karyotype using 25-color FISH and multicolor banding (MCB). *Cytogenet. Cell Genet.* 93, 242–248.

Müller, S., Rocchi, M., Ferguson-Smith, M. A., and Wienberg, J. (1997). Toward a multicolour chromosome bar code. *Hum. Genet.* 100, 271–278.

Müller, S., O'Brien, P. C., Ferguson-Smith, M. A., and Wienberg, J. (1998). Cross-species colour segmenting: a novel tool in human karyotype analysis. *Cytometry* 33, 445–452.

Nederlof, P. M., van der Flier, S., Wiegant, J., Raap, A. K., Tanke, H. J., Ploem, J. S., and van der Ploeg, M. (1990). Multiple fluorescence in situ hybridization. *Cytometry* 11, 126–131.

Nederlof, P. M., van der Flier, S., Vrolijk, J., Tanke, H. J., and Raap, A. K. (1992). Fluorescence ratio measurements of double-labeled probes for multiple in situ hybridization by digital imaging microscopy. *Cytometry* 13, 839–845.

Nietzel, A., Rocchi, M., Starke, H., Heller, A., Fiedler, W., Wlodarska, I., Loncarevic, I. F., Beensen, V., Claussen, U., and Liehr, T. (2001). A new multicolor-FISH approach for the characterization of marker chromosomes: centromere-specific multicolor-FISH (cenM-FISH). *Hum. Genet.* 108, 199–204.

Pardue, M. L., and Gall, J. G. (1969). Molecular hybridization of radioactive DNA to the DNA of cytological preparations. *Proc. Natl. Acad. Sci. USA.* 64, 600–604.

Pinkel, D., Segraves, R., Sudar, D., Clark, S., Poole, I., Kowbel, D., Collins, C., Kuo, W. L., Chen, C., Zhai, Y., Dairkee, S. H., Ljung, B. M., Gray, J. W., and Albertson, D. G. (1998). High resolution analysis of DNA copy number variation using comparative genomic hybridization to microarrays. *Nat. Genet.* 20, 207–211.

Pinkel, D., Straume, T., and Gray, J. W. (1986). Cytogenetic analysis using quantitative, high-sensitivity, fluorescence hybridization. *Proc. Natl. Acad. Sci. USA* 83, 2934–2938.

Pollack, J. R., Perou, C. M., Alizadeh, A. A., Eisen, M. B., Pergamenschikov, A., Williams, C. F., Jeffrey, S. S., Botstein, D., and Brown, P. O. (1999). Genome-wide analysis of DNA copy-number changes using cDNA microarrays. *Nat. Genet.* 23, 41–46.

Rautenstrauß, B. W., and Liehr, T. (2002). *FISH Technology*, Springer Lab Manual, Springer Verlag, Berlin, Heidelberg.

Rigby, P. W., Dieckmann, M., Rhodes, C., and Berg, P. (1977). Labeling deoxyribonucleic acid to high specific activity in vitro by nick translation with DNA polymerase I. *J. Mol. Biol.* 113, 237–251.

Rogan, P. K., Cazcarro, P. M., and Knoll, J. H. (2001). Sequence-based design of single-copy genomic DNA probes for fluorescence in situ hybridization. *Genome Res.* 11, 1086–1094.

Saccone, S., De Sario, A., Della Valle, G., and Bernardi, G. (1992). The highest gene concentrations in the human genome are in telomeric bands of metaphase chromosomes. *Proc. Natl. Acad. Sci. USA* 89, 4913–4917.

Schinzel, A. (2001). *Catalogue of unbalanced chromosome aberrations in man*, 2nd ed., de Gruyter, Berlin, New York.

Schröck, E., du Manoir, S., Veldman, T., Schoell, B., Wienberg, J., Ferguson-Smith, M. A., Ning, Y., Ledbetter, D. H., Bar-Am, I., Soenksen, D., Garini, Y., and Ried, T. (1996). Multicolor spectral karyotyping of human chromosomes. *Science* 273, 494–497.

Schröck, E., and Padilla-Nash, H. (2000). Spectral karyotyping and multicolor fluorescence in situ hybridization reveal new tumor-specific chromosomal aberrations. *Sem. Hematol.* 37, 334–347.

Schwarzacher, T., and Heslop-Harrison, P. (2000). *Practical in situ hybridisation*, BIOS Scientific Publishers Limited, Oxford.

Seong, D. C., Kantarjian, H. M., Ro, J. Y., Talpaz, M., Xu, J., Robinson, J. R., Deisseroth, A. B., Champlin, R. E., and Siciliano, M. J. (1995). Hypermetaphase fluorescence in situ hybridization for quantitative monitoring of Philadelphia chromosome-positive cells in patients with chronic myelogenous leukemia during treatment. *Blood* 86, 2343–2349.

Shaffer, L. G., and Lupski, J. R. (2000). Molecular mechanisms for constitutional chromosomal rearrangements in humans. *Annu. Rev. Genet.* 34, 297–329.

Shay, J. W., Werbin, H., and Wright, W. E. (1994). Telomere shortening may contribute to aging and cancer: a perspective. *Mol. Cell. Differ.* 2: 1–21.

Snijders, A. M., Nowak, N., Segraves, R., Blackwood, S., Brown, N., Conroy, J., Hamilton, G., Hindle, A. K., Huey, B., Kimura, K., Law, S., Myambo, K., Palmer, J., Ylstra, B., Yue, J. P., Gray, J. W., Jain, A. N., Pinkel, D., and Albertson, D. G. (2001). Assembly of microarrays for genome-wide measurement of DNA copy number. *Nat. Genet.* 29, 263–264.

Solinas-Toldo, S., Lampel, S., Stilgenbauer, S., Nickolenko, J., Benner, A., Dohner, H., Cremer, T., and Lichter, P. (1997). Matrix-based comparative genomic hybridization: biochips to screen for genomic imbalances. *Genes Chromosomes Cancer* 20, 399–407.

Speicher, M. R., Ballard, S. G., and Ward, D. (1996). Karyotyping human chromosomes by combinatorial multi-fluor FISH. *Nat. Genet.* 12, 368–375.

Tanke, H. J., Wiegant, J., van Gijlswijk, R. P., Bezrookove, V., Pattenier, H., Heetebrij, R. J., Talman, E. G., Raap, A. K., and Vrolijk, J. (1999). New strategy for multi-colour fluorescence in situ hybridisation: COBRA: COmbined Binary RAtio labelling. *Eur. J. Hum. Genet.* 7, 2–11.

Teixeira, M. R. (2002). Combined classical and molecular cytogenetic analysis of cancer. *Eur. J. Cancer* 38, 1580–1584.

Telenius, H., Carter, N. P., Nordenskjold, M., Ponder, B. A., and Yunnacliffe, A. (1992). Degenerate oligonucleotide-primed PCR: general amplification of target DNA by a single degenerate primer. *Genomics* 13, 718–725.

Tharapel, A. T., Kadandale, J. S., Martens, P. R., Wachtel, S. S., and Wilroy, R. S. Jr. (2002). Prader Willi/Angelman and DiGeorge/velocardiofacial syndrome deletions: diagnosis by primed in situ labeling (PRINS). *Am. J. Med. Genet.* 107, 119–122.

Tijo, J. H., and Levan, A. (1956). The chromosomes of man. *Hereditas* 42, 1–6.

Tönnies, H., Stumm, M., Wegner, R. D., Chudoba, I., Kalscheuer, V., and Neitzel, H. (2001). Comparative genomic based strategy for the analysis of different chromosome imbalances detected in conventional cytogenetic diagnostics. *Cytogenet. Cell. Genet.* 93, 188–194.

Tönnies, H. (2002). Modern molecular cytogenetic techniques in genetic diagnostics. *Trends Mol. Med.* 8, 246–250.

Tönnies, H., Neumann, L. M., Gruneberg, B., and Neitzel, H. (2003). Characterization of a supernumerary ring chromosome 1 mosaicism in two cell systems by molecular cytogenetic techniques and review of the literature. *Am. J. Med. Genet.* 121A, 163–167.

Veltman, J. A., Schoenmakers, E. F., Eussen, B. H., Janssen, I., Merkx, G., van Cleef, B., van Ravenswaaij, C. M., Brunner, H. G., Smeets, D., and van Kessel, A. G. (2002). High-throughput analysis of subtelomeric chromosome rearrangements by use of array-based comparative genomic hybridization. *Am. J. Hum. Genet.* 70, 1269–1276.

Verma, R. S., and Babu, A., eds. (1989). *Human chromosomes: manual of basic techniques*. Pergamon Press, New York.

Wang, N. (2002). Methodologies in cancer cytogenetics and molecular cytogenetics. *Am. J. Med. Genet.* 115, 118–124.

Warburton, D. (2000). *Cytogenetics of reproductive wastage: from conception to birth*. In H. F. L. Mark, ed., Medical Cytogenetics. pp. 213–246, Marcel Dekker, New York, USA.

Ward, B. E., Gersen, S. L., Carelli, M. P., McGuire, N. M., Dackowski, W. R., Weinstein, M., Sandlin, C., Warren, R., and Klinger, K. W. (1993). Rapid prenatal diagnosis of chromosomal aneuploidies by fluorescence in situ hybridization: clinical experience with 4500 specimens. *Am. J. Hum. Genet.* 52, 854–865.

Weber-Matthiesen, K., Pressl, S., Schlegelberger, B., and Grote, W. (1993). Combined immunophenotyping and interphase cytogenetics on cryostat sections by the new FICTION method. *Leukemia* 7, 646–649.

Wegner, R. D. (1999). *Diagnostic Cytogenetics*, Springer Lab Manual, Springer Verlag, Berlin, Heidelberg.

Wells, D., Escudero, T., Levy, B., Hirschhorn, K., Delhanty, J. D., and Munne, S. (2002). First clinical application of comparative genomic hybridization and polar body testing for preimplantation genetic diagnosis of aneuploidy. *Fertil. Steril.* 78, 543–549.

Wessendorf, S., Fritz, B., Wrobel, G., Nessling, M., Lampel, S., Goettel, D., Kuepper, M., Joos, S., Hopman, T., Kokocinski, F., Dohner, H., Bentz, M., Schwaenen, C., and Lichter, P. (2002). Automated screening for genomic imbalances using matrix-based comparative genomic hybridization. *Lab. Invest.* 82, 47–60.

Wiegant, J., Kalle, W., Mullenders, L., Brookes, S., Hoovers, J. M., Dauwerse, J. G., van Ommen, G. J., and Raap, A. K. (1992). High-resolution in situ hybridization using DNA halo preparations. *Hum. Mol. Genet.* 1, 587–591.

CHAPTER 13

Detection of Genomic Duplications and Deletions

GRAHAM R. TAYLOR[1] AND LORYN SELLNER[2]

[1] *Regional Genetics Laboratory & CR-UK Mutation Detection Facility, St James' University Hospital, DNA Laboratory, Leeds, Yorkshire LS9 7TF, UK;*

[2] *Princess Margaret Hospital, Roberts Road, SUBIACO WA 6008, Australia*

TABLE OF CONTENTS

13.1 INTRODUCTION

Gene and chromosome duplications have been implicated as fundamental evolutionary mechanisms (Ohno *et al.*,1968; Li and Gojobori, 1983; Schughart *et al.*, 1989). For the purpose of this chapter, duplications and deletions are defined as those too large to be routinely detected by sequencing or simple PCR-based methods (>100 bp), but too small to be detected by conventional cytogenetic metaphase analysis (<10^6 bp). Unlike other types of rearrangement, for example LINE insertions or balanced translocations, a duplication or deletion will be associated with a direct change in gene dosage. This four-order size range includes small fragments of genes, exons, entire genes, and multiple genes, and the same methods can, of course, be used to measure changes in chromosome count.

Historically, supernumerary chromosomes are the earliest "gene" duplications reported, revealed by cytogenetic techniques (Lejeune, 1960; Patau *et al.*, 1961). Soon afterward, evidence for pathogenic α-globin gene deletions was reported (Ose and Bush, 1962), although more than 10 years elapsed before direct confirmation was achieved (Ottolenghi *et al.*, 1974). β-globin deletions were reported soon afterward (Kan *et al.*, 1975), followed by gene deletions in other hemoglobinopathies and culminating in the use of Southern blotting for the prenatal diagnosis of globin deletions (Orkin *et al.*, 1978). Gene dosage changes were next reported in the immunoglobulin genes (Rabbitts *et al.*, 1980; van Loghem *et al.*, 1980) and the first deletion of a tumor suppressor gene (in retinoblastoma) was reported soon afterward (Junien *et al.*, 1982). The identification of repetitive DNA associated with a gene deletion was described in a form of hereditary persistence of fetal hemoglobin (Jagadeeswaran *et al.*, 1982). A pathogenic role was attributed to repetitive DNA elements by Hess and colleagues (1983), who suggested that DNA insertion elements may disrupt gene correction processes in the two duplication units containing α2- and α1-globin genes. Although the widespread use of Southern blotting (Southern, 1975) from the mid 1970s until the late 1980s may have facilitated the detection of deletions and duplications, the application of the polymerase chain reaction (Mullis *et al.*, 1986) may unintentionally have produced an ascertainment bias away from them, the analysis of dystrophin gene deletions in males being a notable exception (Beggs *et al.*, 1990). This is because a typical PCR is not designed for quantitative analysis, but for optimal purity and yield. However, the observation of germ line deletions in a wide range of genetic conditions has required the development of techniques that can detect gene dosage changes in hemizygotes. These techniques include Southern blotting and PCR modifications, as well as newer methods. Techniques that measure gene dosage can also be adapted to quantify the somatic mosaicism and PCR failure (allele dropout).

13.2 MECHANISMS

Deletions and duplications can be mediated by homologous recombination involving recombinogenic elements, for example *PMP22* (Inoue *et al.*, 2001); or gene duplications, for example the type IV collagen genes *COL4A5* and *COL4A6*, paired head-to-head on chromosome Xq22 deletions in X-linked Alport syndrome. *BRCA1* gene deletions involving a head-to-head of a partial pseudogene have been reported (Brown *et al.*, 2002), as well as *Alu*-mediated intragenic deletions (Puget *et al.*, 1997; Rohlfs *et al.*, 2000). A 26-bp core sequence in two out of five α^0-thalassaemia deletions has been reported (Harteveld *et al.*, 1997), supporting the idea that *Alu* repeats stimulate recombination events not only by homologous pairing, but also by providing binding sites for recombinogenic proteins. Deletion of subtelomeric repeats has been implicated in facioscapulohumeral muscular dystrophy (FSHD) where a remnant fragment, the result of a deletion of tandemly arrayed 3.3 kb repeat units (D4Z4) on 4q35 can be detected (Lemmers *et al.*, 1998). Nonhomologous recombination also has been implicated in gene deletions (Hu and Worton, 1992; Summinaga *et al.*, 2000).

13.3 PATHOLOGICAL CONSEQUENCES

13.3.1 Supernumerary Chromosomes

In practice, when considering alterations in autosomal chromosome copy number, only supernumerary chromosomes need to be addressed, as autosomal monosomy is not compatible with life (unless the individual has monosomy mosaicism). Monosomy of the X chromosome can occur (Turner syndrome), although it is estimated that only 1 in 300 conceptuses with monosomy X survive (Kajii *et al.*, 1980). Triploidy and trisomies of other autosomes are also a common cause of spontaneous abortion. From a clinical perspective, prenatal detection of triploidy and trisomies 13 and 18 are important as fetuses may survive until birth, but most die soon after due to congenital malformations. Trisomies that have less severe consequences are trisomy 21 (Down syndrome), trisomy X (rarely show any physical abnormalities), and Klinefelter syndrome (XXY). Supernumerary chromosomes are readily detectable by conventional cytogenetics (see also previous chapter); however, molecular techniques may be faster and less expensive.

13.3.2 Microdeletions

Pathogenic partial deletions and duplications have been widely reported for all chromosomes, and cause a range of symptoms. Some deletions are easily detectable by conventional cytogenetics, others are smaller (microdeletions) and can be difficult or impossible to detect by conventional cytogenetics, and other cytogenetic techniques such as FISH or

molecular techniques are required. Some of the more common microdeletions with associated syndromes (summarized in Table 12.1) are 22q11.2 (Di George syndrome), 15q11.2/q12 (Prader-Willi and Angelman syndromes), 17p (Miller-Dieker and Smith-Magenis syndromes), 4p16.3 (Wolf syndrome), and 5p15 (crit du chat syndrome).

13.3.3 Subtelomeric Deletions

Several syndromes that are caused by microscopically visible chromosomal deletions and duplications, including the subtelomeric region have been known to be associated with mental retardation (e.g., 4p, 5p, 9p). More recent studies using molecular methods such as detection of loss of hypervariable DNA polymorphisms or microsatellite markers and FISH that detect submicroscopic subtelomeric deletions have shown that these can account for up to 5% of cases with mental retardation (Flint *et al.*, 1995; Knight *et al.*, 1999; Rio *et al.*, 2002). Subtelomeric deletion analysis of all chromosomes can be performed rapidly and cost effectively using new molecular techniques such as MAPH (Sismani *et al.*, 2001).

13.3.4 Gene Deletions or Duplications

A proportion of some types of cancers are caused by inherited germline mutations in tumor suppressor genes, and full or partial gene deletions account for a significant number of these mutations. There are a range of other single gene disorders where deletion or duplication of part or all of the gene accounts for a significant proportion of detected mutations, such as Duchenne muscular dystrophy, spinal muscular atrophy, Charcot-Marie-Tooth disease, Fanconi anemia, congenital adrenal hyperplasia, and rare metabolic disorders such as non-ketotic hyperglycinaemia. Some examples of genes for which an estimate has been obtained for the proportion of germline mutations that are deletions or duplications are shown in Table 13.1.

Since the majority of mutations identified in the tumor suppressor genes lead to the production of a truncated product (Couch and Weber, 1996), until recently most studies performed mutation analysis on genomic DNA using PCR-based techniques such as sequencing, heteroduplex analysis, or the protein truncation test (PTT). Deletions (Petrij-Bosch *et al.*, 1997; Puget *et al.*, 1997; Swensen *et al.*, 1997) or duplications (Puget *et al.*, 1999) within the BRCA1 gene, for example, would have been missed by conventional PCR-based methods and were detected by either reverse transcriptase PCR (RT-PCR) or Southern blotting. Although both PCR and Southern blotting have been adapted to provide quantitative data, PCR has become the method of choice for genetic testing. Estimates of gene dosage typically have been based on comparisons with a reference standard. Other approaches, including the study of junction fragments or microsatellite inheritance and more recently long PCR (Coulter-Mackie *et al.*, 1998), FISH (Voskova-

TABLE 13.1 Proportion (%) of mutations due to deletions or insertions in various genes.

Gene	Disorder	Deletions or duplications (%)	References
BRCA1	Familial breast cancer	4–27	Puget *et al.*, 1999; Gad *et al.*, 2002; Hogervorst *et al.*, 2003
VHL	Von-Hippel Lindau disease	20–47	Shuin *et al.*, 1995; Vortmeyer *et al.*, 2002
DMD	Duchenne muscular dystrophy	60	Koenig *et al.*, 1989
MSH2/ MLH1	Hereditary nonpolyposis colorectal cancer	27–54.8	Wijnen *et al.*, 1998; Gille *et al.*, 2002; Wagner *et al.*, 2002
RB1	Retinoblastoma	14	Richter *et al.*, 2003

Goldman *et al.*, 1997), and array-CGH (Bruder *et al.*, 2001) have also been employed.

The overall contribution of deletions and duplications recorded in the May 2000 Human Gene Mutation database was 5.5% of reported mutations. As of September 2003 the proportion had increased to 6.3% (Stenson *et al.*, 2003). Given the greater technical difficulties in identifying deletions and duplications, this is still likely to be an underestimate. This emphasizes the importance of including the measurement of gene dosage in any comprehensive mutation scan.

Numerous polymorphisms have been described involving deletion or duplication of large chromosomal segments, sometimes involving entire genes; for example, the common deletion polymorphisms of the cytochrome P450 gene CYP2D6 (Meyer and Zanger, 1997), the theta-class glutathione S-transferase gene GSTT1 (Wiencke *et al.*, 1995), and the mu-type gene GSTM1 (Brockmoller *et al.*, 1992).

Regions close to telomeres are especially prone to interchromosomal rearrangements that can lead to different forms of presence/absence polymorphism, including the multiallelic variation in the structure of the 16p telomere (Wilkie *et al.*, 1991), and the deletion polymorphism near the 12q telomere (Baird *et al.*, 2000). The assembly of the human genome sequence is complicated by the presence of duplication/deletion polymorphisms, either because only one form of a region of variable structure is recorded in the sequenced chromosome (Siniscalco *et al.*, 2000) or because the presence of unsuspected polymorphism for tandem duplication can lead to misassembly (Bailey *et al.*, 2001; 2002).

13.4 DIAGNOSTIC TECHNIQUES

Classical cytogenetic genetic techniques have been valuable in identifying supernumerary chromosomes and large deletions. However, the resolution of these methods is limited to several megabases and they would miss many known submicroscopic deletions. Fluoresence *in situ* hybridization (FISH) techniques have greatly expanded the capabilities of cytogenetics (see also Chapter 12); single copy probes allow detection of microdeletions, which would be impossible to detect by conventional cytogenetics; however, due to the specialized equipment and expertise required and the low throughput, detection of submicroscopic deletions has largely become the realm of specialized molecular genetic diagnostics.

13.4.1 Southern Blotting

Southern blotting still has a role in a comprehensive diagnostic service, both for gene dosage measurement and for the measurement of allele expansions. Southern blots can detect deletions or duplications by the identification of novel restriction fragments created by the rearrangement or by measurement of band intensity compared to a control fragment.

Fragment size changes are seen because the region of genomic DNA created by a deletion or duplication may have gained or lost restriction enzyme sites. In those cases, a Southern blot of a genomic digest using an appropriate restriction enzyme will detect novel fragments. There is a risk that some altered fragment sizes may simply be due to restriction fragment length polymorphisms, although the risk can be reduced by performing separate digestions with a different enzyme. Direct visualization of fragment sizes has long been used for detecting α-globin duplications and deletions (Orkin *et al.*, 1978), and the increased size range available in PFGE allows detection of deletions and duplications even in genes as large as the dystrophin gene (Kenwrick *et al.*, 1987).

Estimation of gene dosage by measuring the intensity of probe hybridization (usually in comparison with a control) has identified several instances of gene deletions (Bonifas and Epstein, 1990), though few studies report details of the dose-response curve (such as its linear range) and internal controls are often lacking. Dystrophin deletion carrier testing is possible by assessing relative band intensity (van Essen *et al.*, 1997). Accurate measurement of band intensity requires a phosphoimager, which gives a linear dose response over a wider range than film emulsion. Figure 13.1 shows the application of Southern blotting to measure the relative gene dosage of CYP21A and CYP21B, which can be deleted in congenital adrenal hyperplasia. The risk of inaccuracies introduced by uneven transfer of DNA to the membrane or incomplete washing of the probe require that an additional probe for a control locus should be included as a standard.

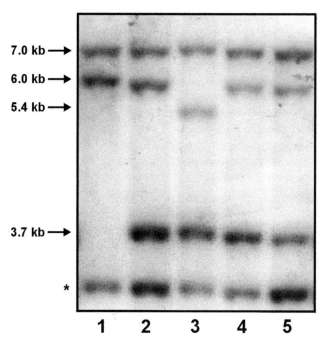

Figure 13.1 Gene dosage in 21 hydroxylase by Southern blotting. The Southern blot shows four major bands: C4A (7 kb), C4B (6 or 5.4 kb), 21-hydroxylase (3.7 kb), and 21-hydroxylase pseudogene (indicated by an asterisk). Sample 1 has a homozygous deletion of 21-OH, sample 2 has equal gene dosage at all four loci. Samples 3 and 4 each have deletions of the 21-hydroxylase pseudogene and C4A, and sample 5 has a deletion of 21-hydroxylase and C4B. (Image courtesy of Dr Kieran Bransfield, Regional Genetics Laboratory, Leeds.)

13.4.2 Microsatellites and SNPs

Microsatellites have been used to detect supernumerary chromosomes (Mansfield, 1993) and large deletions or duplications (Brice *et al.*, 1992). This approach is limited by the fact that semi-quantitative PCR methods do not work well using microsatellites, so only cases in which the allele lengths of the microsatellites differ over chromosomal region of interest can be interpreted unambiguously. Peak heights or areas of microsatellites can be hard to interpret because of preferential amplification of the smaller allele and the stuttering effect that produces a series of minor peaks immediately adjacent to the major peak. The difficulties with multiplex quantitative fluorescent PCR (QF-PCR) are reduced by selection of tetranucleotide repeat markers and is finding increasing use in the rapid diagnosis of common aneuploidies (Donaghue *et al.*, 2003). Loss of heterozyogosity of SNP polymorphisms using high throughput Affymetrix HuSNP arrays has been proposed as a rapid means of identifying allele imbalance caused by genomic deletions in tumor cells (Hoque *et al.*, 2003).

13.4.3 Polymerase Chain Reaction

Neubauer and colleagues (1990) described the coamplification of two amplicons followed by quantitation on the basis of staining intensity as a means of estimating gene dosage. Differential PCR-based methods are semi-quantitative: they determine relative concentrations of two amplicons, but not absolute molar amounts. Taking advantage of the increased sensitivity of fluorescent detection methods Yau *et al.*, (1996) described a multiplex fluorescent PCR that was able to detect female carriers of deletions or duplications in the dystrophin gene. Using peak areas, they were able to give statistical estimates of their assay, likely to be essential for diagnostic applications. Similar methods have been used for testing for hereditary motor and sensory neuropathy (HMSN) duplications (Rowland *et al.*, 2001) and APC deletions (Flintoff *et al.*, 2001). Tosi and colleagues (2002) refined the basic multiplex fluorescent PCR method by selecting shorter fragments, tagging the primers with common tags of 16 nucleotides, and using a modified PCR buffer containing DMSO. The method was termed quantitative multiplex PCR of short fluorescent fragments (QMSF-PCR) and enables rapid design of amplicons to fine-map the limits of a deletion or duplication (Casilli *et al.*, 2002). Although differential PCR methods offer high orders of multiplicity (e.g., 8–15 amplicons per reaction), end point multiplex PCR assays rely on the equivalence of amplification of each fragment in the multiplex in the test sample as well as a control sample. This equivalence may be lost if the starting template DNA concentration is too variable, since different fragments within the multiplex may be amplified with different efficiencies. In some situations it may be necessary to choose a more robust, lower order multiplex base on real-time PCR. Real-time PCR provides a means for continuous detection of product throughout the amplification process, and as such can dispense with a gel separation stage and operate in a closed system (see also Chapter 10). The accumulation of PCR product is monitored by staining using interchelating dyes (e.g., Sybr Green) or by dual-labeled probes such as TaqMan (Laurendeau *et al.*, 1999), molecular beacons (Tyagi and Kramer, 1996), or other fluorescent detection systems. Real-time PCR is becoming widely used as a method for measuring gene dosage (Feldkotter *et al.*, 2002; Bertin *et al.*, 2003; Covault *et al.*, 2003; Gaikovitch *et al.*, 2003; Kim *et al.*, 2003).

13.4.4 MAPH and MLPA

The Multiplex Amplifiable Probe Hybridization (MAPH) method for copy number measurement (Armour *et al.*, 2000) combines hybridization as the primary step to detect copy number, with end point multiplex PCR to amplify the hybridized probes. Sets of short probes corresponding to the segments to be tested, each flanked by the same primer-binding sites, are hybridized with the test genomic DNA immobilized on a solid support. After washing, each specifically bound probe will be present in an amount proportional to its copy number. All probes can then be amplified simultaneously with a single primer pair, and quantified after electrophoretic separation (see Fig. 13.2). This study demonstrated the simultaneous assessment of copy number in a set of 40 human loci, including detection of deletions causing Duchenne Muscular Dystrophy and Prader-Willi/Angelman syndromes. The high order multiplex achieved may be limited only by the need to have probes of varying size for electrophoretic separation. MAPH probes are generated by cloning the target sequences into a plasmid vector, amplifying the cloned sequence using primers directed to the vector with the result that all probes are then flanked with the same sequence. Probes that are intended to be multiplexed must be of sufficient size difference to be resolved by electrophoresis. The membranes are then washed rigorously to remove unbound probe, and the remaining specifically bound probe will be present in an amount proportional to its target copy number. The probes are then stripped from the membrane by boiling, and amplified simultaneously with the universal primer pair. Products are then separated by electrophoresis, and a relative comparison is made between the peak heights. Reduced peak heights compared to internal control probes indicate a reduction in gene copy number (deletion) and an increase in gene copy number (duplication) produces increased peak heights. The assay can be completed in two to three days, requiring one overnight hybridization followed by membrane washing, a PCR step, and a product detection step. The system works well, but the manipulation of small nylon

filters presents some difficulties in sample handling and labeling.

A similar technique called Multiplex Ligatable Probe Amplification (MLPA) avoids the use of filters by using single-stranded ligatable probes and a thermo-stable DNA ligase to produce the amplifiable target (Schouten *et al.*, 2002; Fig. 13.2). In the MLPA technique, genomic DNA is hybridized in solution to probe sets, each of which consists of two halves. One half consists of a target-specific sequence (20–30 nucleotides) flanked by a universal primer sequence, and can be generated synthetically. The other half also has a target-specific sequence at one end (25–43 nucleotides) and a universal primer sequence at the other, but has a variable length stuffer fragment in between (19–370 nucleotides) to generate the size differences necessary in the probes to allow electrophoretic resolution. This larger probe part is generated by cloning the target-specific sequence into M13 derived vectors that already contain the variable length fragments; single-stranded DNA is then purified from the phage particles and made double stranded at two sites by annealing short oligonucleotides in order that the desired probe fragment be liberated by restriction enzyme digestion. The two probe halves are designed such that the target-specific sequences bind adjacently to the target DNA, and can then be joined by use of a ligase. This generates a contiguous probe flanked by universal primer binding sites that can then be amplified by PCR, whereas unbound probe halves cannot be amplified, and hence eliminates the need for removal of excess probe by washing. The amounts of ligated probe produced will be proportional to the target copy number, and after PCR amplification the relative

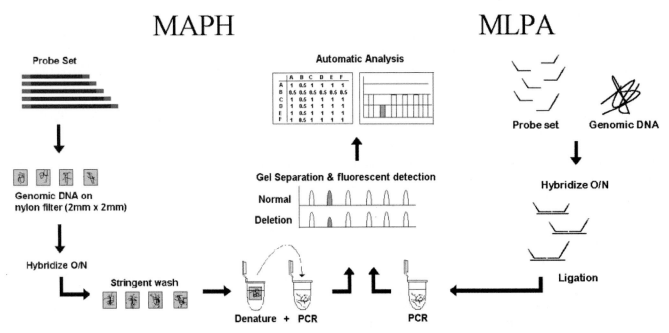

FIGURE 13.2 Comparison of MLPA and MAPH. (Diagram by Donna Sheard, Regional Genetics Laboratory, Leeds.)

peak heights indicate deletion or duplication of target sequence.

In diagnostic use, applications of these high order multiplex methods include the detection of aneuploidies, unbalanced cryptic translocations and whole or partial gene duplications or deletions. Commercial kits based on MAPH and MLPA are available. The high multiplicity of MAPH and MLPA, plus their use of standard genetic laboratory apparatus, make it highly likely that these techniques will become widely used. Both techniques offer a rapid means of scanning up to 40 loci for gene dosage, and are likely to be used widely in research and diagnostic settings. MAPH represents a conceptual breakthrough for the analysis of gene dosage, but the handling of small filter discs is difficult and could pose sample tracking problems in routine medium throughput settings. The liquid-phase solution to sample handling offered by MLPA and the ready availability of robust commercial kits has led to rapid acceptance by diagnostic laboratories worldwide. MLPA uses M13 (single-stranded) probes that are more technically challenging to construct than MAPH probes, which can be made from PCR products.

The kits supplied by MRC-Holland have left a size gap in their probe series to enable users to add up to four extra synthetic probes between 90 and 110 bases. It would be useful to be able to construct additional probes to enable fine mapping and identification of the junction fragments. One solution is to adapt MLPA to an array detection setting, so that fragments were identified by sequence rather than size. This would enable all MLPA probes to be chemically synthesized. Array adaptations of both MAPH and MLPA might enable much higher order multiplexes than are currently possible. However even 40-plex assays represent a substantial gain in the multiplex order typically available in PCR. MAPH or MLPA could provide competition for the array-based CGH approaches: a 96-well array of 40 probes could interrogate over 3,000 loci, representing a better than 1cM coverage of the entire human genome or much higher single chromosome resolution.

13.4.5 Long PCR

Long PCR (Barnes, 1992) is a modified PCR protocol that includes a proof-reading polymerase and short denaturation times to enable the size of amplicons to be increased from 3–5 kb to beyond 30 kb. This opens the possibility of screening for deletions of this scale by direct PCR, with the advantage of producing the junction fragment for further analysis. Long PCR has been used successfully to identify deletions including mitochondrial DNA (Fromenty et al., 1996), C4 gene deletions in the MHC complex (Grant et al., 2000), the CYP2D6 deletion allele, and LDL deletions (Kim et al., 1999). Despite the simplicity in principle of a long PCR approach, in practice it is often difficult to design a robust assay. Direct detection of duplications by amplification of

entire duplicated regions is likely to prove difficult in diploid genomes since the unduplicated allele will have a significant advantage during the amplification process. However, full characterization of a deletion or duplication requires the sequence of the junction fragments. This can be achieved either by using long PCR or by fine mapping, for example using QMSF-PCR followed by conventional PCR and sequencing.

13.4.6 Array-Based CGH

Microarrays have become widely used tools for gene expression studies, and now encouraging developments are taking place in array-based comparative genome hybridization (Array-CGH) as a tool for measuring gene duplications and deletions. Comparative genomic hybridization originally used metaphase chromosomes as targets for differentially labeled probes (e.g., Cy5, Cy3) from control and test samples. Gene dosages changes could be detected by variation of the relative intensity of the two labels (Kallioniemi et al., 1992). By replacing the metaphase spread with microarrayed BAC DNA, Pinkel and colleagues (1998) reported the use of array-based CGH analysis to investigate chromosome 20 gene dosage alterations in breast cancer. DNA purified from BAC clones spaced at approximately 3Mb intervals along the entire chromosome together with some X chromosome controls were arrayed on nitrocellulose membranes as 200–400 micron spots in duplicate. Test DNA from breast cancer cell lines was labeled by nick translation with fluorescein, control DNA with Texas Red, and the spots were counterstained with DAPI. Scanning used custom-built mercury-arc illumination and CCD detection. Other studies have used different dyes (with Cy3/Cy5 being a popular choice) and omitted counterstaining. Many detection systems use confocal laser scanners, which, although more restricted in fluorochrome options, are brighter and have fewer problems with light scattering.

Growing high numbers ($>10^4$) of cloned DNA is expensive and labor-intensive. To circumvent large scale cultures and DNA purifications, array features have been made using degenerate oligonucleotide PCR (DOP-PCR) of BAC templates using 5′ amine–linked primers (Hodgson et al., 2001). Fiegler and colleagues (2003) improved the standard DOP-PCR to reduce nonspecific host and vector amplicons from BAC clones, improving the signal-to-noise ratio. A ligation-mediated PCR BAC labeling method has also been reported that improved signal-to-noise ratios (Snijders et al., 2001). Array CGH is still at an early stage of technical development (see also Chapter 15), but strong commercial and academic interest are likely to result in an increase in research and diagnostic applications in the near future. The supply of robust Cot1 DNA has been a major problem reported by many array-CGH users (Carter et al., 2002). Competitor (Cot1) DNA is used to block the hybridization of repetitive DNA in BAC-derived amplicons. Buckley and colleagues

(2002) constructed a comprehensive microarray representing a human chromosome for analysis of DNA copy number variation. The chromosome 22 microarray covered 34.7 Mb with an average resolution of 75 kb using a sequence-defined, repeat-free, and nonredundant strategy for array preparation. This enabled an increase in array resolution and eliminated the need for Cot1 DNA. Array targets were made using phi29 DNA polymerase synthesis. As the technology becomes more widely used, commercial suppliers are starting to emerge; for example, Spectral Genomics (http://www.spectralgenomics.com/) and genome centers, for example, Leiden University in Holland and the Sanger Centre in the UK are using their resources to manufacture genome or chromosome specific arrays.

13.7 NOMENCLATURE

Recently the Human Genome Variation Society (www.hgvs.org) has taken on the responsibility of establishing a standardized nomenclature for mutation nomenclature that includes deletions and duplications. A nomenclature reference is maintained at this site by Dr. Johan Den Dunnen of Leiden University. The current recommendation for exon or multiexon deletion nomenclature varies, depending on whether or not the breakpoint has been identified. If the breakpoints are not sequenced (e.g., detected on Southern blot or by MLPA), exonic deletions are described as c.88-?_923+?del, indicating a deletion starting at an unknown position in the intron 5′ of cDNA nucleotide 88 and ending at an unknown position in the intron 3′ of cDNA nucleotide 923. A genomic or cDNA reference (e.g., Genbank, EMBL, DDJB) should be cited, including the version number. If the junction fragments are known, then the form g.390_1458del or (g.390_1458del1069) should be used for a genomic reference sequence or c.13-23_301-143del (c.13-23_301-143del1069) for a cDNA reference sequence.

13.8 GENE DOSAGE APPLICATIONS IN TUMOR PROFILING

Partial chromosomal losses occur relatively frequently in a large number of tumor types, and this is readily detectable by demonstrating loss of heterozygosity (LOH) of polymorphic microsatellite markers. Such techniques have been widely employed to investigate various aspects of tumor development:

1. Clonality. Loss of chromosomal regions is essentially an irreversible event in tumor cell development, and all subsequent subclones would be expected to demonstrate the same deletions, and may have accumulated more. LOH analysis of multiple tumors from the same patient can be used to determine whether tumors, either synchronous or metachronous, are clonal in origin, and in the latter case may be used to determine whether subsequent tumors are recurrences or metastases of previous tumors or independent primary tumors. For example, in patients with multiple synchronous lung tumors it is important for treatment decisions to discriminate multicentric lung cancers from intrapulmonary metastases, and this can be aided by clonality studies (Shimizu *et al.*, 2000).

2. Identification of tumor suppressor genes. It is hypothesized that the reason that tumors frequently undergo LOH, and that LOH of particular regions is commonly associated with specific tumor types, is that the regions lost harbor tumor suppressor genes. Identification of such tumor suppressor genes may provide useful tools for diagnosis, prediction of prognosis, and possible therapies, and hence numerous studies have been undertaken to identify regions of LOH specific to individual tumor types and the genes present in these regions. For example, chromosome 9 is the most frequently deleted chromosome in transitional cell carcinoma of the bladder, and candidate genes have been identified in three of the four regions of minimal deletion (Knowles, 1999). In prostate cancer, gains at Xq and 18q are among the most common chromosomal alterations, and amplification of the AR (Xq12), MYC (8q24), and EIF3S3 (8q23) genes have been found in a large fraction of hormone-refractory prostate cancers (Nupponen and Visakorpi, 2000).

3. Diagnosis. LOH analysis has been employed both in detection of the presence of a tumor, as well as differential diagnosis and typing of tumors. For example, LOH has been reliably demonstrated in DNA extracted from urine sediments in bladder cancer patients, and may have a role in the noninvasive diagnosis of bladder cancer (Linn *et al.*, 1997; Berger *et al.*, 2002). Differential diagnosis between renal oncocytomas and renal cell carcinomas can be difficult due to morphological similarities, but is important due to their different prognoses. LOH analysis has shown that they can be differentiated on the basis of spectrum of chromosomal loss (Herbers *et al.*, 1998) LOH analysis has also been used to characterize the aggressive intraductal carcinoma of the prostate, to differentiate it from the less aggressive high-grade dysplasia (prostatic intraepithelial neoplasia, PIN) and to provide evidence that it does not represent invasion of Gleason grade 3 cancers into the ductal/acinar system (Dawkins *et al.*, 2000).

4. Prognosis. In several tumor types, LOH of specific regions have been hypothesized to be key events in tumor evolution and progression. LOH analysis of tumor specimens has been used to determine their value as prognostic markers. It has been found in colorectal

cancers that high level LOH correlated with earlier onset and lymphatic invasion, and hence a poorer prognosis, and low level LOH was more common in earlier stage disease and predicted a more favorable outcome (Choi *et al.*, 2002). In breast cancers, specific chromosomal regions have been identified for which LOH is a significant predictor of lymph-node metastasis and hence may serve as a negative prognostic indicator (Nagahata *et al.*, 2002).

13.9 SUMMARY AND FUTURE DEVELOPMENTS

Gene dosage is a significant contributor to the overall burden of the germ-line and somatic mutations in man. Any comprehensive mutation screen therefore should include measurement of gene dosage. The permeation of genomics into medical practice will increase the demand for mutation screening with applications in diagnosis, predictive testing, and treatment. This, in turn, will encourage the development of more highly automated approaches, the introduction of robust statistical analysis, and quality control of results that may be used in a diagnostic context. Those techniques most easily adaptable to robust laboratory processes and automated data handling will have a competitive advantage. Ease of handling, hence more robust results, perhaps explains the greater uptake of MLPA compared with MAPH. Both MAPH and MLPA produce data using automated DNA sequencers that is spreadsheet-ready. Statistical parameters then may be easily applied as quality measures (Taylor *et al.*, 2003). The same is true of array scanner output and real-time PCR systems, in contrast to classical genetic techniques like Southern blotting or cytogenetics. Whereas MLPA, real time PCR, QMSF-PCR and MAPH approaches are focused around relatively low numbers of targets, the array methods offer very high orders of multiplicity, the so-called "hypothesis-free" approach. It seems likely that the two approaches will converge as arrays become increasingly targeted toward regions or functional clusters of interest and the PCR-based approaches move toward an array-like format, either in capillary arrays or high density microtitre arrays.

References

Armour, J. A., Sismani, C., Patsalis, P. C., and Cross, G. (2000). Measurement of locus copy number by hybridisation with amplifiable probes. *Nucleic Acids Res.* 28, 605–609.

Bailey, J. A., Gu, Z., Clark, R. A., Reinert, K., Samonte, R. V., Schwartz, S., Adams, M. D., Myers, E. W., Li, P. W., and Eichler, E. E. (2002). Recent segmental duplications in the human genome. *Science* 297, 1003–1007.

Bailey, J. A., Yavor, A. M., Massa, H. F., Trask, B. J., and Eichler, E. E. (2001). Segmental duplications: organization and impact within the current human genome project assembly. *Genome Res.* 11, 1005–1017.

Baird, D. M., Coleman, J., Rosser, Z. H., and Royle, N. J. (2000). High levels of sequence polymorphism and linkage disequilibrium at the telomere of 12q: implications for telomere biology and human evolution. *Am. J. Hum. Genet.* 66, 235–250.

Barnes, W. M. (1992). The fidelity of Taq polymerase catalyzing PCR is improved by an N-terminal deletion. *Gene* 112, 29–35.

Beggs, A. H., Koenig, M., Boyce, F. M., and Kunkel, L. M. (1990). Detection of 98% of DMD/BMD gene deletions by polymerase chain reaction. *Hum. Genet.* 86, 45–48.

Berger, A. P., Parson, W., Stenzl, A., Steiner, H., Bartsch, G., and Klocker, H. (2002). Microsatellite alterations in human bladder cancer: detection of tumor cells in urine sediment and tumor tissue. *Eur. Urol.* 41, 532–539.

Bertin, R., Acquaviva, C., Mirebeau, D., Guidal-Giroux, C., Vilmer, E., and Cave, H. (2003). CDKN2A, CDKN2B, and MTAP gene dosage permits precise characterization of mono- and bi-allelic 9p21 deletions in childhood acute lymphoblastic leukemia. *Genes Chromosomes Cancer* 37, 44–57.

Bonifas, J. M., and Epstein, E. H., Jr. (1990). Detection of carriers for X-linked ichthyosis by Southern blot analysis and identification of one family with a de novo mutation. *J. Invest. Dermatol.* 95, 16–19.

Brice, A., Ravise, N., Stevanin, G., Gugenheim, M., Bouche, P., Penet, C., and Agid, Y. (1992). Duplication within chromosome 17p11.2 in 12 families of French ancestry with Charcot-Marie-Tooth disease type 1a. The French CMT Research Group. *J. Med. Genet.* 29, 807–812.

Brockmoller, J., Gross, D., Kerb, R., Drakoulis, N., and Roots, I. (1992). Correlation between trans-stilbene oxide-glutathione conjugation activity and the deletion mutation in the glutathione S-transferase class mu gene detected by polymerase chain reaction. *Biochem. Pharmacol.* 43, 647–650.

Brown, M. A., Lo, L. J., Catteau, A., Xu, C. F., Lindeman, G. J., Hodgson, S., and Solomon, E. (2002). Germline BRCA1 promoter deletions in UK and Australian familial breast cancer patients: Identification of a novel deletion consistent with BRCA1:psiBRCA1 recombination. *Hum. Mutat.* 19, 435–442.

Bruder, C. E., Hirvela, C., Tapia-Paez, I., Fransson, I., Segraves, R., Hamilton, G., Zhang, X. X., Evans, D. G., Wallace, A. J., Baser, M. E., Zucman-Rossi, J., Hergersberg, M., Boltshauser, E., Papi, L., Rouleau, G. A., Poptodorov, G., Jordanova, A., Rask-Andersen, H., Kluwe, L., Mautner, V., Sainio, M., Hung, G., Mathiesen, T., Moller, C., Pulst, S. M., Harder, H., Heiberg, A., Honda, M., Niimura, M., Sahlen, S., Blennow, E., Albertson, D. G., Pinkel, D., and Dumanski, J.P. (2001). High resolution deletion analysis of constitutional DNA from neurofibromatosis type 2 (NF2) patients using microarray-CGH. *Hum. Mol. Genet.* 10, 271–282.

Buckley, P. G., Mantripragada, K. K., Benetkiewicz, M., Tapia-Paez, I., Diaz De Stahl, T., Rosenquist, M., Ali, H., Jarbo, C., De Bustos, C., Hirvela, C., Sinder Wilen, B., Fransson, I., Thyr, C., Johnsson, B. I., Bruder, C. E., Menzel, U., Hergersberg, M., Mandahl, N., Blennow, E., Wedell, A., Beare, D. M., Collins, J. E., Dunham, I., Albertson, D., Pinkel, D., Bastian, B. C., Faruqi, A. F., Lasken, R. S., Ichimura, K., Collins, V. P., and Dumanski, J. P. (2002). A full-coverage, high-resolution human chromosome 22 genomic microarray for clinical and research applications. *Hum. Mol. Genet.* 11, 3221–3229.

Carter, N. P., Fiegler, H., and Piper, J. (2002). Comparative analysis of comparative genomic hybridization microarray technologies: report of a workshop sponsored by the Wellcome Trust. *Cytometry* 49, 43–48.

Casilli, F., Di Rocco, Z. C., Gad, S., Tournier, I., Stoppa-Lyonnet, D., Frebourg, T., and Tosi, M. (2002). Rapid detection of novel BRCA1 rearrangements in high-risk breast-ovarian cancer families using multiplex PCR of short fluorescent fragments. *Hum. Mutat.* 20, 218–226.

Choi, S. W., Lee, K. J., Bae, Y. A., Min, K. O., Kwon, M. S., Kim, K. M., and Rhyu, M. G. (2002). Genetic classification of colorectal cancer based on chromosomal loss and microsatellite instability predicts survival. *Clin. Cancer. Res.* 8, 2311–2322.

Couch, F. J. and Weber, B. L. (1996). Mutations and polymorphisms in the familial early-onset breast cancer (BRCA1) gene. Breast Cancer Information Core. *Hum. Mutat.* 8, 8–18.

Coulter-Mackie, M. B., Applegarth, D. A., Toone, J. R., and Gagnier, L. (1998). A protocol for detection of mitochondrial DNA deletions: characterization of a novel deletion. *Clin. Biochem.* 31, 627–632.

Covault, J., Abreu, C., Kranzler, H., and Oncken, C. (2003). Quantitative real-time PCR for gene dosage determinations in microdeletion genotypes. *Biotechniques* 35, 594–596, 598.

Dawkins, H. J., Sellner, L. N., Turbett, G. R., Thompson, C. A., Redmond, S. L., McNeal, J. E., and Cohen, R. J. (2000). Distinction between intraductal carcinoma of the prostate (IDC-P), high-grade dysplasia (PIN), and invasive prostatic adenocarcinoma, using molecular markers of cancer progression. *Prostate* 44, 265–270.

Donaghue, C., Roberts, A., Mann, K., and Ogilvie, C. M. (2003). Development and targeted application of a rapid QF-PCR test for sex chromosome imbalance. *Prenat. Diagn.* 23, 201–210.

Feldkotter, M., Schwarzer, V., Wirth, R., Wienker, T. F., and Wirth, B. (2002). Quantitative analyses of SMN1 and SMN2 based on real-time lightCycler PCR: fast and highly reliable carrier testing and prediction of severity of spinal muscular atrophy. *Am. J. Hum. Genet.* 70, 358–368.

Fiegler, H., Carr, P., Douglas, E. J., Burford, D. C., Hunt, S., Scott, C. E., Smith, J., Vetrie, D., Gorman, P., Tomlinson, I. P., and Carter, N.P. (2003). DNA microarrays for comparative genomic hybridization based on DOP-PCR amplification of BAC and PAC clones. *Genes Chromosomes Cancer* 36, 361–374.

Flint, J., Wilkie, A. O., Buckle, V. J., Winter, R. M., Holland, A. J., and McDermid, H. E. (1995). The detection of subtelomeric chromosomal rearrangements in idiopathic mental retardation. *Nat. Genet.* 9, 132–140.

Flintoff, K. J., Sheridan, E., Turner, G., Chu, C. E., and Taylor, G. R. (2001). Submicroscopic deletions of the APC gene: a frequent cause of familial adenomatous polyposis that may be overlooked by conventional mutation scanning. *J. Med. Genet.* 38, 129–132.

Fromenty, B., Manfredi, G., Sadlock, J., Zhang, L., King, M. P., and Schon, E. A. (1996). Efficient and specific amplification of identified partial duplications of human mitochondrial DNA by long PCR. *Biochim. Biophys. Acta.* 1308, 222–230.

Gad, S., Caux-Moncoutier, V., Pages-Berhouet, S., Gauthier-Villars, M., Coupier, I., Pujol, P., Frenay, M., Gilbert, B., Maugard, C., Bignon, Y. J., Chevrier, A., Rossi, A., Fricker, J. P., Nguyen, T. D., Demange, L., Aurias, A., Bensimon, A., and Stoppa-Lyonnet, D. (2002). Significant contribution of large BRCA1 gene rearrangements in 120 French breast and ovarian cancer families. *Oncogene* 21, 6841–6847.

Gaikovitch, E. A., Cascorbi, I., Mrozikiewicz, P. M., Brockmoller, J., Frotschl, R., Kopke, K., Gerloff, T., Chernov, J. N., and Roots, I. (2003). Polymorphisms of drug-metabolizing enzymes CYP2C9, CYP2C19, CYP2D6, CYP1A1, NAT2 and of P-glycoprotein in a Russian population. *Eur. J. Clin. Pharmacol.* 59, 303–312.

Gille, J. J., Hogervorst, F. B., Pals, G., Wijnen, J., Van Schooten, R. J., Dommering, C. J., Meijer, G. A., Craanen, M. E., Nederlof, P. M., De Jong, D., McElgunn, C. J., Schouten, J. P., and Menko, F. H. (2002). Genomic deletions of MSH2 and MLH1 in colorectal cancer families detected by a novel mutation detection approach. *Br. J. Cancer* 87, 892–897.

Grant, S. F., Kristjansdottir, H., Steinsson, K., Blondal, T., Yuryev, A., Stefansson, K., and Gulcher, J. R. (2000). Long PCR detection of the C4A null allele in B8-C4AQ0-C4B1-DR3. *J. Immunol. Methods* 244, 41–47.

Harteveld, K. L., Losekoot, M., Fodde, R., Giordano, P. C., and Bernini, L. F. (1997). The involvement of Alu repeats in recombination events at the alpha-globin gene cluster: characterization of two alphazero-thalassaemia deletion breakpoints. *Hum. Genet.* 99, 528–534.

Herbers, J., Schullerus, D., Chudek, J., Bugert, P., Kanamaru, H., Zeisler, J., Ljungberg, B., Akhtar, M., and Kovacs, G. (1998). Lack of genetic changes at specific genomic sites separates renal oncocytomas from renal cell carcinomas. *J. Pathol.* 184, 58–62.

Hess, J. F., Fox, M., Schmid, C. and Shen, C. K. (1983). Molecular evolution of the human adult alpha-globin-like gene region: insertion and deletion of Alu family repeats and non-Alu DNA sequences. *Proc. Natl. Acad. Sci. USA* 80, 5970–5974.

Hodgson, G., Hager, J. H., Volik, S., Hariono, S., Wernick, M., Moore, D., Nowak, N., Albertson, D. G., Pinkel, D., Collins, C., Hanahan, D., and NS Gray, J. W. (2001). Genome scanning with array CGH delineates regional alterations in mouse islet carcinomas. *Nat. Genet.* 29, 459–464.

Hogervorst, F. B., Nederlof, P. M., Gille, J. J., McElgunn, C. J., Grippeling, M., Pruntel, R., Regnerus, R., van Welsem, T., van Spaendonk, R., Menko, F. H., Kluijt, I., Dommering, C., Verhoef, S., Schouten, J. P., van't Veer, L. J., and Pals, G. (2003). Large genomic deletions and duplications in the BRCA1 gene identified by a novel quantitative method. *Cancer Res.* 63, 1449–1453.

Hoque, M. O., Lee, J., Begum, S., Yamashita, K., Engles, J. M., Schoenberg, M., Westra, W. H., and Sidransky, D. (2003). High-throughput molecular analysis of urine sediment for the detection of bladder cancer by high-density single-nucleotide polymorphism array. *Cancer Res.* 63, 5723–5726.

Hu, X., and Worton, R. G. (1992). Partial gene duplication as a cause of human disease. *Hum. Mutat.* 1, 3–12.

Inoue, K., Dewar, K., Katsanis, N., Reiter, L. T., Lander, E. S., Devon, K. L., Wyman, D. W., Lupski, J. R., and Birren, B. (2001). The 1.4-Mb CMT1A duplication/HNPP deletion genomic region reveals unique genome architectural features and provides insights into the recent evolution of new genes. *Genome Res.* 11, 1018–1033.

Jagadeeswaran, P., Tuan, D., Forget, B. G. and Weissman, S. M. (1982). A gene deletion ending at the midpoint of a repetitive DNA sequence in one form of hereditary persistence of fetal haemoglobin. *Nature* 296, 469–470.

Junien, C., Despoisse, S., Turleau, C., Nicolas, H., Picard, F., Le Marec, B., Kaplan, J. C., and de Grouchy, J. (1982). Retinoblastoma, deletion 13q14, and esterase D: application of gene dosage effect to prenatal diagnosis. *Cancer Genet. Cytogenet.* 6, 281–287.

Kajii, T., Ferrier, A., Niikawa, N., Takahara, H., Ohama, K., and Avirachan, S. (1980). Anatomic and chromosomal anomalies in 639 spontaneous abortuses. *Hum. Genet.* 55, 87–98.

Kallioniemi, A., Kallioniemi, O. P., Sudar, D., Rutovitz, D., Gray, J. W., Waldman, F., and Pinkel, D. (1992). Comparative genomic hybridization for molecular cytogenetic analysis of solid tumors. *Science* 258, 818–821.

Kan, Y. W., Holland, J. P., Dozy, A. M., Charache, S., and Kazazian, H. H. (1975). Deletion of the beta-globin structure gene in hereditary persistence of foetal haemoglobin. *Nature* 258, 162–163.

Kenwrick, S., Patterson, M., Speer, A., Fischbeck, K., and Davies, K. (1987). Molecular analysis of the Duchenne muscular dystrophy region using pulsed field gel electrophoresis. *Cell* 48, 351–357.

Kim, S. H., Bae, J. H., Chae, J. J., Kim, U. K., Choe, S. J., Namkoong, Y., Kim, H. S., Park, Y. B., and Lee, C. C. (1999). Long-distance PCR-based screening for large rearrangements of the LDL receptor gene in Korean patients with familial hypercholesterolemia. *Clin. Chem.* 45, 1424–1430.

Kim, S. W., Lee, K. S., Jin, H. S., Lee, T. M., Koo, S. K., Lee, Y. J., and Jung, S. C. (2003). Rapid detection of duplication/deletion of the PMP22 gene in patients with Charcot-Marie-Tooth disease Type 1A and hereditary neuropathy with liability to pressure palsy by real-time quantitative PCR using SYBR Green I dye. *J. Korean Med. Sci.* 18, 727–732.

Knight, S. J., Regan, R., Nicod, A., Horsley, S. W., Kearney, L., Homfray, T., Winter, R. M., Bolton, P., and Flint, J. (1999). Subtle chromosomal rearrangements in children with unexplained mental retardation. *Lancet* 354, 1676–1681.

Knowles, M. A. (1999). Identification of novel bladder tumour suppressor genes. *Electrophoresis* 20, 269–279.

Koenig, M., Beggs, A. H., Moyer, M., Scherpf, S., Heindrich, K., Bettecken, T., Meng, G., Muller, C. R., Lindlof, M., Kaariainen, H., de la Chapelle, A., Kiuru, A., Savontaus, M.-L., Gilgenkrantz, H., Recan, D., Chelly, J., Kaplan, J.-C., Covone, A. E., Archidiacono, N., Romeo, G., Liechti-Gallati, S., Schneider, V., Braga, S., Moser, H., Darras, B. T., Murphy, P., Francke, U., Chen, J. D., Morgan, G., Denton, M., Greenberg, C. R., van Ommen, G. J., and Kunkel L. M. (1989). The molecular basis for Duchenne versus Becker muscular dystrophy: correlation of severity with type of deletion. *Am. J. Hum. Genet.* 45, 498–506.

Laurendeau, I., Bahuau, M., Vodovar, N., Larramendy, C., Olivi, M., Bieche, I., Vidaud, M., and Vidaud, D. (1999). TaqMan PCR-based gene dosage assay for predictive testing in individuals from a cancer family with INK4 locus haploinsufficiency. *Clin. Chem.* 45, 982–986.

Lejeune, J. (1960). Mongolism, regressive trisomy. *Ann. Genet.* 2, 1–38.

Lemmers, R. J., van der Maarel, S. M., van Deutekom, J. C., van der Wielen, M. J., Deidda, G., Dauwerse, H. G., Hewitt, J., Hofker, M., Bakker, E., Padberg, G. W., and Frants, R. R. (1998). Inter- and intrachromosomal sub-telomeric rearrangements on 4q35: implications for facioscapulohumeral muscular dystrophy (FSHD) aetiology and diagnosis. *Hum. Mol. Genet.* 7, 1207–1214.

Li, W. H., and Gojobori, T. (1983). Rapid evolution of goat and sheep globin genes following gene duplication. *Mol. Biol. Evol.* 1, 94–108.

Linn, J. F., Lango, M., Halachmi, S., Schoenberg, M. P., and Sidransky, D. (1997). Microsatellite analysis and telomerase activity in archived tissue and urine samples of bladder cancer patients. *Int. J. Cancer* 74, 625–629.

Mansfield, E. S. (1993). Diagnosis of Down syndrome and other aneuploidies using quantitative polymerase chain reaction and small tandem repeat polymorphisms. *Hum. Mol. Genet.* 2, 43–50.

Meyer, U. A., and Zanger, U. M. (1997). Molecular mechanisms of genetic polymorphisms of drug metabolism. *Annu. Rev. Pharmacol. Toxicol.* 37, 269–296.

Mullis, K., Faloona, F., Scharf, S., Saiki, R., Horn, G., and Erlich, H. (1986). Specific enzymatic amplification of DNA in vitro: the polymerase chain reaction. *Cold Spring Harb. Symp. Quant. Biol.* 51 Pt 1, 263–273.

Nagahata, T., Hirano, A., Utada, Y., Tsuchiya, S., Takahashi, K., Tada, T., Makita, M., Kasumi, F., Akiyama, F., Sakamoto, G., Nakamura, Y., and Emi, M. (2002). Correlation of allelic losses and clinicopathological factors in 504 primary breast cancers. *Breast Cancer* 9, 208–215.

Neubauer, A., Neubauer, B., and Liu, E. (1990). Polymerase chain reaction based assay to detect allelic loss in human DNA: loss of beta-interferon gene in chronic myelogenous leukemia. *Nucleic Acids Res.* 18, 993–998.

Nupponen, N. N., and Visakorpi, T. (2000). Molecular cytogenetics of prostate cancer. *Microsc. Res. Tech.* 51, 456–463.

Ohno, S., Wolf, U., and Atkin, N. B. (1968). Evolution from fish to mammals by gene duplication. *Hereditas* 59, 169–187.

Orkin, S. H., Alter, B. P., Altay, C., Mahoney, M. J., Lazarus, H., Hobbins, J. C., and Nathan, D. G. (1978). Application of endonuclease mapping to the analysis and prenatal diagnosis of thalassemias caused by globin-gene deletion. *N. Engl. J. Med.* 299, 166–172.

Ose, T., and Bush, O. B., Jr. (1962). Erythroblastosis fetalis. Report of cases, with one caused by gene deletion. *Bibl. Haematol.* 13, 290–291.

Ottolenghi, S., Lanyon, W. G., Paul, J., Williamson, R., Weatherall, D. J., Clegg, J. B., Pritchard, J., Pootrakul, S., and Boon, W. H. (1974). The severe form of alpha thalassaemia is caused by a haemoglobin gene deletion. *Nature* 251, 389–392.

Patau, K., Therman, E., Smith, D. W., and Demars, R. I. (1961). Trisomy for chromosome No. 18 in man. *Chromosoma* 12, 280–285.

Petrij-Bosch, A., Peelen, T., van Vliet, M., van Eijk, R., Olmer, R., Drusedau, M., Hogervorst, F. B., Hageman, S., Arts, P. J., Ligtenberg, M. J., Meijers-Heijboer, H., Klijn, J. G., Vasen, H. F., Cornelisse, C. J., van 't Veer, L. J., Bakker, E., van Ommen, G. J., and Devilee, P. (1997). BRCA1 genomic deletions are major founder mutations in Dutch breast cancer patients. *Nat. Genet.* 17, 341–345.

Pinkel, D., Segraves, R., Sudar, D., Clark, S., Poole, I., Kowbel, D., Collins, C., Kuo, W. L., Chen, C., Zhai, Y., Dairkee, S. H., Ljung, B. M., Gray, J. W., and Albertson, D. G. (1998). High resolution analysis of DNA copy number variation using comparative genomic hybridization to microarrays. *Nat. Genet.* 20, 207–211.

Puget, N., Torchard, D., Serova-Sinilnikova, O. M., Lynch, H. T., Feunteun, J., Lenoir, G. M., and Mazoyer, S. (1997). A 1-kb Alu-mediated germ-line deletion removing BRCA1 exon 17. *Cancer Res.* 57, 828–831.

Puget, N., Sinilnikova, O. M., Stoppa-Lyonnet, D., Audoynaud, C., Pages, S., Lynch, H. T., Goldgar, D., Lenoir, G. M., and Mazoyer, S. (1999). An Alu-mediated 6-kb duplication in the BRCA1 gene: a new founder mutation? *Am. J. Hum. Genet.* 64, 300–302.

Rabbitts, T. H., Forster, A., Dunnick, W., and Bentley, D. L. (1980). The role of gene deletion in the immunoglobulin heavy chain switch. *Nature* 283, 351–356.

Richter, S., Vandezande, K., Chen, N., Zhang, K., Sutherland, J., Anderson, J., Han, L., Panton, R., Branco, P., and Gallie, B. (2003). Sensitive and efficient detection of RB1 gene mutations enhances care for families with retinoblastoma. *Am. J. Hum. Genet.* 72, 253–269.

Rio, M., Molinari, F., Heuertz, S., Ozilou, C., Gosset, P., Raoul, O., Cormier-Daire, V., Amiel, J., Lyonnet, S., Le Merrer, M., Turleau, C., de Blois, M. C., Prieur, M., Romana, S., Vekemans, M., Munnich, A., and Colleaux L. (2002). Automated fluorescent genotyping detects 10% of cryptic subtelomeric rearrangements in idiopathic syndromic mental retardation. *J. Med. Genet.* 39, 266–270.

Rohlfs, E. M., Puget, N., Graham, M. L., Weber, B. L., Garber, J. E., Skrzynia, C., Halperin, J. L., Lenoir, G. M., Silverman, L. M., and Mazoyer, S. (2000). An Alu-mediated 7.1 kb deletion of BRCA1 exons 8 and 9 in breast and ovarian cancer families that results in alternative splicing of exon 10. *Genes Chromosomes Cancer* 28, 300–307.

Rowland, J. S., Barton, D. E., and Taylor, G. R. (2001). A comparison of methods for gene dosage analysis in HMSN type 1. *J. Med. Genet.* 38, 90–95.

Schouten, J. P., McElgunn, C. J., Waaijer, R., Zwijnenburg, D., Diepvens, F., and Pals, G. (2002). Relative quantification of 40 nucleic acid sequences by multiplex ligation-dependent probe amplification. *Nucleic Acids Res.* 30, e57.

Schughart, K., Kappen, C., and Ruddle, F. H. (1989). Duplication of large genomic regions during the evolution of vertebrate homeobox genes. *Proc. Natl. Acad. Sci. USA* 86, 7067–7071.

Shimizu, S., Yatabe, Y., Koshikawa, T., Haruki, N., Hatooka, S., Shinoda, M., Suyama, M., Ogawa, M., Hamajima, N., Ueda, R., Takahashi, T., and Mitsudomi, T. (2000). High frequency of clonally related tumors in cases of multiple synchronous lung cancers as revealed by molecular diagnosis. *Clin. Cancer Res.* 6, 3994–3999.

Shuin, T., Kondo, K., Kaneko, S., Sakai, N., Yao, M., Hosaka, M., Kanno, H., Ito, S., and Yamamoto, I. (1995). Results of mutation analyses of von Hippel-Lindau disease gene in Japanese patients: comparison with results in United States and United Kingdom. *Hinyokika Kiyo* 41, 703–707.

Siniscalco, M., Robledo, R., Orru, S., Contu, L., Yadav, P., Ren, Q., Lai, H., and Roe, B. (2000). A plea to search for deletion polymorphism through genome scans in populations. *Trends Genet.* 16, 435–437.

Sismani, C., Armour, J. A., Flint, J., Girgalli, C., Regan, R., and Patsalis, P. C. (2001). Screening for subtelomeric chromosome abnormalities in children with idiopathic mental retardation using multiprobe telomeric FISH and the new MAPH telomeric assay. *Eur. J. Hum. Genet.* 9, 527–532.

Snijders, A. M., Nowak, N., Segraves, R., Blackwood, S., Brown, N., Conroy, J., Hamilton, G., Hindle, A. K., Huey, B., Kimura, K., Law, S., Myambo, K., Palmer, J., Ylstra, B., Yue, J. P., Gray, J. W., Jain, A. N., Pinkel, D., and Albertson, D. G. (2001). Assembly of microarrays for genome-wide measurement of DNA copy number. *Nat. Genet.* 29, 263–264.

Southern, E. M. (1975). Detection of specific sequences among DNA fragments separated by gel electrophoresis. *J. Mol. Biol.* 98, 503–517.

Stenson, P. D., Ball, E. V., Mort, M., Phillips, A. D., Shiel, J. A., Thomas, N. S., Abeysinghe, S., Krawczak, M., and Cooper, D. N. (2003). Human Gene Mutation Database (HGMD): 2003 update. *Hum. Mutat.* 21, 577–581.

Suminaga, R., Takeshima, Y., Yasuda, K., Shiga, N., Nakamura, H., and Matsuo, M. (2000). Non-homologous recombination between Alu and LINE-1 repeats caused a 430-kb deletion in the dystrophin gene: a novel source of genomic instability. *J. Hum. Genet.* 45, 331–336.

Swensen, J., Hoffman, M., Skolnick, M. H., and Neuhausen, S. L. (1997). Identification of a 14 kb deletion involving the promoter region of BRCA1 in a breast cancer family. *Hum. Mol. Genet.* 6, 1513–1517.

Taylor, C. F., Charlton, R. S., Burn, J., Sheridan, E., and Taylor, G. R. (2003). Genomic deletions in MSH2 or MLH1 are a frequent cause of hereditary non-polyposis colorectal cancer: identification of novel and recurrent deletions by MLPA. *Hum. Mutat.* 22, 428–433.

Tyagi, S., and Kramer, F. R. (1996). Molecular beacons: probes that fluoresce upon hybridization. *Nat. Biotechnol.* 14, 303–308.

van Essen, A. J., Kneppers, A. L., van der Hout, A. H., Scheffer, H., Ginjaar, I. B., ten Kate, L. P., van Ommen, G. J., Buys, C. H., and Bakker, E. (1997). The clinical and molecular genetic approach to Duchenne and Becker muscular dystrophy: an updated protocol. *J. Med. Genet.* 34, 805–812.

van Loghem, E., Sukernik, R. I., Osipova, L. P., Zegers, B. J., Matsumoto, H., de Lange, G., and Lefranc, G. (1980). Gene deletion and gene duplication within the cluster of human heavy-chain genes. Selective absence of IgG sub-classes. *J. Immunogenet.* 7, 285–299.

Vortmeyer, A. O., Huang, S. C., Pack, S. D., Koch, C. A., Lubensky, I. A., Oldfield, E. H., and Zhuang, Z. (2002). Somatic point mutation of the wild-type allele detected in tumors of patients with VHL germline deletion. *Oncogene* 21, 1167–1170.

Voskova-Goldman, A., Peier, A., Caskey, C. T., Richards, C. S., and Shaffer, L. G. (1997). DMD-specific FISH probes are diagnostically useful in the detection of female carriers of DMD gene deletions. *Neurology* 48, 1633–1638.

Wagner, A., van der Klift, H., Franken, P., Wijnen, J., Breukel, C., Bezrookove, V., Smits, R., Kinarsky, Y., Barrows, A., Franklin, B., Lynch, J., Lynch, H., and Fodde, R. (2002). A 10-Mb paracentric inversion of chromosome arm 2p inactivates MSH2 and is responsible for hereditary nonpolyposis colorectal cancer in a North-American kindred. *Genes Chromosomes Cancer* 35, 49–57.

Wiencke, J. K., Pemble, S., Ketterer, B., and Kelsey, K. T. (1995). Gene deletion of glutathione S-transferase theta: correlation with induced genetic damage and potential role in endogenous mutagenesis. *Cancer Epidemiol. Biomarkers Prev.* 4, 253–259.

Wijnen, J., van der Klift, H., Vasen, H., Khan, P. M., Menko, F., Tops, C., Meijers Heijboer, H., Lindhout, D., Moller, P., and Fodde, R. (1998). MSH2 genomic deletions are a frequent cause of HNPCC. *Nat. Genet.* 20, 326–328.

Wilkie, A. O., Higgs, D. R., Rack, K. A., Buckle, V. J., Spurr, N. K., Fischel-Ghodsian, N., Ceccherini, I., Brown, W. R., and Harris, P. C. (1991). Stable length polymorphism of up to 260 kb at the tip of the short arm of human chromosome 16. *Cell* 64, 595–606.

Yau, S. C., Bobrow, M., Mathew, C. G., and Abbs, S. J. (1996). Accurate diagnosis of carriers of deletions and duplications in Duchenne/Becker muscular dystrophy by fluorescent dosage analysis. *J. Med. Genet.* 33, 550–558.

CHAPTER **14**

Analysis of Human Splicing Defects Using Hybrid Minigenes

FRANCO PAGANI AND FRANCISCO E. BARALLE

Molecular Pathology Group, International Centre for Genetic Engineering and Biotechnology (ICGEB), Trieste, Italy

---

TABLE OF CONTENTS

14.1 INTRODUCTION

The availability of fast sequencing protocols is at the base of an ongoing genetic diagnostics revolution. It is now possible to scan at least the coding sequences of any gene for pathological variation. Although identifying mutations that produce amino acid changes, stop codons, or frame shift of the reading frame is straightforward, the identification of DNA variations that cause aberrant splicing is not as simple. The unexpected complexity of the splicing process, which correctly selects the coding sequences, the exons, from the more abundant noncoding sequences, the introns, has revealed in the last years the existence of new splicing regulatory elements difficult to identify exclusively by sequence inspection. Variations in these new elements both in coding and noncoding regions may result in an unexpected deleterious effect on the precursor (pre) mRNA splicing. As a result, distinguishing between benign and disease-causing sequence substitutions is a challenge for medical geneticists. In this chapter, the hybrid minigenes assay as a tool for the study and characterization of the effect of human DNA variations on the pre-mRNA splicing process will be described.

14.2 BASIC PRE-mRNA SPLICING PROCESS AND ALTERNATIVE SPLICING

Pre-mRNA splicing is a fundamental process in gene expression and in the generation of proteome diversity. In the nucleus, the splicing process acts on the pre-mRNA, recognizing the protein coding exonic sequences from the more abundant intronic ones, and joins them together forming the mature mRNA that is then transported to the cytoplasm and translated into a protein. The correct splicing process needs intact *cis* acting elements such as the 5′ and 3′ splice sites, the poly-pyrimidine tract, and the branch site (see Fig. 14.1). The 5′ splice site marks the exon/intron junction at the 5′ end on the intron and its hallmark is the invariant GU dinucleotide. At the other hand of the exon, the 3′ splice site region has three conserved elements: the branch site, the poly-pyrimidine tract, followed by the terminal conserved AG dinucleotide. The splicing cut-and-paste catalytic reaction consists of two trans-esterification steps (see Fig. 14.1).

The multicomponent splicing complex, known as spliceosome, that operates the splicing reaction is composed of five small ribonucleoproteins (snRNP) and more than 100 proteins (see Fig. 14.2a). The essential splicing signals on transcribed pre-mRNA are rather short and degenerate sequences at the intron-exon border (see Fig. 14.2a). The first event in the splicing reaction is the correct recognition of these classic splicing signals from the abundant and never-used pseudo-splice sites. For a detailed description of the splicing process see Burge and colleagues (1999).

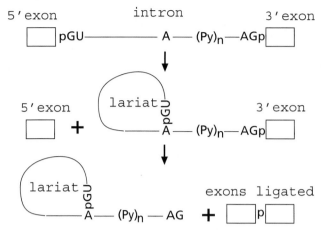

FIGURE 14.1   Overview of the splicing process. Essential splicing signals are the GU/AG dinucleotides at the exon-intron and intron-exon junctions (5′ and 3′ splice sites), respectively, the polypyrimidine tract (Py) and the A nucleotide of the branch site. Splicing takes places in two transesterification steps. In the first step the 2′-hydroxyl group of the A residue at the branch site attacks the phosphate at the GU 5′ splice site. This leads to cleavage of the 5′ exon from the intron and the formation of lariat intermediate. In the following step the two exons are ligated by a second transesterification reaction that involves the phosphate at the 3′ end of the intron and the 3′ hydroxyl of the detached exon. This releases the intron, still in the form of a lariat.

In a typical gene the majority of exons are constitutive; that means they are always included in the final mRNAs. However, alternative usage of different splice sites is a common event in human cells. It may occur in different manners, such as usage of alternative splice sites or mutually exclusive exons, exon skipping/inclusion, or intron retention (Caceres and Kornblihtt, 2002; Black, 2003). The resulting mRNA variable segments in the mature transcript can insert or remove amino acids, shift the reading frame or introduce a termination codon, or even modify regulatory elements for translation or mRNA stability or localization. Alternative splicing results in the production of multiple isoforms from the same transcription unit and is responsible for a substantial part of the complexity of the proteome. It has been estimated that up to 59% of human genes generates multiple mRNAs, an observation that can explain the low unexpected number of genes found in the human genome. A large fraction of alternative spliced transcripts has a restricted developmental and/or cell-type regulation or responds to a variety of external stimuli, and this complexity cannot be explained simply by considering the composition of classical splicing signals.

14.3   NOVEL *cis*-ACTING ELEMENTS INVOLVED IN SPLICING REGULATION

The correct splice site differentiation from the abundance of pseudo-splice sites and the fine-tuning regulation of the alternative splicing process requires auxiliary *cis*-acting elements on the pre-mRNA that assist the spliceosome in the selection of the splice sites. These elements, based on their effect on splicing, have been schematically divided in enhancers and silencers (Cartegni *et al.*, 2002). Enhancers and silencers, respectively, act by stimulating or inhibiting the splicing reactions and can be located either directly in the exons (ESE and ESS), and thus overlap with the selection of the amino acid sequence and codon usage, or in introns (ISE and ISS) even at very long distance from the splice sites (see Fig. 14.2a).

Enhancers and silencers are involved both in constitutive and alternative splicing, and in the majority of cases they lack a well-defined consensus sequence. Furthermore these elements are not always univocally defined and their functions may overlap. In fact in some systems it may be more appropriate to talk about composite exonic regulatory elements of splicing (CERES) as it has been recently described for CFTR exons 9 and 12 (Pagani *et al.*, 2003a; 2003b). Several *trans*-acting splicing factors can interact with enhancers and silencers and accordingly they have been divided into two major groups: members of the serine/arginine-rich (SR) proteins and heterogeneous nuclear ribonucleoproteins (hnRNPs; Fig. 14.2a). In general, but not exclusively, SR protein binding at ESE facilitates the exon recognition whereas hnRNPs are inhibitory. These proteins, in common with the majority of RNA binding proteins, have a modular structure, which consists of one or more RNA-binding domains associated to an auxiliary domain that often is involved in protein-protein interactions.

14.4   HUMAN GENETIC DEFECTS INVOLVING pre-mRNA SPLICING

Genome sequences from different individuals may reveal sequence variations. A difficult task is to correctly distinguish between benign polymorphisms from disease-causing variations. The identification of disease-causing mutations is based primarily on linkage of the mutation with the disease phenotype and on the effect of the mutations on gene expression. This effect generally is assumed to depend on the location of the nucleotide variants. In several cases, the correct interpretation of the molecular nature of the substitution may not be immediately evident (see Fig. 14.3).

The deleterious effect on gene expression of large deletions, nucleotide changes at the consensus GU/AG splice sites, exonic variations that produce stop codons or radical amino acid changes, and significant promoter variations are relatively easy to predict. Variations in nucleotides flanking the canonical GU/AG dinucleotides may be more difficult to assess, because the consensus of these regions shows some variability (Zhang, 1998). Exonic sequence variations overlap with coding sequences and for this reason they frequently are considered only for their effect at the protein

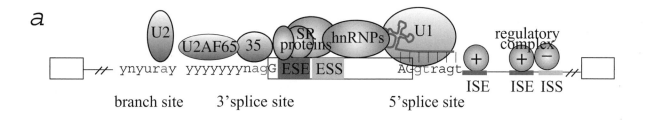

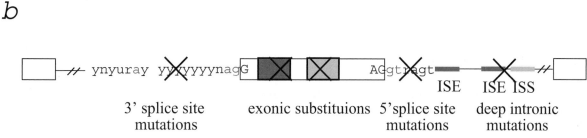

FIGURE 14.2   Regulatory elements in pre-mRNA splicing and mutations that can affect them. a. The essential splicing signals that define the exon boundaries are relatively short and poorly conserved sequences. Only the AG and the GU dinucleotides that directly flank the exon (at the 5′ and 3′ ends, respectively) and the branch point adenosine (all underlined) are always conserved. In most cases there is also a polypyrimidine tract of variable length (a pyrimidine base—cytosine or thymine—is represented by the consensus symbol "y") upstream of the 3′ splice site. The 5′ splice site binds to the U1 RNA by complementarity. The branch point typically is located 18–40 nucleotides upstream from the polypyrimidine tract. Additional enhancer and silencer elements in the exons (ESE; ESS) and/or introns (ISE; ISS) allow the correct splice sites to be distinguished from the many cryptic splice sites that have identical signal sequences. *Trans*-acting splicing factors can interact with enhancers and silencers and accordingly can be subdivided into two major groups: members of the serine arginine (SR) family of proteins and heterogeneous nuclear ribonuclear particles (hnRNPs). In general, but not exclusively, SR protein binding at ESE facilitates the exon recognition whereas hnRNPs are inhibitory. b. Mutations can result in aberrant splicing by affecting different splicing regulatory elements. Exonic variations in ESE or ESS can either change the amino acid or result in synonymous substitutions. Mutations in the intron might occur at the splice sites or deep in the introns.

Functional genomics and pre-mRNA processing alterations

Identification of mutations by systematic genome sequencing

Clear functional alteration   Polymorphism, synonymous variations
                                     atypical and "orphan" mutations

large deletions
splicing defects at consensus splice sites
nonsense mutations   ⤏ splicing defects?
missense mutations
promoter defects

                                     hybrid minigene splicing assay

FIGURE 14.3   Functional genomics and pre-mRNA processing alterations.

function level. Missense mutations modify the amino acid composition and synonymous variations are largely ignored as potentially deleterious. In the same way, deep intronic variations are not evaluated for their potential effect on splicing. Recent evidence from many laboratories has now

indicated that the primary mechanism of disease in a significant fraction of disease-causing mutations is catastrophic splicing abnormalities that disrupt previously unrecognized splicing regulatory elements (Cartegni *et al.*, 2002; Faustino *et al.*, 2003). Given the abundance of *cis*- and *trans*-acting factors involved in the splicing process, it was not completely surprising that a great proportion of human defects are related to the splicing process. Genetic analysis in *NF1* and *ATM* genes has shown that, in about 50% of the patients, mutations affecting splicing were involved (Teraoka *et al.*, 1999; Ars *et al.*, 2000). Of these mutations, 13% and 11% respectively, would have been erroneously classified as frameshift, missense, or nonsense mutations, if the analysis had been limited to genomic sequences. Studies on the mRNA are disclosing that many of them are really exon skipping mutations. Interestingly, most of the splicing mutations identified in these genes did not involve the conserved essential splice sites. Therefore, in most studies attempting to identify disease-causing mutations from the DNA sequence alone, single nucleotide polymorphisms (SNPs) such as these would be overlooked or not correctly classi-fied as producing splicing errors. These sequence variations may affect those auxiliary enhancer and silencer elements that, having a very loose consensus sequence and in several cases overlapping with coding regions, are difficult to identify.

The direct analysis of the processed transcript, mainly in the affected tissue, is the best way to establish with certainty if a particular DNA substitution affects splicing. However, samples obtained for clinical diagnosis are almost always leukocytes from which DNA is prepared. RNA samples from the affected individuals are not always available or not available at all for some tissue-specifically expressed genes, such as those expressed in the brain or the heart. The devel-opment of a functional splicing assay is of utmost impor-tance to determine if a particular sequence variation represents a benign polymorphism or a disease-causing mutation and to study the mechanism involved in normal and aberrant splicing.

14.5 GENERAL STRATEGY OF THE HYBRID MINIGENE ASSAY FOR THE IDENTIFICATION OF SPLICING DEFECTS

The basic principle of hybrid minigene splicing assay is shown in Fig. 14.4. Any genomic region of interest (i.e., exon and short intronic flanking regions), which is impli-cated to a splicing defect because it contains an "orphan" mutation, can be amplified from normal and affected indi-viduals and cloned into the minigene. The minigene plasmid is then transiently transfected in appropriate cell line where it will be transcribed by RNA polymerase II and the result-ing pre-mRNA processed to obtain a mature mRNA. The mRNA splicing pattern is analyzed mainly by RT-PCR with primers specifically designed to amplify processed tran-scripts derived from the minigene to distinguish from endogenous transcripts. The use of RT-PCR does not allow an exact quantification of the amount of each processed tran-script but of relative proportion of each splicing variant. To better quantify the absolute amounts of splicing variants other methods should be used such as RNAse protection assay or real-time PCR (see also Chapter 10).

The length and exon/intron composition of the gene portion to be cloned in the minigene have to be selected carefully and depends on the location of the presumed defect. Given the number and diversity of regulatory splic-ing signals, it is worth evaluating the gene structure as com-pletely as possible. However, for practical reasons, in the majority of cases, sequence substitutions in constitutive included exons are analyzed for exon skipping and require at minimum the cloning of the exon itself along with por-tions of the flanking introns. The same approach can be used for testing variations near the splice sites that can induce

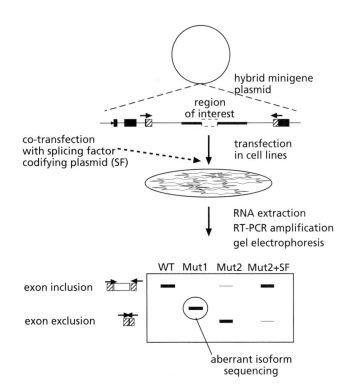

FIGURE 14.4   Schematic drawing of the hybrid minigene splic-ing assay. Study of human variations involved in aberrant splicing using the hybrid minigene. A typical hybrid minigene is a plasmid that contains a simplified version of the gene that will be evaluated for pre-mRNA splicing. It contains at the 5′ end an α-globin gene promoter and SV40 enhancer sequences indicated by a long arrow, to allow polymerase II transcription in the transfected cell lines. This is followed by a series of exonic and intronic sequences (indi-cated as boxes and lines, respectively) that may derive from a reporter gene or from the gene context itself. In this case the reporter gene is composed of α-globin (black boxes) and fibronectin exons (gray boxes); at the 3′ end a functionally com-petent polyadenylation site, derived from the α-globin gene is present. The genomic DNA region of interest that contains a puta-tive orphan splicing mutation is introduced in the minigene in a unique restriction site (*NdeI*). Thick line represents intronic sequences and white dashed box the presumed exonic sequences. The minigenes (normal and mutated) are transfected in cell lines, followed by RNA extraction, RT-PCR (using forward and reverse primers, complementary to the α-globin and fibronectin genes, respectively), and gel electrophoresis. In the example, the analysis of the transcript derived from the wild-type minigene shows the correct inclusion of an exon. Mut1 shows an aberrant spliced form whose identity can be verified by sequence analysis. The Mut2 causes significant exon skipping. In this case, cotransfection of a positive regulatory splicing factor (SF) induces exon inclusion restoring normal splicing.

exon skipping and/or activation of cryptic splice sites. However, in the presence of short introns, it is better to include them along with the nearby exons in the minigene as in these cases intron retention may be the observed splic-ing defect. In some cases, the presence of multiple exons may be necessary (see Section 14.6.3), and if the introns are too long, they can be internally deleted. In any case, comparison between transcripts derived from normal and mutated minigenes is always necessary to identify the disease-causing role of orphan splicing variations.

An alternative method useful for studying splicing that will not be described in detail here is the *in vitro* splicing assay. In *this* assay, labeled, preformed RNA molecules transcribed with bacterial polymerases are incubated in the presence of nuclear extracts and the resulting spliced products resolved on polyacrylamide denaturing gel. With this assay the intermediates of the splicing reactions, such as the lariat formation, can be evaluated. However, in contrast to the hybrid minigene, *in vitro* assay allows the study of relative short sequences, frequently containing a reduced single intron. In addition, and most importantly, *in vitro* assay does not take into account the fact that transcription and splicing are intimately connected in the cell. For a description of the mechanisms that couple transcription and pre-mRNA processing, see recent reviews (Caceres *et al.*, 2002; Maniatis and Reeds, 2002).

14.6 APPLICATIONS OF THE HYBRID MINIGENE ASSAY

14.6.1 3′ Splice Site Variations

Cystic fibrosis is caused by mutations in the CF transmembrane regulator (CFTR) gene and is characterized by pathological features of variable severity at the level of lungs, pancreas, sweat glands, testis, ovaries, and intestine. Some patients have evidence of a clinical disease in only a subgroup of the organ systems. These nonclassic CF forms include late-onset pulmonary disease, male sterility due to congenital bilateral absence of vas deference, and idiopathic pancreatitis. These nonclassic forms have been found associated with a peculiar allele at the polymorphic CFTR intron 8-exon 9 junction. At this locus a variable number of dinucleotide UG repeats (from 9 to 13) followed by a U repeat (U_5, U_7 or U_9) can be found in the normal population. The U_5 allele is considered a disease mutation with incomplete penetrance as it can be found both in normal individuals and in affected patients. The pathologic effect of the U_5 allele has been associated to the alternative splicing of the CFTR exon 9. This exon encodes part of the functionally important first nucleotide-binding domain, and its skipping produces a nonfunctional CFTR protein. The $UG_{13}U_3$ allele has been found in an individual with classic CF (Buratti *et al.*, 2001).

The hybrid minigene experiments shown in Fig. 14.5 indicate the importance of the UG_mU_n polymorphism in determining the amount of transcripts that contain exon 9. Exon 9 sequences, along with part of flanking introns, were introduced in hybrid minigenes, and different polymorphic variants were studied. Transfection experiments showed that

the $UG_{11}U_7$ allele found in normal individuals produced about 85% of normal exon 9 plus mRNAs. The $UG_{11}U_5$ allele, variably associated to nonclassic CF, reduces the amount of normal transcript to about 65% and the $UG_{13}U_3$ allele induces significant exon skipping (only 15% include the exon). The residual amount of normal exon 9 inclusion correlates with the severity of the phenotype.

14.6.2 Nucleotide Substitutions in Exonic Regulatory Elements Involving Splicing

14.6.2.1 MISSENSE AND SILENT MUTATIONS IN COMPOSITE EXONIC REGULATORY ELEMENT OF SPLICING

Skipping of CFTR exon 12 removes a highly conserved region encoding part of the first nucleotide-binding fold of CFTR, rendering the protein nonfunctional. In this exon, some nucleotide substitutions that cause a change in coding sequence did not show a clear association with loss of protein functionality or disease phenotype. This is the case for two interesting and enigmatic missense mutations, D565G and G576A (the latter having previously considered a neutral polymorphism; Fig. 14.6a).

Using the hybrid minigene shown in Fig. 14.6b, it has been shown that the D565G and G576A mutations induce a variable extent of exon 12 skipping that leads to reduced levels of normal transcripts. Interestingly, another missense mutation, the Y577F, which has been reported in a patient with severe CF phenotype, did not cause exon skipping; on the contrary, it increased the amount of transcript with the exon (see Fig. 14.6c). This study was reinforced by the opportunity to study the splicing pattern found in patient's derived cells (see Fig. 14.6d). This analysis showed a perfect concordance between the pattern observed with the minigene system and the one observed in the cells harboring the natural mutation (as it can be seen comparing Figs. 14.6c and d).

To study the nature of the exonic regulatory elements involved, a systematic site directed mutagenesis near the natural substitutions was performed (Pagani *et al.*, 2003b). These experiments clearly showed an overlapping enhancer and silencer functions (see Fig. 14.7).

In fact, nearby mutations or even variations at the same position may result in an enhancing or silencing effect. Due to their peculiar behavior, these elements were named Composite Exonic Regulatory Element of Splicing (CERES) (Pagani *et al.*, 2003b). Interestingly, some site-directed mutants that affect splicing (see Fig. 14.7a, c; underlined) are at the third position of the codon usage and do not modify the amino acid code; hence they would be labeled as neutral variations, if found in a classical genome scanning analysis. This fact indicates that neutral variations, frequently considered not to have pathological consequences, on the contrary may induce a severe splicing defect.

a   *b*

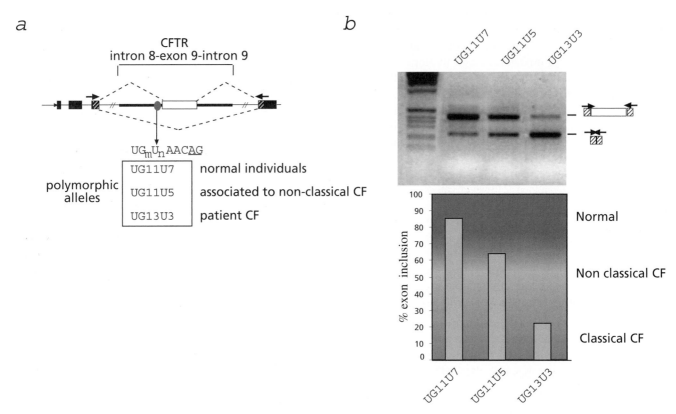

FIGURE 14.5 3′ splice site variations. The example illustrates the use of the minigene approach to study a polymorphic region near the 3′ splice site in CFTR, which causes aberrant exon 9 skipping. a. The polymorphic UG_mU_n locus at the end of CFTR intron 9 contains a variable number of dinucleotide UG repeats (from 9 to 13) followed by a U repeat (U_3, U_5, U_7 or U_9) variably present in the human population. Variants at the locus (mainly the U_5 allele) have been associated to phenotypes of different severity, ranging from classic to non-classic forms of CF, and in some cases found also in normal individuals. In the figure, three representative alleles are shown. The three variant minigenes that contain the indicated number of UG and U near the AG 3′ splice site (underlined) were analyzed. b. Hybrid minigene transient transfection assay showing the effect of the polymorphic variants. The variations induce variable skipping of exon 9 that correlates with the severity of the phenotype. Two processed transcripts are evident in the agarose gel—a normal one that contains the exon 9, and an aberrant one without the exon. Several splicing factors modulate the amount of aberrant exon 9 skipping (not shown), thus possibly influencing the phenotypic expression (Pagani *et al.*, 2000; Buratti *et al.*, 2001).

14.6.2.2 THE IMPORTANCE OF THE GENOMIC CONTEXT IN MINIGENE ANALYSIS: THE EXAMPLE OF NF1 EXON 37

Neurofibromatosis type 1 (NF1) is a common autosomal dominant genetic disorder with a prevalence of approximately 1 in 3,000 individuals and is one of the most common single gene disorders influencing neurological function in humans. The NF1 gene maps to chromosome 17q11.2 and is thought to be a tumor suppressor gene because loss of heterozygosity is associated with the occurrence of benign and malignant tumours in neural crest derived tissues as well as myeloid malignancies. It spans a region of about 350 kb of genomic DNA and contains 60 exons. The NF1 gene transcribes several mRNAs in the size range 11–13 kb; these are markedly expressed in neurons, oligodendrocytes, and non-myelinating Schwann cells. The most common transcript codes for a polypeptide of 2,818 amino acids called neurofibromin. About 50% of the mutations identified result in

aberrant splicing, and most of these defects did not involve the conserved dinucleotides at the splice sites (Ars *et al.*, 2000). The literature describes a very interesting set of mutations that introduce stop codons in exon 37 and may produce significant alteration of the splicing patterns (Hoffmeyer *et al.*, 1998). One of these nonsense mutations, a C-to-A substitution, was studied in detail in different minigene contexts (see Fig. 14.8) and gave interesting information on the regulatory exonic sequences involved. In exon 37, substitutions at the same position may code for a stop codon (TAA and TAG) or are synonymous (TAT). The three substitutions induced the same aberrant splicing pattern in the minigenes causing multiple exon skipping, indicating that this aberrant splicing was not necessarily linked to the presence of a stop codon (see Fig. 14.8c). In addition, this experiment highlights the importance of the appropriate genomic context in determining the effect of the DNA variations on splicing. If only exon 37 and part of its flanking introns is introduced

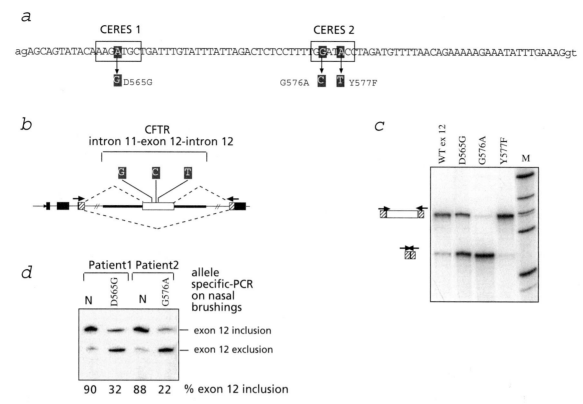

FIGURE 14.6 Missense mutations in Composite Regulatory Elements of Splicing (CERES) in CFTR exon 12. a. Nucleotide sequence of the CFTR exon 12 showing the position of the two CERES (boxed) and natural missense variations. b. Schematic representation of the three hybrid minigenes analysed. These minigenes contain the wild-type exon 12 (WTex12) or the mutations. c. Hybrid minigenes transient transfection assay showing the effect of CFTR exon 12 mutants. The two bands correspond to inclusion (upper) or exclusion (lower) of CFTR exon 12 in mature transcripts. Wild-type exon 12 shows not complete exon inclusion as found *in vivo*. D565G and G576A induce significant exon skipping. On the contrary, Y577F increases the percentage of exon 12 inclusion. d. Allele-specific RT-PCR on RNA extracted from nasal epithelial cells form patient 1 (heterozygote for D565G) and patient 2 (heterozygote for G576A). The RNAs were amplified with allele specific primers that can detect the alleles with the missense variations (D565G or G576A) or the other (normal, N) allele. Note the perfect correlation between the natural amount of exon inclusion (this panel) and the one obtained with the minigene system (panel c).

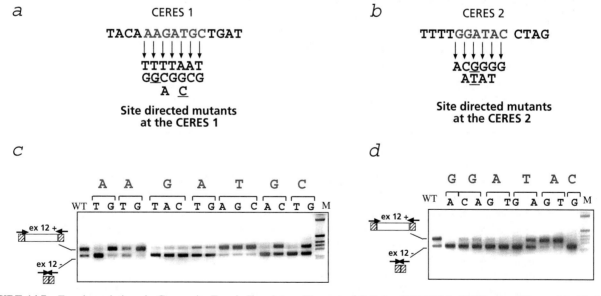

FIGURE 14.7 Exonic variations in Composite Exonic Regulatory Element of Splicing (CERES) in CFTR exon 12. a and b. Nucleotide sequence of the two CERES in CFTR exon 12 and site-directed variations. The site-directed mutants at the third position of the codon usage that do not change the amino acid code are underlined. The different site-directed mutant minigenes were transfected into cells and the percentage of exon 12 inclusion was analysed by RT-PCR. c and d. Hybrid minigenes transient transfection assay showing the effect of natural CFTR exon 12 mutants. Agarose gel electrophoresis of RT-PCR products, obtained from the splicing assay from the site-directed mutants in the two CERES, showing the relative amount of transcripts, with or without exon 12.

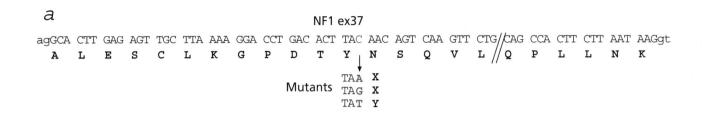

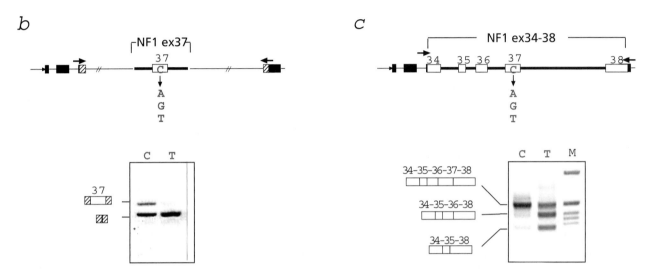

FIGURE 14.8   The example of NF1. a. Nucleotide sequence of part of the NF1 exon 37 showing the three different variants at the same position. The natural C-to-A and C-to-G mutants are nonsense substitutions. The site-directed C-to-T substitution is a silent mutation maintaining the tyrosine codon. The normal and mutant exon 37 variants were introduced in two different minigene contexts. In panel b exon 37 is flanked only by nearby intronic sequences; in panel c exon 37 variants are embedded in the entire genomic region from exon 34 to 38. b. The upper part shows a schematic view of the hybrid minigene containing only the NF1 exon 37 and part of flanking introns. In the lower panel, the resulting patterns of splicing are evident with the exon 37 inclusion and exclusion bands. It is important to note that in the normal C minigene, only a low amount of the transcripts contains exon 37 and this differs from the natural context where exon 37 is constitutively included. The three variants induce complete exon skipping. c. The upper part shows the hybrid minigene that contains the genomic region of NF1 from exon 34 to 38. Lower panel shows the processed transcripts. Contrary to results shown in panel b, the wild-type constructs normally is processed as in the natural context and includes all the five NF1 exons present in the minigene. The three variations induce two aberrant spliced forms: one with skipping of exon 37 alone and one where both exon 36 and 37 are missing. This example clearly shows the importance of the context, indicating that the minigene system can be used only when the wild type sequence gives the same splicing pattern as the chromosomal gene (see also CFTR exon 12 example in Figs. 14.6c and d).

in the minigene, even the wild type exon 37 is not fully recognized by the splicing machinery (see Fig. 14.8b, lane C). These data indicate that additional genomic sequences extending beyond the flanking introns are necessary for correct processing of this exon. Indeed, the minigene that contains the entire normal genomic sequences from exon 34 to exon 38 showed the complete inclusion of wild type exon 37 (see Fig. 14.8c, lane C). Interestingly, in this minigene, the substitutions induced a complex splicing defect that extended beyond the recognition of exon 37. In fact, exon 37 mutants induced not only its skipping but also some transcripts lacking both exon 36 and 37 (see Fig. 14.8c, lane T).

Thus, to study appropriately the effect of human DNA variations on splicing the first step is to ensure that there is a good correlation between the wild type splicing pattern deriving from the chromosomal gene and the minigene con-

struct. If this correlation does not exist the context of the minigene should be appropriately modified.

14.6.2.3 A SILENT MUTATION IN SMN2

Spinal muscular atrophy is a pediatric neurodegenerative disorder caused by homozygous loss of function of the survival motor neuron 1 (*SMN1*) gene. *SMN1* is duplicated in the human genome and its highly homologous copy is called *SMN2*; both genes are transcribed. The *SMN2* gene is present in all the patients but is not able to compensate for the *SMN1* gene defect. *SMN2* differs from *SMN1* by five nucleotides. These variations, either intronic or exonic, are translationally silent. One of these variations, a translationally silent C→T substitution in exon 7, has been shown using a minigene system to be the cause of the inability of *SMN2* to compensate for *SMN1* (Lorson *et al.*, 1999; 2000).

Two minigenes were prepared, SMN1 and SMN2, that differ only for the C→T substitution. Genomic SMN1 and SMN2 DNA including exons 6–8 were cloned into a mammalian expression plasmid, downstream of the constitutively expressing cytomegalovirus promoter. In SMN2 the synonymous substitution causes alternative splicing with skipping of exon 7 in the majority of transcripts (Lorson *et al.*, 1999; 2000; Fig. 14.9).

The truncated transcripts skipping exon 7 encode a protein lacking the 16 C-terminal residues, which is unstable and nonfunctional. Surprisingly, the exact mechanism by which the silent substitution causes exon skipping is not yet clear. Two models have been recently proposed to explain the effect of this substitution: the first model suggests that the mutation causes the inactivation of an ESE binding the SR protein SF2/ASF (Cartegni and Krainer, 2002) and the creation of a new ESS binding hnRNPA1 (Kashima and Manley, 2003). The elucidation of the basic mechanisms is important as the splicing reactivation of the dormant *SMN2* paralogue might represent a new strategic therapy in patients with spinal muscular atrophy (see comments in Buratti *et al.*, 2003; Khoo *et al.*, 2003).

14.6.3 Variations at the 5′ Splice Site

The effect of variations near the invariant AG/GU dinucleotides at the splice sites are not always easy to evaluate. An exon 3 + 5G→C substitution in Neurofibromatosis type 1 gene (NF1) identified during a NF1 genomic sequence clinical screen represents a diagnostic challenge (Baralle *et al.*, 2003). The proximity of the substitution to the 5′ splice site suggests that it may interfere with splicing (see Fig. 14.10a). However there are examples of wild-type 5′ splice sites similar to the exon 3 + 5G→C, in which the corresponding exon is spliced efficiently. For example, in normal *NF-1*, introns 1 and 7 share identical −1 to +5 sequence with the mutated intron 3 and intron 37 carries a T instead of a G at the +5 position (see Fig. 14.10a). It is obvious that sequence analysis alone cannot predict if the nucleotide substitution is pathogenic and, like in most clinical genetics situations, RNA was not available for diagnosis. A region comprising exon 3 and its flanking intronic sequences (see Fig. 14.10b) of the potentially abnormal *NF-1* gene was amplified from the available genomic DNA and inserted into the hybrid minigene constructs. Following transfection and

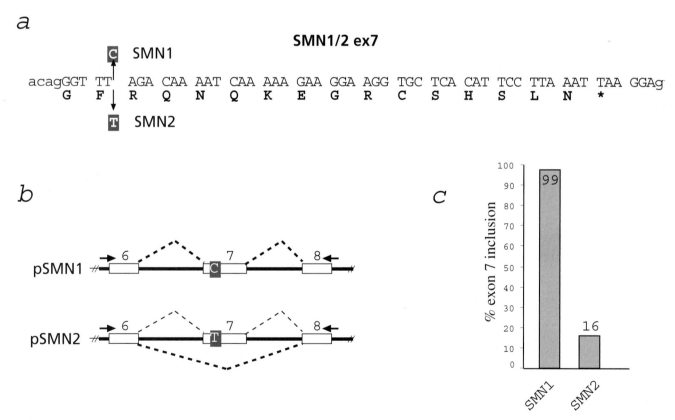

FIGURE 14.9 A silent variation in SMN2 causes exon skipping. a. Nucleotide sequence of the exon 7 of SMN genes showing the position of the synonymous C→T substitution. In exon 7, the SMN1 gene contains a C whereas SMN2 contains a T. b. Two minigenes were prepared, SMN1 and SMN2, that differ only for the C-to-T variant. Genomic SMN1 and SMN2 DNAs including exons 6–8 were cloned into a mammalian expression plasmid, downstream of the constitutively expressing cytomegalovirus promoter. The indicated hybrid minigenes were used to demonstrate that the T variant in SMN2 induces exon skipping resulting in a nonfunctional protein. c. Percentage of exon 7 inclusion derived from transient transfection experiments. Data are from Lorson and colleagues (2000).

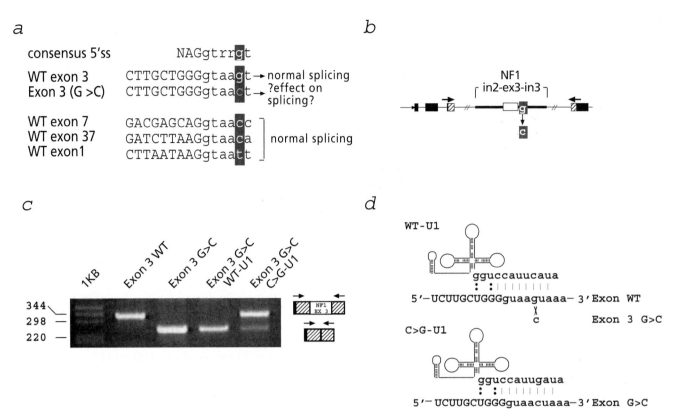

FIGURE 14.10 Identification of 5′ splice site variation in NF1 and correction with U1 snRNA interaction. a. Nucleotide sequence of the 5′ splice sites in wild-type and mutant exon 3 compared to the consensus and with the 5′ splice site of other NF1 exons. The G→C substitution in position +5 in exon 3 deviates from the consensus. However, the 5′ splice site of exons 7, 37, and 1 also deviates from the consensus at the same position, but normally are spliced *in vivo*. b. Hybrid minigenes containing normal and mutated NF1 exon 37 sequences. c. Hybrid minigenes transient transfection assay showing the effect of the exon 3 mutant and the effect of cotransfection of modified U1 snRNAs. The two bands correspond in inclusion (upper) or exclusion (lower) of NF1 exon 3 in the final transcripts. Wild-type exon 3 is included in the mRNA, and the exon 3 G→C substitution causes complete exon skipping. Co-transfection of WT-U1 has no effect on the splicing pattern of mutant exon 3. On the contrary, cotransfection of C→G-U1 induces inclusion of exon 37, indicating that the aberrant splicing defect is corrected in the presence of complementary U1 snRNA. d. Base pairing homology between the 5′ end of normal (WT-U1) and modified (C→G-U1) U1 snRNAs and the exon 3 5′ splice site sequences. The position of the G to C change is indicated. In the upper panel, the G→C substitution disrupts the complementarity with the normal U1. In the lower panel the modified U1 restores the base pair complementarity with the mutant. In other contexts such as the exons 1, 7, and 37 (panel a), such an extensive complementarity with U1 snRNA seems not to be necessary. This example highlights the difficulties found in predicting the effect of this type of mutation.

expression of the constructs, the mRNA produced was analyzed for splicing pattern by RT-PCR. As shown in Fig. 14.10c, the exon 3 + 5G→C mutation dramatically affects pre-mRNA processing, causing exon 3 to be completely skipped. One of the splicing factors that could be involved in the pathogenesis of this defect is U1 snRNP. This ribonucleoprotein particle contains several proteins and a unique RNA, the U1 snRNA.

The U1 snRNA is involved in the recognition of the 5′ splice site by base pair complementarity. The substitution of the guanosine by a cytidine in position + 5 of IVS 3 lessens the degree of U1-snRNA base pairing with the 5′ splice site (see Fig. 14.10d), although not to an extent to render the 5′ splice site nonfunctional as shown by the sequences compared in Fig. 14.10a. To demonstrate that U1 is involved in the splicing defect, a modified version of U1 snRNA was

prepared in which the complementary cytidine to guanosine in the U1 snRNA restore normal base pairing with the mutant (see Fig. 14.10d). Co-expression of this modified U1 snRNA with the minigene carrying the mutation resulted in rescue of exon 3 splicing (see Fig. 14.10c). This experiment proves that the exon 3 + 5G→C variation is a disease-causing mutation that induces aberrant skipping of exon 3 by interfering with the recognition of the 5′ splice site that in this context is sensitive to a shortening of the complementary region.

14.6.4 Identification of Deep Intronic Mutations: The Example of the ATM Gene

In the Ataxia Telengectasia (ATM) gene a novel type of mutation causes a splicing processing defect affecting an

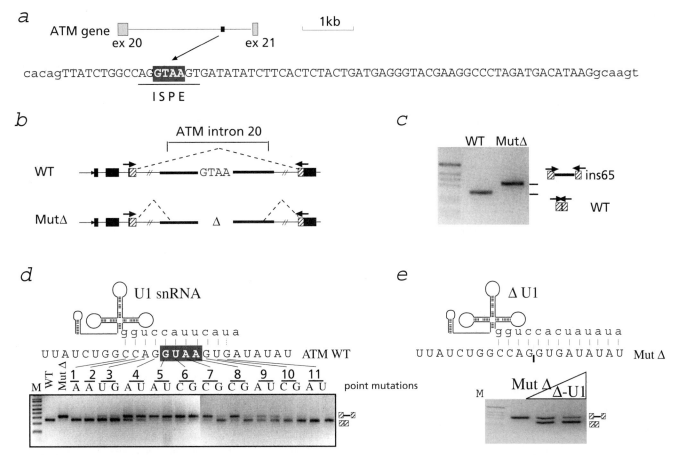

FIGURE 14.11   A splicing processing defect in ATM intron. a. Identification of the splicing defect in the affected patient. A deletion was identified by sequence analysis of the ATM gene and of the mRNA containing the cryptic exon shown in panel a. The sequence of the cryptic exon is shown in uppercase, intronic sequences in lowercase, and the splice sites used are underlined. The 5′ splice site did not have the almost universal GU donor site but a weak GC variant instead. The mutated sequences detected in patient's cDNA and genomic DNA result from the deletion of the four intra-exonic bases GTAA (boxed). b. The 4 bp deletion causes aberrant cryptic exon inclusion in hybrid minigene experiments. Two ATM hybrid minigenes were prepared; one contains wild-type intronic sequences (pWT), the other the ATM variant with the patient's genomic sequences with the 4 bp GTAA deletion (pMutΔ). The primers used in the RT-PCR assay are shown as superimposed arrows. c. Hybrid minigenes were transfected in Hep3B cells and the splicing products analyzed with the specific primers. The agarose gel electrophoresis shows the processed transcripts that correspond to complete intron removal (lower band) or inclusion of the 65 bp long cryptic exon (ins65). d. Base pairing homology between ATM ISPE and the 5′ end of wild type U1 snRNA. It is clear that the ATM ISPE base pairing to U1 snRNA regulates intron splicing processivity. Single point mutations were introduced, one at time, in the ATM minigenes at the ISPE in the entire U1 snRNA complementarity region between position 1 and 11. Hybrid minigene experiments with the different site-directed mutants showed that mutations between position 4 and 9 induced cryptic exon inclusion. Interestingly, mutations 7G and 8G did not induce the aberrant splicing products, but 7C and 8C did. This is because 7G and 8G variants preserve the complementarity to U1 by G:U "wobble" base pairing. e. Restoring U1 snRNAs-ISPE complementarity induces *in vivo* normal intron processivity. The effect of mutant U1 complementary to Mut Δ variant (Δ-U1) on the cryptic exon inclusion was evaluated. The upper panel shows the base pairing homology between the 5′ end of the mutant U1 snRNA and the Mut Δ minigene. The position of the deletion (Δ) in the ATM mRNAs is indicated. Lower panel shows Mut Δ minigene variant cotransfected with increasing amounts of the indicated Δ-U1 snRNA. The aberrant splicing defect is corrected in the presence of complementary U1.

intronic element that is important for correct intron removal (Pagani *et al.*, 2002).

This last example illustrates the importance of intronic sequences in the splicing process, even if they are located far away from the 5′ and 3′ splice sites, pointing to a far more complex RNA processing mechanism than that accepted up to now. In the affected patient, sequence analysis revealed in genomic DNA a 4 bp deletion (GTAA) in

intron 20, about 3 kb from exon 20 and 0.8 kb from exon 21 (see Fig. 14.11a). To test if this deletion was a benign polymorphism or a disease-causing mutation two hybrid minigenes were constructed, one containing the normal ATM intron 20 and the other with the intron deletion (see Fig. 14.11b). Transfection experiments showed that normal ATM sequences were recognized properly as an intron by the splicing machinery and accordingly excluded in the mature transcript. On the contrary, the GTAA deletion caused the activation of a 65 bp cryptic exon (see Fig. 14.11c). Cryptic or pseudo exons normally are activated by intronic mutations that create or strengthen splice sites or create a new branch site. Interestingly, the GTAA deletion did not involve directly the splice sites of the cryptic exon (see Fig. 14.11a). To study the mechanism involved in this new pathogenic mutation, extensive site-directed mutagenesis was performed leading to the identification of an intronic regulatory sequence that was named Intronic Splicing Processing Element (ISPE). The ISPE is complementary to U1 snRNA, which was directly related to the activation of the cryptic exon (see Fig. 14.11d). A clearly functional interaction between U1 snRNP and the ISPE was obtained by complementation experiments with mutant U1 snRNAs, engineered to bind specifically at Δ–ISPE (see Fig. 14.11e). Cotransfection with increasing amounts of Δ-U1 snRNAs with pATMΔ showed the progressive disappearance of the cryptic exon, restoring normal intron processivity (see Fig. 14.11e). These results indicate that the strength of the complementarity of the ISPE to U1 snRNA functionally is related to the splicing defect and modulates the efficiency of intron removal. Thus a new type of intronic U1 snRNP binding site (ISPE) performs an essential function for accurate intron removal. The function of the U1 snRNP-ISPE interaction is apparently quite different from the well-established interaction for the 5′ splice site initial recognition. Deletion of this sequence is involved directly in a splicing processing defect in a human disease that again could not have been predicted from simple DNA sequence inspection.

14.7 CONCLUSIONS

The genomic diversity and genomic pathology data currently available has revealed the extent of ignorance of the basic molecular mechanisms underlying the pre-mRNA splicing process (Teraoka *et al.*, 1999; Ars *et al.*, 2000; Cartegni *et al.*, 2002). Even more worrying is the fact that a lot of clinically relevant mutations may be slipping through the net because their effect on the splicing process is not even considered. An increasing amount of evidence indicates that single nucleotide substitutions in both coding and noncoding sequences might have unexpected deleterious effects on the splicing of the gene transcript (Ars *et al.*, 2000; Jiang *et al.*, 2000; Cartegni *et al.*, 2002; Pagani *et al.*, 2002; 2003a; 2003b; Fernandez-Cadenas *et al.*, 2003; Jiang *et al.*, 2003; Kashima and Manley, 2003). The hybrid minigenes assay described in this chapter represents a useful tool both for the diagnosis of human splicing defects and for the identification of new basic splicing mechanisms. In fact in the diagnostic field, the hybrid minigene assay can be used to distinguish between benign polymorphisms from disease-associated splicing mutations. Furthermore, human splicing errors identified by hybrid minigene assays represent a flag

put on new splicing modulatory elements; this knowledge gives important insight into the basic molecular mechanism of pre mRNA splicing.

Acknowledgements

The financial support of Telethon-Italy (grant n° GGP02453), AIRC and Italian research Ministry (FIRB RBNE01W9PM) is gratefully acknowledged.

References

Ars, E., Serra, E., Garcia, J., Kruyer, H., Gaona, A., Lazaro, C., and Estivill, X. (2000). Mutations affecting mRNA splicing are the most common molecular defects in patients with neurofibromatosis type 1. *Hum. Mol. Genet.* 9, 237–247.

Baralle, M., Baralle, D., De Conti, L., Mattocks, C., Whittaker, J., Knezevich, A., Ffrench-Constant, C., and Baralle, F. E. (2003). Identification of a mutation that perturbs NF1 agene splicing using genomic DNA samples and a minigene assay. *J. Med. Genet.* 40, 220–222.

Black, D. L. (2003). Mechanisms of alternative pre-messenger RNA splicing. *Annu. Rev. Biochem.* 72, 291–336.

Buratti, E., Baralle, F. E., and Pagani, F. (2003). Can a "patch" in a skipped exon make the pre-mRNA splicing machine run better? *Trends Mol. Med.* 9, 229–232.

Buratti, E., Dork, T., Zuccato, E., Pagani, F., Romano, M., and Baralle, F. E. (2001). Nuclear factor TDP-43 and SR proteins promote in vitro and in vivo CFTR exon 9 skipping. *EMBO J.* 20, 1774–1784.

Burge, C., Tuschl, T., and Sharp, P. (1999). Splicing of precursors to mRNAs by the spliceosomes. In R. Gesteland, T. Cech, and J. Atkins. *The RNA World*, pp. 525–560. Cold Spring Harbor, NY, Cold Spring Harbor Press.

Caceres, J. F., and Kornblihtt, A. R. (2002). Alternative splicing: multiple control mechanisms and involvement in human disease. *Trends Genet.* 18, 186–193.

Cartegni, L., Chew, S. L., and Krainer, A. R. (2002). Listening to silence and understanding nonsense: exonic mutations that affect splicing. *Nat. Rev. Genet.* 3, 285–298.

Cartegni, L., and Krainer, A. R. (2002). Disruption of an SF2/ASF-dependent exonic splicing enhancer in SMN2 causes spinal muscular atrophy in the absence of SMN1. *Nat. Genet.* 30, 377–384.

Faustino, N. A. and Cooper, T. A. (2003). Pre-mRNA splicing and human disease. *Genes Dev.* 17, 419–437.

Fernandez-Cadenas, I., Andreu, A. L., Gamez, J., Gonzalo, R., Martin, M. A., Rubio, J. C., and Arenas, J. (2003). Splicing mosaic of the myophosphorylase gene due to a silent mutation in McArdle disease. *Neurology* 61, 1432–1434.

Hoffmeyer, S., Nurnberg, P., Ritter, H., Fahsold, R., Leistner, W., Kaufmann, D., and Krone, W. (1998). Nearby stop codons in exons of the neurofibromatosis type 1 gene are disparate splice effectors. *Am. J. Hum. Genet.* 62, 269–277.

Kashima, T., and Manley, J. L. (2003). A negative element in SMN2 exon 7 inhibits splicing in spinal muscular atrophy. *Nat. Genet.* 34, 460–463.

Khoo, B., Akker, S. A., and Chew, S. L. (2003). Putting some spine into alternative splicing. *Trends Biotechnol.* 21, 328–330.

Lorson, C. L., and Androphy, E. J. (2000). An exonic enhancer is required for inclusion of an essential exon in the SMA-determining gene SMN. *Hum. Mol. Genet.* 9, 259–265.

Lorson, C. L., Hahnen, E., Androphy, E. J., and Wirth, B. (1999). A single nucleotide in the SMN gene regulates splicing and is responsible for spinal muscular atrophy. *Proc. Natl. Acad. Sci. USA.* 96, 6307–6311.

Maniatis, T., and Reed, R. (2002). An extensive network of coupling among gene expression machines. *Nature* 416, 499–506.

Pagani, F., Buratti, E., Stuani, C., Romano, M., Zuccato, E., Niksic, M., Giglio, L., Faraguna, D., and Baralle, F. E. (2000). Splicing factors induce cystic fibrosis transmembrane regulator exon 9 skipping through a nonevoluntionary conserved intronic element. *J. Biol. Chem.* 275, 21041–21047.

Pagani, F., Buratti, E., Stuani, C., Bendix, R., Dork, T., and Baralle, F. E. (2002). A new type of mutation causes a splicing defect in ATM. *Nat. Genet.* 30, 426–429.

Pagani, F., Buratti, E., Stuani, C., and Baralle, F. E. (2003a). Missense, nonsense, and neutral mutations define juxtaposed regulatory elements of splicing in cystic fibrosis transmembrane regulator exon 9. *J. Biol. Chem.* 278, 26580–26588.

Pagani, F., Stuani, C., Tzetis, M., Kanavakis, E., Efthymiadou, A., Doudounakis, S., Casals, T., and Baralle, F. E. (2003b). New type of disease causing mutations: the example of the composite exonic regulatory elements of splicing in CFTR exon 12. *Hum. Mol. Genet.* 12, 1111–1120.

Teraoka, S. N., Telatar, M., Becker-Catania, S., Liang, T., Onengut, S., Tolun, A., Chessa, L., Sanal, O., Bernatowska, E., Gatti, R. A., and Concannon, P. (1999). Splicing defects in the ataxia-telangiectasia gene, ATM: underlying mutations and consequences. *Am. J. Hum. Genet.* 64, 1617–1631.

Zhang, M. Q. (1998). Statistical features of human exons and their flanking regions. *Hum. Mol. Genet.* 7, 919–932.

CHAPTER **15**

DNA Microarrays and Genetic Testing

LARS DYRSKJØT, CLAUS L. ANDERSEN, MOGENS KRUHØFFER, NIELS TØRRING, KAREN KOED, FRIEDRIK WIKMAN, JENS L. JENSEN, AND TORBEN F. ØRNTOFT
Department of Clinical Biochemistry, Molecular Diagnostic Laboratory, Aarhus University Hospital, Skejby, Aarhus N, Denmark

15.1 INTRODUCTION

The completion of the Human Genome Project and the simultaneous advances in microarray technology has had a large impact on the field of molecular biology and especially on cancer research. It is now possible to make systematic genome-wide searches for genes being differentially expressed, amplified, or deleted during cancer development. Earlier this search was limited to only small numbers of candidate targets using, for example, northern blotting assays. The use of microarray-based assays has boosted the assay capacity to include several thousand measurements. The genome-wide searches have identified several genes that can serve as diagnostic or prognostic molecular disease markers or as potential therapeutic targets.

The use of DNA microarray technology has revolutionized molecular biology since its introduction in the mid 1990s. Using DNA microarray analysis, it is now possible to make complete genome-wide screenings e.g., molecular disease markers. Especially in the field of cancer disease classification and outcome prediction, notable findings have been reported by large-scale gene expression profiling.

Basically, a DNA-microarray is an orderly arrangement of usually thousands of defined DNA molecules immobilized on a small glass surface. By convention the immobilized DNA molecules are called the probes and the sample under investigation is the target. Several microarray types exist for monitoring gene expression, single nucleotide polymorphisms (SNPs), and loss or gain of genomic material. Furthermore, microarrays for resequencing of known gene sequences also exists together with small interfering RNA (siRNA) microarrays for high-throughput functional testing.

This chapter introduces the most commonly used microarray types and platforms, with emphasis on DNA-microarrays for gene expression profiling. Furthermore, the most novel findings using microarrays are described together with a summary of the future potentials in the use of microarrays in basic research and in a clinical setting.

15.2 DNA MICROARRAYS AND GENE EXPRESSION PROFILING

Microarrays for measuring gene expression levels are the most commonly used microarrays, and several important findings using this technology recently have been reported. The following sections describe the technology and analysis methods used for generating useful data from microarray experiments. Furthermore, the most important findings in the field of cancer research are described together with descriptions of the potential of using expression profiling for unravelling the affected molecular pathways, for example, during disease progression.

15.2.1 DNA Microarray Technology

Several technologies and protocols exist for monitoring gene expression using microarray technology. In principle, labeled transcripts isolated from biological samples are hybridized to the DNA-microarray probes for determination of the transcript abundance or relative expressions. In conventionally used protocols, total RNA from the biological sample is extracted and reverse transcribed into cDNA. Subsequently, an *in vitro* transcription of the cDNA is carried out with incorporation of modified nucleotides for later coupling with fluorescent molecules. In other protocols, modified nucleotides are incorporated directly into the cDNA product. This procedure, however, is limited in the amount of generated target due to the lack of an amplification step, which may be a problem when working with small amounts of starting material. The labeled target is hybridized to the DNA-microarray slide for several hours to allow hybridization of the sample target to the microarray probes. Following the hybridization procedure and intensive washing of the slide to remove excess target molecules, scanning of the DNA microarray identifies the DNA probe hybridization levels, which reflects the gene expression levels in the samples investigated.

The DNA probes used for expression microarrays are either oligonucleotides or long PCR products amplified from cDNA clones. The advantage of using oligonucleotide probes is that they can be designed to ensure minimal cross-hybridization to other transcripts and furthermore, that each gene will be covered with several probes. In commercially available systems (Affymetrix GeneChips) each gene is covered by up to 20–25mer oligonucleotides and 20–25mer oligonucleotides with a mismatched base in the middle position for measuring the amount of nonspecific hybridization. The use of more than one probe for representing each gene reduces the problem with the nonfunctional probes, which can be a major obstacle when using only a single oligonucleotide probe per gene.

When using PCR products from cDNA clones as probes it is more difficult to minimize cross-hybridization to other transcripts because of the probe length, which may be as long as 500–800 bp. Oligonucleotide and PCR probes are spotted directly on a glass surface using precision robotics and very accurate spotting pins, which are able to apply very small amounts of probe solution to a large number of glass slides. However, it is also possible to synthesize oligonucleotide probes *in situ* on the glass surface (Affymetrix GeneChip technology).

Spotting of the probes introduces probe morphology variation between different microarray slides, and consequently these arrays typically are hybridized with both the biological sample under investigation and a common reference sample. Each sample is labeled with different fluorescent molecules (Cy3 and Cy5 are often used) and the microarray slide is scanned at two different wavelengths to obtain measurements of the transcript hybridization abundance of each sample. The use of a common reference sample makes comparisons between different slides with different probe morphology possible. The two-sample hybridization technique provides relative expression measures (ratios), which reflect on the relative expression of the genes in the two samples. However, when using Affymetrix GeneChip microarrays or noncontact printed microarray slides the probes have highly uniform shapes and sizes and it is possible to hybridize a single sample to each microarray slide and in this way obtain direct measures of the gene expression levels. Figure 15.1 illustrates the difference between the one and two sample microarray systems.

15.2.2 Normalization Strategies

The raw data obtained from a microarray experiment consists of an image of light intensities where some of the pixels are the spots (foreground) to which the labeled transcripts are hybridized, and where the remaining pixels between the spots constitute the background. The abundance of a particular transcript thus is not measured directly and is the result of a translation of the image of intensities into a number. Many sources of variation come into play when trying to evaluate the certainty of an expression level.

At the production stage of a DNA microarray, variations in the spots (morphology) cannot be avoided and the exact position of a spot is not known. The Affymetrix GeneChip microarray differs on this point by having spots of highly uniform size and shape. Even when two spots are labeled with the same DNA probe on the same slide there will be variations. At least the same amount of variation will be seen for slides produced in the same batch and more variation comes into play when using slides from different batches. Also the spots on a slide are produced by a set of pins and these may have different characteristics. The first step in the analysis is to find the spots on the slide and for each spot to subtract a background value. This is necessary since a light intensity of zero can never be found in the situation where the true expression level is zero. There are several competing packages for performing this part of the analysis (SPOT, QuantArray, etc.). The background subtraction is a very delicate process and a wrong correction can lead to nonlinear effects in the later analysis.

Preparation of a sample for hybridization involves several steps and therefore also introduces a source of vari-

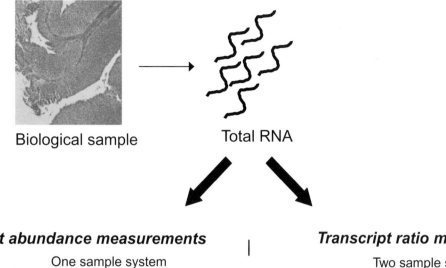

Biological sample Total RNA

Transcript abundance measurements

One sample system

Test sample

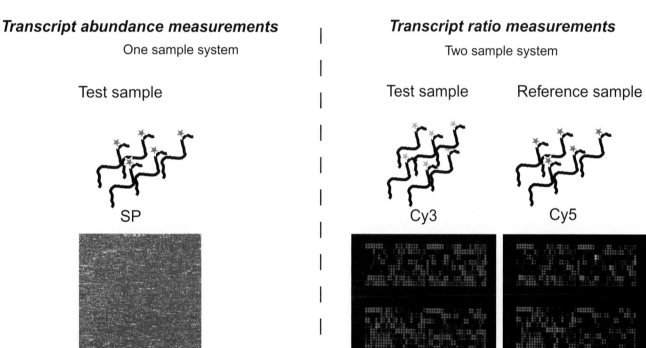

SP

Transcript ratio measurements

Two sample system

Test sample Reference sample

Cy3 Cy5

FIGURE 15.1 One-sample and two-sample microarray systems. In both systems total RNA is extracted from the biological sample under investigation. In one-sample systems like the Affymetrix GeneChip systems (left side of the figure) the total RNA is reverse transcribed into cDNA, which is transcribed into cRNA with incorporation of biotinylated nucleotides. These biotinylated nucleotides are hybridized to the array and subsequently coupled to streptavidin phycoerythrin (SP) conjugates. In this way, direct measurement of transcript abundance is performed. In traditional two sample systems, the test and reference RNA is reverse transcribed into cDNA. The cDNA is transcribed into cRNA with the incorporation of aminoallyl-linked nucleotides. The test cRNA is coupled to Cy3 and reference cRNA is coupled to Cy5. Fluorescence measurements at the two different wavelengths gives the transcript ratios between the test and the references samples. (See color plate)

ation. It is believed, however, that these effects are mainly proportional and can be corrected by scaling all the measurements by the same number. The essence is that normalization between the Cy3 and Cy5 measurements is needed before relative abundances can be calculated. Based on the experience gathered so far the normalization is performed for each pin separately. Contrary to the belief mentioned previously, some authors find that a scaling is not sufficient for making the two channels comparable and they perform instead a nonlinear scaling based on fitting a general curve

to a plot of log differences against log averages (Yang *et al.*, 2002). Normalization of the sample under investigation against the reference sample presumes that the two samples behave similarly for a large fraction of the genes being spotted to the array. Thus, if most of the genes spotted to the array are believed to be involved in the process under study, a normalization cannot be based on the behavior of a large fraction of the genes. Instead the study needs to include control genes that can be assumed to be similarly expressed in the two samples.

Having made a (successful) normalization between the biological sample under investigation and the common reference sample, relative abundance (as a measure of the expression level) can be compared directly between several arrays. For the Affymetrix GeneChip microarray, where only one sample is hybridized to the array, it is necessary to make a normalization between the arrays. The principles for doing this are similar to those discussed earlier for the normalizations in the two-sample system.

On top of the experimentally related error terms described earlier, one has the biological variation for a homogeneous population. This biological variation is often large and influences the sample sizes required for a reliable observation of a differential expression between two groups.

15.2.3 Data Analysis Methods

Gene expression data analysis can be divided into two approaches: unsupervised and supervised methods (Golub *et al.*, 1999). Unsupervised analysis aims at identifying previously unknown relationships between samples. For this purpose hierarchical cluster analysis has been used extensively to group tumors according to similarity in expression profiles. This technique can also be applied to group genes with similar expression profiles across the tumor samples analyzed and in this way identify possible covarying and/or functionally related genes. Hierarchical cluster analysis is a powerful tool for visualization of the expression patterns (Eisen *et al.*, 1998). Other unsupervised methods used are self-organizing maps (SOM), principal-component analysis (PCA), and relevance networks. The unsupervised methods are suited for finding novel relationships between samples (or genes) based on similarities in the gene expression patterns. However, the possibility to identify nonrelevant groups exists because of the complexity in large-scale gene expression data.

Supervised analysis techniques are used for identifying differentially expressed genes between groups of samples; for example, where the histopathological stage or clinical outcome (survival, metastasis, recurrence, progression, etc.) is known. Optimal expression signatures for disease classification or outcome prediction are in this way generated from a set of training samples and the significance of the identified expression signature is tested using independent test samples. Standard statistical tests, such as the student's t-test, are used to identify the genes that show the largest differential expression between the groups. Mathematical methods used for classifying samples based on the optimal gene expression signatures include maximum-likelihood estimates, k-nearest neighbors, support vector machines, and weighted voting schemes, among others (Dudoit *et al.*, 2000).

The use of supervised methods in the analysis of gene expression data, however, involves the risk of oversimplify-ing the clinical groups under investigation. A single clinical class (e.g., poor outcome) may contain several molecular subclasses of tumors with distinct expression patterns, which may obscure and make the selection of differentially expressed genes between the clinical groups analyzed virtually impossible. When using supervised methods for selection of differentially expressed genes the significance of the selected genes can be accessed by permutation analysis. This usually is done by multiple permutations of the sample labels, followed by generation of statistics on the ability to select good differentially expressed genes in random classes, compared to the real data. Such tests determine the likelihood of obtaining the observed expression patterns by chance, which is a real problem when performing thousands of statistical tests.

15.2.4 Disease Diagnosis and Prognosis by Microarray Expression Profiling Analysis

A huge number of published studies involve microarray analysis for gene expression profiling, and several of these studies have demonstrated a potential clinical use of microarrays for diagnosis and for predicting the clinical outcome of patients with various types of cancers. This section reviews some of the most promising clinical studies published to date.

15.2.4.1 BREAST CANCER

Microarray studies of tumours from patients with breast cancer have identified several very interesting disease characteristics as well as intriguing disease outcome predictions. In one of the first studies by Perou and colleagues (2000), the gene expression patterns in 65 breast tumors were studied using microarrays with probes for 8,102 human genes. By the use of unsupervised hierarchical cluster analysis the authors identified molecular distinct subclasses of breast tumors. The subclasses were characterized by *Erb-B2* over-expression and by similarities to basal epithelial cells, luminal epithelial cells, and adipose enriched/normal breast cells. Additional studies of the identified subclasses of breast cancer revealed a subclass of the luminal epithelial cell-like tumors, which demonstrated a significant difference in disease outcome compared to the other groups.

One of the most interesting and clinically promising microarray expression profiling studies identified a 70-gene expression signature for predicting if patients diagnosed with breast cancer would develop disease metastases (van 't Veer *et al.*, 2002). The expression patterns were found using a cohort of 78 young patients that had not received any treatment, and supervised learning methods were used for identification of the optimal genes for predicting disease outcome. The microarrays used in this study contained oligonucleotides probes for approximately 25,000 human genes. This clinically promising prognosis classifier was later validated on an additional tumor set consisting of 234 breast tumors using the same microarray platform (van de Vijver *et al.*, 2002). The validation confirmed a strong correlation between the gene expression signature for a poor prognosis and disease outcome. These microarray-based studies of breast cancers hold much promise for tailoring personalized treatment regimens to patients suffering from breast cancer, as only the patients predicted to develop metastatic disease should be treated with adjuvant chemotherapy.

15.2.4.2 Bladder Cancer

For the study of bladder cancer development and progression, microarray gene expression profiling has also been applied with success. In one of the first studies using clinical material, Affymetrix Genechip with probes for approximately 5,000 human genes and ESTs were used to identify gene expression pattern changes between superficial and invasive tumours (Thykjaer *et al.*, 2001). The identified genes encoded oncogenes, growth factors, proteinases, and transcription factors together with proteins involved in cell cycle, cell adhesion, and immunology. This was the first study to identify genes that separated superficial from invasive bladder tumors.

A more recent microarray-based study on bladder tumors showed advances in disease classification and outcome prediction (Dyrskjot *et al.*, 2003). The authors of this study also used the GeneChips with probes for approximately 5,000 genes and ESTs for identification of a 32-gene expression pattern using 40 tumor samples for classifying tumors according to disease stage. This stage classifier was successfully validated on an independent test set consisting of 68 bladder tumors analyzed on a different array platform. The stage classifier did not only reproduce histopathological staging but added important information regarding subsequent disease progression. This molecular stage classifier holds much promise for the identification of early-stage bladder cancer patients with a poor prognosis. In this study a prognosis classifier for identifying patients with a short interval to superficial disease recurrence was also reported. Tumors from a cohort of 31 patients were used for this purpose to generate the 26-gene disease recurrence expression signature. The clinical significance of the reported prognosis classifier was estimated using cross-validation tests of the 31 samples used (75% correctly classified). However, validation of the disease recurrence signature using independent samples does not exist at present.

15.2.4.3 Lung Cancer

Improvements in classification and prognosis have also been reported in studies of lung cancer by microarray expression profiling of 12,600 transcripts in 186 samples (Bhattacharjee *et al.*, 2001). Unsupervised hierarchical cluster analysis grouped the samples according to established histological classes. In addition, the adenocarcinomas were found to contain four subclasses, and one of these had a less favorable outcome compared to the other subclasses of the adenocarcinomas.

Another interesting microarray-based finding was reported by Beer and colleagues (2002), who studied 86 primary early stage adenocarcinomas from patients with lung cancer. A 50-gene risk index was determined for identifying patients with a poor prognosis. The 50-gene risk index was validated using the data reported previously by Bhattacharjee and colleagues (2001). The authors classified the adenocarcinomas as low and high risk and found a close correlation to disease outcome, and concluded that high-risk patients may benefit from adjuvant treatment.

15.2.4.4 Multicancer Studies

Microarray expression profiling has also been used for studying expression profile changes between different cancers. Ramaswamy and colleagues (2001) used microarrays with probes for 16,063 genes to analyse transcriptional changes in 218 tumor samples and 90 normal samples for generating a multiclass classifier. The classifier showed high classification accuracy, especially for low-grade tumors.

Recently, a signature for predicting metastasis in solid tumors in general has been reported (Ramaswamy *et al.*, 2003). The authors compared the expression profiles from 12 metastatic adenocarcinomas from a wide range of organs with 64 primary adenocarcinomas and identified a 17-gene expression signature through analysis of 279 solid tumors of different origins.

15.2.4.5 Diffuse Large B-Cell Lymphoma

The use of microarray expression profiling in the study of diffuse large B-cell lymphoma (DLBCL) has identified several very interesting disease characteristics, and the power of microarray expression profiling for delineating this clinically very heterogeneous disease with widely different outcomes has been obvious. In one of the first large-scale microarray studies a specialized microarray with 17,856 probes for genes preferentially expressed in lymphoid cells, immunological processes, and cancer was constructed (Alizadeh *et al.*, 2000). A large number of B-cell malignancies was studied using the specialized cDNA microarray platform and two clinically distinct forms of DLBCL were identified using hierarchical cluster analysis—germinal center B-like DLBCL and activated B-like DLBCL. The authors found that the germinal center B-like DLBCL group showed a more favorable outcome. The findings later were validated by expression profiling of DLBCL from 240 patients and a 17-gene expression signature was generated for predicting survival after chemotherapy treatment (Rosenwald *et al.*, 2002).

Instead of the unsupervised approaches just described, Shipp and colleagues (2002) used supervised methods for predicting outcome for patients with DLBCL. A 13-gene expression signature was generated for predicting disease outcome with a very close correlation to clinical outcome and the model was validated by the data from Alizadeh and colleagues (2000). Only three of the 13 optimal genes were in common between the two microarray platforms used and two of these showed a significant correlation to outcome.

15.2.5 Detection of Prevailing Molecular Pathways in Tumors Using Expression Microarrays

Analysis of differential expression in tumor tissues may provide new information about biological pathways involved in a process. The establishment of a high through-put gene expression platform has moved the bottleneck from data production to data analysis. Gene expression profiling studies often contain hundreds of samples resulting in several million data points that must be coped with. Such data are increasingly becoming available for downloading from the Internet, leading to even larger masses of data. Many scientific research groups and bio-informatic compa-nies are working continuously on improving software for a wide variety of data analysis. Many research groups are making these programs freely available on the Internet.

Several cancer studies have compared tumors with normal tissue or tumors at different stages in order to iden-tify new tumor or stage markers. Often a thousand genes or more for which the functions are unknown are reported as potential markers. Instead, the differentially expressed genes are compared to data from the literature for that particular disease or other similar diseases as an initial confirmation of the validity of data. Because of the molecular heterogeneity of most cancers, dimension-reduction methods like single value decomposition or various clustering methods often are applied as the next analysis step. Such analysis helps visualize patterns in the data that may help to define new molecular subclasses in a disease.

In order to generate hypotheses concerning pathways in oncogenesis and disease progression, differentially expressed genes must also be put into a biological context. This may be done by extracting specific information related to gene interplay in signalling networks, called pathways. Once a hypothesis has been generated, specific genes may be chosen for further manipulations in a suited model system in order to test the hypothesis. A common choice for this is cell lines that are as similar as possible to the tumor cells under investigation. In such systems gene expression can be manipulated by various methods such as transfections for upregulation, or antisense or RNAi techniques for down regulation.

A helpful tool for pathway analysis is GenMAPP (Gene MicroArray Pathway Profiler), which is a free computer application (www.genmapp.org). This program allows the direct coupling of gene expression data to a database con-taining information about gene function according to anno-tation from the Gene Ontology Consortium. Once the data has been coupled to the database, the user may extract infor-mation about which functional groups of genes has been affected in the experiment. The application also allows the visualization of expression of genes in pathways drawn as a network with indication of the mode of interaction between the proteins. A large number of premade pathways, called MAPPs, are provided with the program but the users may also design new pathways for automatic coupling to the expression data. The user can set GenMAPP to color all genes up-regulated in an experiment, green, and all genes down-regulated in an experiment, red. Since genes are colored and viewed dynamically, the user can easily switch criteria. For example, the user may wish to see which genes change with different fold cut-offs or combine fold change with other criteria, for example like *Increase* or *Decrease* calls as used in the Affymetrix GeneChip software. Criteria can also be set to view the changes in gene expression on a MAPP at time points during an experiment or at different disease stages. This kind of visualization of gene expression data on MAPPs representing biological pathways and groupings of genes is very helpful to get an overview of the affected pathways in an intuitive way. Figure 15.2 shows an example of a GenMAPP pathway.

15.3 APPLICATION OF CGH ARRAYS FOR STUDIES OF THE MALIGNANT CELL GENOME

Comparative genomic hybridization (CGH, see also Chap-ters 12 and 13) is a technique that detects and maps changes in the copy number of DNA sequences (Kallioniemi *et al.*, 1992). In CGH, DNA from a test (e.g., from a tumor) and reference genome (genomic DNA from a healthy/normal individual) are differentially labeled and hybridized to a rep-resentation of the genome. Originally the representation was a metaphase chromosome spread, but over the past several years, microarray-based representations increasingly have been used. The fluorescence ratios of the test and reference hybridization signals are determined at different positions (the array elements) along the genome and provide infor-mation on the relative copy number of sequences in the test genome compared to the reference (a normal diploid) genome. The array representations can provide a number of advantages over the use of chromosomes, including higher resolution and dynamic range, direct mapping of the copy number changes to the genome sequence, and high-throughput.

A variety of chromosomal aberrations underlie inherited diseases and cancer (see also Table 12.1). Aberrations leading to changes in DNA copy number can be detected by CGH and include interstitial deletions and duplications, nonreciprocal translocations, and gene amplifications. It is important to note that CGH does not provide information on ploidy or location of the rearranged sequences responsible for the copy number change. Furthermore the capability of

TGF Beta Signaling Pathway

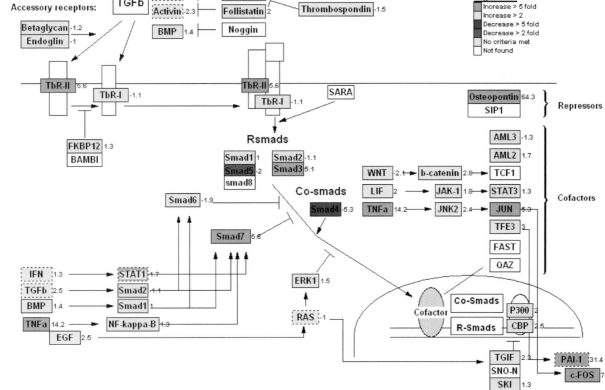

FIGURE 15.2   Visualization of the TGF signalling pathway (provided with GenMAPP). The number next to the gene boxes indicate the fold change of expression, and the color of the boxes indicates which of the criteria as indicated in the Legend are met. If a gene box has two colors (e.g., STAT), this gene is represented by two different probes or probe sets that meet different criteria.

a CGH array to detect aberrations spanning small genomic regions depends on both the size and genomic-spacing of the elements (clones) on the array.

Different applications of CGH arrays impose different performance requirements, so that certain approaches may be suitable for a particular application, whereas others are not. Accordingly, unique CGH arrays have been designed for various applications; for example, for screening of telomeric regions, screening of selected genomic regions known to be frequently involved in cancer, and genome-wide screening. Three different types of array elements have been employed in the various approaches: bacterial artificial chromosomes (BACs), cDNA clones, and oligonucleotides. An analysis typically requires several hundred nanograms of genomic DNA from the specimen when using some BAC arrays (Snijders *et al.*, 2001), or one or more micrograms for cDNA arrays (Pollack, *et al.*, 1999; Monni *et al.*, 2001). Whole genome amplification procedures are likely to reduce substantially the amount of required specimen.

The use of large insert clones like BACs for CGH arrays provides sufficiently intense signals such that single copy

changes affecting individual clones on the array can be detected and aberration boundaries can be located within a fraction of a BAC length. However, propagating and printing BACs can be problematic. BACs are single copy vectors, hence the yield of BAC DNA is low and solutions of the high molecular weight DNA can be viscous, making it difficult to print. Since growing and processing large bacterial cultures is not practical for arrays of thousands of elements, a number of methods have been devised to generate representations of BACs from minute cultures, which can be set up in an industrial fashion.

DNA copy number measurements can also be made using arrays containing elements made from cDNA or oligonucleotides. Typically these arrays initially were produced to measure gene expression. The advantages of these arrays are that they often are readily available and contain large numbers of elements because they were produced to comprehensively assess the transcriptome. The disadvantages are that only large copy number gains provide sufficient signal to determine the boundaries of the amplicons with high resolution.

The detection of low level copy number changes (e.g., losses) requires calculating the running average of multiple clones (5–10 clones) along the genome, and frequently entails discarding measurements on a substantial fraction of elements because they do not provide adequate signals. Thus, the actual genomic resolution of the boundaries of single copy changes and the ability to detect focal single copy changes is considerably less than implied by the average genomic spacing between the clones on the array.

A number of studies have taken advantage of the higher resolution afforded by CGH arrays to more precisely map the boundaries of genomic copy number alterations. Once these are defined, candidate tumor suppressors and oncogenes mapping within the region can readily be identified from the genome sequence databases. Investigation of expression levels of these candidate genes in tumors and model systems (cell lines, etc.) can then be used to determine which of the candidates are most likely to contribute to the disease phenotype and to be the "driver gene(s)" for the copy number alteration. This approach to candidate gene identification is likely to be fruitful, since a correspondence between copy number alterations and changes in gene expression was demonstrated in breast cancer as measured by genome-wide CGH arrays and global expression profiling on the same cDNA array (Hyman *et al.*, 2002; Pollack *et al.*, 2002).

Tumor classification based on copy number profiles obtained using CGH arrays has been reported for several cancer types, including liposarcomas (Fritz *et al.*, 2002), renal cell carcinomas (Wilhelm *et al.*, 2002) and gastric adenocarcinomas (Weiss *et al.*, 2003). However, in a recent study of bladder tumors no significant relationship between copy number alterations and tumor stage and grade was found (Veltman *et al.*, 2003). Instead, an intriguing link among particular genomic loci was revealed; that is, copy number alteration at particular loci occurred together (e.g., gain of CCND1 and deletion of TP53), whereas alterations in the copy number of loci, harboring genes that function in the same pathway such as gains of CCND1 and E2F3 were found to be "complementary" as they did not occur in the same tumors. Together, the studies cited indicate that CGH arrays will be increasingly important in understanding tumor biology over the coming years.

15.4 APPLICATION OF SNP ARRAYS FOR STUDIES OF THE MALIGNANT CELL GENOME

Cancer is a genetic disease that develops from the accumulation of multiple genetic alterations in oncogenes and tumor suppressor genes during tumourigenesis. Identification of the genetic alterations that are responsible for the development of each tumor type is important not only for the basic understanding of the pathogenesis, but also for the development of new principles for pathological classification, disease prevention, and treatment.

Two basic approaches are used to identify the genetic alterations in disease development. The first approach is to identify genetic changes within the tumor. Alterations such as homozygous or heterozygous deletions are thought to be among the most common genetic abnormalities in carcinomas and using loss of heterozygosity (LOH) analysis has become a preferred approach to identify genomic regions involved in tumourigenesis. LOH analysis determines if both alleles are present in a cancer cell by the analysis of polymorphic genetic markers (Primdahl *et al.*, 2002).

A second approach is to define the association between specific polymorphic loci in the genome with the risk of disease development. This is performed either using leukocyte DNA from a group of patients with cancer compared to DNA from a control group, or as linkage studies in families with hereditary cancer. An example of a cancer predisposition caused by a specific polymorphic locus is the [T/A] polymorphism at nucleotide 3,920 in the Adenomatous Polyposis Coli (APC) gene found in 6% of individuals of Ashkenazic Jewish origin. This polymorphism gives a roughly two-fold increase in the lifetime risk of colorectal cancer (Laken *et al.*, 1997). The A allele results in an extended mononucleotide tract in the coding region of the APC gene, which is a much more frequent target for somatic mutations than the normal APC sequence.

Although micro-satellite markers have gained success for LOH analysis and association studies, the instability of these markers in tumor tissue, and the fact that it is difficult to upscale and automate micro-satellite analysis, have made it difficult to use in genome wide LOH-analysis. Attention therefore has been turned to the use of single nucleotide polymorphisms (SNPs) and high-density SNP arrays for LOH analysis, association, and linkage studies.

SNPs are single-based variations with a unique physical location within the genome and with an estimated frequency of one SNP for every 300 base pairs throughout the genome. Considerable effort has been committed to the registration of SNPs by the SNP Consortium (Holden, 2002). The latest version (November 4, 2003) of the National Center of Biotechnology information (NCBI) dbSNP database (available online at http://www.ncbi.nlm.nih.gov/SNP/) holds approximately 5.8 million entries, of which 2.36 million are validated, and 288,265 includes frequency data.

Although SNPs are very abundant, they often have a low average heterozygosity rate compared to micro-satellites, and approximately three times the number of SNPs is required for a resolution equivalent to a micro-satellite.

A variety of techniques have been used to identify SNPs, including conformation based mutation screening, gel electrophoresis based genotyping methods, and fluorescent dye based genotyping technologies (Shi, 2001). The demand for high throughput methods for the parallel analysis of thousands of SNPs in a single experiment has produced a variety of DNA chip systems (Heller, 2002). Microarrays commonly are synthesized by attaching thousands of oligonucleotides to a solid surface. Each oligonucleotide in the array acts as an allele specific probe. The DNA sequences including the SNPs of interest are amplified by PCR and labeled, and then hybridized to the array. A variety of spotting technologies are being developed, such as pin-based fluid, piezo-inkjet fluid-transfer system, and other methods including the use of photolithography for light-directed synthesis of large numbers oligonucleotides on solid surfaces as developed by Affymetrix.

15.4.1 Affymetrix GeneChip® Human Mapping 10K Array

Since the first SNP analysis on high-density oligonucleotide arrays was developed by Affymetrix, the number of SNPs represented on each array have increased from around 600 (Wang, *et al.*, 1998; Mei, *et al.*, 2000) to more than 10,000 SNPs. The latest Affymetrix SNP array with 10,000 SNPs (GeneChip® Human Mapping 10K Array) has a mean inter-marker distance of about 210 kb. The mean genetic gap distance is 0.32 cM and the average heterozygosity of the SNP positions is 0.37. For interrogating each SNP, this 10K array contains at least 40 different 25 bp oligonucleotides. These oligonucleotides differ according to perfect match and mismatch in the SNP position (all four nucleotides are represented at this position), and likewise there are both mismatch and perfect match represented in the flanking sequences around the SNP. This probe redundancy improves the confidence of the genotype calls.

Whereas the early versions of the Affymetrix SNP arrays used multiplex PCR with specific primers for each of the PCR products to amplify the genomic DNA, the 10K SNP array reduces the complexity by ligating a PCR specific adapter (see Fig. 15.3) and consequently only one primer pair is used. The specific adapter sequences are ligated to the 4 bp overhangs created by a prior XbaI restriction enzyme cleavage. The assay is optimized to amplify sequences between 250–1000 bp, so SNPs situated in XbaI fragments below or above this size will not be well represented on the array.

In principle, the Affymetrix SNP arrays can be used for conducting linkage analysis studies, but so far has been used mostly to study LOH. With the new GeneChip® Human Mapping 10K Array, it will be feasible to detect even small deletions and the decreased linkage interval will surely lead to increased use in linkage analysis.

15.4.2 Custom-Made SNP Arrays

Whereas the Affymetrix SNP array system is designed to cover the whole genome, the custom-made SNP arrays are designed to determine the frequency of specific SNPs, or sets of SNPs, of interest in specific genes. One of the most used examples of SNP determination on custom-array is the method described by Fan and colleagues (2000) and Lindroos and colleagues (2002). There are slight differences in the exact method, but the method described here is an adaptation used in the author's laboratory (Christensen and Koed, unpublished results).

Genomic DNA covering the SNPs of interest is amplified in a multiplex PCR amplification (see Fig. 15.4a). Surplus

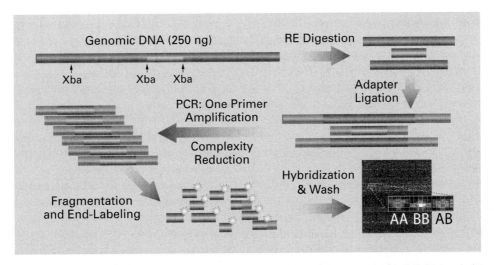

FIGURE 15.3 GeneChip® Human Mapping 10K Array assay overview. Genomic DNA is digested with the restriction enzyme XbaI and ligated to adapters that recognize the cohesive four bp overhangs. A primer that hybridizes to the adapter sequence subsequently is used to amplify the DNA fragments in a PCR reaction. The PCR products are then fragmented, labeled, and hybridized to the GeneChip® Human Mapping 10K Array. Adapted from Affymetrix's datasheet on the GeneChip® Human Mapping 10K Array. (See color plate)

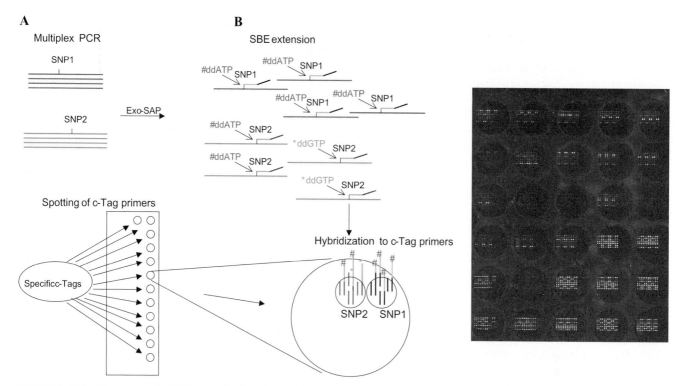

FIGURE 15.4 Custom-made SNP array. A. Overview of the SNP array. First genomic DNA is copied by multiplex PCR (the authors' laboratory routinely runs a multiplex reaction with as many as 10 different primer pairs; Christensen and Koed, unpublished results), primers and nucleotides are degraded by Exonuclease 1 and Shrimp Alkaline Phosphatase (Exo-SAP), Single-base extension (SBE) is carried out in the presence of differentially fluorescence labeled ddNTPs and specific single-base extension primers with specific Tags in the 5' end. The Tags hybridize to complementary Tags (c-Tags) bound to glass slides in a definite pattern. B. Example of a spotted slide hybridized to labeled SBE primers (Christensen and Koed, unpublished results). The same c-Tags were spotted in each well, but each well represents the results from different multiplex PCR reactions. (See color plate)

primers and nucleotides from the PCR amplification are degraded in order to avoid interference in the following single base-pair extension (SBE) reaction. A specific SBE primer that terminates one base upstream to the SNP is hybridized to the single-stranded PCR product and extended one base in the SBE reaction using differentially fluorescence labeled dideoxy nucleoside triphosphates. The SBE primer has a Tag sequence in the 5' end. Complementary Tag (c-Tag) sequences are spotted on a glass slide with the help of a 3'aminogroup. Subsequently, in a silicone-mold, the ddNTP-labeled SBE-Tag primers are hybridized to the c-Tag spotted on the slide (see example in Fig. 15.4b). In each well the material from individual patients can be applied and each individual's composition of SNPs can be determined. Alternatively, pooled DNA from many individuals can be used in the PCR amplification and thereby the frequency of different SNPs can be determined.

Each glass slide contains 80 wells and in each well 100–200 c-Tags can be spotted. If a single bp extension reaction is performed in both the sense and the antisense direction, approximately 50–100 SNPs can be analyzed in each well.

15.5 SEQUENCING MICROARRAYS: p53 STUDIES IN BLADDER CANCER

Testing tumors for mutations in oncogenes and tumor suppressor genes is a valuable tool for predicting the disease course in many cancers, but is limited primarily by time and cost of the analysis. Microarray technology is therefore a very interesting option as it is a very rapid method compared to traditional sequencing. Affymetrix launched the p53 GeneChip in 1999, and in cancer types such as bladder cancer and colon cancer where p53 mutations are frequent, it could be a cost effective screening method, alone or in combination with other methods.

Compared to expression monitoring, where the oligonucleotide probes can be chosen to have suitable hybridization properties, sequencing by hybridization has the inherent drawback that the oligonucleotide probes must cover all positions of the sequence that is tested. Another problem is that the array must be designed for the type of mutations that are tested. The p53 GeneChip was designed to test for all possible base substitutions and single base deletions (see Fig. 15.5).

This means that deletions involving more than one base, and all types of insertions, could not be detected. One can also imagine the difficulties involved in testing two mutations close to each other, or a mutation close to a frequent polymorphism, where one oligonucleotide probe would have several mismatches.

15.5.1 The p53 GeneChip

To test the p53 GeneChip, 140 previously sequenced samples from bladder tumors, and two cell lines with two different homozygotic mutations have been used (Wikman *et al.*, 2000). A dilution experiment where DNA from the two cell lines were mixed in different proportions showed the chip to be able to detect mutated DNA when this constituted less than 2% of total DNA, probably the most sensitive sequencing method known.

The 140 bladder tumor samples were tested on the p53 GeneChip blindly; that is, without knowing the results from the manual sequencing. The results from the report generated by the software were in the form of a score that indicated the size of the deviation from the wild type sequence together with the sequence change. The score was an arbitrary number in the range of 1 to 37. Many samples were reported to have many mutations simultaneously, some with a very low score, and these mutations often occurred at certain bases in the sequence. This indicated that some GeneChip positions were more prone to noise problems than others, which prompted to look for a method to lower the degree of false positives, while maintaining the high sensitivity. A first approach was to introduce a cut-off value to the scores, but it became clear that this not only would reduce the number of false positives, as expected, but also greatly increase the number of false negatives, as some of the mutations were represented by a fairly low score, although in positions without noise (Ahrendt *et al.*, 1999). To overcome this problem, the scores from all experiments were combined into a spreadsheet for statistical analysis. Using this approach, each GeneChip position corresponding to an analysed base in the sequence was treated as a separate entity, with its own noise and threshold characteristics.

The average (A) and standard deviation (S) for the background noise were calculated for each GeneChip position where a score was found. Mutations confirmed by manual sequencing were not included in the calculation of the background statistics. Using the values A + 2S as a new cut-off value drastically improved the specificity of the mutation calling, giving a much better concordance with the results obtained by traditional sequencing (see Table 15.1).

15.5.2 Data Analysis with Neural Networks

In another approach, a neural network was used to predict the sequence from the raw data from the p53 GeneChip (Spicker *et al.*, 2002). The prediction was not more accurate than the GeneChip method, but interestingly, the neural network found inconsistencies in some samples, and by further analysis, using RFLP and sequencing, it was found that some of the inconsistencies were due to a well-known polymorphism (nucleotide number 143 G/C) that the Affymetrix software for some reason did not report.

15.5.3 Benefits and Limitations of Microarray Sequencing

It is a well-known problem, that when sequencing tissue samples, the tumor DNA is mixed with normal DNA from the surrounding tissue, to a lower or higher extent. Furthermore, subclones of the tumor cells may exist. Traditional sequencing methods usually have a threshold of about 30%, as lower contents are obscured by the background noise. Therefore traditional sequencing will underestimate the number of missense mutations, whereas the microarray method is much more sensitive.

FIGURE 15.5 An enlarged part of the p53 GeneChip after hybridization. This is an example of the standard tiling where the sequence can be read visually. Each column corresponds to a base in the sequence, and the five rows correspond to each of the four bases and to a single deletion. The column marked with an asterisk is from a column with alignment controls. (See color plate)

TABLE 15.1   Performance of p53 GeneChip sequencing compared to traditional sequencing. Insertions and deletions are not included in the table.

Method	Ch+/S+[a]	Ch+/S−	Ch−/S+	Ch−/S−	Sensitivity	Specificity
GeneChip Report	65	126	6	31	0.92	34
Fixed cut-off = 10	55	77	16	45	0.77	41
Calculated cut-off	60	10	11	65	0.84	86

[a] Ch: chip sequencing; S: traditional sequencing.

The microarrays did not detect any of the five frameshift mutations present in the 140 bladder tumor samples (Wikman *et al.*, 2000). Four of them were insertions or large deletions, a type of mutation that the microarray is not designed to detect, but one single base deletion should have been detected. The reason for this is not clear, but these positions may generally be noise prone. As many mutations in cancer are frameshift mutations, it is not satisfactory that the microarray does not detect this kind of mutation. However, to benefit from the sensitivity of the microarray, a setup where array sequencing is used in combination with a technique such as DHPLC that effectively detects insertions and deletions would be a cost effective solution for p53 mutation screening in tumors.

The microarray sequencing method is clearly most effective in testing base substitutions. This has been used successfully in SNP detection, both in Affymetrix HuSNP chip, with 1,496 SNPs on one microarray, and with the second generation SNP array, the 10k Mapping Array with more than 10 thousand SNPs typed on one chip (Kennedy *et al.*, 2003).

15.6 RNAi ARRAYS—HIGH THROUGHPUT FUNCTIONAL TESTING

Inhibition of gene expression using the RNA interference (RNAi) pathway mediated by small interfering RNAs (siRNA) has rapidly become the method of choice for studying gene function in mammalian cells. This technology allows the use of RNAi to study gene function in mammalian model systems in which classical methods are often limited and costly. One of the advantages of RNAi is that it is highly specific and that it can knock down a gene by 70–100%. RNAi does not rely on interference with the translation-like antisense oligonucleotides, but rather leads to degradation of the mRNA through the cells' own system. Although the treatment of cells can lead to unwanted induction of the interferon inflammatory pathway, much of this can be avoided by maintaining RNAi concentrations at a reasonable level.

One limitation of the technique has been the uncertainty in predicting the efficacy of siRNAs in silencing a gene. To overcome this problem, siRNA sequences may be monitored by their ability to reduce the expression of cognate target-reporter fusions with easily quantified readouts for the rapid and efficient identification of the most effective siRNA against any gene (Kumar *et al.*, 2003). A combination of this approach with microarray-based cell transfections has an unlimited potential in high-throughput screens for identifying effective siRNA probes for silencing genes in mammalian systems. This method facilitates the development of large-scale siRNA libraries for large-scale functional genomic studies.

The microarray technology often identifies a large number of genes of interest for further functional studies. Recently, the development of a new exciting tool has carried the RNAi technology into the microarray world, allowing highly parallel analysis used on a genome-wide scale (Mousses *et al.*, 2003). In this technology, siRNA probes are dissolved in a transfection matrix and spotted on poly-L-lysine glass slides very much similar to cDNA microarrays. With a spot diameter of 100–500 μm and a center-center distance of 300–1,000 μm, there is space for 2,000 to 15,000 spots on a standard glass slide. Subsequently the slide is overlaid with a monolayer of adherent cells in a cell tray and incubated to allow reverse transfection. For assessment of the effects of gene knock-down a highly magnified digital image is recorded with a charge coupled device (CCD) camera. The gene silencing may be combined with other experiments, such as drug treatment, starvation, and heat shock, and scored with the only limitation that the effect must induce a visible change in phenotype.

15.7 FUTURE ASPECTS OF THE USE OF MICROARRAYS

The development of nano-technologies may bring the use of microarrays to a new level. As is stands today the whole human genome can be included in one microarray, looking at gene expression (see also next chapter). However, there are still unmet needs in terms of analysis of the variation occurring in the human genome and how this may alter cellular behavior in relation to physiological and disease processes. The number of splice variants per gene has been estimated to be at least 5–10 per gene and the total number of SNPs to be more than 5 million. To be able to analyze these very large numbers of variables, new and denser array formats are needed. It also poses a statistical problem on the analysis in terms of the requirements to the materials to be analyzed. Large tissue banks will have to be developed, for example if disease related variations in the genome have to be identified.

Recent years have demonstrated that microarrays could have an important role in parallel analysis of cellular parameters in combination with drugs, genes, viruses, RNAi, and so on. The micro- or nano-format makes it possible to screen for a number of variables simultaneously, for example, during treatment with a new drug candidate. The endpoints have to be measured in parallel in hundreds or thousands of small cell cultures. This is a challenge, as functional assays will have to be developed for this purpose such as apoptosis assays, cell cycle assays, cell kinetic assays, and so on. Proteomic arrays have not been dealt with in this chapter but are, of course, another array area where much is expected in the future (see also Section 18.5).

Some assays based on the function of molecules at DNA or peptide level are promising, such as those for phosphorylation of peptides or binding assays in which binding of a transcription factor to DNA is analyzed with respect to binding inhibitors. Such biochemical process arrays are very promising as they make it possible to screen large libraries of compounds for very specific end-points.

Most array work has aimed at identifying changes in gene expression in clinical samples or *in vitro* in cell lines. The pathways and genes that change their expression are identified and this can be followed by *in vitro* testing of the effect of up or down regulation of the expression of the gene. This effect can again be analyzed by microarrays in a time course study of the gene expression and shifts in pathways.

This situation in which a single parameter is changed and the effect is measured is quite straightforward. However, in the future more than one parameter will have to be altered, such as inhibiting one pathway and promoting another to get more information on the dynamic possibilities in cellular responses. To study such changes with appropriate controls and many time points is costly, and a single factor that will promote this development is the lower cost of arrays. A reduced microarray cost is needed to be able to use multiple arrays for analysis of complicated cellular networks, for example by using Bayesian network modelling.

Apart from such increased use of arrays in defining cellular networks, it is expected that a clinical use of microarrays gradually will be introduced, especially in relation to cancer classification. In breast, lymphoma, leukemia, and bladder malignancies classifiers already have been published and are in clinical testing. They will make it possible to select the right treatment for the individual patient and to develop individual predictors of the disease course needed for individual follow-up programs. Such a use of microarrays may lead to great benefit for the society and to reduced spending on ineffective treatments, and selection of tailored and more effective treatments.

References

Ahrendt, S. A., Halachmi, S., Chow, J. T., Wu, L., Halachmi, N., Yang, S. C., Wehage, S., Jen, J., and Sidransky, D. (1999). Rapid p53 sequence analysis in primary lung cancer using an oligonucleotide probe array. *Proc. Natl. Acad. Sci. USA* 96, 7382–7387.

Alizadeh, A. A., Eisen, M. B., Davis, R. E., Ma, C., Lossos, I. S., Rosenwald, A., Boldrick, J. C., Sabet, H., Tran, T., Yu, X., Powell, J. I., Yang, L., Marti, G. E., Moore, T., Hudson, J., Jr., Lu, L., Lewis, D. B., Tibshirani, R., Sherlock, G., Chan, W. C., Greiner, T. C., Weisenburger, D. D., Armitage, J. O., Warnke, R., Levy, R., Wilson, W., Grever, M. R., Byrd, J. C., Botstein, D., Brown, P. O., and Staudt, L. M. (2000). Distinct types of diffuse large B-cell lymphoma identified by gene expression profiling. *Nature* 403, 503–511.

Beer, D. G., Kardia, S. L., Huang, C. C., Giordano, T. J., Levin, A. M., Misek, D. E., Lin, L., Chen, G., Gharib, T. G., Thomas, D. G., Lizyness, M. L., Kuick, R., Hayasaka, S., Taylor, J. M.,

Iannettoni, M. D., Orringer, M. B., and Hanash, S. (2002). Gene-expression profiles predict survival of patients with lung adenocarcinoma. *Nat. Med.* 8, 816–824.

Bhattacharjee, A., Richards, W. G., Staunton, J., Li, C., Monti, S., Vasa, P., Ladd, C., Beheshti, J., Bueno, R., Gillette, M., Loda, M., Weber, G., Mark, E. J., Lander, E. S., Wong, W., Johnson, B. E., Golub, T. R., Sugarbaker, D. J., and Meyerson, M. (2001). Classification of human lung carcinomas by mRNA expression profiling reveals distinct adenocarcinoma subclasses. *Proc. Natl. Acad. Sci. USA* 98, 13790–13795.

Dudoit, S., Fridlyand, J., and Speed, T. P. (2000). Comparison of Discrimination Methods for the Classification of Tumors Using Gene Expression Data. Technical Report, Berkeley University.

Dyrskjot, L., Thykjaer, T., Kruhoffer, M., Jensen, J. L., Marcussen, N., Hamilton-Dutoit, S., Wolf, H., and Orntoft, T. F. (2003). Identifying distinct classes of bladder carcinoma using microarrays. *Nat. Genet.* 33, 90–96.

Eisen, M. B., Spellman, P. T., Brown, P. O., and Botstein, D. (1998). Cluster analysis and display of genome-wide expression patterns. *Proc. Natl. Acad. Sci. USA* 95, 14863–14868.

Fan, J. B., Chen, X., Halushka, M. K., Berno, A., Huang, X., Ryder, T., Lipshutz, R. J., Lockhart, D. J., and Chakravarti, A. (2000). Parallel genotyping of human SNPs using generic high-density oligonucleotide tag arrays. *Genome Res.* 10, 853–860.

Fritz, B., Schubert, F., Wrobel, G., Schwaenen, C., Wessendorf, S., Nessling, M., Korz, C., Rieker, R. J., Montgomery, K., Kucherlapati, R., Mechtersheimer, G., Eils, R., Joos, S., and Lichter, P. (2002). Microarray-based copy number and expression profiling in dedifferentiated and pleomorphic liposarcoma. *Cancer Res.* 62, 2993–2998.

Golub, T. R., Slonim, D. K., Tamayo, P., Huard, C., Gaasenbeek, M., Mesirov, J. P., Coller, H., Loh, M. L., Downing, J. R., Caligiuri, M. A., Bloomfield, C. D., and Lander, E. S. (1999). Molecular classification of cancer: class discovery and class prediction by gene expression monitoring. *Science* 286, 531–537.

Heller, M. J. (2002). DNA microarray technology: devices, systems, and applications. *Annu. Rev. Biomed. Eng.* 4, 129–153.

Holden, A. L. (2002). The SNP consortium: summary of a private consortium effort to develop an applied map of the human genome. *Biotechniques Suppl.* 22–24, 26.

Hyman, E., Kauraniemi, P., Hautaniemi, S., Wolf, M., Mousses, S., Rozenblum, E., Ringner, M., Sauter, G., Monni, O., Elkahloun, A., Kallioniemi, O. P., and Kallioniemi, A. (2002). Impact of DNA amplification on gene expression patterns in breast cancer. *Cancer Res.* 62, 6240–6245.

Kallioniemi, A., Kallioniemi, O. P., Sudar, D., Rutovitz, D., Gray, J. W., Waldman, F., and Pinkel, D. (1992). Comparative genomic hybridization for molecular cytogenetic analysis of solid tumors. *Science* 258, 818–821.

Kennedy, G. C., Matsuzaki, H., Dong, S., Liu, W. M., Huang, J., Liu, G., Su, X., Cao, M., Chen, W., Zhang, J., Liu, W., Yang, G., Di, X., Ryder, T., He, Z., Surti, U., Phillips, M. S., Boyce-Jacino, M. T., Fodor, S. P., and Jones, K. W. (2003). Large-scale genotyping of complex DNA. *Nat. Biotechnol.* 21, 1233–1237.

Kumar, R., Conklin, D. S., and Mittal, V. (2003). High-throughput selection of effective RNAi probes for gene silencing. *Genome Res.* 13, 2333–2340.

Laken, S. J., Petersen, G. M., Gruber, S. B., Oddoux, C., Ostrer, H., Giardiello, F. M., Hamilton, S. R., Hampel, H., Markowitz,

A., Klimstra, D., Jhanwar, S., Winawer, S., Offit, K., Luce, M. C., Kinzler, K. W., and Vogelstein, B. (1997). Familial colorectal cancer in Ashkenazim due to a hypermutable tract in APC. *Nat. Genet.* 17, 79–83.

Lindroos, K., Sigurdsson, S., Johansson, K., Ronnblom, L., and Syvanen, A. C. (2002). Multiplex SNP genotyping in pooled DNA samples by a four-colour microarray system. *Nucleic Acids Res.* 30, e70.

Mei, R., Galipeau, P. C., Prass, C., Berno, A., Ghandour, G., Patil, N., Wolff, R. K., Chee, M. S., Reid, B. J., and Lockhart, D. J. (2000). Genome-wide detection of allelic imbalance using human SNPs and high-density DNA arrays. *Genome Res.* 10, 1126–1137.

Monni, O., Barlund, M., Mousses, S., Kononen, J., Sauter, G., Heiskanen, M., Paavola, P., Avela, K., Chen, Y., Bittner, M. L., and Kallioniemi, A. (2001). Comprehensive copy number and gene expression profiling of the 17q23 amplicon in human breast cancer. *Proc. Natl. Acad. Sci. USA* 98, 5711–5716.

Mousses, S., Caplen, N. J., Cornelison, R., Weaver, D., Basik, M., Hautaniemi, S., Elkahloun, A. G., Lotufo, R. A., Choudary, A., Dougherty, E. R., Suh, E., and Kallioniemi, O. (2003). RNAi microarray analysis in cultured mammalian cells. *Genome Res.* 13, 2341–2347.

Perou, C. M., Sorlie, T., Eisen, M. B., van de Rijn, M., Jeffrey, S. S., Rees, C. A., Pollack, J. R., Ross, D. T., Johnsen, H., Akslen, L. A., Fluge, O., Pergamenschikov, A., Williams, C., Zhu, S. X., Lonning, P. E., Borresen-Dale, A. L., Brown, P. O., and Botstein, D. (2000). Molecular portraits of human breast tumours. *Nature* 406, 747–752.

Pollack, J. R., Perou, C. M., Alizadeh, A. A., Eisen, M. B., Pergamenschikov, A., Williams, C. F., Jeffrey, S. S., Botstein, D., and Brown, P. O. (1999). Genome-wide analysis of DNA copy-number changes using cDNA microarrays. *Nat. Genet.* 23, 41–46.

Pollack, J. R., Sorlie, T., Perou, C. M., Rees, C. A., Jeffrey, S. S., Lonning, P. E., Tibshirani, R., Botstein, D., Borresen-Dale, A. L., and Brown, P. O. (2002). Microarray analysis reveals a major direct role of DNA copy number alteration in the transcriptional program of human breast tumors. *Proc. Natl. Acad. Sci. USA* 99, 12963–12968.

Primdahl, H., Wikman, F. P., von der Maase, H., Zhou, X. G., Wolf, H., and Orntoft, T. F. (2002). Allelic imbalances in human bladder cancer: genome-wide detection with high-density single-nucleotide polymorphism arrays. *J. Natl. Cancer Inst.* 94, 216–223.

Ramaswamy, S., Tamayo, P., Rifkin, R., Mukherjee, S., Yeang, C. H., Angelo, M., Ladd, C., Reich, M., Latulippe, E., Mesirov, J. P., Poggio, T., Gerald, W., Loda, M., Lander, E. S., and Golub, T. R. (2001). Multiclass cancer diagnosis using tumor gene expression signatures. *Proc. Natl. Acad. Sci. USA* 98, 15149–15154.

Ramaswamy, S., Ross, K. N., Lander, E. S., and Golub, T. R. (2003). A molecular signature of metastasis in primary solid tumors. *Nat. Genet.* 33, 49–54.

Rosenwald, A., Wright, G., Chan, W. C., Connors, J. M., Campo, E., Fisher, R. I., Gascoyne, R. D., Muller-Hermelink, H. K., Smeland, E. B., Giltnane, J. M., Hurt, E. M., Zhao, H., Averett, L., Yang, L., Wilson, W. H., Jaffe, E. S., Simon, R., Klausner, R. D., Powell, J., Duffey, P. L., Longo, D. L., Greiner, T. C., Weisenburger, D. D., Sanger, W. G., Dave, B. J., Lynch, J. C.,

Vose, J., Armitage, J. O., Montserrat, E., Lopez-Guillermo, A., Grogan, T. M., Miller, T. P., LeBlanc, M., Ott, G., Kvaloy, S., Delabie, J., Holte, H., Krajci, P., Stokke, T., and Staudt, L. M. (2002). The use of molecular profiling to predict survival after chemotherapy for diffuse large-B-cell lymphoma. *N. Engl. J. Med.* 346, 1937–1947.

Shi, M. M. (2001). Enabling large-scale pharmacogenetic studies by high-throughput mutation detection and genotyping technologies. *Clin. Chem.* 47, 164–172.

Shipp, M. A., Ross, K. N., Tamayo, P., Weng, A. P., Kutok, J. L., Aguiar, R. C., Gaasenbeek, M., Angelo, M., Reich, M., Pinkus, G. S., Ray, T. S., Koval, M. A., Last, K. W., Norton, A., Lister, T. A., Mesirov, J., Neuberg, D. S., Lander, E. S., Aster, J. C., and Golub, T. R. (2002). Diffuse large B-cell lymphoma outcome prediction by gene-expression profiling and supervised machine learning. *Nat. Med.* 8, 68–74.

Snijders, A. M., Nowak, N., Segraves, R., Blackwood, S., Brown, N., Conroy, J., Hamilton, G., Hindle, A. K., Huey, B., Kimura, K., Law, S., Myambo, K., Palmer, J., Ylstra, B., Yue, J. P., Gray, J. W., Jain, A. N., Pinkel, D., and Albertson, D. G. (2001). Assembly of microarrays for genome-wide measurement of DNA copy number. *Nat. Genet.* 29, 263–264.

Spicker, J. S., Wikman, F., Lu, M. L., Cordon-Cardo, C., Workman, C., Orntoft, T. F., Brunak, S., and Knudsen, S. (2002). Neural network predicts sequence of TP53 gene based on DNA chip. *Bioinformatics* 18, 1133–1134.

Thykjaer, T., Workman, C., Kruhoffer, M., Demtroder, K., Wolf, H., Andersen, L. D., Frederiksen, C. M., Knudsen, S., and Orntoft, T. F. (2001). Identification of gene expression patterns in superficial and invasive human bladder cancer. *Cancer Res.* 61, 2492–2499.

van de Vijver, M. J., He, Y. D., van't Veer, L. J., Dai, H., Hart, A. A., Voskuil, D. W., Schreiber, G. J., Peterse, J. L., Roberts, C., Marton, M. J., Parrish, M., Atsma, D., Witteveen, A., Glas, A., Delahaye, L., van der Velde, T., Bartelink, H., Rodenhuis, S., Rutgers, E. T., Friend, S. H., and Bernards, R. (2002). A gene-expression signature as a predictor of survival in breast cancer. *N. Engl. J. Med.* 347, 1999–2009.

van't Veer, L. J., Dai, H., van de Vijver, M. J., He, Y. D., Hart, A. A., Mao, M., Peterse, H. L., van der Kooy, K., Marton, M. J., Witteveen, A. T., Schreiber, G. J., Kerkhoven, R. M., Roberts, C., Linsley, P. S., Bernards, R., and Friend, S. H. (2002). Gene expression profiling predicts clinical outcome of breast cancer. *Nature* 415, 530–536.

Veltman, J. A., Fridlyand, J., Pejavar, S., Olshen, A. B., Korkola, J. E., DeVries, S., Carroll, P., Kuo, W. L., Pinkel, D., Albertson, D., Cordon-Cardo, C., Jain, A. N., and Waldman, F. M. (2003). Array-based comparative genomic hybridization for genome-wide screening of DNA copy number in bladder tumors. *Cancer Res.* 63, 2872–2880.

Wang, D. G., Fan, J. B., Siao, C. J., Berno, A., Young, P., Sapolsky, R., Ghandour, G., Perkins, N., Winchester, E., Spencer, J., Kruglyak, L., Stein, L., Hsie, L., Topaloglou, T., Hubbell, E., Robinson, E., Mittmann, M., Morris, M. S., Shen, N., Kilburn, D., Rioux, J., Nusbaum, C., Rozen, S., Hudson, T. J., Lipshutz, R., Chee, M., and Lander, E. S. (1998). Large-scale identification, mapping, and genotyping of single-nucleotide polymorphisms in the human genome. *Science* 280, 1077–1082.

Weiss, M. M., Kuipers, E. J., Postma, C., Snijders, A. M., Siccama, I., Pinkel, D., Westerga, J., Meuwissen, S. G., Albertson, D. G., and Meijer, G. A. (2003). Genomic profiling of gastric cancer predicts lymph node status and survival. *Oncogene* 22, 1872–1879.

Wikman, F. P., Lu, M. L., Thykjaer, T., Olesen, S. H., Andersen, L. D., Cordon-Cardo, C., and Orntoft, T. F. (2000). Evaluation of the performance of a p53 sequencing microarray chip using 140 previously sequenced bladder tumor samples. *Clin. Chem.* 46, 1555–1561.

Wilhelm, M., Veltman, J. A., Olshen, A. B., Jain, A. N., Moore, D. H., Presti, J. C., Jr., Kovacs, G., and Waldman, F. M. (2002). Array-based comparative genomic hybridization for the differential diagnosis of renal cell cancer. *Cancer Res.* 62, 957–960.

Yang, Y. H., Dudoit, S., Luu, P., Lin, D. M., Peng, V., Ngai, J., and Speed, T. P. (2002). Normalization for cDNA microarray data: a robust composite method addressing single and multiple slide systematic variation. *Nucleic Acids Res.* 30, e15.

CHAPTER 16

Human Genome Microarray in Biomedical Applications

UTE WIRKNER,[1] CHRISTIAN MAERCKER,[2] JAN SELIG,[1] MECHTHILD WAGNER,[1] HEIKO DRZONEK,[1] ANJA KELLERMANN,[3] ALEXANDRA ANSORGE,[1] IOANNIS AMARANTOS,[1] JOHANNES MAURER,[3] CHRISTIAN SCHWAGER,[1] JONATHON BLAKE,[1] BERNHARD KORN,[2] WOLFGANG WAGNER,[4] ANTHONY D. HO,[4] AMIR ABDOLLAHI,[5] PETER HUBER,[5] UWE RADELOF,[3] AND WILHELM ANSORGE[1]

[1] *European Molecular Biology Laboratory, Biochemical Instrumentation Programme, Heidelberg, Germany;*

[2] *German Genome Resource Center (RZPD), Heidelberg, 69120 Germany;*

[3] *German Genome Resource Center (RZPD), Berlin, 14059, Germany;*

[4] *Department of Medicine V, University of Heidelberg, Heidelberg, Germany;*

[5] *Department of Radiation Oncology, German Cancer Research Center (DKFZ) and Department of Radiation Oncology, University of Heidelberg Medical School, Heidelberg, Germany*

TABLE OF CONTENTS

16.1 INTRODUCTION

Molecular diagnostic technologies will play a significant role in the practice of medicine, public health, and the pharmaceutical industry. Together with genomics and proteomics techniques, the field of pharmacogenomics will develop, with the goal of personalization of diagnosis and therapy, for optimal efficiency and reduced toxicity, contributing to integrated healthcare (see also Chapter 20).

Among the significant technologies in the hands of biological and biomedical researchers for assessing gene expression are the DNA microarrays (summarized in www.nature.com/reviews/focus/microarrays). On glass slides, thousands of known DNA sequences are spotted or synthesized. Test mRNA is isolated from samples, eventually converted to cDNA or amplified, labeled, and then hybridized on the slides. The array on the slide is scanned to measure the amount of sample bound to a particular spot. Subsequent computer analysis generates gene expression profiles of samples, the test results leading to hypotheses and conclusions. Microarrays are providing insights into such areas as study of tumor classification, vaccine development, graft rejection, aging process, circadian gene expression, and stress response (see also previous chapter). Studies are carried out to verify and improve the reproducibility and accuracy of the microarray techniques, as well as to introduce standards in microarray data formats submitted to databases (Brazma *et al.*, 2001).

With the completion of the Human Genome sequence analysis, the interest turned to the production of a complete Human Genome DNA microarray as one of the most powerful and desirable tools for molecular diagnostics. The Human Genome Array, completed at the European Molecular Biology Laboratory (EMBL) by the group of Wilhelm Ansorge, in collaboration with the German Resource Centre (RZPD) team, led by Uwe Radelof and reported in October 2002 (EMBL press release, October 2002; Wirkner *et al.*, 2004), was the first large whole human genome array reported worldwide. This microarray contains the best-characterized and largest cDNA clone selection at the time, with what is believed to be the entire Human Genome with about 52,000 clones of known and predicted human genes. In contrast to "topic-specific" microarrays that contain only a limited number of known genes, this is a tool that can perform a genome-wide differential gene expression profile from any investigation on human samples, and identify new genes, which may play an important role in the investigated test characterized by the samples.

During the years 2003 and 2004, this complete Human Genome microarray was validated in collaborations on several biology and clinical projects (angiogenesis, stem cells, cholesterol traffic, testicular cancer diagnosis, gene deletions, study of cellular receptors, comparative genome hybridization in evolution studies comparing humans with chimpanzees and mice). The DNA microarrays, in combination with the proteomics techniques, like protein and antibody arrays (Maercker *et al.*, 2004; see also Chapter 18), enable the analysis and diagnostics of cellular signalling pathways, similar to the work on endostatin pathway in angiogenesis (Abdollahi *et al.*, 2004a). A promising product of the genome array techniques will be the finding of biomarkers for reliable diagnostics of diseases, as was demonstrated already in many applications; for example, in the study of testicular cancer (Almstrup *et al.*, 2004).

The complete Human Genome microarray is a tool that can be used to compare genome-wide gene expression analysis in different types of cells, with an enormous experimental potential for biologists and clinicians. It will be possible to select for each biomedical application a suitable, most complete smaller subset of genes relevant in the concrete application, and to produce smaller dedicated chips for fast and less expensive analysis.

By the end of 2003 and early 2004, several well-known commercial companies have announced the availability of similar comprehensive DNA arrays with short oligonucleotides (22–60 bp long), among them, Affymetrix, Agilent, Applied Biosystems, and Amersham. In the meantime, the start-up company Nimblegen Systems (http://www.nimblegen.com) offers the whole human genome on an array it sells as a service, with clients sending in the test sample mRNA, the company synthesizes the oligonucleotide array designed by the customer, performs the tests and analysis, then delivers the results to the customers.

16.2 PRODUCTION OF THE HUMAN GENOME MICROARRAY

The authors group has produced in collaboration with the German Resource Centre (RZPD) a whole genome cDNA microarray. Based on the NCBI UniGene clustering (www.ncbi.nlm.nih.gov/Uni-Gene), cDNA clones were selected from the Image clone collection representing 51,145 clusters. The microarray technology developed by the authors group at EMBL was used in the production. The Human Genome microarray was the first large human array completed worldwide (reported in October 2002), containing the best characterized and largest clone selection at the time, nearly the entire Human Genome with about 52,000 clones of known and predicted human genes.

For the production of DNA microarrays, the authors group at EMBL has developed and set up new and improved methods for robot spotting of probes, novel surface chemistry for nucleic acid (and protein) glass attachment and chemistry for labeling of DNA, reliable hybridization conditions, as well as setting up high resolution fluorescence detection systems. Standardization and quality controls are implemented in several steps; for example, quality of total RNA using micro-electrophoresis lab-on-a-chip technology, fluorescent labeling, selection criteria for clones used in normalization and standardization of microarrays, automated analysis of PCR products, and surface chemistry control by electrospray mass spectroscopy. Bioinformatics tools developed included software packages for analysis and evaluation of microarrays, data normalization and standardization, and for databases, central repositories for microarray data and images.

16.2.1 Design and Protocols for Production of the Human Genome Microarray

16.2.1.1 UNIGENESETRZPD3

The UnigeneSetRZPD3 (Radelof *et al.*, 2002) is composed of 51,145 human cDNA clones from the image clone collection (www.ncbi.nlm.nih.gov/Uni-Gene), and NCBI Unigene clustering was taken as a basis for the selection of clones. One representative clone per cluster was selected following these criteria: location as close as possible to the 3′ end of the transcript, insert length between 500 and 1,500 bp. All clones were resequenced and verified.

16.2.1.2 FULL-LENGTH HUMAN cDNA SEQUENCING AND AMPLIFICATION

The importance of the full-length cDNA sequences has been confirmed for analysis of genes, as well as for the future analysis of proteomes. These projects have contributed greatly toward improving the quality of DNA microarrays. For example, in Germany several research groups have joined in a project to sequence many thousands of full-length cDNA clones funded by a grant from the German Federal Ministry of Research (BMBF).

Colony PCR amplification was performed using two different vector primer pairs, which were modified at their 5′ end with a C12 Amino group. Cycling was performed in 96 well plates in a thermocycler. The PCR products were purified on a Biomek FX robot (Beckman) using the Macherey-Nagel Nucleospin kit based on a silica matrix. Since a stacker was added to the robot, 4 × 96 well PCR plates could be purified in one run, in 90 min without manual interference. Items from the stacker are loaded on the robot deck by a shuttle, and bar codes on source and destination (elution) plates were detected and stored during the deck loading.

16.2.1.3 QUALITY CONTROL AND DOCUMENTATION OF PCR PRODUCTS

For quality control, 1 µl from each PCR product was analyzed on a 1.2% ready-to-run agarose gel (Amersham). Sample preparation and gel loading was done on a Biomek 2000 robot.

For quantitative and qualitative evaluation of the gels a software tool—the RTR-skipper—was developed. With this tool, PCR bands are identified and quantified relative to the amount of marker loaded. Multiple bands, missing bands, and low amount of bands are flagged as low quality. A picture of the gel together with a quality table is stored and combined in a purification results table. This table gives an overview on success rate of PCR and purification and a link to the individual gel pictures/quality tables.

The quality and quantity of PCR products were analyzed and documented. DNA fragments were detected and quantified by integration. The evaluated gel picture and a table with the result for each individual PCR fragment were stored in a file and linked to a project table, where results of all purified plates are summarized. Successful amplification and purification resulted a single band detected with a concentration of DNA between 30 and 60 ng/µl.

16.2.1.4 QUALITY CONTROL OF SPOTTED ARRAY BY DNA STAINING

As an initial test of a newly spotted batch of slides, one slide from each batch is stained with Syto-61 to see the spotting efficiency and the DNA attachment before doing an expensive hybridization. The staining gives signals proportional to the amount of DNA spotted. Attached DNA on slides was stained using Syto-61 (Molecular Probes). Slides were incubated in prehybridization solution, stained with Syto-61 solution, rinsed, and scanned as Cy5 at 625 nm.

16.2.1.5 RNA ISOLATION, AMPLIFICATION, AND LABELING FOR HIGHLY SENSITIVE DETECTION

Of particular importance for reproducible protocols is the RNA isolation, and the methods for the preparation of fluorescent cDNA probes that should quantitatively reflect the abundance of different mRNAs in the two samples to be compared (see also previous chapter). Protocols for RNA isolation, amplification, and high sensitivity detection were established and applied.

Total RNA was analyzed on a 2100 Bioanalyser (Agilent) and quantified by optical density OD260 measurement. Five µg or 100 ng of total RNA were amplified using the RiboAmp RNA Amplification Kit (Arcturus) in one or two rounds of amplification, respectively. The amplified RNA was labeled using the Atlas Glas fluorescent labelling kit with Cy3- or Cy5-Monoreactive dye (Amersham Pharmacia). The protocol allows labeling of as low amount of total RNA as 20 ng (RNA amount that can be obtained from only about 50,000 cells). This protocol is based upon a controlled two-round T7 linear RNA amplification and subsequent direct incorporation of fluorescently labeled nucleotides in the synthesized DNA. T7 RNA amplification with subsequent direct labeling was found by the authors laboratory to be the most sensitive and reproducible protocol for low amounts of starting input RNA, in comparison to other available methods (Richter *et al.*, 2002).

The authors have systematically evaluated and compared different published and commercial principles for the synthesis of fluorescently labeled probes for microarray analysis, and observed that individual labeling methods can significantly influence the expression pattern obtained in a microarray experiment. Benefits and limitations of each method were determined (Richter *et al.*, 2002).

16.2.1.6 LOW-COST (FACE-TO-FACE) HYBRIDIZATION

A new modified technique for hybridization of glass slide microarrays has been developed and tested, which reduces by more than half the volume of the hybridization probe needed, cutting the cost of microarray experiments. Following prehybridization, 65 µl of hybridization solution were placed on the spotted side of one slide and instead of a cover slip the slide with the second part of clones was placed with the spotted side down on the first slide. Hybridization was performed 20 h at 42°C. After washing, slides were dried immediately by spinning in a slide centrifuge and scanned in an Axon scanner at 587 nm and 635 nm excitation wavelength. Scanned images were analyzed using the ChipSkipper software (see Section 16.3).

16.2.2 Robotics and Automation

The laboratory automation consists of robots and a microarray spotter. The work involves workflow and protocols optimization, and setting up and implementing databases for data storage and analysis. Higher throughput is achieved on the pipetting robot (96 channels) with implementation of Stacker Carousel (allowing storage of several plates, and improving not attended operation) and optimization of disposables, and in the automated method for PCR cleaning and preparation of spotting plates.

16.2.2.1 PREPARATION OF SPOTTING PLATES AND CONTROLS

PCR products are transferred into 384-well microtiter plates in 2xSSC spotting buffer. By using a Biomek FX with a stacker, four spotting plates could be prepared in one run lasting 40 min. Samples were spotted in 2xSSC and 0.1 M betaine. Spotting plates were prepared on a Biomek FX robot. Consumables, source plates (96-well plates with purified PCR products), and destination plates (386-well spotting) were loaded in a stacker attached to the robot. Items from the stacker were transported to the robot deck by a shuttle and barcode-labeled source and destination plates were scanned during this loading process.

16.2.2.2 NOVEL SPOTTING SYSTEM AND BARCODE

Spotting is performed on an OmniGrid (GeneMachines) spotter, with a new system developed in the group, eliminating missing spots. This is a particularly critical factor for large-scale production of microarrays containing many thousands of spots.

Amino-silane coated slides, with 48 SMP3 split needle pins (from Array It company) were used for spotting at 50% humidity and 20°C. After 384-well spotting plates had been prepared, the PCR fragments were spotted on two slides, the array consisting of 51,145 cDNAs with a spot distance of 180 μm and 24 spots in a row. In each subgrid, labeled oligonucleotides, spike-in controls (known samples added to test samples to facilitate and verify correct analysis; see also previous chapter), and labeled primers were spotted, for positive as well as negative control. After the spotting, the DNA was attached to the slide surface by incubation for 4 h at 65°C. Higher throughput on the microarray spotter was achieved with an added stacker, made functional with lidded plates.

To follow the steps during the process all sample plates have a specific barcode label, which is monitored and combined with the documentation. The barcode control of source and destination plates during purification of PCR products and production of spotting plates helped to verify the right order of samples and to detect a manual turning or mixing of microtiter plates.

16.2.3 Surface Chemistry, Robot for Automated Coating of Glass Slides

Several chemistries were developed and tested for the attachment of DNA and proteins. An automated robotic system was designed, constructed, and tested for high throughput in production of chemically coated glass slides for microarray production. The optimum properties of the surface chemistry invented by the group were verified both for DNA and protein chips, in an academic as well as industrial environment.

16.3 DATA ANALYSIS, DATABASES, AND DATA MANAGEMENT

16.3.1 Microarray Image Analysis, Quantitation, and Evaluation: The Chipskipper Software

Very large data sets are obtained from the whole genome DNA microarrays containing many thousands of spots. Such large amounts of data can be analyzed only with computer systems. There are several software packages for microarray data analysis available on the market from different companies and institutes (www.nature.com/reviews/focus/microarrays). The principles of operation of the microarray data analysis will be shown here in some applications from the ChipSkipper software package.

ChipSkipper, a user-friendly and efficient microarray image analysis software, was extended to meet the requirements for routine production of large genome-wide microarrays. A brief summary of the ChipSkipper software as an example of microarray data analysis software package, and the steps used in the analysis, are described next.

Analysis of microarray scans can be divided into three steps:

1. Image analysis. The overall images are evaluated for irregularities in signal distribution. These may be caused by dust particles or scratches on the glass, uneven slide coating, global flares from sample hybridization and washing, and wrong scanning parameters.
2. Spot quantitation. Integration grids, defining the spot positions are placed onto the microarray images, and underlying spots are numerically analyzed and integrated. As a result, spot intensity values as well as quality values are created.
3. Spot evaluation. The deduced integration values are analyzed with the help of various scatterplots, ratio distribution histograms, or gene specific views.

The general rule in computer science, also valid for microarray analysis, is "garbage in—garbage out." The introduction of low quality data at the beginning of the data evaluation queue may completely scramble any analyses at later stages of data analysis; for example, during clustering. Therefore, in case the microarrays scans are problematic or inconclusive, one should consider repeating the entire experiment.

16.3.1.1 IMAGE ANALYSIS

A big problem in image display from microarray images is information reduction. Single images are scanned with 65,000 levels of intensity. In two-color image overlay on computer displays this is reduced to 256 intensity levels; the human eye probably can distinguish between roughly 100 intensity levels only. Furthermore, the human eye is more sensitive to the green color of the commonly used red/green overlay images. To detect and use most of the information from the microarray scans besides the standard two-color overlay images, additional analysis tools can be used:

• Global/local intensity distribution histograms
• Global/local pixel scatterplots (see Figs. 16.1a, b)
• Line scans across the images
• 3D contour plots (see Fig. 16.1c)

16.3.1.2 SPOT QUANTITATION

In this step of microarray analysis, intensity values for each spot are created. The spot signals are placed on a more or less continuous background signal created by scanner electronics, light scattering, and background fluorescence from slides, slide coating, dust particles, and scratches on the glass, as well as from background fluorescence originating from not completely washed hybridization samples, or even nonincorporated dyes from DNA staining procedures. An image processing based local background detection eliminates these backgrounds for each spot individually. Such segmented signals are integrated and spot intensity values calculated, which are the amount of bound RNA on the spot. Furthermore, signal-to-noise ratios, as well as spot shape qualifiers are generated, which may be used to flag low quality spots; that is, those that have intensities hardly distinguishable from the surrounding background, or those that simply do not look like a spot. Final integration values may be fed into subsequent normalization or clustering programs.

16.3.1.3 IMAGE EVALUATION

The integration values should be analyzed with the help of direct scatterplots (an x/y plot where spot intensities from red/green image are displayed on x/y axes for each spot; see Fig. 16.2), ratio ranked intensity plots (the red/green ratio from each spots is plotted against the ranked; i.e., sorted from low to high average intensity for each spot), or ratio distribution histograms (plots how often a certain up or down regulation occur). Additionally free 2D or 3D scatterplots may be used to display various integration results one against the other to detect additional experimental biases.

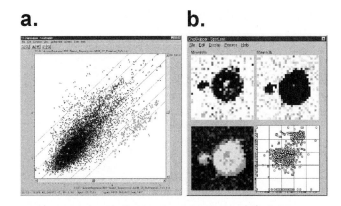

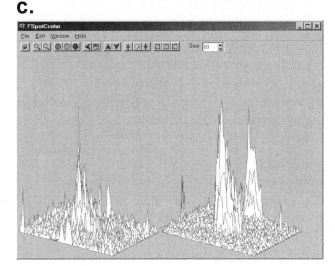

FIGURE 16.1 Visualization of images and integration of results from microarray images. a. "Spot puzzle" shows an artificial pathway-specific microarray image extracted from a genome wide microarray. Spot-thumbnail images from a selected subset of genes are rearranged to create this view. b. "Spotlens" works like a magnification lens into the microarray image. A small region is enlarged and displayed as greyscale images, two-color overlay, and pixel scatterplot. c. "Contour plot" displays a region of the microarray image as pseudo 3D landscape-like map. (See color plate)

These different displays can show irregularities in the data distribution. Some of the irregular patterns may be compensated by normalization techniques. In cases where this seems not to be feasible, it would be recommended to repeat the experiment.

For simple screening experiments it is already possible to generate, at this stage, gene lists of clearly differentially expressed genes. Tracer genes (genes with known behavior under the specific treatment conditions) may be searched and used as quality markers.

A variety of commercial and academic programs are available, offering some or all of the mentioned tasks. Not in all cases and under all circumstances, the software packages coming with microarray scanners are the best analysis tools for the microarray images.

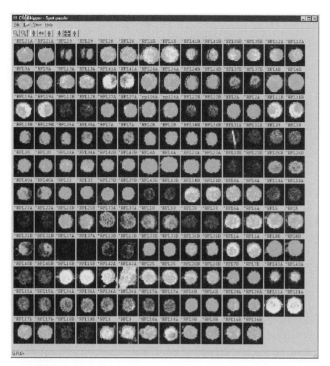

FIGURE 16.2 Scatterplot of spot intensity values. On x/y axes integrated spot intensity values in red/green image are displayed. The large cloud of black-colored spots represents the majority of nondifferentially expressed genes. The green side-cloud represents the strongly differentially expressed genes under the chosen treatment. (See color plate)

16.3.2 Microarray Databases and Data Management

There is an urgent need to set up a central database for microarray data, and a central storage of the image files. The system should allow easy access to large datasets produced over several years, as well as easy data exchange between collaborating groups. The ArrayExpress universal database for storage of any microarray platform, organism, and experimental conditions was developed at the European Molecular Biology Laboratory (EMBL) and its European Bioinformatics Institute (EBI, www.ebi.ac.uk). Annotation tool MiamExpress allows web-based annotation of microarray experiments according to the internationally agreed MIAME standard (Brazma *et al.*, 2001), which was developed and established by a worldwide group of academic and industrial laboratories. Also, Expression Profiler is a collection of clustering tools that offers a variety of algorithms for analysis of microarray experiment series, and can directly interact with ArrayExpress database (see www.ebi.ac.uk for further details).

Due to its nature, MiamExpress covers all presently required information about microarray experiments, protocols, arrays, array layouts, and array data. Future extensions of MIAME will be implemented by the EBI. The use of

MiamExpress guarantees straightforward data submission to ArrayExpress, which will be a prerequisite for publications concerning microarray experiments in the future.

MiamExpress can be extended toward a microarray Laboratory Information and Management System (LIMS) and data mining system. The first aim is to add functionality in such a way that researchers can use MiamExpress like an electronic lab book to

- Annotate all experimental steps with MiamExpress while experiments are done
- Submit experimental protocols to MiamExpress as they are developed and validated
- Store microarray data in MiamExpress directly after image analysis
- Store image files on MiamExpress server

MiamExpress can be also used for implementation of different user levels for

- Selective sharing of data within groups
- An interface to slide production LIMS, quality control, and quality assessment
- The setting up of processing pipelines for data preprocessing, filtering, clustering, and extended analysis
- Querying MiamExpress data for experiments, treatments, protocols, spotted genes or even integration data

MiamExpress is based on publicly available platforms, such as Linux operating system, MySQL database engine, and Apache web server. Thus there are no set-up costs (except the server hardware) and the system can be extended freely.

At present, several prototypes have been implemented for

- Hybridization centered query system
- Tool for batch uploading of microarray hybridization data
- Spot annotation / re-annotation system
- Automated generation of ADF spot tracking files, merged from the annotation database, the slide production LIMS, and generic spotter deconvolution files
- General hybridization table preprocessing and filtering tool

Also, during evaluation of the authors' group collaborative microarray experiments, several publicly available data sources (e.g., Human Protein Reference Database, John Hopkins University, http://www.hprd.org) have been explored and included into the data annotation pipelines mentioned earlier. Contributions may be made by the extensive experimental and computer analysis of data obtained in the various biological and medical collaborative projects with the large comprehensive Human Genome microarrays.

16.4 APPLICATIONS OF THE HUMAN GENOME MICROARRAY

The whole Human Genome microarray was validated during one year, in several collaborative biology and clinical projects (study of angiogenesis, search of marker genes relevant in clinical cancer research studies, testicular cancer diagnosis, gene deletions, stem cells, transcriptome evolution studies in human vs. chimpanzee vs. mouse).

The technology, techniques, and software developed was tested and applied to projects in molecular medicine in collaboration with professional clinical teams, as well as in biology. The tools are used to compare gene activity in different types of cells, with an enormous experimental potential for biologists and clinicians. It will be possible to select for each biomedical application a suitable, most complete smaller subset of genes relevant in the concrete application, and to produce smaller dedicated chips for fast and less expensive analysis. Production and work with microarrays involve also maintenance of the produced array platforms and optimization and improvements in the processing protocols.

There is a large effort in the early diagnosis of cervical cancer (the most prevalent cause of cancer-related deaths among women in the third world), as well as in analysis of genetic alterations in breast cancer cell lines (with the Comparative Genome Hybridization (CGH) technique; see also Chapters 12 and 13).

Among other relevant applications of the Human Genome microarrays were investigations of induced premature senescence in fetal lung fibroblasts cells (expressing telomerase), molecular characterization and specific gene expression patterns in cardiovascular diseases, analysis of gene expression relevant in the study of cellular receptors, examination of cholesterol cellular pathway, and changes in the transcriptome after silencing of p53 gene (analysis with RNA interference technique). By differential gene expression and DNA microarray techniques, a novel protein (the ankyrin repeat containing SOCS box protein 5, a known suppressor of cytokine signaling, and associated with initiation of arteriogenesis), has been identified (Boengler *et al.*, 2003). Microarray techniques (both DNA and protein microarrays) are used in the identification of proteins associated with chemo-resistance of tumors, and to analyze their functional and diagnostic relevance (Huber *et al.*, 2002).

In addition, many other diagnostics and clinical applications have been carried out with the commercially produced and available Human Genome arrays (Affymetrix, Roche, Agilent, Qiagen, Applied Biosystems, Nimblegen, and others).

16.4.1 Adult Human Stem Cells Division Study by Genome-Wide Analysis

The differentiation potential of adult human stem cells was analyzed using the Human Genome microarray (Wagner *et al.*, 2004). The molecular mechanisms that regulate asymmetric divisions of hematopoietic progenitor cells (HPC) are not yet understood. The slow dividing fraction (SDF) of HPC is associated with primitive function and self-renewal, whereas the fast dividing fraction (FDF) predominantly proceeds to differentiation. $CD34^+/CD38^-$ cells of human umbilical cord blood were separated into SDF and FDF. Genome-wide gene expression analysis of these populations was determined using the newly developed Human Transcriptome Microarray containing 51,145 cDNA clones of the Unigene Set-RZPD3. In addition, gene expression profiles of $CD34^+/CD38^-$ cells were compared with those of $CD34^+/CD38^-$ cells. Among the genes showing the highest expression levels in the SDF were the following: CD133, erg, cyclin g2, MDR1, osteopontin, clqr1, ifi16, jak3, fzd3, and hoxa9, a pattern compatible with their primitive function and self-renewal capacity. Furthermore, morphologic differences between SDF and FDF were determined. Cells in the SDF have more membrane protrusions and CD133 is located on these lamellipodia. The majority of cells in the SDF are rhodamine-123^{dull}. These results provide molecular evidence that the SDF is associated with primitive function and serves as a basis for a detailed understanding of asymmetric division of stem cells.

16.4.2 Tumor Angiogenesis

Gene and protein expression patterns relevant to tumor angiogenesis (Abdollahi *et al.*, 2004a) have been analyzed, in particular after physical and pharmaco-therapies (irradiation, ultrasound; Abdollahi *et al.*, 2004b).

It was demonstrated that the set of gene expressions underlying the angiogenic balance in tissues could be molecularly reset en masse by a single protein. Using genome-wide expression profiling, coupled with RT-PCR and phosphorylation analysis, it has been shown that the endogenous angiogenesis inhibitor endostatin down-regulates many signalling pathways in human microvascular endothelium associated with pro-angiogenic activity. Simultaneously, endostatin was found to up regulate many anti-angiogenic genes. The result is a unique alignment between the direction of gene regulation and angiogenic status. Further profiling revealed the regulation of genes not heretofore associated with angiogenesis. Analysis of coregulated genes has shown complex interpathway communications in an intricate signalling network that both recapitulates and extends on current understanding of the angiogenic process. Insights into the nature of genetic networking from the cell biologic as well as therapeutic perspectives were revealed (Abdollahi *et al.*, 2004a).

16.4.3 Embryonic Stem Cell-Like Features of Testicular Carcinoma *in Situ*

Using genome-wide gene expression profiling, more than 200 genes highly expressed in testicular carcinoma *in situ*

(CIS) have been identified, including many never reported in testicular neoplasms (Almstrup *et al.*, 2004). CIS is the common precursor of testicular germ cell tumours (TGCT), which in recent decades have markedly increased and now are the most common malignancy of young adult men. Expression was further verified by semi-quantitative RT-PCR and *in situ* hybridization. Among the highest expressed genes were NANOG and POU5F1, and RT-PCR revealed changes in their stoichiometry upon progression into embryonic carcinoma. CIS expression profile has been compared with patterns reported in embryonic stem cells (ESC), which revealed a substantial overlap and a striking resemblance between CIS cells and ESC. Also, an over-representation of expressed genes has been demonstrated in regions of 17q and 12, reported to be unstable in human ESC. The close similarity between CIS and ESC explains the pluripotency of CIS and supports an early prenatal origin of testicular germ cell tumors (Almstrup *et al.*, 2004).

Some of the highly expressed genes identified in this study are promising candidates for new diagnostic markers for CIS and or TGCTs.

16.4.4 A Model for Transcriptome Evolution

The Human Genome microarray also has been used to analyze a model for transcriptome differences observed in primate and murine brains. It is not known if the majority of changes of gene expression fixed between species and between various tissues during evolution are caused by Darwinian selection or by stochastic processes (Khaitovich *et al.*, 2004). Studied were differences in gene expression levels in the prefrontal cortex of six humans, five chimpanzees, one orangutan, five rhesus macaque, and five crab-eating macaque, also three mouse species. Used in the studies were both commercially available oligonucleotide arrays and the large cDNA microarray, developed in the author's institute. Due to the probe length on the latter microarray, they are much less sensitive to DNA sequence differences and therefore can be used to compare gene expression in species as remote as human and rhesus macaque. The analysis has shown that:

1. Gene expression differences accumulate approximately linearly with time between species.
2. Expression variation within a species correlates positively with expression divergence between species.
3. Rates of expression divergence between species do not differ significantly between intact genes and expressed pseudogenes.
4. Expression differences between brain regions accumulate approximately linearly with time (Khaitovich *et al.*, 2004).

16.5 SUMMARY AND PERSPECTIVES

Initial applications of molecular diagnostics were in the field of infections, but are now expanding in the areas of genetic disorders, genetic screening, cancer, tropical infectious diseases, and others. In the near future, molecular diagnostic technologies will be involved in the development of personalized medicine, based on pharmacogenetics and pharmacogenomics, analyzing DNA markers for predicting patient responses to many common drugs. Sequencing of important DNA fragments by new very fast methods, or resequencing of gene regions by dedicated microarrays, allowing detecting single nucleotide polymorphism markers (SNPs), or mutations may also be possible. As examples of special interest, among others, are techniques for defining the molecular fingerprint of tumors and identification of their subtypes, identification of acquired genetic changes, and for diagnosing congenital genetic defects.

The power, speed, sensitivity, and decreasing cost will be multiplied in combination with the novel techniques of labeling, miniaturization, nanotechnology, and computing.

16.5.1 Oligonucleotide-Based DNA Microarrays, Alternative Splicing

Oligonucleotide-based DNA microarrays are becoming increasingly useful for the analysis of gene expression and SNPs. A systematic study of the sensitivity, specificity, and dynamic range of microarray signals and their dependence on the labeling and hybridization conditions, as well as on the length, concentration, attachment moiety, and purity of the oligonucleotides was carried out (Relogio *et al.*, 2002). Both a controlled set of *in vitro* synthesized transcripts and RNAs from biological samples were used in these experiments. An algorithm was developed that allows efficient selection of oligonucleotides able to discriminate a single nucleotide mismatch. One of the conclusions about the longer oligonucleotides (60mers) is that they can provide significantly better sensitivity than 25 or 30mers, but their specificity in complex mixtures of RNA is significantly lower than that obtained for 25mers (Relogio *et al.*, 2002). These data will facilitate the design and standardization of custom-made microarrays applicable to gene expression profiling, greatly expanding analysis of alternative splicing and its increasing significance in diagnostics, as well as in sequencing analyses.

As mentioned earlier, several well-known commercial companies are producing and distributing oligonucleotide-based DNA microarrays (Affymetrix, Applied Biosystems, Nimblegen, Qiagen, Amersham), which are being applied successfully in numerous biomedical applications.

16.5.2 Molecular Tools for Microarray-Based Genomic Analysis, Protein, and Antibody Arrays

Techniques for both protein and antibody microarrays have been established, and applied in routine projects to verify the functionality and in diagnostics (http://www.function-algenomics.org.uk; see also Chapter 18). A key application, particularly in medical field, requiring highly specific protein ligands or antibodies, will be the microarrays for sensitive detection of presence of certain proteins, for example bacterial antigens. The techniques will be valuable tools in the proteomic research, in applications for monitoring of protein expression levels, protein-protein, and protein-DNA interactions. The work is in progress to develop organ- and disease-specific protein arrays for diagnostics. Another application is the microarray-based analysis of protein expression in tumor cells, defining cancer subtypes by protein expression profile signature.

An indispensable part of these techniques will be the wide availability of specific protein ligands—for example, antibodies—enabling the specific recognition and detection even of low abundance proteins.

The authors have established a novel method for high throughput screening of monoclonal antibody producing hybridomas (De Masi *et al.*, 2004). This method, based on protein microarrays technology, allows for the simultaneous parallel analysis of up to potentially 40,000 hybridomas against one antigen, on a single microarray. The novel method for monoclonal antibody screening, two orders of magnitude less expensive and about ten times faster than the standard techniques, could become a valuable tool in diagnostics projects, clinical biology, as well as in large-scale proteomics analysis.

16.5.3 Cell Arrays, Microinjection, RNA Interference

The cell microarray undoubtedly will be another useful tool, greatly contributing to the field of molecular diagnostics, with its promise to allow studying and determining function of genes in living cells, from various tissues and organs.

The development of DNA microarrays has allowed systematic analysis of gene expression of different organisms. But genes are only one step in the response of an organism to stimulation; the proteins play an important role in this response. The presence or absence of proteins, the protein-protein, and DNA-protein interactions are vital for the understanding of behavior and response of cells.

The cell microarray, or transfected-cell microarray, is a development of the last few years (Ziauddin and Sabatini, 2001). In a cell microarray, one slide contains thousands of cell clusters transfected with a defined DNA, resulting in the cellular expression of a specific protein. This system thus allows studying the presence and localization of proteins inside the cells. Proteins fused with GFP fluorescent reporter protein have been constructed and spotted on glass slides. The cells grow on the slides and they are transfected with plasmids containing the tested protein with the GFP-fusion reporter. The protein is expressed by the cells and allows localization of the protein in the cell by fluorescence microscopy. Each cluster of transfected cells growing above one DNA spot on the slides expresses the corresponding protein. The technique is applied to silencing of genes by the RNA interference. The effects of gene silencing are then analyzed by digital image analysis at a single cell level, in a high throughput cell-based analysis.

References

Abdollahi, A., Hahnfeldt, P., Maercker, C., Gröne, H. J., Debus, J., Ansorge, W., Folkman, J., Hlatky, L., and Huber, P. E. (2004a). Endostatin's antiangiogenic signalling network. *Mol. Cell* 13, 649–663.

Abdollahi, A., Domhan, S., Jenne, J. W., Hallaj, M., Dell Aqua, G., Richter, A., Mueckenthaler, M., Martin, H., Debus, J., Ansorge, W. J., Hynynen, K., and Huber, P.E. (2004b). Apoptosis signals in lymphoblasts induced by focused ultrasound. *FASEB J.* 18, 1413–1414.

Almstrup, K., Hoei-Hansen, C. E., Wirkner, U., Blake, J., Schwager, C., Ansorge, W., Nielsen, J. E., Skakkebæk, N. E., Rajpert-De Meyts, E., and Leffers, H. (2004). Embryonic stem cell-like features of testicular carcinoma in situ revealed by genome-wide gene expression profiling. *Cancer Res.* 64, 4736–4743.

Boengler, K., Pipp, F., Fernandez, B., Richter, A., Schaper, W., and Deindl, E. (2003). The ankyrin repeat containing SOCS box protein 5: a novel protein associated with arteriogenesis. *Biochem. Biophys. Res. Commun.* 302, 17–22.

Brazma, A., Hingamp, P., Quackenbush, J., Sherlock, G., Spellman, P., Stoeckert, C., Aach, J., Ansorge, W., Ball, C. A., Causton, H. C., Gaasterland, T., Glenisson, P., Holstege, F. C., Kim, I. F., Markowitz, V., Matese, J. C., Parkinson, H., Robinson, A., Sarkans, U., Schulze-Kremer, S., Stewart, J., Taylor, R., Vilo, J., and Vingron, M. (2001). Minimum information about a microarray experiment (MIAME)-toward standards for microarray data. *Nat. Genet.* 29, 365–371.

De Masi, F., Chiarella, P., Wilhelm, H., Massimi, M., Bullard, B., Ansorge, and Sawyer, A. (2004). High-throughput mouse monoclonal antibodies using antigen chips. *Nat. Biotechnol.* Submitted.

Huber, P. E., Abdollahi, A., Weber, K. J., Rastert, R., Trinh, T., Krempien, R., Ansorge, W., Wannenmacher, M., and Debus, J. (2002). Expression profiling of irradiated human lung endothelial cells using a large DNA chip. *Int. J. Rad. Oncol. Biol. Phys.* 54, 24–25.

Khaitovich, P., Weiss, G., Lachmann, M., Hellmann, I., Enard, W., Muetzel, B., Wirkner, U., Ansorge, W. and Pääbo, S. (2004). A neutral model of transcriptome evolution. *PLoS Biol.* 2, E152.

Maercker, C., Abdollahi, A., Rutenberg, C., Ridinger, H., Paces, O., Wang, J., Ansorge, W. J., Korn, B., and Huber, P. E. (2004). Antibody chips as a versatile tool for functional analysis of genes involved in tumor angiogenesis. In *Proceedings of the HUGO meeting*, Berlin, p. 59.

Radelof, U., Maurer, J., and Korn, B. (2002). Human UnigeneSet RZPD3—ein Klonsatz für das gesamte genom. *Transkript*, 11, 67.

Relogio, A., Schwager, C., Richter, A., Ansorge, W., and Valcarcel, J. (2002). Optimization of oligonucleotide-based DNA microarrays. *Nucleic Acids Res.* 30, e51.

Richter, A., Schwager, C., Hentze, S., Ansorge, W., Hentze, M. W., and Muckenthaler, M. (2002). Comparison of fluorescent tag DNA labeling methods used for expression analysis by DNA microarrays. *Biotechniques* 33, 620–630.

Wagner, W., Ansorge, A., Wirkner, U., Eckstein, V., Schwager, C., Blake, J., Miesala, K., Selig, J., Saffrich, R., Ansorge, W., and Ho, A. D. (2004). Molecular evidence for stem cell function of the slow dividing fraction among human hematopoietic progenitor cells by genome wide analysis. *Blood*, 104, 675–686.

Wirkner, U., Maercker, C., Abdollahi, A., Wagner, M., Selig, J., Drzonek, H., Kellermann, A., Ansorge, A., Maurer, J., Schwager, C., Blake, J., Korn, B., Wagner, W., Ho, A. D., Huber, P., Radelof, U., and Ansorge, W. J. (2004). Human Genome on a Chip. In *Proceedings of the HUGO meeting*, Berlin, p. 61.

Ziauddin, J., and Sabatini, D. M. (2001). Microarrays of cell expressing defined cDNAs. *Nature* 411, 107–110.

CHAPTER 17

Use of High Throughput Mass Spectrographic Methods to Identify Disease Processes

WILLIAM E. GRIZZLE,[1] O. JOHN SEMMES,[2] WILLIAM L. BIGBEE,[3] GUNJAN MALIK,[2] ELIZABETH MILLER,[1] BARKHA MANNE,[1] DENISE K. OELSCHLAGER,[1] LIU ZHU,[1] AND UPENDER MANNE[1]

[1] *Department of Pathology, University of Alabama at Birmingham, Birmingham, AL;*
[2] *Department of Microbiology and Molecular Cell Biology, Eastern Virginia Medical School, Norfolk, VA;*
[3] *Molecular Biomarkers Group, University of Pittsburgh Cancer Institute, Pittsburgh, PA*

TABLE OF CONTENTS

17.1 INTRODUCTION

The bodily fluids of patients can provide extensive information with respect to the presence or absence of diseases. In the case of neoplastic diseases, one of the major sources of the information in bodily fluids is from tumor cells that are dying and undergoing lysis at a rate that typically is much higher than the rate of death and lysis of normal cells. Also, the locations of the cells undergoing lysis are important because where the contents of dying cells are released affects the uptake of the cellular contents by the vascular-lymphatic system. In most cases, dying malignant cells leak their contents into the interstitial space and subsequently these contents are picked up together with interstitial fluids by the vascular-lymphatic system. In contrast, many normal cells that have a high rate of death (e.g., colorectal epithelial cells) frequently release their contents into lumina of organs; thus, these contents may not be absorbed directly into the vascular system.

One of the best indications of the selective uptake of the contents of tumor cells into the vascular system/urine includes the presence, in serum/urine of cancer patients, of mutated forms of DNAs, of differential levels of oncofetal proteins or products of proto-oncogenes and of auto-antibodies to the products of tumor cells, as well as the development of autoimmune paraneoplastic syndromes (Woolas *et al.*, 1995; Myers *et al.*, 1996; Preuss *et al.*, 2002; Scanlan *et al.*, 2002; Mintz *et al.*, 2003; Wang *et al.*, 2004).

Patients with prostatic adenocarcinomas in whom the prostatic specific antigen (PSA) levels in their sera are higher than the PSA in the sera of many individuals without cancer represent a case in point. This is the basis for the use of PSA as a screening test for prostate cancer. Of importance is that PSA is *not* a tumor-specific molecule, but actually is expressed more strongly in normal prostatic cells than in tumor cells of prostate cancer. The PSA is higher in patients with cancer than in individuals without cancer because dying cells release PSA into the interstitial space and the PSA is picked up by the blood-lymphatic systems; in contrast, dying as well as viable normal and benign prostatic cells release PSA into glandular lumina and thus PSA is contained within the prostatic ducts and leaves the body with the ejaculate (see Fig. 17.1a). In patients with benign prostatic hyperplasia (BPH), PSA also may be elevated in sera; this is likely due to blockage of prostate ducts by the nodules of hyperplasia or concretions of prostatic products and subsequent leakage of the PSA through the walls of partially blocked or inflamed ducts (see Fig. 17.1b).

A

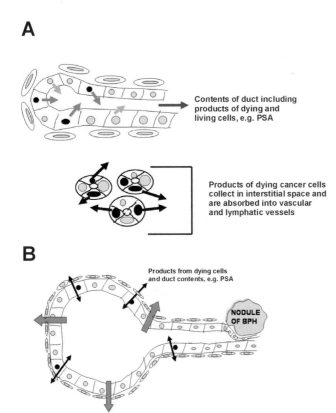

Contents of duct including products of dying and living cells, e.g. PSA

Products of dying cancer cells collect in interstitial space and are absorbed into vascular and lymphatic vessels

B

Products from dying cells and duct contents, e.g. PSA

NODULE OF BPH

FIGURE 17.1 A. Schematic drawing indicating the way that tumors affect the patterns of proteins/protein metabolites detected in bodily fluids and the reason that increased levels of PSA in serum may serve as a molecular marker of prostatic adenocarcinoma. B. Increased levels of PSA may be caused by benign prostatic hyperplasia and similar conditions.

As cancerous cells die, they release not only proteins specific for cancers (e.g., oncofetal tumor markers such as carcinoembryonic antigen (CEA)), but also increased levels of cellular proteins in general as well as proteins characteristic of the cells of origin from which the tumor developed "tissue type proteins" (e.g., PSA). Similarly, metabolites of the cellular proteins are released or are produced proteolytically in blood. From the standpoint of metabolites of proteins released from tumors, the actual pattern of local metabolic cleavage may aid in the diagnosis of disease. Thus, the blood, urine, and other bodily fluids of patients with cancer should be expected to have an increased input of all types of proteins and other molecules from cancer cells. As is the case of PSA, this increased input of proteins into blood can be useful in early detection of disease even if the protein is usually present in normal cells.

Thus, in patients with cancer, it would be expected that all proteins and their metabolites directly released from tumor cells would be selectively and differentially increased. Because cancer cells may contain many oncoproteins and tissue type proteins, and each oncoprotein and

tissue type protein may have multiple metabolites, the mass spectra of proteins from blood that are characteristic of a tumor may be very complex, may spread throughout a large range of masses ($\leq 200,000$ kDa), and be composed of hundreds of informative peaks (Adam *et al.*, 2002; Sorace and Zhan, 2003); however, even some informative peaks may provide redundant information.

In patients with tumors, another source, which can produce an atypical pattern of proteins/peptides/metabolites in the bodily fluids, is the area surrounding deposits of tumor. Molecules produced by such areas can be considered to be components of an "epi-tumor" reaction. Thus, the cells of the tumor as well as the products the tumor secreted into the interstitium act to create an inflammatory and/or other reactive environment in the vicinity of the tumor. Proteins and peptides of this potentially unique reactive area such as cytokines, along with their metabolites may be absorbed in lymphatics and capillaries and ultimately enter the blood circulation. Each classification or subtype of tumor may create both unique and "in-common" products, where in-common products would include factors produced by many subtypes of cancers and/or even by local inflammatory processes unassociated with cancers. Such markers and their metabolites may include cytokines, acute phase reactants, and growth factors associated with repair. In contrast, unique products would be created by a reaction to some unique tumor product or an environment created uniquely by the tumor (e.g., desmoplasia).

Alternatively, molecules secreted or leaked by the tumor or molecules induced by the environment of the tumor are collected in the bloodstream and might induce or inhibit the secretion of other molecules from distant organs/tissues. A similar situation might occur when the body senses illness and enters into a modified metabolic state. An example would be the sick thyroid syndrome or starvation, in which the metabolic rates of certain individuals are reduced greatly in response to these chronic conditions. In such a case, a protein/peptide/molecule may be present in the blood of patients without tumors but may be absent in the spectra of a patient with a tumor. Therefore, molecules and their metabolites that may be useful in separating cancer from noncancer may be numerous, complex, and come from many sources (see Figs. 17.2 and 17.3).

In addition to blood, other bodily fluids may contain molecular features that separate patients with a specific disease process from patients who do not have that disease. As would be expected, other bodily fluids that are filtrates of blood/interstitial fluids such as urine/saliva would be expected to have a characteristic pattern of smaller molecules that are characteristic of blood from a patient with the disease. Other bodily materials such as feces in the colorectum also might be expected to have specific protein patterns or protein fingerprints or signatures characteristic of disease of the gastrointestinal system; similarly, vaginal fluids/mucous might carry a pattern of proteins characteris-

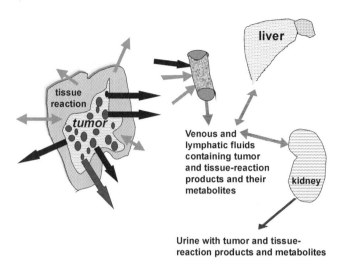

FIGURE 17.2 Modulation of the proteomic spectra of serum obtained from patients with specific types of cancer, by tumors and their epi-phenomenon.

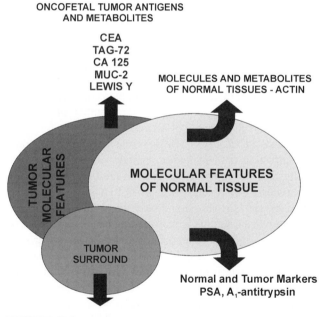

FIGURE 17.3 Contribution of different tissues, both normal and neoplastic, to the proteomic spectra, which separate disease from nondisease.

tic of diseases of the ovary, endometrium, cervix, or vagina (Miller *et al.*, 2003; Chhieng *et al.*, 2003).

Molecular markers in blood that have been used to identify patients with specific cancers have been limited to those proteins known to be overexpressed. The detection of mutations in p53, APC, and K-ras in fecal products has previously been reported to be useful in the detection of colorectal neoplasia (Sidransky *et al.*, 1992; Caldas *et al.*, 1994). With the advent of high throughput methods to screen complex bodily fluids/substances in order to identify molecular

patterns characteristic of specific diseases, new approaches have been developed for the early detection of disease based upon protein patterns or fingerprints found in bodily fluids.

17.2 MASS SPECTROGRAPHIC APPROACHES TO IDENTIFY PROTEIN PATTERNS ASSOCIATED WITH DISEASES

Surface enhanced laser desorption/ionization time of flight mass spectroscopy (SELDI-TOF-MS) is a method that can be used to identify patterns of peptides/proteins (PPs) in complex mixtures of PPs and other molecular species and related molecules based on the biochemical characteristics of molecules of interest and by the molecular weights of these molecules. Typical complex mixtures of PPs and other molecules that may be analyzed by SELDI-TOF-MS include serum, plasma, urine, saliva, effusions, and extracts of feces, tissue, mucous, and cellular preparations including urinary sediment, PAP smears, fine needle aspirations, sputum, and cerebrospinal fluid. PPs characteristics important in identification of peaks for specific PPs by SELDI-TOF-MS include both the molecular weights of the PPs and the binding of the PPs to specific surfaces.

17.2.1 Principle and Design of SELDI-TOF-MS

The SELDI binding surfaces are designed and constructed on metal chips using surface biochemical characteristics to which specific types of PPs bind selectively. An example of such a binding surface is one that binds selectively PPs whose primary structures are mostly hydrophobic (i.e., the C16 hydrophobic interaction chip from Ciphergen, Freemont, CA). Ciphergen constructs these binding areas as "2 mm diameter wells" or "spots" on metal Protein Chip Arrays (PCA) and each currently marketed PCA has either 8 or 16 spots; 1 to 16 individual aliquots of samples or controls can be applied to a PCA (see Fig. 17.4). The binding surfaces of the C16 chip allow for reverse-phase type interactions based on the hydrophobic/hydrophilic properties of PPs and subsequent washing removes unreacted molecules. In theory, after the binding step, the C16 surface can be washed selectively to release differentially PPs based on hydrophobicity. Other PCAs contain neutral, ionic, metal affinity, or other biochemical environments permitting differential binding to spots of a wide range of proteins. Even specific antibodies as well as other bait or receptor molecules can be attached to spots permitting ligand-receptor assays to be performed with the now-customized PCAs (Xiao *et al.*, 2001).

Samples should be analyzed using duplicates or triplicates and aliquots may be applied as undiluted samples or after dilution of each aliquot with, for example, normal physiological saline. For some chips, dilution of aliquots may be very important so the effects of dilution on spectral

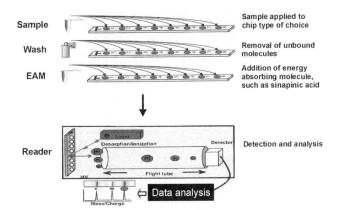

FIGURE 17.4 Illustration of SELDI-TOF Mass Spectrometry. The sample is applied onto a protein chip (which can be hydrophobic, anionic, cationic, or metal binding). Subsequently, the chip is washed to remove all contaminants, unbound proteins, and salt, and finally, the energy-absorbing matrix is added. Following its preparation, the metal chip is loaded into the protein chip reader, which consists of a pulsed nitrogen laser and a time of flight mass spectrophotometer as a detection system. The output of the reader is displayed on a computer output as a peak with a height that is proportional to the total number of molecules hitting the detector per unit time and hence per mass/charge unit (see Section 17.2 for details).

consistency including peak height should be determined before running the actual samples. For example, if fluid samples from extracted tissues are to be run, several dilutions of an equivalent extraction should be tested (e.g., undiluted, and dilutions of 1/2, 1/5, 1/10, 1/15, 1/25, 1/50, and 1/100 should be tested). Each diluted sample should be run in eight identical aliquots, and the peak heights, mass accuracy, and peak resolutions of the spectra should be evaluated. For example, for aliquots of plasma applied to a WCX_2 chip, a 1/10 dilution produces stable spectra but spectra from an undiluted sample of plasma may not be stable.

During an assay, the PPs are released from the binding surface by the addition of an energy-absorbing matrix (EAM) that incorporates proteins bound to the surface of the spot into an energy absorbing crystal matrix; this matrix continues to be localized in the spot of the PCA. The choice of the energy-absorbing molecule varies depending on the molecular weight/charge distribution of the PPs to be analyzed. For PPs in the range of 2 to 20 kDa, α-cyano-4-hydroxycinnamic acid frequently is used. Of note is that spectral peaks, of less than 2 kDa, may arise from components of EAM or from nonprotein molecules or may represent output variance; therefore, evaluation of spectral events in the molecular range of less than 2 kDa may be problematic.

17.2.2 Parameters Affecting Sensitivity

The sensitivity of the SELDI system requires that errors due to preparing the sample for analysis and applying the sample

and EAM to the chip be minimized. In the authors' experience, this requires that a robotic sample processor such as the Beckman Biomek 2000 be coupled to the bioreactor of the SELDI system. Such robotic processing has been compared to hand processing by very experienced technologists at the University of Alabama at Birmingham and has been shown to be greatly superior.

For proper discrimination of disease versus nondisease, the system should be calibrated at least weekly using a group of known PPs that react with the binding surface of the spot being utilized and that are appropriate for the molecular weights being studied and matrix being used. For lower molecular weights (2 to 20 kDa), the use of such calibrators results in a mass accuracy of about ±0.2% for the subsequent assays in this molecular weight range. Also, a control (e.g., a standard sample of a well-characterized serum) should be analyzed weekly and should be run on each chip at a random or nonbiased spot location. In such controls, standard peaks are evaluated as to their amplitude, mass accuracy, and resolution as an index of the assay being run. Most samples are run in triplicate and it is critical that each of the triplicates be loaded on spots of multiple chips in a nonbiased pattern. Specifically, to minimize bias, triplicate should be loaded on a different location and chip and each triplicate should be run on a different day. Obviously, the cases of disease should be randomly mixed with nondisease to avoid biases associated with chips or assays. Also, samples coming from different sites should be such that the number of controls approximates the number of disease cases.

Because SELDI-TOF-MS is so sensitive, various conditions affecting patients and/or samples may affect SELDI results. For example, after four freeze-thaw cycles, the amplitude of prominent peaks in SELDI spectra begin to decrease (see Figs. 17.5a and b). Changes also may occur between when a sample is drawn and when a component of the sample such as serum is frozen. Changes in protein spectra on storage of samples of serum for over five months in a −20°C nonself-defrost freezer also have been documented (see Fig. 17.6). Because sample processing and storage conditions may be variable between sites and between archival collections of sera, SELDI may identify differences between sites or archived storage collections. Similarly, different spectra may be obtained from different populations of patients because of different mixtures of races and ethnic groups. Therefore, it is critical that specimens be analyzed so that biases are not introduced. One approach would be if 10 disease specimens come from site A, then 10 nondisease specimens should also come from the same site. For example, if 10 samples of disease in Hispanics from site A are selected, then 10 samples from Hispanics with nondisease should be selected from site A.

Proteins and peptides in the blood may vary substantially with many patient variables including diurnal and postprandial variations. Changes in peptides also may occur secondary to dietary content, stress, and various

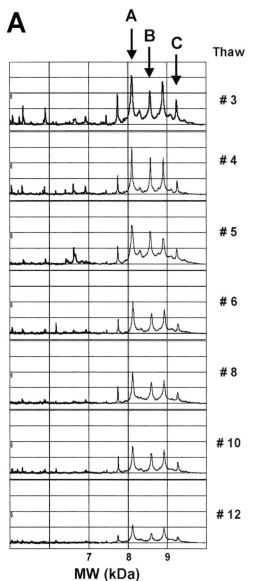

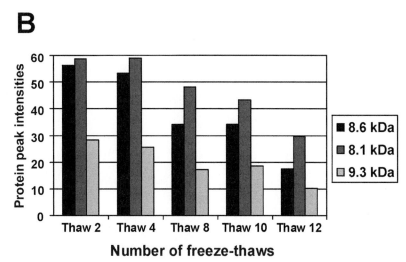

FIGURE 17.5 A. Decrease in protein intensity in human serum due to repetitive freeze-thaws. The spectra display the decrease in protein intensities in three different proteins, labeled A (8.1 kDa), B (8.6 kDa), and C (9.3 kDa) in one human serum sample due to repetitive freeze-thaw processes (indicated as a number next to each panel). Observe the decline in the three peaks (indicated with the arrows), as they are freeze-thawed repeatedly. B. Graph indicating the decrease in protein peak intensities in the human serum sample demonstrated in Fig. 17.5a.

pathophysiological states (e.g., local infections). Such changes will act as noise in SELDI-TOF-MS analysis unless a specific bias is introduced between disease and control groups. For example, if samples of blood from the disease group were drawn randomly with respect to eating but the control group consisted of a group of samples drawn after a glucose challenge of a diabetic workup, a marked bias would be introduced.

17.2.3 Detection and Data Analysis

Following its preparation, the metal chip is loaded into the Protein Chip Reader, which consists of a pulsed nitrogen laser and a time of flight mass spectrophotometer as a detection system (see Fig. 17.4). The laser is fired in multiple

single pulses and these pulses interact with the EAM and metal chip to ionize the PPs and the matrix and components of both are released in the time of flight mass spectrometer. The mass/charge ratio of the PP controls the speed at which the ionized PPs move through the MS chamber with smaller PPs of the same charge moving faster than larger PPs. Each molecule with the same mass/charge ratio moves at the same rate and strikes the detector at approximately the same time. The output is measured as an ion current that varies in magnitude with the total number of molecules with the same specific mass/charge ratio that strike the detector concomitantly. Thus, the intensity of the pulse is proportional to the number of molecules of a type that are ionized, the proportion of molecules of the same type bound to the spot, and factors that affect binding to the spot (e.g., other molecules that may partially block binding or that may bind the molecule of interest and make its ionization less efficient) and factors that affect the release of the PP from the crystal matrix of the EAM.

The output of the reader (ion current and time of flight) is collected by software and displayed on a computer output as a peak that has a height that is proportional to the total number of molecules hitting the detector per unit time and hence per mass/charge unit. For analysis, the spectra at the same molecular weight ranges from multiple laser pulses on

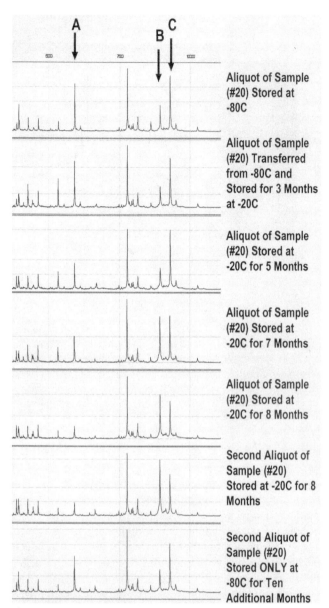

FIGURE 17.6 Change of the protein spectra in serum upon storage at −20°C (nonself-defrost freezer). Observe that the ratio of peaks A-to-B-to-C and A-to-C after storage at −20°C are relatively stable after three (2nd spectra) and five (3rd spectra) months, but the ratios change markedly after seven (4th spectra) and eight (5th and 6th spectra) months of storage at −20°C. In contrast, the spectra of samples stored at −80°C for 10 months (bottom spectra) are very similar to the initial aliquot (top spectra) stored for only a few weeks at −80°C. These data suggest that the proteomic spectra may change on long-term storage at −20°C. Although the samples at −80°C appear stable after 10 months of storage, it is likely that changes would begin to occur after several years.

the same specimen spot are combined. For most studies, the spectra from up to 50 individual laser pulses hitting on one sample spot on a chip are combined to form one spectrum to reduce noise. Molecules with very similar molecular weights and charges, which bind to the same spots on a chip,

usually are detected as shoulders on peaks or peaks that are wider. However, as with other proteomic methods (see also next chapter), the peaks of some proteins, especially those that are present in large amounts, may mask proteins present in smaller amounts.

It is best to collect and analyze separately spectra within limited molecular weight ranges (i.e., 2 to 10 kDa, 10 to 15 kDa, 15 to 30 kDa, 30 to 100 kDa, and 100 to 200 kDa). For each weight range a specific EAM is chosen and molecular weight standards should be run using the matrix selected for that molecular weight range. Thus, triplicates of the same sample should be run separately for each EAM and calibrated area. SELDI-TOF-MS currently works best for identifying PPs at molecular weights of less than 100 kDa and its detection of PPs deteriorates with increasing molecular weights up to 200 kDa, beyond which this method currently has not been demonstrated to be very useful.

Once spectra have been collected, they may be analyzed using several methods. First, just like gene chip arrays or protein arrays, data analysis is done primarily by comparing two or more different conditions or disease states (e.g., the sera of patients with epithelial ovarian cancers versus patients without ovarian epithelial cancers, versus patients with benign epithelial masses, etc.). Thus, key questions include the following:

1. Are the protein patterns in serum different among these groups?
2. Can the spectra separate reproducibly patients with these conditions?
3. Can the method detect low-stage tumors in addition to high-stage tumors?
4. Can low-grade tumors be separated from high-grade tumors?
5. What are the proteins making up the differential spectra?
6. Which peaks are independent?
7. Which spectra from specific types of lesions should be grouped together in the analysis of cancer versus no cancer? For example, in ovarian cancer, studies have grouped spectra from all types of ovarian epithelial cancers together (e.g., ovarian serous adenocarcinomas and ovarian endometrioid adenocarcinomas). This may change in future assays.

Analytical approaches to the analysis of SELDI-TOF-MS have not been developed fully but two somewhat different approaches have been utilized. The first approach, directed, focuses on identifying specific peaks and their amplitudes in a limited spectrum of molecular weight ranges associated with patients who do not have a cancer and identifying differences in spectral peaks in patients who have a specific cancer (e.g., prostatic adenocarcinoma). A decision tree approach is used frequently in such an approach and there usually will be a consistent difference between cancer and noncancer patients of 10 or fewer peaks in a selected molecular weight range (e.g., 2 to 20 kDa) spectrum. The spectra of patients with prostate cancer (PCa) may differ from the spectra of men with benign prostatic hyperplasia (BPH), either by the presence or absence of specific peaks in their spectra.

In all approaches of SELDI-TOF-MS, a learning set with and without the diseases or conditions of interest is used to identify the most useful peaks to separate the disease and nondisease state. For example, using an unblended set of specimens, eight protein peaks in the weight range 2 to 30 kDa may be identified by multiple interactions that separate normal individuals from disease A and normal individuals from disease B, and these same peaks separate disease A from disease B. These eight peaks are now utilized in analysis of a test set to assess whether these peaks are effective in separating diseases A and B from nondisease and disease A from disease B. In this test analysis, samples are analyzed blindly and the test set is run only once.

One of the potential problems with the analysis relying on this approach is the relatively low resolution of the time of flight mass spectroscopy system bundled by Ciphergen with the SELDI system. Between assays, the same protein in one run may be at 1,999 Da and in another run it may be at 2,002 Da. Thus, associating specific peaks to a control or diseased population may be problematic, especially if multiple peaks are present in the ±0.2% range. Thus, extensive effort is devoted to first identifying what constitutes a peak, subtracting the baseline and matching peaks so that all peaks for a specific protein are matched and single peaks within approximately ±0.2% mass are considered one peak for purposes of analysis.

Another approach to analysis, undirected, does not concentrate on identifying and matching specific spectral peaks; instead the amplitude of the spectra at each mass/charge (M/Z) ratio is determined for all areas of the spectra (e.g., 2 to 20 kDa). These values are plotted and analyzed in multidimensional space and clusters of M/Z spectral values of patients with disease A and separate clusters of M/Z and spectral amplitudes of patients without disease A are identified. An unblinded learning set is analyzed by interaction in multidimensional space and areas of space are identified that can be used to identify patients with disease A or without disease A based on groupings of M/Z ratio versus amplitude. A blinded test set is then analyzed once using the same methods. A disadvantage of this method is that it is very sensitive to variations of peak location.

17.3 THEORETICAL CONSIDERATIONS OF SELDI-TOF-MS

SELDI technology relies on differential absorption of proteins/peptides and other molecules from complex mixtures of these molecules, such as in a sample of serum. Each SELDI sample spot is limited in its ability to bind specific molecules. Thus, if protein A with a molecular weight of 20 kDa and a concentration of 10,000 Ci binds avidly to a sample spot of a chip, then protein B of molecular weight of 10 kDa and concentration of 10 Ci, which binds similarly but less avidly, is unlikely to be bound because all binding sites may be occupied by protein A.

Similarly, if peptide C with molecular weight of 5,000 Da and concentration of 1,000 Ci binds to a spot on a chip and protein D with a molecular weight of 5,005 Da and concentration 10 Ci binds similarly, it is unlikely that protein D will be detected. First, given that it ionizes similarly to peptide C, the peak of peptide D will fall under the peak at 5,000 Da and thus may not be detected by the SELDI reader, which has a resolution of 5,000 ± 10 Da. However, such a peak may be identified if the SELDI is coupled to a mass spectrometer of high resolution (e.g., Q STAR). Also, some PPs released from tumors or from areas of the epitumor reaction may be present in too low a concentration to be detected by the SELDI system. The SELDI system tends to detect proteins present in relatively large concentrations.

When the laser is focused on one tiny area of a spot as target for the laser (usually divided into 100 laser targets per spot), the laser will not yield many ionized molecules after more than 50 laser pulses. Thus, not all molecules initially bound to the spot may be ionized by the laser. This brings into question whether or not some molecular species may be bound so strongly that only a small fraction is ionized by the laser and hence is available for analysis. Similarly, molecules may be ionized with components of the EAM attached. Thus, because other molecules and their concentrations may affect the binding of any other specific molecule and once bound the specific molecule may not be freely ionized with laser pulses, the SELDI-TOF-MS as currently used is not a quantitative instrument. However, if the basic concentration of most molecules does not change greatly in a series of samples to be quantitated and molecular traps, such as an antibody are used, semi-quantitative assays may be possible (Xiao *et al.*, 2001).

Tissues may be very heterogeneous so proteins/peptides may vary with site of the tissue removed for analysis. This may be corrected by combining multiple aliquots of the tissue into a single sample. Similarly, bringing specific components of a tissue into solution (e.g., membrane proteins) may add additional problems.

Thus, there are multiple potential problems that may occur in assays using SELDI-TOF-MS of which users and those analyzing SELDI data should be aware. These include the following:

• Low molecular weight peaks (<2 kDa) may not be proteins, but may be components of the crystal matrix, especially the energy absorbing molecule, complex lipids, other molecules such as plastics from sample containers, or instrumental noise.

- Proteins with strong affinities to specific surfaces that are present in patients in large amounts may saturate binding at spots.
- Proteins present in large amounts such as albumin, immunoglobulins, acute phase reactants, and so on, may block the display and subsequent identification of proteins with similar M/Z ratios.
- Some proteins, which are present at very low concentrations, may not be detected by this technology. In order to successfully display other proteins present in moderate concentrations, proteins present at high concentrations (e.g., albumin) may have to be removed from the sample prior to analysis. However, removing binding proteins, such as albumin, may remove informative peptides carried by the binding protein (Mehta *et al.*, 2003–2004).
- SELDI-TOF-MS is very sensitive to variables in sample processing and robotics is required for consistent assays. The robotics adds considerably to the cost of the SELDI-TOF-MS systems.

There have been several criticisms of prior studies of the early detection of cancers using SELDI-TOF-MS. In a recent point-counterpoint discussion (Diamandis, 2003; Petricoin and Liotta, 2003), two arguments were advanced against SELDI-TOF-MS studies of the early detection of PCa. One issue raised was that the SELDI-TOF-MS did not detect PSA as one of the informative peaks in the SELDI-TOF-MS spectra, probably because it is present in a low concentration (Diamandis, 2004). In fact, PSA was not detected in most spectra, as has been the case for other tumor markers in the early detection of other cancers (CA125 in ovarian cancer). It may be that PSA does not interact efficiently with the chemistries of the chips that were used and/or its low concentration may prevent its detection due to the sensitivity of the method. This and related issues of sensitivity deserve additional study. Another major issue was the failure of different groups studying the early detection of PCa to identify the same informative peaks; that is, none of the peaks identified by Petricoin and colleagues (2002) were identified by Adam and colleagues (2002) as being useful in the early detection of PCa.

The fact that different laboratories using dramatically different chip chemistries and assay methods did not identify the same peaks actually is not surprising. For example, consider a case in which there were at least 200 peaks separating PCa from non-PCa. If eight peaks were selected by laboratory A for early detection of PCa, the likelihood of even one of the same eight peaks being identified by laboratory A on a slightly different analysis would be less than 30% (Qu *et al.*, 2002). Of relevance is that Sorace and Zhan (2003) on analysis of the data of Petricoin and colleagues on the early detection of ovarian cancer identified over 450 peaks in the molecular weight range 2,000 to 14,125 Da, which separated ovarian cancer from nonovarian cancer with P values of 10^{-6} or less. Sorace and Zhan (2003) also

identified a spectral peak of very small molecular weight, 2.79 Da, which separated ovarian cancers from noncancer. They proposed that this peak was nonbiologic, and its identification was reported to be indicative of a bias in the set of data. Once more, it should be noted that the molecular weights of peaks below 2 kDa are inaccurate due to an absence of standardization in this area of molecular weights. However, as discussed, there must be a vigorous design of all SELDI-TOF-MS studies both in sample selection and the placement of aliquots of samples in the analysis. Indeed, methodological analysis of any new analytical method such as SELDI-TOF-MS is warranted to uncover sources of variability.

17.4 APPLICATION OF SELDI-TOF-MS IN THE EARLY DETECTION OF CANCER

17.4.1 Early Detection of Prostate Cancer

Efficient clinical management of prostate cancer can be achieved through its detection at its earliest and most curable stages. Although a number of biomarkers for the detection of PCa have been discovered over the past few years, none of them have been clinically validated. Even PSA, which is routinely used in screening of men for PCa, has not been validated in a prospective longitudinal clinical study (Hankey *et al.*, 1999). Indeed, there is growing concern that PSA does not fill the clinical need for a diagnostic biomarker (Wirth *et al.*, 1993; Birtle *et al.*, 2003; Platz *et al.*, 2004). This limited utility of even the most highly touted single protein biomarker underscores failure of single markers and the need for multiprotein biomarker panels. Therefore, there is an increasing emphasis on rapid high throughput multiparametric discovery tools for identification of biomarkers for diagnostics. SELDI-TOF has provided an excellent platform for analysis of complex biological mixtures of proteins. In this section, the major contributions to the application of SELDI to detection and diagnosis of PCa will be discussed.

The precedent work in this field was reported by the authors' laboratories, demonstrating the successful discrimination of PCa (early T1/T2 and advanced T3/T4), benign prostatic hyperplasia (BPH), and healthy age-matched male controls (Adam *et al.*, 2002; Qu *et al.*, 2002). In this study, sera were analyzed on copper activated IMAC 3 ProteinChips. A set of 167 PCa, 77 BPH, and 82 normal individuals were used to train and develop a classification algorithm. This classification tool used seven specific protein peaks from a total of 124 differentially expressed peaks to classify each group. The algorithm then was challenged with a blinded test set of sera from 30 PCa, 15 BPH, and 15 healthy men. A sensitivity of 83% and specificity of 97% were retained through a validated training and test set. A second study by Petricoin and colleagues (2002a) confirmed these earlier results by using SELDI analysis on a separate study cohort and employing significantly different protocols such as alternative chip types and classification approaches. This group obtained sensitivity of 95% in a blinded set of 38 patients with prostate cancer, and a specificity of 78% in a blinded set of 228 patients with benign conditions. Subsequently, a third group also has reported the successful discrimination of PCa with SELDI (Banez *et al.*, 2003). By combining the data obtained from two kinds of ProteinChip arrays, the group achieved 85% sensitivity and 85% specificity for the detection of PCa. Thus, studies carried out by separate groups support the utility of this technique in PCa diagnostics, although each of these groups used different SELDI-based methods and different computational approaches for analysis of the output data. The development of robust classification algorithms and indeed an optimal computational approach is a major effort in our groups (Adam *et al.*, 2001; Qu *et al.*, 2002). The promise of SELDI-based diagnostics for PCa currently is being evaluated and validated by the NCI/EDRN in a multi-institutional trial (Grizzle *et al.*, 2004).

Thus, SELDI-based protein profiling promises to fulfill the need for multiparametric, highly accurate, and sensitive molecular diagnostic tools. The development of this platform and the introduction of novel technologies that can continue down the protein profiling trail may provide significant improvement to clinical decision making.

17.4.2 Detection of Breast Cancer

Two large studies have applied SELDI-TOF-MS serum expression profiling to identify cancers of the breast. Li and colleagues (2002) described the results of a SELDI-TOF-MS profiling study utilizing Ciphergen IMAC3-Ni ProteinChip® arrays for the analysis of 169 serum samples comprising 103 samples from breast cancer patients; 25 samples from women with benign breast diseases; and 41 samples from healthy control women. They reported a classification performance of 93% sensitivity for all cancer patients (stage 0-III), and a specificity of 91% for all non-cancer controls. Vlahou and colleagues (2003) utilizing IMAC3-Cu and WCX$_2$ ProteinChip® arrays analyzed 134 pretreatment serum samples from 42 women with breast cancer, 42 with benign breast diseases, and 47 healthy controls. When profiles from both chip surfaces were included in the analysis, results of 90% sensitivity and 93% specificity were reported. Neither of these studies of breast cancer reported that they were validated against an independent test set. Other studies have studied nipple aspirates using SELDI (Pawletz *et al.*, 2001).

17.4.3 Identification of Patients with Head and Neck Cancer

Wadsworth and colleagues (2004) reported the SELDI-TOF-MS analysis of serum samples from 99 patients with squa-

mous cell carcinoma of the head and neck (SCCHN) together with samples from 25 smoking and 102 nonsmoking controls. IMAC3-Cu ProteinChip® profiles yielded a sensitivity of 83% and a specificity of 90% in discriminating SCCHN patients and the combined controls. In addition, one of the candidate diagnostic features, with a M/Z ratio of 10,068 Da, was provisionally identified as a known biomarker, metallopanstimulin-1 (MPS-1).

17.4.4 Detection of Renal Cancer

Won and colleagues (2003) described a SELDI-TOF-MS WCX$_2$ ProteinChip® array study of 36 serum samples, 15 drawn from renal cell carcinoma (RCC) patients prior to radical nephrectomy, 15 from patients with other benign urologic diseases, and 6 from healthy controls. Correct overall classification of 30/35 profiles included 13/15 of the RCC samples (sensitivity 87%), 12/15 of the nonRCC cases, and 5/5 of the healthy controls (specificity 85%).

17.4.5 Proteomic Profiling of Hepatocellular Carcinomas

Utilizing initial anion exchange chromatography followed by IMAC3-Cu and WCX$_2$ ProteinChip arrays, Poon and colleagues (2003) used SELDI-TOF-MS in the serum profiling of 38 pretreatment samples from patients with histologically confirmed hepatocellular carcinoma (HCC) and 20 samples from patients with chronic liver disease. The SELDI-TOF-MS discrimination between HCC and CLD proteomic profiles resulted in detection of HCC with a sensitivity of 92% and specificity of 90%.

17.4.6 Detection of Lung Cancer

SELDI-TOF-MS profiling on WCX$_2$ ProteinChip® arrays of serum samples from 30 lung cancer patients and 51 age- and sex-matched healthy controls was reported by Xiao and colleagues (2004). Based on a training set, the SELDI-TOF-MS analysis of blinded serum samples from 15 lung cancer patients and 31 controls in a test set yielded a sensitivity of 93% and a specificity of 97%.

17.4.7 Identification of Ovarian Cancer

A serum profiling study of ovarian cancer by Petricoin and colleagues (2002b) using C16 hydrophobic ProteinChip® arrays and SELDI-TOF-MS generated proteomic profiles in a training set of serum samples from 50 patients with ovarian cancer and 50 unaffected women. The classifier developed was used to analyze an independent set of 116 blinded serum samples, 50 from women with ovarian cancer and 66 from unaffected women. Analysis of the blinded test set yielded a correct classification of 46/49 samples from unaffected women and 16/17 samples from women with benign gynecologic disease and nongynecologic inflammatory disease. Remarkably, all 50 of the ovarian cancer case profiles were classified correctly, including 18/18 from women with stage I disease.

Kozak and colleagues (2003) analyzed 109 serum samples from patients with ovarian cancer, 19 from patients with benign tumors, and 56 from healthy women using SAX2 ProteinChip® arrays and SELDI-TOF-MS. A training set of 140 samples yielded three biomarker panels, comprising a total of 14 candidate features with M/Z values ranging from 4.4 to 106.7 kDa. These features correctly classified 21/22 as malignant ovarian neoplasias (10/11 early-stage (I/II)), 6/6 low malignant potential (LMP), 5/6 benign tumors, and 9/10 normal control.

17.4.8 Serum Diagnosis of Pancreatic Cancer

Koopmann and colleagues (2004) used IMAC3-Cu and WCX$_2$ ProteinChip® arrays and SELDI-TOF-MS following anion exchange fractionation to profile preoperative serum samples from 60 patients with resectable adenocarcinoma of the pancreas, 60 samples from age- and sex-matched patients with benign pancreatic diseases including pancreatitis (N = 26), and 60 matched healthy controls. Using the two most significant features identified from the WCX$_2$ profiles with m/z of 3,146 and 12,861 Da, pancreatic cancer patients and controls were discriminated with a sensitivity of 78% and specificity of 97%; when combined with CA19-9, a small improvement of classification accuracy was obtained indicating that the SELDI-identified biomarkers and CA19-9 provided some complementary diagnostic information.

17.5 SUMMARY

The use of SELDI-TOF-MS in the analysis of bodily fluids and solid tissues other than human serum is in its early stages and is beyond the scope of this chapter. However, we note that SELDI-TOF-MS has been used in multiple applications including in the separation of malignant from benign effusions (Chhieng *et al.*, 2003), in the analysis of cervical mucus to separate normal patients from patients with cervical dysplasia (Miller *et al.*, 2003), in the analysis of archival cytologic material (Fetsch *et al.*, 2002), and secretions from cell lines (Diamond *et al.*, 2003; Shiwa *et al.*, 2003).

In these studies the approaches to developing algorithms has varied greatly as have the methods of identification and alignment of peaks and the statistical approaches to separating informative from noninformative peaks. The issues in mathematical/statistical analysis of SELDI-TOF-MS as well as related data (SELDI/QSTAR–TOF-MS and MALDI-TOF-MS) are complex and controversial (Ball *et al.* 2002; Zhu *et al.*, 2003; Baggerly *et al.*, 2003; 2004; Coombes

et al., 2003; 2004). Discussions of these issues are beyond the scope of this chapter.

In summary, the general use of SELDI-TOF-MS in the early detection of disease has proved to be very encouraging. As with any type of new and novel technology, there are many controversies including the use of low resolution mass spectroscopy, issues related to the chromatographic surface binding and laser release of high and low concentration proteins, ability to detect proteins at low concentrations, and the methods of analysis of SELDI-TOF-MS raw data. New issues with respect to SELDI-TOF-MS analysis are being identified daily and some problems (e.g., inter-laboratory standardization and identification of peptides making up peaks) have been already solved. It is therefore safe to predict that such proteomic approaches will have a great impact on the early detection of disease.

Acknowledgements

This manuscript was supported by separate grants from the Early Detection Research Network (EDRN, CA86359-04 to W.E.G., CA85067 to O.J.S., and CA84968 to W.L.B.). The authors thank Ms. Libby Chambers for her secretarial assistance.

References

Adam, B. L., Qu, Y., Davis, J. W., Ward, M. D., Clements, M. A., Cazares, L. H., Semmes, O. J., Schellhammer, P. F., Yasui, Y., Feng, Z., and Wright, G. L. Jr. (2002). Serum protein fingerprinting coupled with a pattern-matching algorithm distinguishes prostate cancer from benign prostate hyperplasia and healthy men. *Cancer Res.* 62, 3609–3614.

Adam, B. L., Viahou, A., Semmes, O. J., and Wright, G. L. Jr. (2001). Proteomic approaches to biomarker discovery in prostate and bladder cancers. *Proteomics* 1, 1264–1270.

Baggerly, K. A., Morris, J. S., Wang, J., Gold, D., Xiao, L.-C., and Coombes, K. R. (2003). A comprehensive approach to the analysis of matrix-assisted laser desorption/ionizatioin-time of flight proteomics spectra from serum samples. *Proteomics* 3, 1667–1672.

Baggerly, K. A., Morris, J. S., and Coombes, K. R. (2004). Reproducibility of SELDI-TOF protein patterns in serum: Comparing data sets from different experiments. *Bioinformatics* 20, 777–785.

Ball, G., Mian, S., Holding, F., Allibone, R. O., Lowe, J., Ali, S., Li, G., McCardle, S., Ellis, I. O., Creaser, C., and Rees, R. C. (2002). An integrated approach utilizing artificial neural networks and SELDI mass spectrometry for the classification of human tumours and rapid identification of potential biomarkers. *Bioinformatics* 18, 395–404.

Banez, L. L., Prasanna, P., Sun, L., Ali, A., Zou, A., Adam, B. L., McLeod, D. G., Moul, J. W., and Srivastava, S. (2003). Diagnostic potential of serum proteomic patterns in prostate cancer. *J. Urol.* 170, 442–446.

Beduschi, M. C., and Osterling, J. E. (1998). Percent free prostate-specific antigen: the next frontier in prostate-specific antigen testing. *Urology* 51, 98–109.

Birtle, A. J., Freeman, A., Masters, J. R., Payne, H. A., and Harland, S. J. (2003). Clinical features of patients who present with metastatic prostate carcinoma and serum prostate-specific antigen (PSA) levels <10 ng/mL: the "PSA negative" patients. *Cancer* 98, 2362–2367.

Caldas, C., Hahn, S. A., Hruban, R. H., Redston, M. S., Yeo, C. J., and Kern, S. E. (1994). Detection of K-ras mutations in the stool of patients with pancreatic adenocarcinoma and pancreatic ductal hyperplasia. *Cancer Res.* 54, 3568–3573.

Chhieng, D. C., Miller, E., Eltoum, I. A., Zhu, L., and Grizzle, W. E. (2003). Protein profile of body cavity fluid by surface enhanced laser desorption/ionization time of flight mass spectrometry (SELDI-TOF). *Mod. Pathol.* 16, 62A.

Coombes, K. R., Fritschie, H. A. Jr., Clarke, C., Chen, J.-N., Baggerly, K. A., Morris, J. S., Xiao, L-C., Hung, M.-C., and Kuterer, H. M. (2003). Quality control and peak finding for proteomics data collected from nipple aspirate fluid by surface-enhanced laser desorption and ionization. *Clin. Chem.* 49, 1615–1623.

Coombes, K. R., Tsavachidis, S., Morris, J. S. Baggerly, K. A., Hung, M.-C., and Kuerer, H. M. (2004). Improved peak detection and quantification of mass spectrometry data acquired from surface-enhanced laser desorption and ionization by denoising spectra with the undecimated discrete wavelet transform. *Biostatistics.* In press.

Diamandis, E. P. (2003). Point Proteomic Patterns in Biological Fluids: Do They Represent the Future of Cancer Diagnosis? *Clin. Chem.* 49, 1272–1275.

Diamandis, E. P. (2004). Commentary: Analysis of serum proteomic patterns for early cancer diagnosis: Drawing attention to potential problems. *J. Natl. Cancer Inst.* 96, 353–356.

Diamond, D. L., Zhang, Y., Gaiger, A., Smithgall, M., Vedvick, T. S., and Carter, D. (2003). Use of ProteinChip array surface enhanced laser desorption/ionization time-of-flight mass spectrometry (SELDI-TOF MS) to identify thymosin beta-4, a differentially secreted protein from lymphoblastoid cell lines. *J. Am. Soc. Mass Spectrom.* 14, 760–765.

Fetsch, P. A., Simone, N. L., Bryant-Greenwood, P. K., Marincola, F. M., Filie, A. C., Petricoin, III, E. F., Liotta, L. A., and Abati A. (2002). Proteomic evaluation of archival cytologic material using SELDI affinity mass spectrometry: potential for diagnostic applications. *Am. J. Clin. Path.* 118, 870–876.

Grizzle, W. E., Adam, B. L., Bigbee, W. L., Conrads, T. P., Carroll, C., Feng, Z., Izbicka, E., Jendoubi, M., Johnsey, D., Kagan, J., Leach, R. J., McCarthy, D. B., Semmes, O. J., Srivastava, S., Thompson, I. M., Thornquist, M. D., Verma, M., Zhang, Z., and Zou, Z. (2003–2004). Serum protein expression profiling for cancer detection: validation of a SELDI-based approach for prostate cancer. *Dis. Markers.* 19, 185–195.

Hankey, B. F., Feuer, E. J., Clegg, L. X., Hayes, R. B., Legler, J. M., Prorok, P. C., Ries, L. A., Merrill, R. M., and Kaplan, R. S. (1999). Cancer surveillance series: interpreting trends in prostate cancer—part I: evidence of the effects of screening in recent prostate cancer incidence, mortality, and survival rates. *J. Natl. Cancer Inst.* 91, 1017–1024.

Koopmann, J., Zhang, Z., White, N., Rosenzweig, J., Fedarko, N., Jagannath, S., Canto, M. I., Yeo, C. J., Chan, D. W., and Goggins, M. (2004). Serum diagnosis of pancreatic adenocarcinoma using surface-enhanced laser desorption and ionization mass spectrometry. *Clin. Cancer Res.* 10, 860–868.

Kozak, K. R., Amneus, M. W., Pusey, S. M., Su, F., Luong, M. N., Luong, S. A., Reddy, S. T., and Farias-Eisner, R. (2003). Identification of biomarkers for cancer using strong anion-exchange ProteinChips: Potential use in diagnosis and prognosis. *Proc. Natl. Acad. Sci. USA* 100, 12343–12348.

Li, J., Zhang, Z., Rosenzweig, J., Wang, Y. Y., and Chan, D. W. (2002). Proteomics and bioinformatics approaches for identification of serum biomarkers to detect breast cancer. *Clin. Chem.* 48, 1296–1304.

Mehta, A. I., Ross, S., Lowenthal, M. S., Fusaro, V., Fishman, D. A., Petricoin, E. F. 3rd, and Liotta, L. A. (2003–2004). Biomarker amplification by serum carrier protein binding. *Dis. Markers* 19, 1–10.

Miller, E. J., Greene, P., Chhieng, D., Partridge, E., Grizzle, W. E., and Eltoum, I. A. (2003). SELDI-TOF protein profiles of cervical mucus in normal vs. dysoplastic epithelium. Abstract #486, presented at the American Association for Cancer Research, 94th Annual Meeting, July 2003, Washington, DC.

Mintz, P. J., Kim, J., Do, K.-A., Wang, W., Zinner, R. G., Cristofanilli, Arap, M. A., Hong, W. K., Troncoso, P., Logothetis, C. L., Pasquanlini, R., and Arap, W. (2003). Fingerprinting the circulating repertoire of antibodies from cancer patients. *Nat. Biotechnol.* 21, 57–63.

Myers, R. B., Brown, D., Oelschlager, D. K., Waterbor, J. W., Marshall, M. E., Srivastava, S., Stockard, C. R., Urban, D. A., and Grizzle, W. E. (1996). Elevated serum levels of p105^{erbB-2} in patients with advanced stage prostatic adenocarcinoma. *Int. J. Cancer* 69, 398–402.

Paweletz, C. P., Trock, B., Pennanen, M., Tsangaris, T., Magnant, C., Liotta, L. A., and Petricoin, III, E. F. (2001). Proteomic patterns of nipple aspirate fluids obtained by SELDI-TOF: potential for new biomarkers to aid in the diagnosis of breast cancer. *Dis. Markers* 17, 301–307.

Petricoin, III, E. F., Ornstein, D. K., Paweletz, C. P., Ardekani, A., Hackett, P. S., Hitt, B. A., Velassco, A., Trucco, C., Wiegand, L., Wood, K., Simone, C. B., Levine, P. J., Linehan, W. M., Emmert-Buck, M. R., Steinberg, S. M., Kohn, E. C., and Liotta, L. A. (2002a). Serum proteomic patterns for detection of prostate cancer. *J. Natl. Cancer Inst.* 94, 1576–1578.

Petricoin, III, E. F., Ardekani, A. M., Hitt, B. A., Levine, P. J., Fusaro, V. A., Steinberg, S. M., Mills, G. B., Simone, C., Fishman, D. A., Kohn, E. C., and Liotta, L. A. (2002b). Use of proteomic patterns in serum to identify ovarian cancer. *Lancet* 359, 572–577.

Petricoin, III, E., and Liotta, L. A. (2003). Counterpoint: The Vision for a New Diagnostic Paradigm. *Clin. Chem.* 49, 1276–1278.

Platz, E. A., DeMarzo, A. M., and Giovannucci, E. (2004). Prostate cancer association studies: Pitfalls and solutions to cancer misclassification in the PSA era. *J. Cell. Biochem.* 91, 553–571.

Poon, T. C. W., Yip, T.-T., Chan, A. T. C., Yip, C., Yip, V., Mok, T. S. K., Lee, C. C. Y., Leung, T. W. T., Ho, S. K. W., and Johnson, P. J. (2003). Comprehensive proteomic profiling identifies serum proteomic signatures for detection of hepatocellular carcinoma and its subtypes. *Clin. Chem.* 49, 752–760.

Preuss K.-D., Zwick, C., Bormann, C., Neumann, F., and Pfreundschuh, M. (2002). Analysis of the B-cell repertoire against antigens expressed by human neoplasms. *Immunol. Rev.* 188, 43–50.

Qu, Y., Adam, B.-L., Yasui, Y., Ward, M. D., Cazares, L. H., Schellhammer, P. F., Feng, Z., Semmes, O. J., and Wright, G. L. Jr. (2002). Boosted decision tree analysis of SELDI mass spectral serum profiles discriminates prostate cancer from non-cancer patients. *Clin. Chem.* 48, 1835–1843.

Scanlan, M. J. U., Gure, A. O., Jungbluth, A. A., Old, L. J., and Chen, Y. T. (2002). Cancer/testis antigens: an expanding family of targets for cancer immunotherapy. *Immunol. Rev.* 188, 22–32.

Shiwa, M., Nishimura, Y., Wakatabe, R., Fukawa, A., Arikuni, H., Ota, H., Kato, Y., and Yamori, T. (2003). Rapid discovery and identification of a tissue-specific tumor biomarker from 39 human cancer cell lines using the SELDI ProteinChip platform. *Biochem. Biophys. Res. Commun.* 309, 18–25.

Sidransky, D., Tokino, T., Hamilton, S. R., Kinzler, K. W., Levin, B., Frost, P., and Vogelstein, B. (1992). Identification of ras oncogene mutations in the stool of patients with curable colorectal tumors. *Science* 256, 102–105.

Sorace, J. M., and Zhan, M. (2003). A data review and re-assessment of ovarian cancer serum proteomic profiling. *BMC Bioinformatics* 4, 24.

Vlahou, A., Schellhammer, P. F., Mentrinos, S., Patel, K., Kondylis, F. I., Gong, L., Nasim, S., and Wright, Jr., G. L. (2001). Development of a novel proteomic approach for the detection of transitional cell carcinoma of the bladder in urine. *Am. J. Pathol.* 158, 1491–1502.

Vlahou, A., Laronga, C., Wilson, L., Gregory, B., Fournier, K., McGaughey, D., Perry, R. R., Wright, Jr., G. L., and Semmes, O. J. (2003). A novel approach toward development of a rapid blood test for breast cancer. *Clin. Breast Cancer* 4, 203–209.

Wadsworth, J. T., Somers, K. D., Stack, B. C. Jr, Cazares, L., Malik, G., Adam, B. L., Wright, G. L. Jr., and Semmes, O. J. (2004). Identification of patients with head and neck cancer using serum protein profiles. *Arch. Otolaryngol. Head Neck Surg.* 130, 98–104.

Wang, M, Block, T. M., Steel, L, Brenner, D. E., and Su, Y. H. (2004). Preferential isolation of fragmented DNA enchances sensitivity of detecting mutation k-ras DNA in circulation. *Clin. Chem.* 50, 211–213.

Wirth, M., Manseck, A., and Heimbach, D. (1993). Value of prostate-specific antigen as a tumor marker. *Eur. Urol.* 24 (Suppl 2), 6012.

Won, Y., Song, H. J., Kang, T. W., Kim, J. J., Han, B. D., and Lee, S. W. (2003). Pattern analysis of serum proteome distinguishes renal cell carcinoma from other urologic diseases and healthy persons. *Proteomics* 3, 2310–2316.

Woolas, R., Conaway, M. R., Xu, F.-J., Jacobs, I. J., Yu, Y., Daly, L., Davies, A. P., O'Briant, K., Berchuck, A., Soper, J. T., Clarke-Pearson, D. L., Rodriguez, G., Oram, D. H., and Bast, Jr., R. C. (1995). Combinations of multiple serum markers are superior to individual assays for discriminating malignant from benign pelvic masses. *Gynecol. Oncol.* 59, 111–116.

Xiao, Z., Adam, B.-L., Cazares, L. H., Clements, M. A., Davis, J. W., Schellhammer, P. F., Dalmasso, E. A., and Wright G. L. Jr. (2001). Quantitation of serum prostate-specific membrane antigen by a novel protein biochip immunoassay discriminates benign from malignant prostate disease. *Cancer Res.* 61, 6029–6033.

Xiao, X., Liu, D., Tang, Y., Guo, F., Xia, L., Liu, J., and He, D. (2003–2004). Development of proteomic patterns for detecting lung cancer. *Dis. Markers* 19, 33–39.

Zhu, W., Wang, X., Ma, Y., Rao, M., Glimm, J., and Kovach, J. S. (2003). Detection of cancer-specific markers amid massive mass spectral data. *Proc. Natl. Acad. Sci. USA* 100, 14666–14671.

Zhukov, T. A., Johanson, R. A., Cantor, A. B., Clark, R. A., and Tockman, M. S. (2003). Discovery of distinct protein profiles specific for lung tumors and pre-malignant lung lesions by SELDI mass spectrometry. *Lung Cancer* 40, 267–279.

CHAPTER 18

The Application of Proteomics to Disease Diagnostics

SAMIR HANASH

Department of Pediatrics, University of Michigan, Ann Arbor, Michigan

TABLE OF CONTENTS

18.1 INTRODUCTION

Proteomics is particularly suited for disease diagnostics as a substantial proportion of diagnostic tests in the clinical laboratory target proteins. Nevertheless, the task of uncovering clinically useful findings from proteomic profiling represents a substantial challenge. Continued improvements in proteomics technology promise to accelerate the pace of discovery of biomarkers and of other proteins of clinical utility, which will positively influence the application of proteomics for disease diagnostics. There is a long history of profiling disease tissue using proteomics. Some three decades ago, two-dimensional polyacrylamide gel electrophoresis (2D PAGE) emerged as a separation technique capable of resolving thousands of cellular proteins in a single gel. It was idealized that 2D gel systems may be capable of displaying all cellular protein constituents simultaneously, thus providing a comprehensive catalogue of proteins that are altered in different disease states. However, the complexity of tissue and cell proteomes and the vast dynamic range of protein abundance have presented a formidable challenge for analysis that no one analytical technique, including 2D PAGE, can overcome. Nevertheless, numerous findings have emerged from disease investigations using 2D PAGE. Moreover, several innovations in two-dimensional separations, coupled with protein tagging and subproteome analysis, have emerged, which offer improved sensitivity and substantially expanded proteome coverage. These technological developments, in combination with mass spectrometry, likely will make an impact on disease related proteomic analysis. Furthermore, there is substantial interest at the present time in

implementing other nongel-based approaches for disease tissue profiling (see Table 18.1). The contributions of various proteomics technologies for disease diagnostics are reviewed in this chapter, with emphasis given in cancer.

18.2 APPLICATIONS OF 2D PAGE TO DISEASE INVESTIGATIONS

During the early years of proteomics and until relatively recently, profiling of protein expression in disease, as well as proteome profiling in general, relied primarily on the use of 2D PAGE, which later was combined with mass spectrometry (Hanash, 2001). These studies generally followed an approach in which a standard cocktail was utilized to solubilize the protein contents of an entire cell population or tissue or biological fluid, followed by separation of the protein contents of the lysate using 2D gels and visualization of the separated proteins, primarily using silver staining. Profiling of disease tissues using this approach and through the display of the most abundant cell proteins has had tangible utility, notably for disease classification, long before the use of DNA microarrays.

Leukemia investigations using proteomics date back to more than two decades ago. These studies demonstrated that leukemias, which represent a heterogeneous group of disorders, could be classified into their different subtypes using 2D PAGE. One study (Hanash *et al.*, 1986a) utilized 2D PAGE to identify polypeptide differences due to lineage between leukemia cells obtained directly from peripheral blood or bone marrow of children with acute leukemia. Among some 400 polypeptides that were analyzed, 12 were detected that could distinguish between the major subtypes of acute lymphoblastic leukemia (ALL) and between ALL and acute myelogenous leukemia (AML). Even more remarkably, polypeptide markers were detected in 2D gels that indicated a myeloid origin of blasts obtained from children who presented with acute leukemia that could not be classified by morphologic, cytochemical, or immune marker analysis (Hanash *et al.*, 1986b). Clinical studies have shown that among children with ALL, age at diagnosis is an

223

TABLE 18.1   Proteomics technologies for disease tissue profiling.

Sample preparations technologies

— Protein enrichment (fractionation, affinity capture)
— Protein tagging

Profiling technologies

Separation based	Nonseparation based
— 2D/MALDI LC/MS	— Protein microarrays — Direct MS

Protein detection strategies

— Staining/fluorescence
— Activity-based assays
— Immune-based assays

TABLE 18.2   Cancer related applications of proteomics.

Protein source	Applications
Tumor lysates	Tumor diagnosis and prognosis
Secreted proteins	Diagnostic markers in serum
Membrane proteins	Diagnostic markers and therapeutic targets
Arrayed tumor derived proteins	Diagnostics based on the presence of antibodies to tumor antigens in serum
Arrayed antibodies	Circulating cancer markers and tumor tissue profiling

important prognostic indicator (Reaman *et al.*, 1985; Crist *et al.*, 1986). Analysis of the polypeptide patterns of leukemia cells of infants and older children with ALL using 2D PAGE showed differences in polypeptide patterns between leukemia cells of infants and older children that otherwise exhibited similar immunological markers (Hanash *et al.*, 1989).

Since the early work aimed at classifying leukemias, proteins with restricted expression in different types of cancer have been identified by numerous groups, including the discovery of mutant proteins (Misek *et al.*, 2002). For example, in a study of breast, ovary, and lung tumors, 20 differentially expressed proteins were identified (Bergman *et al.*, 2000), and in other studies, polypeptides were uncovered that were associated with different histopathological features of lung cancer (Hirano *et al.*, 1995; Schmid *et al.*, 1995). In studies of bladder cancer, Celis' group has investigated more than 1,000 fresh specimens that included tumors, normal biopsies, and cystectomies. These studies have revealed several protein markers for tumor progression (Celis *et al.*, 1996a), and a marker, psoriasin, that is excreted into the urine (Celis *et al.*, 1996b). Their work has led to novel strategies for the identification of tumor heterogeneity among low-grade tumors (Celis *et al.*, 2002) as well as early metaplastic lesions (Celis *et al.*, 1999).

A large number of studies involving lung cancer have been performed in the author's laboratory. At the protein level, these studies have resulted in over 1,000 samples related to lung cancer, which have been processed using 2D gels and for which information has been recorded in the Lung Protein Database. It is of interest to contrast the contributions of DNA microarray profiling of lung cancer with proteomic profiling. In a study of lung adenocarcinomas, transcriptional profiles that predict patient survival have

been identified (Beer *et al.*, 2002). In a parallel study of lung adenocarcinoma using 2D gels, a number of proteins that were also significantly associated with patient survival were identified (Chen *et al.*, 2003). A total of 682 individual protein spots were quantified in 90 lung adenocarcinomas. Thirty-three of 46 survival-associated proteins were identified using mass spectrometry. A leave-one-out cross-validation procedure using the top 20 survival-associated proteins identified by Cox modeling, indicated that protein profiles as a whole can predict survival in stage I tumor patients (P = 0.01). Expression of 12 candidate proteins was confirmed as tumor-derived with immunohistochemical analysis and tissue arrays. Transcriptional profiling from both the same tumors and from an independent study showed mRNAs significantly associated with survival for 11 of 27 encoded genes. Remarkably, combined analysis of protein and mRNA data revealed 11 components of the glycolysis pathway as significantly associated with poor survival. Among these candidates, phosphoglycerate kinase 1 was associated with survival in the protein study, both mRNA studies and in an independent validation set of 117 adenocarcinomas and squamous lung tumors using tissue arrays. These studies indicated that protein expression profiles can predict the outcome of patients with early stage lung cancer and identified new prognostic biomarkers. An overview of the cancer-related applications of proteomics is provided in Table 18.2.

18.3 NOVEL GEL AND NONGEL STRATEGIES FOR DISEASE INVESTIGATIONS

The hopes of displaying all or most cellular proteins in a standard 2D gel have yet to materialize. It has become abundantly clear that the several thousand cellular proteins that may be displayed in a standard 2D gel of a tissue or cell lysate represented a relatively small proportion of the totality of the proteins expressed. This is because many of the proteins detectable in 2D gels of whole cell lysates represent modified forms of a limited numbers of proteins. Thus, 2D PAGE of whole cell or tissue lysates, using the current standard formats, allows analysis of a limited repertoire of cellular proteins representing mostly abundant cytosolic proteins.

To improve the yield of low-abundance proteins in 2D gels, various schemes have been implemented for sample preparation prior to 2D gel analysis in part to reduce sample complexity through fractionation, to allow the detection of low-abundance proteins. For example, liquid-phase isoelectric focusing (IEF) has been utilized to prefractionate in a nongel medium complex mixtures based on their pI. Herbert and Righetti (2000) proposed a protein sample prefractionation approach to isolate proteins into several groups according to their pI, within multicompartment electrolyzers (MCE), that are delimited by immobilized isoelectric membranes with pH values of 3.0, 4.0, 5.0, 6.0, and 10.5. They applied this liquid-phase IEF method to prefractionate human plasma. Proteins in each fraction subsequently were separated by narrow pH 2D PAGE. In plasma separations, since albumin was concentrated within the membranes between pH 5.6 and pH 6.1, both the acidic and basic chambers were free of albumin, resulting in an increase in the number of highly acidic and basic proteins in the fractionated sample compared to whole plasma.

Zuo and Speicher (2002) developed a microscale solution IEF (μsol-IEF) device consisting of six to seven separation chambers bound by the immobilized isoelectric membranes to prefractionate serum sample into a series of well-defined pools prior to subsequent analysis with 2D PAGE. After IEF, each chamber contained only proteins with pI between the pH of the boundary membranes of that chamber. This prefractionation method fractionated complex protein samples into very narrow ranges (<0.5 pH units) with enhanced ability to analyze low-abundance proteins. They demonstrated that 6- to 30-fold greater protein loads were practical for nonalbumin fractions in the subsequent narrow pH range 2D gels, which in turn increased the dynamic range of protein analysis.

Liquid-phase IEF in the Rotofor system (Bio-Rad Laboratory) with 20 isoelectric focusing cells has been utilized to prefractionate samples prior to 2D PAGE (Davidsson *et al.*, 2002). Low-abundance proteins were identified in prefractionated but not in unfractionated samples. These improvements in resolution and sensitivity are promising. They largely stem from the ability to apply greater amounts of proteins. There are numerous other modifications in the 2D gel procedure that have some merit. However, a full presentation of these procedures is outside of the scope of this chapter.

Protein tagging provides an alternative strategy to enhance the detection of low-abundance proteins as illustrated by studies of cell surface proteins (Shin *et al.*, 2003). The cell surface membrane is a cellular compartment of substantial interest. However, a global proteomic analysis of this compartment has represented quite a challenge because of intrinsic features such as high-hydrophobicity and low-expression levels. An intact protein-based strategy was implemented, incorporating the capture of cell surface member proteins from various types of cancer cells by biotin-avidin affinity chromatography and the display of captured proteins using 2D PAGE (see Fig. 18.1). This strategy has been successfully applied to the global profiling of the cell surface proteome of cancer cells (Shin *et al.*, 2003). The surface proteins of viable, intact cells are subjected to biotinylation and the solubilized biotinylated membrane proteins are affinity-captured and purified through the use of a monomeric avidin column. The enriched biotinylated proteins are displayed in 2D gels. Known as well as novel cell surface membrane proteins have been uncovered in our studies. Sabarth and colleagues (2002) applied a similar biotinylation approach for the identification of surface membrane proteins recovered from Helicobacter pylori.

Conventional methods for the comparison of 2D gel images from different samples have relied on the analysis of separate gels prepared for each sample. Due to variability between gels, detection and quantification of protein differences can be problematic. Unlu and colleagues (1997) developed a differential in-gel electrophoresis (DIGE) technique involving tagging of pairs of samples with different fluorescent dyes (Cy3 and Cy5), followed by mixing of the paired samples and their analysis of one 2D gel. Post-run fluorescence imaging of the gel into two images, allows detection of differentially expressed proteins. This method improves the reproducibility and reliability of differential protein expression analysis between samples, and is particularly applicable to comparative analysis of paired affected and nonaffected diseased tissues or controls.

In view of the current interest in the analysis of serum and other biological fluids for the discovery of biomarkers (Adkins *et al.*, 2002; Anderson and Anderson, 2002), several groups have investigated various ways for depleting high-abundance proteins in plasma/serum that interfere with the proteomic analysis of low-abundance proteins. Therefore, it is advantageous to specifically remove those high-abundance proteins in a sample prefractionation step prior to protein separation by either 2D PAGE or other techniques. Specific removal of high-abundance proteins will deplete approximately 85% to 90% of the total protein mass from human plasma/serum. Pieper and colleagues (2003) reported an elegant quantitative strategy for the selective profiling of the low-abundance proteins in human plasma by utilizing a multicomponent immunoaffinity chromatography approach, based on antibody-antigen interactions, to deplete 10 highly abundant proteins from plasma. This selective immunodepletion provided an enriched pool of low-abundance proteins for subsequent 2D PAGE and mass spectrometry analysis. About 350 additional lower abundance proteins were visualized when the high-abundance protein-depleted

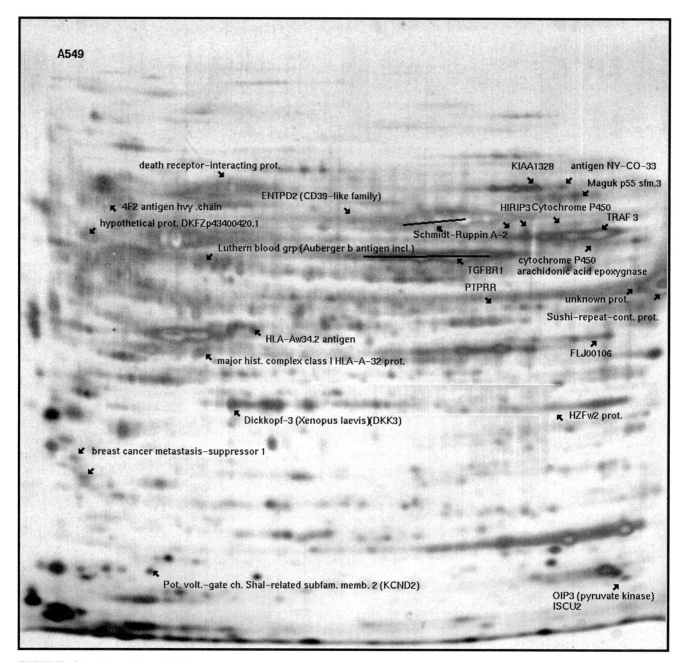

FIGURE 18.1   2D gel blot of biotinylated surface membrane proteins in lung cancer.

plasma samples were resolved by 2D PAGE. Other methods for depletion of high-abundance proteins in serum have also been utilized (Rothemund *et al.*, 2003; Wang and Hanash, 2003).

There is currently a good deal of interest in gel-free systems for protein analysis that have the potential for multiplexing to increase throughput (Liu *et al.*, 2002; Wang and Hanash, 2003). Multimodular combinations of HPLC, liquid phase IEF, and capillary electrophoresis (CE) provide various options for developing high resolution orthogonal 2D liquid phase-based strategies for the separation of

complex mixtures of proteins. A major advantage of liquid separations is that proteins are maintained in solution, which allows online intact protein characterization by mass spectrometry (MS) as well as protein recovery.

18.4 DIAGNOSTIC APPLICATIONS OF MASS SPECTROMETRY

MS, in conjunction with proteomics, has been utilized primarily for protein identification. However, it is possible to profile tissues and biological fluids directly using MS. The potential of MS to yield comprehensive profiles of peptides and proteins in biological fluids without the need to first carry out protein separations has attracted interest. In principle, such an approach would be highly suited for clinical applications because of reduced sample requirements and high throughput. This approach currently is popularized, particularly for serum analysis, by the technology referred to as surface-enhanced laser-desorption ionization (SELDI, Petricoin *et al.*, 2002). Proteins from a patient sample are captured by various types of surfaces with different properties including adsorption, partition, electrostatic interaction, or affinity chromatography. Although such surfaces are referred to as chips, they should not be confused with microarrays as they do not involve any type of arraying. Aside from the use of SELDI, the direct analysis of tissues or biological fluids may simply be accomplished using standard matrix assisted laser desorption ionization (MALDI) without the use of proprietary surfaces.

Some quite noteworthy findings have been reported using SELDI. They include the ability to accurately diagnose ovarian, prostate, breast, and other types of cancer with minimal sample requirement and with high throughput. A study of ovarian cancer that has attracted considerable attention demonstrated the ability of SELDI in combination with an algorithm, to correctly identify all cancer patients, including those with limited stage I disease (Albert *et al.*, 1993).

MALDI MS has been utilized in an innovative fashion to profile tissues *in situ*. A recent study utilized this approach to classify lung tumors based on their proteomic profile (Yanagisawa *et al.*, 2003). Proteomic spectra were obtained for 79 lung tumors and 14 normal lung tissues. More than 1,600 protein peaks were detected from histologically selected 1 mm diameter regions of single frozen sections from each tissue. Class-prediction models based on differentially expressed peaks enabled the classification of lung cancer histologies, distinction between primary tumors and metastases to the lung from other sites, and classification of nodal involvement with 85% accuracy.

The major drawbacks of direct analysis of tissues or biological fluids by MALDI or SELDI are the preferential detection of proteins with a lower molecular mass and the difficulty in determining the identity of proteins whose masses are measured because of lack of correspondence between the masses detected and those predicted for corresponding proteins, due to post-translational modifications. Occasionally, the masses observed match precisely the predicted masses of specific proteins. This was the case in a study of proteins secreted by stimulated CD8 T cells, which led to the identification of the small proteins alpha-defensin 1, 2, and 3 as contributing to the anti-HIV-1 activity of CD8 antiviral factor (Zhang *et al.*, 2002).

There has been some concerns regarding the significance of the diagnostic patterns uncovered using SELDI because the molecules monitored in serum using this approach are likely to be present at concentrations many fold higher than traditional cancer biomarkers. Such markers, therefore, are unlikely to originate from the tumor and thus are considered to be epiphenomena of cancer produced by other organs in response either to the presence of cancer or to a generalized condition of the cancer patient such as debilitation, or acute-phase reaction (Diamandis, 2003). Thus, the role of SELDI and SELDI surfaces in profiling biological fluids remains to be determined.

18.5 CLINICAL APPLICATIONS OF PROTEIN MICROARRAYS

The remarkable achievements in the field of DNA microarrays in the past decade, including the demonstration of the utility of such arrays for analyzing clinical samples, has contributed to the current interest in protein microarrays that allow the systematic analysis of thousands of proteins and that can be used for a variety of clinical applications. Unlike DNA microarrays that provide one measure of gene expression, namely RNA levels (see also Chapter 15), there is a need to implement protein microarray strategies that address the many different features of proteins that have clinical relevance, including on the one hand determination of protein levels in biological samples, and on the other, determination of their functional state, as may be deduced, for example, from their phosphorylation state. This section addresses the utility of protein microarrays for clinical applications. Additional information about protein microarrays may be found in several review articles that have been published in various journals (Cahill and Nordhoff, 2003; Cutler, 2003; Liotta *et al.*, 2003; Phizicky *et al.*, 2003). Furthermore, a recent issue of the journal *Proteomics* (November, 2003) was devoted to protein microarrays and contains both review and original research articles.

The clinical applications of protein microarrays cover a wide range from their use to detect disease or assess response to therapy based on profiling of biological fluids, to their use for profiling of disease tissue to determine disease subtypes and decide on the most appropriate therapy. Although the field of protein microarrays is still evolving, there are currently two broad classes of protein arrays. One class consists of arrays that contain protein capture agents such as antibodies, which are utilized to assay the abundance of corresponding antigens in biological samples, and another class consists of arrays that contain proteins or peptides to be interrogated using cells, tissues, biological fluids, or single agents to uncover their interactions with specific arrayed proteins or peptides.

Arrays containing capture agents may be used in a similar manner as sandwich assays in which an immobilized antibody or any other type of capture agent is utilized to capture proteins of interest in a biological sample and a second labeled antibody or capture agent is used to detect and determine the abundance of the protein (Osada *et al.*, 1998). Such an approach therefore requires two capture agents that recognize different epitopes that need to be available for binding in order to assay any given protein(s). This dual requirement creates constraints for interrogating the abundance of large numbers of proteins simultaneously in a biological sample. Alternatively, a label-free detection method, such as MS or surface plasmon resonance, may be utilized.

Yet another alternative is to tag biological samples before they are hybridized to immobilized capture agents. The levels of the proteins in the biological sample, for which corresponding capture agents are arrayed, is determined based on quantifying the amount of tag bound to the immobilized capture agents. A potentially quite useful approach to profiling disease tissue and biological fluids is to determine not only the levels of specific proteins but also their post-translational modifications such as phosphorylation. To that effect, a dual detection system in which captured proteins are assessed with respect to their phosphorylation status has been implemented by Steinberg and colleagues (2003). This group applied a novel phosphoprotein dye technology suitable for the fluorescent detection of phosphoserine-, phosphothreonine-, and phosphotyrosine-containing proteins to the determination of protein kinase and phosphatase substrate preference (Steinberg *et al.*, 2003).

The utility of the fluorescent dye technology was demonstrated using phosphoproteins and phosphopeptides as well as with protein kinase reactions performed in miniaturized microarray assay format (Martin *et al.*, 2003). A small fluorescent probe was employed as a sensor of protein phosphorylation status. The detection limit for phosphoproteins on a variety of different commercially available protein array substrates was found to be 312–625 fg, depending upon the number of phosphate residues (Martin *et al.*, 2003). The development of reagents that allow assessment of protein modification such as phosphorylation, glycosylation, or other functionally relevant protein changes have substantial utility for clinical applications.

Antibody microarrays with a relatively limited content as compared to DNA microarrays, have become commercially available. For example, the BD Clontech Ab Microarray 500 consists of 500 distinct antibodies that detect a variety of human proteins either in circulation or in cell and tissue extract. These correspond to a broad range of functional classes involved in signal transduction, cell growth and proliferation, apoptosis, inflammation, and the immune response. A complete list of the antibodies on this array is available at http://atlasinfo.clontech.com/abinfo/array-list-action.do. Other commercially available antibody microarrays that interrogate a more limited or a particular class of proteins are also available. The content and variety of commercially available microarrays is likely to continue to expand. A potential limiting factor for the use of commercial microarrays is their cost.

The compelling need for developing capture agents with the prerequisite specificity has led numerous biotechnology companies to devise novel strategies that include aptamers (SomaLogic, http://www.somalogic.com/), ribozymes (Archemix, http://www.archemix.com/), partial-molecule imprints (Aspira Biosystems, http://www.aspirabio.com), and modified binding proteins (Phylos, http://www.phylos.com).

For assays of protein interactions, arrays that contain either peptides or proteins are being produced. Peptides may be synthesized in very large numbers directly on the chip (Pellois *et al.*, 2002). Alternatively, recombinant proteins may be arrayed. There is substantial interest at the present time to assemble large sets of purified recombinant proteins for various applications including microarray construction, as has been done with the yeast (Zhu *et al.*, 2001; Schweitzer *et al.*, 2003). Yet another type of protein array consists of arraying lysates prepared individually from multiple clinical samples such as tumors. Such arrays are produced in batches in which each array contains the full complement of clinical samples. Each array is interrogated with a single antibody to determine the level or post-translational modification of a particular protein (Paweletz *et al.*, 2001). A high degree of sensitivity, precision, and linearity has been demonstrated using such arrays, making it possible to quantify the phosphorylation status of signal proteins in human tissue cell subpopulations.

Using this approach, Paweletz and colleagues (2001) have longitudinally analyzed pro-survival checkpoint proteins at the transition stage from patient matched histologically normal prostate epithelium to prostate intraepithelial neoplasia and to invasive prostate cancer. Cancer progression was associated with increased phosphorylation of Akt, suppression of apoptosis pathways, as well as decreased phosphorylation of ERK. At the transition from histologically normal epithelium to intraepithelial neoplasia, a statistically significant increase in phosphorylated Akt and a concomitant suppression of downstream apoptosis pathways preceding the transition into invasive carcinoma were observed.

Disease tissue profiling studies that have utilized protein microarrays are beginning to emerge, particularly in the cancer field. Belov and colleagues (2001; 2003) have used a microarray of 60 antibodies directed against clusters of differentiation (CD) antigens to immunophenotype various types of leukemia. This group utilized whole cells to determine expression of antigens expressed on the surface. Based on the expression pattern of CD antigens, a fingerprint was defined for chronic lymphocytic leukemia.

As a model to better understand how patterns of protein expression shape the tissue microenvironment, Knezevic and colleagues (2001) analyzed protein expression in tissue derived from squamous cell carcinomas of the oral cavity through an antibody microarray approach for high-throughput proteomic analysis. Utilizing laser capture microdissection to procure total protein from specific cell populations, this group demonstrated that quantitative, and potentially qualitative, differences in expression patterns of multiple proteins within epithelial cells reproducibly correlated with oral cavity tumor progression. Differential expression of multiple proteins was found in stromal cells surrounding and adjacent to regions of diseased epithelium that directly correlated with tumor progression of the epithelium. Most of the proteins identified in both cell types were involved in signal transduction pathways, leading therefore to the hypothesis that extensive molecular communications involving complex cellular signaling between epithelium and stroma play a key role in driving oral cavity cancer progression.

Cytokines have become a popular target for analysis using microarrays. Schweitzer and colleagues (2002) printed 75 cytokine antibodies on chemically derivatized glass slides and investigated cytokine secretion from human dendritic cells induced with lipopolysaccharide or tumor-necrosis factor-α. This group used isothermal rolling-circle amplification (Lizardi *et al.*, 1998) to increase the sensitivity of their assay. Some of the detection antibodies were found to cross-react with other cytokines on the array. Therefore, they divided their collection of capture antibodies into two groups that were spotted on a different portion of the same glass slide separated by a Teflon barrier.

A clinically relevant application of protein microarrays is the identification of proteins that induce an antibody response in autoimmune disorders (Robinson *et al.*, 2002). Microarrays were produced by attaching several hundred proteins and peptides to the surface of derivatized glass slides. Arrays were incubated with patient serum, and fluorescent labels were used to detect autoantibody binding to specific proteins in autoimmune diseases, including systemic lupus erythematosus and rheumatoid arthritis. In a more recent study (Robinson *et al.*, 2003), a myelin proteome microarray was developed to profile the evolution of autoantibody responses in experimental autoimmune encephalomyelitis, a model for multiple sclerosis. Increased diversity of autoantibody responses predicted a more severe clinical course. Chronic experimental autoimmune encephalomyelitis was associated with previously undescribed extensive intra- and intermolecular epitope spreading of autoreactive B-cell responses. Proteomic monitoring of autoantibody responses provided a useful approach to monitor autoimmune disease and to develop and tailor disease- and patient-specific tolerizing DNA vaccines.

An important consideration in protein microarrays is that proteins undergo numerous post-translational modifications that may be highly important for their functions and disease development. However, these modifications are generally not captured using either recombinant proteins or antibodies that do not distinctly recognize specific forms of a protein. One approach for comprehensive analysis of proteins in their modified forms is to array proteins directly

isolated from cells and tissues following protein fractionation schemes (Madoz-Gurpide *et al.*, 2001). Fractions that react with specific probes are within the reach of chromatographic and gel-based separation techniques for resolving their individual protein constituents and of mass spectrometric techniques for identification of their constituent proteins.

A productive approach for cancer marker identification has been the analysis of serum for autoantibodies against tumor proteins. There is increasing evidence for an immune response to cancer in humans, demonstrated in part by the identification of autoantibodies against a number of intracellular and surface antigens detectable in sera from patients with different cancer types (Hanash, 2003). The identification of panels of tumor antigens that elicit an antibody response may have utility in cancer screening, diagnosis, or in establishing prognosis. Such antigens may also have utility in immunotherapy against the disease.

There are several approaches for the detection of tumor antigens that induce an immune response (Hanash, 2003). For most antigenic proteins that induce an antibody response in cancer identified using proteomics, post-translational modifications contributed to the immune response. For example, in a study of lung cancer, sera from 60% of patients with lung adenocarcinoma, and 33% of patients with squamous cell lung carcinoma, but none of the noncancer controls, exhibited IgG-based reactivity against proteins identified as glycosylated annexins I and II (Brichory *et al.*, 2001).

Microarrays that contain proteins derived from tumor cells have the potential of substantially accelerating the pace of discovery of tumor antigens and yielding a molecular signature for immune responses directed against protein targets in different types of cancer (Haab, 2003; Nam *et al.*, 2003). In a study of colon cancer (Nam *et al.*, 2003) microarrays printed with 1,760 separate protein fractions, isolated from the LoVo colon adenocarcinoma cell line, were hybridized with individual sera. A fraction that exhibited IgG based reactivity with 9/15 colon cancer sera was found to contain Ubiquitin C-terminal hydrolase L3 (UCH-L3) by tandem MS (ESI-Q-TOF). The highest levels of UCH-L3 mRNA among the 329 tumors of different types analyzed by DNA microarrays were found in colon tumors. Independent validation by Western blotting demonstrated UCH-L3 antibodies in 19/43 sera from patients with colon cancer, and in 0/54 sera from subjects with lung cancer, colon adenoma, or otherwise healthy. These data point to the utility of microarrays printed with natural proteins for the identification of cancer markers.

18.6 CONCLUSIONS

2D PAGE has provided a robust platform for the analysis of protein constituents of cells, tissues, and biological fluids. Although the reach of 2D gels has been limited to relatively abundant proteins, numerous improvements have been introduced that expand the reach of 2D gels. Nevertheless, it is likely that other, nongel-based approaches, such as liquid-based separations and the use of microarrays that contain protein capture agents notably antibodies, will provide an alternative for expression proteomics. However the reach of such approaches at the present time is rather limited, maintaining 2D PAGE as a gold standard for proteome analysis.

References

Adkins, J. N., Varnum, S. M., Auberry, K. J., Moore, R. J., Angell, N. H., Smith, R. D., Springer, D. L., and Pounds, J. G. (2002). Toward a human blood serum proteome: analysis by multidimensional separation coupled with mass spectrometry. *MCP* 1, 947–955.

Albert, A. S., Thorburn, A. M., Shenolikar, S., Mumby, M. C., and Feramisco, J. R. (1993). Regulation of cell cycle progression and nuclear affinity of the retinoblastoma protein by protein phosphatases. *Proc. Natl. Acad. Sci. USA* 90, 388–392.

Anderson, N. L., and Anderson, N. G. (2002). The human plasma proteome. *Mol. Cell. Proteomics.* 1, 845–867.

Beer, D. G., Kardia, S. L. R., Huang, C.-C., Giordano, T. J., Levin, A. M., Misek, D. E., Lin, L., Chen, G., Gharib, T. G., Thomas, D. G., Lizyness, M. L., Kuick, R., Hayasaka, S., Taylor, J. M. G., Iannettoni, M. D., Orringer, M. B., and Hanash, S. (2002). Gene-expression profiles predict survival of patients with lung adenocarcinomas. *Nat. Med.* 8, 816–824.

Belov, L., de la Vega, O., dos Remedios, C. G., Mulligan, S. P., and Christopherson, R. I. (2001). Immunophenotyping of leukemias using a cluster of differentiation antibody microarray. *Cancer Res.* 61, 4483–4489.

Belov, L., Huang, P., Barber, N., Mulligan, S. P., and Christopherson, R. I. (2003). Identification of repertoires of surface antigens on leukemias using an antibody microarray. *Proteomics* 3, 2147–2154.

Bergman, A. C., Benjamin, T., Alaiya, A., Waltham, M., Sakaguchi, K., Franzen, B., Linder, S., Bergman, T., Auer, G., Appella, E., Wirth, P. J., and Jornvall, H. (2000). Identification of gel-separated tumor marker proteins by mass spectrometry. *Electrophoresis* 21, 679–686.

Brichory, F. M., Misek, D. E., Yim, A. M., Krause, M. C., Giordano, T. J., Beer, D. G., and Hanash, S. M. (2001). An immune response manifested by the common occurrence of annexins I and II autoantibodies and high circulating levels of IL-6 in lung cancer. *Proc. Natl. Acad. Sci. USA* 98, 9824–9829.

Cahill, D. J., and Nordhoff, E. (2003). Protein arrays and their role in proteomics. *Adv. Biochem. Eng. Biotechnol.* 83, 177–187.

Celis, J. E., Ostergaard, M., Basse, B., Celis, A., Lauridsen, J. B., Ratz, G. P., Andersen, I., Hein, B., Wolf, H., Orntoft, T. F., and Rasmussen, H. H. (1996a). Loss of adipocyte-type fatty acid binding protein and other protein biomarkers is associated with progression of human bladder transitional cell carcinomas. *Cancer Res.* 56, 4782–4790.

Celis, J. E., Rasmussen, H. H., Vorum, H., Madsen, P., Honore, B., Wolf, H., and Orntoft, T. F. (1996b). Bladder squamous cell carcinomas express psoriasin and externalize it to the urine. *J. Urol.* 155, 2105–2112.

Celis, J. E., Celis, P., Ostergaard, M., Basse, B., Lauridsen, J. B., Ratz, G., Rasmussen, H. H., Orntoft, T. F., Hein, B., Wolf, H., and Celis, A. (1999). Proteomics and immunohistochemistry define some of the steps involved in the squamous differentiation of the bladder transitional epithelium: a novel strategy for identifying metaplastic lesions. *Cancer Res.* 59, 3003–3009.

Celis, J. E., Celis, P., Palsdottir, H., Ostergaard, M., Gromov, P., Primdahl, H., Orntoft, T. F., Wolf, H., Celis, A., and Gromova, I. (2002). Proteomic strategies to reveal tumor heterogeneity among urothelial papillomas. *MCP* 1, 269–279.

Chen, G., Charib, T., Prescott, M., Huang, C., Shedden, K., Taylor, J., Thomas, D., Greenson, J., Kardia, S., Beer, D., Rennert, G., Cho, K., Gruber, D., Fearon, E., and Hanash, S. (2003). Protein expression profiles predictive of survival in lung adenocarcinomas. *Proc. Natl. Acad. Sci. USA* 100, 13537–13542.

Crist, W., Pullen, J., Boyett, J., Falletta, J., van Eys, J., Borowitz, M., Jackson, J., Dowell, B., Frankel, L., Quddus, F., Ragab, A., and Vietti, T. (1986). Clinical and biologic features predict a poor prognosis in acute lymphoid leukemias in infants: A pediatric oncology group study. *Blood* 67, 135–140.

Cutler, P. (2003). Protein arrays: the current state of-the-art. *Proteomics* 3, 3–18.

Davidsson, P., Folkesson, S., Christiansson, M., Lindbjer, M., Dellheden, B., Blennow, K., and Westman-Brinkmalm, A. (2002). Identification of proteins in human cerebrospinal fluid using liquid-phase isoelectric focusing as a prefractionation step followed by two-dimensional gel electrophoresis and matrix-assisted laser desorption/ionization mass spectrometry. *Rapid Commun. Mass Spectrom.* 16, 2083–2088.

Diamandis, E. P. (2003). Point: Proteomic patterns in biological fluids: do they represent the future of cancer diagnostics? *Clin. Chem.* 49, 1272–1275.

Haab, B. B. (2003). Methods and applications of antibody microarrays in cancer research. *Proteomics* 3, 2116–2122.

Hanash, S. M., Baier, L. J., McCurry, L., and Schwartz, S. (1986a). Lineage related polypeptide markers in acute lymphoblastic leukemia detected by two-dimensional electrophoresis. *Proc. Natl. Acad. Sci. USA* 83, 807–811.

Hanash, S. M., Baier, L. J., Neel, J. V., and Niezgoda, W. (1986b). Genetic analysis of thirty three platelet polypeptides detected in two-dimensional polyacrylamide gels. *Am. J. Hum. Genet.* 38, 352–360.

Hanash, S. M., Kuick, R., Strahler, J. R., Richardson, B. C., Reaman, G., Stoolman, L., Hanson, C., Nichols, D., and Tueche, J. (1989). Identification of a cellular polypeptide that distinguishes between acute lymphoblastic leukemia in infants and in older children. *Blood* 73, 527–532.

Hanash, S. (2001). 2D or not 2D—is there a future for 2D gels in proteomics? Insights from York proteomic meeting. *Proteomics* 1, 635–637.

Hanash, S. (2003). Harnessing immunity for cancer marker discovery. *Nat. Biotechnol.* 21, 37–38.

Herbert, B., and Righetti, P. G. (2000). A turning point in proteome analysis: Sample prefractionation via multicompartment electrolyzers with isoelectric membranes. *Electrophoresis* 21, 3639–3648.

Hirano, T., Franzen, B., Uryu, K., Okuzawa, K., Alaiya, A. A., Vanky, F., Rodrigues, L., Ebihara, Y., Kato, H., and Auer, G. (1995). Detection of polypeptides associated with the histopathological differentiation of primary lung carcinoma. *Br. J. Cancer* 72, 840–848.

Knezevic, V., Leethanakul, C., Bichsel, V. E., Worth, J. M., Prabhu, V. V., Gutkind, J. S., Liotta, L. A., Munson, P. J., Petricoin III, E. F., and Krizman, D. B. (2001). Proteomic profiling of the cancer microenvironment by antibody arrays. *Proteomics* 1, 1271–1278.

Liotta, L. A., Espina, V., Mehta, A. I., Calvert, V., Rosenblatt, K., Geho, D., Muns, P. J., Young, L., Wulfkuhle, J., and Petroicoin 3rd, E. (2003). Protein microarrays: meeting analytical challenges for clinical applications. *Cancer Cell* 3, 317–325.

Liu, H., Berger, S. J., Chakraborty, A. B., Plumb, R. S., and Cohen, S. A. (2002). Multidimensional chromatography coupled to electrospray ionization time-of-flight mass spectrometry as an alternative to two-dimensional gels for the identification and analysis of complex mixtures of intact proteins. *J. Chromatogr. B Analyt. Technol. Biomed. Life Sci.* 782, 267–289.

Lizardi, P. M., Huang, X., Zhu, Z., Bray-Ward, P., Thomas, D. C., and Ward, D. C. (1998). Mutation detection and single-molecule counting using isothermal rolling circle amplification. *Nat. Genet.* 19, 225–232.

Madoz-Gurpide, J., Wang, H., Misek, D. E., Brichory, F., and Hanash, S. M. (2001). Protein based microarrays: A tool for probing the proteome of cancer cells and tissues. *Proteomics* 1, 1279–1287.

Martin, K., Steinberg, T. H., Cooley, L. A., Gee, K. R., Beechem, J. M., and Patton, W. F. (2003). Quantitative analysis of protein phosphorylation status and protein kinase activity on microarrays using a novel fluorescent phosphorylation sensor dye. *Proteomics* 7, 1244–1255.

Misek, D., Chang, C., Kuick, R., Hinderer, R., Giordano, T., Beer, D., and Hanash, S. (2002). Transforming properties of a Q18ÆE mutation of the microtubule regulator Op18. *Cancer Cell* 2, 217.

Nam, M. J., Madoz-Gurpide, J., Wang, H., Lescure, P., Schmalbach, C. E., Zhao, R., Misek, D. E., Kuick, R., Brenner, D. E., and Hanash, S. M. (2003). Molecular profiling of the immune response in colon cancer using protein microarrays: Occurrence of autoantibodies to ubiquitin C-terminal hydrolase L3. *Proteomics* 3, 2108–2115.

Osada, M., Ohba, M., Kawahara, C., Ishioka, C., Kanamaru, R., Katoh, I., Ikawa, Y., Nimura, Y., Nakagawara, A., Obinata, M., and Ikawa, S. (1998). Cloning and functional analysis of human p51, which structurally and functionally resembles p53. *Nat. Med.* 4, 839–843.

Paweletz, C. P., Charboneau, L., Bichsel, V. E., Simone, N. L., Chen, T., Gillespie, J. W., Emmert-Buck, M. R., Roth, M. J., Petricoin III, E. F., and Liotta, L. A. (2001). Reverse phase protein microarrays which capture disease progression show activation of pro-survival pathways at the cancer invasion front. *Oncogene* 20, 1981–1989.

Pellois, J. P., Zhou, X., Srivannavit, O., Zhou, T., Gulari, E., and Gao, X. (2002). Individually addressable parallel peptide synthesis on microchips. *Nat. Biotechnol.* 20, 922–926.

Petricoin, E. F., Zoon, K. C., Kohn, E. C., Barrett, J. C., and Liotta, L. A. (2002). Clinical proteomics: translating benchside promise into bedside reality. *Nat. Rev. Drug Discov.* 1, 683–695.

Phizicky, E., Bastiaens, P. I., Zhu, H., Snyder, M., and Fields, S. (2003). Protein analysis on a proteomic scale. *Nature* 422, 208–215.

Pieper, R., Su, Q., Gatlin, C. L., Huang, S. T., Anderson, N. L., and Steiner, S. (2003). Multi-component immunoaffinity subtraction chromatography: an innovative step towards a comprehensive survey of the human plasma proteome. *Proteomics* 3, 422–432.

Reaman, G., Zeltzer, P., Bleyer, W. A., Amendola, B., Level, C., Sather, H., and Hammond, D. (1985). Acute lymphoblastic leukemia in infants less than one year of age: a cumulative experience of the Children's Cancer Study Group. *J. Clin. Oncol.* 3, 1513–1521.

Robinson, W. H., DiGennaro, C., Hueber, W., Haab, B. B., Kamachi, M., Dean, E. J., Fournel, S., Fong, D., Genovese, M. C., de Vegvar, H. E., Skriner, K., Hirschberg, D. L., Morris, R. I., Muller, S., Pruijin, G. J., van Verooij, W. J., Smolen, J. S., Brown, P. O., Steinman, L., and Utz, P. J. (2002). Autoantigen microarrays for multiplex characterization of autoantibody responses. *Nat. Med.* 8, 295–301.

Robinson, W. H., Fontoura, P., Lee, B. J., de Vegvar, H. E., Tom, J., Pedotti, R., DiGennaro, C. D., Mitchell, D. J., Fong, D., Ho, P. P., Ruiz, P. J., Maverakis, E., Stevens, D. B., Bernard, C. C., Martin, R., Kuchroo, V. K., van Noort, J. M., Genain, C. P., Amor, S., Olsson, T., Utz, P. J., Garren, H., and Steinman, L. (2003). Protein microarrays guide tolerizing DNA vaccine treatment of autoimmune encephalomyelitis. *Nat. Biotechnol.* 21, 1033–1039.

Rothemund, D. L., Locke, V. L., Liew, A., Thomas, T. M., Wasinger, V., and Rylatt, D. B. (2003). Depletion of the highly abundant protein albumin from human plasma using the Gradiflow. *Proteomics* 3, 279–287.

Sabarth, N., Lamer, S., Zimny-Arndt, U., Junglut, P. R., Meyer, T. F., and Bumann, D. (2002). Identificaiton of surface proteins of Helicobacter pylori by selective biotinylation, affinity purification, and two-dimensional gel electrophoresis. *J. Biol. Chem.* 277, 27896–27902.

Schmid, H. R., Schmitter, D., Blum, P., Miller, M., and Vonderschmitt, D. (1995). Lung tumor cells: a multivariate approach to cell classification using two-dimensional protein pattern. *Electrophoresis* 16, 1961–1968.

Schweitzer, B., Predki, P., and Snyder, M. (2003). Microarrays to characterize protein interactions on a whole-proteome scale. *Proteomics* 3, 2190–2199.

Schweitzer, B., Roberts, S., Grimwade, B., Shao, W., Wang, M., Fu, Q., Shu, Q., Laroche, I., Zhou, Z., Tchernev, V. T., Christiansen, J., Velleca, M., and Kingsmore, S. F. (2002). Multiplexed protein profiling on microarrays by rolling-circle amplification. *Nat. Biotechnol.* 20, 359–365.

Shin, B. K., Wang, H., Yim, A. M., Le Naour, F., Brichory, F., Jang, J. H., Zhao, R., Puravs, E., Tra, J., Michael, C. W., Misek, D. E., and Hanash, S. M. (2003). Global profiling of the cell surface proteome of cancer cells uncovers an abundance of proteins with chaperone function. *J. Biol. Chem.* 278, 7607–7616.

Steinberg, T. H., Agnew, B. J., Gee, K. R., Leung, W. Y., Goodman, T., Schulenberg, B., Hendrickson, J., Beechem, J. M., Haugland, R. P., and Patton, W. F. (2003). Global quantitative phosphoprotein analysis using multiplexed proteomics technology. *Proteomics* 3, 1128–1144.

Unlu, M., Morgan, M. E., and Minden, J. S. (1997). Difference gel electrophoresis: A single gel method for detecting changes in protein extracts. *Electrophoresis* 18, 2071–2077.

Wang, H., and Hanash, S. (2003). Multi-dimensional liquid based separations in proteomics. *J. Chromatogr. B Analyt. Technol. Biomed. Life Sci.* 787, 11–18.

Yanagisawa K, Shyr Y, Xu, B. J., Massion, P. P., Larsen, P. H., White, B. C., Roberts, J. R., Edgerton, M., Gonzalez, A., Nadaf, S., Moore, J. H., Caprioli, R. M., and Carbone, D. P. (2003). Proteomic patterns of tumour subsets in non-small-cell lung cancer. *Lancet* 362, 433–439.

Zhang, L., Yu, W., He, T., Yu, J., Caffrey, R. E., Dalmasso, E. A., Fu, S., Pham, T., Mei, J., Ho, J. J., Zhang, W., Lopez, P., and Ho, D. D. (2002). Contribution of human alpha-dfensin 1,2 and 3 to the anti-HIV-1 activity of CD8 antiviral factor. *Science* 298, 995–1000.

Zhu, H., Bilgin, M., Bangham, R., Hall, D., Casamayor, A., Bertone, P., Lan, N., Jansen, R., Bidlingmaier, S., Houfek, T., Mitchell, T., Miller, P., Dean, R. A., Gerstein, M., and Snyder, M. (2001). Global analysis of protein activities using proteome chips. *Science* 293, 2101–2105.

Zuo, X., and Speicher, D. W. (2002). Comprehensive analysis of complex proteomes using microscale solution isoelectrofocusing prior to narrow pH range two-dimensional electrophoresis. *Proteomics* 2, 58–68.

SECTION II

Applications of Molecular Diagnostics and Related Issues

CHAPTER **19**

Identification of Genetically Modified Organisms

FARID E. AHMED

Department of Radiation Oncology, Leo W. Jenkins Cancer Center, The Brody School of Medicine, LSB 014, East Carolina University, Greenville, NC, USA

TABLE OF CONTENTS

19.1 INTRODUCTION AND HISTORICAL PERSPECTIVE

Adopting modern biotechnology, including genetic transformation, has developed new plant varieties. Genetically modified (GM) products contain an additional trait encoded by an introduced gene(s), which generally produce a protein(s) that confers the trait of interest. Raw material (e.g., grains) and processed products (e.g., foods) derived from GM crops might thus be identified by testing for the presence of introduced DNA, or by detection of expressed novel protein(s) encoded by the genetic material. Both qualitative (i.e., those that give a yes/no answer) and quantitative diagnostic methods are available (Ahmed, 2002).

Most developed countries over the last few years have established mechanisms for adjudicating on the safety of novel food before it is marketed (Moseley, 1999). GM foods have not gained worldwide acceptance because of unmollified consumer suspicion resulting from earlier food and environmental concerns, transparent regulatory oversight, and mistrust in government bureaucracies, all factors that fueled debates about the environmental and public health safety issues of introduced genes; for example, potential gene flow to other organisms, the destruction of agricultural diversity, allergenicity, antibiotic resistance, gastrointestinal problems (Gaskell *et al.*, 1999; Hasslberg, 2000). Social acceptance of such novel foods or ingredients are not uniform in developed countries. Consumer concerns can be based on ethical considerations (i.e., scientists playing God) or safety worries (i.e., more testing needs to be done; Moseley, 1999). Other economical and ethical issues pertaining to intellectual property rights came into play (Serageldin, 1999) with the realization that inadvertent contamination of non-GM seeds with genetically modified organisms (GMOs) is likely. These factors induced countries, exemplified by the European Union (EU), either to restrict the import of bioengineered foods or to introduce legislation requiring mandatory labeling of GMO foods or food ingredients containing additives and flavorings that have been genetically modified or have been produced from GMOs (Commission Regulation, 2000a; 2000b). EU regulations mandate labeling of food containing GMOs (Council Regulation, 1997; 1998). Norway and Switzerland, which are not members of the EU, demand the labeling of GMOs in their food (Hardegger *et al.*, 1999). Moreover, in 1998, the EU introduced a *de facto* moratorium on the import and production of GM foods. In March 2003, the European Commission upheld the moratorium and is standing firm on its decision that any food containing more than 0.9% of a GM product would carry a label (Hellemans, 2003). In the USA, recent legislation did not stipulate mandatory labeling of GM foods but has instead recommended a voluntary labeling of bioengineered foods and requested that companies must notify the US Food and Drug Administration (FDA) of their intent to market GM foods at least 120 days in advance of launch (Editorial, 2000).

It appears that insufficient attention has been given to the following issues:

- The introduction of the same gene into different types of cells can produce distinct proteins.
- The introduction of any gene (either from the same or different species) can significantly change overall gene expression, and thus the phenotype of the recipient cells.
- Enzymatic pathways introduced to synthesize micronutrients may interact with endogenous pathways leading to production of novel metabolically-active molecules.

Consequently, as with secondary modifications, it is possible that any or all of these perturbations result in unpredictable outcome (Shubert, 2002). Although short- and long-term toxicity and metabolic studies could be carried out to address concerns, it is probably not feasible that they will detect relevant changes unless extensive safety testing is carried out on GM crops, and more importantly provided the effect(s) in question are known so that they can be targeted and carefully studied.

Assessing the risk that a GM food poses to human health has been based on comparing its chemical composition to that of unmodified food; if the composition is similar, it is considered safe for human consumption. This concept, called substantial equivalence, has been based on relatively easy and inexpensive tests (Kuuiper *et al.*, 2001). However, some scientists in Europe criticize the methodologies currently used and contend that such tests do not include all biological, toxicological, and immunological aspects of GM food (Hellemans, 2003).

The first generation of "input traits" of GM crops (e.g., traits with purely agronomic benefits) is entering its ninth year, with the large majority derived from these leading crops in North America, namely canola (rapeseed), corn, and soybean. These products harbor traits that serve agronomic purpose (i.e., benefiting farmers, but not necessarily the consumers). They entered the North American market with minimum regulations and without segregation, and have been judged by regulators as substantially equivalent to existing varieties. Second generation crops, which involve output modifications (traits with health and nutritional benefits; Agius *et al.*, 2003), most likely will not be cleared unless their purity is assured; this is a problematic prospect given the current difficulties of attaining gene containment.

Third generation crops with new industrial, nutriceutical, or pharmaceutical properties most likely will require an effective gene control system so that they may be allowed to enter the markets (Kleter *et al.*, 2001). Regardless of how effective regulations are, some producers (either deliberately or inadvertently) will misappropriate these technologies, creating risks and liabilities. Moreover, many plant species are sexually promiscuous, creating natural gene flow to related species, and leading to the following liability issues:

- The potential of volunteer seeds inadvertently left in the field to germinate the following year(s)
- The potential for pollen flown from GM crops to non-GM crops
- The potential of co-mingling of GM and non-GM crops, which could jeopardize the value of both crops and product lines if transgene remain undetected before processing
- The potential for environmental risks associated with uncontrolled gene flow from GM varieties into related plants, which will impede export of GM varieties to countries not willing to adopt the new technologies

These liability issues have resulted in some disastrous consequences and have imposed significant cost on the food industry (Smyth *et al.*, 2002). About 36 countries have legislation making it imperative that governments, the food industry, testing laboratories, and crop producers develop ways to accurately quantitate GMOs in crops, foods, and food ingredients to assure compliance with threshold levels of GM-products (Butler and Peichardt, 1999; Kuiper, 1999; US National Academy of Sciences, 2000). The aim of this chapter is to detail diagnostic test methods, their potential, and their limitations.

19.2 SAMPLING PLANS

Both sample size and sampling procedures are important issues for testing GMOs in raw material and food ingredients if one is to avoid problems of nonhomogeneity. The sampling plan should be performed in a manner that ensures that the sample is statistically representative, and the sample size must be sufficient to allow adequate sensitivity, because the statistical significance achievable with a small sample size is weak (Gilbert, 1999). When a sample is used to represent the content of a lot, its content is likely to deviate from the actual content of the lot. Sampling errors create risks for both the buyer and the seller in a transaction. The buyer may get a lot with a higher concentration than desired. The seller may also have a lot that meets a contract rejected. These two types of risks are referred to as seller's and buyer's risk, respectively, and the relationship between them is seen in Fig. 19.1. Probability can be used to estimate a likely range that a sample may deviate from the true lot content, as long as the sample is properly taken (Whitaker *et al.*, 2001).

Results from theoretical and simulation research have shown that GMOs distribution in bulk commodities is more likely to be heterogeneous than homogenous, which poses a serious limit to the unconditional acceptance of the assumption of random distribution of GM material and to the use of binomial distribution to estimate producer and consumer risks (Paoletti *et al.*, 2003). Due to the heterogenous distribution of biotech seeds (or beans or kernels) within a load,

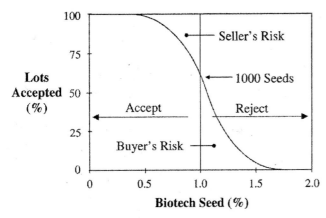

FIGURE 19.1 Operating characteristic curve illustrating buyer's and seller's risk. From Whitaker and colleagues (2001); with permission.

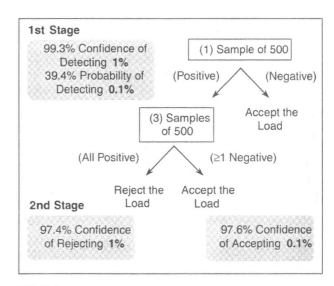

FIGURE 19.2 Threshold testing protocol providing high confidence that 1% will be rejected and 0.1% will be accepted. From Remund and colleagues (2001); with permission.

there is a probability that a sample from the load will contain a concentration that is higher or lower than the true concentration. Using a sample size of 1000 seeds, the probability of accepting a load containing various percentages of biotechs is defined by the operating characteristic (OC) curve shown in Fig. 19.1. The buyer's and seller's risk for a load containing a true concentration of 1% biotech units, randomly distributed in a load, is indicated by the area of the figure between the QC curve and the line representing 1%. Using these statistic principles to manage marketing risks, different sampling and testing strategies have been developed to meet the demand of the market.

In situations where biotech grains have been approved by government authorities, appreciable concentration of biotech grains in a load may be tolerated. In this case, it is possible to imagine a transaction where the buyer wants high confidence that the beans being purchased contain less than 1% Roundup Ready™ (RR) soybeans to avoid labeling. At the same time, the seller wants reasonable confidence that if the load contains less than 1% RR, there is a high probability that the load will be accepted. Figure 19.2 illustrates an example of a sampling and testing protocol, called threshold testing, that uses lateral flow strips for qualitative determination of novel protein, which could be useful under such a scenario to provide acceptable assurances to both parties (Remund *et al.*, 2001). A limitation of the use of this technology is the difficulty of testing all available biotech events using a single sample. To use the threshold testing protocol, it is necessary to limit the maximum number of kernels in a sample so that the presence of only a single biotech kernel will always give a positive response. When performing threshold testing for multiple biotech events in a single sample, it is necessary to limit the number of kernels so that the lowest expressing event will always be detected, and such a constrain significantly limits the sensitivity of the methods for high expressing events. This constrain on sensitivity, and the fact that there is a large and ever-changing

list of biotech events, have curtailed efforts to develop methods to detect multiple biotech events in a single sample of corn grain (Stave, 2004).

As raw material often comes from different suppliers, and given that industrial activities are structured in space and time, one can expect a portion of the original chronological order always to be present in the spatial structure of any lot. Under this assumption, a systematic sampling approach is recommended over a random sampling approach. Systematic sampling should take place when lots are being unloaded (i.e., when there is the option of continuous sampling of the entire consignment). This is preferable to the situation of large batches in silos or trucks, where it is difficult to access remote parts even when employing sampling probes (Paoletti *et al.*, 2003).

It is difficult to make clear recommendations on the number of increments used to produce the bulk sample, because the number of increments required to minimize the sampling error will depend upon the heterogeneity of the lot under investigation. The lack of data on the expected distributions of real lots makes it impossible to establish objective criteria to address this problem. When defining the number of increments to be sampled, it should be taken into account that even modest levels of heterogeneity will compromise sampling reliability when 30 to 50 increments are used to produce the bulk sample (Paoletti *et al.*, 2003).

All the sampling steps necessary to produce final samples of suitable working size from large bulk samples are needed. However, assuming random distribution to estimate the errors associated with each of these secondary sampling steps is not going to pose a problem as long as grinding and mixing the material is properly carried out. Sampling size reduction should be attempted only when all the sampled material is reduced to the same particle size as the smaller and more uniform this is after milling, the more successful the mixing will be in ensuring homogeneity in the population, and ultimately in minimizing sampling error. Sampling size is selected to best meet the needs of the buyer and seller, and often involves a compromise between precision and cost (Paoletti *et al.*, 2003).

19.3 CERTIFIED REFERENCE MATERIAL

Appropriate reference materials for positive and negative controls provide the basis for the validation of analytical procedures and for assessing the performance of methods and laboratories. Reference material should be independent of the analytical methods and should be focused on raw material or base ingredients rather on finished foods (Ahmed, 2002). Various types of material can be used as a certified reference material (CRM) for the detection of GMOs: matrix based material (produced from seeds), pure DNA, or protein standards. Pure DNA standards can consist of either genomic DNA extracted from plant leaves or plasmid DNA in which a short sequence of a few hundred base pairs (bp) has been cloned and amplified in a suitable vector. Pure protein can be extracted from ground seed, or produced by recombinant DNA technology. Different requirements have to be met by CRM depending on their use for identification and/or quantitation of GMOs, and each has advantages and disadvantages (see Table 19.1). In contrast to protein detection methods, in which a single standard can be settled on relatively easy, DNA-based methods are better served through combinations of several positive controls. The availability of reference materials currently is limited owing to concerns over intellectual property rights

and costs (Serageldin, 1999). The Institute of Reference Materials and Measurements at the Joint Research Center in Geel, Belgium offer through Fluka (Buchs, Switzerland) a limited number of reference materials for modified soya, corn, and maximizer maize (MM) (Ahmed, 2002).

19.4 PROTEIN-BASED TESTING METHODS

Immunoassay technologies using antibodies are ideal for qualitative and quantitative detection of many types of proteins in complex matrices when the target analyte is known (Brett *et al.*, 1999). Both monoclonal (highly specific) and polyclonal (often more sensitive) antibodies can be used depending on the amounts needed and the specificity of the detection system (e.g., antibodies to whole protein, or to specific peptide sequences) depending on the particular application, time allotted for testing, and cost. On the basis of typical concentrations of transgenic material in plant tissue (>10 µg/tissue), the detection limits of protein immunoassays can predict the presence of modified proteins in the range of 1% GMOs (Stave, 1999). Immunoassays in which antibodies are attached to a solid phase have been used in two formats:

- A competitive assay in which the detector and analyte compete to bind with capture antibodies
- A two-site (double antibody sandwich) assay in which target analyte sandwich between the capture antibody and the detector antibody; this assay is deemed preferable

19.4.1 Western Blot

The Western blot is a highly specific method that provides qualitative results suitable for determining whether a sample

TABLE 19.1   Advantages and disadvantages of various types of GMO CRMs. Adapted from Trapmann and colleagues (2004); with permission.

Type of CRM	Advantages	Disadvantages
Matrix GMO CRMs	• suitable for protein- and DNA-based methods • extraction covered • commutability	• different extractability (?) • large production needed • low availability of raw material due to restricted use of seeds • degradation • variation of the genetic background
Genomic DNA CRMs	• good calibrant • less seeds needed • commutability	• large production needed • low availability of raw material due to restricted use of seeds • variation of the genetic background • long-term stability (?)
Pure protein DNA CRMs	• less seeds needed	• commutability (?)
Plasmidic DNA CRMs	• easy to produce in large quantities • broad dynamic range	• plasmid topology • discrepancies • commutability (?)

contains the target protein below or above a predetermined threshold level (Lipton *et al.*, 2000), and is particularly useful for the analysis of insoluble protein (Brett *et al.*, 1999). Further, because electrophoretic separation of protein is carried out under denaturing conditions, any problems of solubilization, aggregation, and coprecipitation of the target protein with adventitious proteins are eliminated (Rogan *et al.*, 1999; Sambrook and Russel, 2000). The detection limits of Western blots vary between 0.25% for seeds and 1% for toasted meal (Smyth *et al.*, 2002). This method, even though it can provide quantitative results and is sensitive, is considered more suited to research applications than to routine testing because it is not amenable to automation. Western blotting generally takes about two days and costs about $150/sample (Ahmed, 2002).

19.4.2 ELISA

ELISA assumes more than one format: a microwell plate (or strip) format, and a coated tube format. The antibody-coated microwell strips, with removable strips of 8 to 12 wells, are quantitative, highly sensitive, economical, provide high throughput, and are ideal for quantitative high volume laboratory analysis, provided that the protein is not denatured. The typical run time for a plate assay is 90 min, and an optical plate reader determines concentration levels in the samples. Detection limits for CP4 EPSPS soybean protein was 0.25% for seeds and 1.4% for toasted meal. The antibody-coated tube format is suited for field-testing, with typical run times ranging from 15 to 30 min, and tubes can be read either visually or by an optical tube reader; results are qualitative. Because there is no quantitative internal standard within the assay, no extra information can be obtained concerning the presence of GMO at the ingredient level in food. ELISA test generally takes about 1 to 2 h, and costs about $5.0/sample on the average (Ahmed, 2002).

19.4.3 Lateral Flow Strip

A lateral flow strip is a single unit device that allows for manual testing of individual samples. Each nitrocellulose strip consists of three components: a reservoir pad on which an antibody coupled to a colored particle such as colloidal gold or latex is deposited, a result window, and a filter cover (see Fig. 19.3). An analyte-specific capture antibody (Ab) is also immobilized on the strip. Inserting the strip in an eppendorf vial containing an extract from plant tissue solution harboring a transgenic protein, leads to the solution moving toward the reservoir pad solubilizing the reported Ab, which binds to target analyte and forms an analyte-Ab complex that flows with the liquid sample laterally along the surface of the strip. When the complex passes over the zone where the capture Ab has been immobilized, it binds to the Ab and produces a colored band. The presence of two bands indicates a positive test for the protein of interest. A single band indi-

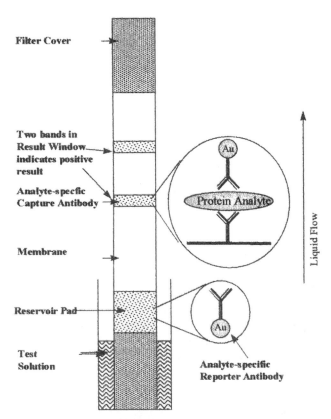

FIGURE 19.3 A schematic view of lateral flow strip assay format illustrating the principles of the assay. From Lipton and colleagues (2000); with permission.

cates that the test was performed correctly, but that there was no protein of interest present in the sample. This test provides a yes/no or threshold (semi-quantitative) determination of the target protein in about 10 min at an average cost of about $2/sample, is appropriate for field or on-site applications, and does not require specialized equipment (Ahmed, 2002). Commercially available lateral flow strips currently are limited to few biotechnology-derived protein-producing GMO products, but strips that can detect multiple proteins simultaneously are being developed (Smyth *et al.*, 2002).

19.4.4 Other Immunoassay Formats

In addition to microplate ELISA and later flow devices, other immunoassay formats use magnetic particles as the solid support surface. The magnetic particles can be coated with the capture antibody and the reaction carried out in a test tube. The particles with bound reactants are separated from unbound reactants in solution using a magnet. Advantages of this format are superior kinetics because the particles are free to move in reaction solutions, and increased precision owing to uniformity of the particles. Other less commonly used formats utilize nesting or combination of two steps in one (Brett *et al.*, 1999; Stave, 1999; Lipton *et al.*, 2000; Ahmed, 2002). In the near future, improvements in immunoassays are expected to occur via advances in antibody technology and improved instruments (Smyth *et al.*, 2002). Recent advances in proteomics have implications for GMOs' diagnosis. However, their sensitivity has to be increased, and instrument cost has to be brought down (Ahmed, 2004).

19.5 DNA-Based Testing Methods

Molecular diagnostic methods for GMOs detect and quantify those DNA sequences that have been introduced into the organism during the process of gene modification. The DNA that has been engineered into a crop consists of several elements that govern its functioning. They are typically a promoter sequence, structural gene, and a stop sequence for the gene. Although several techniques are available, two are commonly used: Southern blot, and particularly, PCR analyses. Microarray-based and real-time biosensor technologies have recently also been applied for diagnosing GMOs.

19.5.1 Southern Blot

The method involves fixing isolated sample DNA onto nitrocellulose or nylon membranes, probing with double-stranded (ds)-labeled nucleic acid probe(s) specific to the organisms whose diagnosis is desired, and detecting hybridization radiographically, fluoremetrically, or by chemiluminescence. Earlier probes were labeled with ^{32}P. However, nonradioactive probe labeling methods employing the indirect or the direct approach have been developed to allow convenient diagnosis and avoid use of radioactive isotopes. Indirect methods relying on fluorescence and chemiluminiscence have been developed (Ross *et al.*, 1999). Direct methods utilizing fluorophores are faster than indirect methods because the antibody conjugates, incubation, and associated blocking and washing steps have been eliminated (Osborne, 2000).

A comparison of the performance of PCR, ELISA, and DNA hybridization for the diagnosis of the causal agent of bacterial ring rot *Calvibacter michiganensis* subsp. *sepedonicus* in crude extracts of field grown potatoes were carried out. Results showed that PCR was slightly better than ELISA, and both PCR and ELISA were superior to DNA hybridization in detection sensitivity. On the other hand, the two DNA-based assays (PCR and DNA hybridization) have the advantage of not relying on an arbitrary positive threshold, and had greater specificity (Drennan *et al.*, 1993; Slack *et al.*, 1996).

Recently, an alternative Southern blot technology has been attempted using near infrared (IR) fluorescent dyes (emissions at approximately 700 and 800 nm) coupled to a carbodiimide reactive group and attach directly to DNA in a 5 min reaction. The signals for both dyes are detected simultaneously (limit in the low zeptomolar range) by two detectors of an IR imager, something not yet possible with conventional radioactive or chemiluminescent detection techniques (Stull, 2001).

Although Southern hybridization can be quantitative, it is mainly used as a research tool for GMOs detection, and is not suitable for routine day-to-day diagnosis because of its low throughput, as it takes on average two days to complete at a cost of $150/sample, and it is unsuitable for automation (Ahmed, 2002).

19.5.2 Polymerase Chain Reaction (PCR)

Because not all GM foods contain expressed protein(s) or having antibodies available to detect them, and the rather low expression levels of transgenic products in tissue used for human consumption (Lipton *et al.*, 2000), more sensitive PCR methods are used for diagnosis. A positive result, however, has to be conformed by a specific assay determining the unique modification (Schreiber, 1999). Two essential prerequisites for the application of PCR-based diagnostic methods are complete knowledge of the foreign gene construct within the GMO to be detected, and the ability to extract significant amounts of amplifiable DNA from the samples to be diagnosed (Ahmed, 1995); whereas the availability of CRM and criteria for standardization are the limiting factors for PCR diagnosis (Wurz *et al.*, 1999).

PCR exploits the specificity of DNA polymerase to allow the selective amplification of specific DNA segments occurring at low frequency in a complex mixture of other DNA sequences. In a standard PCR test, two pairs of primers are used: forward, sense or $5' \rightarrow 3'$; and reverse, antisense or $3' \rightarrow 5'$. These primers are designed to hybridize on opposite strands of the sequence of interest, and through a series of repetitive cycles (of 2–3 thermal steps) amplify the sequence between the primers millions of times (see also Fig. 1.2). Amplified fragments can be subjected to agarose gel electrophoresis to separate amplified DNA according to size, although other separation methods such as high performance liquid chromatography (HPLC) and capillary electrophoresis (CE) are used (Ahmed, 1995; De Palma, 2001).

PCR diagnosis of GMOs involves four critical steps:

1. Sampling and sample preparation.
2. DNA purification and aliquot size.
3. PCR amplification and detection of reaction products.
4. Interpretation of results.

Sampling issues have been discussed in a previous section. For routine purposes, a sample size in 2.5 to 3 kg range is recommended (Fagan, 2004). The field samples are ground and homogenized, and duplicate subsamples (between 1–2 g) are taken in a manner that ensures that the analytical sample is representative of field samples in accordance with recommendations of the ISO/CEN working group on GMO testing (Whitaker *et al.*, 2001). DNA is then isolated independently from the duplicate analytical subsample for later PCR analysis. Purification procedures not producing DNA that is free from PCR inhibitors or DNA degradation must be minimized, and DNA yield must be sufficient for reliable analysis. There are two basic procedures commonly used for isolating DNA from food and ingredient products for GMO diagnosis: the CTAB method, based on incubating food sample in the presence of the detergent cetyltrimethylammonium bromide; and the Wizard method, employing DNA-binding silica resins (Promega Corp., Madison, Wisconsin). Both methods produce satisfactory DNA isolation without unacceptable DNA degradation and are cost effective. Extraction methods for GMO diagnosis employing several commercial kits recently have been reviewed in detail (Terry *et al.*, 2002).

Food and agricultural products contain numerous compounds that are inhibitory to PCR such as polysaccharides, caramelized sugar, proteins, fats, cocoa extracts, phenols, Ca^{++}, Fe^{++}, and other secondary metabolites and trace compounds. Although not easy to separate these inhibitors, an extraction kit (e.g., QI Amp DNA Stool Kit™ (QIAGEN Inc., Valencia, CA)) was reported to effectively remove PCR inhibitory substances such as cocoa from highly processed foods containing GMOs (Tengel *et al.*, 2001).

Factors such as excessive heat, nuclease activity, and low pH (quite common in food processing) also contribute to DNA degradation. This is most likely for products with long shelf lives, such as prepared meatballs in tomato sauce and beef burgers. Data suggest that the critical minimum average DNA size for successful PCR analysis is approximately 400 bp (Meyer, 1999). Remedies such as increasing the amount of sample DNA added, selecting primer sets whose recognition sites within the fragment DNA are very close together resulting in amplicons approximately 100 bp long, have been recommended (Fagan, 2004).

Most currently available GMOs in the EU contain any of three genetic elements: the cauliflower mosaic virus (CaMV) 35S promoter, the nopalin synthase (NOS) terminator, or the kanamycin-resistance marker gene (nptII), and others (Ahmed, 2002). These elements also occur naturally in some plants and soil microorganisms, and can thus be detected using PCR giving false positive results. If the PCR assay gives a positive result, product-specific PCR methods that have been developed for a range of different GM foods, can be carried out. These product-specific PCR methods are based on the use of a primer pair set that spans the boundary of two adjacent genetic elements (e.g., promoters, target genes, and terminators), or that are specific for detection of the altered target gene sequence. Detection limits are in the range of 20 pg to 10 ng of target DNA and 0.0001–1% of the mass fraction of GMOs (Smyth *et al.*, 2002).

Different methods can be used to confirm the PCR results:

- Specific cleavage of the amplified product by restriction endonuclease digestion (Meyer, 1999)
- Hybridization with a DNA probe specific for a target sequence
- Direct sequencing of the PCR product (Sambrook and Russel, 2000)
- Nested PCR (Bouw *et al.*, 1998), in which two sets of primer pairs bind specifically to the amplified target sequence

19.5.2.1 QUALITATIVE PCR

The first method for GMO identification in foodstuffs developed to identify the Flavr Savr™ tomatoes was a qualitative PCR application assay because the genetic modification did not produce a protein in the plant (Meyer, 1995). This tomato contains—in addition to the polygalacturonase (PG) gene, which degrades pectin in the cell wall—the Kan^r gene, conferring resistance to kanamycin and the cauliflower mosaic virus promoter CaMV 35S. PCR detection was achieved by designing two pairs of primers: one pair amplified a 173 bp fragment for Kan^r, and the second pair amplified a 427 bp fragment that contains part of the promoter sequence (Lüthy, 1999). Other methods for detection of RR soybean, containing the genetic element from the crown gall causing bacterium *Agrobacterium tumifaciens* producing the enzyme 3-enolpyruvyl-shikimate-5-phosphate-synthase (EPSPS) that makes the plant resistant to the herbicide glyphosate, and of maximizer maize™ (MM) containing the synthetic endotoxin cryIA (b) gene employing qualitative PCR, were developed (Lüthy, 1999).

Later on, a nested PCR method was applied to the detection of EPSPS gene in Soya meal pellets and flour, as well as a number of processed complex products such as infant formulas, tofu, tempeh, soy-based desserts, bakery products, and meal-replacing products (Bouw *et al.*, 1998). In this two-step method, an outer primer was used to amplify a 352 bp fragment, followed by an inner primer set to amplify a 156 bp. This resulted in improved selectivity and sensitivity of the PCR reaction. RR bean DNA could be detected at 0.02%, but processed products (e.g., candy, biscuits, lecithins, cocoa-drink powder, and vegetarian paste) were undetectable by PCR due to DNA breakdown as a result of heating, and low pH, which resulted in increased nuclease activity leading to depurination and hydrolysis (Meyer, 1999; Wurz *et al.*, 1999). The presence of inhibitory components and low amounts of DNA in some material (e.g., lecithin, starch derivatives and refined Soya oil) makes it difficult to develop a single reliable method for detection of all products (Lüthy, 1999).

19.5.2.2 LIMITING DILUTION PCR

A semiquantitative method for RR detection based on the limited dilution method has been the official method for detection of GM foods in Germany (German Food Act LMBG § 35, Jankiewciz *et al.*, 1999). This method is based on optimization of the PCR so that amplification of an endogenous control gene will take place on an all-or-none fashion occurring at the terminal plateau phase of the PCR, and the premise that one or more targets in the reaction mixture (e.g., GMO) will give rise to a positive result. Accurate quantitation is achieved by performing multiple replicates at serial dilutions of the material(s) to be assayed.

At the limit of dilution, where some endpoints are positive and some are negative, the number of targets present can be calculated from the proportion of negative endpoints by using Poisson statistics (Sykes *et al.*, 1998). In this method, two measurements are used for setting limits for the GMOs content of foods: a theoretical detection limit ($L_{Theoret}$), defined as the lowest detectable amplification determined from the serial dilution of target DNA with/without background DNA; and a practical detection limit (L_{Pract}), defined as the lowest detectable amplicon determined by examining certified reference material CRM containing different mass fraction of GM and non-GM organisms. The $L_{Theoret}$ for both RR and MM (0.0005%) is generally two or more orders of magnitude lower than L_{Pract} (0.1%, Sykes *et al.*, 1998). An advantage of this method is that it does not require coamplification of added reporter DNA. However, caution should be exercised when using this technique because of potential contamination of PCR reactions due to various dilutions and manipulations (Hupfer *et al.*, 1998).

19.5.2.3 QUANTITATIVE ENDPOINT PCR

An important aspect in GMO food analysis is quantitation, since maximum limits of GMO in food are the basis for labeling in countries like the EU, and the fast increasing number of GM foods on the market demands the development of more advanced multidetection systems (Schreiber, 1999). Therefore, adequate quantitative PCR detection methods were developed. Quantitative competitive (QC)-PCR was first applied for the determination of the 35S-promoter in RR and MM in Switzerland as described by Studer and colleagues (1998). In this method, an internal DNA standard was coamplified with target DNA (Hübner *et al.*, 1999). The standard was constructed using linearized plasmids containing a modified PCR amplicon, which was an internal insert of 21 or a 22 bp deletion in case of RR and MM DNA, respectively. QC-PCR consists of four steps:

1. Coamplification of standard and target DNA in the same reaction tube.
2. Separation of the products by an appropriate method such as agarose gel electrophoresis and staining the gel by ethidium bromide.
3. Analysis of the gel densitometrically.
4. Determining the relative amounts of target and standard DNA by regression analysis.

At the equivalent point, the starting concentration of internal standard and target are equal. In the QC-PCR, the com-petition between the amplification of internal standard DNA and target DNA generally leads to loss of detection sensitivity. Nevertheless, the method allows for the detection of as little as 0.1% GMO DNA (Hübner *et al.*, 1999).

19.5.2.4 QUANTITATIVE REAL-TIME PCR

To overcome some of the limitations of conventional quantitative endpoint PCR, a real-time PCR was introduced that provided a large dynamic range of amplified molecule, allowing for higher sample throughput, decreased labor, and increased fluorescence (Ahmed, 2002).

Several commercially available real-time PCR thermal cyclers, although different in design and operation, all automate the analytical procedure and allow cycle-by-cycle monitoring of reaction kinetics; this mode of monitoring permits calculating the concentration of target sequence. Several formats are used to estimate the amount of PCR product:

- The ds-DNA-binding dye SYBR Green I
- Hybridization probes, or fluorescence resonance energy transfer (FRET) probes
- Hydrolysis probes (TaqMan® technology)
- Molecular beacons (see also Chapter 10)

These systems also permit differentiation between specific- and nonspecific-PCR products (such as primer-dimer) by the probe hybridization or by using melt curve analysis of PCR products, as nonspecific products tend to melt at a much lower temperature than the longer specific products (Ahmed, 2000).

19.5.2.5 QUALITY ASSURANCE ISSUES

There are generally two sources of sample-to-sample variability in PCR experiments: differences caused by variation in the quantity or quality of the samples (e.g., partial degradation or the presence of contaminants), and random sample-to-sample variation (which includes user-induced ones). Unfortunately, random variability is a fact in PCR; the best way to minimize it is to run duplicate samples and average the data. Variability caused by operator error can be minimized by making a cocktail of reagents or master mix (Fagan, 2004).

PCR is a very sensitive amplification process. Therefore, extreme care must be taken to prevent contamination of primers and PCR reagents with cDNAs, cRNAs, RNase, or DNase. PCR reactions should be set up in a laminar flow hood, employing the same precautions used in aseptic procedures, in a laboratory separate from the laboratory where the reaction will be run. Examples of these procedures include cleaning all work surfaces with DNA Away (Molecular Bio-Products, San Diego, CA), wearing gloves, using aerosol-resistant pipette tips, using dedicated pipettors, routinely cleaning the pipettes with DNA Away, using DEPC-treated water for all dilutions, and adding primers and/or DNA to the master mixes after all component tubes have been closed to prevent cross contamination (Ahmed, 1995).

Changes made in reagents by manufacturers often produce different results, or may change the level of sensitivity of the assays. Collecting and storage procedures were found to influence detectable levels of soluble markers; even using the same kits, studies demonstrated differences among results obtained by various laboratories suggesting that factors relating to assay performance may be responsible. It is important that a laboratory be attuned to quality control and good laboratory practice issues in testing and sample storage (see also Chapter 34); assessing analytical sensitivity for minimum detection level of markers under consideration; intra- and interassay variability; problems inherent in use of different suppliers—and sometimes different batches—for the same reagents; and the need to establish median, mean SD, and 5th and 95th percentiles for references' range (Ahmed, 2000).

A method for preamplification inactivation of amplified DNA that allows for the selective destruction of previously amplified DNA can be used, in which deoxynucleotide 5'-uridine triphosphate (dUTP) is substituted for deoxynucleotide 5'-thymine triphosphate (dTTP) in the master mix. All amplified DNA therefore will contain U instead of T. When a new amplification reaction is set up, the master mix is supplemented with the enzyme uracil-DNA-glycosylase (UDG) prior to the start of the temperature cycling. This enzyme catalyzes the excision of uracil from ss- and ds-DNA (but not RNA). During the first denaturation cycle of the PCR program, there is strand scission of the aglycosidic linkage, destroying contaminating templates. UDG is inactive at most temperatures used during PCR cycling (i.e., above 55°C), so newly synthesized amplicons will not be degraded prior to detection. The level of effective sterilization approaches 10^6 copies of input DNA, which is much more than should be necessary to control aerosol carryover contamination. Moreover, the presence of UDG at the start of the amplification appears to enhance the specificity of the amplification reaction. Because residual UDG activity can potentially degrade dU-PCR products under certain conditions (i.e., prolonged incubation of the product at either 4 or 25°C) following thermal cycling, the samples are kept at 72°C to protect amplified dU products. The UDG inhibitor protein Ugi is used to inactivate residual UDG activity and protect dU-containing products during benchtop manipulations (Ahmed, 2003).

Optimization experiments for PCR can be carried out by attempting various cycles (from 25 to 40); changing annealing temperatures (in 3°C increments above or below calculated annealing temperature), employing hot start Taq DNA polymerase, and varying $MgCl_2$ and dNTPs concentrations. In case of amplifying regions of high GC content, denaturants such as formamide and DMSO may be tried. Both positive and negative controls should be employed. To detect inhibitors of Taq DNA polymerase, additional validation procedures, or a method referred to as quality control PCR need to be employed (Ahmed, 2003).

Maintaining uniformly high standards of performance in the laboratory carrying out PCR diagnosis includes adherence to a thorough system of standardized operating procedures, assessment of laboratory performance from within, and through external performance assessment schemes, accreditation, and availability of international standardized and validated testing methods (Fagan, 2004).

19.5.3 DNA Microarray Technology

The study of gene expression by microarray technology is important because changes in the physiology of an organism or a cell are accompanied by changes in patterns of gene expression (Jordan, 1998). The technology is based on immobilization of cDNA or oligonucleotides on solid support. The major advances of DNA microarray technology results from the small size of the array, permitting more information to be packed into the chip, thereby allowing for higher sensitivity, enabling the parallel screening of large number of genes, and providing the opportunity to use smaller amounts of starting material (see also Chapter 15). Mainly the introduction of differently labeled fluorescent probes for control and test samples made possible miniaturization of arrays, and because microarrays can be produced in series, this facilitates comparative analysis of a number of samples. Microarray technology has been employed for diagnosing GMOs in plants and foods (Van Hal *et al.*, 2000). The principles of the technique are illustrated in Fig. 19.4.

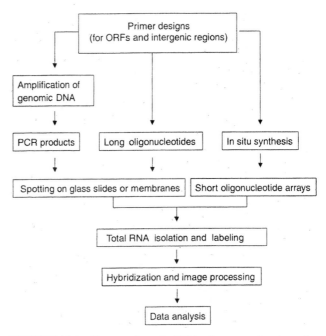

FIGURE 19.4 Construction of a DNA microarray showing principles of gene expression analysis. From Ye and colleagues (2001); with permission.

Like any analytical technique, quality assurance issues (e.g., sensitivity, specificity, ruggedness, test performance) must be rigorously tested and evaluated. The sensitivity of microarray detection is an area where research has been carried out to improve its performance. Recent reports have indicated sensitivities to be in the order of fg of purified DNA in complex samples containing potential PCR inhibitors and competing target DNAs (Wilson *et al.*, 2002). If carried out correctly, high-density oligonucleotide arrays were shown in some cases to confirm the identity of predicted sequence changes within the gene with about 99% accuracy covering a minimum of 97% of the sequence under study (Hacia *et al.*, 1998). The hindrances to utilization of the microarray technique on a wide scale in diagnosis include lack of standardization, and the initial expense of the technology. A typical basic experiment involving 50 or so pairwise hybridizations would cost between $5,000 and $10,000 in reagents. An investment of about one million dollars will bring into existence a DNA microarray facility (Ahmed, 2000). Attempts have been made to reduce cost by employing a visualization system based on an enzyme-linked system to generate a dye visible to the naked eye, and increasing the size of spots on the arrays where probes are attached form few microns to approximately 1 to 2 mm in diameter, thus allowing signal visualization by the naked eye, instead of using sophisticated confocal microscopy (Grohmann, 2002).

19.5.4 DNA-Based Biosensors

DNA biosensors are based on the immobilization of a DNA probe, and on monitoring and recording the variation of a transducer signal when the complementary target in solution interacts with the probe, forming a stable complex. There are currently two types of real-time biosensing transducers for GMO diagnosis: piezoelectric, in the form of quartz crystal microbalance (QCM), and surface plasmon resonance (SPR); both the QCM and SPR biosensors have been used by immobilizing specific 25-mer oligonucleotide probes using different sensing surfaces (e.g., a screen-printed electrode (or chip), the piezoelectric crystal, and the Biacore™ sensor (M5) dextran-coated chip). These sequences are complementary to the sequences of the CaMV 35S-promoter and the NOS terminator (Minunni *et al.*, 2001).

The piezoelectric and SPR sensors require biotinylation of the probes, but not for the electrochemical one. The PCR probe must first be denatured to produce a single-stranded (ss) DNA capable of hybridization. For QCM sensing, denaturation at 95°C for 5 min followed by incubation on ice was reported to be adequate. For SPR sensing using the Biacore X instrument (Upsala, Sweden), the thermal treatment did not allow for an adequate amount of ss-DNA to reach the sensor surface and react with the probe. However, when magnetic beads were used, the problem was overcome. Figure 19.5 shows QCM signals for three different denaturing methods: thermal, enzymatic, and magnetic using CaMV 35S PCR-amplified plasmid DNA. Signals obtained from GMO samples for different foods (e.g., dietetic snacks and drinks, certified reference material (CRM)) were treated only thermally. The system has been optimized using synthetic complementary oligos (25-mer), and the specificity of the system (which relies on the immobilized probe sequence) was tested with a noncomplementary oligo (23-mer). The hybridization study was performed also with DNA samples from CRM soybean powder con-

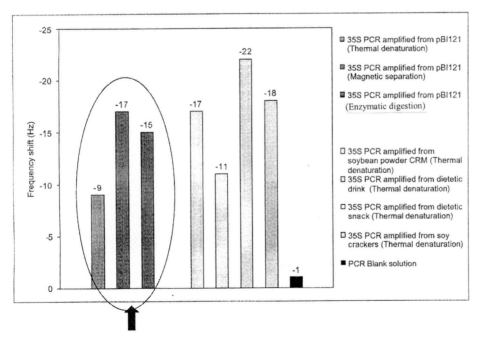

FIGURE 19.5 QCM sensor. Results obtained with different PCR amplified samples: plasmid DNA from pBI121 (containing CaMV 35S) treated with three different denaturation methods (thermal, enzymatic, and magnetic particles); transgenic CRM, and dietetic snacks and light drinks (also containing CaMV 35S). From Minunni and colleagues (2003); with permission.

taining 2% GMO and amplified by PCR. Nonamplified genomic or plasmid DNA were also used. The amplified CaMV 35S resulted in a fragment 195 bp long (Minunni, 2003).

These affinity systems are attractive for DNA sensing since their versatility often is associated with probability and with absence of any labeling. Additionally, many analyses can be performed on the same sensor surface with the possibility of reuse of the QCM and SPR devices up to 30 times and more than 100 times, respectively. Future prospects in this rapidly developing area will result—within few years—in new equipment and formats that will enhance the detection sensitivity of these methods.

The BIACORE X instrument is priced at $112,000. The chip employed for SPR analysis (e.g., dextran modified) cost about $200 per chip, with one chip capable of performing 100 analyses of PCR-amplified samples. A system based on piezo sensing, however, is much cheaper, averaging about $13,000 (including software). The sensing element (i.e., piezoelectric crystal) cost approximately $25, allowing up to 25 analyses per surface, and doubling the number of reactions with both surfaces used. Reagents for both methods cost nearly the same (about $70 for biotinylated probes immobilized on the transducer surface). Usually about 35 probes are needed for the analysis, making a probe cost per chip in the range of $1. Synthetic oligonucleotides are needed to generate a calibration curve, at a cost of $3 per chip.

Based on affinity biosensing, GeneScan Europe recently has introduced a test kit for diagnosing of GMOs in food products, which allows a multiplex PCR for the specific detection of DNA sequences from plant species and GM traits using a biosensor chip and a biochip reader. The detection limit for the GMO Chip kit is in the range of 250 copies of each of the target DNA sequences in the PCR (GeneScan Europe, 2001).

19.6 NEAR-INFRARED (NIR) TECHNOLOGY

NIR transmittance spectroscopy has been widely used by grain handlers in elevators in the United States and in most of the world for nondestructive analysis of whole grains for the prediction of parameters including moisture, protein, oil, fiber, and starch. The technique has recently been used to distinguish RR from conventional soybean, but unfortunately, the sensitivity has been quite low (Rousel *et al.*, 2001). Attempts are being made to increase the sensitivity of NIR systems by accumulating an adequate database from which to glean a common NIR "signature" for a given GMO, and linking the systems to sophisticated computers to increase signal detection, but the capacity has not yet been realized (Roussel and Cogdill, 2004).

TABLE 19.2   Summary of methods that specifically detect recombinant deoxyribonucleic acid (rDNA) products produced by GM foods. Modified from Ahmed (2002); with permission.

Parameter	Protein-based			DNA-based					
	Western blot	ELISA	Lateral flow strip	Southern blot	Qualitative[@] PCR	QC-PCR and Limiting dilution	Real-time PCR	DNA Microarrays	DNA sensors
Needs special equipment	Difficult Yes	Moderate Yes	Simple No	Difficult Yes	Difficult Yes	Difficult Yes	Difficult Yes	Difficult Yes	Difficult Yes
Sensitivity	High	High	High	Moderate	Very high	High	High	High	Low
Duration[#]	2 d	30–90 min	10 min	6 h*	1.5 d	2 d	1 d	2 d	2 d
Cost/sample	US$150	US$5	US$2	US$150	US$250	US$350	US$450	US$600	US$200
Provides quantitative results	No	Yes[$]	No	No	No	Yes	Yes**	Yes	No
Suitable for field test	No	Yes[$]	Yes	No	No	No	No	Yes	Yes
Employed mainly in	Academic labs	Test facility	Field testing	Academic labs	Test facility	Test facility	Test facility	Academic* labs	Academic* labs

◆ NIR detects structural changes (not DNA or protein), is fast (<1 min) and inexpensive (~$1).
@ Including nested PCR and GMO Chip.
# Excluding time allotted for sample preparation.
* When nonradioactive probes are used; otherwise 30 h with ^{32}P-labeled probes.
$ As in the antibody-coated tube format.
** With high precision.
* In development.

19.7 CONCLUSIONS

The commonly used methods for diagnosing GMOs in foods are presented in Table 19.2. To respond to global regulations requiring food labeling, a tiered approach may be employed using first qualitative PCR for GMOs detection. If no GMOs were detected using a validated qualitative method, the product(s) would be evaluated for the presence of protein. If no protein is detected, the product is presumed not detectable. If the qualitative PCR showed a positive result, the product is considered as "nonapproved GMO," and a validated real-time PCR is used to detect the level of GMO. If the level is above an established threshold, the product also is considered nonapproved GMO, but if below the threshold, the product need not be labeled (Ahmed, 2002). The high sensitivity and specificity of PCR methods and their ability to be applied to different food matrices make them suitable for diagnosing GMOs at low thresholds in various foods. Developments in DNA microarrays, biosensor devices, and proteomics promises to increase diagnostic approaches for GMOs detection dramatically (Ahmed, 2004).

The greatest uncertainty of using DNA-based assays, as for protein-based methods, is that not all products derived from GM foods (e.g., refined oil) contain enough DNA. In addition, heating and other processes associated with finished food production can degrade DNA. Similarly, if GMO is expressed on a relative basis (i.e., % GMO), it is important to know whether the estimate is to be based on total DNA from all sources, or on the basis of analyzed product DNA. This approach, known as genetic equivalence, which is sound and pragmatic, was correlated with results of studies where GMO content was expressed on a percent weight basis. Quantitative PCR might best be applied at the early stages in the food production chain. Using the genome equivalent approach to assess the GMO content of food ingredients and tracking the ingredients used should allow for accurate estimation of GMOs. This approach is also consistent with current EU food labeling regulations that focus on ingredients, and is also applicable to finished products containing more than one GMO-derived ingredient (Ahmed, 2002).

References

Agius, F., Gonzalez-Lamothe, R., Cabarello, J. L., Manoz-Blanco, J., Botella, M. A., and Valpuesta, V. (2003). Engineering increased vitamin C levels in plants by overexpression of a D-galacturonic acid reductase. *Nat. Biotechnol.* 21, 177–181.

Ahmed, F. E. (1995). Applications of molecular biology to biomedicine and toxicology. *J. Env. Sci. Health* C11, 1–51.

Ahmed, F. E. (2000). Molecular markers for early cancer detection. *J. Env. Sci. Health* C18, 75–125.

Ahmed, F. E. (2002). Detection of genetically modified organisms in foods. *Trends Biotechnol.* 20, 215–223.

Ahmed, F. E. (2003). Colon cancer: prevalence, screening, gene expression and mutation, and risk factors and assessment. *J. Env. Sci. Health* C21, 65–131.

Ahmed, F. E. (2004). Other methods for GMOs detection & overall assessment of the risks. In *Testing of Genetically Modified Organisms in Food*, F. E. Ahmed, ed., pp. 285–313, Haworth Press, Binghamton, NY.

Bouw, E., Keijer, J., Kok, E. J., Kuiper, H. A., and Van Hoef, A. M. A. (1998). Development and application of a selective detection method for genetically modified soy and soy-derived products. *Food Addit. Contam.* 15, 767–774.

Brett, G. M., Chambers, S. T., Huang, L., and Morgan, M. R. A. (1999). Design and development of immunoassays for detection of proteins. *Food Control* 10, 401–406.

Butler, D., and Reichardt, T. (1999). Long-term effect of GM crops serves up food for thought. *Nature* 398, 651–656.

Commission Regulation (EC) No 50/2000, 10 January 2000a, on the labeling of foodstuffs and food ingredients containing additives and flavorings that have been genetically modified or have been produced from genetically modified organisms. *Official J. Eur. Communities: Legislation* 6, 15–17.

Commission Regulation (EC) No 49/2000, 10 January 2000b, amending Council Regulation (EC) No 1139/98 concerning the compulsory indication on the labeling of certain foodstuffs produced from genetically modified organisms of particulars other than those provided for in Directive 79/112/EEC, *Official J. Eur. Communities: Legislation* 6, 13–14.

Council Regulation (EC) No 285/97, 27 January 1997, concerning Novel Foods and Novel Food Ingredients. *Official J. European Communities*, No. L43, 1–5.

Council Regulation (EC) No 1139/98, 26 May 1998, concerning the compulsory indication of the labelling of certain foodstuffs produced from genetically modified organisms, *Official J. Eur. Communities: Legislation* 159, 4–7.

De Palma, A. (2001). Capillary electrophoresis. *Gen. Eng. News* 21, 21–22, 81.

Drennan, J. L., Westra, A. A. G., Slack, S. A., Delserone, L. M., and Collmer, A. (1993). Comparison of a DNA hybridization probe and ELISA for the detection of *Calvibacter michiganensis* subsp. *Sepodonicus* in field-grown potatoes. *Plant Dis.* 77, 1243–1247.

Editorial (2000). Overseeing Biotech Foods. *Genet. Engineer.* 20, 1, 37, 83.

Fagan, J. (2004). DNA-based methods for detection and quantification of GMOs. In *Testing of Genetically Modified Organisms in Food*, F. E. Ahmed, ed. Haworth Press, pp. 163–220, Binghamton, NY.

Gaskell, G., Bauer, M. W., Durant, J., and Allum, N. C. (1999). Worlds apart? The reception of genetically modified food in Europe and the U.S. *Science* 285, 384–386.

GeneScan Europe (2001). GMO Chip: Test Kit for the Detection of GMOs in Food Products, Cat. No. 5321300105, Bremen, Germany.

Gilbert, J. (1999). Sampling of raw materials and processed foods for the presence of GMOs. *Food Control* 10, 363–365.

Grohmann, L. (2002). GMO chip internal validation results. Symposium—GMO Analytik Heute, Frankfurt.

Hacia, J. G., Makalowski, W., Edgemon, K., Erdos, M. R., Robbins, C. M., Fodor, S. P. A., Brody, L. C., and Collins, F. S. (1998). Evolutionary sequence comparison using high-density oligonucleotide array. *Nat. Genet.* 18, 155–158.

Hardegger, M., Brodmann, P., and Hermann, A. (1999). Quantitative detection of the 35S promoter and the NOS terminator using quantitative competitive PCR. *Eur. Food Res. Technol.* 209, 83–87.

Haslberger, A. G. (2000). Monitoring and labeling for genetically modified products. *Science* 287, 431–432.

Hellemans (2003). A Consumers fear cancels European GM research. *The Scientist* 17, 52–54.

Hemmer, W. (1997). Foods Derived from Genetically Modified Organisms and Detection Methods. *BATS Report* 2/97.

Hübner, P., Studer, E., and Lüthy, J. (1999). Quantitative competitive PCR for the detection of genetically modified organisms in food. *Food Control* 10, 353–358.

Hupfer, C., Hotzel, H., Sachse, K., and Engel, K-H. (1998). Detection of genetic modification in heat treated products of Bt maize by polymerase chain reaction. *Z. Lebensum. Unters. Forsch.* 206, 203–207.

Jankiewicz, A., Broll, H., and Zagon, J. (1999). The official method for the detection of genetically modified soybeans (German Food Act LMBG 35): a semi-quantitative study of sensitivity limits with glyphosate-tolerant soybeans (Roundup Ready) and insect resistant Bt maize (Maximizer). *Eur. Food. Res. Technol.* 209, 77–82.

Jordan, B. R. (1998). Large-scale expression measurement by hybridization methods: from high-density membranes to "DNA chips." *J Biochem.* 124, 251–258.

Kleter, G. A., van der Krieken, W. M., Kok, E. J., Bosch, D., Jordi, W., and Gilissen, L. J. W. J. (2001). Regulation and exploitation of genetically modified crops. *Nat. Biotechnol.* 19: 1105–1110.

Kuiper, H. A. (1999). Summary report of the ILSI Europe Workshop on detection methods for novel foods derived from genetically modified organisms. *Food Control* 10, 339–349.

Kuiper, H. A., Kleter, G., Noteborn, H. P. J. M., and Kok, E. (2001). Assessment of the food safety issues related to genetically modified foods. *Plant J.* 27: 503–528.

Lipp, M., Brodmann, P., Pietsch, K., Pauwles, J., Anklam, E., Borchers, T., Braunschweiger, G., Busch, U., Eklund, E., Eriksen, F. D., Fagan, J., Fellinger, A., Gaugitsch, H., Hayes, D., Hertel, C., Hortner, H., Joudrier, P., Kruse, L., Meyer, R., Miraglia, M., Muller, W., Phillipp, P., Popping, B., Rentsch, R., and Wurtz, A. (1999). IUPAC collaborative trial study of a method to detect genetically modified soybeans and maize in dried powder. *J. AOAC Int.* 82, 923–928.

Lipton, C. R., Dautlick, X., Grothaus, G. D., Hunst, P. L., Magin, K. M., Mihaliak, C. A., Ubio, F. M., and Stave, J. W. (2000). Guidelines for the validation and use of immunoassays for determining of introduced proteins in biotechnology enhanced crops and derived food ingredients. *Food Agric. Immunol.* 12, 153–164.

Lüthy, J. (1999). Detection strategies for food authenticity and genetically modified foods. *Food Control* 10, 359–361.

Meyer R. (1995). Nachweis gentechnisch der Flavr Savr™-Tomate. *Z. Lebensum. Unters. Forsch.* 201, 583–586.

Meyer, R. (1999). Development and application of DNA analytical methods for the detection of GMOs in food. *Food Control* 10, 391–399.

Minunni, E. (2003). Biosensors based on nucleic acid interaction. *Spectroscopy* 17, 613–625.

Minunni, M., Tombelli, S., Mariotti, E., Mascini, M., and Mascini, M. (2001). Biosensors as a new analytical tool for detection of genetically modified organisms (GMOs). *Fres. J. Anal. Chem.* 369, 589–593.

Moseley, B. E. B. (1999). The safety and social acceptance of novel foods. *Int. J. Food Microbiol.* 50, 25–31.

Osborne, J. (2000). A review of radioactive and non-radioactive based techniques used in life science applications—Part I: Blotting techniques. *Life Sci News (Amersham Biosciences)* 6, 1–4.

Paoletti, C., Donatelli, M., Kay, S., and Van der Eede, G. (2003). Simulating kernel lot sampling: the effect of heterogeneity on the detection of GMO contamination. *Seed Sci. Technol.* 31, 629–638.

Remund, K., Dixon, D. A., Wright, D. L., and Holden, L. R. (2001). Statistical considerations in seed purity testing for transgenic traits. *Seed Sci. Res.* 11, 101–119.

Rogan, G. J., Dudin, Y. A., Lee, T. C., Magin, K. M., Astwood, J. D., Bhakta, N. S., Leach, J. N., Sanders, P. R., and Fuchs, R. L. (1999). Immunodiagnostic methods for selection of 5-enolpyruvyl shikimate-3-phosphate synthase in Roundup Ready® soybeans. *Food Control* 10, 407–414.

Ross, R., Ross, X-L, Rueger, R. B., Laengin, T., and Reske-Kunz, A. B. (1999). Nonradioactive detection of differentially expressed genes using complex RNA or DNA hybridization probes. *BioTechniques* 26, 150–155 .

Roussel, S., and Cogdill, R. P. (2004). Near infrared spectroscopic methods. In *Testing of Genetically Modified Organisms in Food*, F. E. Ahmed, ed., pp. 255–284, Haworth Press, Binghamton, NY.

Roussel, S. A., Hardy, C. L., Hurburgh, C. R. Jr., and Rippke, G. R. (2001). Detection of Roundup Ready™ soybean by near-infrared spectroscopy. *Applied Spectroscopy* 55, 1425–1430.

Sambrook, J., and Russel, D. (2000). *Molecular Cloning: A Laboratory Manual*, 3rd edition, Cold Spring Harbor Laboratory Press.

Schreiber, G. A. (1999). Challenges for methods to detect genetically modified DNA in foods. *Food Control* 10, 351–352.

Serageldin, I. (1999). Biotechnology and food security. *Science* 285, 387–389.

Shubert, D. (2002) A different perspective on GM food. *Nat. Biotechnol.* 20, 696.

Slack, S. A., Drennan, J. L., and Westra, A. A. G. (1996). Comparison of ELISA and hybridization for the detection of *Calvibacter michiganensis* subsp. *Sepodonicus* in field-grown potatoes. *Plant Dis.* 80, 519–524.

Smyth, S., Khachatourians, G. G., and Phillips, P. W. B. (2002). Liabilities and economics of transgenic crops. *Nat. Biotechnol.* 20, 537–541.

Stave, J. W. (1999). Detection of new or modified proteins in novel foods derived from GMO-future needs. *Food Control* 10, 367–374.

Stave, J. W. (2004). Protein-based methods: case studies for GMO detection. In *Testing of Genetically Modified Organisms in Food*, F. E. Ahmed, ed., pp. 147–161, Haworth Press, Binghamton, NY.

Studer, E., Rhyner, C., Lüthy, J., and Hübner, P. (1998). Quantitative competitive PCR for the detection of genetically modified soybean and maize. *Z Lebensum Unters Forsch* 207, 207–213.

Stull, D. (2001). A feat of fluorescence. *The Scientist* 15, 20–21.

Sykes, P. J., Neoh, S. H., Brisco, M. J., Hughes, E., Condon, J., and Morley, A. A. (1998). Quantitation of targets for PCR by use of limiting dilution. In *The PCR Technique: Quantitative PCR*, J. W. Larrick, ed., pp. 81–93, Eaton Publishing Co., Natick, Massachusetts.

Tengel, C., Scuber, P., Setzke, E., Balles, J., and Sprenger-HauBels, M. (2001). PCR-based detection of genetically modified soybean and maize in raw and highly processed food stuff. *BioTechniques* 31, 426–429.

Terry, C. F., Harris, N., and Parkes, H. C. (2002). Detection of genetically modified crops and their derivatives: critical steps in sample preparation and extraction. *J. AOAC Inter.* 85, 768–774.

Trapmann, S., Corbisier, H., and Schimmel, H. (2004). Reference materials and standards. In *Testing of Genetically Modified Organisms in Food*, F. E. Ahmed, ed., pp. 101–115, Haworth Press, Binghamton, NY.

US National Academy of Sciences (2000). *Genetically Modified Pests-Protected Plants: Science and Regulations*, National Academy Press.

Van Hal, N. L. W., Vorst, O., van Houwelingen, A. M. M. L., Kok, E. J., Peijumenburg, A., Aharoni, A., van Tunen, A. J., and Keijer, J. (2000). The application of DNA microarrays in gene expression analysis. *J. Biotechnol.* 78, 271–280.

Whitaker, T. B., Freese, L. F., Giesbrecht, F. G., and Slate, A. B. (2001). Sampling grain shipments to detect genetically modified seed. *J. AOAC Inter.* 84, 1941–1946.

Wilson, W. J., Strout, C. L., DeSantis, T. Z., Stilwell, J. L., Carrano, A. V., and Andersen, G. L. (2002). Sequence-specific identification of 18 pathogenic microorganisms using microarray technology. *Mol. Cell. Probes* 16, 119–127.

Wurz, A., Bluth, A., Zeltz, P., Pfeifer, C., and Willmund, R. (1999). Quantitative analysis of genetically modified organisms (GMO) in processed food by PCR-based methods. *Food Control* 10, 385–389.

Ye, R. W., Wang, T., Bedzyk, L., and Croker, K. M. (2001). Application of DNA microarrays in microbial systems. *J. Microbiol.* 47, 257–272.

CHAPTER 20

Pharmacogenetics and Pharmacogenomics: Impact on Drug Discovery and Development

KLAUS LINDPAINTNER

Roche Genetics, F. Hoffmann-La Roche, Ltd, Basel, Switzerland

TABLE OF CONTENTS

20.1 INTRODUCTION

Genetics and genomics today are widely proclaimed as about to revolutionize the face of medicine. In a more measured assessment, their use provides significant opportunities, based on a more fundamental understanding of cell biology and molecular disease pathology, and raises significant challenges with regard to work yet to be done. The implementation of molecular genetics and biology will continue to provide, as it has done already, better ways to diagnose and treat illnesses, but it will do so at a stepwise and evolutionary pace, based on an improved understanding of the nature of disease, allowing more specific treatments, better risk prediction, and the implementation of preventive strategies. As such, future progress in biomedicine will travel the same well-treaded paths of improved differential diagnosis and risk prediction along which it has advanced for the last decades and centuries. So, although meaningful biomedical research today, by and large, depends on the use of the newly developed tools of genetics and genomics and the insights gained through them, it is unlikely to fundamentally change the direction of medical progress.

Advances made over the last 30 years in molecular biology, molecular genetics and genomics, and in the development and refinement of associated methods and technologies have had a major impact on the understanding of biology, including the action of drugs and other biologically active xenobiotics. The tools that have been developed to allow these advances, and the knowledge of fundamental principles underlying cellular function thus derived, have become quintessential and indeed indispensable for almost any kind and field of biological research, including future progress in biomedicine and health care.

One aspect in particular of the broad scope across which progress in biology has been achieved, namely the understanding of genetics and especially, sequencing of the human genome, has uniquely captured the imagination of both scientists and the public. Given the austere beauty of Mendel's laws of inheritance, the compelling aesthetics of the double helix structure, and the awe-inspiring accomplishment of cataloging billions of base pairs, and last, but not least, a public relations campaign unprecedented in its scope in the history of scientific achievement, this reaction is quite understandable. However, the high expectations raised, regarding the degree and timeframe of impact that these technologies will have on the practice of health care, are almost certainly unrealistic. Situated at the interface between pharmacology and genetics/genomics, pharmacogenetics and pharmacogenomics (usually without any further definition of what these terms mean) are commonly touted as heralding a revolution in medicine.

It is important to realize that, with regard to pharmacology and drug discovery, accomplishments in basic biology—starting sometime in the last third or quarter of the last century—have indeed already led to what may well be considered a rather fundamental, perhaps paradigmatic shift from the chemical paradigm to a biological paradigm: historically, drug discovery was driven by medicinal chemistry, with biology serving an almost secondary, ancillary role that examined new molecules for biological function. The ability to comprehend cell biology and function, based on a newly developed set of tools to investigate the physiological effects of biomolecules and pathways on their basic, molecular level, has since reversed this directionality: the biologist is now driving the process, requesting from the chemist compounds that modulate the function of these biomolecules or pathways, with the expectation of a more predictable impact on physiological function and the correction of its pathological derailments.

Indeed, as pointed out earlier, the major change in how drugs are discovered, from the chemical to the biological paradigm, has already occurred some time ago; what the current advances, in due time, promise to allow researchers to do is to move from a physiology-based to a (molecular) pathology-based approach toward drug discovery, promising the advancement from a largely palliative to a more cause/contribution-targeting pharmacopoeia.

This chapter is intended to provide a necessarily somewhat subjective view of what the disciplines of genetics and genomics stand to contribute, and how they actually have contributed for many years, to drug discovery and development, and more broadly to the practice of health care. Particular emphasis will be placed on examining the role of genetics, acquired or inherited variations at the level of DNA-encoded information, in real life—that is, with regard to common complex disease—a realistic understanding of this role is absolutely essential for a balanced assessment of the impact of genetics on health care in the future. Definitions for some of the terms that are in wide and often unreflected use today, almost always sorely missing from both academic and public policy-related documents on the topic, will be provided, with an understanding that much of the field is still in flux, and that these may well change.

With regard to the actual opportunities and challenges genetics and genomics provide in the field of health care, a perspective staged by time windows is helpful. Thus,

- The aspect that holds the greatest promise, successfully targeting newly recognized, causally relevant targets with innovative drugs based on a more fundamental and functional understanding of disease causation or contribution on the molecular level, is also the one that lies farthest in the future, and commensurately little can be said about it lest we were to lose ourselves in pure speculation.
- The more mid-term impact of these technologies, applicable to the earlier stages of drug discovery, will be covered in some detail, as here, too, the true relevance of various applications of genomic technologies have yet to be fully established.
- The most imminent application to medicines that are either on the market or in clinical trials (i.e., pharmacogenetics) will receive the major emphasis, for the obvious reasons that these applications are in part already being implemented today in clinical trials. Here, a more systematic classification than generally found will be attempted.

It is important to remain mindful that what will be discussed is, to a large extent, still uncharted territory, so by necessity many of the positions taken, reasoned on today's understanding and knowledge, must be viewed as somewhat tentative in nature. Where appropriate and possible, select examples will be provided, although it should be pointed out that much of the literature in the area of genetic epidemiology and pharmacogenetics lacks the stringent standards normally applied to peer-reviewed research, and replicate data are generally absent.

20.2 DEFINITION OF TERMS

There is widespread indiscriminate use of, and thus confusion about, the terms pharmacogenetics and pharmacogenomics. Although no universally accepted definition exists, there is an emerging consensus on the differential meaning and use of the two terms (see Fig. 20.1).

20.2.1 Pharmacogenetics

The term *–genetics* relates etymologically to the presence of individual properties, and interindividual differences in

- **Pharmacogenetics:**

  o Differential effects of a drug, *in vivo*, in different patients, dependent on the presence of inherited gene variants

  o Assessed primarily by **genetic (SNP) and genomic (expression) approaches**

  o A concept to provide more patient/disease-specific health care

  o **One** drug – **many** genomes (i.e., different patients)

  o Focus: **patient** variability

- **Pharmacogenomics:**

  o Differential effects of compounds, *in vivo* or *in vitro*, on gene expression, among the entirety of expressed genes

  o Assessed by **expression profiling**

  o A tool for compound selection/drug discovery

  o **Many** drugs (i.e., early-stage compounds) – **one** genome [i.e., normative genome (database, technology platform)]

  o Focus: **compound** variability

FIGURE 20.1 Terminology of pharmacogenetics and pharmacogenomics.

these properties, as a consequence of having inherited (or acquired) them. Thus, the term *pharmacogenetics* describes the interactions between a drug and an individual's (or perhaps more accurately, groups of individuals) characteristics as they relate to differences in the DNA-based information. Pharmacogenetics, therefore, refers to the assessment of clinical efficacy and/or the safety and tolerability profile, the pharmacological, or response-phenotype, of a drug in groups of individuals that differ with regard to certain DNA-encoded characteristics and tests the hypothesis that these differences, if indeed associated with a differential response-phenotype may allow prediction of individual drug response. The DNA-encoded characteristics are most commonly assessed based on the presence or absence of polymorphisms at the level of the nuclear DNA, but may be assessed at different levels where such DNA variation translates into different characteristics, such as differential mRNA expression or splicing, protein levels or functional characteristics, or even physiological phenotypes, all of which would be seen as surrogate, or more integrated markers of the underlying genetic variant. It should be noted that some authors continue to subsume all applications of expression profiling under the term pharmacogenomics, in a definition of the term that is more driven by the technology used rather than by functional context.

20.2.2 Pharmacogenomics

In contrast, the terms pharmacogenomics, and its close relative, toxicogenomics, are etymologically linked to genomics, the study of the genome and of the entirety of expressed and nonexpressed genes in any given physiologic state. These two fields of study are concerned with a comprehensive, genome-wide assessment of the effects of pharmacological agents, including toxins/toxicants on gene expression patterns. Pharmacogenomic studies are thus used to evaluate the differential effects of a number of chemical compounds, in the process of drug discovery commonly applied to lead selection with regard to inducing or suppressing the expression of transcription of genes in an experimental setting. Except for situations in which pharmacogenetic considerations are front-loaded into the discovery process, interindividual variations in gene sequence usually are not taken into account in this process.

In contrast to pharmacogenetics, pharmacogenomics therefore does not focus on differences among individuals with regard to the drug's effects, but rather examines differences among several (prospective) drugs or compounds with regard to their biological effects using a generic set of expressed or nonexpressed genes. The basis of comparison is quantitative measures of expression, using a number of more or less comprehensive gene-expression-profiling methods, commonly based on microarray formats. By extrapolation from the experimental results to theoretically desirable patterns of activation or inactivation of expression of genes in the setting of integrative pathophysiology this approach is hoped to provide a faster, more comprehensive, and perhaps even more reliable way to assess the likelihood of finding an ultimately successful drug than previously available schemes involving mostly *in vivo* animal experimentation.

Thus, although both pharmacogenetics and pharmacogenomics refer to the evaluation of drug effects using (primarily) nucleic acid markers and technology, the directionalities of their approaches are distinctly different: pharmacogenetics represents the study of *differences among a number of individuals with regard to clinical response to a particular drug (one drug, many genomes)*, whereas pharmacogenomics represents the study of *differences among a number of compounds with regard to gene expression response in a single (normative) genome/expressome (many drugs, one genome)*. Accordingly, the fields of intended use are distinct: the former will help in the clinical setting to find the medicine most likely to be optimal for a patient (or the patients most likely to respond to a drug); the latter will aid in the setting of pharmaceutical research to find the best drug candidate from a given series of compounds under evaluation.

20.3 LONG-TERM TIMEFRAME: CAUSATIVE TARGETS ADDRESSING DERANGED FUNCTION DIRECTLY

20.3.1 Palliative and Causative Acting Drugs

By far the largest fraction of today's pharmacopoeia does not target disease at its cause, as these causes are largely unknown, but by modulating a pathway that affects the disease-relevant phenotype or function. Such drugs are referred to as symptomatic or palliative agents. The pathways they target are known from more than a century of physiological, biochemical, and pharmacological research. The pathways they modulate are disease-phenotype-relevant (albeit not disease-cause-relevant), and though they are not dysfunctional, their modulation can be used effectively to counterbalance the effect of a dysfunctional, disease-causing pathway. Thus signs and symptoms of the disease can be alleviated, often with striking success, notwithstanding the fact that the real cause of the disease remains untouched. A classic example of such an approach is the acute treatment of thyrotoxicity with β-adrenergic blocking agents: even though the sympathetic nervous system in this case does not contribute causally to tachycardia and hypertension, dampening even its baseline tonus through this class of rapidly acting drugs can quickly and successfully relieve the cardiovascular symptoms and signs of this condition, and may well prevent a heart attack if the patient has underlying coro-

nary disease, before the causal treatment (in this case available through partial chemical ablation of the hyperactive thyroid gland) can take effect.

20.3.2 The Two-Fold Challenge of Common Complex Disease

It stands to reason, of course, that a drug that addresses the actual cause of the disease should provide superior treatment. However, finding these deranged functions is not trivial, even with the aid of all the molecular biological, genetic, and genomic tools that are presently available.

There is an emerging consensus that all common complex diseases (the health problems that are by far the main contributors to society's disease burden as well as to public and private health spending) are multifactorial in nature; that is, that they are brought upon by the coincidence of certain intrinsic (inborn or acquired) predispositions and susceptibilities on the one hand, and extrinsic, environment-derived influences on the other, with the relative importance of these two influences varying across a broad spectrum. In some diseases external factors appear to be more important, whereas in others intrinsic predispositions prevail.

However, it is important to note, as it is commonly neglected in discussions on this topic, that in the majority of common complex diseases, genetic factors are less important than environmental and lifestyle factors, as exemplified by heritability coefficients generally between 0.2 and 0.5 for these conditions. Thus, it must be recognized from the outset that by targeting the genetic aspects of these diseases it is expected to make a minor contribution to curing or preventing them.

The complex nature of these diseases renders the discovery of genetic variants that contribute causally to any of them a major challenge. The complexity confronted occurs on two levels: Several inherent predisposing susceptibility traits, and generally more than one environmental or lifestyle risk factor coincide in any one individual for the disease to occur; thus any of the genetic variants present provides only a modest contribution to overall disease causation, and thus investigational hurdles of discovering it. This intraindividual complexity is further accentuated by interindividual diversity based on the fact that any one clinical diagnosis is bound to be etiologically heterogeneous at the level of molecular pathology. Thus, consensus exists that the same conventional clinical diagnosis given to different individuals is quite likely to reflect the outcome of different constellations of inborn susceptibility factors and/or of environmental and lifestyle-related risks. So, it is expected that, both on the level of an individual patient and even more so on the population level, where all relevant genetic-epidemiological studies need to be conducted, the disease-causing (or better, -contributing) effect of any one intrinsic, genetically encoded characteristic will be, by and large, quite modest, and likely drowned out by noise, unless very large (and very expensive) studies are conducted. Carrying out such studies, which will attach function and clinical outcomes to the human genome sequence and its variations, is a huge task that looms many times larger than the sequencing effort itself; however, it is the *sine-qua-non* without which the whole genome sequencing effort essentially will remain inconsequential.

After large-scale efforts based primarily on the sib-pair linkage study design have failed, with few exceptions, to yield disease-gene variants in common complex disease, it appears that populations with large, genealogically well-characterized families and a certain degree of genetic isolation may provide the most likely successful approach toward characterization of these disease susceptibility gene variants. Not many such populations exist; examples are the Utah Mormons, the Icelandic nation, and the French settlers in Northern Quebec, where initial work has proved to be promising. Of course, results obtained in such isolated populations will need to be validated in other, more mainstream populations, if the new target is to be pursued strictly as a causative target.

Common, complex diseases, and thus the vast majority of what is to be clinically applied genetics behave almost fundamentally different from rare, classic, monogenic, Mendelian diseases: whereas in the latter the impact of the genetic variant is typically categorical in nature, that is, deterministic, in the former case the presence of a disease-associated genetic variant is merely of probabilistic value, raising (or lowering) the likelihood of disease occurrence to some extent, but never predicting it in a black-and-white fashion.

Communicating this difference to a public that has long been misled into a perception of everything genetic being of deterministic, Mendelian quality, represents a second, no less important and difficult challenge. Unless this effort is successful, by engaging in a true dialogue with all stakeholders, in providing the basis for informed discourse and sensible decision-making on the societal level, the full potential of our deepening understanding of biology and of these technological advances will not be recognized.

20.3.3 Applicability of Genetic Targets/Medicines

Will such newly discovered causative targets, and the medicines that may eventually be developed to modulate them, be indeed applicable only to that fraction of the population with the clinical diagnosis in whom the targeted mechanism contributes materially to the disease? It is too early to tell. Cleary, in the latter group such medicines may be expected to be particularly effective, and sometimes they will be exclusively effective in these patients; however, by uncovering what essentially will commonly be a new, previously not recognized mechanistic pathway, such drugs may be of value as palliative medicines also in those individuals in whom the mechanism in question is actually not dysfunctional. The former case is illustrated by Herceptin (see later); an example for the latter is the finding that glucokinase activators, which correct the molecular defect in the rare MODY (mature onset diabetes of the young) type 2 patients in whom the enzyme is dysfunctional, can raise the activity of normal glucokinase and thus lower glucose in just about everyone.

20.3.4 Acceleration of Drug Discovery/Development Through Better Targets

Hopes have been nurtured by some that the implementation of genetics and genomics would smarten up the drug discovery process and thus potentially accelerate it and reduce its cost. Quite on the contrary, it appears for now that the opposite is happening: Latencies from inception of a project to the launch of a new drug have lengthened to almost 15 years, and the average cost per successful launch has gone up to almost $900 million (information available at http://www.bcg.com/publications). Interestingly, about two-thirds of the costs incurred are now apparently spent in the preclinical phase, according to this report.

Although it is unclear to what extent this data truly reflect reality, it actually stands to reason that the preclinical phase, in particular, would be more lengthy and expensive if one tackles a completely novel target, about which, other than the association with disease that the genetic approach provides, virtually nothing is known. This may be complicated by the fact that such targets may not belong to the classic few druggable target families, such a target may not be chemically tractable at all, or may encounter additional hurdles due to the bias of most chemical libraries in favor of conventional target families. These formidable challenges are somewhat counterbalanced by the expectation that targets selected based on causative disease contribution, overall, may have a somewhat higher likelihood of success; thus, if today's attrition rate of 9 per 10 compounds that are introduced into clinical testing could be reduced by as little as one, to 8 per 10, this would already translate, over time, into a doubling of productivity.

20.4 MID-TERM TIMEFRAME— PHARMACOGENOMICS/ TOXICOGENOMICS: FINDING NEW MEDICINES QUICKER AND MORE EFFICIENTLY

Once a screen (assay) has been set up in a drug discovery project and lead compounds are identified, the major task becomes the identification of an optimized clinical candidate molecule among the many compounds synthesized by medicinal chemists. Conventionally, such compounds are screened in a number of animal or cell models for efficacy and toxicity, experiments that, while having the advantage of being conducted in the *in vivo* setting, commonly take significant amounts of time and depend entirely on the similarity between the experimental animal condition/setting and its human counterpart; that is, the validity of the model.

Although such experiments will never be replaced entirely by expression profiling on either the nucleic acid (genomics) or the protein (proteomics) level, this technique offers powerful advantages and complimentary information. First, efficacy and profile of induced changes can be assessed in a comprehensive fashion (within the limitations—primarily sensitivity and completeness of transcript representation) of the technology platform used. Second, these assessments of differential efficacy can be carried out much more expeditiously than in conventionally used, (patho-)physiology-based animal models. Third, the complex pattern of expression changes revealed by such experiments may provide new insights into possible biological interactions between the actual drug target and other biomolecules, and thus reveal new elements, or branchpoints of a biological pathway that may be useful as surrogate markers, novel diagnostic analytes, or as additional drug targets. Fourth, increasingly important, these tools serve to determine specificity of action among members of gene families that may be highly important for both efficacy and safety of a new drug. It must be borne in mind that any and all such experiments are limited by the coefficient of correlation with which the expression patterns determined are linked to the desired *in vivo* physiological action of the compound.

A word of caution regarding microarray-based expression profiling would appear to be in order: the power of comprehensive (almost) genome-wide assessment of expression patterns has led to what may justly be described as somewhat of an infatuation with this technology that at times leaves a certain degree of critical skepticism to be desired. In particular, the pair-wise comparison algorithms used in much of this work (competition staining of a case and a control sample on the same physical array) raise a number of questions regarding selection bias, which take on particular significance since the overall sample sizes are commonly (very) small. Biostatistical analytic approaches are commonly less than sophisticated, if at all used. Additionally, it is important to remain aware of the fact that all microarray expression data are of only associative character, and must be interpreted mindful of this limitation.

As a subcategory of this approach, toxicogenomics is increasingly evolving as a powerful adjuvant to classic toxicological testing. As pertinent databases are being created from experiments with known toxicants, revealing expression patterns that may potentially be predictive of longer-term toxic liabilities of compounds, future drug discovery efforts should benefit by insights allowing earlier killing of compounds likely to cause such complications.

When using these approaches in drug discovery, even if implemented with proper biostatistics and analytical rigor,

it is imperative to understand the probabilistic nature of such experiments. A promising profile on pharmacogenomic and toxicogenomic screens will enhance the likelihood of having selected an ultimately successful compound, and will achieve this goal quicker than conventional animal experimentation, but will do so only with a certain likelihood of success. The less reductionist approach of the animal experiment will still be needed. It is to be anticipated, however, that such approaches will constitute an important, time- and resource-saving first evaluation or screening step that will help to focus and reduce the number of animal experiments that will ultimately need to be conducted.

20.5 SHORT-TERM TIMEFRAME— PHARMACOGENETICS: MORE TARGETED, MORE EFFECTIVE MEDICINES

20.5.1 Genes and Environment

It is common knowledge that today's pharmacopoeia—in as much as it represents enormous progress compared with what physicians had only 15 or 20 years ago—is far from perfect. Many patients respond only partially, or fail to respond altogether, to the drugs they are given, and others suffer adverse events that range from unpleasant to serious and life threatening.

If a pharmacological agent is regarded as one of the extrinsic, environmental factors in a common complex disease scenario, with a potential to affect the health-status of the individual to whom it is administered, then individually differing responses to such an agent would, under the multifactorial and heterogeneous paradigm of common complex disease elaborated upon earlier, be regarded as the expression of differences in the intrinsic characteristics of these patients. As long as variation in the exposure to the drug is excluded, this is important, as in clinical practice nonadherence to prescribed regimens of administration, or drug-drug interactions interfering with bioavailability of the drug, are by far the most likely culprits when such differences in response-phenotype are observed. The influence of such intrinsic variation on drug response may be predicted to be more easily recognizable and more relevant the steeper the dose-response curve of a given drug is.

The argument for the particular likelihood of observing environmental factor/gene interactions with drugs among all other environmental influences goes along the same lines. Among all these environmental factors that people are exposed to, drugs might be particularly likely to interact specifically and selectively with the genetic properties of a given individual, as their potency and (compared, for example, to foodstuffs) narrow therapeutic window makes interactions with innate individual susceptibilities that affect the interaction with drugs more likely.

Clearly, a better, more fundamental and mechanistic understanding of the molecular pathology of disease in general and of the role of intrinsic, biological properties regarding the predisposition to contract such diseases, as well as of drug action on the molecular level, will be essential for future progress in health care. Current progress in molecular biology and genetics has indeed provided with some of the prerequisite tools that should help to reach the goal of such a more refined understanding.

20.5.2 An Attempt at a Systematic Classification of Pharmacogenetics

Two conceptually quite different scenarios of interindividually differential drug response may be distinguished on the basis of the underlying biological variance (see Fig. 20.2).

* In the first case, the underlying biological variation is in itself not disease-causing or -contributing, and becomes clinically relevant only in response to the exposure to the drug in question (classic pharmacogenetics).
* In the second case, the biological variation is directly disease-related, is per se of pathological importance, and represents a subgroup of the overall clinical disease/diagnostic entity. The differential response to a drug is

thus related to how well this drug addresses, or is matched to, the presence or relative importance of the pathomechanism it targets, in different patients; that is, the molecular differential diagnosis of the patient (disease-mechanism-related pharmacogenetics).

Although these two scenarios are conceptually rather different, they result in similar practical consequences with regard to the administration of a drug, namely stratification based on a particular, DNA-encoded marker. Therefore it seems legitimate to subsume both under the umbrella of pharmacogenetics.

20.5.3 Classic Pharmacogenetics

This category includes differential *pharmacokinetics* and *pharmacodynamics*. Pharmacokinetic effects due to interindividual differences in absorption, distribution, metabolism (with regard to both activation of pro-drugs, inactivation of the active molecule, and generation of derivative molecules with biological activity), or excretion of the drug. In any of these cases, differential effects observed are due to the presence at the intended site of action either of inappropriate concentrations of the pharmaceutical agent, or of inappropriate metabolites, or of both, resulting either in

* **Classical Pharmacogenetics**
  * **Pharmacokinetics**
    * Absorption
    * Metabolism
      * Activation of pro-drugs
      * Deactivation
      * Generation of biologically active metabolites
    * Distribution
    * Elimination
  * **Pharmacodynamics**
    * Palliative drug action (modulation of disease-symptoms or disease-signs by targeting physiologically relevant systems, without addressing those mechanism that cause or causally contribute to the disease)
* **Molecular differential-diagnosis-related pharmacogenetics**
  * Causative drug action (modulation of actual causative of contributory mechanisms

FIGURE 20.2 Systematic classification of pharmacogenetics.

TABLE 20.1 Chronology of pharmacogenetics.

Pharmacogenetic phenotype	Described in	Underlying gene/mutation	Identified in
Sulphonal-porphyria	1890	Porphobilinogen-deaminase	1985
Suxamethonium-hypersensitivity	1957–1960	Pseudocholinesterase	1990–1992
Primaquin hypersensitivity; favism	1958	G-6-PD	1988
Long QT-Syndrome	1957–1960	*Herg* etc	1991–1997
Isoniazid slow/fast acetylation	1959–1960	N-acetyltranferase	1989–1993
Malignant hyperthermia	1960–1962	Ryanodine receptor	1991–1997
Fructose-intolerance	1963	Aldoalse B	1988–1995
Vasopressin insensitivity	1969	Vasopressin receptor 2	1992
Alcohol-susceptibility	1969	Aldehyde-dehydrogenase	1988
Debrisoquine-hypersensitivity	1977	CYP2D6	1988–1993
Retinoic acid resistance	1970	PML-RARA fusion-gene	1991–1993
6-Mercaptopurin-toxicity	1980	Thiopurine-methyltransferase	1995
Mephenytoin resistance	1984	CYP2C19	1993–1994
Insulin-insensitivity	1988	Insulin receptor	1988–1993

lack of efficacy or toxic effects. Pharmacogenetics, as it relates to pharmacokinetics, has been recognized as an entity for more than 100 years, going back to the observation, commonly credited to Archibald Garrod that a subset of psychiatric patients treated with the hypnotic, sulphonal, developed porphyria. Since then, the underlying genetic causes for many of the previously known differences in enzymatic activity have been elucidated, most prominently with regard to the P450 enzyme family, and these have been the subject of recent reviews (Evans and Relling, 1999; Dickins and Tucker, 2001; see also http://www.imm.ki.se/CYPalleles; Table 20.1). However, such pharmacokinetic effects are also seen with membrane transporters, such as in the case of differential activity of genetic variants of MDR-1 that affects the effective intracellular concentration of antiretrovirals (Fellay *et al.*, 2002), or of the purine-analogue-metabolizing enzyme, thiomethyl-purine-transferase (Dubinsky *et al.*, 2000).

Notably, despite the widespread recognition of isoenzymes with differential metabolizing potential since the middle of the 20th century, the practical application and implementation of this knowledge has been minimal so far. This may be the consequence, on one hand, of the irrelevance of such differences in the presence of relatively flat dose-effect-curves (i.e., a sufficiently wide therapeutic window), as well as, on the other hand, the fact that many drugs are subject to complex, parallel metabolizing pathways, where in the case of underperformance of one enzyme, another one may compensate. Such compensatory pathways may well have somewhat different substrate affinities, but allow plasma levels to remain within therapeutic concentrations. Thus, the number of such polymorphisms that have found practical applicability is rather limited and, by and large, restricted to determinations of the presence of functionally deficient variants of the enzyme, thiopurine-methyl-tranferase, in patients prior to treatment with purine-analogue chemotherapeutics (see Table 20.2).

TABLE 20.2 Pharmacogenetics systematics.

Enzymes	Testing substance
Phase I enzyme	
Aldehyd-dehydrogenase	Acetaldehyd
Alcohol-dehydrogenase	Ethanol
CYP1A2	Caffeine
CYP2A6	Nicotine, coumarin
CYP2C9	Warfarin
CYP2C19	Mephenytoin, omeprazole
CYP2D6	Dextromethorphan, dbrisoquine, sparteine
CYP2E1	Chloroxazone, caffeine
CYP3A4	Erythromycin
CYP3A5	Midazolam
Serum cholinesterase	Benzoylcholine, Butrylcholine
Paraoxonase/arylesterase	Paraoxon
Phase II enzyme	
Acetyltransferase (NAT1)	*Para*-aminosalizylsäure
Acetyltransferase (NAT2)	Isoniazid, sulfamethazine, caffeine
Dihydropyrimidin-dehydrogenase	5-fluorouracil
Glutathione-transferase (GST-M1)	*Trans*-stilbene-Oxid
Thiomethyltransferase	2-mercaptoethanol, D-penicillamine, captopril
Thiopurine-methyltransferase	6-mercaptopurine, 6-thioguanine, 8-azathioprine
UDP-glucuronosyl-transferase (UGT1A)	Bilirubin
UDP-glucuronosyl-transferase (UGT2B7)	Oxazepam, ketoprofen, estradiol, morphine

Pharmacodynamic effects, in contrast, may lead to interindividual differences in a drug's effects despite the presence of appropriate concentrations of the intended active (or activated) drug compound at the intended site of action. Here, DNA-based variation in how the target molecule or another (downstream) member of the target molecule's mechanistic pathway can respond to the medicine modulates the effects of the drug. This will apply primarily to palliatively working medicines, as discussed earlier.

Figure 20.3 helps to clarify these somewhat complex concepts, in which a hypothetical case of a complex trait/disease is depicted where excessive, dysregulated function of one of the trait-controlling/contributing pathways (see Fig. 20.3; lanes A and B) causes symptomatic disease. The example used refers to blood pressure as the trait, and hypertension as the disease in question, respectively (for the case of Fig. 20.3, defective or diminished function of a pathway, an analogous schematic could be constructed, and again for a deviant function). A palliative treatment would be one that addresses one of the pathways that, though not dysregulated, contributes to the overall deviant physiology (see Fig. 20.3; lane F); the respective pharmacogenetic/pharmacodynamic scenario would occur if this particular pathway, due to a genetic variant, was not responsive to the drug chosen (see Fig. 20.3; lane G). A palliative treatment may also be ineffective if the particular mechanism targeted by the palliative drug is due to the presence of a molecular variant provides less than the physiologically expected baseline contribution to the relevant phenotype (see Fig. 20.3; lane H). In such a case, modulating an *a priori* unimportant pathway in the disease scenario will not yield successful palliative treatment results (see Fig. 20.3; lanes I and J).

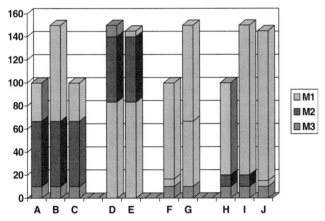

FIGURE 20.3 Three molecular mechanisms (M1, M2, M3) contribute to a trait. **Lane A**: Normal physiology. **Lane B**: Diseased physiology D1: derailment (cause/contribution) of molecular mechanism 1 (M1). **Lane C**: Diseased physiology D1: causal treatment T1 (aimed at M1). **Lane D**: Diseased physiology D3: derailment (cause/contribution) of molecular mechanism 3 (M3). **Lane E**: Diseased physiology D3, treatment T1: treatment does not address cause. **Lane F**: Diseased physiology D1, palliative treatment T2 (aimed at M2). **Lane G**: Diseased physiology D1, palliative treatment T2; T2-refractroy gene variant in M2. **Lane H**: Normal physiology variant: differential contribution of M1 and M2 to normal trait. **Lane I**: Diseased physiology D1 variant: derailment of mechanism M1. **Lane J**: Diseased physiology D1 variant: treatment with T2. Solid bars: Normal function; stippling pathological dysfunction; hatching therapeutic modulation (from Lindpaintner, 2003; with permission).

Several of the most persuasive examples that have been accumulated to date for such palliative-drug-related pharmacogenetic effects have been observed in the field of asthma. The treatment of asthma relies on an array of drugs aimed at modulating different generic pathways, thus mediating bronchodilation or anti-inflammatory effects, often without regard to the possible causative contribution of the targeted mechanism to the disease. One of the mainstays of the treatment of asthma is activation of the β-2-adrenoceptor by specific agonists, which leads to relaxation of bronchial smooth muscles and, consequently, bronchodilation. Recently, several molecular variants of the beta-2-adrenoceptor have been shown to be associated with differential treatment response to such beta-2-agonists (Martinez *et al.*, 1997; Tan *et al.*, 1997). Individuals carrying one or two copies of a variant allele that contains a glycine in place of arginine in position 16 were found to have a 3- and 5-fold reduced response to the agonist, respectively.

This was shown in both *in vitro* (Green *et al.*, 1994; 1995) and *in vivo* (Green *et al.*, 1995) studies to correlate with an enhanced rate of agonist-induced receptor receptor-down-regulation, but not with any difference in transcriptional or translational activity of the gene, or with agonist binding. In contrast, a second polymorphism affecting position 19 of the beta upstream peptide was shown to affect translation (but not transcription) of the receptor itself, with a 50% decrease in receptor numbers associated with the variant allele, which happens to be in strong linkage disequilibrium with a variant allele position 16 in the receptor. The simultaneous presence of both mutations thus would be predicted to result in low expression and enhanced down-regulation of an otherwise functionally normal receptor, depriving patients carrying such alleles of the benefits of effective bronchodilation as a palliative (i.e., noncausal) countermeasure to their pathological airway hyper-reactivity. Importantly, there is no evidence that any of the allelic variants encountered are associated with the prevalence or incidence, and thus potentially the etiology of the underlying disease (Reihsaus *et al.*, 1993; Dewar *et al.*, 1998). This would reflect the scenario depicted in Fig. 20.3 (lane H).

Inhibition of leukotriene synthesis, another palliative approach toward the treatment of asthma, proved clinically ineffective in a small fraction of patients who carried only non-wild-type alleles of the 5-lipoxygenase promoter region (Drazen *et al.*, 1999). These allelic variants previously had been shown to be associated with decreased transcriptional activity of the gene (In *et al.*, 1997). Consistent with the clinical observations, it stands to reason that in the presence of already reduced 5-lipoxygenase activity, pharmacological inhibition may be less effective (see Fig. 20.3; lanes H, I, and J). Of note, again, there is no evidence for a primary, disease-causing or -contributing role of any 5-lipoxygenase variants; they all were observed at equal frequencies in disease-affected and nonaffected individuals (In *et al.*, 1997).

Pharmacogenetic effects may account not only for differential efficacy, but may also contribute to differential occurrence of adverse effects. An example of this scenario is provided by the well-documented pharmacogenetic association between molecular sequence variants of the 12S rRNA, a mitochondrion-encoded gene, and aminoglycoside-induced ototoxicity (Fischel-Ghodsian, 1999). Intriguingly, the mutation that is associated with susceptibility to ototoxicity renders the sequence of the human 12S rRNA similar to that of the bacterial 12S rRNA gene, and thus effectively turns the human 12S rRNA into the (bacterial) target for aminoglycoside drug action, presumably mimicking the structure of the bacterial binding site of the drug (Hutchin and Cortopassi, 1994). As in the other examples, presence of the 12S rRNA mutation per se has no primary, drug-treatment-independent pathologic effect.

One may speculate that, analogously, such molecular mimicry may occur within one species: adverse events may arise if the selectivity of a drug is lost because a gene that belongs to the same gene-family as the primary target, loses its identity vis-à-vis the drug and attains, based on its structural similarity with the principal target, similar or at least increased affinity to the drug. Depending on the biological role of the imposter molecule, adverse events may occur, even though the variant molecule, again, may be quite silent with regard to any contribution to disease causation. Although we currently have no obvious examples for this scenario, it is certainly imaginable for various classes of receptors and enzymes.

20.5.4 Pharmacogenetics as a Consequence of Molecular Differential Diagnosis

As alluded to earlier, there is general agreement today that any of the major clinical diagnoses in the field of common complex disease, such as diabetes, hypertension, cancer, and so on, are composed of a number of etiologically (i.e., at the molecular level) more or less distinct subentities. In the case of a causally acting drug this may imply that the agent will be appropriate, or will work best, only in that fraction of all the patients who carry the (all-inclusive and imprecise) clinical diagnosis in whom the dominant molecular etiology, or at least one of the contributing etiological factors, matches the biological mechanism of action that the drug in question modulates (see Fig. 20.3; lane C). If the mechanism of action of the drug addresses a pathway that is not disease-relevant, perhaps already down-regulated as an appropriate physiologic response to the disease, then the drug may logically be expected not to show efficacy (see Fig. 20.3; lanes D and E).

Thus, unrecognized and undiagnosed disease heterogeneity, disclosed indirectly by presence or absence of response to a drug targeting a mechanism that contributes only to one of several molecular subgroups of the disease, provides an important explanation for differential drug response and likely represents a substantial fraction of what

today is somewhat indiscriminately subsumed under the term pharmacogenetics.

Currently, the most frequently cited example for this category of pharmacogenetics is trastuzamab (HERCEPTIN®), a humanized monoclonal antibody directed against the her-2-oncogene. This breast cancer treatment is prescribed based on the level of her-2-oncogene expression in the patient's tumor tissue. Differential diagnosis at the molecular level not only provides an added level of diagnostic sophistication, but also actually represents the prerequisite for choosing the appropriate therapy. Because tastuzamab specifically inhibits a gain-of-function variant of the oncogene, it is ineffective in the two-thirds of patients who do not over-express the drug's target, whereas it significantly improves survival in the third of patients that constitute the subentity of the broader diagnosis breast cancer in whom the gene is expressed (Baselga *et al.*, 1996). Some have argued against this being an example of pharmacogenetics because the parameter for patient stratification (i.e., for differential diagnosis) is the somatic gene expression level rather than a particular genotype data (Haseltine, 1998). This is a difficult argument to follow, since in the case of a treatment-effect-modifying germ-line mutation it would obviously not be the nuclear gene variant per se, but also its specific impact on either structure/function or on expression of the respective gene/gene product that would represent the actual physiological corollary underlying the differential drug action. Conversely, an *a priori* observed expression difference is highly likely to reflect an as yet undiscovered sequence variant. Indeed, as pointed out earlier, there are a number of examples in the field of pharmacogenomics where the connection between genotypic variant and altered expression already has been demonstrated (In *et al.*, 1997; McGraw *et al.*, 1998).

Another example, although still hypothetical, of how proper molecular diagnosis of relevant patho-mechanisms will significantly influence drug efficacy, is in the evolving class of anti-AIDS/HIV drugs that target the CCR5 cell-surface receptor (Huang *et al.*, 1996; Dean *et al.*, 1996; Samson *et al.*, 1996). These drugs would be predicted to be ineffective in those rare patients, who carry the delta-32 variant, but who nevertheless have contracted AIDS or test HIV-positive (most likely due to infection with an SI-virus phenotype that utilizes CXCR4; O'Brien *et al.*, 1997; Theodorou *et al.*, 1997).

It should be noted that the pharmacogenetically relevant molecular variant need not affect the primary drug target, but may equally well be located in another molecule belonging to the system or pathway in question, both up- or down-stream in the biological cascade with respect to the primary drug target.

20.5.5 Different Classes of Markers

Pharmacogenetic phenomena, as pointed out previously, need not be restricted to the observation of a direct association between allelic sequence variation and phenotype, but may extend to a broad variety of indirect manifestations of underlying, but often (as yet) unrecognized sequence variation. Thus, epigenetic modifications, such as differential methylation of the promoter-region of O6-methylguanine-DNA-methylase has recently been reported to be associated with differential efficacy of chemotherapy with alkylating agents. If methylation is present, expression of the enzyme that rapidly reverses alkylation and induces drug-resistance is inhibited, and therapeutic efficacy is greatly enhanced (Esteller *et al.*, 2000).

20.5.6 Complexity Is to Be Expected

In the real world, it is likely that not only one of the scenarios depicted, but a combination of several may affect how well a patient responds to a given treatment, or how likely it is that he or she will suffer an adverse event. Thus, a fast-metabolizing patient with poor-responder pharmacodynamics may be particularly unlikely to gain any benefit from taking the drug in question, whereas a slow-metabolizing status may counterbalance in another patient the same inopportune pharmacodynamics, and a third patient, who is a slow metabolizer and displaying normal pharmacodynamics, may be more likely to suffer adverse events. In all of them, both the pharmacokinetic and pharmacodynamics properties may result from the interaction of several of the mechanisms described earlier. In addition, it is known, of course, that coadministration of other drugs or even the consumption of certain foods may affect and further complicate the picture for any given treatment.

20.6 INCORPORATING PHARMACOGENETICS INTO DRUG DEVELOPMENT STRATEGY

20.6.1 Diagnostics First, Therapeutics Second

It is important to note that despite the public hyperbole and the high-strung expectations surrounding the use of pharmacogenetics to provide personalized care these approaches are likely to be applicable only to a fraction of medicines that are being developed. Further, if and when such approaches will be used, they will represent no radical new direction or concept in drug development but simply a stratification strategy as it has been used all along.

An increasingly sophisticated and precise diagnosis of disease, arising from a deeper, more differentiated understanding of pathology at the molecular level, that will increasingly subdivide today's clinical diagnoses into molecular subtypes, will foster medical advances, which, if considered from the viewpoint of today's clinical diagnosis, will appear as pharmacogenetic phenomena, as described earlier. However, the sequence of events that is today often presented as characteristic for a pharmacogenetic scenario, namely, exposing patients to the drug, recognizing a differ-

ential (i.e., (quasi-) bimodal-) response pattern, discovering a marker that predicts this response, and creating a diagnostic product to be comarketed with the drug, henceforth, is likely to be reversed. Rather, in the case of pharmacogenetics due to a match between drug action and dysregulation of a disease-contributing mechanism, researchers likely will search for a new drug specifically, and *a priori*, based on a new mechanistic understanding of disease causation or contribution (i.e., a newly found ability to diagnose a molecular subentity of a previously more encompassing, broader, and less precise clinical disease definition). Thus, pharmacogenetics will not be as much about finding the right medicine for the right patient, but instead about finding the right medicine for the disease (-subtype), as physicians have aspired to do all along throughout the history of medical progress. This is, in fact, good news: the conventional pharmacogenetic scenario would invariably present major challenges from both a regulatory and a business development and marketing standpoint, as it would confront development teams with a critical change in the drug's profile at a very late point during the development process. In addition, the timely development of an approvable diagnostic in this situation is difficult at best, and its marketing as an add-on to the drug a less than attractive proposition to diagnostics business. Thus, the practice of pharmacogenetics will, in many instances, be marked by progress along the very same path that has been one of the main avenues of medical progress for the last several hundred years: differential diagnosis first, followed by the development of appropriate, more specific treatment modalities.

Thus, the sequence of events in this case may well involve, first, the development of an *in vitro* diagnostic test as a stand-alone product that may be marketed on its own merits, allowing the physician to establish an accurate, state-of-the-art diagnosis of the molecular subtype of the patient's disease. Sometimes such a diagnostic may prove helpful even in the absence of specific therapy by guiding the choice of existing medicines and/or of nondrug treatment modalities such as specific changes in diet or lifestyle. Availability of such a diagnostic, as part of the more sophisticated understanding of disease, will undoubtedly foster and stimulate the search for new, more specific drugs; and once such drugs are found, availability of the specific diagnostic will be important for carrying out the appropriate clinical trials. This will allow a prospectively planned, much more systematic approach toward clinical and business development, with a commensurate greater chance of actual realization and success.

20.6.2 Probability, Not Certainty

In practice, some extent of guesswork will remain, due to the nature of common complex disease. First, all diagnostic approaches, including those based on DNA analysis in common complex disease, as stressed earlier, will ultimately only provide a measure of probability, not of certainty; thus, although the variances of drug response among patients who do (or do not) carry the drug-specific subdiagnosis will be smaller, there will still be a distribution of differential responses. Although by-and-large the drug will work better in the responder group, there will be some patients among this subgroup who will respond less or not at all, and conversely, not everyone belonging to the nonresponder group will completely fail to respond, depending perhaps on the relative magnitude with which the particular mechanism contributes to the disease. It is important to bear in mind, therefore, that even in the case of fairly obvious bimodality, patient responses will still show distribution patterns, and that all predictions as to responder or nonresponder status will have only a certain likelihood of being indeed accurate (see Fig. 20.4). Therefore, the terms responder and nonresponder, as applied to groups of patients stratified based on a DNA marker, represent Mendelian-thinking-inspired misnomers that should be replaced by more appropriate terms that reflect the probabilistic nature of any such classification; for example, likely (non)responder.

In addition, based on the current understanding of the polygenic and heterogeneous nature of these disorders, even in an ideal world where all possible susceptibility gene variants for a given disease are known and treatments exist for them, it will be possible only to exclude, in any one patient, those that do not appear to contribute to the disease, and therefore deselect certain treatments. However, there will most likely be a small number, perhaps two to four, of potentially disease-contributing gene-variants, whose relative contribution to the disease will be very difficult, if not impossible, to rank in an individual patient. Likely then, trial and error, and this great intangible quantity, physician experience, will still play an important role, albeit on a more limited and subselective basis.

The alternative scenario, where differential drug response and/or safety occurs as a consequence of a pathologically not relevant, purely drug-response related pharmacogenetics scenario is more likely to present greater difficulty in planning and executing a clinical development program because, presumably, it will be more difficult to anticipate or predict differential responses *a priori*. When such a differential response occurs, it will also potentially be more difficult to find the relevant marker(s), unless it happens to be among the obvious candidate genes implicated in the disease physiopathology or the treatment's mode of action.

Although screening for molecular variants of these genes, and testing for their possible associations with differential drug response, is a logical first step, if unsuccessful, it may be necessary to embark on an unbiased genome-wide screen for such a marker or markers. Despite recent progress in high-throughput genotyping, the obstacles that will have to be overcome on the technical, data-analysis, and cost levels are formidable. They will limit the deployment of such programs, at least for the foreseeable future, to select cases in which there are very solid indications for doing so, based on clinical data showing a near-categorical (e.g., bimodal) distribution of treatment outcomes. Even then, it is expected to encounter for every success, which will be owed to a favorably strong linkage-disequilibrium across considerable genomic distance in the relevant chromosomal region, as many or more failures, in cases where the culpable gene variant cannot be found due to the higher recombination rate or other characteristics of the stretch of genome on which it is located.

20.7 REGULATORY ASPECTS

At this writing, regulatory agencies in both Europe and the United States are beginning to show keen interest in the potential role that pharmacogenetic approaches may play in the development and clinical use of new drugs, and the potential challenges that such approaches may present to the regulatory approval process. Although no formal guidelines have been issued, the pharmaceutical industry has already been reproached (albeit in a rather nonspecific manner) for not being more proactive in the use of pharmacogenetic markers. It will be of key importance for all concerned to engage in an intensive dialogue at the end of which, it is hoped, will emerge a joint understanding that stratification according to DNA-based markers is fundamentally nothing new, and not different from stratification according to any other clinical or demographic parameter, as has been used all along.

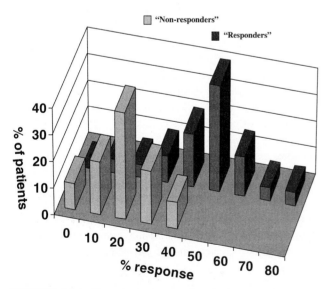

FIGURE 20.4 Hypothetical example of bimodal distribution according to marker that indicates nonresponder or responder status. Note that in both cases a distribution is present, with overlaps. Thus the categorization into responders or nonresponders based on the marker must be understood to convey only the probability of belonging to one or the other group (from Lindpaintner, 2003; with permission).

Still, based in the perception (in the case of common complex diseases scientifically unjustified) that DNA-based markers represent a different class of stratification parameters, a number of important questions will need to be addressed and answered, hopefully always in analogy to conventional stratification parameters, including those referring to ethical aspects. Among the most important ones are questions concerning:

- The need and/or ethical justification (or lack thereof) to include likely nonresponders in a trial for the sake of meeting safety-criteria, which, given the restricted indication of the drug, may indeed be excessively broad
- The need to carry out conventional-size safety trials in the disease-stratum eligible for the drug (if the stratum represents a relatively small fraction of all patients with the clinical diagnosis, it may be difficult to amass sufficient numbers and/or discourage companies from pursuing such drugs to the disadvantage of patients)
- The need to use active controls if the patient/disease stratum is different from that in which the active control was originally tested
- The strategies to develop and gain approval for the applicable first-generation diagnostic, as well as for the regulatory approval of subsequent generations of tests to be used to determine eligibility for prescription of the drug
- A number of ethical-legal questions relating to the unique requirements regarding privacy and confidentiality for genetic testing that may raise novel problems with regard to regulatory audits of patient data (see Section 20.9)

A concerted effort to avoid what has been termed genetic exceptionalism, the differential treatment of DNA-based markers as compared with other personal medical data, should be made so as to not further unnecessarily complicate the already very difficult process of obtaining regulatory approval. This seems justified based on the recognized fact that in the field of common complex disease DNA-based markers are not at all different from conventional medical data in all relevant aspects, namely specificity, sensitivity, and predictive value.

20.8 PHARMACOGENETIC TESTING FOR DRUG EFFICACY VS SAFETY

20.8.1 Greater Efficacy: Likely

In principle, pharmacogenetic approaches may be useful both to raise efficacy and to avoid adverse events, by stratifying patient eligibility for a drug according to appropriate markers. In both cases, clinical decisions and recommendations must be supported by data that have undergone rigorous biostatistical scrutiny. Based on the substantially different prerequisites for and opportunities to acquiring such data, and to applying them to clinical decision-making, the use of pharmacogenetics for enhanced efficacy is expected to be considerably more common than for the avoidance of adverse events.

The likelihood that adequate data on efficacy in a subgroup may be generated is reasonably high, given the fact that unless the drug is viable in a reasonably sizeable number of patients, it will probably not be developed for lack of a viable business case, or at least only under the protected environment of orphan drug guidelines. Implementation of pharmacogenetic testing to stratify for efficacy, provided that safety in the nonresponder group is not an issue, will primarily be a matter of physician preference and sophistication, and potentially of third-party payer directives, but would appear less likely to become a matter of regulatory mandate, unless a drug has been developed selectively in a particular stratum of the overall indication (in which case a contraindication label for other strata is likely to be issued). Indeed, an argument can be made against depriving those who carry the likely nonresponder genotype regarding eligibility for the drug, but who individually, of course, may respond to the drug with a certain, albeit lower probability. From a regulatory aspect, use of pharmacogenetics for efficacy, if adequate safety data exist, appears largely unproblematic—the worst-case scenario (a genotypically inappropriate patient receiving the drug) would result in treatment without expected beneficial effect, but with no increased odds to suffer adverse consequences; that is, much of what one would expect under conventional paradigms.

20.8.2 Avoidance of Serious Adverse Effects: Less Likely, with Exceptions

The utility and clinical application of pharmacogenetic approaches toward improving safety, in particular with regard to serious adverse events, will meet with considerably greater hurdles and therefore is less likely expected to become reality. A number of reasons are cited for this: first, in the event of serious adverse events associated with the use of a widely-prescribed medicine, withdrawal of the drug from the market usually is based almost entirely on anecdotal evidence from a rather small number of cases, in accordance with the Hippocratic mandate *primum non nocere*. If the sample size is insufficient to demonstrate a statistically significant association between drug exposure and event, as is typically the case, it will most certainly be insufficient to allow meaningful testing for genotype-phenotype correlations; the biostatistical hurdles become progressively more difficult as many markers are tested and the number of degrees of freedom applicable to the analysis for association continues to rise. Therefore, the fraction of attributable risk shown to be associated with a given at-risk (combination of) genotype(s) would have to be very substantial for regulators to accept such data. Indeed, the low prior probability of the adverse event, by definition, can be expected to yield an equally low positive (or negative) predictive value. Second, the very nature of safety issues raises the hurdles substantially because in this situation the worst-case scenario—administration of the drug to the wrong patient—will result in higher odds to cause harm to the patient.

Therefore, it is likely that the practical application of successfully investigating and applying pharmacogenetics toward limiting adverse events will likely be restricted to the more exceptional case of diseases with dire prognosis, where a high medical need exists, where the drug in question offers unique potential advantages (usually bearing the characteristics of a life-saving drug), and where, therefore, the tolerance even for relatively severe side effects is much greater than for other drugs. This applies primarily to areas like oncology or HIV/AIDS, for which the recently reported highly specific and acceptably sensitive association between the MHC gene variant, *HLA B5701*, and occurrence of a severe hypersensitivity reaction is a prime example (Mallal *et al.*, 2002; Hetherington *et al.*, 2002).

In most other indications, the sobering biostatistical and regulatory considerations discussed represent barriers that are unlikely to be overcome easily; and the proposed, conceptually highly attractive, routine deployment of pharmacogenetics as generalized drug surveillance or pharmaco-vigilance practice following the introduction of a new pharmaceutical agent faces these scientific as well as formidable economic hurdles (Roses, 2000).

20.9 CHALLENGE—GENETICS AND SOCIETY: ETHICAL, LEGAL, AND SOCIETAL ISSUES

20.9.1 Data Protection Needs to be Matched by Person Protection

Whereas public attitudes toward genetics and genomics span a broad spectrum from enthusiastic approval to total rejection, it is fair to say that among the most outspoken voices from the lay-community are many who are skeptical, critical, or outright negative about the pursuit of genetic research and the possible implementations of some of its outcomes in society. There is a widespread public sentiment of fear of the consequences, which centers primarily on two issues: genetic engineering and the lack of control over one's private medical-genetic data. Both concerns are thoroughly understandable, if not always justified, since any new and powerful technology carries the innate risk of being abused by unscrupulous individuals, to the detriment of others. The following sections, as is appropriate for the scope of this chapter, shall focus on the latter of these concerns regarding the fate of genetic information.

Much of the discussion about ethical and legal issues relating to pharmacogenetics is centered on the issue of genetic testing, a topic that recently has also been the focus of a number of guidelines, advisories, white papers, and so on, issued by a number of committees in both Europe (information available at http://www.eshg.org/fiorireport.htm) and the United States. It is interesting to note that the one characteristic that virtually all these documents share is an almost studious avoidance of defining what exactly a genetic test is. Where definitions are given, they tend to be very broad, including not only the analysis of DNA but also of transcription and translation products affected by inherited variation. Inasmuch as the most sensible solution to this dilemma ultimately will hopefully be a consensus to treat all personal medical data in a similar fashion regardless of the degree to which DNA-encoded information affects it (noting that there really isn't any medical data that are not to some extent affected by intrinsic patient properties), it may, for the time being, be helpful to let the definition of what constitutes genetic data be guided by the public perception of genetic data, inasmuch as the whole discussion of this topic is prompted by these public perceptions.

In the public eye, genetic test usually is understood as either:

- Any kind of test that establishes the diagnosis of or predisposition for one of the classic monogenic, heritable disease
- Any kind of test based on structural nucleic acid analysis (sequence)

This includes the (non-DNA-based) Guthrie test for phenylketonuria and forensic and paternity testing, as well as a DNA-based test for Lp(a), but not the plasma-protein-based test for the same marker (even though the information derived is identical). Since monogenic disease, in effect, is excluded from this discussion, it makes sense to restrict the definition of genetic testing to the analysis of (human) DNA sequence.

It is clear that DNA-based structural data (genetic data) must never be procured without the explicit permission and informed consent of the individual (see also Chapter 31). It is equally clear that once such information is created, it should be guarded by the most reliable data protection systems that are available to prevent any of this information getting into the hands of third parties not intended to have access to it. At the same time, to provide the expected benefits in terms of guiding medical management, this information will need to be shared with a more or less extensive number of participants in the patient's health care, some of which may potentially use it in ways contrary to the patient's interests. This is true for any drug approved in conjunction with a DNA-based test: mere prescription of such a medicine discloses implicitly the outcome of the test to a wide number of stakeholders, among them certainly the patient's health insurer, who may gain from this information added knowledge about future health risks of this patient and take, accordingly, certain actions; whether these are scientifically-actuarially justified is irrelevant in this discussion.

Since data protection, in real life, therefore is always limited or compromised, additional measures are needed to protect the individual from disadvantages based on his or her data. Thus, personal protection, in addition to data protection must be provided. This requires a framework of regulations or laws that govern the use of personal medical information—the latter term is preferred to restricting such protection to DNA-derived data because many conventional medical data carry similar or larger information content, particularly in the area of common complex disease. Such a framework can, in democratically governed systems, arise only as a consensus among all stakeholders involved, among them patients, physicians, insurers, employers, and so on, that represents the optimal compromise, which maximizes the benefits for both the individual and society. Such a framework will define which uses of the information are endorsed by society as lawful, and which are shunned as illegal. Given such a framework, much of the anxiety that abounds today with regard to the possible leakage of private medical data will be laid to rest, because it is primarily the potential (ab)use of information to their detriment that patients fear (which is not to trivialize the loss of autonomy that is, of course, also of concern).

20.9.2 Public Education and Information

The essential requirement to reach such a consensus that will allow the use of genetics and genomics in the best interest of all concerned is an informed dialogue among the various stakeholders, which can begin to take place only once (mis)perceptions are replaced by objective and neutral information as the basis to form informed opinions. Progress in the fields of genetic and genomics has been rapid and substantial over the last few decades, and it has been accompanied by a great deal of hyperbole in the popular press.

Meanwhile, geneticists have continued to cultivate an arcane and forbidding vernacular that adds only to the appearance of purposeful secrecy that the public is reacting to, instead of having made extra efforts to reach out to the public in a concerted educational campaign. As part of the Human Genome Project, substantial amounts of funding have been provided to work in the area of bioethics, and much progress has been achieved there. Similar or even greater efforts need to be undertaken in the area of public information and education, which surely will go a long way toward resolving some of the fears as well as unrealistic hopes the public currently associates with genetics and genomics. The author of this communication and his colleagues have assembled an interactive CD-ROM-based educational program that is distributed freely upon request from http://www.rochegenetics.com.

20.9.3 Ethical-Societal Aspects of Pharmacogenetics

Based on the perceived particular sensitivity of genetic data, institutional review boards commonly apply a specific set of rules to granting permission to test for DNA-based markers

in the course of drug trials or other clinical research, including (variably) separate informed consent forms, the anonymization of samples and data, specific stipulations about availability of genetic counseling, provision to be able to withdraw samples at any time in the future, and so on.

Arguments have been advanced that genotype determinations for pharmacogenetic characterization (Roses, 2000), in contrast to genetic testing for primary disease risk assessment, are less likely to raise potentially sensitive issues with regard to patient confidentiality, the misuse of genotyping data or other nucleic-acid-derived information, and the possibility of stigmatization. Although this is certainly true when pharmacogenetic testing is compared to predictive genotyping for highly penetrant Mendelian disorders, it is not apparent why in common complex disorders issues surrounding predictors of primary disease risk would be any more or less sensitive than those pertaining to predictors of likely treatment success/failure. Both can be expected to provide, in most cases, a modicum of better probabilistic assessment, based on the modest degree of sensitivity, specificity, and positive/negative predictive value observed with tests for pharmacogenetic interactions. If, however misguided this would be given the anticipated quite limited information content of such tests, such information was to be used against the patient, then two lines of reasoning may actually indicate an increased potential for ethical issues and complex confrontations among the various stakeholders to arise from pharmacogenetic data.

First, even though access to genotyping and other nucleic acid-derived data related to disease susceptibility can be strictly limited, the very nature of pharmacogenetic data calls for a rather more liberal position regarding use: if this information is to serve its intended purpose, that is, improving the patients chance for successful treatment, then it is essential that it is shared among at least a somewhat wider circle of participants in the health care process. Thus, the prescription for a drug that is limited to a group of patients with a particular genotype will inevitably disclose the receiving patient's genotype to any one of a large number of individuals involved in the patient's care at the medical and administrative level. The only way to limit this quasi-public disclosure of this patient's genotype data would be if he or she were to sacrifice the benefits of the indicated treatment for the sake of data confidentiality.

Second, patients profiled to carry a high disease probability along with a high likelihood for treatment response may be viewed, from the standpoint of, for example, insurance risk, as quite comparable to patients displaying the opposite profile, for example, a low risk to develop the disease, but a high likelihood not to respond to medical treatment if the disease indeed occurs. For any given disease risk, then, patients less likely to respond to treatment would be seen as a more unfavorable insurance risk, particularly if nonresponder status is associated with chronic, costly illness rather than with early mortality, the first case having much more far-reaching economic consequences. The pharmacogenetic profile may thus, under certain circumstances, even become a more important (financial) risk-assessment parameter than primary disease susceptibility, and would be expected, in as much as it represents but one stone in the complex-disease mosaic, to be treated with similar weight, or lack thereof, as other genetic and environmental risk factors.

Practically speaking, the critical issue is not only, and perhaps not even predominantly, the real or perceived sensitive nature of the information, and how it is, if at all, disseminated and disclosed, but how and to what end it is used. Obviously, generation and acquisition of personal medical information must always be contingent on the individual's free choice and consent, as must be all application of such data for specific purposes. Beyond this, however, there is today an urgent need for the requisite dialogue and discourse among all stakeholders within society to develop and endorse a set of criteria by which the use of genetic, indeed of all personal medical information, should occur.

It will be critically important that society as a whole endorses, in an act of solidarity with those less fortunate— that is, at higher risk of developing disease, or less likely to respond to treatment—rules that guarantee the beneficial and legitimate use of the data in the patient's interest while prohibiting their use in ways that may harm the individual, personally, financially, or otherwise. As long as the political decision processes to reflect societal consensus is trustworthy, and as long as such consensus reflects the principles of justice and equality, the resulting set of principles should ensure such proper use of medical information. Indeed, both aspects—data protection and patient/subject protection—are seminal components of the mandates included in the World Health Organization's "Proposed International Guidelines on Ethical Issues in Medical Genetics and Genetic Services" (available at http://www.who.int/ncd/hgn/hgnethic.htm), which mandate autonomy, beneficence, nonmaleficence, and justice.

20.10 SUMMARY

Genetics and genomics, in their various implementations, will represent an important new avenue toward understanding disease pathology and drug action, and will offer new opportunities of stratifying patients to achieve better treatment success. As such, these approaches represent a logical, consequent step in the history of medicine, evolutionary, rather than revolutionary. The implementation of genetic approaches will take time, and will not apply to all diseases and all treatments equally. Pharmacogenetic information will be probabilistic and relative, not deterministic or absolute. Application of genetics and genomics to the drug discovery and development process, as well as to medical practice will provide help, but no simple solutions, and will not be a panacea. Importantly, society at large will need to find ways to sanction the proper use of private medical information, thus allowing and protecting its unencumbered use for the benefit of patients while safeguarding them from unintended use. Increased efforts to inform and educate the general public, leading to a more realistic assessment of the actual potential of genetic approaches to provide benefit or cause harm, will be key to quell both the exalted hopes and exaggerated fears that are so often associated with the topic.

Acknowledgements

This communication represents the author's personal views and not necessarily those of any of his institutional or corporate affiliations, including F. Hoffmann-La Roche.

References

Baselga, J., Tripathy, D., Mendelsohn, J., Baughman, S., Benz, C. C., Dantis, L., Sklarin, N. T., Seidman, A. D., Hudis, C., Moore, J., Rosen, P. P., Twaddell, T., Henderson, I. C., and Norton, L. (1996). Phase II study of weekly intravenous recombinant humanized anti-p185(HER2) monoclonal antibody in patients with HER2/neu-overexpressing metastatic breast cancer. *J. Clin. Oncol.* 14, 737–744.

Dean, M., Carrington, M., Winkler, C., Huttley, G. A., Smith, M. W., Allikmets, R., Goedert, J. J., Buchbinder, S. P., Vittinghoff, E., Gomperts, E., Donfield, S., Vlahov, D., Kaslow, R., Saah, A., Rinaldo, C., Detels, R., and O'Brien, S. J. (1996). Genetic restriction of HIV-1 infection and progression to AIDS by a deletion of the CKR5 structural gene. *Science* 273, 1856–1862.

Dewar, J. C., Wheatley, A. P., Venn, A., Morrison, J. F. J., Britton, J., and Hall, I. P. (1998). β2 adrenoceptor polymorphisms are in linkage disequilibrium, but are not associated with asthma in an adult population. *Clin. Exp. Allergy* 28, 442–448.

Dickins, M., and Tucker, G. (2001). Drug disposition: To phenotype or genotype. *Int. J. Pharm. Med.* 15, 70–73; see also: http://www.imm.ki.se/CYPalleles.

Drazen, J. M., Yandava, C. N., Dube, L., Szczerback, N., Hippensteel, R., Pillari, A., Israel, E., Schork, N., Silverman, E. S., Katz, D. A., and Drajesk, J. (1999). Pharmacogenetic association between ALOX5 promoter genotype and the response to anti-asthma treatment. *Nat. Genet.* 22, 168–170.

Dubinsky, M., Lamothe, S., Yang, H. Y., Targan, S. R., Sinnett, D., Theoret, Y., and Seidman, E. G. (2000). Pharmacogenomics and Metabolite Measurement for 6-Mercaptopurine Therapy in Inflammatory Bowel Disease. *Gastroenterology* 118, 705–713.

Esteller, M., Garcia-Foncillas, J., Andion, E., Goodman, S. N., Hidalgo, O. F., Vanaclocha, V., Baylin, S. B., and Herman, J. G. (2000). Inactivation of the DNA-repair gene mgmt and the clinical response of gliomas to alkylating agents. *N. Engl. J. Med.* 343, 1350–1354.

Evans, W. E., and Relling, M. V. (1999). Pharmacogenomics: Translating functional genomics into rational therapies. *Science* 206, 487–491; see also http://www.sciencemag.org/feature/data/1044449.shl.

Fellay, J., Marzolini, C., Meaden, E. R., Back, D. J., Buclin, T., Chave, J. P., Decosterd, L. A., Furrer, H. J., Opravil, M., Pantaleo, G., Retelska, D., Ruiz, L., Schinkel, A. H., Vernazza, P., Eap, C. B., and Telenti, A. (2002). Response to antiretroviral treatment in HIV-1-infected individuals with allelic variants of the multidrug resistance transporter 1: a pharmacogenetics study. *Lancet* 359, 30–36.

Fischel-Ghodsian, N. (1999). Genetic factors in aminoglycoside toxicity. *Ann. N Y Acad. Sci.* 884, 99–109.

Green, S. A., Turki, J., Innis, M., and Liggett, S. B. (1994). Amino-terminal polymorphisms of the human beta 2-adrenergic receptor impart distinct agonist-promoted regulatory properties. *Biochemistry* 33, 9414–9419.

Green, S. A., Turki, J., Bejarano, P., Hall, I. P., and Liggett, S. B. (1995). Influence of beta 2-adrenergic receptor genotypes on signal transduction in human airway smooth muscle cells. *Am. J. Respir. Cell Mol. Biol.* 13, 25–33.

Haseltine, W. A. (1998). Not quite pharmacogenomics. *Nat. Biotechnol.* 16, 1295.

Hetherington, S., Hughes, A. R., Mosteller, M., Shortino, D., Baker, K. L., Spreen, W., Lai, E., Davies, K., Handley, A., Dow, D. J., Fling, M. E., Stocum, M., Bowman, C., Thurmond, L. M., and Roses, A. D. (2002). Genetic Variations in *HLA-B* region and hypersensitivity reactions to abacavir. *Lancet* 359, 1121–1122.

Huang, Y., Paxton, W. A., Wolinsky, S. M., Neumann, A. U., Zhang, L., He, T., Kang, S., Ceradini, D., Jin, Z., Yazdanbakhsh, K., Kunstman, K., Erickson, D., Dragon, E., Landau, N. R., Phair, J., Ho, D. D., and Koup, R. A. (1996). The role of a mutant CCR5 allele in HIV-1 transmission and disease progression. *Nat. Med.* 2, 1240–1243.

Hutchin, T., and Cortopassi, G. (1994). Proposed molecular and cellular mechanism for aminoglycoside ototoxicity. *Antimicrob. Agents Chemother.* 38, 2517–2520.

In, K. H., Asano, K., Beier, D., Grobholz, J., Finn, P. W., Silverman, E. K., Silverman, E. S., Collins, T., Fischer, A. R., Keith, T. P., Serino, K., Kim, S. W., De Sanctis, G. T., Yandava, C., Pillari, A., Rubin, P., Kemp, J., Israel, E., Busse, W., Ledford, D., Murray, J. J., Segal, A., Tinkleman, D., and Drazen, J. M. (1997). Naturally occurring mutations in the human 5-lipoxygenase gene promoter that modify transcription factor binding and reporter gene transcription. *J. Clin. Invest.* 99, 1130–1137.

Lindpaintner, K. (2003). Pharmacogenetics and the future of medical practice. *J. Mol. Med.* 81, 141–153.

Mallal, S., Nolan, D., Witt, C., Masel, G., Martin, A. M., Moore, C., Sayer, D., Castley, A., Mamotte, C., Maxwell, D., James, I., and Christiansen, F. T. (2002). Association between presence of *HLA-B*5701*, *HLA-DR7*, and *HLA-DQ3* and hypersensitivity to HIV-1 reverse-transcriptase inhibitor abacavir. *Lancet* 359, 727–732.

Martinez, F. D., Graves, P. E., Baldini, M., Solomon, S., and Erickson, R. (1997). Association between genetic polymorphisms of the beta2-adrenoceptor and response to albuterol in children with and without a history of wheezing. *J. Clin. Invest.* 100, 3184–3188.

McGraw, D. W., Forbes, S. L., Kramer, L. A., and Liggett, S. B. (1998). Polymorphisms of the 5′ leader cistron of the human beta2-adrenergic receptor regulate receptor expression. *J. Clin. Invest.* 102, 1927–1932.

O'Brien, T. R., Winkler, C., Dean, M., Nelson, J. A. E., Carrington, M., Michael, N. L., and White, G. C. 2nd. (1997). HIV-1 infection in a man homozygous for CCR5 delta 32. *Lancet* 349, 1219.

Reihsaus, E., Innis, M., MacIntyre, N., and Liggett, S. B. (1993). Mutations in the gene encoding for the beta 2-adrenergic receptor in normal and asthmatic subjects. *Am. J. Respir. Cell Mol. Biol.* 8, 334–349.

Roses, A. (2000). Pharmacogentics and future drug development and delivery. *Lancet* 355, 1358–1361.

Samson, M., Libert, F., Doranz, B. J., Rucker, J., Liesnard, C., Farber, C. M., Saragosti, S., Lapoumeroulie, C., Cognaux, J., Forceille, C., Muyldermans, G., Verhofstede, C., Burtonboy, G., Georges, M., Imai, T., Rana, S., Yi, Y., Smyth, R. J., Collman, R. G., Doms, R. W., Vassart, G., and Parmentier, M. (1996). Resistance to HIV-1 infection in Caucasian individuals bearing mutant alleles of the CCR-5 chemokine receptor gene. *Nature* 382, 722–725.

Tan, S., Hall, I. P., Dewar, J., Dow, E., and Lipworth, B. (1997). Association between beta 2-adrenoceptor polymorphism and susceptibility to bronchodilator desensitisation in moderately severe stable asthmatics. *Lancet* 350, 995–999.

Theodorou, I., Meyer, L., Magierowska, M., Katlama, C., and Rouzioux, C. (1997). HIV-1 infection in an individual homozygous for CCR5 delta 32. Seroco Study Group. *Lancet* 349, 1219–1220.

CHAPTER **21**

Molecular Diagnostic Applications in Forensic Science

BRUCE BUDOWLE,[1] JOHN V. PLANZ,[2] ROWAN CAMPBELL,[2] AND ARTHUR J. EISENBERG[2]

[1] *Federal Bureau of Investigation, Laboratory Division, Quantico, VA;*
[2] *DNA Identification Laboratory, University of North Texas Health Science Center, Ft. Worth, TX*

TABLE OF CONTENTS

21.1 INTRODUCTION

Forensic human identity testing is focused primarily on determining the source of biological samples found at crime scenes. The biological samples encountered in criminal cases or in mass disasters can be compromised by the presence of environmental contaminants, they can be degraded, and typically they are limited in quantity. Furthermore, the samples are diverse, including such materials as bloodstains, saliva stains, semen stains, bone, teeth, hair, or muscle tissue, and are found on a variety of substrates. Therefore, there are a number of challenges to consider when developing and implementing analytical methods that genetically characterize biological evidence. These include:

- Extracting DNA of sufficient quantity and quality to enable analysis
- Developing robust analytical technology so that reliable results can be obtained
- Establishing reliable interpretation guidelines
- Conveying the results in a court of law

Historically, polymorphic protein genetic markers were used to potentially differentiate individuals (Murch and Budowle, 1986; Saferstein, 1995; Budowle and Brown, 2001). The classic protein polymorphic system is the ABO blood group system, which had been used for decades for identity testing. There are four common ABO blood types: A, B, O, and AB. For example, if a bloodstain found at a crime scene is typed as B, then only B-type individuals can be possible sources of the sample. Those who have blood types A, O, and AB (slightly more than 90% of the population) are excluded as possible sources of the bloodstain. Other protein genetic marker systems included serum and red cell proteins, such as group-specific component, transferrin, hemoglobin, phosphoglucomutase-1, acid phosphatase, esterase D, and others. However, protein-based genetic systems were limited in use because the stability or persistence is low for these proteins in biological samples exposed to the environment, the discrimination power of the systems is low, and the protein markers did not exist at typeable levels in all tissues of an individual. Typing genetic polymorphisms at the DNA level, rather than the protein level, obviates these limitations to a much greater degree.

The ability to type DNA from biological evidence is one of the most important developments in forensic science since the advent of latent fingerprint analysis. Any biological material that contains nucleated cells, including blood, semen, saliva, hair, bones, and teeth, potentially can be typed for DNA polymorphisms. The human DNA typing techniques and the battery of available genetic markers are more sensitive, more specific, and better resolving than the classic protein genetic markers just described. Most importantly, DNA technology affords the forensic scientist the ability to exclude from consideration individuals who have been falsely associated with a biological sample and to reduce the number of potential contributors of the sample to a few (if not one) individuals. The molecular biology methods available include a wide range of genetic markers and a variety of typing strategies. The methods that have been or are used routinely for human identity testing include restriction fragment length polymorphism (RFLP) typing of variable number of tandem repeat (VNTR) loci (Wyman and White, 1980; Jeffreys *et al.*, 1985a; 1985b; Budowle and Baechtel, 1990), and amplification of target DNA molecules by the polymerase chain reaction (PCR, Saiki *et al.*, 1985) with subsequent typing of specified genetic markers (Saiki *et al.*, 1989; Kasai *et al.*, 1990; Budowle *et al.*, 1991; 1996a; 1996b; Comey and Budowle, 1991; Edwards *et al.*, 1991; Sullivan *et al.*, 1991a; 1991b; Holland *et al.*, 1993; Wilson *et al.*, 1995a; 1995b). A subclass of VNTR loci, known as short tandem repeat (STR) or microsatellite loci, is the primary class of genetic markers used for human identity testing. Such loci are highly abundant in the human genome, and the forensically informative ones, residing on autosomal chromosomes, are hypervariable. The STR loci are composed of tandemly repeated sequences, each of which is two to seven base pairs (bp) in length. Loci containing repeat sequences consisting of four bp (or tetranucleotides) are used routinely for human identification (Edwards *et al.*, 1991; Budowle *et al.*, 1998; Chakraborty *et al.*, 1999). Some pentanucleotide repeat loci are also used (Budowle *et al.*, 2001c). The number of common alleles at a single forensically important STR locus ranges usually from three to eight. Because the region containing the repeats is generally quite small (a STR region with 15 copies of a 4 bp repeat is 60 bp in length), they are amenable to amplification by the PCR. Amplified alleles, however, are somewhat larger— generally less than 350 bp—due to the incorporation, into the amplicon, of regions that flank the repeat sequence and spacing requirements for multiplex analysis. Because of the use of the PCR, less than 1 ng of human template DNA is needed for STR analysis, with less than 200 pg of template DNA (an equivalent of 33 human diploid cells) at times yielding reliable typing results. The main benefit of amplifying small-sized amplicons is that many degraded DNA samples can be successfully amplified at the STR loci. Another advantage is that a number of STR loci can be amplified simultaneously in a multiplex PCR. To facilitate analysis, by reducing labor and consumption of evidence, 5 to 15 STR loci can be amplified in a single PCR and subsequently simultaneously typed. Commercial kits are available to assist in typing multiple STR loci (Lins *et al.*, 1998; Micka *et al.*, 1999; Moretti *et al.*, 2001; Budowle *et al.*, 2001a; Holt *et al.*, 2002; Krenke *et al.*, 2002; 2004). The process for typing the amplified STRs entails separating the fragments, usually by capillary electrophoresis (CE; see also Section 7.5.4), and detecting fluorescently labeled products in real time. Profile patterns from different samples, that is, evidence and reference samples, are compared and interpretations of match, exclusion, or inconclusive are made. The multiple locus STR profiles (typically 13 loci in the United States) can be very informative.

Although the battery of 13 autosomal STR loci offer a high degree of discrimination power, such that only one or a few individuals may carry a particular multilocus profile, there are some samples from forensic cases where these core autosomal STR loci may not be informative. Forensic biological samples are often found at scenes of violent crimes and typically these are composed of more than one donor.

Identification of genotypes at STR loci from a male perpetrator may not be possible in a mixed sample composed of low levels of male DNA. When there is a large background of female DNA in a mixed male/female sample, the female DNA competes for the reagents during STR amplification so that little or no male contribution is observed. By amplifying loci on male-specific DNA, for example, Y-chromosome STR loci, the competition for reagents as occurs with autosomal markers is substantially reduced. Vaginal swabs, fingernail scrapings, diluted stains, and the like may provide Y-specific genetic profiles where the autosomal STR loci from the portion of the samples derived from a male perpetrator may be too low to detect or at a level too low to interpret reliably (de Knijff *et al.*, 1997; Jobling *et al.*, 1997; Prinz *et al.*, 1997; Kayser *et al.*, 1997a; 1997b; 1998; Honda *et al.*, 1999; Corach *et al.*, 2001; Dekairelle and Hoste, 2001; Gusmão *et al.*, 2001; Ballantyne and Hall, 2003; Budowle *et al.*, 2003b; Sinha *et al.*, 2003a; 2003b).

The Y-chromosome genetic markers are the most recent class of DNA-based markers to gain interest as tools for human identity testing (Chakraborty, 1985; de Knijff *et al.*, 1997; Jobling *et al.*, 1997; Kayser *et al.*, 1997a; 1998; Roewer *et al.*, 1996; 2000; 2001). Because of their more recent history in forensic analyses, this chapter describes the general characteristics of Y-STRs in forensic analyses, some analytic procedures for typing Y-STR loci, and general interpretation issues to consider. Although only Y-chromosome and briefly mitochondrial DNA (mtDNA) markers are described, you will gain a better appreciation of forensic science as it applies to human identification.

21.2 GENERAL CHARACTERISTICS OF Y-CHROMOSOME MARKERS

Y-chromosome DNA resides in the nucleus, but markers residing on the Y-chromosome differ in some respects from the autosomal loci (de Knijff *et al.*, 1997; Jobling *et al.*, 1997; Gill *et al.*, 2000; Budowle *et al.*, 2003b; Kayser, 2003). The Y-chromosome is the smallest chromosome of the human genome (about 60 million bp), and unlike the autosomal chromosomes, it is transmitted solely paternally, from father to son. Most of the DNA in the Y-chromosome is nonrecombinant, with only the most distal portions of the chromosome able to recombine with the X-chromosome. Therefore, barring mutation, the Y-linked DNA types are identical for all paternal relatives, including male siblings. This characteristic can be helpful in human identity cases, such as those involving the analysis of the remains of a missing person, where known paternal relatives can provide reference samples for direct comparison to the unknown Y-linked DNA types. Moreover, reference samples from living individuals can be used to aid in the identification of paternal relatives that are several generations removed. Such was the case with the study of the paternal lineage of Thomas

Jefferson and that of male offspring of Sally Hemmings (Foster *et al.*, 1998). However, due to its mode of inheritance, all paternal relatives of an accused male suspect will share the same Y-marker profile and this must be considered when evaluating and presenting evidence.

Both Y-STR loci and single nucleotide polymorphisms (SNPs) have been used in evolutionary studies and are available for identity testing. SNP-based Y-markers will not be discussed further; these Y-markers have yet to be used significantly for forensic analyses and have been used predominately for deciphering prehistoric human migration routes of global dispersal of modern humans (Underhill *et al.*, 2001; Wells *et al.*, 2001; Cruciani *et al.*, 2002; Lell *et al.*, 2002; Mountain *et al.*, 2002). Y-STR polymorphic loci are being used more routinely, and commercially available kits are available for typing forensically informative loci.

The Y-STR loci have certain characteristics that make them useful for forensic casework analyses. These include:

1. Y-markers can be useful in DNA analysis of violent crime samples, because males are involved in the majority of violent crimes (Bureau of Justice Statistics, 2001).
2. More than 100 Y-linked STR loci have been discovered (Kayser, 2003), and many are sufficiently polymorphic for forensic applications.
3. Y-STR loci reside on the nonrecombinant portion of the Y-chromosome. The Y-markers are transmitted from father to male progeny as a haplotype. A haplotype is a specific array of alleles/loci observed in an individual. Because there is no independent assortment of the Y-loci, they can be very informative in multigenerational paternal lineage studies.
4. Because of the haploid nature of Y-STR loci, for most loci, only one allele per locus is displayed per individual, as opposed to two alleles for autosomal loci. Thus interpretation of profiles in mixed samples is simplified. For example, fewer alleles per locus may make it easier to elucidate the number of male contributors in a multimale contributor sample.
5. Those Y-STR loci selected for forensic applications are male-specific. DNA from a female victim does not contribute to the Y-linked DNA profile. Thus, again interpretation in mixture cases is simplified by having fewer alleles to evaluate.
6. For autosomal STR typing, about 200 pg – 1 ng of template DNA are used. When the male portion of the mixed sample is too low, for example 0.3–10% of the total DNA, using 200 pg – 1 ng of template DNA in the PCR will not provide meaningful information for autosomal STRs. The amount of male DNA is too low to render a detectable profile. However, much larger amounts of total female/male DNA can be placed in a PCR when typing Y-STR loci. For example, suppose that 400 ng of a sample of DNA were composed of 399 ng from a female contributor and 1 ng from a male contributor. If a total of 1 ng of DNA from this sample was used for analysis, then only the female's type would be detected (if autosomal STRs were analyzed). However, putting all 400 ng into a PCR and typing for male-specific loci would enable genetic identity profiling of the male perpetrator. Quantification assays have been developed so that relative quantities of Y-chromosomal DNA (and hence male DNA) in a cocktail of nuclear DNA can be estimated.
7. Stochastic affects due to too few template molecules in the PCR are less of an issue for interpretation of Y-STR loci than for autosomal loci. For autosomal loci, a heterozygote can appear as a "pseudohomozygoe" if too few template molecules are amplified during PCR (Budowle *et al.*, 2001b). For Y-STR loci, this is less of an issue. Too few copies of template DNA will result in only either the allele being amplified or the locus yielding no result. Thus, the minimum template requirements for the PCR of Y-STR loci can be lower than for autosomal STR loci.
8. The analytic tools and apparatus for typing Y-STR loci are the same as those used for typing autosomal loci. Therefore, no new equipment needs to be acquired in a forensic laboratory to implement Y-STR typing.

The International Y-STR User Group has recommended use of nine Y-STR loci for identity testing. These are the loci: DYS19, DYS385a, DYS385b, DYS389I, DYS389II, DYS390, DYS391, DYS392, and DYS393, defined as the European Minimal Haplotype (EMH) (Kayser *et al.*, 1997b; Roewer *et al.*, 2001). In the United States, the Scientific Working Group on DNA Analysis Methods (SWGDAM) also recommended the use of these nine loci as well as the loci DYS438 and DYS439 (Ayub *et al.*, 2000). So that the forensic community could analyze these 11 loci in a single multiplex system, two multiplex kits have been developed. The PowerPlex® Y System kit (Promega, Madison, WI) enables amplification of these 11 loci and the additional locus DYS437 (Ayub, 2000; Krenke, 2003) for a total 12 Y-STR loci. The Y-PLEX™ 12 kit (Reliagene, New Orleans, LA) also enables amplification of the SWGDAM 11 loci plus the amelogenin locus, which is used routinely in forensic DNA typing for gender determination (Budowle *et al.*, 1996a). These 11 loci provide a considerable power of discrimination, but much less than that of the battery of autosomal STR loci (Roewer *et al.*, 1996; Kayser *et al.*, 1997b; 2000; 2002; Budowle *et al.*, 2003b). Other candidate loci that may increase haplotype gene diversity values are GATA A7.1, GATA A7.2, GATA A10, GATA C4, and GATA H4 (White *et al.*, 1999; Ayub *et al.*, 2000; Beleza *et al.*, 2003; Quintans *et al.*, 2003).

21.3 METHODOLOGY

Forensic biological specimens often are limited in quality and quantity. Sometimes only one analysis may be possible.

Undue consumption of evidence may leave nothing for retesting. To be effective, techniques that are developed for crime laboratories should be sufficiently robust so practitioners can use them readily. Protocols should minimize unnecessary steps and improve, if possible, performance time, reliability, and detection sensitivity. Moreover, the procedures used in forensic applications usually are subjected to extensive validation. Validation is a process used to assess the ability of defined procedures to reliably obtain desired results, to define conditions that are required to obtain the result, to determine the limitations of the analytical procedure, and to identify aspects that must be monitored and controlled (Budowle *et al.*, 2000). Validated methods are an essential requirement for routine work in the crime laboratory.

Although not discussed in detail in this chapter, quality assurance and quality control practices are requisite for ensuring high quality results. Adherence to using tested reagents, calibrated equipment, and known control samples is necessary to obtain reliable results with confidence from standard operating protocols. Often when problems arise while analyzing good quality samples, it is likely that quality assurance and quality control practices were not followed. A quality assurance program should be implemented to demonstrate personnel, equipment, and reagents perform as expected (Budowle *et al.*, 2000; see also Chapter 34).

The following discussion addresses a subset of potential issues to consider when typing Y-STR loci in forensic samples. Crime scene samples can be degraded and contaminated with materials that inhibit analytical processes, particularly the PCR. The practices employed and issues considered in forensic analyses could be useful for scientists in other disciplines where high quality, as well as challenging, samples may be encountered.

21.3.1 Extraction

The success of DNA typing relies on the isolation of DNA of sufficient quantity, quality, and purity. DNA can be extracted from most types of biological material found at a crime scene. Depending on the typing procedure, the required quantity (i.e., the amount of retrievable DNA) and quality (i.e., the length of the fragmented molecules) of DNA can vary widely. There will be forensic samples where the quantity is too low or the sample has been subjected to environmental insults such that they are too degraded to analyze; no extraction procedure will improve the ability to type such samples. However, with the sensitivity afforded by the PCR, many small-sized, degraded samples can be typed. Purity of the DNA extract refers to a quality of cleanliness, such that subsequent analytical assays can be carried out effectively. Generally, DNA extraction protocols that overcome, remove, or dilute PCR inhibitors are sought. Although many extraction protocols are similar, tremendous

variation in the specific details are possible, and yet comparably effective results can be obtained.

Ideally, extraction protocols should be simple and inexpensive to perform. Procedures should be suitable for extraction of DNA from small liquid blood samples, bloodstains, and other fluids and tissues. Because of space, very effective procedures such as salting out (Grimberg *et al.*, 1989), organic extraction (Sambrook *et al.*, 1989), or use of chaotropic reagent (Scherczinger *et al.*, 1997; Shedlock *et al.*, 1997; Sinclair and McKechnie, 2000; Iudica *et al.*, 2001) will not be described here. These well-known procedures are compatible with both RFLP and PCR-based procedures, because the isolated DNA is double-stranded. Most molecular biologists are familiar with these procedures. Instead, two procedures used more commonly in forensic applications are presented. These are the use of Chelex-100 (Singer-Sam *et al.*, 1989; Walsh *et al.*, 1991) and the use of FTA paper (Del Rio *et al.*, 1996; Burgoyne and Rogers, 1997).

Chelex-100 (BioRad, Hercules, CA) is an ion-exchange resin that binds cations that can inhibit the PCR. The Chelex protocol entails placing a small sample in 5% (w/v) Chelex-100 and incubating the sample at 56°C for 30 minutes. After incubation, the sample is boiled and centrifuged. A portion of the supernatant is used (Budowle *et al.*, 2000). Boiling denatures the DNA, rendering it unsuitable for restriction enzyme digestion (i.e., RFLP typing). However, chelex-extracted DNA is compatible with current forensic DNA quantitation assays and with the PCR. The DNA extract used for the PCR should not contain residual Chelex, because the material will chelate magnesium ion (essential for polymerase activity). Chelex-100 can be used on a variety of tissues sources.

The other method of DNA purification, particularly useful for reference samples, is the washing away of cellular debris from DNA immobilized in FTA paper (Fitzco, Minneapolis, MN; Whatman, Clifton, NJ). Blood is spotted on FTA paper, the cells are automatically lysed, the DNA released from the cells is immobilized in the FTA matrix, and the sample is dried. Heme, or other cellular debris that may inhibit enzymatic activity in subsequent assays is simply washed away without loss of DNA. The washed, immobilized DNA, housed in a circular punch (approximately 1.2 mm in diameter), can be used directly as a DNA template source.

Blood deposited on FTA paper is an excellent source of genomic DNA. The paper facilitates handling, prevents degradation of the DNA, and enables storage of the material at ambient temperature. In addition to blood, epithelial cells from buccal swabs may be transferred and stored within the FTA matrix, facilitating the processing of reference samples. Buccal cells may be applied directly by a sponge applicator swab to the FTA paper during the sample collection phase, or the cells may be eluted from swabs and then placed on the FTA paper. Other cell-containing materials, such as urine, may be stored and processed with the FTA system.

The FTA system also provides several safety and quality assurance benefits. The matrix is impregnated with several reagents that are deleterious to bacteria, viruses, and fungi, as well as serving as a free radical trap to prevent DNA damage. Therefore handling and transport are simplified. The materials stored in FTA paper present fewer biohazard concerns. Moreover, the materials can be stored at ambient temperature without concern regarding degradation; thus, refrigerated storage is unnecessary. The paper matrix is available in several formats from single spot cards to 96-positions per sheet for both manual and robotic processing. The DNA is immobilized in the matrix and experiments have demonstrated no cross contamination during multiple sample sheet processing.

The evidence from rape cases, such as vaginal swabs and stained clothing, most often contains nucleated cells from the male contributor (i.e., predominately sperm) and the female victim (i.e., epithelial cells). Sperm cells can be separated from other cells during extraction, because their cell membranes contain thiol-rich proteins and are resistant to cell lysis in the absence of a reducing agent (Gill *et al.*, 1985; Giusti *et al.*, 1986). The absence of a reducing agent results in a preferential lysis of nonsperm cells. The nonsperm DNA is released into the supernatant, whereas the sperm cells remain intact and are pelleted. After removal of the supernatant, the intact sperm can be lysed in the presence of a reducing agent. Many different extraction procedures have been modified to enable this differential extraction of sperm and nonsperm DNA.

21.3.2 Typing

The PCR process is well known and will not be described here. Commercial kits are available that contain the reagents necessary to amplify at least the 11 core SWGDAM Y-STR loci.

After the quantity of recovered DNA is determined (Waye *et al.*, 1989; Walsh *et al.*, 1992; Budowle *et al.*, 1995; Giusti and Budowle, 1995), typically 150 pg to 1 ng of male template DNA are placed into the PCR. Because Y-STR loci are only being typed, female DNA can be present in much greater quantities and not impact on the male-specific results. For example, samples containing up to and beyond 1000 ng of female DNA, along with the male DNA, have provided successful Y-STR results (Sinha *et al.*, 2003a; 2003b). Until recently, forensic DNA quantitation methods were human- (and higher primate) specific, but not gender-specific. Recently, a real-time PCR method was developed that can estimate the amount of male DNA in a sample. The Quantifiler™ Y Human DNA Quantification Kit (Applied Biosystems, Foster City, CA) has a dynamic range of at least 23 pg to 50 ng, is human- (and higher primate) specific, and is commercially available.

The loci for analysis have been determined by the SWGDAM, giving direction to commercial manufacturers to develop requisite loci containing kits. The PowerPlex® Y System kit (Promega, Madison, WI) and the Y-PLEX™ 12 kit (Reliagene, New Orleans, LA) contain reagents needed for amplification, to include primer sets, which are specific for the various loci, PCR buffer, allelic ladders, and an internal lane standard. AmpliTaq Gold® DNA polymerase (Applied Biosystems, Foster City, CA) is purchased separately. Allelic ladders contain the common alleles in the general population for each locus and are used to normalize electrophoretic migration for the PCR products analyzed (see Fig. 21.1). The basic criteria for the 11 core Y-STR loci are displayed in Table 21.1. The primers are designed to amplify each locus specifically and tested for cross reactivity with other regions of the human genome, as well as with the other primers in the multiplex. The primer sets contained within each kit consist of both unlabeled and those labeled with one of three distinctive fluorescent dyes. One primer of each primer pair per Y-STR locus is labeled with a fluorescent dye so that PCR products can be detected during elec-

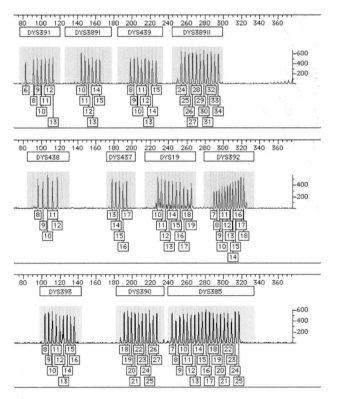

FIGURE 21.1 Profile of allelic ladder provided with the Power-Plex® Y System kit (Promega, Madison, WI). The common alleles at each locus are included. The values 80–360 above the peaks are scan units that relate to elution time. The numbers on the Y-axis (200, 400, 600) are relative fluorescent units and are correlated to relative yield of amplified product. Kindly provided by B. Krenke of the Promega Corporation.

TABLE 21.1 General criteria of Y-STR loci in the two commercial multiplex kits.*

Locus	Dye (Promega)	PCR product size (bases) (Promega)	Dye (Reliagene)	PCR product size (bases) (Reliagene)	Allele range	Repeat motif	GenBank accession #
DYS392	JOE	294–327	FAM	103–139	6–18	TAT	G09867
DYS390	TMR	191–227	FAM	163–207	17–28	TCTA / TCTG	G09611 AC011289
DYS385a/b	TMR	243–315	FAM	220–288	7–25	GAAA	Z93950
DYS393	TMR	104–136	JOE	100–136	8–17	AGAT	G09601
DYS389I	FL	148–168	JOE	179–207	10–17	TCTG / TCTA	G09600 AF140635
DYS391	FL	90–118	JOE	230–262	6–14	TCTA	G09613
DYS389II	FL	256–296	JOE	292–332	24–34	TCTG / TCTA	G09600 AF140635
DYS437	JOE	183–199	—	—	13–17	TCTA / TCTG	AC002992
DYS19	JOE	232–268	NED	174–210	10–19	TAGA	X77751
DYS439	FL	203–231	NED	230–258	8–15	GATA	AC002992
DYS438	JOE	101–121	NED	292–327	6–14	TTTTC	AC002531
Amelogenin	—	—	NED	104–110	X,Y	—	M55418 and M55419

* Data are from PowerPlex® Y System Technical Manual No. D018, (Promega Corporation, Madison, WI) and Shewale and colleagues (2004).

trophoretic separation. For Y-STR typing, fluorescent tags are covalently bound to the 5′ end of one of the primers of a primer pair per locus. Thus, a fluorescent molecule is incorporated into PCR products and detected using laser-induced fluorescence. The internal lane standard is labeled with another distinctive fluorescent dye. The use of multi-color dyes permits the analysis of loci with overlapping size ranges.

The polymorphism of STR loci is based on different numbers of repeats contained within the locus (or amplified product) among individuals in the population. Thus, alleles are determined by the different size of amplified products, differing by repeat increments or portions of repeats (i.e., micro-variants). The amplified fragments are separated according to size by CE using, for example, the ABI Prism™ 310 or 3100 Genetic Analyzer (Applied Biosystems, Foster City, CA). CE, rather than polyacrylamide slab gel electrophoresis, is used to fractionate the amplified products for allele designation, primarily because of increased automation and reduction of the need for manual skills in gel preparation by the analyst (see also Chapter 7). With CE, manual gel pouring and sample loading are eliminated; loading a sieving medium into the capillary and sample injection is achieved by automatic means. Because the charge-to-mass ratio of different size DNA molecules is the same, mobility in free solution is independent of fragment size. Therefore, a sieving medium is required to separate DNA fragments. Use of a soluble medium is advantageous, because the buffer and the polymer can be treated in a similar fashion to a free solution separation and readily pumped into the capillary. Therefore, the injection of the sieving medium can be automated, and fresh sieving medium can be used with each CE

run. Refilling capillaries also reduces the chance of contamination from previous samples that were analyzed in the same capillary.

The high surface area-to-volume ratio of a capillary enables electrophoretic separations to be carried out at very high field strengths by increasing heat dissipation. Faster separation times are thereby achieved, and resolution may be improved compared with some slab gel electrophoresis methods. Furthermore, real time detection is performed with CE. No post-electrophoresis detection is required. Results are stored directly in the computer, thus facilitating subsequent data manipulations and analyses.

Injection of DNA samples is performed electrokinetically. The amount of the DNA in the sample is dependent on the mobility and concentration of the other ions in the sample. The preferential injection of higher charge-to-mass molecules (e.g., Cl^{-1} ions) with an electrokinetic injection affects, to a degree, the quantity of sample that can be injected into the capillary. Dialysis of samples can remove such competing ions. For most forensic protocols, the samples do not need removal of competing ions.

21.3.3 Interpretation of Results

For the application of Y-STR loci, interpretation issues need to be considered. Many are the same as those for autosomal STR loci. These include: a quality evaluation of the data, whether the sample is composed of multiple donors or a single source sample, and the significance of matching data.

Figure 21.2 shows a Y-STR profile from a single male donor. Each locus displays one allele, except the DYS385 locus, which can present two alleles, due to a tandem dupli-

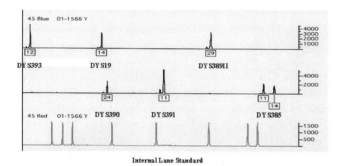

FIGURE 21.2 A six-locus Y-STR profile from a single male donor, generated using the Y-PLEX™ 6 kit (Reliagene, New Orleans, LA). The internal lane standard is a concoction of non-human DNA fragments of known mass; the standard is placed in every sample after PCR and prior to electrophoresis so DNA fragment sizes can be estimated and normalized. Kindly provided by S. Sinha of Reliagene.

cation of the locus (Kayser *et al.*, 1997b). A minimum amount of DNA is required to evaluate the profile. This threshold interpretation level is determined in-house during validation and is set at a minimum relative fluorescent units (rfus) level. When two or more alleles are observed at multiple loci, then the sample is likely a mixture. In addition to the alleles, there are four artifact classes to consider before rendering an interpretation of a mixture. The first are stutter peaks. These are amplification products typically one repeat smaller in size than the true allele. Stutter is inherent in the PCR and most likely due to repeat slippage (Levinson and Gutman, 1987; Schlotterer and Tautz, 1992). Most stutter products range from not detectable to about 15% of the associated allele (Sinha *et al.*, 2003a; 2003b). The specified Y-STR loci were selected because stutter products are nominal. STR loci with smaller size repeats, particularly dinucleotide repeats and homopolymeric stretches, tend to have high levels of stutter and would substantially complicate interpreting mixtures. The stutter peaks are considered more so, when interpreting mixtures samples where there is a minor contributor (displaying allelic products comparable in yield to that of the stutter peaks).

Terminal nucleotide addition is another artifact generated during the PCR (Smith *et al.*, 1995; Brownstein *et al.*, 1996; Magnuson *et al.*, 1996), but is more controllable than generation of stutter. The enzyme *Taq* polymerase has a tendency to add a nucleotide, generally adenine, to the 3′ end of the amplicon without the presence of a template. Addition is termed +A and the true amplicon size is termed −A. The addition of the terminal nucleotide is dependent on temperature and the sequence adjacent to the extension site. There are two approaches to address addition of the nucleotide: either design the assay to minimize addition or design the assay to promote addition. It is easier to drive the reaction toward adenine addition. By adding an extension step at 60°C for 30 minutes to one hour after the PCR cycling, almost all of the amplified products achieve the +A state (Moretti *et al.*, 2001). Allelic ladders provided with

commercial kits contain alleles in the +A state as well. In some casework analyses, sample profiles may display significant −A products. These are indicative of too much template DNA in the PCR or the presence of PCR inhibitors in the sample. To eliminate the presence of −A products, one can dilute the sample and subject it to another PCR. Alternatively, after the PCR, fresh *Taq* polymerase can be added to the PCR and only the extension step at 60°C is carried out. AmpliTaq Gold® DNA polymerase should not be used in the latter approach. To activate AmpliTaq Gold® DNA polymerase, a heat activation step is required, which will denature the DNA duplex. Terminal nucleotide addition requires double-stranded DNA.

Another potential artifact is due to fluorophore synthesis by-products (or by leaching of bound tags). These products are not DNA derived and can appear in negative control samples. Often these artifacts are not observed until very small quantity samples are analyzed. The fluorescent products are present but cannot be seen in good quantity samples due to the scale. Because these by-products are present in negative controls, they can be readily considered when interpreting results.

The last artifact to consider is cross-reactivity of Y-STR primers with female DNA. The PCR primers are designed to not generate products with female DNA (and particularly with loci on the X-chromosome). However, when excessive amounts of female DNA are in the PCR, nonspecific products may be observed. Sinha and colleagues (2003a) observed two TAMRA-labeled amplified products of sizes 255 bp and 448 bp when the quantity of female DNA in the PCR was 10 ng or greater. Thus, when an evidence sample dictates that large amounts of female DNA are placed in a PCR, it is imperative to run a positive control from, for example, the female victim with similar amounts of template DNA.

If the negative control samples show no product, the positive control sample types correctly, and the evidence and reference profiles are determined to be of sufficient quality, DNA profiles from evidentiary sample(s) are compared with profiles from known exemplar(s) to determine whether or not they are similar. There are three general interpretations that can be rendered:

1. Inclusion, or match—the Y haplotypes from the two samples are sufficiently similar and potentially could have originated from the same source.
2. Exclusion—the Y-haplotype profiles are dissimilar and could not have originated from the same source.
3. Inconclusive—there are insufficient data to render an interpretation (Budowle *et al.*, 2003b).

Exclusions and inconclusive results require no further analysis. However, when a Y-haplotype obtained from a forensic specimen matches that of a suspect or victim (or cannot be excluded as arising from a biological relative), it is common practice to place some significance upon the likelihood of such an occurrence. Accordingly, statistics derived from population data are applied with the intent to provide an estimate of how common or rare a DNA profile is in the relevant population(s). The mode of Y-STR inheritance must be considered to decide the statistical approach to use when placing weight on an observed Y-STR haplotype. As described earlier, the Y-STR loci reside on the nonrecombinant portion of the Y-chromosome. Thus, the rarity of a multilocus Y-STR profile cannot be estimated as the combined product of the allele frequencies at each locus, as is done for the autosomal STR loci (Roewer *et al.*, 1996; 2000; 2001; de Knijff *et al.*, 1997; Jobling *et al.*, 1997; Kayser *et al.*, 1997b; 2002; Gill *et al.*, 2000; Budowle *et al.*, 2003b). They must be evaluated as a haplotype of loci (i.e., inherited as a block). This can be achieved by evaluating the rarity of the Y-STR haplotypes in reference to which the observation of match is made (Budowle *et al.*, 2003b).

The counting method is one method employed (and the most likely to be used in the United States) to convey an estimate of the rarity of the Y-haplotype (Budowle *et al.*, 2003b). Basically, the number of times a particular haplotype is observed in a reference database(s) is counted and divided by the number of profiles in the data set. Then, a correction for sampling error is applied. It is a very conservative approach, because although all loci are physically linked, they are not all in disequilibrium to the same extent (Budowle *et al.*, 2003b), likely due to the relatively high mutation rate of STR loci (Kayser *et al.*, 2000).

The reference population database(s) used in forensics is composed of convenience samples (NRC II Report 1996); that is, they are collected from such sources as paternity test laboratories and blood banks. The databases generally are divided into major population groups of the potential contributors of the evidence. For forensic applications the relevance and representativeness of these databases are considered and for autosomal loci the populations are relatively homogeneous within a major population group, such as Caucasians, African Americans, and Hispanics. Because of the lack of recombination and smaller effective population size of Y-markers compared with autosomal markers, the degree of heterogeneity is expected to be greater among populations within a major population group. Y-haplotype data demonstrate population substructure more so than the autosomal loci (de Knijff *et al.*, 1997; Jobling *et al.*, 1997; Kayser *et al.*, 1997a; Kayser, 2003; Budowle *et al.*, 2003b). Therefore, correcting for effects of substructure when estimating the rarity of the profile need to be instituted. Such practices already are applied to calculations of the rarity of autosomal genotypes. The inbreeding coefficients will be larger for Y-STR loci. Data support that within Europe, interpopulation variability for Y-STR loci tends to be low within a geographic region. Common haplotypes appear throughout Europe. The ϕ_{st} values, similar to Wright's Fst

values, are typically <0.01, such as those observed in central Europe. However, across Europe different regional affiliations are evident. The ϕ_{st} values can be as high as 0.10. Not surprisingly, distances for European to non-European range from 0.2–0.5 (Roewer *et al.*, 1996; 2000). The United States is somewhat more amalgamated than Europe and people tend to migrate more so. Kayser and colleagues (2003) analyzed African American, United States Caucasian, and Hispanic Y-STR data (on European minimal haplotype) and did not find significant geographic heterogeneity for Y-STR haplotypes within the major United States population groups. For the major populations in the United States the effects of substructuring for the Y-chromosome are shown to be less than those observed across Europe.

21.4 CASE EXAMPLES

There are many cases that have been analyzed and myriad examples that could be presented to gain an appreciation of the application of Y-STR analysis in identity testing. Two cases are described generally: a paternity case and an assault case.

In a paternity test, a woman accused a man of being the father of her child and the man denied the allegation. A paternity test was undertaken using 15 autosomal STR markers. After accounting for the maternally inherited alleles, there were two loci in the child that had alleles (paternal in origin) that were not observed in the alleged biological father. Because of the relatively high mutation rate for STR loci (approximately at 10^{-3}), one or two exculpatory loci are not sufficient for excluding the man as the biological father. However, the paternity index is substantially reduced because of these two loci. Further testing is required to attain a sufficient paternity index. The 12-locus Y-haplotype for the child and the alleged father were identical. Thus, the genetic data could not exclude the alleged father as the biological father and because of the Y-STR typing a paternity index greater than 1,000 was obtained.

A woman was attacked at night and her identification of the assailant was questioned, but the alibi of the suspect was weak. However, she did manage to scratch the arm of her assailant. The material under her fingernails was removed by scraping. There was very little tissue from her male assailant. In fact, most of the tissue obtained derived from the victim. An attempt was made to type the extracted DNA with the 13 core STR loci and the result yielded a single-source profile consistent with a reference sample from the victim. Subsequently, the total recovered DNA (about 200 ng) was typed for six Y-STR loci. A profile was generated equivalent to about 750 pg of male template DNA, and the Y-haplotype was different than that derived from the reference sample of the accused. If not for the special qualities afforded by Y-specific DNA testing, the suspect would not have been excluded as a possible perpetrator.

21.5 MITOCHONDRIAL DNA MARKERS ANALYSIS IN FORENSIC SCIENCE

Although the focus of this chapter has been on male lineage Y-STRs, it is worth mentioning briefly the forensic value and application of the female lineage equivalent; that is, mitochondrial (mt) DNA. Mitochondria are subcellular organelles that contain an extrachromosomal genome, separate and distinct from the nuclear genome. Human mtDNA differs from nuclear DNA in the following ways.

- It exists as a closed circular, rather than linear, genome.
- The genome is substantially smaller, consisting of approximately 16.5 kilobases.
- The molecule predominately consists of coding sequences (for 2 ribosomal RNAs, 22 transfer RNAs, and 13 proteins), and only one notable noncoding region approximately 1,100 bp long, called the displacement loop (D-loop) or control region.
- All mtDNA molecules are inherited maternally.
- There are many copies per cell.
- Similar to the Y chromosome, mtDNA does not undergo recombination (for review see Budowle *et al.*, 2003a).

The main advantage of analyzing mtDNA, compared with nuclear DNA, is the high copy number of mtDNA (Copenhagen and Clayton, 1974). Although each set of nuclear chromosomes is present in only two copies per cell, hundreds to thousands of mtDNA molecules reside within a cell. For forensic cases, where the evidence is particularly challenging, such as old bones, severely decomposed or charred remains, or single hair shafts, there may be insufficient quantity and/or quality nuclear DNA for analysis. However, because of sheer copy number, the probability of obtaining a successful typing result from mtDNA is higher than that from nuclear DNA. One can also hypothesize that the circular structure of the mtDNA may make the molecule more resistant to nuclease activity than nuclear DNA. Thus, mtDNA is likely to persist longer than nuclear DNA under environmentally difficult conditions.

Instead of variation in the number of repeats being exploited for forensic identification, the variation among individuals is at the sequence of the nucleotides in the mtDNA. The greatest concentration of genetic variation resides in two hyper variable regions in the control region designated HV1 and HV2 (Greenberg *et al.*, 1983). These are sequenced using the Sanger method (Sanger *et al.*, 1977), where four different fluorescently labeled dideoxyribonucleoside triphosphate analogs (ddATP, ddCTP, ddGTP, and ddTTP) are placed in the chain synthesis reaction. The ddNTP in the sequencing reaction is incorporated into the growing chain by complementary base pairing to the template, as are dNTPs, and therefore competes with its dNTP analog for incorporation. However, chain elongation is terminated at the point where the ddNTP is incorporated since,

unlike dNTPs, ddNTPs do not have a 3′ hydroxyl group, which is necessary for chain elongation. Thus, the extended fragments are labeled by incorporation of a ddNTP with a fluorescent dye attached. The labeled fragments are separated and detected by CE as is done with Y-STRs.

Because the mtDNA molecule does not undergo recombination, the mtDNA markers, barring mutation, are identical for all maternal relatives, which includes siblings. Thus, mtDNA typing can be useful in cases, such as those involving the analysis of the remains of a missing person, where known maternal relatives, separated by several generations, can provide meaningful reference samples for direct comparison to the questioned mtDNA type (Ginther *et al.*, 1992; Sullivan *et al.*, 1992; Holland *et al.*, 1993). For example, if one were attempting to identify the remains of a child using mtDNA analysis, reference samples could be obtained from the alleged mother, siblings, maternal grandmother, or children if the missing person is female. One of the first and most noted cases applying mtDNA typing was the identification of the remains of Tsar Nicholas II. The maternal reference samples were obtained from Countess Xenia Cheremeteff-Sfiri and the Duke of Fife (Gill *et al.*, 1994; Ivanov *et al.*, 1996).

Although bones and teeth are typical sources for identification in such cases as Tsar Nicholas II and in the identification of human remains in general, an important forensic application of mtDNA sequencing is in the analysis of hair shafts. Individual hairs contain very limited quantities of nuclear DNA, such that no molecular analysis of nuclear DNA markers is possible. However, mtDNA sequences can be obtained from as little as 1–2 cm of a single hair shaft (Wilson *et al.*, 1995a; 1995b; see Fig. 21.3). Thus, hair, a biological material often found at crime scenes, can be genetically characterized routinely to assist in resolving criminal cases.

21.6 LEGAL ADMISSIBILITY

DNA evidence, in itself, does not prove guilt or innocence. It conveys information about who may have contributed the sample and those who could not be the source of the evidence sample. Other evidence about the crime is considered by the fact finder to determine whether or not an accused individual is guilty. It is important to appreciate that the forensic field has a particular constraint not routinely encountered in other scientific disciplines, and that is, the results are often presented in a nonscience forum, which is the courtroom. Although many genetic typing technologies have been demonstrated to be robust and reliable, the law determines the admissibility of the technology. The court carries out a process, called an admissibility hearing, to ensure that there is some level of confidence in the evidence before it may be presented in a court proceeding. The federal standard for admissibility, Daubert versus Merrell Dow

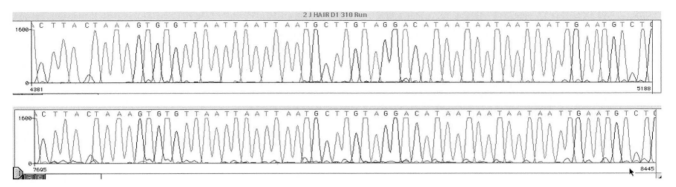

FIGURE 21.3   A sequence of a portion of the HV2 region of the human mtDNA genome. The top panel is from mtDNA derived from a hair shaft. The bottom panel is from mtDNA derived from a whole blood sample.

Pharmaceuticals, Inc. (1993), places the trial judge as a gate-keeper for determining that an expert's testimony both rests on a reliable foundation and is relevant to the issue(s). Factors to consider in evaluating subsequent expert testimony include:

- The testability of the science
- Peer review
- The error rate of the method
- Existence of standards
- General acceptance by those who are familiar with it

The court's role is to attempt to screen out inappropriate and misleading (i.e., termed "junk") science.

Science and the law do not obtain the truth in the same manner. In science, beliefs and findings are questioned continuously using the scientific method. Hypotheses are proposed and experiments are carried out to test hypotheses. If the data do not refute a hypothesis, the hypothesis gains more support and through constructive incremental steps the hypothesis becomes grounded and accepted as reasonable and reliable. In the legal setting, an adversarial approach is used. Admissibility challenges to scientific evidence are classic examples of bias in ascertainment. Rarely is exculpatory evidence challenged. Inculpatory evidence is what is typically challenged for admissibility. Therefore, results that support the prosecution's hypothesis of guilt are the likely ones to be challenged and adversarially criticized. Even if the result is objectively obtained, it will support one side's beliefs over that of the other. Typically, a defense attorney is placed in the position or has the responsibility to vigorously create doubt, regardless of his/her personal beliefs of the client's guilt. However, at times, a prosecuting attorney may want to create doubt. In the courtroom, one may exploit the standard practice of science "to question" as lack of consensus, even if most, if not all, agree that the approach is reliable. Because the challenges are adversarial in the courtroom, the scientific arena and testability have been and are a better forum for determining the validity of forensic analytical methods.

21.7   CONCLUSIONS

The advent of molecular biology has been a boon to human forensic identification cases. Small, environmentally insulted samples can be genetically characterized and, in many cases, to such resolution that only one or a few individuals could be the source of the samples. In fact, DNA typing is one of the best exculpatory tools available when analyzing biological evidentiary materials. The technology and applications have become well established and routinely used. Now methods are being developed to address more challenging samples. One set of markers, those residing on the Y-chromosome, enable the forensic scientist to analyze samples that were not possible previously, such as mixtures composed of minute amounts of male DNA amidst a large background of female DNA. The Y-STRs were presented to exemplify some of the issues to consider when implementing DNA typing methods into the forensic laboratory. Core sets of Y-STR loci have been defined and commercial kits have been developed. For methods to be used routinely, they must be validated, quality assurance measures must be implemented, interpretation guidelines instituted, and methods should be developed for placing statistical weight on an evidence profile that matches a reference sample profile.

Acknowledgements

This is publication number 04-03 of the Laboratory Division of the Federal Bureau of Investigation. Names of commercial manufacturers are provided for identification only, and inclusion does not imply endorsement by the Federal Bureau of Investigation.

References

Ayub, Q., Mohyuddin, A., Qamar, R., Mazhar, K., Zerjal, T., Mehdi, S. Q., and Tyler-Smith, C. (2000). Identification and characterization of novel human Y-chromosomal microsatellites from sequence database information. *Nucleic Acids Res.* 28, e8.

Ballantyne, J., and Hall, A. (2003). Strategies for the design and assessment of Y-short tandem repeat multiplexes for forensic uses. *Forensic Sci. Rev.* 15, 137–149.

Beleza, S., Alves, C., Gonzalez-Neira, A., Lareu, M., Amorim, A., Carracedo, A., and Gusmao, L. (2003). Extending STR markers in Y chromosome haplotypes. *Int. J. Legal Med.* 117, 27–33.

Brownstein, M. J., Carpten, J. D., and Smith, J. R. (1996). Modulation of non-templated nucleotide addition by Taq DNA polymerase: primer modifications that facilitate genotyping. *Biotechniques* 20, 1004–1010.

Budowle, B., and Baechtel, F. S. (1990). Modifications to improve the effectiveness of restriction fragment length polymorphism typing. *Appl. Theor. Electrophoresis* 1, 181–187.

Budowle, B., Chakraborty, R., Giusti, A. M., Eisenberg, A. J., and Allen, R. C. (1991). Analysis of the VNTR locus D1S80 by the PCR followed by high-resolution PAGE. *Am. J. Hum. Genet.* 48, 137–144.

Budowle, B., Baechtel, F. S., Comey, C. T., Giusti, A. M., and Klevan, L. (1995). Simple protocols for typing forensic biological evidence: chemiluminescent detection for human DNA quantitation and RFLP analyses and manual typing of PCR amplified polymorphisms. *Electrophoresis* 16, 1559–1567.

Budowle, B., Koons, B. W., and Errera, J. D. (1996a). Multiplex amplification and typing procedure for the loci D1S80 and amelogenin. *J. Forensic. Sci.* 41, 660–663.

Budowle, B., Koons, B. W., Keys, K. M., and Smerick, J. B. (1996b). Methods for typing the STR triplex CSF1PO, TPOX, and HUMTHO1 that enable compatibility among DNA typing laboratories. In *Advances in Forensic Haemogenetics* 6, A. Carracedo, B. Brinkmann, W. Bar, eds., pp. 107–114, Springer-Verlag, Berlin.

Budowle, B., Moretti, T. R., Niezgoda, S. J., and Brown, B. L. (1998). CODIS and PCR-based short tandem repeat loci: Law enforcement tools. In *Second European Symposium on Human Identification* 1998, Promega Corp., Madison, Wisconsin, pp 73–88.

Budowle, B., Smith, J. A., Moretti, T., and DiZinno, J. (2000). *DNA Typing Protocols: Molecular Biology and Forensic Analysis*, BioTechniques Books, BioForensic Sciences Series, Eaton Publishing, Natick, MA.

Budowle, B., Collins, P. J., Dimsoski, P., Ganong, C. K., Hennessy, L. K., Leibelt, C. S., Rao-Coticone, S., Shadravan, F., and Reeder, D. J. (2001a). Population data on the STR loci D2S1338 and D19S433. Forensic Science Communications 3(2), July 2001. Available: http://www.fbi.gov/hq/lab/fsc/backissu/july2001/index.htm.

Budowle, B., Hobson, D. L., Smerick, J. B., Smith, J. A. L. (2001b). Low copy number—consideration and caution. In *Twelfth International Symposium on Human Identification 2001*, Promega Corp., Madison, WI. http://www.promega.com/ussymp12proc/default.htm.

Budowle, B., Masibay, A., Anderson, S. J., Barna, C., Biega, L., Brenneke, S., Brown, B. L., Cramer, J., De Groot, G. A., Douglas, D., Duceman, B., Eastman, A., Giles, R., Hamill, J., Haase, D. J., Janssen, D. W., Kupferschmid, T. D., Lawton, T., Lemire, C., Llewellyn, B., Moretti, T., Neves, J., Palaski, C., Schueler, S., Sgueglia, J., Sprecher, C., Tomsey, C., and Yet, D.

(2001c). STR primer concordance study. *Forensic. Sci. Int.* 124, 47–54.

Budowle, B., and Brown, B. L. (2001). The use of DNA typing for forensic identification. *Forensica* 1, 9–37.

Budowle, B., Allard, M. W., Wilson, M. R., and Chakraborty, R. (2003a) Forensics and mitochondrial DNA: applications, debates, and foundations. *Annu. Rev. Genomics Hum. Genetics* 4, 119–141.

Budowle, B., Sinha, S. K., Lee, H. S., and Chakraborty, R. (2003b). Utility of Y chromosome STR haplotypes in forensic applications. *Forensic Sci. Rev.* 15, 153–164.

Bureau of Justice Statistics (2001). Criminal Victimization in the United States, Table 38, Library of Congress, NCJ 197064, available at http://www.ojp.usdoj.gov/bjs/pub/pdf/cvus01.pdf.

Burgoyne, L., and Rogers, C. (1997). Bacterial typing: Storage and processing of stabilized reference bacteria for polymerase chain reaction without preparing DNA—an example of an automatable procedure. *Anal. Biochem.* 247, 223–227.

Chakraborty, R. (1985). Paternity testing with genetic markers: Are Y-linked genes more efficient than autosomal ones? *Am. J. Med. Genet.* 21, 297–305.

Chakraborty, R., Stivers, D. N., Su, B., Zhong, Y., and Budowle, B. (1999). The utility of STR loci beyond human identification: Implications for the development of new DNA typing systems. *Electrophoresis* 20, 1682–1696.

Comey, C. T., and Budowle, B. (1991). Validation studies on the analysis of the HLA-DQ alpha locus using the polymerase chain reaction. *J. Forens. Sci.* 36, 1633–1648.

Copenhagen, D., and Clayton, D. A. (1974). The number of mitochondrial deoxyribonucleic acid genomes in mouse L and human HeLa cells. *J. Biol. Chem.* 249, 7791–7795.

Corach, D., Risso, L. F., Marino, F., Penancino, G., and Sala, A. (2001). Routine Y-STR typing in forensic casework. *Forensic Sci. Int.* 118, 131–135.

Cruciani, F., Santolamazza, P., Shen, P., Macaulay, V., Moral, P., Olckers, A., Modiano, D., Holmes, S., Destro-Bisol, G., Coia, V., Wallace, D. C., Oefner, P. J., Torroni, A., Cavalli-Sforza, L. L., Scozzari, R., and Underhill, P. A. (2002). A back migration from Asia to sub-Saharan Africa is supported by high-resolution analysis of human Y-chromosome haplotypes. *Am. J. Hum. Genet.* 70, 1197–1214.

Daubert versus Merrell Dow Pharmaceuticals, Inc. 509 U.S. 579, 1993.

De Knijff, P., Kayser, M., Caglià, A., Corach, D., Fretwell, N., Gehrig, C., Graziosi, G., Heidorn, F., Herrmann, S., Herzog, B., Hidding, M., Honda, K., Jobling, M., Krawczak, M., Leim, K., Meuser, S., Meyer, E., Oesterreich, W., Pandya, A., Parson, W., Penacino, G., Perez-Lezaun, A., Piccinini, A., Prinz, M., Schmitt, C., Schneider, P. M., Szibor, R., Teifel-Greding, J., Weichhold, G., and Roewer, L. (1997). Chromosome Y microsatellite: population genetic and evolutionary aspects. *Int. J. Legal Med.* 110, 134–149.

Dekairelle, A. F., and Hoste, B. (2001). Application of Y-STR pentaplex PCR (DYS19, DYS389I and II, DYS390, and DYS393) to sexual assault cases. *Forensic Sci. Int.* 118, 122–125.

Del Rio, S., Marino, M. A., and Belgrader, P. (1996). Reusing the same bloodstained punch for sequential DNA amplifications and typing. *BioTechniques* 20, 970–974.

Edwards, A., Civitello, A., Hammond, H. A., and Caskey, C. T. (1991). DNA typing and genetic mapping with trimeric and tetrameric tandem repeats. *Am. J. Hum. Genet.* 49, 746–756.

Foster, E. A., Jobling, M. A., Taylor, P. G., Donnelly, P., de Knijff, P., Mieremet, R., Zerjal, T., and Tyler-Smith, C. (1998). Jefferson fathered slave's last child. *Nature* 396, 27–28.

Gill, P., Jeffreys, A. J., and Werrett, D. J. (1985). Forensic application of DNA fingerprints. *Nature* 318, 577–579.

Gill, P., Ivanov, P. L., Kimpton, C., Piercy, R., Benson, N., Tully, G., Evett, I., Hagelberg, E., and Sullivan, K. (1994). Identification of the remains of the Romanov family by DNA analysis. *Nat. Genet.* 6, 130–135.

Gill, P., Brenner, C., Brinkman, B., Budowle, B., Carracedo, A., Jobling, M. A., de Knijff, P., Kayser, M., Krawczak, M., Mayr, W. R., Morling, N., Olaisen, B., Pascali, V., Prinz, M., Roewer, L., Schneider, P. M., Sajantila, A., and Tyler-Smith, C. (2000). DNA Commission of the International Society of Forensic Genetics: recommendations on forensic analysis using Y-chromosome STRs. *Int. J. Legal Med.* 114, 305–309.

Ginther, C., Issel-Tarver, L., and King, M. C. (1992). Identifying individuals by sequencing mitochondrial DNA from teeth. *Nat. Genet.* 2, 135–138.

Giusti, A. M., Baird, M., Pasquale, S., Balasz, I., and Glassberg, G. (1986). Application of deoxyribonucleic acid (DNA) polymorphisms to the analysis of DNA recovered from sperm. *J. Forensic. Sci.* 31, 409–417.

Giusti, A. M., and Budowle, B. (1995). Chemiluminescence-based detection system for human DNA quantitation and restriction fragment length polymorphism (RFLP) analysis. *Appl. Theor. Electrophor.* 5, 89–98.

Greenberg, B. D., Newbold, J. E., and Sugino, A. (1983). Intraspecific nucleotide sequence variability surrounding the origin of replication in human mitochondrial DNA. *Gene* 21, 33–49.

Grimberg, J., Nawoschik, S., Belluscio, L., McKee, R., Turck, A., and Eisenberg, A. (1989). A simple and efficient nonorganic procedure for the isolation of genomic DNA from blood. *Nucleic Acids Res.* 17, 8390.

Gusmão, L., Alves, C., and Amorium, A. (2001). Molecular characterisation of four human Y-specific microsatellites (DYS434, DYS437, DYS438, DYS439) for population and forensic studies. *Ann. Hum. Genet.* 65, 285–291.

Holland, M. M., Fisher, D. L., Mitchell, L. G., Rodriguez, W. C., Canik, J. J., Merril, C. R., and Weedn, V. W. (1993). Mitochondrial DNA sequence analysis of human skeletal remains: Identification of remains from the Vietnam War. *J. Forensic Sci.* 38, 542–553.

Holt, C. L., Buoncristiani, M., Wallin, J. M., Nguyen, T., Lazaruk, K. D., and Walsh, P. S. (2002). TWGDAM validation of AmpFlSTR PCR amplification kits for forensic DNA casework. *J. Forensic Sci.* 47, 66–96.

Honda, K., Roewer, L., and de Knijff, P. (1999). Male DNA typing from 25-year-old vaginal swabs using Y chromosomal STR polymorphisms in retrial request case. *J. Forensic Sci.* 44, 868–872.

Iudica, C. A., Whitten, W. M., and Williams, N. H. (2001). Small bones from dried mammal museum specimens as a reliable source of DNA. *BioTechniques* 30, 732–734.

Ivanov, P. L., Wadhams, M. J., Roby, R. K., Holland, M. M., Weedn, V. W., and Parsons, T. J. (1996). Mitochondrial DNA sequence heteroplasmy in the Grand Duke of Russia Georgij

Romanov establishes the authenticity of the remains of Tsar Nicholas II. *Nat. Genet.* 12, 417–420.

Jeffreys, A. J., Wilson, V., and Thein, S. L. (1985a). Hyper variable minisatellite regions in human DNA. *Nature* 314, 67–73.

Jeffreys, A. J., Wilson, V., and Thein, S. L. (1985b). Individual-specific fingerprints of human DNA. *Nature* 316, 76–79.

Jobling, M. A., Pandya, A., and Tayler-Smith, C. (1997). The Y chromosome in forensic and paternity testing. *Int. J. Legal Med.* 110, 118–124.

Kasai, K., Nakamura, Y., and White, R. (1990). Amplification of a variable number of tandem repeat (VNTR) locus (pMCT118) by the polymerase chain reaction (PCR) and its application to forensic science. *J. Forensic Sci.* 35, 1196–1200.

Kayser, M., de Knijff, P., Dieltjes, P., Krawczak, M., Nagy, M., Zerjal, T., Pandya, A., Tyler-Smith, C., and Roewer, L. (1997a). Applications of microsatellite-based Y chromosome haplotyping. *Electrophoresis* 18, 1602–1607.

Kayser, M., Cagliá, A., Corach, D., Fretwell, N., Gehrig, C., Graziosi, G., Heidorn, F., Herrmann, S., Herzog, B., Hidding, M., Honda, K., Jobling, M., Krawczak, M., Leim, K., Meuser, S., Meyer, E., Oesterreich, W., Pandya, A., Parson, W., Penacino, G., Perez-Lezaun, A., Piccinini, A., Prinz, M., Schmitt, C., Schneider, P. M., Szibor, R., Teifel-Greding, J., Weichhold, G., de Knijff, P., and Roewer, L. (1997b). Evaluation of Y-chromosomal STRs: a multicenter study. *Int. J. Legal Med.* 110, 125–133.

Kayser, M., Kruger, C., Nagy, M., Geserick, G., and Roewer, L. (1998). Y-chromosomal experiences and recommendations. In B. Olaisen, B. Brinkmann, P. J. Lincoln, eds., in *Progress in Forensic Genetics* 7, pp. 494–496, Elsevier Science Amsterdam.

Kayser, M., Roewer, L., Hedman, M., Henke, L., Henke, J., Brauer, S., Kruger, C., Krawczak, M., Nagy, M., Dobosz, T., Szibor, R., de Knijff, P., Stoneking, M., and Sajantila, A. (2000). Characterization and frequency of germline mutations at microsatellite loci from the human Y chromosome, as revealed by direct observation in father/son pairs. *Am. J. Hum. Genet.* 66, 1580–1588.

Kayser, M., Brauer, S., Willuweit, S., Schadlich, H., Batzer, M., Zawacki, J., Prinz, M., Roewer, L., and Stoneking, M. (2002). Online Y-chromosomal short tandem repeat haplotype reference database (YHRD) for US populations. *J. Forensic Sci.* 47, 513–519.

Kayser, M. (2003). The Human Y-chromosome—introduction into genetics and applications. *Forensic Sci. Rev.* 15, 77–89.

Kayser, M., Brauer, S., Schadlich, H., Prinz, M., Batzer, M., Zimmerman, P. A., Boatin, B. A., and Stoneking, M. (2003). Y chromosome STR haplotypes and the genetic structure of U.S. populations of African, European, and Hispanic Ancestry. *Genome Res.* 13, 624–634.

Krenke, B. E., Tereba, A., Anderson, S. J., Buel, E., Culhane, S., Finis, C. J., Tomsey, C. S., Zachetti, J. M., Masibay, A., Rabbach, D. R., Amiott, E. A., and Sprecher, C. J. (2002). Validation of a 16-locus fluorescent multiplex system. *J. Forensic Sci.* 47, 773–785.

Krenke, B. (2003). The PowerPlex® Y system. *Profiles in DNA* 6, 6–9.

Krenke, B., Viculis, L., Richard, M. L., Prinz, M., Milne, S. C., Ladd, C., Gross, A. M., Gornall, T., Frappier, R., Eisenberg, A., Barna, C., Aranda, X., Adamowicz, M. S., and Budowle, B.

(2005). Validation of a male-specific, 12-locus fluorescent short tandem repeat (STR) multiplex. *Forensic Sci. Int.*; 148, 1–14.

Lell, J. T., Sukernik, R. I., Starikovskaya, Y. B., Su, B., Jin, L., Schurr, T. G., Underhill, P. A., and Wallace, D. C. (2002). The dual origin and Siberian affinities of Native American Y chromosomes. *Am. J. Hum. Genet.* 70, 192–206.

Levinson, G., and Gutman, G. A. (1987). Slipped strand mispairing: a major mechanism for DNA sequence evolution. *Mol. Biol. Evol.* 4, 203–221.

Lins, A. M., Micka, K. A., Sprecher, C. J., Taylor, J. A., Bacher, J. W., Rabbach, D. R., Bever, R. A., Creacy, S. D., and Schumm, J. W. (1998). Development and population study of an eight-locus short tandem repeat (STR) multiplex system. *J. Forensic Sci.* 43, 1168–1180.

Magnuson, V. L., Ally, P. S., Nylund, S. J., Karanjuwala, Z. E., Rayman, J. B., Knapp, J. I., Lowe, A. L., Ghosh, S., and Collins, F. S. (1996). Substrate nucleotide-determined non-templated addition of adenine by Taq DNA polymerase: implications for PCR-based genotyping. *BioTechniques* 21, 700–709.

Micka, K. A., Amiott, E. A., Hockenberry, T. L., Sprecher, C. J., Lins, A. M., Rabbach, D. R., Taylor, J. A., Bacher, J. W., Glidewell, D. E., Gibson, S. D., Crouse, C. A., and Schumm, J. W. (1999). TWGDAM validation of a nine-locus and a four-locus fluorescent STR multiplex system. *J. Forensic Sci.* 44, 1243–1257.

Moretti, T. R., Baumstark, A. L., Defenbaugh, D. A., Keys, K. M., and Budowle, B. (2001). Validation of short tandem repeats (STRs) for forensic usage: Performance testing of fluorescent multiplex STR systems and analysis of authentic and simulated forensic samples. *J. Forensic Sci.* 46, 647–660.

Mountain, J. L., Knight, A., Jobin, M., Gignoux, C., Miller, A., Lin, A. A., and Underhill, P. A. (2002). SNPSTRs: empirically derived, rapidly typed, autosomal haplotypes for inference of population history and mutational processes. *Genome Res.* 12, 1766–1772.

Murch, R., and Budowle, B. (1986). Applications of isoelectric focusing in forensic serology. *J. Forensic Sci.* 31, 869–880.

National Research Council: The Evaluation of Forensic DNA Evidence; National Academy Press: Washington, DC; 1996.

Prinz, M., Boll, K., Baum, H., and Shaler, B. (1997). Multiplexing of Y chromosome specific STRs and performance of mixed samples. *Forensic Sci. Int.* 85, 209–218.

Quintans, B., Beleza, S., Brion, M., Sanchez-Diz, P., Lareu, M., and Carracedo, A. (2003). Population data of Galicia (NW Spain) on the new Y-STRs DYS437, DYS438, DYS439, GATA A10, GATA A7.1, GATA A7.2, GATA C4 and GATA H4. *Forensic Sci. Int.* 131, 220–224.

Roewer, L., Kayser, M., Dieltjes, P., Nagy, M., Bakker, E., Krawczak, M., and de Knijff, P. (1996). Analysis of molecular variance (AMOVA) of Y chromosome specific microsatellites in two closely related human populations. *Hum. Mol. Genet.* 7, 1029–1033.

Roewer, L., Kayser, M., de Knijff, P., Anslinger, K., Cagliá, A., Corach, D., Füredi, S., Henke, L., Hidding, M., Kärgel, H. J., Lessig, R., Nagy, M., Pascali, V. L., Parson, W., Rolf, B., Schmitt, C., Szibor, R., Teifel-Greding, J., and Krawczak, M. (2000). A new method for the evaluation of matches in non-recombining genomes: application to Y-chromosomal short tandem repeat (STR) haplotypes in European males. *Forensic Sci. Int.* 114, 31–43.

Roewer, L., Krawczak, M., Willuweit, S., Nagy, M., Alves, C., Amorim, A., Anslinger, K., Augustin, C., Betz, A., Bosch, E., Cagliá, A., Carracedo, A., Corach, D., Dekairelle, A. F., Dobosz, T., Dupuy, B. M., Füredi, S., Gehrig, C., Gusmaõ, L., Henke, J., Henke, L., Hidding, M., Hohoff, C., Hoste, B., Jobling, M. A., Kärgel, H. J., de Knijff, P., Lessig, R., Liebeherr, E., Lorente, M., Martínez-Jarreta, B., Nievas, P., Nowak, M., Parson, W., Pascali, V. L., Penacino, G., Polski, R., Rolf, B., Sala, A., Schmidt, U., Schmitt, C., Schneider, P. M., Szibor, R., Teifel-Greding, J., and Kayser, M. (2001). Online reference database of European Y-chromosomal short tandem repeat (STR) haplotypes. *Forensic Sci. Int.* 118, 106–113.

Saferstein, R., ed. *Criminalistics—An introduction to Forensic Science*, 5e. New Jersey: Prentice Hall Education; 1995.

Saiki, R. K., Scharf, S., Faloona, F., Mullis, K. B., Horn, G. T., Erlich, H. A., and Arnheim, N. (1985). Enzymatic amplification of beta-globin genomic sequences and restriction analysis for diagnosis of sickle cell anemia. *Science* 230, 1350–1354.

Saiki, R. K., Walsh, P. S., Levenson, C. H., and Erlich, H. A. (1989). Genetic analysis of amplified DNA with immobilized sequence-specific oligonucleotide probes. *Proc. Natl. Acad. Sci. USA* 86, 6230–6234.

Sambrook, J., Fritsch, E. F., and Maniatis, T. (1989). *Molecular Cloning: A Laboratory Manual*. 2e. Cold-Spring Harbor Laboratory, Cold-Spring Harbor Laboratory Press, New York.

Sanger, F., Nicklen, S., and Coulson, A. R. (1977). DNA sequencing with chain-terminating inhibitors. *Proc. Natl. Acad. Sci. USA* 74, 5463–5468.

Scherczinger, C. A., Bourke, M. T., Ladd, C., and Lee, H. C. (1997). DNA extraction from liquid blood using QIAamp. *J. Forensic Sci.* 42, 893–896.

Schlotterer, C., and Tautz, D. (1992). Slippage synthesis of simple sequence DNA. *Nucleic Acids Res.* 20, 211–215.

Shedlock, A. M., Haygood, M. G., Pietsch, T. W., and Bentzen, P. (1997). Enhanced DNA extraction and PCR amplification of mitochondrial genes from formalin-fixed museum specimens. *BioTechniques* 22, 394–400.

Shewale, J. G., Nasir, H., Schneida, E., Gross, A. M., Budowle, B., and Sinha, S. K. (2004). Y-chromosome STR system, Y-PLEX™ 12, for forensic casework: development and validation. *J. Forensic Sci.*; 49, 1278–1290.

Sinclair, K., and McKechnie, V. M. (2000). DNA extraction from stamps and envelope flaps using QIAamp and QIAshredder. *J. Forensic Sci.* 45, 229–230.

Singer-Sam, J., Tanguay, R. L., and Riggs, A. (1989). Use of Chelex to improve the PCR signal from a small number of cells. *Amplifications: A Forum for PCR Users* 3, 11.

Sinha, S. K., Budowle, B., Arcot, S. S., Richey, S. L., Chakraborty, R., Jones, M. D., Wojtkiewicz, P. W., Schoenbauer, D. A., Gross, A. M., Sinha, S. K., and Shewale, J. G. (2003a). Development and validation of a multiplexed Y-chromosome STR genotyping system, Y-PLEX™6, for Forensic Casework. *J. Forensic Sci.* 48, 93–103.

Sinha, S. K., Nasir, H., Gross, A. M., Budowle, B., and Shewale, J. G. (2003b). Development and validation of the Y-PLEX™5, a Y-chromosome STR genotyping system for forensic casework. *J. Forensic Sci.* 48, 985–1000.

Smith, J. R., Carpten, J. D., Brownstein, M. J., Ghosh, S., Magnuson, V. L., Gilbert, D. A., Trent, J. M., and Collins, F. S. (1995). Approach to genotyping errors caused by nontemplated nucleotide addition by Taq DNA polymerase. *Genome Res.* 5, 312–317.

Sullivan, K. M., Hopgood, R., and Gill, P. (1991a). Identification of human remains by amplification and automated sequencing of mitochondrial DNA. *Int. J. Legal Med.* 105, 83–86.

Sullivan, K. M., Hopgood, R., Lang, B., and Gill, P. (1991b). Automated amplification and sequencing of human mitochondrial DNA. *Electrophoresis* 12, 17–21.

Sullivan, K. M., Hopgood, R., and Gill, P. (1992). Identification of human remains by amplification and automated sequencing of mitochondrial DNA. *Int. J. Legal Med.* 105, 83–86.

Underhill, P. A., Passarino, G., Lin, A. A., Shen, P., Mirazon Lahr, M., Foley, R. A., Oefner, P. J., and Cavalli-Sforza, L. L. (2001). The phylogeography of Y chromosome binary haplotypes and the origins of modern human populations. *Ann. Hum. Genet.* 65, 43–62.

Walsh, S. Metzger, D. A., and Higuchi, R. (1991). Chelex 100 as a medium for simple extraction of DNA for PCR-based typing from forensic material. *BioTechniques* 10, 506–513.

Walsh, P. S., Varlaro, J., and Reynolds, R. (1992). A rapid chemiluminescent method for quantitation of human DNA. *Nucleic Acids Res.* 20, 5061–5065.

Waye, J., Presley, L., Budowle, B., Shutler, G. G., and Fourney, R. M. (1989). A simple and sensitive method for quantifying human genomic DNA in forensic specimen extracts. *Biotechniques* 7, 852–855.

Wells, R. S., Yuldasheva, N., Ruzibakiev, R., Underhill, P. A., Evseeva, I., Blue-Smith, J., Jin, L., Su, B., Pitchappan, R., Shanmugalakshmi, S., Balakrishnan, K., Read, M., Pearson, N. M., Zerjal, T., Webster, M. T., Zholoshvili, I., Jamarjashvili, E., Gambarov, S., Nikbin, B., Dostiev, A., Aknazarov, O., Zalloua, P., Tsoy, I., Kitaev, M., Mirrakhimov, M., Chariev, A., and Bodmer, W. F. (2001). The Eurasian heartland: a continental perspective on Y-chromosome diversity. *Proc. Natl. Acad. Sci. USA* 98, 10244–10249.

White, P., Tatum, O., Deaven, L., and Longmire, J. (1999). New, male-specific microsatellite markers from the Y-chromosome. *Genomics* 57, 433–437.

Wilson, M. R., DiZinno, J. A., Polanskey, D., Replogle, J., and Budowle, B. (1995a). Validation of mitochondrial DNA sequencing for forensic casework analysis. *Int. J. Legal Med.* 108, 68–74.

Wilson, M. R., Polanskey, D., Butler, J., DiZinno, J. A., Replogle, J., and Budowle, B. (1995b). Extraction, PCR amplification, and sequencing of mitochondrial DNA from human hair shafts. *Biotechniques* 18, 662–669.

Wyman, A. R., and White, R. (1980). A highly polymorphic locus in human DNA. *Proc. Natl. Acad. Sci. USA* 77, 6754–6758.

CHAPTER 22

Molecular Diagnostics and Comparative Genomics in Clinical Microbiology

ALEX VAN BELKUM

Erasmus MC, Department of Medical Microbiology and Infectious Diseases Unit Research and Development, Rotterdam, The Netherlands

TABLE OF CONTENTS

22.1 INTRODUCTION

Invasive infections can be caused by a wide variety of protozoan, fungal, microbial, and viral pathogens. This renders microbiological and viral diagnosis a complicated task, especially since many of these pathogens generate similar and sometimes even identical clinical syndromes. As a further complicating factor, clinical samples submitted for culture-based diagnostic procedures frequently are contaminated by microorganisms naturally colonizing the epithelial lining of the human body. Moreover, the specimens also may contain substances that limit the chances for pathogen's survival and, thereby, cultivation success. On the whole, microbial diagnostics usually is performed on a wide variety of clinical samples and the microbial agents requiring identification may differ in prevalence with age, sex, habits, and several other personal characteristics of the individual patient. The prevalence of infectious agents also may be subject to strong seasonal and geographic variation. In all, culture-oriented clinical microbiology seems to be an art rather than a precise scientific approach. With respect to correctness of a diagnosis, much depends on the expertise of the clinical microbiologist or infectious disease (ID) specialist: during the diagnostic procedures several important decisions must be taken and these may be biased due to personal preferences rather than scientific appropriateness.

The choice of microbiological laboratory procedure usually is dictated by personal experiences: doctors ask for a specific type of diagnostic assay, but this may not always be the most obvious one. After clinical parameters have been assessed at bedside, the first step in microbial diagnosis normally consists of a straightforward (gram-) staining procedure. This segregates gram-positive from gram-negative organisms and already, in part, may lead to direct species-identification of the pathogen involved. Malaria parasites, for instance, show very distinct morphological features upon simple Giemsa staining of a blood smear (Iqbal *et al.*, 2003). Subsequently, the clinical specimen may be further analyzed by cultivation-based tests. In addition, concomitant serum samples may be analyzed for antibodies raised by the host's immune system against the pathogen involved. Microbial antigens rather than the complete and viable organism itself can be searched for as well. Finally, when investigations are finished and a pathogen has been identified beyond reasonable doubt, rational therapy can be implemented, if available. Therapy usually is preceded by assessment of the antimicrobial susceptibility profile and/or virulence characteristics of the microorganism involved; that is, in case of successful cultivation, of course. It goes without saying that speed and quality of the diagnostic procedures in the end determine the clinical impact and efficacy of antibiotic therapies.

Since some pathogens may be transmitted easily between humans, not only diagnosis is important. It is also relevant to keep track of the spreading of certain pathogens, both in the hospital and in the open population. Classic microbiology uses methods such as phagetyping, serotyping, or antibiogram comparison to assess epidemiological relatedness among strains of a given pathogenic species. The diagnostic and epidemiological methods just mentioned have been

Molecular Diagnostics
Patrinos and Ansorge

in place for already quite some time. Seeing the continuing impact of infectious diseases on modern society (i.e., the AIDS pandemic, biological warfare, emergence of multidrug resistant bacteria, the recent SARS outbreak), there is a persisting need for continuous improvement of diagnostic microbiology and virology (Demain, 1998). Molecular microbiology has provided the laboratory with additional tools that will help significantly to improve the quality of microbial detection and (subspecies) identification (see Fig. 22.1).

Molecular microbiological techniques initially were developed during the 1960s and 1970s in fundamental research laboratories. As a direct result, over the past one or two decades, the identification of the organisms that are infectious to humans has been greatly facilitated by the development and application of specific molecular hybridization (probe) tests (Ksiazek *et al.*, 2003). In addition, the availability of (real-time) nucleic acid amplification methods has been instrumental in the development of another, more recent category of direct, highly sensitive diagnostic assays. The polymerase chain reaction (PCR) is among the most popular of these methods (Wolk *et al.*, 2001). The currently available molecular assays enable both the direct detection of antigens (DNA and RNA) of putative pathogens in clinical material and the genetic identification (also know as DNA fingerprinting or comparative genomics) of microorganisms obtained by culture.

The availability of molecular diagnostics initially was considered a panacea, but replacement of conventional tests

for detection and identification of microorganisms by molecular procedures turned out to be a slow and cautious process. However, several of the innovative nucleic acid identification tests are currently just beyond or slightly ahead of their breakthrough. For all the clinically relevant microorganisms one or more molecular tests for detection and identification, even below the species level, have now become available, although commercially developed tests are for sale for only the most prevalent infectious disease agents. High-throughput use of these tests, however, is still somewhat restricted to the larger laboratories possessing adequate technical equipment and analytical expertise.

In this chapter, the current state-of-the-art of molecular-diagnostics and comparative genomics in medical microbiology will be described, together with the technological advances that have been made recently. In addition, some of the problems remaining to be solved prior to general acceptance of nucleic acid-mediated detection and identification of microbial pathogens will be also reviewed. This chapter highlights the success of the novel applications by providing examples of modern molecular diagnostic approaches in the field of bacterial infections. Viral diagnosis will not be discussed in depth, but a short section describing the current state of affairs within virology will be provided later in this chapter.

22.2 TECHNOLOGICAL IMPROVEMENTS

Essentially, molecular diagnostics of infectious diseases is simple and straightforward. A clinical sample (or a not yet identified organism derived from a microbiological culture experiment) is provided and the use of a broad-spectrum nucleic acid isolation procedure generates material to be included as template in nucleic acid hybridization or amplification assays. These reactions proceed and positive or negative results are produced, interpreted, and translated into a diagnostic result. However, this simple scenario has been compromised by a variety of obvious and emerging problems. Most of these have been solved, but some remain and are continuously investigated in search of elegant solutions. Some examples of recent technological successes follow.

22.2.1 Optimization of the Isolation of Template Nucleic Acid Molecules

During the past decade, the guanidinium isothiocyanate/Celite affinity procedure was accepted as the gold standard in many molecular diagnostic laboratories (Boom *et al.*, 1990). Its main downside was the fact that the manual version of this procedure was and still is quite time-consuming and laborious. Fortunately, alternative systems have become available. Extraction simplification and automation have progressed well and after the commercial

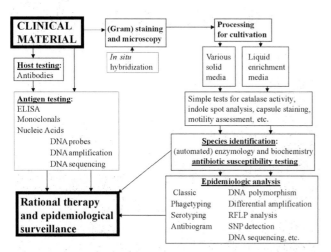

FIGURE 22.1 Schematic representation of sample routing in clinical microbiology. The classic methods are depicted in standard letter type, whereas the novel molecular diagnostic tools are depicted in italic and bold lettering. The arrows indicate the flow of activities; note that epidemiological investigations can be guided by data from both the antigen detection and culture based test systems (ELISA: enzyme-linked immuno-sorbent assay; RFLP: restriction fragment length polymorphism; SNP: single nucleotide polymorphism).

availability of handy spin column assays (e.g., the Qiagen kits; McOrist *et al.*, 2002), fully automated nucleic acid isolation systems have been developed. Organon, recently acquired by bioMérieux, has automated the Celite affinity procedure through the Nuclisense Extractor machine. Roche Molecular Systems is marketing the MagnaPure DNA isolation robot, which allows for parallel DNA extractions for hundreds of samples per working day, invoking limited hands-on time only (Van Doornum *et al.*, 2003). Further improvements in the nucleic acid yield and sample throughput are anticipated. The currently separate DNA extraction and amplification machines will soon be combined in a single apparatus (Cockerill and Smith, 2002).

To monitor the extraction process for loss of target or contamination by inhibitory compounds, internal process controls should be introduced (see also Section 22.4. for more details). This methodology identifies deleterious effects during extraction (loss of sample) and amplification (inhibition), and enables one to be more confident on both positive and negative results generated. It has to be stated that molecular microbiologists have been putting in enormous efforts to demonstrate the efficacy of nucleic acid-mediated testing over the past years. Far more than classic microbiology, DNA/RNA testing has been subjected to extremely detailed and stringent intercentre quality control studies. Interestingly, most of the new generations of tests withstood the comparisons with classic testing quite easily, thereby confirming their superiority over the widely accepted classic test systems (Harmsen *et al.*, 2001; Niesters, 2002).

22.2.2 Prevention of Carry-Over Contamination

The implementation of PCR and other amplification tests in routine laboratories requires logistic adaptation and training programs for the personnel involved. For one, the various stages in PCR diagnostics (i.e., nucleic acid purification, preparation of enzyme master mixes, collating the complete PCR samples, amplification, and finally, the analysis of the amplified material) should preferably be performed in physically separate laboratories. The most important adaptations are those required to prevent contamination and subsequent false positive results. The current generation of commercial assays uses intelligent systems such as the UNG glycosylase approach (Meier *et al.*, 1993). This results in pre-PCR destruction of amplimers generated in prior tests; this system has proven to be quite robust (Rys and Persing, 1993). There are several other measures and precautions to be taken that essentially prevent carry over contamination from occurring. Replacement of DNA-contaminated disposable articles by clean ones, avoiding pipette abuse (aerosol formation!), adequate routing of patient materials, and the use of molecular-grade PCR ingredients are a few of these alternative options. A comprehensive discussion on most of the important

factors was recently published by Millar and colleagues (2002).

22.2.3 Real-Time PCR Systems

The initial molecular tests for infectious diseases were strictly qualitative in nature: they provided a straightforward yes-or-no answer to questions relating to the absence or presence of certain pathogens. With recent improvements of the nucleic acid-mediated technology quantitative aspects also became addressable. A variety of real-time PCR machines have been developed. These include, among others, the GeneAmp 5700 and Prism 7700 by Applied Biosystems, the BioRad iCycler, the Roche LightCycler, the Cepheid Smartcycler and GeneXpert, the MX400 by Stratagene and the Rotor Gene by Corbette Research (see also Chapter 10). All these machines share a high assay speed (between 20 minutes and 2 hours per run) and sufficient sample capacity (from 16 to 384 samples per run). Real-time technologies combining PCR and TaqMan hydrolysis probes (Lunge *et al.*, 2002), NASBA and molecular beacons (Weusten *et al.*, 2002) or hybridization probes (Bidet *et al.*, 2003) helped to improve the tedious process of real-time detection and identification. It has been well established that the new technologies have the reproducible ability to detect (small amounts of) an infectious agent in all sorts of clinical specimens.

The question whether nucleic acid quantification may have an added value for clinical infectious disease management in distinct groups of patients should and must be answered in the near future. The main issue here is whether larger amounts of pathogens present may be an indication of the severity of (underlying) disease. If this is the case (as suggested by some virology studies; Schutten and Niesters, 2001) this implies that quantitative tests could also be used successfully to monitor the efficacy of a treatment protocol.

22.2.4 Broad Spectrum PCR and PCR Multiplexing

In some cases, species of microbial pathogens capable of invoking an infectious syndrome may be large in number. Infection of the lower and upper airways, for instance, can be caused by a variety of bacterial species. These include, in random order, *Streptococcus pneumoniae*, *Moraxella catarrhalis*, *Haemophilus influenzae*, *Bordetella pertussis*, *Mycoplasma pneumoniae*, *Legionella pneumophila*, *Chlamydia pneumoniae*, *Staphylococcus aureus*, *Klebsiella pneumoniae*, and others covering a wide variety of viruses, fungi, and parasites as well. It would be very cost-effective if these pathogens, involved in similar, clinically manifest diseases, could be detected by a single assay. Broad spectrum PCR coupled to species-specific nucleic acid probing assays could provide such an opportunity (Roth *et al.*, 2000), whereas multiplex PCRs have been described that are capable of achieving a similar feat (Ko *et al.*, 2003). This type of assay will become increasingly important over the coming years, especially with the introduction of quantitative variants of these tests. The downside of multiplexing is that the multiplex PCR usually is considered to be less sensitive than the simplex approach.

The detection of ill-defined or even previously unrecognized bacterial pathogens can be pursued by broad spectrum PCR, a method already alluded to in the previous section (Relman, 1998). The method usually involves PCR targeting of universally conserved, eubacterial PCR priming sites. A well-known example of such targets are the ribosomal genes, although successful broad spectrum amplification of genes encoding DNA polymerase, DNA gyrase, or protein elongation factors has been described as well. Broad spectrum PCR has recently been used to try and resolve a variety of clinical syndromes. These include, for instance, periodontal diseases (Kumar *et al.*, 2003) and nonchlamydial, nongonococcal urethritis in men (Riemersma *et al.*, 2003). The latter study elegantly demonstrated that in infectious diseases not only the presence of a pathogen may be important, but also that the absence of certain apparently healthy components of the resident flora may be helpful in diagnosing disease. There may be, however, fundamental obstructions to the diagnostic implementation of broad spectrum PCR. For instance, there is an intense discussion still ongoing on the presence of native, nonpathogenic commensal bacteria in human blood (Nikkari *et al.*, 2001; McLaughlin *et al.*, 2002). Such phenomena may in the end completely frustrate the application of broad spectrum PCR in nucleic acid extracts of these matrices.

22.3 PERSISTING PROBLEMS WITH MOLECULAR DIAGNOSTICS

In clinical microbiology, companies focus on a limited number of economically attractive microbes for the development of well-standardized, commercially available assays. From this perspective, it is deemed unlikely that the number and scope of actively marketed infectious disease tests will increase rapidly over the coming years: all of the economically interesting diagnostic areas now seem to be covered and primarily the competition between diagnostic companies may increase significantly when different tests aiming at the same agent become more widely available. However, it should be emphasized that there is a large and still growing list of less prevalent microbial pathogens for which the commercial availability of nucleic acid test systems would be much appreciated by medical microbiologists and infectious disease specialists. For most of these minor pathogens, molecular diagnosticians have to rely on tests that have been developed in house; a major drawback is the fact that these tests generally are not accepted and widely used. These tests usually suffer from restricted

quality control only (see also Chapter 34). Hopefully, the small niche markets will raise the interest of start-up or other small companies.

Besides the current lack of specific test availability, another bottleneck is due to the lack of convenient and reproducible nucleic acid isolation systems to be used in combination with the diversity of clinical specimens received by the clinical microbiology laboratory. Although several open systems are available, there still is a clear need for versatile, highly reliable automated systems. When available, these machines may significantly enhance full implementation of nucleic acid detection in the day-to-day laboratory setting. Furthermore, this will enable the laboratory to generate results within a short turn-around time, which is essential for infectious disease management. In addition, only limited amounts of information are available on the (dis)functioning of the present generation of robots and other instruments in use for diagnostic analyses. This scant availability has thus far frustrated detailed intercenter comparisons on the machines' performance.

Lack of standardization and quality control programmes is another persisting problem area in molecular diagnostics (see also Chapter 34). It has been known since the early 1990s that one of the most significant hurdles to be overcome by molecular technologies is false-positivity due to contamination and false negativity due to the large differences in sensitivity between various test systems (Schuurman *et al.*, 1996). European intercenter quality control research programs have indeed shown the clear need for standardized reference materials and the general desire to participate in quality control programs, for instance the ones organized by the Quality Control for Molecular Diagnostics (QCMD) organization (Van Vliet *et al.*, 2001). This latter organization, initially funded by the European Union, provides quality control schemes for an increasing number of viral and bacterial targets.

Besides the need for well-defined qualitative quality control programs, there is also a lack of standardization of (semi) quantitative assays. External standards should be developed in a format that shows maximum identity to the samples of interest. The conclusion is that at present absolute quantification is very hard to achieve, a problem that may persist for years to come. Also, in the field of subspecies identification of clinical bacterial isolates, big problems have been noted. Many of the molecular technologies used for fingerprinting the genomes of microbes failed to be reproducibly implemented even in different but closely collaborating laboratories (Van Belkum *et al.*, 1995; Van Belkum *et al.*, 1998a). Although incidental successes have been reported, there is no current gold standard procedure for microbial typing (De Lencastre *et al.*, 1996; Murchan *et al.*, 2003). The technology most likely to survive in the end and already now providing useful tools for detailed international microbial dissemination studies probably will be multilocus sequence typing (MLST), a technology that can now be applied using Affymetrix DNA chip technology as well (Maiden *et al.*, 1998; Van Leeuwen *et al.*, 2003). This technology will be discussed in a later section of this chapter.

22.4 MOLECULAR VIRUS DETECTION

One has to realize that the translation from experimental research tools into routine molecular diagnostics is still not completed in many cases. In virology this is mainly hampered by the fact that many of the new testing systems are really new: there are no historic alternatives for the sensitive detection procedure by non-nucleic acid methodologies, with the detection of the hepatitis virus, the human metapneumovirus, or the coronavirus causing SARS as recent examples (Su *et al.*, 2002; Poon *et al.*, 2003; Van den Hoogen *et al.*, 2001). This renders comparative quality assessment difficult if not impossible and both laboratory scientists and clinicians need to put their trust in a test that stands largely unverified. However, the powerful versatility and reliability of nucleic acid testing has already convinced most investigators with clinical responsibilities.

The availability of unprecedented tests in virology has accelerated the development of process and quality controls (Oberste *et al.*, 2000; Van Elden *et al.*, 2001; Zaaijer *et al.*, 1993). One of the means to measure and validate the performance of diagnostic testing is the use of external quality assessment (EQA) control programs. The QCMD network again supplies standardized reference sample panels to be used for monitoring of development and implementation of nucleic acid detection technologies, both in a qualitative and a quantitative manner (Quint *et al.*, 1995; Valentine-Thon *et al.*, 2001; Wallace, 2003). The use of EQA has until recent years been focusing almost exclusively to blood-borne viruses, for which standardized tests are now commercially available. However, the need for EQA relating to the large panel of clinically relevant but commercially less interesting targets is growing (Espy *et al.*, 2000; Savolainen *et al.*, 2002).

Besides external quality control, virology has also excelled in the development of internal control procedures. One of these elegant control systems concerns a complete nonhuman seal herpes DNA virus (Van Doornum *et al.*, 2003). A real time and quantitative TaqMan assay was developed for this Phocine Herpes Virus (PhHV). The virus can be grown relatively easily in cell culture and is available in sheer unlimited amounts. Also, a universal RNA virus was introduced as an internal control; the Phocine Distemper Virus (PDV) could serve the same purpose as PhHV, be it for reverse transcriptase (RT) PCR applications (personal communication, Dr. Bert Niesters, Virology, Erasmus MC, Rotterdam, The Netherlands). The assumption here is that an intact virus, when used as universal internal control, behaves more similarly in the extraction procedure as compared to target viruses of interest, in contrast to using, for example, a plasmid as internal control. A low and fixed amount of this virus (equal to an amount giving a cut-off (Ct) value in the real-time assay of approximately 30–33 cycles of amplification), needs to be added to each clinical sample before the extraction procedure starts. In virology this mostly involves serum or plasma, although analysis of sputum and throat samples has also been successful. The internal control virus genomes are co-extracted and subsequently amplified in a quantitative manner in (currently) a separate tube.

22.5 EXAMPLES FROM BACTERIOLOGY

In all protocols, molecular diagnostic bacteriology starts with the purification of DNA or RNA from the bacteria involved. In other words, procedures aimed at the detection of a certain specific pathogen can be adapted simply to the detection of other pathogens in general. The commercially available molecular procedures suited for the detection of the sexually transmitted, intracellular pathogen *Chlamydia trachomatis* will be highlighted in the following pages. In addition, methods suited for the detection and subspecific identification of the gram-positive microorganism *Staphylococcus aureus*, often based on in-house testing, will be discussed in detail.

22.5.1 Commercial Test Systems for *Chlamydia trachomatis* Diagnosis

C. trachomatis is a microbe capable of causing two major clinical syndromes. When infecting the eye it can cause trachoma, which in the end leads to blindness. When infecting the fallopian tubes it can cause pelvic inflammatory disease (PID), a major cause of infertility in women (for a review see Schachter, 1985). It is clear that diagnosis of *C. trachomatis* infection, in Western countries usually disseminated by sexual contacts or through vertical transmission during birth, requires some priority. Classical *C. trachomatis* diagnosis depended on the organism's capability to infect certain receptor cell lines. Cervical swabs were taken, transported to the laboratory in a specialized conservation medium and the suspected material was inoculated into the cell culture system. The appearance of a cytopathogenic effect was indicative for the presence of *C. trachomatis* (Suchland *et al.*, 2003). This test system was insensitive, laborious, expensive, and required high levels of analytical expertise. Hence, the availability of an initial DNA hybridization-based culture confirmation test spurred enthusiasm from the diagnostic community (Tenover, 1993).

In addition to the extended culture diagnosis, the first direct nucleic acid tests became available in the early 1980s of the last century. These commercial tests manufactured by GenProbe (San Diego, CA) were initially aimed at direct ribosomal RNA detection, the second generation of tests was also able to generate limited ribosomal RNA amplification (Verkooyen *et al.*, 2003). Ultimately, various amplification mediated tests were developed based on technologies such as bacteriophage Qβ RNA polymerase mediated amplification (Stefano *et al.*, 1997), the ligase chain reaction or LCR (Blocker *et al.*, 2002) and, of course, the PCR (see Verkooyen *et al.*, 2003 and references therein). These tests all underwent extensive comparisons with the culture-based assay and among each other, and in the end the PCR tests prevailed, although the LCR tests demonstrated adequate sensitivity and specificity (Pannekoek *et al.*, 2003).

With the apparent preference of many researchers for the PCR, commercial developments took various quantum leaps. Roche Molecular Systems is now selling an integral PCR-based detection system called COBAS AMPLICOR. This system facilitates automated *C. trachomatis* diagnostics, not only based on the use of cervical swabs but also on less invasive, more patient-friendly urine samples (Leslie *et al.*, 2003). In conclusion, in a mere 15 years a drastic diagnostic change has been observed concerning the detection of *C. trachomatis*: from the cumbersome cell culture systems, to be combined with cervical swabs, we now have a new gold standard technology: the detection of *C. trachomatis* in urine can be performed with a machine generating an adequate result within half a working day. So, added to the increased sensitivity and specificity of the test system, speed and throughput also have improved significantly over the past decade (for a concise meta-analysis of some of the available data, see Table 22.1). Although detailed cost-effectiveness studies have not been published thus far, preliminary explorations already indicated that in populations with a prevalence of *C. trachomatis* infections over 3.9% the direct costs of PCR mediated screening programs are low, as compared to symptom-driven analyses (Paavonen *et al.*, 1998).

22.5.2 Molecular Detection and Identification of *Staphylococcus aureus*

Classical *S. aureus* culture is still amenable to improvement. Novel selective growth media providing excellent yield and specificity are still strong competitors to diagnostic DNA testing. In the case of *S. aureus*, especially the detection of the methicillin-resistant version of *S. aureus* (MRSA), it is an important driving factor behind diagnostic improvement. Colonization of patients or medical personnel with this particularly antibiotic-multiresistant bug predisposes to dangerous, difficult-to-treat infections. Colonization and subsequent infection, or spread of MRSA in the hospital setting, must therefore be prevented, hence the continuous need for methods for MRSA detection and genetic identification (see Table 22.2 for a short list of currently available methods). MRSA, like methicillin susceptible *S. aureus* (MSSA) has its prime ecological niche in the vestibulum nasi, the foremost compartment of the nose.

Diagnostic procedures often are focused on the analysis of nasal swabs. An example of a recently described, new medium is the Oxacillin Resistance Screening Agar developed by Blanc and colleagues (2003). This medium provides adequate results within 48 hours after inoculation. The most extensive comparison of culture-based assays was recently presented by Safdar and colleagues (2003), who compared 32 different media. Their conclusion was that optimal samples are taken with standard rayon swabs, the material should be enriched overnight in salt-containing trypticase soy broth and the pre-enrichment culture should subsequently be inoculated on lipovitellin-containing mannitol salt agar containing oxacillin. Addition of an oxacillin disk on the agar surface further increased the specificity.

It has been shown that the quality of the clinical specimen is one of the prime determinants of diagnostic efficacy. For instance, nasal swabs are to be preferred over perianal or throat swabs for assessing the colonization status of a person (Singh *et al.*, 2003). However, PCR is still considered to be the future gold standard diagnostic tool, especially for the detection and identification of MRSA. Many of the *S. aureus* PCR tests in one way or another include the detection of the methicillin resistance gene encoding the penicillin binding protein PBB2' (Miyamoto *et al.*, 2003). The first mecA specific test was published as early as 1991 (Murakami *et al.*, 1991) and many followed over the past decade.

Testing for the presence of MRSA is clinically important. For instance, Hallin and colleagues (2003) demonstrated that PCR diagnosis was instrumental in the modification and optimization of antibiotic therapy in 7/28 (25%) of patients included in their clinical study. Also, among cardiac surgery patients the prevention of post-surgical *S. aureus* wound infections was better guided by PCR than by culture (Shrestha *et al.*, 2003). Besides PCR, other molecular amplification methodologies are also applied to the detection of *S. aureus*. A new method, isothermal signal amplification (ISA), allows for the detection of at least 2×10^5 MRSA cells (Levi *et al.*, 2003). Although the method may have its advantages, the fact that in general a time-consuming enrichment culture is required prior to ISA may in the end be a significant obstacle. The latest development is that of real-time quantitative PCR (Borg *et al.*, 2003; Fang and Hedin, 2003; Gueudin *et al.*, 2003).

Many other nucleic acid-controlled technologies are applied in order to reach further refinement of molecular diagnostics. This includes genome sequencing and computerized or experimental comparative genomics in order to increase the number of adequate target sequences (Baba *et al.*, 2002). Transcription profiling using DNA chips is useful for enhancing our understanding of stage- or stimulus-specific gene expression in *S. aureus* (Mongodin *et al.*, 2003).

A very important new feature of the diagnosis of *S. aureus* infection is the possibility to simultaneously detect the presence of certain virulence genes. For instance, French

TABLE 22.1 Meta-analysis of *Chlamydia trachomatis* molecular diagnostics versus culture. Various diagnostic procedures are listed in the column on the left, some information on each and every system is indicated in the other columns. This is not a complete survey, this is meant for highlighting some of the more common technologies and their respective performance in clinical diagnostics.

Procedure	Target sequence	Specificity	Sensitivity	Number of samples	References
Radioactive probe hybridization	Cryptic 7kbp plasmid	91–94%	87–90%	1214 conjunctival samples	Dean et al., 1989
PCR	16S rRNA	No cross-reactivity with DNA from 13 different bacterial species	One bacterial cell in 100.000 host cells	DNA extracts from cell lines and various bacterial species	Pollard et al., 1989
PCR and radioactive RNA hybridization	Conserved plasmid	100%	100%	200 vaginal swabs	Griffais and Thibon, 1989
Probe hybridization	16S rRNA	Not discussed	10–100 picograms	Variable numbers from different sources, mainly conjunctival swabs	Cheema et al., 1991
PCR-RFLP analysis	Plasmid and outer membrane protein 1 (omp1)	Not relevant	30 culture positive, 38 PCR positive	209 cervical scrapes	Lar et al., 1993
LCR	Major Outer Membrane Protein MOMP	All serovars positive, no cross reaction, not even with C. psittaci and C. pneumoniae	Three elementary bodies	Dilution series of serovar L2 DNA	Dille et al., 1993
LCR	Plasmid and MOMP	99.8–100% (40–85% when culture efficacy was assessed)	93–98% for plasmid PCR, 68% for the MOMP PCR	1500 urines and urethral swabs	Chernesky et al., 1994
Qβ replicase	16S rRNA	85 out of 88 culture negatives were also PCR negative	1000 molecules, five out of six culture positive ones	94 urogenital samples	Shah et al., 1994
Qβ replicase and PCR	Both 16S rRNA	Both methods added three positives to the culture positives	Five elementary bodies for both methods	94 endocervical samples	An et al., 1995
PCR	16S rRNA and MOMP for discordance analysis	PCR can give positive signals for two weeks after antibiotic treatment	92.7 for the PCR, 79.1% for culture after discordance analysis	1110 cervical swabs	Goessens et al., 1995
Capture PCR	C1Q capture and endogenous plasmid	Not discussed	90–95% respectively for low and high positive cultures; capture helps concentrate bacteria	71 cervical swabs including samples with high and low positivity in the culture test	Herbrink et al., 1995
Transcription-mediated amplification TMA, LCR, and COBAS AMPLICOR PCR	TMA: 16S rRNA LCR: plasmid Amplicor: plasmid	TMA: 98% LCR: 99% Amplicor: 99% versus culture	TMA: 90% LCR: 91% Amplicor: 96% versus culture	First void urine and urethral / cervical scrapes for 544 males and 456 females, respectively	Goessens et al., 1997
TMA and COBAS AMPLICOR PCR	TMA: 16S rRNA Amplicor: plasmid	TMA: 100% Amplicor: 99%	TMA: 85% Amplicor: 97%	First void urine of 320 males and 338 females	Pasternack et al., 1997

TABLE 22.2 Examples of MRSA specific test systems. Four categories of test systems are included. Again, this is not a complete survey, this is meant for highlighting some of the more common technologies and their respective performance in clinical diagnostics.

Test principle	Data output	References
Culture-based tests		
The **BBL Crystal MRSA ID test** assesses viability by means of biochemical detection of oxygen consumption through a fluorescent compound.	Fluorescence, commercially available multiwell system for assessing growth in the presence of various antibiotics.	Qadri *et al.*, 1994; Zambardi *et al.*, 1996; Kubina *et al.*, 1999
The **Soft Salt Mannitol Agar Cloxacillin test** consists of tubes containing this medium; can be used at bedside for direct swab inoculation.	An indicator substance induces a color change upon the presence of MRSA.	Mir *et al.*, 1998
CHROMagar Staph aureus and Oxa Resistance Screen Agar are chromogenic plate media; addition of 4 ug/ml renders the media suitable for MRSA identification.	*S. aureus* colonies on the agar show a distinct purple color on Chromagar; some problems have been noted with test sensitivity.	Merlino *et al.*, 2000; Apfalter *et al.*, 2002
Lipovitellin Salt Mannitol Agar is a selective medium; lipovitellin is a glycosylated, lipid binding protein present in the yolk of egg-laying animals, showing homology to human apolipoprotein B.	Growth or not.	Verghese *et al.*, 1999
Vitek technology is developed by bioMerieux and represents automated biochemical and anti-microbial screening of bacterial isolates.	Biochemical and antimicrobial profiles are produced with limited hands on time.	Knapp *et al.*, 1994
Oxacillin Resistant Screen Agar is another chromogenic, selective culture medium.	Blue stained colonies, does require follow-up identification reactions.	Blanc *et al.*, 2003
Protein detection systems		
230 kiloDalton S. aureus surface protein was found to be useful in the agglutination of nonagglutinable strains of MRSA.	Experimental set up using purified antibodies reacting with the protein.	Kuusela *et al.*, 1994
MRSA Screen facilitates the detection of PBP-2a in crude cell preparations.	Latex agglutination system.	Van Leeuwen *et al.*, 1999
Nucleic acid detection systems		
EVIGENE probe hybridization can be used for non-PCR mediated etection of MRSA.	Staining and spectrophotometry at 405 nm.	Levi and Towner, 2003
Velogene Rapid MRSA Identification Assay is a probe-mediated procedure, which uses a chimeric probe for cycling mediated recognition of MecA gene.	Fluorescence value based on fluorescein, this procedure is intermediate between probe mediated and nucleic acid amplification mediated testing.	Louie *et al.*, 2000; Van Leeuwen *et al.*, 2001; Arbique *et al.*, 2001
Nucleic acid amplification		
Multiplex PCR for nuc, Tnase and MecA facilitates detection of MRSA.	Home-brew PCR, analysis of products by gel electrophoresis.	Brakstadt *et al.*, 1993
Multiplex PCR for femA and mecA facilitates detection of MRSA.	Home-brew PCR, analysis of products by gel electrophoresis. Has also been combined with immunoassay for product identification, including mupirocin resistance.	Vannuffel *et al.*, 1998; Towner *et al.*, 1998; Perez-Roth *et al.*, 2001; Jonas *et al.*, 2002
Combined Immunomagnetic enrichment of S. aureus followed by TaqMan PCR using the SmartCycler.	Fluorescence measurements using the SmartCycler hardand software.	Francois *et al.*, 2003
TaqMan PCR for the S. aureus nuc gene in combination with selective culture-based pre-enrichment.	Fluorescence measurement.	Fang and Hedin, 2003

studies convincingly showed that the presence of the gene for the Panton-Valentine leukocidin (PVL) confers strong disease invoking capacities upon the staphylococcus. PVL-positive strains have been implicated as causal agents in very severe cases of necrotising pneumonia, a disease that is leading to significant mortality (Gillet *et al.*, 2002). In addition, it has been shown by a Dutch research group that the presence of one or more virulence genes may lead to enhancement of impetigo in children. It was shown that the presence of the exfoliative toxin B (ETB) gene and also the PVL gene led to an increase in the number and the overall size of the impetigo lesions (Koning *et al.*, 2003).

With the availability of complete inventories of putative virulence genes, as based on whole genome comparisons, the possibilities for predictive diagnosis will increase in the future: the virulence gene repertoire of a colonizing *S. aureus* strain can be assessed by molecular diagnostics, and depending on its virulence gene profile it may be decided that the strain needs to be eliminated prior to a patient undergoing surgical treatment, which reduces the risk of post surgical wound infection.

Finally, the detection of subspecies genetic polymorphism is important for fingerprinting staphylococcal isolates and, hence, facilitates epidemiologic studies into the dissemination of clones of MSSA and MRSA. In the past, epidemiologic studies essentially were based on a strain's antimicrobial resistance profile, biochemical characteristics, or its phage type (Weiss and Nitzkin, 1971). Phage typing determines a strain's susceptibility toward infection by a large panel of lytic *S. aureus* specific bacteriophages. Essentially, a binary code is developed consisting of alternating sensitivity or resistance toward given phages (Weller, 2000). Forty years ago, phage typing was the epidemiological gold standard and used even for successful and informative nationwide analysis of the dissemination of identical *S. aureus* phagetypes. This clonal dissemination revealed interesting staphylococcal colonization and infection dynamics and set the stage for the development of more stable and reproducible typing systems. Again, molecular microbiology provided most of the alternative possibilities. Phage typing has, for instance, been compared in detail with random amplification of polymorphic DNA, a PCR method generating DNA fingerprints consisting of nonidentified DNA molecules (Van Belkum *et al.*, 1993). This showed that the DNA mediated procedure was better in the sense that it appeared to be more reproducible and that its resolving power was clearly enhanced.

Over the past years a relatively large set of alternative strategies for typing *S. aureus* has become available. This includes the assessment of mutation in a variable number of tandem repeat (VNTR) loci (Sabat *et al.*, 2003), a method aiming at the detection of unit number variation in certain genetic loci of fast-evolving repetitive DNA. The mechanism behind this form of DNA polymorphism is slipped strand mispairing: during replication the DNA polymerase

skips or adds a repeat unit as a consequence of the complex tertiary structure of the repetitive DNA domain (Van Belkum *et al.*, 1998b). Specific for MRSA, various methods for fine-typing of the PBP-encoding gene and its neighboring sequences have been developed (Ito *et al.*, 2001; Oliveira and De Lencastre, 2002); these methods usually depend on the selective PCR mediated amplification of locus specific sequence elements (see Fig. 22.2).

Although nearly all DNA typing methods appear to be useful for epidemiologic analysis of MRSA and MSSA (Van Belkum, 2000), pulsed field gel electrophoresis (Stranden *et al.*, 2003) and multilocus sequence typing (MLST; Feil *et al.*, 2003) are the two methods that are currently best appreciated. Hundreds of papers have been published describing the use of PFGE for epidemiological comparisons of sets of MRSA and MSSA strains (Van Belkum, 2000; see Fig. 22.2 for examples). These research efforts all strongly con-

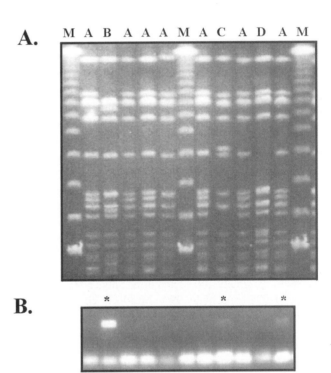

FIGURE 22.2 Gel electrophoretic analysis of genetic diversity and the presence of a methicillin resistance gene in strains of *Staphylococcus aureus*. A. Fingerprints generated by pulsed field gel electrophoresis (PFGE) of DNA macrorestriction fragments. Four different characteristic banding patterns are observed (indicated by the letters A to D, respectively, above each lane); these serve the purpose of unequivocal determination of relatedness between strains. Lanes marked M contain concatemeric phage lambda genomes, the smallest one being 50,000 bp in size. B. PCR amplification of part of the mecA gene, the gene encoding the penicillin binding protein 2A, the product of which shows diminished affinity toward the antibiotic; hence its causal involvement in resistance. Three positive reactions are observed (marked with an asterisk).

tributed to the current awareness of the nature of international MRSA dissemination (Aires de Sousa, 2003). MLST is relatively new and defines single nucleotide polymorphisms (SNPs) in housekeeping genes. Each gene sequence is translated into an allele code and the accumulation of all (generally 7 or 8) alleles leads to an allelic profile. This can be redefined into a single digit Sequence Type (ST). Its resolving power is not as strong as that of PFGE but its biggest advantage is that sequencing data are extremely portable. Data can be put in a single large database and each and every individual researcher can compare his strains according to MLST type with the database entries. This facilitates worldwide comparisons on staphylococcal genotypes to be made and a number of really seminal studies on the population structure of MRSA and MSSA recently have been published (Day *et al.*, 2001; Fitzgerald *et al.*, 2001; Enright *et al.*, 2002; Feil *et al.*, 2003). Overall, five or six major MRSA clones have travelled the world and many more minor types, restricted to certain locations, have been identified.

22.6 FUTURE PERSPECTIVES

Nucleic acid-based tests are now slowly being introduced into routine clinical microbiology laboratories (Check, 2001). Part of the obvious delay is due to the inadequately perceived and prejudiced shortcomings of nucleic acid testing; it is laborious, expensive, and requires high levels of laboratory expertise (Vaneechoutte and Van Eldere, 1997). However, these historic objections are slowly being taken apart, and accelerated introduction of molecular diagnostics should be pursued in many cases. PCR testing for *Legionalla pneumophila*, the agent of Legionnaires' Disease, for instance, (sensitivity 80–100%, specificity more than 90%) improves significantly over culture (sensitivity 10–80%, specificity 100%) and urine antigen testing (sensitivity 70–90%, specificity 99%, Murdoch, 2003). The same is true for *Trichomonas vaginalis*, an example of a sexually transmitted parasite. Several PCR tests showing increased sensitivity and excellent specificity have been described (Van Der Schee *et al.*, 1999), even using urine instead of an invasive swab as clinical specimen. Why are these tests not implemented immediately and massively in clinical microbiology? The most frequently perceived comments in this respect still concern the supposed levels of complexity and elevated costs of the molecular tests. It is not generally known what the cost-effectiveness of the molecular technology is and as long as costs only are appreciated more than improvement in the sensitivity and specificity of clinical testing, swift introduction of new tests into clinical practice will remain utopia.

Another important future application of molecular testing will be in the realm of host susceptibility toward infections (Relman, 2002). When genetic profiling tests for humans become more widely available (and their implications better understood) all infectious disease specialists will ultimately profit from these new molecular services. Many examples of genes important in host defense against infections have been described already (Santos *et al.*, 2002; Leveque *et al.*, 2003; and many, many more!!). It will be interesting to see where these exciting developments will lead over the coming decade. Whether or not this type of genetic data in the end should be collected for each and every individual is already now a matter of intense, ethical debate.

Genomics and proteomics, not to forget metabolomics, transcriptomics and all of the other "omics" sciences, have started to dominate the microbiology field over the past five years. The availability of complete genome sequences for both bacteria and the larger viruses has opened new avenues of fundamental and applied research (Cummings *et al.*, 2002). From the diagnostic perspective, genomics has facilitated the characterization of novel diagnostic and epidemiologically intriguing target sequences, whereas proteomics has facilitated the identification of complete protein profiles expressed by pathogens under different environmental conditions. The use of high-throughput "omics" methods has enhanced our understanding of infectious diseases significantly and the emphasis for the coming years should be on functional "omics" approaches, where microbiological "omics" data will be linked to microbial phenotypes or features of the host pathogen interaction.

22.7 CONCLUDING REMARKS

It needs to be emphasized that in principle the issues covered in this chapter for some very specific pathogens can be extrapolated to species and isolates of each and every other microbial infectious disease agent. This exemplifies the beauty of molecular diagnostics: the focused, nearly universally applicable technology is not only suited for the detection of pathogens; the subsequent genetic profiling of pathogens and assessment of their virulence potential is enabled by the same technology. Moreover, homologous technology also can be used to measure host response and in the end predict host susceptibility toward infectious diseases. The outcome of molecular diagnostic studies will continue to improve our understanding of infectious disease over the coming years. Ours are exciting times!!

References

Aires de Sousa, M., and De Lencastre, H. (2003). Evolution of sporadic isolates of methicillin resistant *Staphylococcus aureus* (MRSA) in hospitals and their similarities to isolates of community acquired MRSA. *J. Clin. Microbiol.* 41, 3806–3815.

An, Q., Liu, J., O'Brien, W., Radcliffe, G., Buxton, D., Popoff, S., King, W., Vera-Garcia, M., Lu, L., and Shah, J. (1995). Comparison of characteristics of Q beta replicase-amplified assay with competitive PCR assay for *Chlamydia trachomatis. J. Clin. Microbiol.* 33, 58–63.

Apfalter, P., Assadian, O., Kalczyk, A., Lindenmann, V., Makristathis, A., Mustafa, S., Rotter, M., and Hirschl, A. M. (2002). Performance of a new chromogenic oxacillin resistance screen medium (Oxoid) in the detection and presumptive identification of methicillin-resistant *Staphylococcus aureus. Diagn. Microbiol. Infect. Dis.* 44, 209–211.

Arbique, J., Forward, K., Haldane, D., and Davidson, R. (2001). Comparison of the Velogene Rapid MRSA identification assay, Denka MRSA Screen assay and BBL Crystal MRSA ID system for rapid identification of methicillin-resistant *Staphylococcus aureus. Diagn. Microbiol. Infect. Dis.* 40, 5–10.

Baba, T., Tajeuchi, F., Kuroda, M., Yuzawa, H., Aoki, K., Oguchi, A., Nagai, Y., Iwama, N., Asano, K., Naimi, T., Kuroda, H., Cui, L., Yamamoto, K., and Hiramatsu, K. (2002). Genome and virulence determinants of high virulence community-acquired MRSA. *Lancet* 359, 1819–1827.

Bidet, P., Brahimi, N., Chalas, C., Aujard, Y., and Bingen, E. (2003). Molecular characterisation of serotype III Group B Streptococcal isolates causing neonatal meningitis. *J. Infect. Dis.* 188, 1132–1137.

Blanc, D. S., Wenger, A., and Bille, J. (2003). Evaluation of a novel medium for screening specimens from hospitalised patients to detect methicillin-resistant *Staphylococcus aureus. J. Clin. Microbiol.* 41, 3499–3502.

Blocker, M. E., Krysiak, R. G., Behets, F., Cohen, M. S., and Hobbs, M. M. (2002). Quantification of *Chlamydia trachomatis* elementary bodies in urine by ligase chain reaction. *J. Clin. Microbiol.* 40, 3631–3634.

Boom, R., Sol, C. J. A., Salimans, M. M. M., Jansen, C. L., Wertheim-van Dillen, P. M. E., and Van der Noordaa, J. (1990). Rapid and simple method for the purification of nucleic acids. *J. Clin. Microbiol.* 28, 495–503.

Borg, I., Rohde, G., Loseke, S., Bittscheidt, J., Schultze-Werninghaus, G., Stephan, V., and Bufe, A. (2003). Evaluation of a quantitative real-time PCR for the detection of respiratory syncytial virus in pulmonary diseases. *Eur. Respir. J.* 21, 944–951.

Brackstad, O. G., Maeland, J. A., and Tveten, Y. (1993). Multiplex PCR for detection of genes for *Staphylococcus aureus* thermonuclease and methicillin resistance and correlation with oxacillin resistance. *APMIS* 101, 681–688.

Check, W. (2001). Nucleic acid-based tests move slowly into clinical labs. *ASM News* 67, 560–565.

Cheema, M. A., Schumacher, H. R., and Hudson, A. P. (1991). RNA-directed molecular hybridization screening: evidence for inapparent chlamydial infection. *Am. J. Med. Sci.* 302, 261–268.

Chernesky, M. A., Lee, H., Schachter, J., Burczak, J. D., Stamm, W. E., McCormack, W. M., and Quinn, T.C. (1994). Diagnosis of *Chlamydia trachomatis* urethral infection in symptomatic and asymptomatic men by testing first-void urine in a ligase chain reaction assay. *J. Infect. Dis.* 170, 1308–1311.

Cockerill III, F. R., and Smith, T. F. (2002). Rapid real-time PCR: a revolution for clinical microbiology. *ASM News* 68, 77–83.

Cummings, L., Riley, L., Black, L., Souvorov, A., Reschenchuk, S., Dondoshka, A., and Tatusova, T. (2002). Genomic BLAST: custom defined virtual databases for complete and unfinished genomes. *FEMS Microbiol. Lett.* 216, 133–138.

Day, N. P., Moore, C. E., Enright, M. C., Berendt, A. R., Smith, J. M., Murphy, M., Peacock, S. J., Spratt, B. G., and Feil, E. J. (2001). A link between virulence and ecological abundance in natural populations of *Staphylococcus aureus. Science* 292, 114–116.

Dean, D., Pant, C. R., and O'Hanley, P. (1989). Improved sensitivity of a modified polymerase chain reaction amplified DNA probe in comparison with serial tissue culture passage for detection of Chlamydia trachomatis in conjunctival specimens from nepal. *Diagn. Microbiol. Infect. Dis.* 12, 133–137.

De Lencastre, H., Severina, E. P., Roberts, R. B., Kreiswirth, B. N., and Tomasz, A. (1996). Testing the efficacy of a molecular surveillance network: MRSA and VREF genotyping in six hospitals in the metropolitan New York area. The Emerging Infectious Diseases Initiative Pilot Group. Bacterial Antibiotic Resistance Surveillance Group. *Microb. Drug Resist.* 2, 343–351.

Demain, A. L. (1999). Achievements in microbial technology. *Biotechnol. Adv.* 8, 291–301.

Dille, B. J., Butzen, C. C., and Birkenmeyer, L. G. (1993). Amplification of *Chlamydia trachomatis* DNA by ligase chain reaction. *J. Clin. Microbiol.* 31, 729–731.

Enright, M. C., Robinson, D. A., Randle, G. E., Feil, E. J., Grundmann, H., and Spratt, B. G. (2002). The evolutionary history of methicillin-resistant *Staphylococcus aureus* (MRSA). *Proc. Natl. Acad. Sci. USA* 99, 7687–7692.

Espy, M. J., Uhl, J. R., Mitchell, P. S., Thorvilson, J. N., Svien, K. A., Wold, A. D., and Smith, T. F. (2000). Diagnosis of herpes simplex virus infections in the clinical laboratory by LightCycler PCR. *J. Clin. Microbiol.* 38, 795–799.

Fang, H., and Hedin, G. (2003). Rapid screening and identification of methicillin-resistant *Staphylococcus aureus* from clinical samples by selective-broth and real-time PCR. *J. Clin. Microbiol.* 41, 2894–2899.

Feil, E. J., Cooper, J. E., Grundmann, H., Robinson, D. A., Enright, M. C., Berendt, T., Peacock, S. J., Maynard Smith, J., Murphy, M., Spratt, B. G., Moore, C. E., and Day, N. P. J. (2003). How clonal is *Staphylococcus aureus*? *J. Bacteriol.* 185, 3307–3316.

Fitzgerald, J. R., Sturdevant, D. E., Mackie, S. M., Gill, S. R., and Musser, J. M. (2001). Evolutionary genomics of *Staphylococcus aureus*: insight into the origin of MRSA and the toxic shock syndrome epidemic. *Proc. Natl. Acad. Sci. USA* 98, 8821–8826.

Francois, P., Pittet, D., Bento, M., Pepey, B., Vaudaux, P., Lew, D., and Schrenzel, J. (2003). Rapid detection of methicillin resistant *Staphylococcus aureus* directly from sterile or non-sterile clinical samples by a new molecular assay. *J. Clin. Microbiol.* 41, 254–260.

Gillet, Y., Issartel, B., Vanhems, P., Fournet, J. C., Lina, G., Bes, M., Vandenesch, J., Piemont, Y., Brousse, N., Floret, D., and Etienne, J. (2002). Association between *Staphylococcus aureus* strains carrying the genes for Panton-Valentine leukocidin and highly lethal necrotising pneumoniae in young immunocompetent patients. *Lancet* 359, 753–759.

Goessens, W. H. F., Kluytmans, J. A., Den Toom, N., Van Rijsoort-Vos, T., Niesters, B. G., Stolz, E., Verbrugh, H. A., and Quint, W. G. V. (1995). Influence of volume of sample processed on detection of *Chlamydia trachomatis* in urogenital samples by PCR. *J. Clin. Microbiol.* 33, 251–253.

Goessens, W. H. F., Mouton, J. W., Van der Meijden, W. I., Deelen, S., Van Rijsoort-Vos, T., Lemmens-den Toom, N., Verbrugh,

H. A., and Verkooijen, R. P. (1997). Comparison of three commercially available amplification assays, AMP CT, LCx, and COBAS AMPLICOR for detection of *Chlamydia trachomatis* in first-void urine. *J. Clin. Microbiol.* 1997, 2628–2633.

Griffais, R., and Thibon, M. (1989). Detection of *Chlamydia trachomatis* by the PCR. *Res. Microbiol.* 140, 139–141.

Gueudin, M., Vabret, A., Petitjean, J., Gouarin, S., Brouard, J., and Freymuth, F. (2003). Quantitation of respiratory syncytial virus RNA in nasal aspirates of children by real-time RT-PCR assay. *J. Virol. Methods* 109, 39–45.

Hallin, M., Maes, N., Byl, B., Jacobs, F., De Gheldre, Y., and Struelens, M. J. (2003). Clinical impact of a PCR assay for identification of *Staphylococcus aureus* and determination of methicillin resistance directly from blood cultures. *J. Clin. Microbiol.* 41, 3942–3944.

Harmsen, D., Singer, C., Rothganger, J., Tonjum, T., De Hoog, G. S., Shah, A. J., and Frosch, M. (2001). Diagnostics of neisseriaeae and moraxellaceae by ribosomal DNA sequencing: ribosomal differentiation of medical microorganisms. *J. Clin. Microbiol.* 39, 936–942.

Herbrink, P., Van den Munckhof, H. A., Niesters, H. G., Goessens, W. H. F., Stolz, E., and Quint, W. G. V. (1995). Solid phase C1q-directed bacterial capture followed by PCR for detection of *Chlamydia trachomatis* in clinical specimens. *J. Clin. Microbiol.* 33, 283–286.

Iqbal, J., Hira, P. R., Al-Ali, F., Khalid, N., and Sher, A. (2003). Modified Giemsa staining for rapid diagnosis of malaria infections. *Med. Princ. Pract.* 12, 156–159.

Ito, T., Katayama, Y., Asada, K., Mori, N., Tsutsumimoto, K., Tiensasitorn, C., and Hiramatsu, K. (2001). Structural comparison of three types of staphylococcal cassette chromosome *mec* integrated in the chromosome of methicillin-resistant *Staphylococcus aureus*. *Antimicrob. Agents Chemother.* 45, 1323–1336.

Jonas, D., Speck, M., Daschner, F. D., and Grundmann, H. (2002). Rapid PCR based identification of methicillin resistant *Staphylococcus aureus* from screening swabs. *J. Clin. Microbiol.* 40, 1821–1823.

Knapp, C. C., Ludwig, M. D., and Washington, D. A. (1994). Evaluation of differential inoculum disk diffusion method and Vitek GPS-SA card for detection of oxacillin resistant staphylococci. *J. Clin. Microbiol.* 32, 433–436.

Ko, K. S., Kim, J. M., Kim, J. W., Jung, B. Y., Kim, W., Kim, I. J., and Kook, Y. H. (2003). Identification of *Bacillus anthracis* by rpoB sequence analysis and multiplex PCR. *J. Clin. Microbiol.* 41, 2908–2914.

Koning, S., Van Belkum, A., Snijders, S., Van Leeuwen, W., Verbrugh, H., Nouwen, J., Op 't Veld, M., Suijlekom-Smit, L. W., Van der Wouden, J., and Verduin, C. (2003). Severity of non-bullous *Staphylococcus aureus* impetigo in children is associated with strains harboring genetic markers for exfoliative toxin B, Panton Valentine leukocidin and the multidrug resistance plasmid pSK41. *J. Clin. Microbiol.* 41, 3017–3021.

Ksiazek, T. G., Erdman, D., Goldsmith, C. S., Zaki, S. R., Peret, T., Emery, S., Urbani, C., Comer, J. A., Lim, W., Rollin, P. E., Dowell, F. S., Ling, A. F., Humar, C. D., Shieh, W. J., Guarner, J., Paddock, C. D., Rota, P., Fields, B., DeRisi, J., Cox, N., Hughes, J. M., LeDuc, J. W., Bellini, W. J., Anderson, L. J., and the SARS Working Group. (2003). A novel coronavirus associated with severe acute respiratory syndrome. *New Engl. J. Med.* 348, 1953–1966.

Kubina, M., Jaulhac, B., Delabranche, X., Lindenmann, C., Piemont, Y., and Monteil, H. (1999). Oxacillin susceptibility testing of staphylococci directly from Bactec Plus blood cultures by the BBL Crystal MRSA ID system. *J. Clin. Microbiol.* 37, 2034–2036.

Kumar, P. S., Griffen, A. L., Barton, J. A., Paster, B. J., and Moeschberger, M. L. (2003). New bacterial species associated with chronic periodontitis. *J. Dent. Res.* 82, 338–344.

Kuusela, P., Hilden, P., Savolainen, K., Vuento, M., Lyytikainen, O., and Vuopio-Varkila, J. (1994). Rapid detection of methicillin-resistant *Staphylococcus aureus* strains not identified by slide agglutination tests. *J. Clin. Microbiol.* 32, 143–147.

Lar, J., Walboomers, J. M., Roosendaal, R., Van Doornum, G. J., MacLaren, D. M., Meijer, C. J., and Van den Brule, A. J. (1993). Direct detection and genotyping of *Chlamydia trachomatis* in cervical scrapes by using PCR and restriction fragment length polymorphism analysis. *J. Clin. Microbiol.* 31, 1060–1065.

Leslie, D. E., Azzato, F., Ryan, N., and Fyfe, J. (2003). An assessment of the Roche Amplicor *Chlamydia trachomatis/Neisseria gonorrhoeae* multiplex PCR assay in diagnostic use on a variety of specimen types. *Commun. Dis. Intell.* 27, 373–379.

Leveque, G., Forgetta, V., Morroll, S., Smith, A. L., Bumstead, N., Barrow, P., Loredo-Osti, J. C., Morgan, K., and Malo, D. (2003). Allelic variation in TLR4 is linked to susceptibility to *Salmonella enterica* serovan Typhimurium infection in chickens. *Infect Immun.* 71, 1116–1124.

Levi, K., Bailey, C., Bennett, A., Marsh, P., Cardy, D. L. N., and Towner, K. J. (2003). Evaluation of an isothermal signal amplification method for rapid detection of methicillin-resistant *Staphylococcus aureus* from patient-screening swabs. *J. Clin. Microbiol.* 41, 3187–3191.

Levi, K., and Towner, K. J. (2003). Detection of methicillin resistant *Staphylococcus aureus* (MRSA) in blood with the EVIGENE MRSA detection kit. *J. Clin. Microbiol.* 41, 3890–3892.

Louie, L., Matsumara, S. O., Choi, E., Louie, M., and Simor, A. E. (2000). Evaluation of three rapid methods for detection of methicillin resistance in *Staphylococcus aureus*. *J. Clin. Microbiol.* 38, 2170–2173.

Lunge, V. R., Miller, B. J., Livak, K. J., and Batt, C. A. (2002). Factors affecting the performance of the 5′ nuclease PCR assay for *Listeria monocytogenes* detection. *J. Microbiol. Methods* 51, 361–368.

Maiden, M. C., Bygraves, J. A., Feil, E., Morelli, G., Russell, J. E., Urwin, R., Zhang, Q., Zhou, J., Zurth, K., Caugant, D. A., Feavers, I. M., Achtman, M., and Spratt, B. G. (1998). Multilocus sequence typing: a portable approach to the identification of clones within populations of pathogenic microorganisms. *Proc. Natl. Acad. Sci. USA* 95, 3140–3145.

McLaughlin, R. W., Vali, H., Lau, P. C. K., Palfree, R. G. E., De Ciccio, A., Sirois, M., Ahmad, D., Villemur, R., Desrosiers, M., and Chan, E. C. S. (2002). Are there naturally occurring pleomorphic bacteria in the blood of healthy humans? *J. Clin. Microbiol.* 40, 4771–4775.

McOrist, A. L., Jackson, M., and Bird, A. R. (2002). A comparison of five methods for extraction of bacterial DNA from human faecal samples. *J. Microbiol. Methods* 50, 131–139.

Meier, A., Persing, D. H., Finken, M., and Bottger, E. C. (1993). Elimination of contaminating DNA within polymerase chain reaction reagents: implications for a general approach to detection of uncultured pathogens. *J. Clin. Microbiol.* 31, 646–652.

Merlino, J., Leroi, M., Bradbury, R., Veal, D., and Harbour, C. (2000). New chromogenic identification and detection of *Staphylococcus aureus* and methicillin resistant *S. aureus*. *J. Clin. Microbiol.* 38, 2378–2380.

Millar, B. C., Xu, J., and Moore, J. E. (2002). Risk assessment models and contamination management: implications for broad-range ribosomal DNA PCR as a diagnostic tool in medical bacteriology. *J. Clin. Microbiol.* 40, 1575–1580.

Mir, N., Sanchez, M., Baquero, F., Lopez, B., Canlderon C., and Canton, R. (1998). Soft salt-mannitol agar cloxacillin test: a highly specific bedside screening test for detection of colonization with methicillin resistant *Staphylococcus aureus*. *J. Clin. Microbiol.* 36, 986–989.

Miyamoto, H., Imamura, K., Kojima, A., Takenaka, H., Hara, N., Ikenouchi, A., Tanabe, T., and Taniguchi, H. (2003). Survey of nasal colonisation by, and assessment of novel multiplex PCR method for detection of biofilm-forming methicillin-resistant staphylococci in healthy medical students, *J. Hosp. Infect.* 53, 215–223.

Mongodin, E., Finan, J., Climo, M. W., Rosato, A., Gill, S. R., and Archer, G. L. (2003). Microarray transcription analysis of clinical *Staphylococcus aureus* isolates resistant to vancomycin. *J. Bacteriol.* 185, 4638–4643.

Murakami, K., Minamide, W., Wada, K., Nakamura, E., Teraoka, A., and Watanabe, S. (1991). Identification of methicillin resistant strains of *Staphylococcus aureus* by polymerase chain reaction. *J. Clin. Microbiol.* 29, 2240–2244.

Murchan, S., Kaufmann, M. E., Deplano, A., De Ryck, R., Struelens, M., Fussing, V., Salmenlinna, S., Vuopio-Varkila, J., El Solh, N., Cuny, C., Witte, W., Tassios, P. T., Legakis, N., Van Leeuwen, W., Van Belkum, A., Vindel, A., Garaizar, J., Haeggman, S., Olsson Lilljequist, B., Ransjo, U., Coombs, J., and Cookson, B. D. (2003). Harmonization of PFGE protocols for epidemiological typing of strains of methicillin resistant *Staphylococcus aureus*: a single approach developed by consensus in ten European laboratories and its application for tracing the spread of related strains. *J. Clin. Microbiol.* 41, 1574–1585.

Murdoch, D. R. (2003). Diagnosis of *Legionella* infection. *Clin. Infect. Dis.* 36, 64–69.

Niesters, H. G. M. (2002). Clinical virology in real time. *J. Clin. Virol.* 125, 3–12.

Nikkari, S., McLaughlin, I. J., Bi, W., Dodge, D. E., and Relman, D. A. (2001). Does blood from healthy human subjects contain bacterial DNA? *J. Clin. Microbiol.* 39, 1956–1959.

Oberste, M. S., Maher, K., Flemister, M. R., Marchetti, G., Kilpatrick, D. R., and Pallansch, M. A. (2000). Comparison of classic and molecular approaches for the identification of untypeable enteroviruses. *J. Clin. Microbiol.* 38, 1170–1174.

Oliveira, D. C., and De Lencastre, H. (2002). Multiplex PCR strategy for rapid identification of structural types and variants of the *mec* element in methicillin-resistant *Staphylococcus aureus*. *Antimicrob. Agents Chemother.* 46, 2155–2161.

Paavonen, J., Puolakkainen, M., Paukku, M., and Sintonen, H. (1998). Cost benefit analysis of first void urine *Chlamydia trachomatis* during a screening program. *Obstet. Gynecol.* 92, 292–298.

Pannekoek, Y., Westenberg, S. M., Eijk, P. P., Repping, S., Van der Veen, F., Van der Ende, A., and Dankert, J. (2003). Assessment of *Chlamydia trachomatis* infection of semen specimens by ligase chain reaction. *J. Med. Microbiol.* 52, 777–779.

Pasternack, R., Vuorinen, P., and Miettinen, A. (1999). Comparison of a transcription-mediated amplification assay and PCR for the detection of *Chlamydia trachomatis* in first void urine. *Eur. J. Clin. Microbiol. Infect. Dis.* 18, 142–144.

Perez-Roth, E., Claverie-Martin, F., Villar, J., and Mendez-Alvarez, S. (2001). Multiplex PCR for simultaneous identification of *Staphylococcus aureus* and detection of methicillin and mupirocin resistance. *J. Clin. Microbiol.* 39, 4037–4041.

Pollard, D. R., Tyler, S. D., Ng, C. W., and Kozee, K. R. (1989). A PCR protocol for the specific detection of *Chlamydia* spp. *Mol. Cell. Probes* 3, 383–389.

Qadri, S. M., Ueno, Y, Imambaccus, H., and Almodovar, E. (1994). Rapid detection of MRSA by Crystal MRSA ID system. *J. Clin. Microbiol.* 32, 1830–1832.

Quint, W. G. V., Heijtink, R. A., Schirm, J., Gerlich, W. H., and Niesters, H. G. M. (1995). Reliability of methods for hepatitis B virus DNA detection. *J. Clin. Microbiol.* 33, 225–228.

Relman, D. A. (1998). Detection and identification of previously unrecognised microbial pathogens. *Emerg. Infect. Dis.* 4, 382–389.

Relman, D. A. (2002). New technologies, human-microbe interactions, and the search for previously unrecognized pathogens. *J. Infect. Dis.* 186, S254–S258.

Riemersma, W. A., Van der Schee, C. J., Van der Meijden, W., Verbrugh, H. A., and Van Belkum, A. (2003). Microbial population diversity in the urethras of healthy men and males suffering from nonchlamydial, nongonococcal urethritis. *J. Clin. Microbiol.* 41, 1977–1986.

Roth, A., Reischl, U., Streubel, A., Naumann, L., Kroppenstedt, R. M., Haber, Z., Fischer, M., and Mauch, H. (2000). Novel diagnostic algorithm for identification of mycobacteria by genus specific amplification of the 16S-23S rRNA gene spacer and application of restriction endonuclease. *J. Clin. Microbiol.* 38, 1094–1104.

Rys, P. N., and Persing, D. H. (1993). Preventing false positives: quantitative evaluation of three protocols for inactivation of PCR amplification products. *J. Clin. Microbiol.* 31, 2356–2360.

Sabat, A., Krzyszto-Russjan, J., Strzalka, W., Filipek, R., Kusowska, K., Hryniewicz, W., Travis, J., and Potemka, J. (2003). New method for typing *Staphylococcus aureus* strains: multiple locus variable number of tandem repeat analysis of polymorphism and gentic relatedness of clinical isolates. *J. Clin. Microbiol.* 41, 1801–1804.

Safdar, N., Narans, L., Gordon, B., and Maki, D. G. (2003). Comparison of culture screening methods for detection of nasal carriage of methicillin-resistant *Staphylococcus aureus*: a prospective study comparing 32 methods. *J. Clin. Microbiol.* 41, 3163–3166.

Santos, A. R., Suffys, P. N., Vanderborght, P. R., Moraes, M. O., Vieira, L. M. M., Cabello, P. H., Bakker, A. M., Matos, H. J., Huizinga, T. W. J., Ottenhoff, T. H. M., Sampaio, E. P., and Sarno, E. N. (2002). Role of tumour necrosis factor-alpha and interleukin 10 promoter gene polymorphisms in leprosy. *J. Infect. Dis.* 186, 1687–1691.

Savolainen, C., Blomqvist, S., Mulders, M. N., and Hovi, T. (2002). Genetic clustering of all 102 human rhinovirus prototype strains: serotype 87 is close to human enterovirus 70. *J. Gen. Virol.* 83, 333–340.

Schachter, J. (1985). Overview of *Chalamydia trachomatis* infection and the requirement for a vaccine. *Rev. Infect. Dis.* 7, 713–716.

Schutten, M., and Niesters, H. G. M. (2001). Clinical utility of viral quantification as a tool for disease monitoring. *Expert. Rev. Mol. Diagn.* 2, 153–162.

Schuurman, R., Descamps, D., Weverling, G. J., Kaye, S., Tijnagel, J., Williams, I., Van Leeuwen, R., Tedder, R., Boucher, C. A., Brun-Vezinet, F., and Loveday, C. (1996). Multicenter comparison of three commercial methods for quantification of human immunodeficiency virus type 1 RNA in plasma. *J. Clin. Microbiol.* 34, 3016–3022.

Shah, J. S., Liu, J., Smith, J., Popoff, S., Radcliffe, G., O'Brien, W. J., Serpe, G., Olive, D. M., and King, W. (1994). Novel ultra-sensitive Q beta-replicase amplified hybridization assay for detection of *Chlamydia trachomatis*. *J. Clin. Microbiol.* 32, 2718–2724.

Shrestha, N. K., Shermock, K. M., Gordon, S. M., Tuohy, M. J., Wilson, D. A., Cwynar, R. E., Banbury, M. K., Longworth, D. L., Isada, C. M., Mawhorter, S. D., and Procop, G. W. (2003). Predictive value and cost-effectiveness analysis of rapid polymerase chain reaction for pre-operative detection of nasal carriage of *Staphylococcus aureus*. *Infect. Control Hosp. Epidemiol.* 24, 327–333.

Singh, K., Gavin, P. J., Vescio, T., Thomson, R. B., Deddish, R. B., Fisher, A., Noskin, G. A., and Peterson, L. R. (2003). Microbiologic surveillance using nasal cultures alone is sufficient for detection of methicillin-resistant *Staphylococcus aureus* in neonates. *J. Clin. Microbiol.* 41, 2755–2757.

Stefano, J. E., Genovese, L., An, Q., Lu, L., McCarty, J., Du, Y., Stefano, K., Burg, J. L., King, W., and Lane, D. J. (1997). Rapid and sensitive detection of *Chlamydia trachomatis* using a ligatable binary RNA probe and Q beta replicase. *Mol. Cell. Probes* 11, 407–426.

Stranden, A., Frei, R., and Widmer, A. F. (2003). Molecular typing of methicillin-resistant *Staphylococcus aureus*: can PCR replace pulsed-field gel electrophoresis? *J. Clin. Microbiol.* 41, 3181–3186.

Suchland, R. J., Geisler, W. M., and Stamm, W. E. (2003). Methodologies and cell lines used for antimicrobial susceptibility testing of *Chlamydia* spp. *Antimicrob. Agents Chemother.* 47, 636–642.

Tenover, F. C. (1993). DNA hybridisation techniques and their application to the diagnosis of infectious diseases. *Infect. Dis. Clin. North Am.* 7, 171–181.

Towner, K. J., Talbot, J. C., Curran, R., Webster, C. A., and Humphreys, H. (1998). Development and validation of a PCR-based immunoassay for the rapid detection of methicillin resistant *Staphylococcus aureus*. *J. Med. Microbiol.* 47, 607–613.

Valentine-Thon, E., Van Loon, A. M., Schirm, J., Reid, J., Klapper, P. E., and Cleator, G. M. (2001). European Proficiency Testing Program for Molecular Detection and Quantitation of Hepatitis B Virus DNA. *J. Clin. Microbiol.* 39, 4407–4412.

Van Belkum, A., Bax, R., Peerbooms, P., Goessens, W., Van Leeuwen, N., and Quint, W. G. V. (1993). Comparison of phage typing and DNA fingerprinting by polymerase chain reaction for discrimination of methicillin-resistant *Staphylococcus aureus*. *J. Clin. Microbiol.* 31, 798–803.

Van Belkum, A., Kluytmans, J., Van Leeuwen, W., Bax, R., Quint, W., Peters, E., Fluit, A., Vandenbroucke Grauls, C., Van den Brule, A., Koeleman, H., Elaichoune, A., Vaneechoutte, M., Tenover, F., and Verbrugh, H. (1995). Multicenter evaluation of arbitrarily primed PCR for typing of *Staphylococcus aureus* strains. *J. Clin. Microbiol.* 33, 1537–1547.

Van Belkum, A., Scherer, S., Van Alphen, L., and Verbrugh, H. (1998a). Short sequence repeats in prokaryotic genomes. *Microbiol. Mol. Biol. Rev.* 62, 275–293.

Van Belkum, A., Van Leeuwen, W., Kaufmann, M. E., Cookson, B. D., Forey, F., Etienne, J., Goering, R., Tenover, F., Steward, C., O'Brien, F., Grubb, W., Legakis, N., Morvan, A., El Solh, N., De Ryck, R., Struelens, M., Salmenlinna, S., Vuopio-Varkila, J., Koositra, M., Witte, W., and Verbrugh, H. (1998b). Assessment of resolution and intercenter reproducibility of genotyping of *Staphylococcus aureus* by pulsed field gel electrophoresis of SmaI macrorestriction fragments: a multi-center study. *J. Clin. Microbiol.* 36, 1653–1659.

Van Belkum, A. (2000). Molecular epidemiology of MRSA strains: state of affairs and tomorrows possibilities. *Microb. Drug Resist.* 6, 173–188.

Van den Hoogen, B. G., De Jong, J. C., Groen, J., Kuiken, T., De Groot, R., Fouchier, R. A., and Osterhaus, A. D. (2001). A newly discovered human pneumovirus isolated from young children with respiratory tract disease. *Nat. Med.* 7, 719–724.

Van der Schee, C. J., Van Belkum, A., Zwijgers, L., Van der Brugge, E., O'Neill, E., Luijendijk, A., Van Rijsoort Vos, T., Van der Meijden, W. I., and Verbrugh H. A. (1999). Improved diagnosis of *Trichomonas vaginalis* infection by PCR using vaginal swabs and urine specimens compared to diagnosis by wet mount microscopy, culture and fluorescent staining. *J. Clin. Microbiol.* 37, 4127–4130.

Van Doornum, G. J., Guldemeester, J., Osterhaus, A. D., and Niesters, H. G. M. (2003). Diagnosing herpesvirus infections by real time amplification and culture. *J. Clin. Microbiol.* 41, 576–580.

Vaneechoutte, M., and Van Eldere, J. (1997). The possibilities and limitations of nucleic acid amplification technology in diagnostic microbiology. *J. Med. Microbiol.* 46, 188–194.

Van Elden, L. J., Nijhuis, J., Schipper, P., Schuurman, R., and Van Loon, A. M. (2001). Simultaneous detection of influenza viruses A and B using real-time quantitative PCR. *J. Clin. Microbiol.* 39, 196–200.

Van Leeuwen, W. B., Van Pelt, C., Luijendijk, A., Verbrugh, H. A., and Goessens, W. H. F. (1999). Rapid detection of methicillin resistance in *Staphylococcus aureus* isolates by the MRSA screen latex agglutination test. *J. Clin. Microbiol.* 37, 3029–3030.

Van Leeuwen, W. B., Kreft, D. E., and Verbrugh, H. A. (2002). Validation of rapid screening tests for the identification of methicillin resistance in staphylococci. *Microb. Drug Resist.* 8, 55–59.

Van Leeuwen, W., Jay, C., Snijders, S., Durin, N., Lacroix, B., Verbrugh, H. A., Enright, M. C., Troesch, A., and Van Belkum, A. (2003). Multilocus sequence typing of *Staphylococcus aureus* with DNA array technology. *J. Clin. Microbiol.* 41, 3323–3326.

Vannuffel, P., Laterre, P. F., Bouyer, M., Gigi, J., Vandercam, B., Reynaert, M., and Gala, J. L. (1998). Rapid and specific molecular identification of methicillin-resistant *Staphylococcus aureus* in endotracheal aspirates from mechanically ventilated patients. *J. Clin. Microbiol.* 36, 2366–2368.

Van Vliet, K. E., Muir, P., Echevarria, J. M., Klapper, P. E., Cleator, G. M., and Van Loon, A. M. (2001). Multicenter proficiency testing of nucleic acid amplification methods for the detection of enteroviruses. *J. Clin. Microbiol.* 39, 3390–3392.

Verghese, S., Padmaja, P., Sudha, P., Vanitha, V., and Mathew, T. (1999). Nasal carriage of MRSA in a cardiovascular tertiary care centre and its detection by lipovitellin salt mannitol agar. *Indian J. Pathol. Microbiol.* 42, 441–446.

Verkooyen, R., Noordhoek, G. T., Klapper, P. E., Reid, J., Schirm, J., Clears, A., Ieven, M., and Hoddevik, G. (2003). Reliability of nucleic amplification methods for detection of *Chlamydia trachomatis* in urine: results of the first international collaborative quality control study among 96 laboratories. *J. Clin. Microbiol.* 41, 3013–3016.

Wallace, P. S. (2003). Linkage between the journal and Quality Control Molecular Diagnostics (QCMD). *J. Clin. Virol.* 27, 211–212.

Weiss, D. L., and Nitzkin, P. (1971). Relationship between lysogeny and lysis in phage *S. aureus* systems. *Am. J. Clin. Pathol.* 56, 593–596.

Weller, T. M. (2000). Methicillin resistant *Staphylococcus aureus* typing methods should be the international standard? *J. Hosp. Infect.* 44, 160–172.

Weusten, J. J., Carpay, W. M., Oosterlaken, T. A., Van Zuijlen, M. C., and Van der Paar, P. A. (2002). Principles of quantitation of viral loads using nucleic acid-sequence based amplification in combination with homogeneous detection using molecular beacons. *Nucleic Acids Res.* 30, e26.

Wolk, D., Mitchell, S., and Patel, R. (2001). Principles of molecular microbiology testing methods. *Infect. Dis. Clin. North Am.* 15, 1157–1204.

Zaaijer, H. L., Cuypers, H. T., Reesink, H. W., Winkel, I. N., Gerken, G., and Lelie, P. N. (1993). Reliability of polymerase chain reaction for detection of hepatitis C virus. *Lancet* 341, 722–724.

Zambardi, G., Fleurette, J., Schito, G. C., Auckenthaler, R., Bergogne-Berezin, E., Hone, R., King, A., Lenz, W., Lohner, C., Makrishtatis, A., Marco, F., Muller-Serieys, C., Nonhoff, C., Phillips, I., Rohner, P., Rotter, M., Schaal, K. P., Struelens, M., and Viebahn, A. (1996). European multicentre evaluation of a commercial system for identification of MRSA. *Eur. J. Clin. Microbiol. Infect. Dis.* 15, 747–749.

CHAPTER **23**

Preimplantation Genetic Diagnosis

KAREN SERMON,[1,3] INGE LIEBAERS,[1,3] AND ANDRÉ VAN STEIRTEGHEM[2,3]

[1] *Centre for Medical Genetics;*
[2] *Centre for Reproductive Medicine;*
[3] *Research Centre Reproduction and Genetics;*
Medical School and University Hospital of the Dutch-speaking Brussels Free University (Vrije Universiteit Brussel, VUB), Brussels, Belgium

TABLE OF CONTENTS

23.1 WHAT IS PREIMPLANTATION GENETIC DIAGNOSIS?

Preimplantation genetic diagnosis (PGD) is a very early form of prenatal diagnosis. Oocytes or preimplantation embryos are obtained *in vitro* and are genetically analyzed, after which only those embryos that are judged to be free of the genetic defect under consideration are transferred. To this end, polar bodies are removed from oocytes, or blastomeres (either at the cleavage stage or at the blastocyst stage) are removed from preimplantation embryos, and these cells are used for the genetic diagnosis. The first report on PGD was by Handyside and colleagues (1990) and described the sexing of embryos in families who were at risk for X-linked disease. The sex of the embryos was established using PCR of repetitive Y-sequences, and resulted in two pregnancies and the birth of healthy female babies.

Initially, PGD was developed to help couples at risk for monogenic diseases. The only technique available to analyze single cells (polar bodies or blastomeres) was Polymerase Chain Reaction (PCR), which was used either to sex the embryos (Handyside *et al.*, 1990) or to detect specific mutations such as the ΔF508 mutation in cystic fibrosis (Handyside *et al.*, 1992; Verlinsky *et al.*, 1992). Later, fluorescent *in situ* hybridization (FISH) was downscaled to the single cell level (Griffin *et al.*, 1992) and this opened possibilities to analyze embryos at the chromosomal level: again, either for sexing (Harper *et al.*, 1994) or for chromosomal aberrations such as Robertsonian or reciprocal translocations (Conn *et al.*, 1999), or, finally, for aneuploidy screening (Verlinsky and Kuliev, 1996).

23.2 INDICATIONS FOR PGD

23.2.1 Monogenic Diseases

Not surprisingly, most PGD cycles for monogenic diseases have been performed for the most frequent monogenic diseases. These are also more or less the monogenic diseases for which most prenatal diagnoses are performed. Examples are, of course, most importantly cystic fibrosis, spinal muscular atrophy, β-globinopathies, Duchenne's muscular dystrophy, fragile X disease, myotonic dystrophy, and Huntington's disease. Cystic fibrosis was the first monogenic disease for which a specific diagnosis at the single cell level was performed, analyzing the most frequent ΔF508 mutation (Handyside *et al.*, 1992), and several reports using slightly different approaches have been published since

(Verlinsky *et al.*, 1992; Liu *et al.*, 1994; Moutou and Viville, 1999; Goossens *et al.*, 2000). More recently, multiplex approaches designed for analysis of single cells, and analyzing a CF mutation with one or more linked markers (Strom *et al.* 1998; Moutou *et al.*, 2002; Goossens *et al.*, 2003), or a panel of linked markers, have been described (Dreesen *et al.*, 2000; Eftedal *et al.*, 2001; Vrettou *et al.*, 2002). Similarly, strategies have been developed for PGD for spinal muscular atrophy, both based on analysis of the most prevalent mutation (a large deletion encompassing several exons, Dreesen *et al.*, 1998; Fallon *et al.*, 1999; Moutou *et al.*, 2001) and on the multiplex analysis of the mutation and linked markers (Moutou *et al.*, 2003).

Mutations in the β-globin gene, both leading to sickle cell anemia and β-thalassemia have received considerable attention from the PGD community, because of their high frequency in certain populations. Because so many different mutations can be found in the β-globin gene (only one of them leading to sickle cell anemia, the others leading to various types of β-thalassemia), the strategies developed for PGD are different for example from spinal muscular atrophy and aim to detect as many different mutations as possible with the same assay. Examples of this are given by Xu and colleagues (1999), on PGD for sickle cell anemia; Vrettou and colleagues (1999), using DGGE for PGD for β-thalassemia; Kuliev and colleagues (1999), on polar body biopsy for PGD for β-thalassemia; De Rycke and colleagues (2001), on PGD for sickle cell anemia and β-thalassemia using fluorescent PCR; and most recently Jiao and colleagues (2003), using reverse dot-blot for PGD for β-thalassemia.

Patients at risk for transmitting an autosomal dominant disease have always been particularly interested in PGD because of the high recurrence risk. Examples of this abound in the literature, and again the more frequent diseases have received the most attention. The authors (Sermon *et al.*, 1999a) and others (Harton *et al.*, 1996; Blaszczyk *et al.*, 1998) have described several different approaches for Marfan's disease. PGD for diseases caused by dynamic mutations have also been reported. Noteworthy here are myotonic dystrophy type 1 (DM1; Sermon *et al.*, 1998a; 2001; Pyiamongkol *et al.*, 2001; Dean *et al.*, 2001) and Huntington's disease (Sermon *et al.*, 1998a; 2001; Stern *et al.*, 2002).

Sexing with FISH has been most frequently used for X-linked diseases such as Duchenne's muscular dystrophy, hemophilia A and B, retinitis pigmentosa, and others (Staessen *et al.*, 1999), but more and more specific DNA diagnoses are developed and used. The advantages of a specific DNA diagnosis are important: first, healthy male embryos are not discarded and second, female carriers can be identified and, according to the patient's wishes and the center's policy, are then not selected for transfer. Examples of tests developed for the specific diagnosis of an X-linked disease are given by Hussey and colleagues (1999), Ray and

colleagues (2001), and Girardet and colleagues (2003) for Duchenne's muscular dystrophy. Sermon and colleagues (1999b) and Apessos and colleagues (2001) described protocols for PGD for fragile X syndrome. Because girls who are carriers of fragile X can be affected, only a specific DNA analysis can be used for PGD.

For more examples of diseases for which PGD has been performed and the appropriate references, see Sermon and colleagues (2004).

23.2.2 Chromosomal Aberrations

Reciprocal translocations are characterized by the exchange of fragments between chromosomes, whereas in Robertsonian translocations a whole acrocentric chromosome is translocated to another one through centromeric fusion. Normal carriers of these translocations are at risk to have children with congenital anomalies and mental retardation due to chromosomal imbalances or more frequently suffer from recurrent miscarriages or infertility (especially if the male is a carrier). This explains the large interest this group of patients has shown for PGD. The first reports (Munné *et al.*, 1998; Iwarsson *et al.*, 1998) described the use of probes that were custom-designed for one specific translocation usually occurring in only one family, which is why the application of this approach remained limited. However, it was only since the widespread availability of fluorescent probes in different colors (Conn *et al.*, 1999; Van Assche *et al.*, 1999; Coonen *et al.*, 2000), that it has been possible to propose PGD to these patients in more than a handful of highly specialized centres. Examples of reports on larger series are those by Munné and colleagues (2000), Durban and colleagues (2001), and Pickering and colleagues (2003).

23.2.3 Aneuploidy Screening

It has been a well-established fact that human embryos carry cytogenetic abnormalities in high proportions: using classical karyotyping to investigate embryos, between 23% and 80% of embryos were found to be aneuploid, the latter number found in embryos of poor quality (Zenzes and Casper, 1992; Pellestor *et al.*, 1994). More detailed information has become available after the advent of FISH. Earlier reports by Delhanty and colleagues (1993), Munné and colleagues (1993), and Harper and colleagues (1995) showed a whole range of abnormalities, such as monosomies, trisomies, triploidies, and combined abnormalities. These authors also reported that 70% of the embryos analyzed were developing abnormally, even if only five chromosomes (X, Y, 13, 18, and 21) were analysed.

In a more recent report on a large number of nonviable cleavage stage embryos, Marquez and colleagues (2000) could show that aneuploidy (from 1.4% in patients between 20 and 34 years to 52.4% in patients between 40 and 47) increases with maternal age, and polyploidy and mosaicism is related to poor embryo morphology. Several authors suggested that, considering the high rate of abnormalities in preimplantation embryos, together with the higher risk for fetal aneuploidy at an advanced maternal age and the fact that 50% to 60% of all spontaneous abortions from clinically recognized pregnancies carry an abnormal karyotype (Boué et al., 1985), embryo selection based on chromosome complement would improve *in vitro* fertilization (IVF) results in groups of patients with poor outcome, as well as avoid the birth of babies with chromosomal defects. The obvious patient groups or cases for whom PGD-aneuploidy screening (PGD-AS) could be beneficial are patients with advanced maternal age (Verlinsky et al., 1999; Munné et al., 2003a), repetitive implantation failure after IVF (Gianaroli et al., 1999) and recurrent miscarriage not due to translocations (Wilton et al., 2002; Rubio et al., 2003).

Comparative Genomic Hybridization (CGH, Voullaire et al., 2000; Wells and Delhanty, 2000; Wilton et al., 2001; Wells et al., 2002) has been used in PGD-AS: the important advantage of CGH over FISH is that a whole karyotype is obtained (see also Chapters 12 and 13). In this way, abnormalities are found in embryos, which would have been missed by FISH. Wilton and colleagues (2003) estimate that FISH for five or nine chromosomes would have missed 38% and 25% of the abnormal blastomeres, respectively. However, the complexity of the CGH, as well as the time currently needed to obtain a karyotype (five days) explain why for the time being, CGH is not as widely applied as FISH. PGD-AS using FISH has now become widely applied in the patient groups mentioned earlier, because of the relative ease of the technique and the large potential patient group. However, the evaluation of the benefit of these treatments awaits the results of large prospective trials, because the PGD-AS data in these studies are not compared to a suitable control group (Wilton et al., 2002). Large multicenter, randomized studies are currently undertaken and will allow in the future to evaluate the efficiency of PGD-AS and to delineate patient groups who will most benefit from PGD-AS.

23.3 TECHNOLOGIES USED IN PGD

23.3.1 Assisted Reproductive Technology

The assisted reproductive technology (ART) used to obtain embryos for PGD will only briefly be outlined, as it lies mostly outside the scope of this chapter.

The first step in a PGD cycle, as in many ART, is controlled ovarian hyper-stimulation, aimed at obtaining as many mature oocytes as possible. Although there is a tendency in regular IVF to reduce the number of oocytes in order to obtain better quality oocytes and to prevent multiple pregnancies (Macklon and Fauser, 2000), in PGD a large

number of oocytes are still required because many embryos will be deemed untransferable because they are affected or carry important aneuploidies (Vandervorst et al., 1998). Since the introduction of gonadotrophin releasing hormone antagonists (GnRHa), the following stimulation protocol is mostly used: follicle stimulating hormone (FSH, usually the recombinant variant) is started on day 2 of the cycle to start follicle maturation (Kolibianakis et al., 2002). At day 6 of the cycle, GnRHa is started to suppress a luteinizing hormone surge, which would cause a spontaneous ovulation. When blood hormone and ultrasound monitoring show that a sufficient number of mature oocytes are present, an injection of human chorionic gonadotrophin (hCG) is given to induce ovulation. Follicles can then be punctured transvaginally and under ultrasound guidance, and oocytes are aspirated.

Intracytoplasmic sperm injection (ICSI) usually is preferred over regular IVF, because the risk for unexpected fertilization failure is reduced. Moreover, when PCR is used for diagnosis, the presence of sperm stuck to the zona pellucida after IVF represent an important source of contamination (Liebaers et al., 1998). The aspirated oocytes are first denuded of the surrounding granulosa cells, either through enzymatical (hyaluronidase) or mechanical (aspiration in and out of a pipette with a small diameter) means, or a combination of both. Concurrently, sperm is washed and prepared for ICSI. During ICSI, the oocyte is held in a holding pipette at nine o'clock, while the sperm is injected with a fine injection pipette at three o'clock (Joris et al., 1998).

After about 16 hours in culture, normal fertilization can be observed if two pronuclei and two polar bodies are present. Nowadays, embryo culture media are complex systems, with special formulations adapted for each step in the embryo's development. Fertilization, early development (up to day three), and development to blastocyst are performed in different formulations. Several of these so-called sequential media are now commercially available (Schoolcraft and Gardner, 2001).

23.3.2 Biopsy Techniques: Pros and Cons

23.3.2.1 POLAR BODY BIOPSY

Just before fertilization, a normal oocyte is at the metaphase II stage of the meiosis; that is, it has extruded the first polar body (PBI). The second polar body (PBII) is then extruded after normal fertilization. These two PBs have no further function in the embryonic development, and can thus be retrieved for analysis. The genetic content of the oocyte is the mirror image of the genetic content of the polar bodies and can thus indirectly be deduced (Verlinsky et al., 1992; 1999). The advantages of PB biopsy are self-evident: the embryo proper is not disturbed and there are thus no detrimental effects of the decrease of embryonic mass as in blastomere biopsy. Another important argument in favor is ethical. Indeed, it is possible to biopsy and analyze the two PBs before syngamy of the male and female pronucleus, and thus before what is legally regarded in several countries (e.g., Germany), as the beginning of life. A third advantage advanced by the proponents of PB biopsy, but which has failed to convince the authors, is that the difficulties raised by mosaicism in the embryo are avoided. Whether this argument is valid depends on the importance imparted on the presence of mosaicism. The important disadvantages have led most centers to prefer cleavage stage embryo biopsy. First, only the maternal contribution is analyzed, thus the technique is not applicable in autosomal dominant diseases or translocations where the father is carrier. It is also not applicable for sexing. Second, if the analysis is to be finished before syngamy, it leaves very little time to complete the diagnosis. Conversely, if ethics are not an issue, more time is available for analysis than in cleavage stage biopsy (see Section 23.3.2.2).

Technically, PB biopsy is quite straightforward. The zona pellucida (ZP) is breached using either mechanical slitting with a fine needle or laser technology. The chemical breaching of the ZP is not a valid alternative as the Acidic Tyrode's solution used for this purpose damages the oocyte. After hatching, a small diameter pipette is introduced into the hole and the two polar bodies are removed (Verlinsky *et al.*, 1999). If a FISH analysis is to be performed (for translocations or PGD-AS), then the two PBs can be removed in one operation because FISH analysis will reveal which cell is PBI and which is PBII. However, if analysis at the DNA level is carried out, then it is mandatory to biopsy the PBI before fertilization and the PBII after fertilization, to be able to follow the segregation of the respective alleles in oocyte and PBs (Rechitsky *et al.*, 1999).

23.3.2.2 CLEAVAGE STAGE BIOPSY

Cleavage stage biopsy is the most widely spread technique for obtaining embryonic material for PGD (Sermon et al, 2005). The biopsy is performed at the morning of day 3 when the embryo is normally at the eight-cell stage. Since human embryos have a quite erratic cleavage pattern, some embryos will be arrested at the four-cell stage, whereas others will be retarded and will have reached only the six-cell stage, and still others will have a lead and will have reached the ten-cell stage. One or two cells are retrieved, and it is generally agreed that the biopsy of embryos below the six-cell stage is of limited benefit. Whether the biopsy of two cells would significantly impair the implantation potential of the embryo is still under debate, although some studies seem to indicate that good quality embryos easily recover from the removal of one quarter of their cell mass (Van de Velde *et al.*, 2000; Parriego *et al.*, 2003). However, there seems to be a consensus that for some applications (e.g., PGD for autosomal dominant diseases) the risk for misdiagnosis on one cell is too important and in these cases two cells should be biopsied, but this risk is often consid-

ered much smaller in other applications (e.g., PGD-AS) and here, biopsy of one cell is sufficient (De Vos and Van Steirteghem, 2001).

Several techniques have been used to breach the ZP and to make the blastomeres accessible. The ZP can be opened mechanically, like for polar body biopsy. Usually, two perpendicular slits are made with a needle, giving rise to flaps in the ZP that can be lifted to allow the introduction of the biopsy pipette (Cieslak *et al.*, 1999). The chemical opening is the most widespread method. A thin stream of Acidic Tyrode's (AT) solution (pH 2.2) is applied with a drilling pipette of about 10–12 μm to the ZP to dissolve it. This technique requires some skill, as the AT can lyse the cells immediately under the ZP. Lately, the opening of the ZP using a noncontact diode infrared laser has found entrance in the PGD lab. Two or exceptionally three pulses of 5–8 ms and with a wavelength of 1.48 μm are applied at a safe distance (more than 8 μm from the nearest blastomere; De Vos and Van Steirteghem, 2001). Joris and colleagues (2003) have found that the number of cells lysed after zona hatching is reduced significantly using the laser as compared to AT, and that the time needed to biopsy the cell(s) is also reduced significantly. Moreover, the laser does not have a detrimental effect on the further development of the embryos.

At about the eight-cell stage, embryos start compaction; that is, the cell-to-cell adhesions increase and the cell boundaries become indistinguishable, a process that is completed at the morula stage. Especially with the introduction of the use of sequential media, more embryos are at the compacted stage at the moment of biopsy and this can lead to important difficulties to retrieve one or two cells without cell lysis. It was shown by Dumoulin and colleagues (1998) that the use of Ca^{++} and Mg^{++}-free medium to de-compact the embryos did not have a detrimental effect on the development of the embryo. This is why most centers performing PGD at the cleavage stage incubate the embryos in Ca^{++} and Mg^{++}-free medium, either only before biopsy, or during the whole biopsy procedure (ESHRE PGD Consortium Steering Committee, 2002).

For the biopsy itself, several techniques have been developed and applied with changing success. Biopsy by flow displacement (where a flow of fluid is applied through one opening in the zona release the blastomere through another hole) or extrusion (where pressure is applied to the embryo to push a blastomere through the opening in the ZP) are just mentioned for completeness. The most widely used method is described extensively in Joris and colleagues (2003). The embryo is held with a holding pipette (100 μm outer, 25–30 μm inner diameter) through the application of a slight negative pressure. A blunt biopsy pipette (inner diameter of 35–40 μm) is introduced through the hole in the ZP and the blastomere is gently aspirated. The blastomere is usually not aspirated completely into the pipette, but is "grabbed" partially and pulled through the hole in the ZP (see Fig. 23.1).

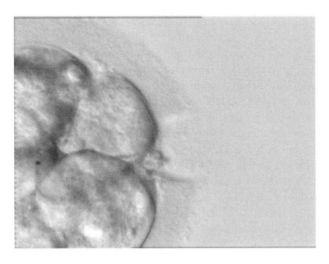
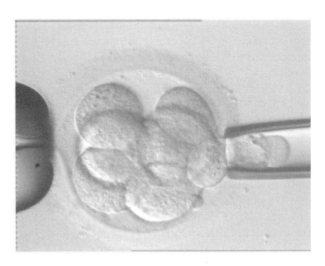

a)   b)

FIGURE 23.1 Example of an embryo biopsy. Panel a) shows the hole in the zona pellucida after application of the laser for zona hatching. Panel b) shows how one blastomere is gently aspirated outside the embryo. Panel c) shows two biopsied blastomeres with a clear nucleus.

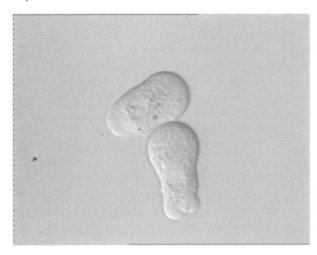

c)

23.3.3 Single Cell PCR: Pitfalls and Their Solution

PCR is the only method that allows the analysis of the DNA from one single cell; that is, 6 pg of DNA. Since the first report on PGD, the methods of analysis of the PCR products have evolved from electrophoresis on simple agarose gels (Handyside *et al.*, 1990), over fragment analysis on automated sequencers (Sermon *et al.*, 1998b), to minisequencing (Fiorentino *et al.*, 2003), and real-time PCR (Rice *et al.*, 2002). Although the refinement of the analysis methods has increased the efficiency and accuracy tremendously, there are still a number of pitfalls that are inherent to single cell PCR and that still cause incorrect diagnosis. These are:

- The (lack of) specificity of the PCR
- Contamination with DNA extraneous to the analyzed cell
- Allele drop-out

23.3.3.1 SPECIFICITY OF THE PCR

From one or two DNA template molecules, it takes about 50 to 60 PCR cycles to visualize the PCR product on simple ethidium bromide-stained agarose gels. However, a-specific products and smears appear long before that, because Taq DNA polymerase incorporates mistakes. This problem was solved in the early 1990s with the introduction of nested PCR (Holding and Monk, 1989). In this method, a first PCR round is performed, followed by a second PCR round in which a small amount of the first round PCR product serves as a template and primers are used that amplify a fragment inside the first fragment. A-specific fragments are thus not amplified in the second PCR round, and a clear, pure PCR product is obtained.

Later, the number of PCR cycles necessary to obtain enough PCR product for analysis was significantly reduced with the introduction of fluorescent PCR. Fluorescent PCR products are obtained using one fluorescently labeled primer, after which the fragments are analyzed on an automated sequencer. Not only is this method much more sensitive than the ethidium bromide staining (approximately 1000-fold), but the resolution of fragments is also much greater. A difference in fragment length of one base pair can be distinguished (Lissens and Sermon, 1997).

The introduction of DNA polymerases with a high proofreading activity, alone or in mixtures with regular Taq polymerase, has reduced the number of cycles necessary for a sufficient yield even more (Sermon *et al.*, 1998b). Finally, the need for the tedious fine-tuning of PCR protocols for single cell analysis was significantly reduced with the introduction of mini-sequencing techniques as applied, for instance, in the SNaPshot kit from Applera (Fiorentino *et al.*, 2003).

23.3.3.2 CONTAMINATION WITH EXTRANEOUS DNA

The introduction of one molecule of foreign DNA in the PCR tube along with the cell that is to be analyzed can lead to a wrong diagnosis. Two types of contamination with extraneous DNA can be distinguished: contamination with DNA from cells (e.g., from the operator, sperm cells, or granulosa cells stuck to the embryo), from extracted DNA (genomic DNA), or from carry-over with PCR fragments from previous DNA amplifications.

A number of measures need to be taken to avoid and/or detect contamination (Lissens and Sermon, 1997).

1. Granulosa cells are meticulously removed from the oocytes and fertilization is achieved through ICSI to avoid contamination with sperm.
2. The pre-PCR area, where all the reactions are set up and the post-PCR area, where the PCR products are analyzed need to be strictly separated. No DNA, except primers and single cell samples or highly diluted DNA, should be brought into the pre-PCR area.
3. The PCR reactions are set up in a laminar flow, that is fully equipped with dedicated pipettes, filtered tips and UV light to be turned on whenever the flow is not in use. In the same room, the products dedicated to pre-PCR work are kept and are not mixed, for example, with primers used in a routine DNA laboratory.
4. The manipulation of the small PCR tubes requires a certain amount of care, and even skill, in order to prevent contamination. Especially ill-fitting gloves can be a source of carry-over contamination.
5. For each sample, a blank containing all PCR components except DNA should be run. However, a contaminated blank is only an indication of a more general problem and does not mean that the corresponding sample is contaminated. A more efficient way to detect (genomic) contamination is the use of linked or unlinked polymorphic markers, amplified in duplex with the locus of interest.

23.3.3.3 ALLELE DROP-OUT AND PREFERENTIAL AMPLIFICATION

Allele drop-out (ADO) is defined as the nonamplification of one allele when starting from a single cell. It can thus be detected only in a heterozygous cell. In preferential amplification, one allele is less amplified than the other. Both are related problems because the allele that is less amplified, but still present, can go undetected if detection methods with a low sensitivity are used (e.g., ethidium bromide-stained agarose gels). ADO can lead to serious misdiagnoses, for

example in the detection of mutations in autosomal dominant diseases. When the affected allele drops out, the cell appears to be unaffected and the corresponding embryo could be transferred. This has been called an unacceptable error (Navidi and Arnheim, 1991). Conversely, in an autosomal recessive situation where both parents carry the same mutation (e.g., ΔF508 in CF) ADO would not lead to the transfer of affected embryos: if ADO occurs in a carrier cell, drop-out of the healthy allele would lead to the diagnosis of an affected embryo, and the embryo would not be transferred, whereas if the affected allele drops out, the embryo would be diagnosed as homozygous healthy instead of carrier, which is still a healthy embryo, and could be transferred. This was termed an acceptable misdiagnosis.

Supposedly, ADO has led to two misdiagnoses in CF (Harper and Handyside, 1994; Verlinsky, 1996). The parents were both carrying different mutations and the embryos in which a misdiagnosis occurred were compound heterozygotes.

A first efficient way to significantly reduce ADO is to use fluorescent PCR. As mentioned earlier, it is 1000-fold more sensitive than detection of PCR products on agarose gels, requires much fewer cycles to obtain an analyzable amount of PCR product, and allows for a clear distinction between preferential amplification and real ADO. Together with the use of more efficient DNA polymerases, fluorescent PCR has been the most important breakthrough so far to reduce the risk for ADO (Sermon *et al.*, 1998b).

ADO cannot be reduced to zero, so detection methods have been devised. The most important is the multiplex PCR of the mutation together with linked markers or a set of linked markers, and has now become the golden standard in PGD for monogenic diseases (Lewis *et al.*, 2001). Usually, microsatellites are used as linked markers because they are highly polymorphic and thus can be used in several families (Rechitsky *et al.*, 1999; Dreesen *et al.*, 2000; Apessos *et al.*, 2001; Dean *et al.*, 2001; Pyiamongkol *et al.*, 2001; Goossens *et al.*, 2003; Moutou *et al.*, 2003). Figure 23.2 shows how multiplex PCR can detect ADO. Simply stated, the results in one locus should concord with the results in another locus.

Several new and promising techniques have emerged that will facilitate the development of new diagnostic tests. As it is, it takes a considerable amount of time and effort to make single cell PCR efficient enough to obtain interpretable results, without too much background or a-specific bands. Minisequencing has been advocated as a means to reliably detect mutations and single nucleotide polymorphisms (SNPs) at the single cell level without a tedious preclinical work-up. Moreover, although SNPs are dimorphic but micro-satellites are polymorphic, SNPs are much more widespread in the genome and easier to introduce in a single cell PCR assay (Fiorentino *et al.*, 2003).

Another newly introduced method in single cell PCR is real-time PCR, which has the advantage that PCR fragments

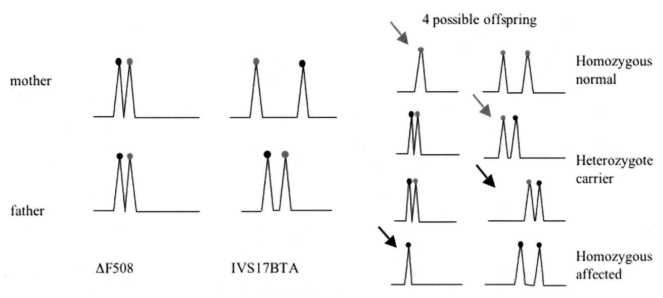

FIGURE 23.2   The principle of a duplex PCR for cystic fibrosis (mutation ΔF508 and IVS17BTA intragenic marker). Both parents are carriers of the ΔF508 mutation (black dot). In the marker, the black dot segregates with the mutation and the grey dot with the normal allele. Four different offspring are possible. If a result is obtained with the combination of the grey arrows (i.e., normal for the mutation and heterozygous for the marker), it can be assumed that ADO of the affected allele occurred in the mutation. This embryo could be transferred because both cells give the same result (transferable embryo). If the two black arrows are found, an ADO of the normal allele can be assumed in the mutation. Nevertheless, this embryo will not be transferred because both loci give opposite results (transferable/not transferable).

are analyzed as they are formed, and not after a complete PCR program. It has been used by several authors with a low rate of ADO (Pierce *et al.*, 2000; Rice *et al.*, 2002). This can probably be explained by the fact that in real-time PCR, small fragments are generated (typically about 50 bp), and it is a well-known fact that ADO increases with longer PCR fragments.

Finally, microarrays are currently pervading all branches of molecular biology, not only for expression studies with cDNA arrays, but also for DNA applications in cytogenetics as well as at the basepair level (Syvänen, 1999). For cytogenetic applications, the array would be covered with oligonucleotides spread over the whole chromosome set (Osborne *et al.*, 2003). For the analysis at the base pair level, a combination of microsequencing and microarray technology would be applied: the array would be covered in oligonucleotides that anneal specifically to PCR fragments just adjacent to the SNP to be analyzed. As in minisequencing, the complementary nucleotide would be added in (Kurg *et al.*, 2000). However, microarrays on a solid support are high-throughput systems whose main advantage is that a large number of sites are analyzed simultaneously. As of now, single-cell analysis is still PCR based, and as long as only a limited number of loci can be amplified in one reaction, to apply microarrays in PGD would be an inefficient use of a powerful tool. A more versatile system than solid support arrays is, for example, the flow cytometry-based minisequencing (Colinas *et al.*, 2000).

23.3.4   Single Cell Cytogenetics

23.3.4.1   FLUORESCENCE IN SITU HYBRIDIZATION

In FISH, fluorescent probes carrying distinct fluorochromes for different chromosomes, are hybridized to cell nuclei spread either in metaphase or in interphase (see also Chapter 12). When embryos are analyzed, the cells are usually in interphase. Careful choosing of the type and location of the probes allows not only enumeration of chromosomes, as in sexing and PGD-AS, but also the diagnosis of chromosome imbalances in structural abnormalities.

Two methods are currently in use for the fixation of the blastomeres. Briefly, the first method (derived from the Tarkowski method to fix embryos) consists of the following steps. First, the blastomere is lysed in hypotonic solution (1% sodium citrate in water) to release the nucleus. Second, the fixative (3:1 acetic acid:methanol) is used to dissolve away the cytoplasm and to fix the nucleus to the glass slide. The second method was described by Coonen and colleagues (2000): the blastomere is brought in spreading solution (0.1 N hydrochloric acid, 0.01% Tween 20), which is then removed and the process repeated until the nucleus is free of cytoplasm and attached to the poly-L-lysine-coated glass slide. Some authors claim that the modified Tarkowski method yields better spread nuclei, so that the different fluorescent dots are better separated and more easily enumerated (Vellila *et al.*, 2002). The Coonen method is easier to use, and gives reasonable spreading results (Staessen *et al.*,

1999), even when five different chromosomes are analyzed concurrently, in the laboratories that are more experienced with this method (Staessen *et al.*, 2003).

Fluorescent probes can be made in-house or, as is more common nowadays, are commercially available. For more widespread uses, such as sexing and PGD-AS, kits are available, containing directly labeled probes for X and Y, 13, 18, and 21 (MultiVysion PGT kit from Vysis) or 13, 16, 18, 21, and 22 (MultiVysion PB kit from Vysis). These chromosomes were chosen either because they are present in liveborn trisomies (trisomies 13, 18, and 21) or because they are frequently present in miscarriages (16 and 22). Because the number of fluorochromes available is limited, the number of chromosomes that are analyzed can be increased either through ratio labeling and computerized analysis, or through the application of two and three hybridization rounds. Munné and colleagues (2003a) showed that in this way, up to nine different chromosomes (X, Y, 13, 15, 16, 17, 18, 21, and 21) could be analyzed for aneuploidy screening, and these authors reported an increase in implantation rate, as compared to the implantation rate obtained after analysis with five probes.

For structural abnormalities, a judicious choice of probes is made depending on the chromosomes involved and the breakpoints present. The first cases were performed using probes that specifically delineate the translocation breakpoints (Munné *et al.*, 1998; Munné *et al.*, 2000). The disadvantage was that appropriate probes first had to be cloned and then shown to be useful in FISH, but the advantage was that balanced translocated karyotypes could be distinguished from normal karyotypes. A more general approach was described by Conn and colleagues (1999), taking advantage of the availability of centromeric and subtelomeric probes for each chromosome. For Robertsonian translocations, a combination of one locus-specific probe centromeric to the breakpoint of one of the chromosomes involved and one on the telomere of the other chromosome involved, is applied. Three FISH probes are used for reciprocal translocations: two on the centromeric side of the breakpoints of the chromosomes involved, and a third one on one of the telomeres on the other side of the breakpoint (see Fig. 23.3). The two or three probes used must each carry a different fluorochrome, and it must be possible to use them together in one assay (Scriven *et al.*, 2000).

23.3.4.2 COMPARATIVE GENOMIC HYBRIDIZATION

CGH was developed to characterize the often complex chromosomal rearrangements present in cancer tissue. Total genomic DNA from the cells to be analyzed (e.g., the cancer cells) is labeled with a green fluorochrome, and a normal DNA sample is labeled with another, for example red, fluorochrome. Both labeled DNA samples are mixed and hybridized to a normal metaphase. If the test sample contains more of a certain sequence, for example, in a duplication, then that sequence will show up more green after

computerized analysis of the metaphase. Conversely, if the test sample contains less of a certain sequence (e.g., a deletion), this sequence will be more red on analysis. CGH has been down-scaled to the single cell level by introducing a first step of whole genome amplification of the DNA in the single blastomere (Voullaire *et al.*, 2000; Wells and Delhanty, 2000; Wilton *et al.*, 2001; Wells *et al.*, 2002).

The most widely used whole genome amplification method is degenerate oligonucleotide primed PCR (or DOP-PCR). At this point, CGH at the single cell level still requires several days for analysis, which is why the groups that have presented clinical application of CGH in PGD either had to recur to polar body analysis (Wells *et al.*, 2002), or to cryopreservation of the embryos (Voullaire *et al.*, 2000; Wilton *et al.*, 2001). Both authors, however, foresee that the introduction of microarrays (M-CGH) to replace the metaphase spread would significantly reduce the time necessary for the CGH analysis, and bring it back well within time to transfer embryos before day 5. The specificity of the sequences in the microarrays would lead to shorter hybridization and computer analysis times, obviating the need for polar body analysis or cryopreservation. Other advantages of M-CGH would be the greater resolution (100–200 kb instead of 2–10 Mb), and the important computerization of the whole procedure would allow more IVF centers to use M-CGH routinely for aneuploidy screening (Wells and Brynn, 2003).

23.4 OUTCOME OF PGD

23.4.1 Accuracy of the Diagnosis

One misdiagnosis for sexing and two for CF are mentioned in early reports (Harper and Handyside, 1994; Lissens and Sermon, 1997), and these were mainly due to the low efficiency of single cell PCR. Further technical developments—that is, FISH for sexing and multiplex fluorescent PCR—have ruled out the reoccurrence of this type of error. Munné and colleagues (1999) reported one misdiagnosis (trisomy 21 after aneuploidy screening) on a total of 57 pregnancies. These authors estimate that the misdiagnosis rate after biopsy of one blastomere was around 7% in a large series; nearly 6% of the embryos were misdiagnosed due to mosaicism in the embryo (Munné *et al.*, 2003b). Pickering and colleagues (2003) reported one misdiagnosis for spinal muscular atrophy out of 18 ongoing pregnancies. The third report of the ESHRE PGD Consortium (2002) mentions eight misdiagnoses: five after PCR (two for sexing, one each for Duchenne's muscular dystrophy, β-thalassemia, CF and myotonic dystrophy) and three after FISH (one each for social sexing, translocation (11;22), and a trisomy 21 after PGD-AS). This gives a figure of 8/265 (3%) if only those fetal sacs for which a control had been carried out are counted. The identification of these misdiagnoses and the

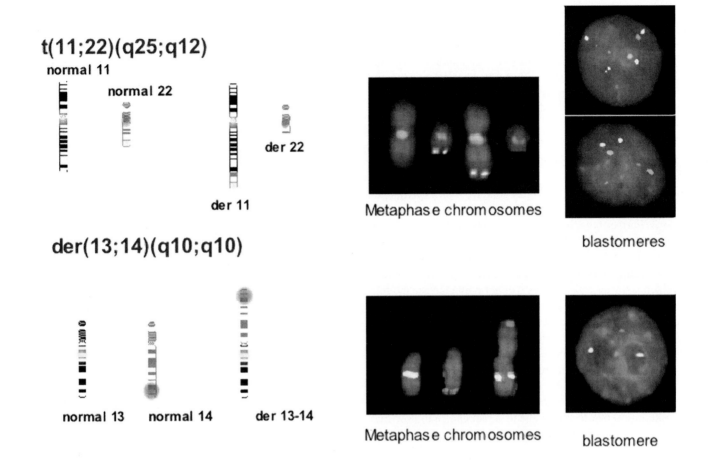

Panel A

Panel B

FIGURE 23.3 Example of the probes chosen for a reciprocal translocation (t(11;22), panel a) and for a Robertsonian translocation (t(13;14), panel b). The results are shown schematically, on lymphocyte metaphase spreads, and on interphase spreads of blastomeres. (See color plate)

reasons why they occurred has lead to the initiation of the drawing up of guidelines for PGD within ESHRE.

23.4.2 Pregnancy Outcome

The European Society for Human Reproduction and Embryology (ESHRE) PGD Consortium has reported on PGD cycles and their outcome (ESHRE PGD Consortium, 2002).

For structural chromosomal abnormalities (among which are translocations), 368 cycles led to 62 clinical pregnancies (17%). For sexing, 254 cycles led to 41 pregnancies (16%). For monogenic diseases, 575 cycles led to 119 (21%) pregnancies. These numbers are lower than can be expected in a regular IVF cycle, but it must be taken into account that a large cohort of embryos is diagnosed as affected or abnormal. Especially in patients carrying reciprocal translocations, as many as 80% of the embryos carry an unbalanced karyotype (Munné *et al.*, 1998; Van Assche *et al.*, 1999; Pickering *et al.*, 2003). This contrasts with the results for

PGD-AS: of the 799 cycles reported, 199 or 25% led to a clinical pregnancy in patient groups with a poor prognosis (advanced maternal age, repetitive IVF failure, and recurrent miscarriages).

The most important reason for morbidity and mortality in the pregnancies conceived after PGD is multiplicity. In this, and many other aspects such as birth weight and congenital malformations, children born after PGD are comparable to children born after ICSI (Bonduelle *et al.*, 2002).

The International Working Group reported that more than 3,000 clinical PGD cycles had been applied by mid-2001, resulting in around a 24% pregnancy rate (The International Working Group, 2001). Close to 700 children have been born following these pregnancies and 4.9% of these were reported to show abnormalities. Taking the four most active groups worldwide together, a total of 2774 PGD cycles, resulting in 2265 transfers and 652 clinical pregnancies (29%) were mentioned. Here too, it was stressed that outcome of pregnancy was comparable to IVF populations.

Another large cohort of children, not included in the ESHRE PGD Consortium report, were reported by Strom and colleagues (2000). A total of 109 infants were described and here, too, the conclusions were that children born after PGD are very comparable with children born after ICSI, and that PGD is a safe method to avoid the birth of children with genetic defects.

23.5 CONCLUSIONS

In just over a decade, PGD has gone from an experimental procedure to a widely accepted alternative for prenatal diagnosis. Furthermore, in many centers it has become an important tool to improve IVF results in certain patient categories. Although it is still labor-intensive and time consuming, introduction of new technologies has made it safer and more reliable. Newer technologies will not only increase the reliability further, but will also make it easier and more accessible to our patients. Nevertheless it is the authors' firm conviction that a close follow-up of the results obtained with PGD, not only in terms of technical safety and prevention of misdiagnosis, but more importantly in the follow-up of the pregnancies and babies born, is mandatory to bring PGD at the same level with prenatal diagnosis. Thus, PGD will make it possible to offer patients a real choice in how they make decisions in their reproductive lives.

References

Apessos, A., Abou-Sleiman, P., Harper, J., and Delhanty, J. (2001). Preimplantation genetic diagnosis of the fragile X syndrome by use of polymorphic linked markers. *Prenat. Diagn.* 21, 504–511.

Blaszczyk, A., Tang, Y., Dietz, H., Adler, A., Berkeley, A., Krey, L., and Grifo, J. (1998). Preimplantation genetic diagnosis of human embryos for Marfan's syndrome. *J. Assist. Reprod. Genet.* 15, 281–284.

Bonduelle, M., Liebaers, I., Deketelaere, V., Derde, M.-P., Camus, M., Devroey, P., and Van Steirteghem, A. (2002). Neonatal data on a cohort of 2889 infants born after ICSI (1991–1999) and of 2995 infants born after IVF (1983–1999). *Hum. Reprod.* 17, 671–694.

Boué, A., Boué, J., and Gropp, A. (1985). Cytogenetics of pregnancy wastage. *Adv. Hum Genet.* 14, 1–57.

Cieslak, J., Ivakhnenko, V., Wolf, G., Sheleg, S., and Verlinsky, Y. (1999). Three-dimensional partial zona dissection for preimplantation genetic diagnosis and assisted hatching. *Fertil. Steril.* 71, 308–313.

Colinas, J., Bellisario, R., and Pass, K. (2000). Multiplexed genotyping of β-globin variants from PCR-amplified newborn blood spot DNA by hybridization with allele-specific oligonucleotides coupled to an array of fluorescent microspheres. *Clin. Chem.* 46, 996–998.

Conn, C., Cozzi, J., Harper, J., Winston, R., and Delhanty, J. (1999). Preimplantation genetic diagnosis for couples at high risk of Down syndrome pregnancy owing to parental translocation or mosaicism. *J. Med. Genet.* 36, 45–50.

Coonen, E., Martini, E., Dumoulin, J., Hollanders-Crombach, H., deDie-Smulders, C., Geraedts, J., Hopman, A., and Evers, J. (2000). Preimplantation genetic diagnosis of a reciprocal translocation t(13;11)(q27.3;q24.3) in siblings. *Mol. Hum. Reprod.* 6, 199–206.

De Rycke, M., Van de Velde, H., Sermon, K., Lissens, W., De Vos, A., Vandervorst, M., Vanderfaeillie, A., Van Steirteghem, A., and Liebaers, I. (2001). Preimplantation genetic diagnosis for sickle-cell anemia and for β-thalassemia. *Prenat. Diagn.* 21, 214–222.

De Vos, A., and Van Steirteghem, A. (2001). Aspects of biopsy procedures prior to preimplantation genetic diagnosis. *Prenat. Diagn.* 21, 767–780.

Dean, N., Tan, S., and Ao, A. (2001). The development of preimplantation genetic diagnosis for myotonic dystrophy using multiplex fluorescent polymerase chain reaction and its clinical application. *Mol. Hum. Reprod.* 7, 895–901.

Delhanty, J., Griffin, D., Handyside, A., Harper, J., Atkinson, G., Pieters, M., and Winston, R. (1993). Detection of aneuploidy and chromosomal mosaicism in human embryos during preimplantation sex determination by fluorescent in situ hybridisation, (FISH). *Hum. Mol. Genet.* 2, 1183–1185.

Dreesen, J., Bras, M., de Die-Smulders, C., Dumoulin, J., Cobben, J., Evers, J., Smeets, H., and Geraedts, J. (1998). Preimplantation genetic diagnosis of spinal muscular atrophy. *Mol. Hum. Reprod.* 4: 881–885.

Dreesen, J., Jacobs, L., Bras, M., Herbergs, J., Dumoulin, J., Geraedts, J., Evers, J., and Smeets, H. (2000). Multiplex PCR of polymorphic markers flanking the CFTR gene: a general approach for preimplantation genetic diagnosis of cystic fibrosis. *Mol. Hum. Reprod.* 6, 391–396.

Dumoulin, J. C., Bras, M., Coonen, E., Dreesen, J., Geraedts, J. P., and Evers, J. L. (1998). Effect of Ca2+/Mg2+-free medium on the biopsy procedure for preimplantation genetic diagnosis and further development of human embryos. *Hum Reprod.* 13, 2880–2883.

Durban, M., Benet, J., Boada, M., Fernandez, E., Calafell, J. M., Lailla, J. M., Sanchez-Garcia, J. F., Pujol, A., Egozcue, J., and Navarro, J. (2001). *Hum. Reprod. Update* 7, 591–602.

Eftedal, I., Schwartz, M., Bendtsen, H., Andersen, A., and Zieve, S. (2001). Single intragenic microsatellite preimplantation genetic diagnosis for cystic fibrosis provides positive allele identification of all CFTR genotypes for informative couples. *Mol. Hum. Reprod.* 7, 307–312.

ESHRE PGD Consortium Steering Committee. (2002). ESHRE Preimplantation Genetic Diagnosis Consortium: data collection III (May 2001). *Hum. Reprod.* 17, 233–246.

Fallon, L., Harton, G. L., Sisson, M. E., Rodriguez, E., Field, L. K., Fugger, E. F., Geltinger, M., Sun, Y., Dorfmann, A., Schoener, C., Bick, D., Schulman, J., Levinson, G., and Black, S. H. (1999). Preimplantation genetic diagnosis for spinal muscular atrophy type I. *Neurology.* 53, 1087–1090.

Fiorentino, F., Magli, M. C., Podini, D., Ferraretti, A. P., Nuccitelli, A., Vitale, N., Baldi, M., and Gianaroli, L. (2003). The minisequencing method: an alternative strategy for preimplantation genetic diagnosis of single gene disorders. *Mol. Hum. Reprod.* 9, 399–410.

Gianaroli, L., Magli, C., Ferraretti, A., and Munné, S. (1999). Preimplantation diagnosis for aneuploidies in patients undergoing in vitro fertilization with a poor prognosis: identification of the categories for which it should be proposed. *Fertil. Steril.* 72, 837–844.

Girardet, A., Hamamah, S., Déchaud, H., Anahory, T., Coubes, C., Hédon, B., Demaille, J., and Claustres, M. (2003). Specific detection of deleted and non-deleted dystrophin exons together with gender assignment in preimplantation genetic diagnosis of Duchenne muscular dystrophy. *Mol. Hum. Reprod.* 9, 421–427.

Goossens, V., Sermon, K., Lissens, W., Vandervorst, M., Vanderfaeillie, A., De Rycke, M., De Vos, A., Henderix, P., Van de Velde, H., Van Steirteghem, A., and Liebaers, I. (2000). Clinical application of preimplantation genetic diagnosis for cystic fibrosis. *Prenat. Diagn.* 20, 571–581.

Goossens, V., Sermon, K., Lissens, W., De Rycke, M., Saerens, B., De Vos, A., Henderix, P., Van de Velde, H., Platteau, P., Van Steirteghem, A., Devroey, P., and Liebaers, I. (2003). Improving clinical preimplantation genetic diagnosis for cystic fibrosis by duplex-PCR using two polymorphic markers or one polymorphic marker in combination with the detection of the ΔF508 mutation. *Mol. Hum. Reprod.* 9, 559–567.

Griffin, D., Wilton, L., and Handyside, A. (1992). Dual fluorescent in situ hybridisation for simultaneous detection of X and Y chromosome-specific probes for sexing of human preimplantation embryonic nuclei. *Hum. Genet.* 89, 18–22.

Handyside, A., Kontogianni, E., Hardy, K., and Winston, R. (1990). Pregnancies from biopsied human preimplantation embryos sexed by Y-specific DNA amplification. *Nature* 344, 768–770.

Handyside, A., Lesko, J., Tarin, J., Winston, R., and Hughes, M. (1992). Birth of a normal girl after in vitro fertilization and preimplantation diagnostic testing for cystic fibrosis. *N. Eng. J. Med.* 327, 905–909.

Harper, J., and Handyside, A. (1994). The current status of preimplantation diagnosis. *Curr. Obst. Gynecol.* 4, 143–149.

Harper, J. C., Coonen, E., Ramaekers, F. C., Delhanty, J. D., Handyside, A. H., Winston, R. M. and Hopman, A. H. (1994). Identification of the sex of human preimplantation embryos in two hours using an improved spreading method and fluorescent in-situ hybridization (FISH) using directly labelled probes. *Hum. Reprod.* 9, 721–724.

Harper, J., Coonen, E., Handyside, A., Winston, R., Hopman, A., and Delhanty, J. (1995). Mosaicism of autosomes and sex chromosomes in morphologically normal, monospermic preimplantation human embryos. *Prenat. Diagn.* 15, 41–49.

Harton, G., Tsipouras, P., Sisson, M., Starr, K., Mahoney, B., Fugger, E., Schulman, J., Kilpatrick, M., Levinson, G., and Black, S. (1996). Preimplantation genetic testing for Marfan syndrome. *Mol. Hum. Reprod.* 2, 713–715.

Holding, C., and Monk, M. (1989). Diagnosis of beta-thalassaemia by DNA amplification in single blastomeres from mouse preimplantation embryos. *Lancet*, 2, 532–535.

Hussey, N., Donggui, H., Froiland, D., Hussey, D., Haan, E., Matthews, C., and Craig, J. (1999). Analysis of five Duchenne muscular dystrophy exons and gender determination using conventional duplex polymerase chain reaction on single cells. *Mol. Hum. Reprod.* 5, 1089–1094.

Iwarsson, E., Richter-Ährlund, L., Inzunza, J., Rosenlund, B., Fridström, M., Hillensjö, T., Sjöblom Nordenskjöld, M., and Blennow, E. (1998). Preimplantation genetic diagnosis of a large pericentric inversion of chromosome 5. *Mol. Hum. Reprod.* 7, 719–723.

Jiao, Z., Zhou, C., Li, J., Shu, Y., Liang, X., Zhang, M., and Zhuang, G. (2003). Birth of healthy children after preimplantation diagnosis of β-thalassemia by whole-genome amplification. *Prenat. Diagn.* 23, 646–651.

Joris, H., Nagy, P., Van de Velde, H., De Vos, A., and Van Steirteghem, A. (1998). Intracytoplasmic sperm injection: laboratory set-up and injection procedure. *Hum. Reprod.* 13, Suppl 1, 76–86.

Kanavakis, E., Vrettou, C., Palmer, G., Tretis, M., Mastrominas, M., and Traeger-Synodinos, J. (1999). Preimplantation genetic diagnosis in 10 couples at risk for transmitting β-thalassemia major: clinical experience including the initiation of six singleton pregnancies. *Prenat. Diagn.* 19, 1217–1222.

Kolibianakis, E., Bourgain, C., Albano, C., Osmanagaoglu, K., Smitz, J., Van Steirteghem, A., and Devroey, P. (2002). Effect of ovarian stimulation with recombinant follicle-stimulating hormone, gonadotropin releasing hormone antagonists, and human chorionic gonadotropin on endometrial maturation on the day of oocyte pick-up. *Fertil. Steril.* 78, 1025–1029.

Kuliev, A., Rechistky, S., Verlinsky, O., Ivakhnenko, V., Cieslak, J., Evsikov, S., Wolf, G., Angostiniotis, M., Kalakoutis, G., Strom, C., and Verlinsky, Y. (1999). Birth of healthy children after preimplantation diagnosis of thalassemias. *J. Assist. Reprod. Genet.* 16, 207–211.

Kurg, A., Tonisson, N., Georgiou, I., Shumaker, J., Tollett, J., and Metspalu, A. (2000). Arrayed primer extension: solid-phase four-color DNA resequencing and mutation detection technology. *Genet. Test.* 4, 1–7.

Lewis, C., Pinel, T., Whittaker, J., and Handyside, A. (2001). Controlling misdiagnosis errors in preimplantation genetic diagnosis: a comprehensive model encompassing extrinsic and intrinsic sources of error. *Hum. Reprod.* 16, 43–50.

Liebaers, I., Sermon, K., Staessen, C., Joris, H., Lissens, W., Van Assche, E., Nagy, P., Bonduelle, M., Vandervorst, M., Devroey, P., and Van Steirteghem, A. (1998). Clinical experience with preimplantation genetic diagnosis and intracytoplasmic sperm injection. *Hum. Reprod.* 13, Suppl 1, 186–195.

Lissens, W., and Sermon, K. (1997). Preimplantation genetic diagnosis: current status and new developments. *Hum. Reprod.* 12, 1756–1761.

Liu, J., Lissens, W., Silber, S., Devroey, P., Liebaers, I., and Van Steirteghem, A. (1994). Birth after preimplantation diagnosis of the cystic fibrosis ΔF508 mutation by polymerase chain reaction in human embryos resulting from intracytoplasmic sperm injection with epididymal sperm. *JAMA* 272, 1858–1860.

Macklon, N., and Fauser, B. (2000). Regulation of follicle development and novel approaches to ovarian stimulation for IVF. *Hum. Reprod. Update* 6, 307–312.

Marquez, C., Sandalinas, M., Bahçe, M., Alikani, M., and Munné, S. (2000). Chromosome abnormalities in 1255 cleavage-stage human embryos. *RBMOnline* 1, 17–26.

Moutou, C., and Viville, S. (1999). Improvement of preimplantation genetic diagnosis (PGD) for cystic fibrosis mutation ΔF508 by fluorescent polymerase chain reaction. *Prenat. Diagn.* 19, 1248–1250.

Moutou, C., Gardes, N., Rongières, C., Ohl, J., Bettahar-Lebugle, K., Wittemer, C., Gerlinger, P., and Viville, S. (2001). Allele-specific amplification for preimplantation genetic diagnosis (PGD) of spinal muscular atrophy. *Prenat. Diagn.* 21, 498–503.

Moutou, C., Gardes, N., and Viville, S. (2002). Multiplex PCR combining delta F508 mutation and intragenic microsatellites of the CFTR gene for pre-implantation genetic diagnosis (PGD) of cystic fibrosis. *Eur. J. Hum. Genet.* 10, 231–238.

Moutou, C., Gardes, N., and Viville, S. (2003). Duplex PCR for preimplantation genetic diagnosis (PGD) of spinal muscular atrophy. *Prenat. Diagn.* 23, 685–689.

Munné, S., Lee, A., Rosenwaks, Z., Grifo, J., and Cohen, J. (1993). Diagnosis of major chromosome aneuploidies in human preimplantation embryos. *Hum. Reprod.* 8, 2185–2191.

Munné, S., Fung, J., Cassel, M., Marquez, C., and Weier, H. (1998). Preimplantation genetic diagnosis of translocations: case-specific probes for interphase cell analysis. *Hum. Genet.* 102, 663–674.

Munné, S., Magli, C., Cohen, J., Morton, P., Sadowy, S., Gianaroli, L., Tucker, M., Marquez, C., Sable, D., Ferraretti, A., Massey, J., and Scott, R. (1999). Positive outcome after preimplantation diagnosis of aneuploidy in human embryos. *Hum. Reprod.* 1, 2191–2199.

Munné, S., Sandalinas, M., Escudero, T., Fung, J., Gianaroli, L., and Cohen J. (2000). Outcome of preimplantation genetic diagnosis of translocations. *Fertil. Steril.* 73, 1209–1218.

Munné, S., Sandalinas, M., Escudero, T., Velilla, E., Walmsley, R., Sadowy, S., Cohen, J., and Sable, D. (2003a). Improved implantation after preimplantation genetic diagnosis of aneuploidy. *Repr. BioMed. Online* 7, 91–97.

Munné, S., Sandalinas, M., Escudero, T., Marquez, C., and Cohen, J. (2003b). Chromosome mosaicism in cleavage-stage human embryos: evidence of a maternal age effect. *Repr. BioMed. Online* 4, 223–232.

Navidi, W., and Arnheim, N. (1991). Using PCR in preimplantation genetic disease diagnosis. *Hum. Reprod.* 6, 836–849.

Osborne, E., Trouson, A., and Cram, D. (2003). Microarray detection of Robertsonian translocations. *Hum. Reprod.* 18, Suppl 1, xviii59.

Parriego, M., Solé, M., Vidal, F., Boada, M., Santalo, J., Egozcue, J. Veiga, A., and Barri, P. N. (2003). One- or two-cell biopsy: does it affect implantation potential in PGD? *Repr. BioMed. Online* 7, Suppl 1, 31.

Pellestor, F., Girardet, A., Andréo, B., Arnal, F., and Humeau, C. (1994). Relationship between morphology and chromosomal constitution in human preimplantation embryo. *Mol. Reprod. Dev.* 39, 141–146.

Pickering, S., Polidoropoulos, N., Caller, J., Scriven, P., Mackie Ogilvie, C., Braude, P., and the PGD Study Group. (2003). Strategies and outcomes of the first 100 cycles of preimplantation genetic diagnosis at the Guy's and St. Thomas' Center. *Fertil. Steril.* 79, 81–90.

Pierce, K., Rice, J., Aquiles Sanchez, J., Brenner, C., and Wangh, L. (2000). Real-time PCR using molecular beacons for accurate detection of the Y-chromosome in single human blastomeres. *Mol. Hum. Reprod.* 6, 1155–1164.

Piyamongkol, W., Harper, J., Sherlock, J., Doshi, A., Serhal, P., Delhanty, J., and Wells, D. (2001). A successful strategy for preimplantation genetic diagnosis of myotonic dystrophy using multiplex fluorescent PCR. *Prenat. Diagn.* 21, 223–232.

Ray, P., Vekemans, M., and Munnich, A. (2001). Single cell multiplex PCR amplification of five dystrophin gene exons combined with gender determination. *Mol. Hum. Reprod.* 7, 489–494.

Rechitsky, S., Strom, C., Verlinsky, O., Amet, T., Ivakhnenko, V., Kukharenko, V., Kuliev, A., and Verlinsky, Y. (1999). Accuracy of preimplantation diagnosis of single-gene disorders by polar body analysis of oocytes. *J. Assist. Reprod. Genet.* 16, 192–198.

Report of the 11th Annual Meeting of International Working Group on Preimplantation Genetics. (2001). Preimplantation genetic diagnosis: experience of 3000 clinical cycles. *Reprod. BioMed. Online* 3, 49–53.

Rice, J., Aquiles Sanchez, J., Pierce, K., and Wangh, L. (2002). Real-time PCR with molecular beacons provides a highly accurate assay for detection of Tay-Sachs alleles in single cells. *Prenat. Diagn.* 2, 1130–1134.

Rubio, C., Simon, C., Vidal, F., Rodrigo, L., Pehlivan, T., Remohi, J., and Pellicer, A. (2003). Chromosomal abnormalities and embryo development in recurrent miscarriage couples. *Hum. Reprod.* 18, 182–188.

Schoolcraft, W., and Gardner, D. (2001). Blastocyst versus D2 or D3 transfer. *Sem. Reprod. Med.* 19, 259–268.

Scriven P., O'Mahony, F., Bickerstaff, H., Yeong, C.-T., Braude, P., and Mackie Ogilvie C. (2000). Clinical pregnancy following blastomere biopsy and PGD for a reciprocal translocation carrier: analysis of meiotic outsomes and embryo quality in two IVF cycles. *Prenat. Diagn.* 20, 587–592.

Sermon, K., Goossens, V., Seneca, S., Lissens, W., De Vos, A., Vandervorst, M., Van Steirteghem, A., and Liebaers, I. (1998a). Preimplantation diagnosis for Huntington's disease (HD): clinical application and analysis of the HD expansion in affected embryos. *Prenat. Diagn.* 18, 1427–1436.

Sermon, K., De Vos, A., Van de Velde, H., Seneca, S., Lissens, W., Joris, H., Vandervorst, M., Van Steirteghem, A., and Liebaers, I. (1998b). Fluorescent PCR and automated fragment analysis for the clinical application of preimplantation genetic diagnosis of myotonic dystrophy (Steinert's disease). *Mol. Hum. Reprod.* 4, 791–795.

Sermon, K., Lissens, W., Messiaen, L., Bonduelle, M., Vandervorst, M., Van Steirtegem, A., and Liebaers, I. (1999a). Preimplantation genetic diagnosis of Marfan syndrome with the use of fluorescent polymerase chain reaction and the Automated Laser Fluorescence DNA Sequencer. *Fertil. Steril.* 71, 163–165.

Sermon, K., Seneca, S., Vanderfaeillie, A., Lissens, W., Joris, H., Vandervorst, M., Van Steirteghem, A., and Liebaers I. (1999b) Preimplantation diagnosis for Fragile X syndrome based on the detection of the non-expanded paternal and maternal CGG. *Prenat. Diagn.* 19, 1223–1230.

Sermon, K., Seneca, S., De Rycke, M., Goossens, V., Van de Velde, H., De Vos, A., Platteau, P., Lissens, W., Van Steirteghem, A., and Liebaers I. (2001). PGD in the lab for triplet repeat diseases—myotonic dystrophy, Huntington's disease and Fragile-X syndrome. *Mol. Cell. Endocrinol.* 183, Suppl 1, 77–85.

Sermon, K., Moutou, C., Harper, J., Geraedts, J., Scriven, P., Wilton, L., Magli, M. C., Michiels, A., Viville, S., De Die, C. (2005) ESHRE PGD Consortium data collection IV: May-December 2001. *Hum Reprod.* 20, 19–34.

Staessen, C., Van Assche, E., Joris, H., Bonduelle, M., Vandervorst, M., Liebaers, I., and Van Steirteghem, A. (1999). Clinical experience of sex determination by fluorescent in situ hybridisation for preimplantation genetic diagnosis. *Mol. Hum. Reprod.* 5, 392–399.

Staessen, C., Tournaye, H., Van Assche, E., Michiels, A., Van Landuyt, L., Devroey, P., Liebaers, I., and Van Steirteghem, A. (2003). PGD in 47, XXY Klinefelter's syndrome patients. *Hum. Reprod. Update* 9, 319–330.

Stern, H., Harton, G., Sisson, M., Jones, S., Fallon, L., Thorsell, L., Getlinger, M., Black, S., and Schulman, J. (2002). Nondisclosing preimplantation genetic diagnosis for Huntington disease. *Prenat. Diagn.* 22, 303–307.

Strom, C., Ginsberg, N., Rechitsky, S., Cieslak, J., Ivakhenko, V., Wolf, G., Lifchez, A., Moise, J., Valle, J., Kaplan, B., White, M., Barton, J., Kuliev, A., and Verlinsky, Y. (1998). Three births after preimplantation genetic diagnosis for cystic fibrosis with sequential first and second polar body analysis. *Am. J. Obstet. Gynecol.* 178, 1298–1306.

Strom, C., Levin, R., Strom, S., Masciangelo, C., Kuliev, A., and Verlinsky, Y. (2000). Neonatal outcome of preimplantation genetic diagnosis by polar body removal: the first 109 infants. *Pediatrics* 106, 650–653.

Syvanen, A. C. (1999). From gels to chips: "minisequencing" primer extension for analysis of point mutations and single nucleotide polymorphisms. *Hum. Mutat.* 13, 1–10.

Thornhill, A. R., De Die-Smulders, C. E., Geraedts, J. P., Harper, J. C., Harton, G. L., Lavery, S. A., Moutou, C., Robinson, M. D., Schmutzler, A. G., Scriven, P. N., Sermon, K. D., Wilton L. (2005) ESHRE PGD Consortium "Best practice guidelines for clinical preimplantation genetic diagnosis (PGD) and preimplantation genetic screening (PGS)". *Hum Reprod.* 20, 35–48.

Van Assche, E., Staessen, C., Vegetti, W., Bonduelle, M., Vandervorst, M., Van Steirteghem, A., and Liebaers, I. (1999). Preimplantation genetic diagnosis and sperm analysis by fluorescence in situ hybridization for the most common reciprocal translocation t(11;22). *Mol. Hum. Reprod.* 5, 682–690.

Van de Velde, H., De Vos, A., Sermon, K., Staessen, C., De Rycke, M., Van Assche, E., Lissens, W., Vandervorst, M., Van Ranst, H., Liebaers, I., and Van Steirteghem, A. (2000). Embryo implantation after biopsy of one or two cells from cleavage-stage embryos with a view to preimplantation genetic diagnosis. *Prenat. Diagn.* 20, 1030–1037.

Vandervorst, M., Liebaers, I., Sermon, K., Staessen, C., De Vos, A., Van de Velde, H., Van Assche, E., Joris, H., Van Steirteghem, A., and Devroey, P. (1998). Succesful preimplantation genetic diagnosis is related to the number of available cumulus-oocyte-complexes. *Hum. Reprod.* 13, 3169–3176.

Velilla, E., Escudero, T., and Munné, S. (2002). Blastomere fixation techniques and risk of misdiagnosis for preimplantation genetic diagnosis of aneuploidy. *Reprod. BioMed. Online* 4, 210–217.

Verlinsky, Y., Rechitsky, S., Evsikov, S., White, M., Cieslak, J., Lifchez, A., Valle, J., Moise, J., and Strom, C. (1992). Preconception and preimplantation diagnosis for cystic fibrosis. *Prenat. Diagn.* 12, 103–110.

Verlinsky, Y. (1996). Preimplantation genetic diagnosis (Editorial). *J. Assist. Reprod. Genet.* 13, 87–89.

Verlinsky, Y., and Kuliev, A. (1996). Preimplantation diagnosis of common aneuploidies in fertile couples of advanced maternal age. *Hum. Reprod.* 11, 2076–2077.

Verlinsky, Y., Cieslak, J., Ivakhnenko, V., Evsikov, S., Wolf, G., White, M., Lifchez, A., Kaplan, B., Moise, J., Valle, J., Ginsberg, N., Strom, C., and Kuliev, A. (1999). Prevention of age-related aneupoidies by polar body testing of oocytes. *J. Assist. Reprod. Genet.* 16, 165–169.

Voullaire, L., Slater, H., Williamson, R., and Wilton, L. (2000). Chromosome analysis of blastomeres from human embryos by using CGH. *Hum. Genet.* 105, 210–217.

Vrettou, C., Palmer, G., Kanavakis, E., Tretis, M., Antoniadi, T., Mastrominas, M., and Traeger-Synodinos, J. (1999). A widely applicable strategy for single cell genotyping of β-thalassemia mutations using DGGE analysis: application to preimplantation genetic diagnosis. *Prenat. Diagn.* 19, 1209–1216.

Vrettou, C., Tzetis, M., Traeger-Synodinos, J., Palmer, G., and Kanavakis, E. (2002). Multiplex sequence variation detection throughout the CFTR gene appropriate for preimplantation genetic diagnosis in populations with heterogeneity of cystic fibrosis mutations. *Mol. Hum. Reprod.* 8, 880–886.

Wells, D., and Delhanty, J. (2000). Comprehensive chromosomal analysis of human preimplantation embryos using whole genome amplification and single cell comparative gnomic hybridisation. *Mol. Hum. Reprod.* 6, 1055–1062.

Wells, D., Escudero, T., Levy, B., Hirschhorn, K., Delhanty, J., and Munné, S. (2002). First clinical application of comparative genomic hybridisation and polar body testing for preimplantation genetic diagnosis of aneuploidy. *Fertil. Steril.* 78, 543–549.

Wells, D., and Brynn, L. (2003). Cytogenetics in reproductive medicine: the contribution of comparative genomic hybridisation (CGH). *BioEssays*, 25, 289–300.

Wilton, L., Williamson, R., McBain, J., Edgar, D., and Voullaire, L. (2001). Birth of a healthy infant after preimlantation confirmation of euploidy by comparative genomic hybridisation. *N. Engl. J. Med.* 34, 1537–1541.

Wilton, L. (2002). Preimplantation genetic diagnosis for aneuploidy screening in early human embryos: a review. *Prenat. Diagn.* 22, 312–318.

Wilton, L., Voullaire, L., Sargeant, P., Williamson, R., and McBain, J. (2003). Preimplantation aneuploidy screening using comparative genomic hybridization or fluorescence in situ hybridization of embryos from patients with recurrent implantation failure. *Fertil. Steril.* 80, 860–868.

Xu, K., Shi, Z., Veeck, L., Hughes, M., and Rosenwaks, Z. (1999). First unaffected pregnancy using preimplantation genetic diagnosis for sickle cell anemia. *JAMA* 281, 1701–1706.

Zenzes, M., and Casper, R. (1992). Cytogenetics of human oocytes, zygotes, and embryos after in vitro fertilization. *Hum. Genet.* 88, 367–375.

CHAPTER **24**

Genetic Monitoring of Laboratory Animals

JEAN-LOUIS GUÉNET[1] AND FERNANDO BENAVIDES[2]

[1] *Unité de Génétique des Mammifères, Institut Pasteur, Paris, France;*
[2] *The University of Texas M. D. Anderson Cancer Center, Science Park-Research Division, Smithville, Texas, USA*

TABLE OF CONTENTS

24.1 INTRODUCTION

The use of laboratory rodents in biological research raises a number of issues. Some of these are related to the ethical aspects of experimenting with living animals, and others are related to the cost of the experiments or the type of facilities required to perform these experiments. A way to alleviate these problems would be to design experiments in a way such that they each yield the largest possible amount of novel and reliable information, thus reducing to a minimum assays leading to inconclusive results (Festing and Altman, 2002). To reach this goal, experiments should be performed with carefully designed protocols and genetically defined animals. In this chapter, different aspects related to the genetic quality of laboratory animal strains and its control will be discussed. This is a very important aspect of experimental medicine that should be considered as a front line priority.

24.2 THE CONTROL OF GENETIC QUALITY

24.2.1 The Different Kinds of Laboratory Animal Strains

Strains of laboratory rodents are of two kinds. The first kind consists of genetically uniform populations, each of these populations representing only a sample among the many allelic forms that exist in the species. The other kind consists of genetically heterogeneous (GH) populations, segregating for a variety of alleles.

Inbred strains belong to the first kind. They are artificial populations, analogous (but not identical) to a clone of mammals with, however, the two sexes. They are very different from human populations but they have the enormous advantage of being extremely uniform from the genetic point of view and they are supposed to remain so, generation after generation, provided they are bred with an appropriate protocol (Festing, 1979; 1996; Hedrich, 1990). Inbred strains have derivatives such as *recombinant inbred strains* (RIS; Taylor, 1996), *congenic strains*, and *recombinant congenic strains* (RCS; Groot *et al.*, 1996) and, according to Grüneberg, the introduction of these strains into biology was "comparable in importance with that of the analytical balance in chemistry" (Morse, 1978a).

Opposed to inbred strains, randombred and outbred stocks are genetically heterogeneous, and in this respect, they are more similar to human populations. *Randombred stocks*, as their name indicates, are bred with no specific protocol: in other words, progenitors at generation F are mated randomly to produce generation F + 1. The genetic constitution of these stocks is unknown and may change from one generation to the next. *Outbred stocks*, on the contrary, are bred with a specific protocol and, although they segregate for a variety of alleles at a number of loci, the frequency of these alleles in the population fluctuates within limits that are determined by the breeding system and monitored (Silver, 1995; Hartl, 2001).

The decision to use an inbred strain rather than an outbred stock in an experimental protocol depends on the experiment and, in particular, on the biological question that is addressed. In fact, both populations are interesting and the reasons for choosing one or the other will become obvious in the following sections, once their genetic constitution and main characteristics are explained.

24.2.2 What Exactly Is an Inbred Strain?

For the International Committee on Genetic Nomenclature a strain can be regarded as inbred if it has been propagated by mating systematically brothers to sisters for 20 or more consecutive generations, and individuals of the strain can be traced to a single ancestral pair at the twentieth or subsequent generation. At this point the genome of each animal within the strain, on the average, will have no more than 1% to 2% residual heterozygosity and members of the same inbred strain can then be regarded, for most purposes, as genetically identical (Davisson, 1996). In practice, most of the mouse or rat inbred strains that are commonly used have undergone several tens of generations of inbreeding and some have even been bred with this system since the middle of the last century, which means, for over 200 generations. This definition of an inbred strain is important and calls for a few comments.

As it has already been mentioned, mice of the same inbred strain are *genetically identical*. They are also homozygous at all loci of their genome and in particular at all loci that were polymorphic in the founder ancestors. After a few tens of generations, one of the alleles that were segregating at a given locus becomes fixed and the others (i.e., up to three) are lost. This process of allele loss (or fixation) is easy to understand considering that, when by chance, an allele that was present at generation F is not represented in at least one of the two mice that are mated to produce generation F + 1, it is then permanently lost. In other words, alleles are lost during the inbreeding process but, except in the rare case of mutations, none is ever introduced from an external source. The sorting of the alleles that are retained in the strain and those that are lost also depends on chance and, if the experiment could be remade from scratch, with the same founder mice, it would lead to a strain with a different genetic constitution after the same 20 generations. This means that an inbred strain represents a unique, although fortuitous, assortment of alleles. To give an idea of the genetic profile of an inbred strain one could imagine a totally artificial scenario where an X-bearing spermatozoa, taken from a wild mouse, fertilizes an oocyte. Then the female pronucleus is removed from the oocyte in question (before it merges with the male pronucleus) and finally the chromosomes that were brought into the oocyte by the spermatozoa are duplicated. This new (n + n = 2n) conceptus would be a female with the two chromosomes of each pair absolutely identical. This is precisely how the genome of all the members of an inbred strain looks like with, of course, the exception of sex chromosomes.

During the process of inbreeding, the progress toward homozygosity is not constant. It is relatively fast during the first few generations then it slows down and, after 20 generations of unrelaxed inbreeding, it is no more than about 1% to 2% of the loci that are still segregating. A mathematical series based on Fibonacci's numbers traditionally is used to model the progress toward homozygosity while inbreeding progress, but it is only an approximation (see Fig. 24.1).

Since complete homozygosity is virtually reached, at all loci, after a few tens of brother to sister matings, it may then come to mind that it is no longer necessary to use such a stringent breeding system to propagate the strain. In fact, this would be hazardous because mutations constantly occur and the interruption of inbreeding would allow an increasing load of new polymorphisms to accumulate. Even if the spontaneous mutation rate is very low, there are so many genes in a mammalian genome that this source of polymorphism won't be negligible. Accordingly, inbreeding must not be relaxed.

All animals of an inbred strain are *isogenic*. This means that they all are absolutely genetically identical. This is an enormous advantage because scientists working with the same inbred strain but in different laboratories or at different periods in time can perform experiments where fluctuations in the experimental results, by definition, would not be the consequence of possible differences in the genetic constitution. Being isogenic also means that one can define in detail the phenotypic characteristics of an inbred strain by

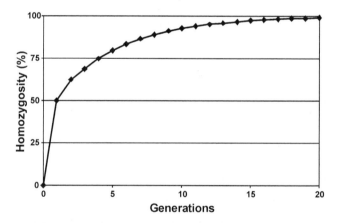

FIGURE 24.1 When a population of laboratory animals is propagated exclusively by mating brothers to sisters, the percentage of loci that are heterozygous decreases regularly. This progression towards homozygosity is modeled quite faithfully by the successive terms of a mathematical series where the numerators are Fibonacci's numbers (1, 1, 2, 3, 5, 8, 13, etc., the term ranked N being the sum of the terms ranked N-1 and N-2) and denominators the successive terms of the series 2^N. From the fifth generation (F5) onward, the level of heterozygosity falls off roughly by 19% at each generation so that, at F + 40, 99.98% of the genome is expected to be homozygous. In practice, considering that most of the loci were already homozygous in the first progenitors (as they are in most mammals, irrespective of the species) and keeping in mind that loci in a mammalian genome are not assorting independently but on the contrary are inherited as chromosomal chunks of various size, one can consider that a strain is totally inbred after 60 generations of inbreeding, not taking into account the continual appearance of spontaneous mutations.

accumulating experimental data concerning this strain from several sources. For example, The Jackson Laboratory has recently developed a program to establish a collection of baseline phenotypic data on the 48 most popular inbred strains of mouse, through a coordinated international effort. Information collected in this program (The Mouse Phenome Database; Paigen and Eppig, 2000) is freely available to the research community through the Internet (http://aretha. jax.org/pub-cgi/phenome/mpdcgi?rtn=docs/phenomelist).

The development of this database, which is regularly updated, was possible only because mice of inbred strains are isogenic. Around 400 different mouse and 200 rat inbred strains are available worldwide and a dozen among these strains, in both species, have become extremely popular.

To finish with the unique characteristics of inbred strains one could mention that they are fundamental tools for all experiments where a high level of standardization is required. They can be used as such, and they can be used for the production of interstrain F1 hybrids, which also represent genetically homogeneous populations and which, when taken by pair, are absolutely comparable to monozygotic twins. Inbred strains can also be used to generate GH populations when, for example, F1 hybrids between strain A and strain B (abbreviated AXBF1) are crossed with similar F1 hybrids between strain C and strain D (CXDF1) to generate a 4-way heterogeneous stock. In this case, although genetically heterogeneous, the basic components of such a stock, the original inbred strains, are perfectly identified and such GH stocks can be produced over and over, at will, when necessary. GH stocks with an even more complex structure (for example 8-way crosses stemming from eight different and unrelated inbred strains) have been produced for research in quantitative genetics.

24.2.3 The Difficulties of Maintaining Strain Purity

For people in charge of a breeding colony, the challenge is to keep each strain as pure as possible and to make sure that it remains true to its specification. This aim could be summarized in a few simple questions: are the mice to be used really from an inbred strain? Are these mice the same sort as the mice previously used? Is the strain used the same as the one somebody else is using for his/her experiments?

Three factors can lead to changes in the genetic constitution of inbred strains: the *genetic drift* due to residual heterozygosity, the occurrence of *mutations*, and the most important of all, the *contamination* by unintentional or accidental outcross with another strain.

Genetic drift due to residual heterozygosity is more a theoretical than an actual threat because most of the strains that are used nowadays stem from a rather small number of primitive strains that were already highly inbred when they began to be distributed. It must be remembered, however, that most of the existing differences among the major sublines of the C3H family (for example, C3H/An from Andervont, C3H/He from Heston, C3H/Bi from Bittner, and C3H/Fg from Fuchs, etc.) are, for the major part, due to incomplete inbreeding before distribution by L.C. Strong of the original progenitors (Morse, 1978b). Accordingly, if genetic drift can now be disregarded as a source of genetic divergence this was not the case at the beginning of last century and must be kept in mind, for a rigorous designation of the various substrains. Neglecting this may lead to serious difficulties, or even inconsistencies, in the interpretation of the experimental results.

Mutations are important to consider for two reasons: first, because their occurrence is beyond the control of the colony manager, and second, because they are very insidious and frequently impossible to detect by simple, superficial phenotypic observation. Based on extensive statistics, the mutations rates have been estimated to be in the range of 10^{-7} to 0.5×10^{-6} per locus per gamete for mutations toward a dominant allele and between 0.6 to 0.8×10^{-6} per locus per gamete for mutations toward a recessive allele (Schlager and Dickie, 1967).

Because these mutation rates are not negligible, the International Committee on Genetic Nomenclature has decided that two strains with the same origin but separated in different colonies by 100 or more generations (for example, 47 in laboratory A and 53 in laboratory B) should be considered as two different substrains and designated differently, in compliance with the rules for nomenclature (Davisson, 1996). Examples are common in the literature where two substrains of the same original inbred strain behave differently because a mutation occurred in one of the two substrains and dramatically altered the standard of the strain in question (Sultzer, 1968; Moisset, 1978; Morse, 1978a; Bulfield *et al.*, 1984). Data recently collected concerning single nucleotide polymorphisms (SNPs) in different C57BL/6 substrains kept independently for a few years at The Jackson Laboratory indicates that the mutation rate generating SNPs is very low (Wade, 2002). If, in addition, we assume that only one SNP out of ten is translated into a functional polymorphism (Beier, 2000), this would suggest that the occurrence of new mutations is not a serious issue in the generation of subline divergence. The main problem, however, is that the consequences of a new mutation are totally unpredictable. Mice of the C57BL/6JOlaHsd substrain for example, are homozygous for a deletion of the *Snca* locus (encoding for α-synuclein) on chromosome 6 (Specht and Schoepfer, 2001). This deletion has modest phenotypic effects (Chen *et al.*, 2002) but might interfere in an unpredictable manner with other mutations if, for example, the C57BL/6JOlaHsd substrain is used as a background strain for making knockout congenics (Wotjack *et al.*, 2003). Similarly, if mice of substrain C3H/HeJ are to be infected experimentally with Gram-negative bacteria they may react very differently from mice of substrain C3H/OuJ. This is explained by the occurrence of a mutation at the *Tlr4* locus (encoding for a *Toll*-like receptor) that occurred in the substrain C3H/HeJ where all mice are homozygous for the defective allele (Poltorak *et al.*, 1998). A very similar comment could be made for mice of the CBA/N substrain, which, unlike mice of all other CBA substrains, are homozygous for an X-linked mutation (*Btk^{xid}*) producing a syndrome of immunodeficiency homologous to the Bruton's disease in man.

The accidental matings of individuals from one inbred strain with another strain is by far the most important source of alteration of the genetic profile of inbred strains. Genetic contaminations of this type, which always result in a sudden and massive exchange of alleles, generally are observed between strains that have similar coat color (i.e., albino (*Tyr^c*/*Tyr^c*), agouti (*A*/*A*), or non-agouti (*a*/*a*)). Even more frequently, these accidental crosses occur between an interstrain F1 and one of the parental strains. Strains A2G and C57BL/Ks are two well-known examples where massive genetic contamination occurred in the past. A2G was considered a substrain of strain A until it was discovered that they became contaminated after an "illegitimate" mating with an unknown partner (a wild mouse?). C57BL/Ks (now C57BLKS) derives from strain C57BL/6 but was contaminated with up to 30% of the DBA/2 genome. This was suspected because C57BL/Ks mice have an *H2* haplotype that is not like the haplotype normally found in C57BL/6 mice (they are *H2^d* instead of *H2^b*) and also because congenic mice for the same obese (*Lep^{ob}*) mutation on these two backgrounds (C57BL/6J and C57BL/Ks) exhibited a totally different phenotype (Herberg and Coleman, 1977). The suspicion of genetic contamination has now been molecularly documented.

All three sources of changes in the genetic constitution of inbred strains must be taken into account by animal breeders. If the genetic drift due to residual heterozygosity is no longer a serious problem and the occurrence of mutation is insidious and unavoidable, the genetic contamination after an accidental cross with another strain is, by far, the most serious threat.

24.3 MONITORING THE GENETIC QUALITY OF INBRED STRAINS

A variety of techniques have been described in the past to assay the genetic quality of inbred strains. All these techniques were based on the postulates that each inbred strain, as previously mentioned, is *a priori* supposed to be homozygous at all loci of its genome and that all animals of the same strain have exactly the same genetic make-up (they are isogenic). These techniques, summarized in Table 24.1, were designed following the progress in genetics of the species and consisted in the analysis of a few phenotypic traits, controlled by a set of specific alleles, and defining a specific pattern for each strain.

TABLE 24.1 Summary of the most representative techniques to assay the genetic quality of inbred strains.

Phenotype-based

- External characteristics
- Reproductive performances
- Skin grafting
- Protein analysis by gel electrophoresis

DNA-based

- RFLP analysis
- PCR amplification of microsatellite markers (SSLP)
- SSCP analysis of 100/250 bp long DNA stretches
- DNA sequencing (detection of SNPs)

Reciprocal skin grafting, for example, has been extensively used in the 1960s because histocompatibility (the complete and permanent acceptation of transplants between any two mice or rats) is controlled by many genes and requires complete genetic identity between the donor and the recipient (Bailey, 1960). Skin grafting was a relatively nonexpensive protocol, fast (in skilled hands!) but, unfortunately, it was often influenced in both directions by environmental factors yielding false positive and false negative results. It is no longer used for the control of strain purity although some refinements in the protocol, using some modern *in vitro* assays, might still be interesting as supplementary tests.

Analysis of the electric charge of enzymatic proteins by electrophoresis in gels became popular in the mid-1970s because the technique was highly reliable and relatively easy to handle (Groen, 1977). However, this technique had the major drawback of being expensive to apply because each test required the use of specific and costly reagents.

Nowadays most of the genetic monitoring techniques applied to inbred strains are based on DNA technology and are extremely powerful. However, it must be kept in mind that the control of genetic purity must be undertaken in a broader context, considering also several simple parameters of very different nature, and not by applying very sophisticated molecular techniques alone. Among these parameters, a careful observation of the members of the same inbred strain, even if it may appear rather subjective, is always a very important source of information.

The first among these parameters is the external phenotype of the animals that must be totally homogeneous. All members of an inbred strain must have exactly the same phenotype and, in pigmented strains, the observation of different coat colors must be considered highly suspicious. A white spot on the belly of C3H mice is, however, common and should even be regarded as a criterion of purity as well as occasional tail kinks or eye abnormalities in C57BL/6.

With experience, a sagacious observer should even be able to recognize subtle characteristics discriminating the five major albino strains of mice (A, AKR, BALB/c, FVB, and SJL) or rats only considering the hair texture and tail shape. Similarly, several traits such as aggressiveness, open field behavior, life span, spontaneous diseases (tumors in particular), as well as a whole range of biochemical, immunological, and physiological characteristics are very much strain specific and are important to consider as criteria of genetic purity. For example, the occurrence of testicular teratomas in some substrains of 129 mice (129S2/SvPas in particular) is a criterion of great value because that sort of tumor is under polygenic control.

Another change that is almost always the direct consequence of a genetic contamination is a dramatic increase in the breeding performance. With a few exceptions, inbred strains are not very prolific and the age of puberty or first pregnancy and the average number of pups weaned per week and per female, are highly heritable characters. An abrupt change in these performances must be considered suspicious.

Apart from these zootechnical parameters, which are very easy to monitor by the mere analysis of the breeding records and a careful analysis of the phenotypes of the animals, other tests are available for the monitoring of genetic purity. Most of these tests are based on the analysis of a few strain-specific DNA sequences, revealed by the routine techniques of DNA structural analysis such as endonucleases-generated restriction polymorphisms (RFLP), mini-satellite-based DNA fingerprinting (Benavides *et al.*, 1998), PCR amplification of microsatellites (SSLP), single strand conformation polymorphism (SSCP, see also Chapter 6), or DNA sequencing to identify various SNPs.

Nowadays, microsatellite markers are very popular because they are extremely easy to type at a very low cost (Montagutelli and Guénet, 1994; Benavides *et al.*, 1999). The technique consists in the amplification of short repeated sequences, in general dinucleotides of the type $(CA)_n$ or $(TA)_n$, with flanking primers using mouse or rat DNA templates. There is an enormous number of microsatellites loci in the mouse and rat genomes (probably around 10^5) and it is generally not a problem to find a set of such molecular markers whose amplification products define a strain specific pattern (see Fig. 24.2). This strain-specific pattern may be assayed on a sample of animals of the strain and compared to a reference pattern that is archived in the laboratory. Routine analysis of DNA samples with microsatellite markers will confirm isogenicity and, provided the markers have been carefully selected, it could also guarantee that the strain whose DNA is assayed indeed corresponds to its designation (see Fig. 24.3).

Microsatellite markers have been a true revolution in the genetic monitoring of laboratory inbred strains because the test is extremely simple, affordable, and highly reliable. It

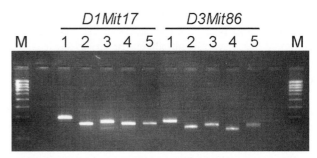

FIGURE 24.2 *Segment of an agarose gel showing the characteristic bands obtained after PCR amplification of SSLP markers. The amplification products were obtained using genomic DNA from five mouse inbred strains: SEG/Pas (1), C57BL/6Pas (2), C3H/HePas (3), DBA/2Pas (4), and BALB/cPas (5). The first five bands represent microsatellite marker D1Mit17, and the other five marker D3Mit86. Note that each inbred strain exhibits only one band (allele) per SSLP locus (indicative of their homozygous state), and that the bands have a standard size for each strain. Using a set of carefully selected microsatellite markers it is possible to define a pattern characteristic of a given strain. M: DNA size marker.*

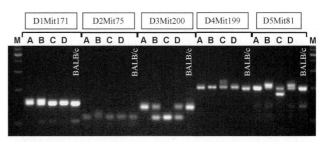

FIGURE 24.3 *Example of genetic contamination demonstrated after a genetic control with PCR-micro-satellites (SSLPs) markers. The analysis was carried out using mouse genomic DNA from mice that supposedly belong to the BALB/c inbred strain. PCR products were separated on a 4% agarose gel and visualized with ethidium bromide under UV transillumination. The figure depicts DNA amplifications for markers D1Mit171, D2Mit75, D3Mit200, D4Mit199, and D5Mit81 with the test samples (A to D) and BALB/c DNA standard. All the markers evidenced contamination with non-BALB/c (foreign) alleles in at least one sample. M: DNA size marker.*

can then be repeated more frequently and on a large sample of animals. A drawback of this test, maybe the only one, is a consequence of the occurrence of SNPs or mutations in the microsatellite sequences. Typing DNA samples from the BXD set of recombinant inbred strains, between the parental strains C57BL/6 and DBA/2, Dallas and colleagues (1992) found that several amplification products were from neither of the parental types and were termed as mutants, accordingly. This occurred in the frequency of 10^{-2} to 10^{-4}, a frequency that is not so low. When such a mutation is discovered, it is then recommended to type a dozen more microsatellites, some of them flanking the mutant allele. In general this is sufficient to clarify the situation. It would also be advisable to type all the breeders in the colony and to eliminate the new mutant allele (or to select it in the homozygous state) but not to keep it segregating in the inbred strain.

Even though microsatellites are remarkable markers for genetic monitoring of inbred strains because of their ease of use and specificity, they are not the only markers available. Sequencing a few strain-specific SNPs is another possibility that is not really expensive and in most research institutions it can be performed, in passing, at the occasion of an unrelated sequencing effort. Petkov and colleagues, from The Jackson Laboratory (Maine, USA) has recently described the allelic distribution of 235 SNPs in 48 mouse strains and selected a panel of 28 SNPs, enough to distinguish the almost 300 inbred, wild-derived, congenic, consomic, and recombinant inbred strains maintained at The Jackson Laboratory (Petkov *et al.*, 2004). This set of markers, encompassing all mouse chromosomes, is an excellent tool (fast, reliable, and economic) for detecting genetic contaminations in mouse facilities by way of automated PCR systems.

Finally, it is worth noting that, regardless the technique used, the occurrence of a new mutation in a coding sequence cannot be detected, except by chance, or if it results in an obvious phenotypic change. The only possibility for detecting such changes would be to resequence the genome of the strain, something that would be prohibitive and probably not worth doing in the end. Techniques are now being developed for the detection of DNA mismatches, which may become useful for the detection of new mutations, but their cost is still high.

24.4 PRESERVING THE GENETIC PURITY OF INBRED STRAINS

As it has been already discussed in the previous paragraphs, there are very efficient techniques to monitor the genetic quality of inbred strains. However, once a strain is recognized contaminated, the situation is irreversible, all individuals of the strain must be discarded, and another strain must be developed from a new set of breeders. The dramatic consequences of a genetic contamination imply that steps should be taken to prevent it from occurring again. Roughly, one can imagine that there are two efficient ways to preserve an inbred nucleus from genetic contamination: embryo freezing and complete physical isolation.

Embryo freezing is, theoretically, the most efficient way of preservation because, once frozen, genomes are insensitive to mutations (DNA does not replicate) and of course contamination cannot occur. Experiments have been performed a few years ago, which indicated that cosmic radiations do not represent a serious risk for deep-frozen embryos (Whittingham *et al.*, 1977). This means that choosing this

procedure for genome preservation is very secure and relatively inexpensive. However, the technique has two serious drawbacks. First, it can be achieved safely only in laboratories where the technical staff is trained for embryo handling and, second, some strains are very difficult to preserve and the percentage of living embryos recovered after thawing is extremely low or null. This drawback, which is inherent to particular strains, cannot be bypassed.

Sperm freezing may, in some circumstances, be used in substitution to embryo freezing, but is of no use when a diploid genome must be preserved.

Complete isolation of a breeding nucleus, for example, into a plastic isolator is a very efficient way to preserve genetic integrity and it is also an elegant way to preserve, at the same occasion, the health status of a rodent colony. Most of the commercial breeders of laboratory rat and mice have chosen this strategy, which combines several advantages at a relatively low cost.

Finally, and as an anecdotal comment, it is interesting to note that any recessive coat color mutation occurring in an inbred nucleus, at least theoretically, can be used as a natural seal of genetic purity since any illegitimate mating will result in the immediate disappearance of the trait. At several occasions, new alleles of the mutations leaden (symbol *ln*—chromosome 1), brown (*Tyrp^b*—chromosome 4), dilute (*Myo5^d*—chromosome 9), or pink-eyed dilution (*p*—chromosome 7) were found segregating in the author's colonies of C3H or C57BL/6, which could have been used to tag those strains with an institutional marker. However, it seems that scientists, when offered these naturally tagged mice are always reluctant to use them; a C57BL/6 inbred mouse must be solid black and any other color is suspicious!

24.5 CONTROLLING THE GENETIC STANDARD OF OUTBRED STOCKS

Outbred stocks are genetically heterogeneous and segregate for several alleles (in general two, sometimes three) at a few loci. They are very useful tools in many areas of experimental biology because the response of this type of stocks depends upon a large pool of genes and not only upon a few artificially selected alleles as in the case of inbred strains. In this respect, they appear complementary to the inbred strains but obviously they are neither homozygous nor isogenic and accordingly cannot be genetically monitored with the same protocols. What is important in this case is to make sure that the percentage of the different alleles in the population does not change much from one generation to the next and it is clear then that the strategies used for the monitoring of these strains are based on sampling and statistics. Here again, microsatellite markers can give an indication and hence this is a safe strategy to test a large sample of presumptive breeders retaining for the next generation a pool of mice or rats in which the frequency of each allelic variant is not statistically different from the frequency observed in generation F-1. Of course, the larger the sample the better it is to type an even greater number of microsatellites. Finally, it must also be kept in mind that, in the case of inbred strains only, one brother and one sister are required to breed generation F + 1, in the case of outbred stocks the larger the pool of breeders the better. Breeding programs have been developed to minimize the fluctuation of allele frequency at each generation (Hartl, 2001) but it is safe, from time to time, to perform a very large survey of the structure of the population and to keep the results as a reference.

4-way and 8-way genetically heterogeneous stocks, which are produced by intercrossing F1 hybrids bred from four or eight independent inbred strains, are an interesting alternative to the outbred stocks. In addition, they have the enormous advantage not to require any genetic monitoring. Like the best champagne wines have a constant taste because they are a blend of several vintages, GH stocks are produced by crossing a small number of highly standardized inbred strains.

24.6 THE CONTROL OF HEALTH STATUS

Although the control of the health status is out of the scope of this chapter, it is worth making a brief comment about it because the health status is a major source of variation in the biological response of laboratory animals and accordingly it has become a serious concern. In fact, though very efficient strategies exist for the control of the genetic quality of laboratory animals it is very difficult (and also very expensive) to perform a thorough control of their microbiological status and even more difficult to keep it constant all year round.

Lists of micro-organisms that are pathogenic for rats and mice have been previously published and a battery of tests exist for the diagnostic of these pathogens that are reliable provided they are applied routinely and carefully performed. Some of these tests are PCR-based and allow the detection of some specific pathogens (*Helicobacter*, mouse hepatitis virus, etc.; see also Chapter 22). One can then guarantee with a high level of confidence that a particular rodent colony is free of specific pathogens (specific pathogen free or SPF). Animals of this type have been available for several decades by most laboratory animals breeding farms. Since these mice are also controlled at regular intervals for their genetic quality one may consider that the quality of laboratory animals is now totally under control. However scientists have recently reported that variations in the biological response of some transgenic mice could be seriously influenced by differences in the population of enteric bacteria even if these bacteria are not registered as pathogenic for the mouse (Sellon *et al.*, 1998). In other words, it looks as if "clean" mice may not be the best material for some investigations in immunology. These observations indicate that, in the future, a standardized micro flora for laboratory rodents may have to be defined and with it, the procedures to monitor it.

24.7 CONCLUSIONS

In this chapter, the problems associated with the maintenance of standardized strains of laboratory rodents and the need to control their genetic quality and health status regularly have been reviewed. These controls, provided they are carefully and regularly performed, will contribute to the optimization of the experimental protocols, which in turn result in reducing the cost of the experiments by minimizing the number of animals needed. It is also of cardinal importance to use and follow with care, in all published works, the rules of nomenclature for inbred strains in order to avoid possible inconsistencies that may result from subtle genetic differences between the many substrains available.

References

Bailey, D. W., and Usama, B. (1960). A rapid method of grafting skin on tails of mice. *Plast. Reconstr. Transplant. Bull.* 25, 424–425.

Beier, D. R. (2000). Sequence-based analysis of mutagenized mice. *Mamm. Genome* 11, 594–597.

Benavides, F., Cazalla, D., and Pereira, C. (1998). Evidence of genetic heterogeneity in a BALB/c mouse colony as determined by DNA fingerprinting. *Lab. Anim.* 32, 80–85.

Benavides, F. J. (1999). Genetic contamination of an SJL/J mouse colony: Rapid detection by PCR-based microsatellite analysis. *Contemp. Top. Lab. Anim Sci.* 38, 54–55.

Bulfield, G., Siller, W. G., Wight, P. A., and Moore, K. J. (1984). X chromosome-linked muscular dystrophy (*mdx*) in the mouse. *Proc. Natl. Acad. Sci. USA* 81, 1189–1192.

Chen, P. E., Specht, C. G., Morris, R. G., and Schoepfer, R. (2002). Spatial learning is unimpaired in mice containing a deletion of the alpha-synuclein locus. *Eur. J. Neurosci.* 16, 154–158.

Dallas, J. F., Wotjak, C. T., and Holter, S. M. (1992). Estimation of microsatellite mutation rates in recombinant inbred strains of mouse. *Mamm. Genome* 3, 452–456.

Davisson, M. T. (1996). In *Genetic variants and strains of the laboratory mouse*, M. F. Lyon, S. Rastan, and S. D. M. Brown, eds. Rules for nomenclature of inbred strains, pp. 1532–1536. Oxford University Press, Oxford.

Festing, M. F. (1979). Inbred strains in biomedical research. The MacMillan Press Ltd.

Festing, M. F. (1996). In *Genetic variants and strains of the laboratory mouse*, M. F. Lyon, S. Rastan, and S. D. M. Brown, eds. Origins and characteristics of inbred strains of mice, pp. 1537–1576. Oxford University Press, Oxford.

Festing, M. F., and Altman, D. G. (2002). Guidelines for the design and statistical analysis of experiments using laboratory animals. *ILAR J.* 43, 244–258.

Glenister, P. H., and Thornton, C. E. (2000). Cryoconservation-archiving for the future. *Mamm. Genome* 11, 565–571.

Groen, A. (1977). Identification and genetic monitoring of mouse inbred strains using biochemical polymorphisms. *Lab. Anim.* 11, 209–214.

Groot, P. C., Moen, C. J. A., Hart, A. A. M., Snoek, M., and Démant, P. (1996). In *Genetic variants and strains of the laboratory mouse*, M. F. Lyon, S. Rastan, and S. D. M. Brown, eds. Recombinant congenic strains—genetic composition, pp. 1660–1670. Oxford University Press, Oxford.

Hartl, D. L. (2001). Genetic management of outbred laboratory rodent populations. Charles River Genetic Literature, http://www.criver.com/techdocs/index.html.

Hedrich, H. J. (1990). In *Genetic monitoring of inbred strains of rats*, H. J. Hedrich, ed. Inbred strains in biomedical research, pp. 1–7. Gustav Fischer Verlag, Stuttgart.

Herberg, L., and Coleman, D. L. (1977). Laboratory animals exhibiting obesity and diabetes syndromes. *Metabolism* 26, 59–99.

Moisset, B. (1978). In *Origins of inbred mice*, H. C. Morse, 3rd, ed. Subline differences in behavioral responses to pharmacological agents, pp. 483–484. Academic Press, New York.

Montagutelli, X., and Guénet, J.-L. (1994). Genetic monitoring of inbred strains by analysis of microsatellite polymorphisms. *Fifth Symposium of the Federation of European Laboratory Animal Science Associations*. Royal Society of Medicine Press, London.

Morse, H. C., 3rd (1978a). In *Origins of inbred mice*, H. C. Morse, 3rd, ed. Introduction, pp. 3–22. Academic Press, New York.

Morse, H. C., 3rd (1978b). In *Origins of inbred mice*, H. C. Morse, 3rd, ed. Differences among sublines of inbred mouse strains, pp. 441–444. Academic Press, New York.

Nomura T., Esaki, K., and Tomita, T. (1984). *ICLAS manual for genetic monitoring of inbred mice*. University of Tokyo Press.

Paigen, K., and Eppig, J. T. (2000). A mouse phenome project. *Mamm. Genome* 11, 715–717.

Petkov, P. M., Cassell, M. A., Sargent, E. E., Donnelly, C. J., Robinson, P., Crew, V., Asquith, S., Vonder Haar, R., and Wiles, M. V. (2004). Development of a SNP genotyping panel for genetic monitoring of the laboratory mouse. *Genomics* 83, 902–911.

Poltorak, A., He, X., Smirnova, I., Liu, M. Y., Van Huffel, C., Du, X., Birdwell, D., Alejos, E., Silva, M., Galanos, C., Freudenberg, M., Ricciardi-Castagnoli, P., Layton, B., and Beutler, B. (1998). Defective LPS signaling in C3H/HeJ and C57BL/10ScCr mice: mutations in Tlr4 gene. *Science* 282, 2085–2088.

Schlager, G., and Dickie, M. M. (1967). Spontaneous mutations and mutation rates in the house mouse. *Genetics* 57, 319–330.

Sellon, R. K., Tonkonogy, S., Schultz, M., Dieleman, L. A., Grenther, W., Balish, E., Rennick, D. M., and Sartor, R. B. (1998). Resident enteric bacteria are necessary for development of spontaneous colitis and immune system activation in interleukin-10-deficient mice. *Infect. Immun.* 66, 5224–5231.

Silver, L. M. (1995). *Mouse Genetics: Concepts and Applications*. Oxford University Press, Oxford (also available via the Internet at http://www.informatics.jax.org).

Specht, C. G., and Schoepfer, R. (2001). Deletion of the alpha-synuclein locus in a subpopulation of C57BL/6J inbred mice. *BMC Neurosci.* 2, 11.

Sultzer, B. M. (1968). Genetic control of leucocyte responses to endotoxin. *Nature* 219, 1253–1254.

Taylor, B. A. (1996). In *Genetic variants and strains of the laboratory mouse*, M. F. Lyon, S. Rastan, and S. D. M. Brown, eds. Recombinant inbred strains, pp. 1597–1599. Oxford University Press, Oxford.

Wade, C. M., Kulbokas, E. J., 3rd, Kirby, A. W., Zody, M. C., Mullikin, J. C., Lander, E. S., Lindblad-Toh, K., and Daly, M. J. (2002). The mosaic structure of variation in the laboratory mouse genome. *Nature* 420, 574–578.

Whittingham, D. G., Lyon, M. F., Glenister, P. H., Wotjak, C. T., and Holter, S. M. (1977). Long-term storage of mouse embryos at 196 degrees C: the effect of background radiation. *Genet. Res.* 29, 171–181.

Wotjak, C. T., Henniger, M. S., and Holter, S. M. (2003). C57BLack/BOX? The importance of exact mouse strain nomenclature. *Trends Genet.* 19, 183–184.

CHAPTER **25**

The Use of Locus-Specific Databases in Molecular Diagnostics

CHRISTOPHE BEROUD

Laboratoire de Génétique Moléculaire, CHU de Montpellier, Institut Universitaire de Recherche Clinique, Montpellier Cedex, France

TABLE OF CONTENTS

25.1 INTRODUCTION

Development of genetics has led to the emergence of a new field, bioinformatics. This domain usually is defined as informatics tools used in molecular biology. It must satisfy the following requirements: huge data size (billions of base pairs), large diversity of data (repeated sequences, protein domains, various species), integration of published references, and so on. If the development of a central database collecting all this information with appropriate analyzing tools is theoretically feasible, practically it is impossible. Then it was decided to split these data in various linked databases allowing the user to find an answer quickly without any apparent limitation. This structure is available today via the Internet. If this system seems to be exhaustive, it is sometimes hard to find the requested information and this is particularly true for mutations.

In the last decade, major advances have been made in the cloning and identification of genes involved in human diseases. Concurrently, advances in technology have led to the identification of numerous mutations in these genes, ranging from point mutations to large rearrangements (see previous chapters). It rapidly has become clear that the knowledge and organization of these alterations in structured repositories will be of great importance not only for diagnosis but also for clinicians and researchers.

The first database collecting mutations from a single gene has been published in 1976. It included 200 mutations

from the globin gene and has led to the HbVar database (Hardison *et al.*, 2002; Patrinos *et al.*, 2004). Meanwhile, Victor McKusick started to include gene mutations in his "Mendelian Inheritance in Man" compendium (OMIM: http://www3.ncbi.nlm.nih.gov/omim/). It was only a decade later that David Cooper began listing mutations in genes to allow him to determine which mutational change was the most common (Cooper *et al.*, 1998). In the mid-1990s the Human Genome Organization-Mutation Database Initiative (HUGO-MDI) was created, in order to organize this new domain of genetics, mutation analysis (Cotton *et al.*, 1998).

Since these early days, two main approaches have been developed: the *core* databases, also known as central databases, and the *locus-specific* databases (LSDB). The core databases have been created to collect published mutations from all genes. For this purpose, various automatic routines have been created to automatically scan a large set of publications for new mutations. The best example is the Human Gene Mutation Database (HGMD), which constitutes a comprehensive core collection of data on germ-line mutations in nuclear genes underlying or associated with human inherited disease (www.hgmd.org). By March 2003, the database contained an excess of 39,415 different lesions detected in 1,516 different nuclear genes.

Experts in specific fields do not maintain these databases, but they integrate data on mutations, their distribution, and references. Only published mutations are collected. Each mutation is entered only once in order to avoid confusion between recurrent and identical-by-descent lesions and the phenotypic description associated to the mutation is very limited, preventing any study on phenotypic variability. These databases frequently are referred as "mile wide and inch deep databases," as they included mutations from many genes but with a limited description. On the contrary, a LSDB collects all published and unpublished mutations from a specific gene. Experts, also known as curators, validate these data. A study performed by Richard Cotton in 2000 demonstrated that, for some LSDBs, up to 50% of data corresponded to unpublished mutations (personal communication). In addition, each mutation can occur more than once in a particular position, and hence appearing more than once in the database, allowing the identification of mutation hotspots. Furthermore, the annotation of each mutant includes a full molecular and phenotypic description. Therefore, these databases are referred to as "inch wide and mile deep databases."

25.2 WHY A LOCUS-SPECIFIC DATABASE IS REQUIRED FOR DIAGNOSIS

25.2.1 Identifying Causative Mutations

In the previous chapters, it has been nicely demonstrated that detection of DNA sequence variation can be very efficiently performed using a plethora of techniques. But how can one decide that it is really a causative mutation? Usually, in diagnostic laboratories, if a missense mutation is detected, additional arguments need to be collected prior to concluding that the mutation in question is in fact the causative mutation in the family. In the absence of a functional test, the segregation of the mutation in the affected family members, the absence of this variation in a panel of at least 100 control samples, the biochemical nature of the substitution, the region where the mutation is located, and the degree of conservation among species are some of the arguments for a causative mutation. This approach is often both time consuming and cost effective. The use of a LSDB can provide researchers with valuable information to help in such a decision process. For example, if the variation has already been reported as a polymorphism, one will have a direct access to its frequency and distribution among various populations (see Fig. 25.1).

If the mutation has been reported as a causative one, the full description of the mutation and the corresponding literature is provided in the LSDB. Furthermore, a good LSDB includes not only the reference sequence of the gene but also the description of structural domains and data about conservation of each residue of the protein among species.

It could also be useful to have direct access from the LSDB to the restriction map modifications associated with a mutation in order to confirm the detected substitution. This option is available only in some LSDBs (see Fig. 25.2; Beroud et al., 2000).

Finally, as many publications include a large set of data, it is often possible to observe typing errors or use of a wrong nomenclature, due to reference to an old sequence (up to 10% of errors for some publications). The use of a LSDB solves these problems, as most of the tools include an automatic naming system, using a reference sequence, and the human genetic code (see Fig. 25.3).

25.2.2 Providing Information About Phenotypic Patterns Associated with a Specific Mutation

LSDBs are far more than just inert repositories, as they include analyzing tools, which exploit the power of computers to answer complex queries, such as phenotypic heterogeneity and genotype/phenotype correlations. It is therefore possible to have a complete phenotypic description of patients with the same mutation. This valuable information could be useful for predictive medicine. A good illustration is given by the distribution of mutations of the FBN1 gene (MIM# 134797) that are associated with Marfan syndrome (MFS) and a spectrum of conditions phenotypically related to MFS, including dominantly inherited ectopia lentis, severe neonatal MFS, and isolated typical features of MFS.

MFS, the founding member of heritable disorders of connective tissue, is a dominantly inherited condition characterized by tall stature and skeletal deformities, dislocation of the ocular lens, and propensity to aortic dissection (Collod-Beroud and Boileau, 2002). The syndrome is characterized by considerable variation in the clinical phenotype between families and also within the same family. Severe neonatal MFS has features of the MFS and of congenital contractural arachnodactyly present at birth, along with unique features such as loose, redundant skin, cardiac malformations, and pulmonary emphysema (Collod-Beroud and Boileau, 2002). A specific pattern of mutations is observed in exons 24 to 26 in association with the neonatal MFS. In fact, 73.1% of mutations are located in this region in the neonatal form of the disease, but only 4.8% of mutations associated with a classical MFS are located in these exons (FBN1 database: http://www.umd.be; Collod et al., 1996; Collod-Beroud et al., 1997; 1998; 2003).

Another example is the identification of a specific type of mutations of the VHL gene, for which the loss of function results both in the predisposition to cancer in the von Hippel-Lindau disease (MM #193300) and in sporadic renal cell carcinoma (RCC). The von Hippel-Lindau disease is a dominantly inherited familial cancer syndrome, with an incidence of 1 in 36,000, predisposing to the development of retinal, cerebellar and spinal hæmangioblastomas, pancreatic cysts, and carcinoma and RCC (Maher and Kaelin, 1997; Richard et al., 1998). It has been shown that the nature of the mutation itself can be considered as a risk factor to develop specific tumors for VHL patients. In fact, patients with germline mutations leading to truncated proteins have a higher risk to develop a RCC than patients with missense mutations (77% versus only 55%, p < 0.05, Gallou et al., 1999). The VHL database is accessible at http://www.umd.be (Beroud et al., 1998).

25.2.3 Defining an Optimal Strategy for Mutation Detection

As LSDBs collect all published and unpublished mutations, they are very useful to define an optimal mutation screening strategy. Therefore, an overview of the distribution of mutations at the exonic level can help to focus on specific exons, where most of the mutations are located. The best example is given by the study of the TP53 gene involved in up to

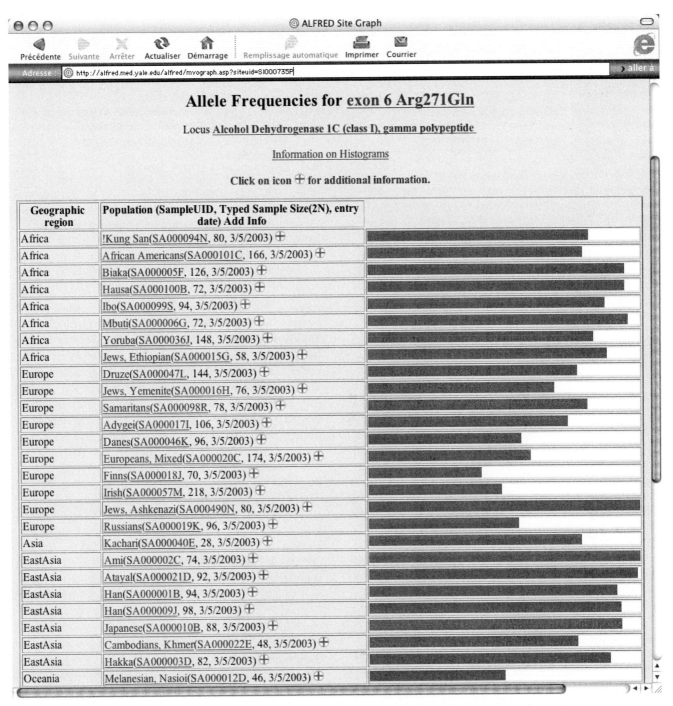

FIGURE 25.1 Distribution of the exon 6 *Arg271Gln* polymorphism from the ADH1C gene in various populations. Dark gray: frequency of the Arg allele; White: frequency of the Gln allele. Data are extracted from the ALlele FREquency Database (http://alfred.med.yale.edu/alfred/index.asp) that includes 830 polymorphisms, 352 populations, and 16,244 frequency tables.

50% of human cancers (Soussi, 2000). This gene is composed of 11 exons from which 10 are transcribed in a 393 amino acids protein. The distribution of the 14,970 mutations reported from 1,409 references in the TP53 database (http://www.umd.be) shows that 30.23% of mutations are located in exon 5, 27.2% in exon 7, 24.29% in exon 8, and 12.95% in exon 6; thus 94.87% of mutations are located in

4 out of the 11 exons of the gene (Beroud and Soussi, 2003; Soussi and Beroud, 2003). This observation has led 39% of research groups to search for mutations only in these exons, whereas 13% performed a complete scanning of the gene (Soussi and Beroud, 2001). Although this strategy is cost-effective, one needs to be careful in case of a negative result and should perform a complete scanning in order to avoid

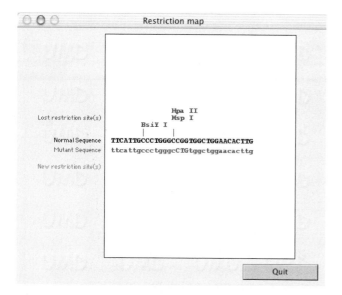

FIGURE 25.2   Restriction map modifications of the 2333G → T mutation in the ATP7B gene (http://www.umd.be). Note that three enzymes will produce two restriction fragments for the wild type allele and one uncut fragment for the mutant allele, allowing the identification or confirmation of the mutation.

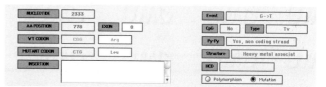

FIGURE 25.3   Input of a mutation in the ATP7B database. The mutation position can be indicated at the nucleotide or amino acid level. The software automatically displays the exon number, the wild type codon, and amino acid. The user indicates the mutation (here, the mutant codon CTG) and the software automatically computes the mutational event, the involvement of a CpG dinucleotide and/or pyrimidine doublet, the mutation type, the location of the mutation in a specific structural domain and a Highly Conserved Domain (HCD), and the international nomenclature: 2333G → T (not shown).

bias, as described in Soussi and Beroud (2001). Similarly, the study of mutational events will help to choose the best technical approach for mutation screening. For example, if most mutations are nonsense mutations, the protein truncation test could be one of the best approaches, whereas sequencing is considered the golden standard for mutation detection. If tools and data are available (this is the case only for few LSDBs today), the mutation pattern associated with an ethnic group/region can give valuable information as shown in Fig. 25.4.

If LSDBs are dealing with mutations and their analysis, they are usually part of a complete web site dedicated to a specific gene or disease. It is then frequent to find additional information about primers and technical conditions to help new research groups or diagnostic laboratories to establish their own diagnostic procedures. A very nice web site is hosted in Leiden (The Netherlands), about various genes involved in muscular dystrophies (http://www.dmd.nl). Figure 25.5 shows a list of primers that can be used for multiplex PCR to detect deletions of the DMD gene involved in Duchenne and Becker myopathies.

25.3 SELECTING THE PROPER LOCUS-SPECIFIC DATABASE

With the development of the Internet, accessing LSDBs has been greatly facilitated. Nevertheless, it can be sometimes difficult to choose the "best" database. For example, if searching for TP53 mutations is required, then one has to access at least six different databases. Two of them are recognized as reference databases (http://p53.curie.fr/ and http://www.iarc.fr/P53/index.html). The most efficient way to directly select the reference database is to use portals from either the Human Genome Variation Society (http://www.hgvs.org/) or either the HGMD (http://archive.uwcm.ac.uk/uwcm/mg/docs/oth_mut.html, Cooper *et al.*, 1998). The human genome is thought to contain about 40,000 genes and presently about 3,000 are known to be implicated in genetic diseases. In the HGMD 39,415 mutations from 1,516 genes are reported, but less than 500 LSDBs are available. It is clear that data are missing or not available for many genes.

The HGVS produced guidelines and recommendations for nomenclature, content, structure, and deployment of mutation databases (Scriver *et al.*, 1999; 2000; den Dunnen and Antonarakis, 2000; 2001; Claustres *et al.*, 2002; den Dunnen and Paalman, 2003). In addition, generic tools are available to help research groups set up their own database (Beroud *et al.*, 2000; Brown and McKie, 2000; Ohtsubo *et al.*, 2003). This user-friendly software will promote the creation of more and better LSDBs, by reducing or eliminating the present requirement for substantial knowledge of computing and bio-informatics to establish an LSDB from scratch. Potential curators will be encouraged to set up LSDBs, with the choice of using these software packages locally on their own workstations or having their databases hosted on a central server or pool of servers on the Internet. Using a system that can be run on any platform will reduce the risk of data becoming "lost" through lack of funding for an individual LSDB. Since data can be transferred directly between platforms or locations they will remain accessible to all even if database curation for some reason is interrupted.

The goal for the next decade is then the creation of freely accessible LSDBs on the Internet for virtually all genes

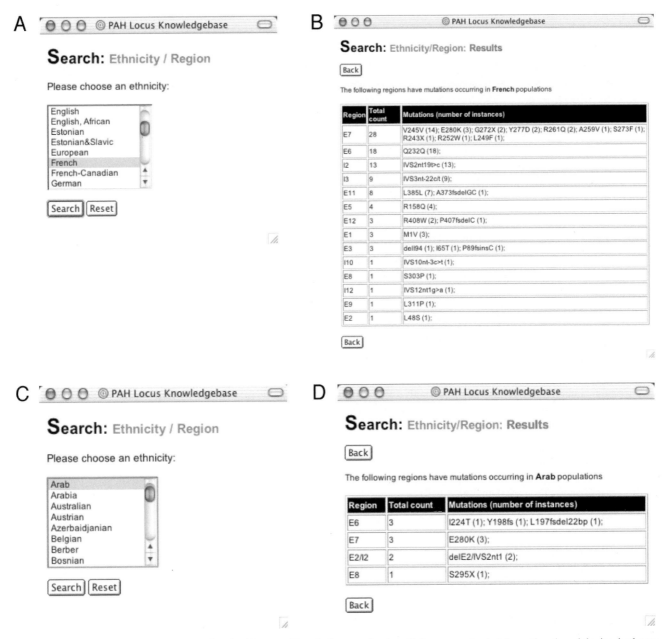

FIGURE 25.4 Mutations patterns associated with a specific ethnic group/region. Data were extracted from the phenylalanine hydroxylase (PAH) gene database (http://www.pahdb.mcgill.ca/). A and C: ethnicity/region selection; B and D: specific pattern of mutations found in French and Arab populations, respectively.

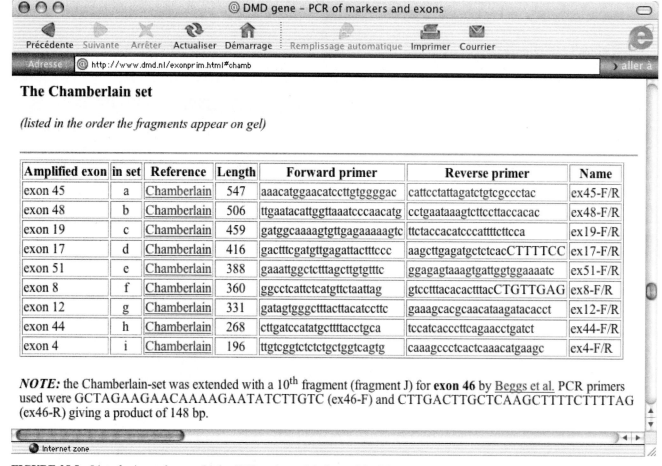

FIGURE 25.5 List of primers for a multiplex PCR to detect deletions of the DMD gene. Data were extracted from the Leiden Duchenne and Becker myopathies database (http://www.dmd.nl).

involved in human genetic diseases. Concomitantly, new tools should be developed to query simultaneously these LSDBs, in order to address new challenges such as the genetic heterogeneity (one disease—several alternative genes). A prototype is already available at http://www.umd.be.

References

Beroud, C., Joly, D., Gallou, C., Staroz, F., Orfanelli, M. T., and Junien, C. (1998). Software and database for the analysis of mutations in the VHL gene. *Nucleic Acids Res.* 26, 256–258.

Beroud, C., Collod-Beroud, G., Boileau, C., Soussi, T., and Junien, C. (2000). UMD (Universal mutation database): a generic software to build and analyze locus-specific databases. *Hum. Mutat.* 15, 86–94.

Beroud, C., and Soussi, T (2003). The UMD-p53 database: new mutations and analysis tools. *Hum. Mutat.* 21. 176–181.

Brown, A. F., and McKie, M. A. (2000). MuStaR and other software for locus-specific mutation databases. *Hum. Mutat.* 15, 76–85.

Claustres, M., Horaitis, O., Vanevski, M., and Cotton, R. G. (2002). Time for a unified system of mutation description and reporting: a review of locus-specific mutation databases. *Genome Res.* 12, 680–688.

Collod, G., Beroud, C., Soussi, T., Junien, C., and Boileau, C. (1996). Software and database for the analysis of mutations in the human FBN1 gene. *Nucleic Acids Res.* 24, 137–140.

Collod-Beroud, G., Beroud, C., Ades, L., Black, C., Boxer, M., Brock, D. J., Godfrey, M., Hayward, C., Karttunen, L., Milewicz, D., Peltonen, L., Richards, R. I., Wang, M., Junien, C., and Boileau, C. (1997). Marfan Database (second edition): software and database for the analysis of mutations in the human FBN1 gene. *Nucleic Acids Res.* 25, 147–150.

Collod-Beroud, G., Beroud, C., Ades, L., Black, C., Boxer, M., Brock, D. J., Holman, K. J., de Paepe, A., Francke, U., Grau, U., Hayward, C., Klein, H. G., Liu, W., Nuytinck, L., Peltonen, L., Alvarez Perez, A. B., Rantamaki, T., Junien, C., and Boileau, C. (1998). Marfan Database (third edition): new mutations and new routines for the software. *Nucleic Acids Res.* 26, 229–233.

Collod-Beroud, G., and Boileau, C. (2002). Marfan syndrome in the third Millennium. *Eur. J. Hum. Genet.* 10, 673–681.

Collod-Beroud, G., Le Bourdelles, S., Ades, L., Ala-Kokko, L., Booms, P., Boxer, M., Child, A., Comeglio, P., De Paepe, A.,

Hyland, J. C., Holman, K., Kaitila, I., Loeys, B., Matyas, G., Nuytinck, L., Peltonen, L., Rantamaki, T., Robinson, P., Steinmann, B., Junien, C., Beroud, C., and Boileau, C. (2003). Update of the UMD-FBN1 mutation database and creation of an FBN1 polymorphism database. *Hum. Mutat.* 22, 199–208.

Cooper, D. N., Ball, E. V., and Krawczak, M. (1998). The human gene mutation database. *Nucleic Acids Res.* 26, 285–287.

Cotton, R. G., McKusick, V., and Scriver, C. R. (1998). The HUGO Mutation Database Initiative. *Science* 279, 10–11.

den Dunnen, J. T., and Antonarakis, S. E. (2000). Mutation nomenclature extensions and suggestions to describe complex mutations: a discussion. *Hum. Mutat.* 15, 7–12.

den Dunnen, J. T., and Antonarakis, S. E. (2001). Nomenclature for the description of human sequence variations. *Hum. Genet.* 109, 121–124.

den Dunnen, J. T., and Paalman, M. H. (2003). Standardizing mutation nomenclature: why bother? *Hum. Mutat.* 22, 181–182.

Gallou, C., Joly, D., Mejean, A., Staroz, F., Martin, N., Tarlet, G., Orfanelli, M. T., Bouvier, R., Droz, D., Chretien, Y., Marechal, J. M., Richard, S., Junien, C., and Beroud, C. (1999). Mutations of the VHL gene in sporadic renal cell carcinoma: definition of a risk factor for VHL patients to develop an RCC. *Hum. Mutat.* 13, 464–475.

Hardison, R. C., Chui, D. H., Giardine, B., Riemer, C., Patrinos, G. P., Anagnou, N., Miller, W., and Wajcman, H. (2002). HbVar: A relational database of human hemoglobin variants and thalassemia mutations at the globin gene server. *Hum. Mutat.* 19, 225–233.

Maher, E. R., and Kaelin, W. G., Jr. (1997). von Hippel-Lindau disease. *Medicine* 76, 381–391.

Ohtsubo, M., Shibuya, K., Kudoh, J., Minoshima, S., and Shimizu, N. (2003). Integrated database for mutations in disease genes: MutationView/KMDB. *Tanpakushitsu Kakusan Koso* 48, 762–769.

Patrinos, G. P., Giardine, B., Riemer, C., Miller, W., Chui, D. H., Anagnou, N. P., Wajcman, H., and Hardison, R. C. (2004). Improvements in the HbVar database of human hemoglobin variants and thalassemia mutations for population and sequence variation studies. *Nucleic Acids Res.* 32, D537–D541.

Richard, S., Beroud, C., Joly, D., Chretien, Y., and Benoit, G. (1998). Von Hippel-Lindau disease and renal cancer: 10 years of genetic progress. GEFVHL (French-Speaking Study Group on von Hippel-Lindau disease). *Prog. Urol.* 8, 330–339.

Scriver, C. R., Nowacki, P. M., and Lehvaslaiho, H. (1999). Guidelines and recommendations for content, structure, and deployment of mutation databases. *Hum. Mutat.* 13, 344–350.

Scriver, C. R., Nowacki, P. M., and Lehvaslaiho, H. (2000). Guidelines and recommendations for content, structure, and deployment of mutation databases: II. Journey in progress. *Hum. Mutat.* 15, 13–15.

Soussi, T. (2000). The p53 tumor suppressor gene: from molecular biology to clinical investigation. *Ann. N. Y. Acad. Sci.* 910, 121–137.

Soussi, T., and Beroud, C. (2001). Assessing TP53 status in human tumours to evaluate clinical outcome. *Nat. Rev. Cancer* 1, 233–240.

Soussi, T., and Beroud, C. (2003). Significance of TP53 mutations in human cancer: a critical analysis of mutations at CpG dinucleotides. *Hum. Mutat.* 21, 192–200.

CHAPTER **26**

Safety Analysis in Retroviral Gene Therapy: Identifying Virus Integration Sites in Gene-Modified Cells

STEPHANIE MAIER-LAUFS,[1] K. ZSUZSANNA NAGY,[1] BERNHARD GENTNER,[1] W. JENS ZELLER,[1]
ANTHONY D. HO,[2] AND STEFAN FRUEHAUF[2]

[1] *Innovative Cancer Diagnostic and Therapy, German Cancer Research Center;*
[2] *Department of Internal Medicine V, University of Heidelberg, Heidelberg, Germany*

TABLE OF CONTENTS

26.1 INTRODUCTION

Oncoretroviral gene transfer is currently the most developed system to achieve transgene expression in hematopoietic cells. Low transduction efficiency of vectors into hematopoietic stem cells (HSC) made this technology inoperative for therapeutic application for a long time. In the early 1990s the transduction efficiency of those vectors hardly reached 1% in different clinical trials. Meanwhile, this situation has changed due to

- The application of adequate hematopoietic growth factors (Drexler, 1996; Petzer *et al.*, 1996; Shah *et al.*, 1996)
- The use of fibronectin fragments in the transduction process (Moritz *et al.*, 1996; Hanenberg *et al.*, 1996)
- Application of pseudo-typed viral particles with increased binding specifity for HSCs (Porter *et al.*, 1996; Movassaqh *et al.*, 1998; Kelly *et al.*, 2000; Gatlin *et al.*, 2001; Kiem *et al.*, 1998, Sanders, 2002)
- The optimization of transgene expression of retroviral vector constructs (Halene and Kohn, 2000; Moritz and Williams, 1998)

These advanced conditions today allow one to achieve an *in vivo* transduction efficiency of 5% to 20% (Abonour *et al.*, 2000) and contributed to the resumption of several clinical trials on gene therapy of monogenetic diseases (Bordignon and Roncarolo, 2002; Grez *et al.*, 2000; Kohn, 2002). For example, the therapeutic benefits in the French trial on X-chromosomal severe combined immunodeficiency (x-SCID), which is due to a defect in the gamma cytokine receptor subunit of interleukin-2, -4, -7, -9, and -15 receptors, showed the potential of the improved gene transfer technology (Cavazzana-Calvo *et al.*, 2000).

The latter trial unfortunately has also shown clearly the risks of gene therapy by appearance of insertion mutagenesis-associated cases of acute leukemia (Bonetta, 2002; Hacein-Bey-Abina *et al.*, 2003; Marshall, 2003). Methods that allow one to recognize insertion sites of retroviral vectors therefore have become indispensable for the safety assessment in clinical trials and experimental studies involving retroviral-mediated gene therapy (Fruehauf *et al.*, 2002; Baum *et al.*, 2003).

Retroviruses are a class of enveloped viruses containing two single-stranded RNA molecules as the genome. Following infection, the viral genome is reverse-transcribed into double-stranded DNA, which integrates into the host genome and is referred to as a provirus. Sequences, known as long terminal repeats (LTRs), are located at each end of the viral genome, which include promoter/enhancer regions and sequences involved with integration. In retroviral vectors the viral genes (*gag, pol,* and *env*) are replaced with the transgene of interest (see Fig. 26.1).

26.2 METHODS USED TO DETECT RETROVIRAL INTEGRATION SITES

The experimental analysis of proviral integration sites in human HSCs is challenging. Stem cell tracking techniques

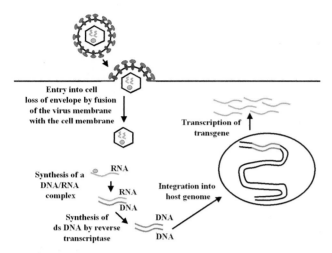

FIGURE 26.1 Retroviral gene transduction. The retroviral vector is packaged into a retrovirus, which consists of envelope and nucleocapsid. The retrovirus interacts with the host cell surface. After fusion of the virus membrane with the cell membrane, the nucleocapsid enters the cytoplasm of the host cell. The viral RNA is reverse transcribed into double-stranded (ds) DNA. After transport into the host cell nucleus, the viral DNA integrates into the host cell genome. The integrated viral DNA is referred to as a provirus. Transgene expression is mediated by the host cell transcription and translation repertoire.

based on detecting the retroviral integration sites and using this as a unique tag traditionally were used in isologous mouse studies (Dick *et al.*, 1985; Jordan and Lemischka, 1990). However, these techniques are not easily applied to xenogenic transplantation models due to lower contents of engrafted cells carrying the proviral tag in this latter setting. Retroviral integration patterns in transduced human cord blood cells transplanted into immune-deficient mice were detected by Southern blotting if engraftment and transduction efficiency were high (Barquinero *et al.*, 2000; Guenechea *et al.*, 2001). Inverse polymerase chain reaction, an alternative technique used by Nolta and colleagues (1996), is more sensitive, but requires clonal preparations as starting material. This approach cannot reliably detect multiple integration sites in one reaction (Kim *et al.*, 2000). An oligo-cassette mediated polymerase chain reaction technique described by Rosenthal and Jones (1990) and modified by Schmidt and colleagues (2001) is a promising approach to detect different integration sites simultaneously. This ligation-mediated PCR (LM-PCR) technique has been optimized for analysis of peripheral blood progenitor cells (PBPCs, Laufs *et al.*, 2003). Further methods to identify retroviral integration sites are fluorescence *in-situ* hybridization (FISH; see also Chapter 12) and arbitrary primer PCR. It has been shown that the integration sites observed with LM-PCR could be confirmed by FISH analysis (Laufs *et al.*, 2003).

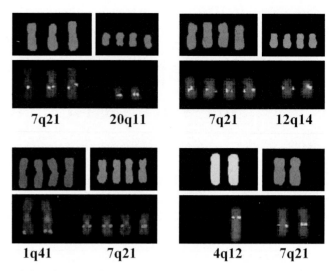

7q21   20q11   7q21   12q14

1q41   7q21   4q12   7q21

FIGURE 26.2 Chromosomal mapping of proviral sequences by FISH in cell lines. FISH analysis using SF1m vector plasmid DNA as probe and subsequently performed multicolor FISH was established to detect proviral inserts in SF1m-transduced HT1080 cell line clones. Chromosomes are identified by different colors. Chromosomes with a hybridization signal are highlighted: the *MDR*1 transgene could be detected on chromosomes 1q41, 4q12, 12q14, and 20q11. The human wild-type *MDR*1 locus (7q21) could also be detected. The highlighted chromosomes (top panel) show no translocations, so that the specific vector signal (bottom panel) can be assigned clearly to the depicted chromosomes. (Adapted from Laufs *et al.*, 2003; with permission.) (See color plate)

26.3 IDENTIFYING VIRUS INTEGRATION SITES BY FLUORESCENCE IN SITU HYBRIDIZATION

Metaphase spreads from vector-transduced cells are hybridized by using a retroviral vector probe (Lichter and Cremer, 1992) followed by whole chromosome painting probes for 24-color FISH (Speicher *et al.*, 1996). For transduced cells, at least 10 to 15 metaphase spreads should be acquired. Subsequently, the Re-FISH protocol (Müller *et al.*, 2002) can be performed, and metaphase chromosomes are hybridized by using the multicolor FISH protocol (Speicher *et al.*, 1996). At least five pools of whole chromosome painting probes are amplified and labeled by degenerate oligonucleotide primed-PCR (DOP-PCR; Telenius *et al.*, 1992) with the use of five spectrally distinguishable fluorochromes (fluorescein isothiocyanate (FITC), Cy3, Cy3.5, Cy5, Cy5.5). Each probe is hybridized in the presence of Cot-1 DNA for 48 hours. For evaluation, metaphase spreads are acquired by using highly specific filter sets and images are processed using the appropriate software, such as the Leica Multicolor Karyotyping (MCK) software (see Fig. 26.2).

26.4 IDENTIFYING VIRUS INTEGRATION SITES BY LIGATION-MEDIATED PCR

The goal of the LM-PCR method is to detect the junction between the integrated provirus and the flanking genomic sequence. The sequence of the LTR is known, but the flanking genomic sequence is unknown. To overcome this difficulty, the genomic DNA is digested with a restriction enzyme, which cuts upstream from the LTR in the genomic sequence and in the proviral sequence, but not within the LTR region. Fragments containing the LTR-genomic DNA junctions are tagged with a biotinylated LTR specific primer. Since the 5′ and 3′ LTR have identical sequence, two types of fragments are tagged: the external band, which contains the unknown flanking sequence; and the internal band, which contains only vector sequence. The biotin-tagged fragments are enriched using streptavidin-coated paramagnetic beads. Thus, other DNA fragments are removed.

In the next step, an adaptor oligonucleotide cassette of known sequence is ligated on the LTR-distant end of fragments allowing PCR amplification. To increase the specificity and the sensitivity of PCR, a second, nested PCR is performed. The resulting PCR-products are excised from agarose gel, separately cloned following DNA extraction, and sequenced (see Fig. 26.3). Sequences are confirmed as integration sites when the following criteria are met:

- Sequence is flanked by LTR and adapter sequences
- Sequence is matched uniquely (>90% identity) to human genome

The LM-PCR method first has been established and optimized on transduced HT1080 cell line clones (Laufs *et al.*, 2003) obtained by single cell sorting. Therefore, two cell line clones with one integration site each and one cell line clone with two integration sites have been analyzed. These cell line clones were mixed together to test for the ability to detect multiple clones in one reaction. Four different integration sites and one internal band were readily coamplified (see Fig. 26.4).

These integration sites obtained by LM-PCR were confirmed by FISH analysis (see Table 26.1).

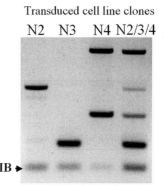

Transduced cell line clones

N2　N3　N4　N2/3/4

IB ▶

Cell line clone	Band-length (kb)	Vector-sequence	Human flanking DNA sequence	GenBank accession number	Chromosomal localization
N2	0.22	..gggctttca	CTTCCCATTCT..	AC121337	12q13
N3	0.33	..gggctttca	TCAGACCCTTC..	AC092015	1q41
N3	0.67	..gggctttca	TATGGAAGCTT..	AC083798	3q21
N4	0.25	..gggctttca	AACCCATTCAC..	AL109824	20q11
IB	0.15	..gggctttca ttccccccttttct			

FIGURE 26.4 Retroviral integrations detected by LM-PCR. Three SF1m-transduced HT1080 cell line clones are analyzed. DNA from each clone is used in separate reactions (N2, N3, and N4) and mixed together in one reaction (N2/3/4). The internal band (IB) originates from the 3′ LTR and is identical for all SF1m vector-transduced cells. The length and the sequence of each band is shown. The mapping result of the detected human flanking sequences is shown as the matched Genbank accession number and as chromosomal localization. (Adapted from Laufs *et al.*, 2003; with permission.)

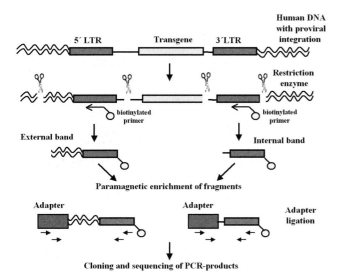

FIGURE 26.3 Schema of ligation-mediated PCR technique. A restriction digest is performed on human genomic DNA containing proviral integrants. Fragments containing LTR-genomic DNA junctions are marked with a biotinylated LTR-specific primer and enriched by streptavidin-coated paramagnetic beads. Other DNA fragments are removed. An adaptor oligonucleotide cassette is ligated to flanking DNA. Solid-phase nested PCR is performed with adapter-specific and LTR-specific primers. PCR bands are excised after gel electrophoresis, separately cloned, and sequenced.

TABLE 26.1 Correlation between flanking DNA sequences mapped to human chromosomes and localization of FISH signals.

Cell line clone	Integration site sequences obtained by LM-PCR	FISH signals
N2	12q13	12q14
N3	1q41	1q41
N4	20q11	20q11

Figure 26.3 labels: 5′ LTR, Transgene, 3′LTR, Human DNA with proviral integration; Restriction enzyme; biotinylated primer; External band; Internal band; Paramagnetic enrichment of fragments; Adapter; Adapter ligation; Cloning and sequencing of PCR-products.

The addition of genomic mouse DNA (as background DNA) to the sample did not change the results, and negative controls with mock-transduced mouse DNA produced no PCR bands on agarose gel. An additional optimization step was the excision of the bands from agarose gel, which increased the number of identified integrations.

The LM-PCR method has been performed on transduced human CD34+ PBPCs, repopulating the bone marrow of NOD/SCID mice (Laufs *et al.*, 2003). The human multidrug resistance 1 (*MDR*1) gene was transferred into human mobilized CD34+ PBPCs by a Friend mink cell focus-forming/murine embryonic stem-cell virus type vector, the SF1m retroviral vector, carrying the *MDR*1 gene (Baum *et al.*, 1995). This vector has been shown to mediate high P-glycoprotein (*MDR*1 gene product) expression, conferring a drug-resistant phenotype in early hematopoietic cells *in vitro* (Baum *et al.*, 1995; Eckert *et al.*, 1996; Hildinger *et al.*, 1998; Hildinger *et al.*, 1999) and permitting sustained transgene expression *in vivo* (Schiedlmeier *et al.*, 2000; Baum *et al.*, 1996). Chimeric mouse bone marrow DNA was digested and LM-PCR was performed. An integration site library was constructed by cloning the resulting PCR products. The cloning procedure allowed sensitive detection of fragments. Subsequently, the cloned PCR products were sequenced to prove the presence of LTR-genomic DNA junctions, and up to 32 different clones were identified from one chimeric bone marrow. Repeated LM-PCR using the same chimeric bone marrow DNA resulted in less than 10% overlap between two separate LM-PCR analyses, pointing to the polyclonality of human hematopoiesis following experimental transplantation.

To investigate the specificity of the method, the LM-PCR was performed on DNA, isolated from bone marrow of three mice that had received transplants of nontransduced human CD34+ cells (mock transduction). The LM-PCR product was concentrated and cloned as a whole, omitting procedures in which PCR fragments could get lost. Not a single LTR-flanking DNA junction could be identified in three mock mice that had received untransduced human PBPCs from different transplantation experiments (Laufs *et al.*, 2003).

26.5 IDENTIFYING VIRUS INTEGRATION SITES USING ARBITRARY PRIMER PCR

This approach is a sensitive and rapid method to characterize retroviral integration sites in small clonal cell samples such as hematopoietic colonies. Methods using arbitary primers (Silver and Keerikatte, 1989; Sørensen *et al.*, 1993) or primers against repetitive genomic DNA sequences (Butler *et al.*, 2001) that hybridize to the unknown flanking DNA regions have been described for the analysis and identification of retroviral integration sites. These methods do not require prior manipulations like DNA restriction enzyme digest, adapter ligation, or circularization of DNA frag-

ments. Therefore, they can potentially be performed with very small amounts of DNA.

To detect proviral integrants in human genomic DNA, a PCR reaction is performed using biotinylated LTR-specific and arbitrary primers. Arbitrary primers consist of a long 5′ tail of known sequence, seven random nucleotides, and five fixed nucleotides at the 3′ end. Due to the fact that these primers are partly degenerate (random nucleotides), they enable amplification of proviral flanking human DNA fragments, without any information on sequence. The biotinylated fragments are enriched by streptavidin-coated paramagnetic beads. A nested PCR is performed using a nested LTR-specific primer and a primer that binds to the 5′ tail sequence of the arbitrary primer. This second PCR enhances the specificity of the amplification. The obtained PCR-products are excised from agarose gel and directly sequenced following DNA extraction (see Fig. 26.5).

An arbitary primer PCR technique (Sørensen *et al.*, 1993) was adapted by Gentner and colleagues (2002), and the results were validated with LM-PCR on retrovirally transduced cell line clones. The applicability of arbitrary primer PCR was shown to analyze retroviral integration sites in colony-forming human CD34+ PBPCs, a cell population that is an indicator for transduction of human stem cells (HSCs). Seven different arbitrary primers were constructed according to a model described by Sørensen and colleagues (1993), which has previously been used for integration analysis of

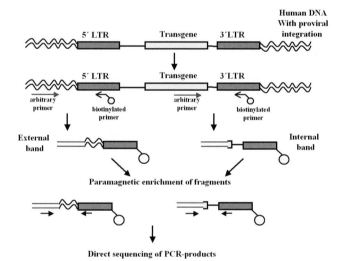

FIGURE 26.5 Outline of arbitrary primer PCR with arbitrary primers. A PCR reaction is performed on human genomic DNA containing proviral integrants using biotinylated LTR-specific and arbitrary primers. Arbitrary primers allow amplification of proviral flanking human DNA fragments without any information on sequence because of their partly degenerate feature. After paramagnetic enrichment of biotinylated fragments, a nested PCR is performed in order to enhance the specificity of amplification. The obtained PCR-products are excised from agarose gel and directly sequenced following DNA extraction. (See color plate)

wild-type retroviruses (Sørensen *et al.*, 1993; Sorensen *et al.*, 1996; see Fig. 26.6).

To identify the most suitable primers for the application, arbitrary primer PCRs were performed with each of those primers in combination with a specific primer for the known proviral LTR segment on two human fibrosarcoma cell line clones (HT1080 clone N2 and N3) that were transduced with the *MDR*1 gene (see Section 26.4). Three arbitrary primers (FP2, FP4 and FP5; see also Fig. 26.6) turned out to be most useful, as they did not avidly bind to proviral sequences close to the 3′ LTR.

FP2, FP4, and FP5 generated PCR products of the LTR-genomic DNA junctions (5′ LTR external fragment) in both of the HT1080 cell line clones, which were subsequently sequenced. In clone N3 two different junctions were detected. The first one was detected by FP2 and FP5, and the second junction was detected by FP4. Importantly, both integrations could be detected in one reaction using a cocktail of the primers FP2 and FP4. Furthermore these HT1080 cell line clones were analyzed with the LM-PCR method in order to validate the results. After extensive studies with three different restriction enzymes to create a large variety of amplification permissive restriction fragment length polymorphisms, a total of three different retroviral integrations in the two HT1080 cell-line clones were found. It appeared

that vector integrations detected with arbitrary primer PCR were identical to those detected by LM-PCR, thus confirming the reliability and sensitivity of both methods.

Next, the optimized arbitrary primer PCR technique was applied in order to detect retroviral vector integration sites in colony-forming human PBPCs. Mobilized CD34⁺ PBPCs were transduced with the SF1m retroviral vector and cultured in semisolid medium in the presence or absence of vincristine. Vincristine was added to the plates in order to select for *MDR*1-expressing (vector positive) colonies. As expected, the percentage of *MDR*1-positive colonies was up to five fold increased in the presence of vincristine.

A total of 182 individual colonies from five different SF1m transduced PBPC donor samples were analyzed for the presence of the *MDR*1 gene. In total, 72 colonies (40%) were positive for the *MDR*1 transgene and thus qualified for integration site analysis. The DNA amount of a *MDR*1-positive colony corresponds to a sensitivity of 75 or more cells per reaction. Of the 72 colonies, 61 yielded a specific PCR product; that is, either an informative band (25 colonies) or an internal 3′ LTR fragment. In most cases, one to five clear, robust bands per colony were amplified. In 10 out of 25 colonies containing external bands, the informative fragment was the only amplicon seen in the gel.

Arbitrary primer PCR has been used successfully to study wild-type retroviral integrations in cell lines (Sørensen *et al.*, 1993), retroviral integrations in tumor DNA from mice (Sørensen *et al.*, 1996), and hepatitis B virus integrations in peripheral blood mononuclear cells (Laskus *et al.*, 1999). It was demonstrated that this simple and fast approach also can be used to study retroviral vector integrations in day 14 colonies of CD34⁺ PBPCs, a setting in which DNA amount is very limited. High-quality sequences of the LTR-genomic DNA junctions could be obtained after only five standard steps (1st PCR, enrichment, 2nd PCR, agarose gel electrophoresis, direct sequencing).

26.6 CONCLUSIONS

To assess the mutagenic risk of retroviral gene therapy, it is important to characterize vector integration sites in individual stem cells and their progeny.

It is well established for wild-type retroviruses that genomic integration is not a completely random process (Holmes-Son *et al.*, 2001). It was shown by some researchers that the central core domain of the retroviral integrase plays an important role in determining the target specificity (Shibagaki and Chow, 1997). The efficiency of chromosomal sites to become a preferred integration target appears to be further affected by several factors, such as transcriptional activity (Weidhaas *et al.*, 2000), DNAse I hypersensitivity (Vijaya *et al.*, 1986), methylation (Kitamura *et al.*, 1992), GC content (Leclercq *et al.*, 2000), nuclear scaffold attachment (Mielke *et al.*, 1996), nucleosome struc-

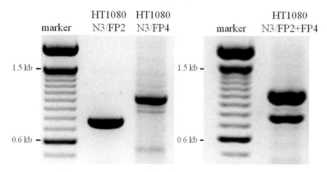

	long 5′ tail	seven random nucleotides	five fixed nucleotides
FP1:	5′-GACTCACTATAGGGCACGCGTGGT	NNNNNNN	GCGCT -3′
FP2:	5′-GACTCACTATAGGGCACGCGTGGT	NNNNNNN	CGCGT -3′
FP3:	5′-GACTCACTATAGGGCACGCGTGGT	NNNNNNN	CCGGT -3′
FP4:	5′-GACTCACTATAGGGCACGCGTGGT	NNNNNNN	GGCCT -3′
FP5:	5′-GACTCACTATAGGGCACGCGTGGT	NNNNNNN	CGCGA -3′
FP6:	5′-GACTCACTATAGGGCACGCGTGGT	NNNNNNN	CCGGA -3′
FP7:	5′-GACTCACTATAGGGCACGCGTGGT	NNNNNNN	GCGCA -3′

FIGURE 26.6 Combination of arbitrary primers allows simultaneous detection of retroviral integration sites. Sequences of arbitrary primers with partly degenerate tail (top panel). Individual SF1m transduced HT1080 cells are selected by single-cell deposition. Arbitrary primer PCR with a primer cocktail allows simultaneous detection of different retroviral integration sites in HT1080 cell line clone N3. Arbitrary primer PCR is performed on HT1080 cell line clone N3 using either arbitrary primers FP2 and FP4 alone or FP2/FP4 in a cocktail.

ture (Pruss *et al.*, 1994), and DNA structure of higher order (Katz *et al.*, 1998).

In a study with turkey embryo cellular DNA, all genomic regions contained integration targets for the wild-type avian leukosis virus DNA, with a frequency that varied from approximately 0.2 to 4 times that expected for random integration. Within regions, the frequency of use of specific sites varied considerably. Integrations in some sites occurred 280 times more frequently than expected for random integrations (Withers-Ward *et al.*, 1994). In the mouse model, ecotropic wild-type viral integration (evi) sites have been described that contribute to lymphomagenesis when located upstream of proliferation-inducing genes (Schmidt *et al.*, 1996). In a recent publication it was reported that global analysis of cellular transcription indicated that active genes were preferential integration targets for lentiviral cDNA (Schroder *et al.*, 2002); another study in cell lines showed that retroviral MLV vector integration occurred preferentially at the transcription start region (Wu *et al.*, 2003).

Using a highly sensitive and specific ligation-mediated PCR (LM-PCR) followed by sequencing of retroviral vector integration sites, it has been demonstrated that multiple transduced human PBPC clones mediated engraftment in the bone marrow of immune-deficient mice. 141 proviral inserts were unambiguously mapped to the human genome. Integrations were found in all chromosomes. Vector integrations occurred with significantly increased frequency into chromosomes 17 and 19 (see Fig. 26.7) and into specific regions of chromosomes 6, 13, and 16 although the majority of chro-

mosomes were targeted. These findings revealed for the first time that retroviral vector integration into human marrow-repopulating cells can be nonrandom (P = 0.00037, Laufs *et al.* 2003). On chromosome 17 and chromosome 19, ten and nine integrations resulting from three donors were found, respectively, and three integrations were expected for both chromosomes when the chromosome size and completed human genome sequences (EMBL genome monitoring table, http://www.ebi.ac.uk/genomes/mot/) were taken into account. These differences are highly significant as evidenced by the RISC-score (retroviral integration estimate into chromosome; see Fig. 26.7).

The integration site analysis and the calculated RISC-score should be a valuable addendum for the safety assessment of current preclinical or clinical stem cell gene therapy protocols. A data bank project "retroviral insertion estimate of chromosomal integration" (RISC) containing more than 200 integration sequences has been set up to recognize critical genomic regions and genes involved with possible transforming capacity. Such data collections will allow further safety type studies and monitoring that are needed to restore confidence in clinical trials involving integrating vectors.

Acknowledgments

The technical assistance of Bernhard Berkus, Hans-Jürgen Engel, Sigrid Heil (German Cancer Research Center, Heidelberg, Germany), and the support of the animal

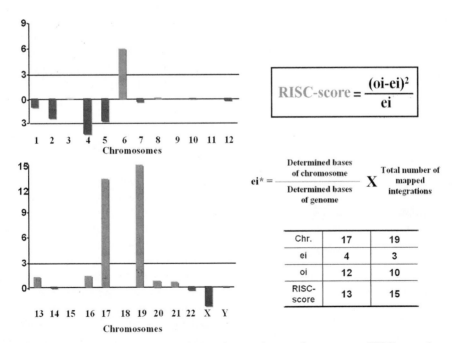

FIGURE 26.7 Schematic distribution of retroviral vector integrations on human chromosomes. RISC-score (retroviral integration estimate into chromosome) is defined from observed integrations (oi) and expected integrations (ei). For calculating the expected integrations, the chromosome size and completed human genome sequences (EMBL genome monitoring table, http://www.ebi.ac.uk/genomes/mot/) are taken into account. In case of nonpreferential integration, a RISC-score of 0 would be expected. For chromosomes 17 and 19 a much higher RISC-score is obtained than for the other chromosomes. *Results based on the human genome sequencing status from 10/24/2002.

facility team of the German Cancer Research Center are gratefully acknowledged. The authors thank Birgit Stähle (EUFETS AG, Idar-Oberstein, Germany) for her help at the start of this project and Christopher Baum (Department of Hematology and Oncology Hannover Medical School, Germany) for providing the retroviral vector. This work was supported in part by grant 10-1294-Ze3 of the Deutsche Krebshilfe/Dr. Mildred-Scheel-Stiftung, by grant M 20.4 of the H.W. & J. Hector-Stiftung.

References

Abonour, R., Williams, D. A., Einhorn, L., Hall, K. M., Chen, J., Coffman, J., Traycoff, C. M., Bank, A., Kato, I., Ward, M., Williams, S. D., Hromas, R., Robertson, M. J., Smith, F. O., Woo, D., Mills, B., Srour, E. F., and Cornetta, K. (2000). Efficient retrovirus-mediated transfer of the multidrug resistance 1 gene into autologous human long-term repopulating hematopoietic stem cells. *Nat. Med.* 6, 652–658.

Barquinero, J., Segovia, J. C., Ramirez, M., Limon, A., Guenechea, G., Puig, T., Briones, J., Garcia, J., and Bueren, J. A. (2000). Efficient transduction of human hematopoietic repopulating cells generating stable engraftment of transgene-expressing cells in NOD/SCID mice. *Blood* 95, 3085–3093.

Baum, C., Hegewisch-Becker, S., Eckert, H. G., Stocking, C., and Ostertag, W. (1995). Novel retroviral vectors for efficient expression of the multidrug resistance (mdr-1) gene in early hematopoietic cells. *J. Virol.* 69, 7541–7547.

Baum, C., Eckert, H. G., Stockschlader, M., Just, U., Hegewisch-Becker, S., Hildinger, M., Uhde, A., John, J., and Ostertag, W. (1996). Improved vectors for hematopoietic stem cell protection and in vivo selection. *J. Hematotherapy* 5, 323–329.

Baum, C., Dullmann, J., Li, Z., Fehse, B., Meyer, J., Williams, D. A., and von Kalle, C. (2003). Side effects of retroviral gene transfer into hematopoietic stem cells. *Blood* 101, 2099–2114.

Bonetta, L. (2002). Leukemia case triggers tighter gene-therapy controls. *Nat. Med.* 8, 1189.

Bordignon, C., and Roncarolo, M. G. (2002). Therapeutic applications for hematopoietic stem cell gene transfer. *Nat. Immunol.* 3, 318–321.

Butler, S. L., Hansen, M. S., and Bushman, F. D. (2001). A quantitative assay for HIV DNA integration in vivo. *Nat. Med.* 7, 631–634.

Cavazzana-Calvo, M., Hacein-Bey, S., de Saint Basile, G., Gross, F., Yvon, E., Nusbaum, P., Selz, F., Hue, C., Certain, S., Casanova, J. L., Bousso, P., Deist, F. L., and Fischer, A. (2000). Gene therapy of human severe combined immunodeficiency (SCID)-X1 disease. *Science* 288, 669–672.

Dick, J. E., Magli, M. C., Huszar, D., Phillips, R. A., and Bernstein, A. (1985). Introduction of a selectable gene into primitive stem cells capable of long-term reconstitution of the hemopoietic system of W/Wv mice. *Cell* 42, 71–79.

Drexler, H. G. (1996). Expression of FLT3 receptor and response to FLT3 ligand by leukemic cells. *Leukemia* 10, 588–599.

Eckert, H. G., Stockschlader, M., Just, U., Hegewisch-Becker, S., Grez, M., Uhde, A., Zander, A., Ostertag, W., and Baum, C. (1996). High-dose multidrug resistance in primary human hematopoietic progenitor cells transduced with optimized retroviral vectors. *Blood* 88, 3407–3415.

Fruehauf, S., Veldwijk, M. R., Berlinghoff, S., Basara, N., Baum, C., Flasshove, M., Hegewisch-Becker, S., Kroger, N., Licht, T., Moritz, T., Hengge, U. R., Zeller, W. J., and Laufs, S. (2002). Gene therapy for sarcoma. *Cells Tissues Organs* 172, 133–144.

Gatlin, J., Melkus, M. W., Padgett, A., Kelly, P. F., and Garcia, J. V. (2001). Engraftment of NOD/SCID mice with human CD34(+) cells transduced by concentrated oncoretroviral vector particles pseudotyped with the feline endogenous retrovirus (RD114) envelope protein. *J. Virol.* 75, 9995–9999.

Gentner, B., Laufs, S., Nagy, K. Z., Zeller, W. J., and Fruehauf, S. (2002). Rapid detection of retroviral vector integration sites in colony-forming human peripheral blood progenitor cells using PCR with arbitrary primers. *Gene Ther.* 10, 789–794.

Grez, M., Becker, S., Saulnier, S., Knoss, H., Ott, M. G., Maurer, A., Dinauer, M. C., Hoelzer, D., Seger, R., and Hossle, J. P. (2000). Gene therapy of chronic granulomatous disease. *Bone Marrow Transplant.* 25 (Suppl 2), S99–104.

Guenechea, G., Gan, O. I., Dorrell, C., and Dick, J. E. (2001). Distinct classes of human stem cells that differ in proliferative and self-renewal potential. *Nat. Immunol.* 2, 75–82.

Hacein-Bey-Abina, S., von Kalle, C., Schmidt, M., Le Deist, F., Wulffraat, N., McIntyre, E., Radford, I., Villeval, J. L., Fraser, C. C., Cavazzana-Calvo, M., and Fischer, A. (2003). A serious adverse event after successful gene therapy for X-linked severe combined immunodeficiency. *N. Engl. J. Med.* 348, 255–256.

Halene, S., and Kohn, D. (2000). Gene therapy using hematopoietic stem cells: Sisyphus approaches the crest. *Hum. Gene Ther.* 11, 1259–1267.

Hanenberg, H., Xiao, X. L., Dilloo, D., Hashino, K., Kato, I., and Williams, D. A. (1996). Colocalization of retrovirus and target cells on specific fibronectin fragments increases genetic transduction of mammalian cells. *Nat. Med.* 2, 876–882.

Hildinger, M., Eckert, H. G., Schilz, A. J., John, J., Ostertag, W., and Baum, C. (1998). FMEV vectors: both retroviral long terminal repeat and leader are important for high expression in transduced hematopoietic cells. *Hum. Gene Ther.* 5, 1575–1579.

Hildinger, M., Schilz, A., Eckert, H. G., Bohn, W., Fehse, B., Zander, A., Ostertag, W., and Baum, C. (1999). Bicistronic retroviral vectors for combining myeloprotection with cell-surface marking. *Gene Ther.* 6, 1222–1230.

Holmes-Son, M. L., Appa, R. S., and Chow, S. A. (2001). Molecular genetics and target site specificity of retroviral integration. *Adv. Genet.* 43, 33–69.

Jordan, C. T., and Lemischka, I. R. (1990). Clonal and systemic analysis of hematopoiesis in the mouse. *Genes Dev.* 4, 220–232.

Katz, R. A., Gravuer, K., and Skalka, A. M. (1998). A preferred target DNA structure for retroviral integrase in vitro. *J. Biol. Chem.* 273, 24190–24195.

Kelly, P. F., Vandergriff, J., Nathwani, A., Nienhuis, A. W., and Vanin, E. F. (2000). Highly efficient gene transfer into cord blood nonobese diabetic/severe combined immunodeficiency repopulating cells by oncoretroviral vector particles pseudotyped with the feline endogenous retrovirus (RD114) envelope protein. *Blood* 96, 1206–1214.

Kiem, H. P., Andrews, R. G., Morris, J., Peterson, L., Heyward, S., Allen, J. M., Rasko, J. E., Potter, J., and Miller, A. D. (1998). Improved gene transfer into baboon marrow repopulating cells using recombinant human fibronectin fragment CH-296 in combination with interleukin-6, stem cell factor, FLT-3 ligand, and megakaryocyte growth and development factor. *Blood* 92, 1878–1886.

Kim, H. J., Tisdale, J. F., Wu, T., Takatoku, M., Sellers, S. E., Zickler, P., Metzger, M. E., Agricola, B. A., Malley, J. D., Kato, I., Donahue, R. E., Brown, K. E., and Dunbar, C. E. (2000). Many multipotential gene-marked progenitor or stem cell clones contribute to hematopoiesis in nonhuman primates. *Blood* 96, 1–8.

Kitamura, Y., Lee, Y. M., and Coffin, J. M. (1992). Nonrandom integration of retroviral DNA in vitro: effect of CpG methylation. *Proc. Natl. Acad. Sci. USA* 89, 5532–5536.

Kohn, D. B. (2002). Adenosine deaminase gene therapy protocol revisited. *Mol. Ther.* 5, 96–97.

Laskus, T., Radkowski, M., Wang, L. F., Nowicki, M., and Rakela, J. (1999). Detection and sequence analysis of hepatitis B virus integration in peripheral blood mononuclear cells. *J. Virol.* 73, 1235–1238.

Laufs, S., Gentner, B., Nagy, K. Z., Jauch, A., Benner, A., Naundorf, S., Kuehlcke, K., Schiedlmeier, B., Ho, A. D., Zeller, W. J., and Fruehauf, S. (2003). Retroviral vector integration occurs into preferred genomic targets of human hematopoietic stem cells with long-term bone marrow repopulating ability. *Blood* 101, 2191–2198.

Leclercq, I., Mortreux, F., Cavrois, M., Leroy, A., Gessain, A., Wain-Hobson, S., and Wattel, E. (2000). Host sequences flanking the human T-cell leukemia virus type 1 provirus in vivo. *J. Virol.* 74, 2305–2312.

Lichter, P., and Cremer, T. (1992). In *Human Genetics—A Practical Approach*, D. E. Rooney, B. H. Czepulkowski BH, eds. Chromosome analysis by non-isotopic in situ hybridization, pp. 157–192. IRL Press, Oxford, United Kingdom.

Marshall, E. (2003). Second child in French trial is found to have leukemia. *Science* 299, 320.

Mielke, C., Maass, K., Tummler, M., and Bode, J. (1996). Anatomy of highly expressing chromosomal sites targeted by retroviral vectors. *Biochemistry* 35, 2239–2252.

Moritz, T., Dutt, P., Xiao, X., Carstanjen, D., Vik, T., Hanenberg, H., and Williams, D. A. (1996). Fibronectin improves transduction of reconstituting hematopoietic stem cells by retroviral vectors: evidence of direct viral binding to chymotryptic carboxy-terminal fragments. *Blood* 88, 855–862.

Moritz, T., and Williams, D. A. (1998). In *Hematopoietic Cell Transplantation*, E. D. Thomas, K. Blume, S. J. Forman, eds. Methods of gene transfer: genetic manipulation of hematopoietic stem cells, pp. 79–86. Blackwell Science, Malden, MA.

Movassagh, M., Desmyter, C., Baillou, C., Chapel-Fernandes, S., Guigon, M., Klatzmann, D., and Lemoine, F. M. (1998). High-level gene transfer to cord blood progenitors using gibbon ape leukemia virus pseudotype retroviral vectors and an improved clinically applicable protocol. *Hum. Gene Ther.* 9, 225–234.

Müller, S., Neusser, M., and Wienberg, J. (2002). Towards unlimited colors for fluorescence in situ hybridization (FISH). *Chromosome Res.* 10, 223–232.

Nolta, J. A., Dao, M. A., Wells, S., Smogorzewska, E. M., and Kohn, D. B. (1996). Transduction of pluripotent human hematopoietic stem cells demonstrated by clonal analysis after engraftment in immune-deficient mice. *Proc. Natl. Acad. Sci. USA* 93, 2414–2419.

Petzer, A. L., Hogge, D. E., Landsdorp, P. M., Reid, D. S., and Eaves, C. J. (1996). Self-renewal of primitive human hematopoietic cells (long-term-culture-initiating cells) in vitro and their expansion in defined medium. *Proc. Natl. Acad. Sci. USA* 93, 1470–1474.

Porter, C. D., Collins, M. K., Tailor, C. S., Parkar, M. H., Cosset, F. L., Weiss, R. A., and Takeuchi, Y. (1996). Comparison of efficiency of infection of human gene therapy target cells via four different retroviral receptors. *Hum. Gene Ther.* 7, 913–919.

Pruss, D., Bushman, F. D., and Wolffe, A. P. (1994). Human immunodeficiency virus integrase directs integration to sites of severe DNA distortion within the nucleosome core. *Proc. Natl. Acad. Sci. USA* 91, 5913–5917.

Rosenthal, A., and Jones, D. S. (1990). Genomic walking and sequencing by oligo-cassette mediated polymerase chain reaction. *Nucleic Acids Res.* 18, 3095–3096.

Sanders, D. A. (2002). No false start for novel pseudotyped vectors. *Curr. Opin. Biotechnol.* 13, 437–442.

Schiedlmeier, B., Kuehlcke, K., Eckert, H. G., Baum, C., Zeller, W. J., and Fruehauf, S. (2000). Quantitative assessment of retroviral transfer of the human multidrug resistance 1 gene to human mobilized peripheral blood progenitor cells engrafted in nonobese diabetic/severe combined immunodeficient mice. *Blood* 95, 1237–1248.

Schmidt, T., Zornig, M., Beneke, R., and Moroy, T. (1996). MoMuLV proviral integrations identified by Sup-F selection in tumors from infected myc/pim bitransgenic mice correlate with activation of the gfi-1 gene. *Nucleic Acids Res.* 24, 2528–2534.

Schmidt, M., Hoffmann, G., Wissler, M., Lemke, N., Muessig, A., Glimm, H., Williams, D. A., Ragg, S., Hesemann, C. U., and von Kalle, C. (2001). Detection and direct genomic sequencing of multiple rare unknown flanking DNA in highly complex samples. *Hum. Gene Ther.* 12, 743–749.

Schroder, A. R., Shinn, P., Chen, H., Berry, C., Ecker, J. R., and Bushman, F. (2002). HIV-1 integration in the human genome favors active genes and local hotspots. *Cell* 110, 521–529.

Shah, A. J., Smogorzewska, E. M., Hannum, C., and Crooks, G. M. (1996). Flt3 ligand induces proliferation of quiescent human bone marrow CD34+CD38-cells and maintains progenitor cells in vitro. *Blood* 87, 3563–3570.

Shibagaki, Y., and Chow, S. A. (1997). Central core domain of retroviral integrase is responsible for target site selection. *J. Biol. Chem.* 272, 8361–8369.

Silver, J., and Keerikatte, V. (1989). Novel use of polymerase chain reaction to amplify cellular DNA adjacent to an integrated provirus. *J. Virol.* 63, 1924–1928.

Sørensen, A. B., Duch, M., Jorgensen, P., and Pedersen, F. S. (1993). Amplification and sequence analysis of DNA flanking integrated proviruses by a simple two-step polymerase chain reaction method. *J. Virol.* 67, 7118–7124.

Sørensen, A. B., Duch, M., Amtoft, H. W., Jorgensen, P., and Pedersen, F. S. (1996). Sequence tags of provirus integration sites in DNAs of tumors induced by the murine retrovirus SL3-3. *J. Virol.* 70, 4063–4070.

Speicher, M. R., Gwyn Ballard S., and Ward D. C. (1996). Karyotyping human chromosomes by combinatorial multi-fluor FISH. *Nat. Genet.* 12, 368–375.

Telenius, H.. Carter, N. P., Bebb, C. E., Nordenskjold, M., Ponder, B. A., and Tunnacliffe, A. (1992). Degenerate oligonucleotide-primed PCR: general amplification of target DNA by a single degenerate primer. *Genomics* 13, 718–725.

Vijaya, S., Steffen, D. L., and Robinson, H. L. (1986). Acceptor sites for retroviral integrations map near DNase I-hypersensitive sites in chromatin. *J. Virol.* 60, 683–692.

Weidhaas, J. B., Angelichio, E. L., Fenner, S., and Coffin, J. M. (2000). Relationship between retroviral DNA integration and gene expression. *J. Virol.* 74, 8382–8389.

Withers-Ward, E. S., Kitamura, Y., Barnes, J. P., and Coffin, J. M. (1994). Distribution of targets for avian retrovirus DNA integration in vivo. *Genes Dev.* 8, 1473–1487.

Wu, X., Li, Y., Crise, B., and Burgess, S. M. (2003). Transcription start regions in the human genome are favored targets for MLV integration. *Science* 300, 1749–1751.

CHAPTER 27

Automated DNA Hybridization and Detection

BERT GOLD

Human Genetics Section, Laboratory of Genomic Diversity, National Cancer Institute at Frederick, Frederick, MD, USA

TABLE OF CONTENTS

27.1 INTRODUCTION

The accuracy and precision of molecular identification has improved considerably as a consequence of the application of robots to the tasks of nucleic acid hybridization. Automation has enabled higher throughput molecular assays and diagnostics. Inspiration for automating these processes has come from the successful effort to map and sequence the human genome in the past ten years. Specific tasks that have been automated have been manually chained together so that nucleic acid extraction, amplification, hybridization, and detection are dovetailed. In a clinical molecular diagnostics laboratory, these processes are monitored, interpreted, and reported by skilled technologists and diagnosticians. In a research environment, chaining these processes has permitted high throughput SNP characterization and voluminous data production.

Several instrument and reagent manufacturers have participated in the automation initiative. The short-term future should witness increasing integration and a diminished manual component for dovetailing automated, high throughput subtasks. As laboratory throughput has grown, the need for biological information handling has grown. As a consequence, database management and result reporting and analysis systems are poised for a period of rapid growth. Since the last broad review of this topic in 1995 (Beck, 1995), the human genome initiative has permitted great strides in automating molecular hybridization and detection technology. Although nonradioactive detection was then a novelty, it has now become commonplace. And, whereas clinical molecular diagnosis was then an exception, it is now the accepted US standard for Cystic Fibrosis (CF) Carrier screening. In fact, American College of Medical Genetics/American College of Obstretrics and Gynecology guidelines promulgate prenatal carrier screening of every Caucasian couple of childbearing age (Grody *et al.*, 2001).

Automated hybridization now encompasses nucleic acid extraction, microarray spotting, colony and plaque hybridization, picking, and many parts of DNA sequencing, primer extension, and single nucleotide polymorphism (SNP) detection. The aforementioned earlier review also subsumed fluorescent *in situ* hybridization (FISH) applications under its rubric, but that topic, comparative genomic hybridization (CGH), and array CGH are beyond the scope of this chapter.

27.2 DNA HYBRIDIZATION

27.2.1 Origins of DNA Hybridization

DNA hybridization was anticipated by Chargaff's rule and enumerated in Watson and Crick's 1953 landmark publications. Spiegelman's contribution to the field was recognizing that it could be used to gauge phage similarity and quantify viruses (Gillespie and Spiegelman, 1965). During the same period, Ron Davis, a graduate student in Norman Davidson's laboratory at the time, developed the heteroduplex method for visualizing deletions in phage genomes with the electron microscope (Davis and Davidson, 1968). Roy Britten, Barbara Hough-Evans, and Eric Davidson began experiments aimed at investigating sequence diversity through examination of hybridization kinetics (Davidson and Hough, 1969). Of course, all these were wet bench experiments carried out with manually purified and quantitated (often sheared) nucleic acids in laboratory glassware or quartz cuvettes.

27.2.2 Southern Blot Hybridization

Ed Southern invented his namesake southern blot in 1975 (Southern, 1975). The original method required the purchase of a qualified batch of nitrocellulose membrane; an explosive, brittle solid support that occasionally dehydrates and loses wettability characteristics required for nucleic acid transfer. The blotting method, as originally described, required osmotic wicking to provide a flow of high-salt transfer solution through the gel and directly onto a properly wet membrane. After transfer, blots were washed in a lower salt buffer and then dried in a vacuum oven. Radioactive probing in the first 10 years of Southern blotting was standard operating procedure and only in the past ten years has given way to nonisotopic techniques (Gold *et al.*, 2000). The Southern blot provided a cumbersome yet reliable method for molecular diagnosis. It permitted direct interrogation of human genomic sequence through cleavage pattern identification using an appropriate restriction endonuclease. Unfortunately, it also often required the use of radioactivity and copious amounts of human genomic DNA.

From 1986, when Lander and Botstein first suggested that restriction fragment length polymorphism could be used for gene mapping (Lander and Botstein, 1986), through the early 1990s, Southern blot characterization of restriction fragment length polymorphism became a central method for gene mapping.

DNA from families affected with genetic disease was collected, digested with various inexpensive lots of restriction endonuclease, and sized by electrophoresis. The Oncor ProbeTech, shown in Fig. 27.1a, was helpful for this procedure, and was purchased by a number of laboratories to help automate Southern blotting. The device, which is no longer in production, consisted of an electrophoresis chamber with pumps for recirculating buffer, and a safety lock to prevent accidental shock to the operator. Recirculation is a requirement to minimize Joule heating and provide high resolution Southern blotting. Without buffer recirculation, ion polarization during electrophoresis distorts the band pattern made by near saturating, high-molecular weight DNA as it runs through agarose pores (Current Protocols in Human Genetics, 2000).

In addition to the recirculation pumps, the device shown in Fig. 27.1 contained convenient gel trays that permitted denaturation and neutralization steps, both preblotting requirements of the Southern blot protocol to be accom-

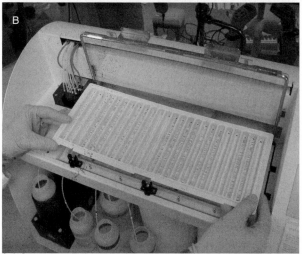

FIGURE 27.1 Oncor Probetech and Tecan ProfiBlot. A. The Oncor probetech was an automated southern blot device that is no longer in production. It was used routinely to produce Southern blots for RFLP studies intended for gene mapping through the 1990s. It was also used extensively for Southern blot hybridization required for characterization of Fragile X syndrome diagnostic confirmation and carrier testing. B. The Tecan ProfiBlot is an automated RDB developing device that provides more reproducible LiPA results. Latter photo courtesy of Roche Molecular Systems.

plished efficiently. The company sold a line of pretested buffers that could be used off the shelf, or diluted for electrophoresis, denaturation, and neutralization. Also, a charged nylon membrane supplied by the company was precut to fit the gel chamber and a vacuum pump built into the device provided suction to transfer sized, genomic DNA from the gel to the membrane. Although production of these devices ceased in the middle 1990s, they were workhorses in a number of molecular diagnostics and gene mapping laboratories, and provide an exemplary first effort at automation.

27.2.3 From Southern Blot to PCR

During the 1990s, as molecular genetic diagnosis came of age, polymerase chain reaction-restriction fragments length polymorphism (PCR-RFLP) analysis became the gold standard for molecular mutation screening. PCR-RFLP was believed by some clinicians to provide greater analytic specificity than spot hybridization techniques (Gold, 2001; 2003). This is because two levels of specificity were required for its proper function. In Southern blotting, both hybridization and sizing form the basis for unambiguous identification of a molecular fragment. Similarly, in PCR-RFLP, both accurate amplification and proper sizing of an amplified fragment are required for unambiguous molecular identification. However, PCR-RFLP is not practical for high mutation spectrum disorders (hereditary disorders with hundreds of known disease-bearing alleles such as CF or β-thalassemia). Direct DNA sequencing would be the most thorough mutation detection approach, and is being tried in some laboratories (Groman *et al.*, 2002; Strom *et al.*, 2003), but it is arduous, capital intensive, time-consuming to analyze, and requires relatively expensive supplies.

Rapid reliable molecular diagnosis of high mutation spectrum disorders requires a unique approach. Because the majority of these hereditary illnesses occur in families with no prior history, a high-sensitivity screening test to identify at-risk carriers is required. Molecular carrier screening tests must be cost-effective to be practical. Commercialization of molecular diagnostics for high mutation spectrum disorders involves special considerations. Among these are low cost, high throughput, and rapid interpretation. Automated printing and processing of standardized reverse dot blots (see Fig. 27.1b) in conjunction with robust, optimized multiplex PCR reagents provides a practical commercial mutation detection method. Technical factors, such as methods of attachment of oligonucleotide probes to solid supports, the nature of the solid supports themselves, and PCR product visualization techniques play an important role in formatting high throughput reverse dot blot diagnostics. Reverse dot blots (RDB) have become popular for clinical molecular disease diagnosis recently (see also Chapter 2); however, newer chip-based techniques and homogeneous assay systems are likely to supplant the reverse dot blot in the not-too-distant future.

27.3 DNA EXTRACTION

In the past five years, three automated platforms for DNA extraction have emerged as field leaders. These are the Roche MagnaPure system, the Qiagen BioRobot 9604, and the Gentra Autopure LS system (pictured in Figs. 27.2a–d). Unique specifications and requirements describe each of these systems. And yet, each automatic method provides sufficient pure genomic DNA product from peripheral blood lymphocytes to perform conventional PCR and restriction endonuclease cleavage for southern blotting.

The extraction principle underlying the Roche Magna-Pure system (see Fig. 27.2a, b) is that glass beads are coated with a magnetic layer held inside reaction tips. Bead bound nucleic acids are cycled in reaction tips through a series of processing steps. Once the sample is loaded, there is no operator intervention until extraction is complete. Roche sells two MagnaPure Systems, of these the larger takes 90 minutes to purify 32 samples with a per sample expendable cost of approximately $2. Typical yield is 10 µg from 200 µL of peripheral blood collected in EDTA anticoagulant. High salt nucleic acid adherent hydrated silica gel tubes provide the basis for Qiagen's nucleic acid extraction technology. Subsequent to lysis, gel adherent cell nucleic acids are extracted by high salt and alcohol buffer washed, dried, and eluted in a low salt buffer. Unfortunately, a final centrifugation step requires pulling the samples off the robot, as the centrifuge has not been interfaced. Qiagen's BioRobot 9604 (see Fig. 27.2c) takes approximately two hours to extract 96 samples, with yields in the range of 4–6 µg from 200 µL of whole blood. New to the marketplace is Qiagen's MDX gantry robot that uses a vacuum manifold to enable faster purification and standardize processing. Gentra System's Puregene kits, which are based on a salting out procedure, have long provided nucleic acid substrate for molecular diagnostic assays. Gentra's Autopure LS instrument (see Fig. 27.2d) provides 35 µg per mL of whole human peripheral blood DNA, with more than 70% to 90% between 100 and 200 kb. The DNA is also consistent with long-term storage characteristics (up to ten years) that have been discussed in the literature (Farkas *et al.*, 1996). Sample sizes up to 10 mL of whole blood are accommodated by the instrument, with a throughput of eight samples in less than one hour, 16 samples in 80 minutes, and 96 samples per 8-hour shift. The instrument accommodates 16 samples at a time.

27.4 QUANTIFYING DNA

Diverging views characterize methods for genomic DNA quantification in preparation for PCR or restriction endonuclease cleavage assays. On the one hand, the optical density at wavelength maxima for the macromolecule (OD 260) compared with the wavelength maxima for phenylalanine and tyrosine, representing protein presence (OD 280) customarily have been compared in a ratio to estimate nucleic acid purity (OD 260/OD 280). Early workers charted ratios to provide rigor to purity judgments. In the current milieu, this has mean using high throughput spectrophotometers, such as Molecular Devices Spectra Max micro-well plate reader to provide wavelength scans for 96 to 384 samples simultaneously (Hilbert *et al.*, 2000). This device rigged into a Beckman FX liquid handling robot is shown in Fig. 27.2e.

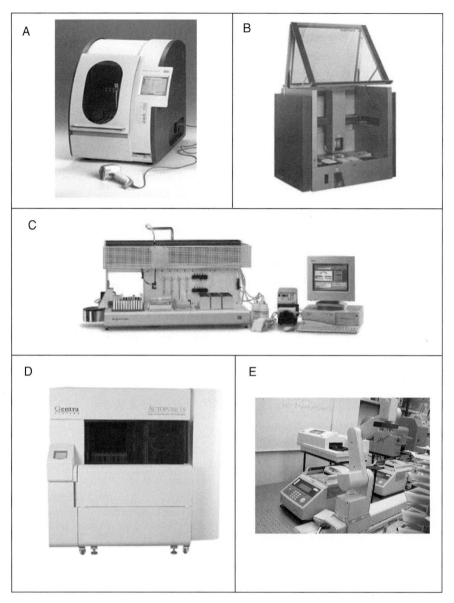

FIGURE 27.2 DNA extraction and one spectrophotometric quantitation system. A. The MagNA Pure Compact System is a small robot intended to automate nucleic acid extraction in low throughput laboratories. The system can simultaneously extract up to eight samples using the same principles as the larger Roche system. B. MagNA Pure LC instrument, which is a table-top robot capable of 32 simultaneous nucleic acid isolations. The system uses silica-coated magnetic core beads to extract nucleic acid in disposable reaction tips. Operating costs and details are discussed in the text. C. Qiagen's BioRobot 9604 uses adherence to a silica bead in high salt conditions to provide a basis for its robotic extraction method. Although a manual centrifugation step is required for DNA purification, a newly designed Qiagen MDX robot (not shown) has an on-board vacuum system that supplants the need for centrifugation. D. Gentra Systems Autopure LS uses a well-characterized salting out method to provide high DNA yield and purity on 16 simultaneous peripheral blood samples with walk-away operation. E. A Molecular Devices SpectraMax Plus 384 micro-well plate scanning spectrophotometer can use specially engineered 384-well quartz cuvette plates to provide accurate optical densities at 230, 260, and 280 nanometers. Here, the spectrophotometer is shown integrated into a Beckman FX based sequencing pipeline constructed by David Munroe and his collaborators at the National Cancer Institute in Frederick, Maryland.

On the other hand, numerous labs (CIDR, Myriad Genetic Laboratories), report more accurate genomic DNA measures using fluorescent quantitation with the intercalating dye Pico Green or one of the Hoechst stains (Hoechst 33258 and its relatives, Hopwood *et al.*, 1997). These are read on a fluorimeter (such as the Molecular Devices Gemini series or Tecan Genios) in 96-well or 384-well format and compared with a standard dilution series of DNAs, compiled on the same instrument at the Emax of the intercalating dye.

27.5 ROBOTICS

In his chapter, Stephan Beck points to several breadboards as providing initial hybridization robotics (Beck, 1995). Among these were Beck's own DNA sequencing device; his hybridization device (Alderton *et al.*, 1994), and George Church's genomic sequencing machine (Church and Gilbert, 1984; Church and Gilbert, 1985). However, the first effort at using off-the-shelf liquid handling technology to successfully automate an ongoing laboratory operation was the application of the Beckman Biomek 1000 to RDB in the Department of Human Genetics of the Baylor College of Medicine (DeMarchi *et al.*, 1994). There, a team began robotically spotting successive nylon membranes with amplification products and then hybridizing these with a variety of probes. It was through this effort that robotization was first attempted for forward dot-blots. It was not until 2001 that a robot procedure was published for spotting reverse dot blots (Lappin *et al.*, 2001). Commercialization and the desire to retain competitive advantage means most automated procedures remain unpublished in detail. In fact, no central repository for automated laboratory robotic codes or procedures has yet been established.

27.5.1 A Revolution in Liquid Handling Robotics

One important force propelling molecular hybridization technology forward is the revolution witnessed in liquid handling technology. Beckman-Coulter's development of a series of disposable tip liquid handling robots (Multimek, Biomek 1000 and 2000, and Beckman FX) has permitted the automation of a variety of recalcitrant laboratory tasks. Nearly simultaneously, companies like Tecan and Packard developed a series of gantry robots using continuous pipetting technology to allow relatively inexpensive liquids (because the robots have significant void volumes) that have long shelf lives to be accurately dispensed. Hamilton Corporation, a micro-syringe manufacturer, built a series of robots that provide small volume accuracy based on positive pressure displacement, and can be programmed for bleach or distilled water wash between solution additions.

This noncomprehensive description of current liquid handling technology has made possible a variety of automated solutions to blotting tasks. However, there is a hidden cost: there has been little standardization of the software required to operate these robots even though some computer engineers have advocated the use of Tcl and and Tk to the task (Ousterhout, 1994). Unfortunately, the use of proprietary robot controller software has flourished, and therefore a true automated laboratory requires an engineering team's support. As a consequence, engineering costs may cause automation to be too expensive for laboratories that would otherwise use it.

27.6 REVERSE DOT BLOT

Details of the use of a liquid handling laboratory robot for printing of RDBs was published by Lappin and colleagues (2001). RDBs most often are arrayed so that the normal oligonucleotide probes sit on a line adjacent to one another while variant probes are not on the same line. This permits rapid interpretation of heterozygotes, where one normal and one variant allele are indicated (see also Chapter 2). However, where two mutations are close to each other and fall within the sequence of the oligonucleotide probe (a situation often observed with respect to hemoglobin Hb A, Hb S, and Hb C), a different pattern is obtained. DNA from individuals homozygous for Hb S that have sickle cell disease will not hybridize to the normal probe at either the S or C position of the blot. Instead, such DNA hybridizes to both the mutant S and mutant C probes.

Similar RDB results are observed in the case of variants neighboring the ΔF508, such as ΔI507 or I506V. Confounding neighboring variants are also present in β-thalassemia diagnosis where IVS I-1 (G→T) and IVS I-6 (T→C) mutations, five nucleotides apart, can each contribute to disease. Nevertheless, even though RDBs possess limitations, they provide a widely used and versatile mutation detection platform.

In early 2000, the American College of Medical Genetics and the American College of Obstetrics and Gynecology recommended CF carrier testing for all US Caucasian couples that anticipate children (Grody *et al.*, 2001). Since that time, two CF line blot assays that provide mutation detection have become popular methods in US Molecular Diagnostics laboratories. The CF LAp, offered by Roche Diagnostics (Indianapolis, IN, USA) and Innogenetics (Gent, Belgium) line probe assay (LiPA), have been well received by molecular diagnostics laboratories. Recently, a Johns Hopkins group has evaluated a 58 allele line probe assay for CF, and discussed its analytic utility in a variety of diagnostic situations (Wang *et al.*, 2002). Limitations, such as individuals who are compound heterozygote at closely spaced loci, occasionally fail to signal the presence of one or the other allele because of interference. As mentioned earlier, it is one characteristic of all sequence specific assays that nucleotide variants within the probed region (usually approximately 17 nucleotides, although some new chemistries permit probes approximately 13 nucleotides and still maintain specificity) affect test accuracy (Current Protocols in Human Genetics, 2000). An example of this kind of interaction is detection of an S549/R553X heterozygote by Wang and colleagues (2002). This failed to hybridize with the G551D wild type probe encompassed by the probe for R553X. Another example is a ΔF508 mutation/I506V polymorphism heterozygote that failed to hybridize against the normal sequence in the region. Confounding results should notify an alert diagnostician that a given test uncertainty bears further investigation.

As mentioned earlier, RDBs are versatile and might be printed for the detection of a variety of genetic lesions. Efforts to commercialize RDB hybridization for CF has motivated many laboratories to engage in home-brew testing via reverse allele-specific oligonucleotides for a variety of hereditary illness and infectious disease typing. In spite of certain limitations, this technology is useful for the reproducible detection of any point mutation or small deletion or insertion in any amplified genomic sample. The general utility of RDBs might allow mechanized production of screening strips for aldolase B mutations, causing hereditary fructose intolerance (Lau and Tolan, 1999), nondeletion α-thalassemia (Chan *et al.*, 1999), or adult onset mitochondrial disorders such as Leber hereditary optic neuropathy (Schollen *et al.*, 1997) or even Hepatitis A contamination in food (Jean *et al.*, 2001). A recently published RDB assay for congenital adrenal hyperplasia has high clinical utility in identifying sexually ambiguous newborns (Yang *et al.*, 2001). As discussed earlier, the RDB is flexible, inexpensive to implement, and uses off-the-shelf commercially available hardware, reagents, and software.

27.6.1 Commercialization of the Reverse Dot Blot

Commercialization of RDB strips required significant effort in ensuring their reproducibility and uniformity. During the period of scale-up at Roche Molecular Systems, it was found that dot-blots themselves often gave equivocal results in the form of a halo around the dots. This had more to do with surface tension considerations when applying amino-conjugated or BSA-conjugated oligonucleotides (printing) than it had to do with any other single consideration. As a consequence, a decision was made to convert the dots to lines of uniform width. Automation for printing the line blots was first developed at Ismeca USA, Inc. (Carlsbad, CA, USA), which never found an adequate market to maintain production of their Bio-Line Dispenser. Commercial RDB printing equipment is currently available from Bio-Dot, Inc. (Irvine, CA, USA). Roche developed a Linear Striper and a Multi-dispenseR2000 Controller (IVEK, N Springfield, VT, USA) to coat nylon-backed sheets. Stephen Lappin, Jeffrey Cahlik, and the author described the techniques used for RDB automated printing at Quest diagnostics during 2001 (Lappin *et al.*, 2001).

Extensive improvements in reproducibility and manufacture accompanied the commercialization process at Roche Molecular Systems and production of strips at Quest. Incubation temperature, temperature equilibration, and uniformity in conjugate and substrate distribution are known to be critical variables for color or luminescence detection. Both Tecan (Maennedorf, Switzerland) and Dynal Biotech (Oslo, Norway) manufacture specially engineered incubators and chemical dispenser devices that automate part of the RDB hybridization and developing process (Tecan's ProfiBlot is

pictured in Fig. 27.1b). RDB assays generally are limited by an inability to detect large or quantitative deletions and an inability to characterize all but modestly expanded repeat sequences (ascertainment of the exact size of an expansion is often required for accurate molecular diagnosis or binning), but these strips can provide a means of accurate and reproducible genotype assignments. Automated spotting or line blotting of RDB strips allows the printing of large numbers of these with a minimum of operator intervention. Automation also permits higher density arraying. Optimum conditions for nucleotide hybridization and development of the RDB strips were devised and automated through the use of Tecan's Profiblot (see Fig. 27.1b) and Dynal's AutoReli. These shaking baths give uniformity to the strip development that is not easily achieved in any other fashion.

27.7 5′ NUCLEOTIDASE (TAQMAN) ASSAYS

David Gelfand and his colleagues invented the 5′ nucleotidase assay on the basis of his observation that Taq polymerase retained some 5′ endonuclease activity even when its $3' \rightarrow 5'$ proofreading activity was blocked, as in Amplitaq Gold (Holland *et al.*, 1991; see also Chapter 10). These researchers took advantage of this activity to modify PCR so that amplification, conducted in the presence of a probe labeled with a label on its 5′ end, permits visualization of a perfectly hybridized sequence. At first, the Gelfand group used a radioactive label, but colleagues at Applied Biosystems and Genentech modified a Fluorescent Energy Transfer method, dual labeling oligonucleotide probes with a 3′ quencher and a 5′ fluorescent reporter to detect perfectly hybridized oligomers (Heid *et al.*, 1996). On a fluorescent energy transfer probe, the fluorescent moieties on the probe differ in their excitation and emission spectra, such that one quenches the other when in close proximity. The fluorescent molecule selected as the quencher remains attached to the oligonucleotide probe before, during, and after the assay. The reporter is liberated during the PCR extension and allows the measurement of increasing fluorescence throughout the assay. By conducting the PCR in optically clear plastic tubes irradiating these during the assay at the excitation wavelength of the reporter, the increasing fluorescence of a positive sample can be visualized during (or shortly after) the test.

A small modification of the assay allows two alleles to be discerned in the test sample. Instead of using only one probe, with one fluorescent and one quencher molecule, two differently labeled oligonucleotides—one with fluorescent moiety A and a one base pair difference with an oligonucleotide with an alternative fluorescent moiety B—can be used simultaneously in the same tube (Morin *et al.*, 1999). One of these detects the mutation and the other detects the normal gene sequence. Through application of this strategy, it is possible to visualize the presence of normal, homozygous mutant, or heterozygous samples in a single tube. Drawbacks of the technique include the difficulty of optimizing assay conditions, the expense of the complicated synthesis of TaqMan probes, and the cost of readers required for visualizing the results. In spite of these disadvantages, this method has been adopted in several high-throughput genotyping facilities.

27.8 CAPILLARY THERMAL CYCLER

The Light-cycler is a circulating hot air thermal cycler equipped with 32 experimental positions that can be excited and monitored in real-time. Thermal cycling is very rapid, with a standard 30-cycle reaction completing in less than half an hour (Wittwer *et al.*, 1989; 1990; Wittwer and Garling, 1991; Wittwer *et al.*, 1997; Ririe *et al.*, 1997). For homogeneous analysis using the 5′ nucleotidase technology, the dynamic range of the laser-exciter diode detection circuit may not be great enough to detect reactions easily quantitated in the TaqMan instruments. However, until recently, the LightCycler was unique in that it permitted production of a melting curve of hybridized fluorescent labeled probe annealed to amplified genomic DNA. Thus, until recently, alleles not easily visualized using 5′ nucleotidase reactions could be resolved on the LightCycler. Recently, the 5′ nucleotidase platforms (ABI 7700; ABI 7900) have permitted visualization of melt curves using a 5′ GC-clamp strategy, with three probes Stoffel Fragment and SybrGreen in the reaction mixes. This permits visualization of alleles through synthesis of relatively inexpensive oligonucleotide primers and a melting strategy.

27.9 ELECTRONIC HYBRIDIZATION

Newer technologies have begun to make their way into the hybridization automation marketplace. Nanogen (San Diego, CA) has recently developed a CF assay on their proprietary, silicon chip-based platform (Sosnowski *et al.*, 1997; Fig. 27.3). The chip consists of a silicon microchip with 100 electronically addressable sites (see Fig. 27.3c). A hydrogel layer overlying the chip is embedded with streptavidin. Electronic addressing is used to concentrate DNA to a particular test site(s) on the microarray, enabling sequential addition of biotinylated oligonucleotides, in the capture down mode, or biotinylated samples, in the amplicon down mode, to different test sites (see Fig. 27.3b). Interrogation and genotyping is accomplished by hybridization of fluorescently labeled reporter probes. Laser excitation and a charge-coupled device (CCD) camera are used to monitor hybridization. The equipment requires a sizable capital investment in specialized hardware and relatively expensive cartridges; however, Nanogen provides ways to make efficient use of each chip.

Unlike other microarrays, Nanogen's microarray can be used for analysis of multiple patient samples. Another unique feature is that, although each test site can be used only once, unused sites on a given chip can be saved for later use. The Nanogen technology is a huge forward leap in automation of DNA hybridization because it permits electronic alteration of hybridization specificity. In addition, the machine provides an environment to explore the effects of a myriad of conjugates, derivatives, and alternative nucleotides on nucleic acid hybridization. Whether the newly developed NanoChip® CF test (Nanogen) is commercially successful or not, it is part of a trend toward more reliable, efficiently automated, hybridization assays, mass produced and capable of technical adjustment at the performance site. The Nanogen instrument has several advantages over conventional laboratory hybridization, not the least of which is that the hybridization condition can be precisely adjusted through the use of an electrical field that runs through the unit.

27.10 HYBRIDIZATION ARRAYS

Affymetrix and its HapMap partner, Perlegen, have been photolithographically attaching oligonucleotides to quartz, enabling the production of dense gene interrogation microarrays for the past ten years (Pease *et al.*, 1994). Quartz is used because of its natural hydroxylation properties. To begin the process, quartz is washed to provide a uniform surface and bathed in silane, which reacts with the hydroxyl groups to form a matrix of covalent linker molecules. The intramolecular distance between silanes determines the probe's packing density. Affymetrix and Perlegen have worked out proprietary methods to permit synthesis of arrays holding over 500,000 probe locations (which they call features) within a 1.28 cm square coverslip. Linker molecules, attached to the silane matrix, provide a surface that may be spatially activated by light. Probe synthesis occurs in parallel, permitting simultaneous nucleotide addition of A, C, T, or G nucleotide to features. Definition of which oligonucleotide chains will receive a nucleotide in each step is accomplished through the use of photolithographic masks, each with 18 to 20 μ^2 windows corresponding to the dimensions of individual features. The use of ultraviolet light to irradiate the mask in the first step of synthesis permits exposed linkers to become deprotected and available for nucleotide coupling. Chrome marks on the wafer ensure accurately masked alignment during the deprotection step. After activation, the wafer surface is flooded with a solution containing a single type of deoxyribonucleotide with a removable protection group. Nucleotides attach to exposed (unmasked) activated linkers.

Occasionally, activated molecules fail to attach the new nucleotide. To prevent these aberrant chains from compro-

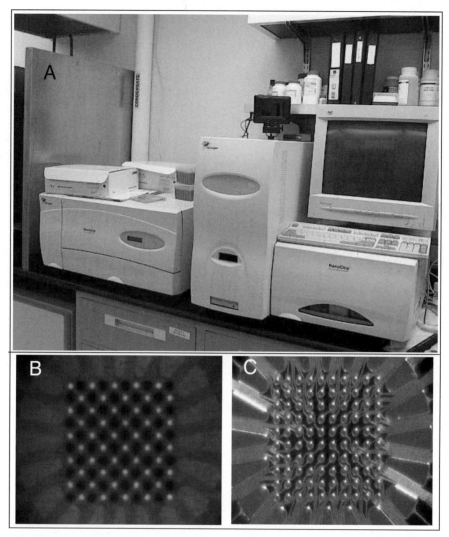

FIGURE 27.3 Nanogen System. A. The entire system as it sits on a benchtop. A personal computer controls the electronic gradient within the NanoChip as well as the Laser reader. B. A 50 addressable position chip with DNA probes gathered at the electrically active interrogation sites within the agarose gel overlay. Visualization of fluorescing signal represents the location of an abundance of newly loaded, biotinylated DNA, subsequent to loading and electric field application. C. Detailed structure of the 100 position NanoChip. This photolithographically constructed silicon wafer permits addressing 100 sites with each added biotinylated oligonucleotide, in the capture down mode, or biotinylated sample, in the amplicon down mode (see text for details).

mising chip quality, a capping step is used to truncate them. In addition, branched oligonucleotide formation is prevented through specific side chain blocking chemistry. Iterative synthesis steps involve masking, deprotection, and coupling. The process is repeated until the probes reach approximately 25 nucleotides. Although each position in the sequence of an oligonucleotide is occupied by 1 of 4 nucleotides, resulting in an apparent need for $25 \times 4 = 100$ different masks per wafer, the synthesis process is optimized to minimize masking requirements. This is done through the use of algorithms that adjust probe growth kinetics and the informatic identification of probe products where the same mask can be used multiple times.

At synthesis conclusion, the wafers are deprotected, diced, and the resulting individual arrays are packaged in flow cell cartridges. Depending on the number of probe features per array, a single 5-inch square quartz wafer can yield between 49 and 400 arrays.

Subsequent to the manufacturing process both Affymetrix and Perlegen do a comprehensive series of quality control tests including sampling arrays from every wafer to run control hybridizations. A quantitative test of hybridization is also performed using standardized control probes.

Application of Affymetrix arrays to the detection of SNPs is accomplished through a series of labeling steps (Fan *et al.*, 2000). First, approximately 250 µg of total genomic

DNA is digested with restriction enzyme XbaI and ligated to adaptors provided by Affymetrix that recognize the cohesive 4 bp 5′ overhang. There is no purification between digestion and adapter ligation so both large and small restriction fragments are ligated with adapters. PCR primers that contain the adapter sequence then are used to amplify the fragments in the 250 to 1000 bp size range. These molecules then are labeled in a reaction that consists of a proprietary biotin-labeled end labeling reagent and terminal deoxynuclotidyl transferase. This incubation produces the labeled molecules that are hybridized to the GeneChip array. Subsequent to hybridization and washing in the instrument that Affymetrix has engineered for this purpose, chips are flooded with phycoerythrin conjugated streptavidin, which binds to the biotinylated amplicons attached to the matrix, and permits visualization in an elaborate scanner supplied by the company.

27.11 MICRO-WELL PLATE ARRAYS

Chemical attachment of amino-conjugated oligonucleotide probes to plastic micro-well plates is now a well-established methodology. This is the system used by Roche for popular micro-well plate assays for HIV and HCV viral load (Mulder *et al.*, 1994). It has also been used by several groups for Factor V Leiden mutation detection (Zehnder *et al.*, 1997; Kowalski *et al.*, 2000). In brief, the method consists of attachment of amino-conjugated oligonucleotides to BSA-coated micro-well plates or plates coated with N-oxysuccinimide esters; such esters react the nucleophilic primary amine conjugates. Once coupling is accomplished, plates are extensively washed before being used for hybridization capture of biotinylated gene-specific amplicons. After post-hybridization washing, subsequent incubation of the plate with a streptavidin conjugated with alkaline phosphatase permits visualization using tetramethylbenzidine or acid-tetramethylbenzidine chemistry. Finally, plates containing visualized amplicons are optically scanned using a micro-well plate reader and interpreted using appropriate software. These assays are efficient and low-cost compared with RFLP-PCR or RFLP-PCR coupled with probe hybridization; however, they have two major drawbacks. These are the extensive manipulation of post-PCR products, because they are nonhomogeneous and the requirement for preparation and validation of appropriate capture plates.

27.12 MICROARRAY PRINTING

Now that microarray printing and visualization has become routine in many laboratories (Schena, 1999), former constraints preventing indel or gene copy number detection via microarray have diminished (Beheshti *et al.*, 2002; Fiegler *et al.*, 2003; Yu *et al.*, 2003). As a consequence, the ability to gauge gene copy number, genomic deletions, or the presence of amplified regions has become possible through the printing of oligonucleotide arrays. These may be printed on glass slides or coverslips through the use of microarray preparation robots. Competitive hybridization between normal and patient genomic DNA labeled with Cy3 and Cy5 through sample amplification permits visualization using a microarray scanner and consequent label ratio determination. Metrics for comparative genomic hybridization, developed by Pinkel, Gray and their collaborators (Pinkel *et al.*, 1998), can then be used to determine regions of comparative genomic amplification or putative deletion. In addition, careful control of hybridization conditions, probe synthesis, and attachment should permit SNPs at large numbers of sites to be scored using this technology. This kind of home-brew microarray manufacture for genetic disease research and diagnosis is still in its infancy, in part because of the expense of oligonucleotide synthesis; however, these methods should provide some healthy competition for the suppliers of still relatively expensive photolithographically manufactured genomic chips discussed earlier.

27.13 SUMMARY

Genotyping assay efficiency and throughput has considerably improved as a consequence of automation. Automation of many of the subtasks of nucleic acid hybridization including amplification, electrophoresis, and homogeneous signal detection has enabled the establishment of higher throughput molecular assays and diagnostics. Process automation has been motivated in part by the successful effort to map and sequence the human genome. Specific subtask automation has been accompanied by manual chaining together so that nucleic acid extraction, amplification, hybridization, and detection are dovetailed. There are incremental gains in clinical laboratory efficiency through automated process monitoring, result interpretation, and reporting afforded by current laboratory management information systems. In a research environment, chaining manual processes with automated subtasks, such as robotic liquid handling, has permitted high throughput genomic sequencing, genotyping, and voluminous data production. Many instrument and reagent manufacturers have participated in the automation initiative. There will undoubtedly be many improvements in automation and detection of DNA hybridization in the short-term future.

Acknowledgments

The author wishes to thank Dr. Michael Dean for providing time for the composition of this manuscript. The content of this publication does not necessarily reflect the views or policies of the Department of Health and Human Services, nor does mention of trade names, commercial products, or organizations imply endorsement by the US Government. This chapter has been funded in whole or in part with federal funds from the National Cancer Institute and the National Institutes of Health.

References

Alderton, R. P., Kitau, J., and Beck, S. (1994). Automated DNA hybridization. *Anal. Biochem.* 218, 98–102.

Beck, S. (1995). Automated DNA Hybridization and Detection. In *Molecular Biology: Current Innovations and Future Trends.* G. H. Griffin AM, ed. Wymondham, UK, Horizon Scientific Press, pp. 83–91.

Beheshti, B., Park, P. C., Braude, I., and Squire, J. A. (2002). Microarray CGH. *Methods Mol. Biol.* 204, 191–207.

Chan, V., Yam, I., Chen, F. E., and Chan, T. K. (1999). A reverse dot-blot method for rapid detection of non-deletion alpha thalassaemia. *Br. J. Haematol.* 104, 513–515.

Church, G. M., and Gilbert, W. (1984). Genomic sequencing. *Proc. Natl. Acad. Sci. USA* 81, 1991–1995.

Church, G. M., and Gilbert, W. (1985). The genomic sequencing technique. *Prog. Clin. Biol. Res.* 177, 17–21.

Davidson, E. H., and Hough, B. R. (1969). High sequence diversity in the RNA synthesized at the lampbrush stage of oogenesis. *Proc. Natl. Acad. Sci. USA* 63, 342–349.

Davis, R. W., and Davidson, N. (1968). Electron-microscopic visualization of deletion mutations. *Proc. Natl. Acad. Sci. USA* 60, 243–250.

DeMarchi, J. M., Richards, C. S., Fenwick, R. G., Pace, R., and Beaudet, A. L. (1994). A robotics-assisted procedure for large scale cystic fibrosis mutation analysis. *Hum. Mutat.* 4, 281–290.

Dracopoli, N. C., Haines, J. L., Korf, B. R., Morton, C. C., Seidman, C. E., Seidman, J. G., and Smith, D. R., eds. (2000). *Current Protocols in Human Genetics,* New York: Wiley.

Fan, J. B., Chen, X., Halushka, M. K., Berno, A., Huang, X., Ryder, T., Lipshutz, R. J., Lockhart, D. J., and Chakravarti, A. (2000). Parallel genotyping of human SNPs using generic high-density oligonucleotide tag arrays. *Genome Res.* 10, 853–860.

Farkas, D. H., Drevon, A. M., Kiechle, F. L., DiCarlo, R. G., Heath, E. M., and Crisan, D. (1996). Specimen stability for DNA-based diagnostic testing. *Diagn. Mol. Pathol.* 5, 227–235.

Fiegler, H., Carr, P., Douglas, E. J., Burford, D. C., Hunt, S., Scott, C. E., Smith, J., Vetrie, D., Gorman, P., Tomlinson, I. P., and Carter, N. P. (2003). DNA microarrays for comparative genomic hybridization based on DOP-PCR amplification of BAC and PAC clones. *Genes Chromosomes Cancer* 36, 361–374.

Gillespie, D., and Spiegelman, S. (1965). A quantitative assay for DNA-RNA hybrids with DNA immobilized on a membrane. *J. Mol. Biol.* 12, 829–842.

Gold, B., Radu, D., Balanko, A., and Chiang, C. S. (2000). Diagnosis of Fragile X syndrome by Southern blot hybridization using a chemiluminescent probe: a laboratory protocol. *Mol. Diagn.* 5, 169–178.

Gold, B. (2001). DNA genotyping. *Adv. Clin. Chem.* 36, 171–234.

Gold, B. (2003). Origin and utility of the reverse dot-blot. *Expert Rev. Mol. Diagn.* 3, 143–152.

Grody, W. W., Cutting, G. R., Klinger, K. W., Richards, C. S., Watson, M. S., and Desnick, R. J. (2001). Laboratory standards and guidelines for population-based cystic fibrosis carrier screening. *Genet. Med.* 3, 149–154.

Groman, J. D., Meyer, M. E., Wilmott, R. W., Zeitlin, P. L., and Cutting, G. R. (2002). Variant cystic fibrosis phenotypes in the absence of CFTR mutations. *N. Engl. J. Med.* 347, 401–407.

Heid, C. A., Stevens, J., Livak, K. J., and Williams, P. M. (1996). Real time quantitative PCR. *Genome Res.* 6, 986–994.

Hilbert, H., Lauber, J., Lubenow, H., and Dusterhoft, A. (2000). Automated sample-preparation technologies in genome sequencing projects. *DNA Seq.* 11, 193–197.

Holland, P. M., Abramson, R. D., Watson, R., and Gelfand, D. H. (1991). Detection of specific polymerase chain reaction product by utilizing the 5′—3′ exonuclease activity of Thermus aquaticus DNA polymerase. *Proc. Natl. Acad. Sci. USA* 88, 7276–7280 .

Hopwood, A., Oldroyd, N., Fellows, S., Ward, R., Owen, S. A., and Sullivan, K. (1997). Rapid quantification of DNA samples extracted from buccal scrapes prior to DNA profiling. *Biotechniques* 23, 18–20.

Jean, J., Blais, B., Darveau, A., and Fliss, I. (2001). Detection of hepatitis A virus by the nucleic acid sequence-based amplification technique and comparison with reverse transcription-PCR. *Appl. Environ. Microbiol.* 67, 5593–5600.

Kowalski, A., Radu, D., and Gold, B. (2000). Colorimetric microwell plate detection of the factor V Leiden mutation. *Clin. Chem.* 46, 1195–1198.

Lander, E. S., and Botstein, D. (1986). Mapping complex genetic traits in humans: new methods using a complete RFLP linkage map. *Cold Spring Harb. Symp. Quant. Biol.* 51 Pt 1, 49–62.

Lappin, S., Cahlik, J., and Gold, B. (2001). Robot printing of reverse dot blot arrays for human mutation detection. *J. Mol. Diagn.* 3, 178–188.

Lau, J., and Tolan, D. R. (1999). Screening for hereditary fructose intolerance mutations by reverse dot-blot. *Mol. Cell. Probes* 13, 35–40.

Morin, P. A., Saiz, R., and Monjazeb, A. (1999). High-throughput single nucleotide polymorphism genotyping by fluorescent 5′ exonuclease assay. *Biotechniques* 27, 538–544.

Mulder, J., McKinney, N., Christopherson, C., Sninsky, J., Greenfield, L., and Kwok, S. (1994). Rapid and simple PCR assay for quantitation of human immunodeficiency virus type 1 RNA in plasma: application to acute retroviral infection. *J. Clin. Microbiol.* 32, 292–300.

Ousterhout, J. K. (1994). *Tcl and the Tk toolkit.* Addison-Wesley: Reading, MA.

Pease, A. C., Solas, D., Sullivan, E. J., Cronin, M. T., Holmes, C. P., and Fodor, S. P. (1994). Light-generated oligonucleotide arrays for rapid DNA sequence analysis. *Proc. Natl. Acad. Sci. USA* 91, 5022–5026.

Pinkel, D., Segraves, R., Sudar, D., Clark, S., Poole, I., Kowbel, D., Collins, C., Kuo, W. L., Chen, C., Zhai, Y., Dairkee, S. H., Ljung, B. M., Gray, J. W., and Albertson, D. G. (1998). High resolution analysis of DNA copy number variation using comparative genomic hybridization to microarrays. *Nat. Genet.* 20, 207–211.

Ririe, K. M., Rasmussen, R. P., and Wittwer, C. T. (1997). Product differentiation by analysis of DNA melting curves during the polymerase chain reaction. *Anal. Biochem.* 245, 154–160.

Schena, M. (1999). *DNA microarrays: a practical approach.* Oxford University Press: Oxford; New York.

Schollen, E., Vandenberk, P., Cassiman, J. J., and Matthijs, G. (1997). Development of reverse dot-blot system for screening of mitochondrial DNA mutations associated with Leber hereditary optic atrophy. *Clin. Chem.* 43, 18–23.

Sosnowski, R. G., Tu, E., Butler, W. F., O'Connell, J. P., and Heller, M. J. (1997). Rapid determination of single base mismatch mutations in DNA hybrids by direct electric field control. *Proc. Natl. Acad. Sci. USA* 94, 1119–1123.

Southern, E. M. (1975). Detection of specific sequences among DNA fragments separated by gel electrophoresis. *J. Mol. Biol.* 98, 503–517.

Strom, C. M., Huang, D., Chen, C., Buller, A., Peng, M., Quan, F., Redman, J., and Sun, W. (2003). Extensive sequencing of the cystic fibrosis transmembrane regulator gene: Assay validation and unexpected benefits of developing a comprehensive test. *Genet. Med.* 5, 9–14.

Wang, X., Myers, A., Saiki, R. K., and Cutting, G. R. (2002). Development and evaluation of a PCR-based, line probe assay for the detection of 58 alleles in the cystic fibrosis transmembrane conductance regulator (CFTR) gene. *Clin. Chem.* 48, 1121–1123.

Wittwer, C. T., Fillmore, G. C., and Hillyard, D. R. (1989). Automated polymerase chain reaction in capillary tubes with hot air. *Nucleic Acids Res.* 17, 4353–4357.

Wittwer, C. T., Fillmore, G. C., and Garling, D. J. (1990). Minimizing the time required for DNA amplification by efficient heat transfer to small samples. *Anal. Biochem.* 186, 328–331.

Wittwer, C. T., and Garling, D. J. (1991). Rapid cycle DNA amplification: time and temperature optimization. *Biotechniques* 10, 76–83.

Wittwer, C. T., Ririe, K. M., Andrew, R. V., David, D. A., Gundry, R. A., and Balis, U. J. (1997). The LightCycler: a microvolume multisample fluorimeter with rapid temperature control. *Biotechniques* 22, 176–181.

Yang, Y. P., Corley, N., and Garcia-Heras, J. (2001). Reverse dot-blot hybridization as an improved tool for the molecular diagnosis of point mutations in congenital adrenal hyperplasia caused by 21-hydroxylase deficiency. *Mol. Diagn.* 6, 193–199.

Yu, W., Ballif, B. C., Kashork, C. D., Heilstedt, H. A., Howard, L. A., Cai, W. W., White, L. D., Liu, W., Beaudet, A. L., Bejjani, B. A., Shaw, C. A., and Shaffer, L. G. (2003). Development of a comparative genomic hybridization microarray and demonstration of its utility with 25 well-characterized 1p36 deletions. *Hum. Mol. Genet.* 12, 2145–2152.

Zehnder, J. L., Benson, R. C., and Cheng, S. (1997). A microplate allele-specific oligonucleotide hybridization assay for detection of factor V Leiden. *Diagn. Mol. Pathol.* 6, 347–352.

CHAPTER 28

The Use of Microelectronic-Based Techniques in Molecular Diagnostic Assays

PIOTR GRODZINSKI,[1] MICHAEL WARD,[1] ROBIN LIU,[2] KATHRYN SCOTT,[3] SAUL SURREY,[4] PAOLO FORTINA[3]

[1] *Bioscience Division, Los Alamos National Laboratory, Los Alamos, NM;*

[2] *Applied Nanobioscience Center, Arizona State University, Tempe, AZ;*

[3] *Center for Translational Medicine, Department of Medicine, Jefferson Medical College, Thomas Jefferson University, Philadelphia, PA;*

[4] *Department of Medicine, Cardeza Foundation for Hematologic Research, Jefferson Medical College, Thomas Jefferson University, Philadelphia, PA*

TABLE OF CONTENTS

28.1 INTRODUCTION

Contemporary on-chip biological assays have benefited from the multitude of techniques developed during the microelectronic revolution of the last three decades. The ability of delivering electric and magnetic signals to the localized portions of the biochip permitted exploitation of fundamental dielectric and magnetic properties of biological molecules and cells. For example, most biological molecules exhibit an electric charge and thus they migrate when exposed to an electric field. The interaction between an external electric field and these molecules can be used for their manipulation and concentration. The differentiation in dielectric properties of cells led to the development of dielectrophoretic techniques allowing for sorting, separation, and isolation of cells from different organisms and tissues. The ability to monitor electron transfer between the molecule and the external measurement system formed the basic principles for electronic detection of DNA.

Similarly, the vast arsenal of microfabrication capabilities developed in the microelectronic industry has been leveraged toward on-chip assay applications, where integration of conductive, heating, and magnetic elements resulted in development of micro-PCR devices, microarray DNA chips containing thermal gradients, and dielectrophoretic and magnetic devices for cell and molecule preconcentration.

This chapter will include a discussion on microfabrication, cell sorting, molecule preconcentration, and sample preparation techniques followed by DNA/RNA amplification on chips, electronic-assisted hybridization assays, and examples of commercially available platforms.

28.2 MICROFABRICATION

The engagement of microfabrication techniques permits for realization of high throughput, integrated analytical systems. Use of microfabrication techniques allows for building of miniature networks of chambers and channels with a dimension resolution <1 μm. Performing assays within these

350   *Molecular Diagnostics*

devices reduces reagent volume and facilitates reduction of assay duration (Kricka, 1998; Freemantle, 1999). In addition, high parallelism of reactions is achieved by placement of many assay paths within the same device; thus, analysis throughput can be increased (Fodor *et al.*, 1991; Simpson *et al.*, 1998). Furthermore, a multitude of assay functions can be integrated into the same device leading to assay automation.

Individual components of a microfluidic device are required to perform a variety of functions including mixing, thermal cycling, valving, and detection, which eventually will need to be integrated into one chip. There are a number of material platforms under consideration for fabrication of such devices including glass (Manz *et al.*, 1992; Harrison *et al.*, 1993), silicon (Mastrangelo *et al.*, 1998; Burns *et al.*, 1998), plastic (Alonso-Amigo, 2000; Anderson *et al.*, 2000; Grodzinski *et al.*, 2001), and hybrid approaches (Burns *et al.*, 1998). Devices made of glass and silicon both take advantage of lithographic techniques developed in the integrated circuit chip industry. They use photolithography, wet etching, metal sputtering, and anodic and thermal bonding. The glass-based platform is attractive due to its well-understood surface chemistry, and it possesses a surface charge, which allows for using electrokinetic pumping methods for fluid transport. Glass is also transparent in the UV region of the spectrum, which simplifies detection.

Silicon, on the other hand, permits for integration of complex complimentary metal-oxide semiconductor (CMOS) electronics for control and data processing within a chip.

Plastics offer an attractive alternative for fabrication of disposable devices due to low cost (particularly important for disposable devices) and a wide range of advantageous physical and chemical properties, including very good biocompatibility. Bulk micromachining of plastics can be achieved through hot embossing, injection molding, and casting. Rapid prototyping techniques taking advantage of direct writing in polymer plates are also available and permit production of complex structures (see Fig. 28.1).

Due to the distinct advantages of various chip fabrication techniques, hybrid approaches with integrated electronics on silicon substrates with glass or plastic channels are also being pursued (Burns *et al.*, 1998).

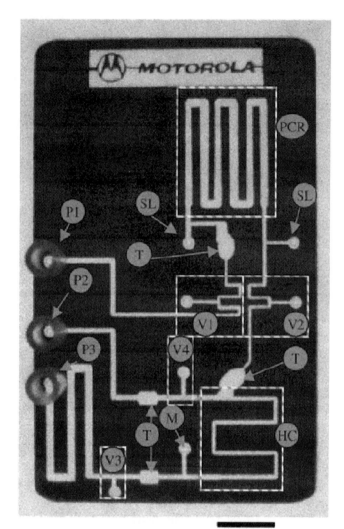

FIGURE 28.1 Monolithic, integrated DNA assay device produced using direct laser writing in 250 mm polycarbonate plates and boned through thermal bonding process. Serpentine PCR channel (PCR), hybridization channel (HC), pluronics valves (V1–4), pluronic traps (T), hydrophobic air permeable membrane (M), PCR reagent loading holes (SL), sample driving syringe pump P1, waste withdrawing syringe pump (P2), and wash syringe pump (P3) are indicated (adapted from Liu *et al.*, 2002).

28.3 CHIPS FOR SAMPLE PREPARATION

Sample preparation represents the most time-consuming and labor-intensive procedure in DNA analysis, and also introduces one of the largest variables in subsequent analyses due to its complexity. Due to these reasons, initial efforts in bringing the assay to the chip level were focused on development of back-end detection schemes resulting in demonstrations of DNA microarray biochips (Fodor *et al.*, 1991; Fortina *et al.*, 2000) and electrophoresis separation chips

(Manz *et al.*, 1992; Harrison *et al.*, 1993; Woolley and Mathies, 1994). However, the quest for a complete, integrated analytical chip system requires the treatment of "real" samples of bodily or environmental fluids. Recent research attempts use methods of separating cells from complex samples. Two of these methods are discussed in the following sections, which in the author's opinion are most promising. These methods utilize dielectrophoresis and immunomagnetic separation.

28.3.1 Use of Dielectrophoresis

Immersion of a biological cell into a medium of different dielectric property, and subsequent application of an external nonuniform electric field causes cell polarization and induction of charges at the cell/medium interface (Wang and Cheng, 2001). The dipole moment develops due to separation of positive and negative charges within the cell. Dielectrophoretic force (DEP) originates from interaction of this dipole moment with an external nonuniform electric field. Depending on the relationship of dipole moment and electric field vectors (directions), the DEP force can be positive (parallel vectors) or negative (anti-parallel vectors), causing motion of the cell in different directions (Pethig *et al.*, 1992). These vector relationships depend on the relative dielectric property of the cell and the medium. The former depends on the cell type and physiological state. Negative effects on development or behavior of cells were not observed either during or after application of electrical fields (Pethig *et al.*, 1992).

The DEP technique does not require cell labeling, and uncomplicated devices containing only interdigitated electrodes can be built. When nonuniform AC electric field is used, cells migrate to locations of minimum dielectric potentials and get immobilized there. Furthermore, differences in dielectric properties between cells allows for tuning of electric field conditions so that only single cell types from a complex mixture are immobilized.

For example, DEP separation techniques, which exploit differential dielectric properties among different biological cells, have been developed. Cultured cervical carcinoma (HeLa) cells and *E. coli* cells were separated and isolated from normal blood cells (Cheng *et al.*, 1998a; 1998b) on a silicon chip (see Fig. 28.2). The NanoChip™, developed by Nanogen, contains a two-dimensional silicon-based array of the platinum electrode sites. These sites can be addressed individually, such that AC electric field can be directed to the site of choice. Cells, bacteria, or other molecules can be moved rapidly and concentrated to the appropriate address in the NanoChip™ electronic microarray. A permeation layer, usually porous agarose or polyacrylamide, allows an electrical contact between the electrode and cells or molecules in solution, while preventing physical contact, which may lead to undesired oxidation or denaturing.

As described in Cheng and colleagues (1998b), separation of HeLa cells from blood cells took approximately 3 min after addition of the mixture. Furthermore, a chip also was designed to integrate subsequent sample processing and DNA/RNA hybridization (Cheng *et al.*, 1998b). The DEP separation of bacteria containing a plasmid from blood cells, electronic lysis of isolated cells, and digestion of proteins were performed. Presence of plasmid sequence was detected by hybridization, and the entire process took approximately 4 min to complete after the mixture was added to the chip. Lysis was performed using a series of electronic pulses, and all positions had a proteinase K mixture added to digest proteins released during lysis. Plasmid and genomic RNA and DNA were released from the cells as a result of the lysis. Hybridization was performed on another chip, requiring removal of the nucleic acids. The hybridization chip contained a permeation layer of streptavidin agarose with biotinylated capture probes at specific addresses. Each target sequence was electronically moved to the separate addresses and bound as expected to the proper fluorescent capture probe.

A diversity of dielectrophoretic techniques resulted in modifications of simple cell arrest approaches and produced more sophisticated operation schemes, such as Field-Flow Fractionation (FFF) DEP and traveling-wave DEP (Gascoyne and Vykoukal, 2002). The former is used to separate cells in the vertical direction within a flow channel and combines variation of flow velocity due to parabolic distribution of the flow front in pressure-driven flow and DEP. The latter involves application of an AC electrical field in a sequential manner to the set of electrodes enabling simultaneous concentration and motion of the cells.

DEP devices also can be used for focusing (Edman *et al.*, 1998), preconcentration, and selective entrapment of DNA fragments (Chou *et al.*, 2002). The latter approach used an electrode-less DEP chip configuration where electrodes are removed to the end of fluidic channels, and DEP force is created due to crowding of electric field lines within the microfabricated channel constriction (see Fig. 28.3). The advantage of this approach lies in prevention of DNA hydrolysis when it is brought to the vicinity of the metal electrode.

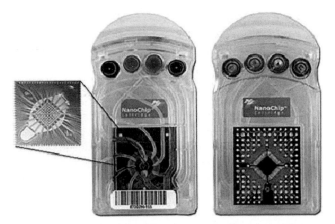

FIGURE 28.2 NanoChip™ Cartridge and Electronic Microarray. The latest generation NanoChip™ Cartridge is used on the automated Molecular Biology Workstation developed by Nanogen. Each microarray has 100 test sites.

28.3.2 Use of Magnetic Devices

Molecule and cell labeling with particles or beads, which could exhibit magnetic properties, are common in molecu-

(A)

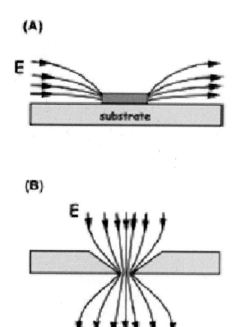

(B)

(C)

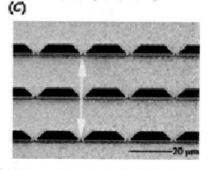

FIGURE 28.3   Schematic of a microfluidic DEP trap. A. A metallic DEP trap is made of microfabricated wire(s) on a substrate. The wire(s) may be either free-floating or connected to a voltage source. B. An electrode-less DEP trap made of dielectric constrictions. The solid lines, indicated by E, are electric field lines. C. A scanning electron micrograph of an electrode-less DEP device consisted of a constriction array etched in quartz. The constrictions are 1 μm wide and 1.25 μm deep. The whole chip measures 1 × 1 cm. The applied electric field direction z is shown by the double-headed arrow (adapted from Chou *et al.*, 2002).

lar biology. The magnetic beads then can be manipulated, transferred, and immobilized in a magnetic field. Resulting molecule or cell separation and sorting then can occur. Incorporating chip elements enabling selective production of such fields allows for localized formation of these fields.

Traditional cell separation protocols rely predominantly on immunological methods utilizing labeling of cells with specific antibodies, which are usually covalently linked to particles, for example, paramagnetic beads (Chalmers *et al.*, 1998a; 1998b). The bead particles can be trapped in an external magnetic field, and a subset of cells bound to the beads can be isolated via an antibody. This method carries substantial promise for separation of rare cells occurring at low concentrations, due to its potential for high selectivity. Thus, developing chips capable of producing magnetic fields and magnetic gradients and enabling isolation of magnetically labeled cells within the chips at predetermined locations is of great importance. A method to effectively produce strong magnetic gradients is important while capturing small (<150 nm) beads, which possess too weak a momentum to be trapped in uniform fields.

Introduction of complex fabrication processes to fabricate on-chip integrated microcoils has been successful, and has facilitated an alternative approach instead of use of external magnets for bead arrest. Meander-type magnetic coils have been demonstrated using a planar magnetic conductor and two-level magnetic core (Ahn and Allen, 1993). Such coils can be used directly to drive magnetic actuators. Hybrid processes, where wire bonding combined with microfabrication techniques have been used, which result in coils (Christenson *et al.*, 1995). Microfluidic structures with manually wound coils used for bead trapping and coil-to-coil bead transport also have been evaluated (Joung *et al.*, 2000). Shallow channels need to be used to maintain gradient strength in the bead flow field.

Methods to produce high magnetic gradients for cell separation have been discussed (Tibbe *et al.*, 1999; 2002; Berger *et al.*, 2001; Ward *et al.*, 2002), and devices of approximately 1 μm wide and 100 nm tall sputtered ferromagnetic lines used to create very high local gradients at the edges of the lines. A cell-tracks device (Immunicon Corp. Hundington Valley, PA) uses nickel magnetized in an external field (Tibbe *et al.*, 1999; 2002), whereas another uses permanently magnetized cobalt-chrome-tantalum materials (Berger *et al.*, 2001). When labeled cells are brought in close proximity (50 μm) to the lines, they can be captured by the magnetic force resulting from the edge gradients. The Immunicon device (see Fig. 28.4) achieves this proximity using a specialized external gradient magnet while the other uses shallow flow channels. The Immunicon device is engineered for optical observation of trapped cells in a CD-like configuration capable of laser scanning; the second is designed for cell sorting and uses hydrodynamic forces in the channel to direct captured cells along the wires to collection points.

Ward and colleagues (2002) developed microfluidic channels with built-in gradients using submillimeter iron or nickel-iron matrix elements fabricated into acrylic and polycarbonate substrate plastic channels. Those metal features geometrically concentrate external magnetic flux to form the necessary magnetic gradients for capture of weakly susceptible particles. Targets are collected at the walls of the channel where they can be washed or otherwise processed in a flow stream. These first generation devices were used to concentrate *E. coli* cells from blood and employed double-sided iron embedded polycarbonate channels 100 μm in height. Plating experiments of *E. coli* inoculated blood demonstrated that large sample volumes could be con-

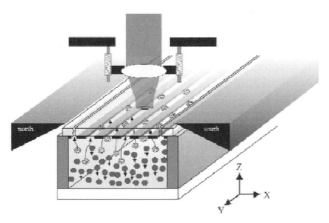

FIGURE 28.4 Schematic of Celltracks device and method. The chamber with microfabricated nickel lines is placed in the external magnet device. Labeled cells in the chamber are drawn upward where they align along the nickel. They can then be analyzed by the scanning optics (adapted from Tibbe *et al.*, 1999).

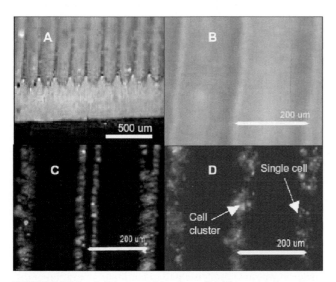

FIGURE 28.5 A. Laser-machined acrylic ridge structure prior to electroplating. B. Ni-Fe ridge device with fluorescein-labeled CD45 antibody coupled to IgG1-specific magnetic beads prior to separation. C. Same device after separation. Fluorescence is focused on the ridges. D. Similar device after separation of white blood cells using the fluorescent anti-CD45 antibodies (adapted from Ward *et al.*, 2002). (See color plate)

centrated in minutes with relatively high recovery rates. About 50% recovery was obtained at the maximum tested average flow rate of approximately 19 mm/sec, which corresponded to an approximately 350 µl/min flow rate. For capture of rare target mammalian cells from blood, alternative, more uniform and more easily manufactured designs using compression-molded, saw-toothed ridges in acrylic, which are coated with soft magnetic material, were also explored (see Fig. 28.5). A 90% cell capture efficiency using a 120 µl/min flow rate was achieved.

28.3.3 Cell Separation by Size

Physical filters relying on separation of cells by size have been demonstrated (Wilding *et al.*, 1998). White blood cells were isolated from whole blood in silicon-glass 4.5 µl microchips containing a series of 3.5 µm feature-sized weir-type filters, formed by an etched silicon dam spanning the flow chamber. Genomic DNA targets, for example, the dystrophin gene, then were directly amplified using PCR from cells isolated on the filters. This dual function microchip provides a means to simplify nucleic acid analyses by integrating in a single device two key steps in the analytical procedure: cell isolation and PCR (Cheng *et al.*, 1996; 1998c; Wilding *et al.*, 1998).

28.4 DNA AND RNA AMPLIFICATION IN MICROCHIP FORMAT

DNA amplification using Polymerase Chain Reaction (PCR) is essential to most genetic analysis applications of integrated microchips. Rapid operation, small sample volume, and parallel amplification of different amplicons within the same chip are among the desired features of amplification in the microchip environment. Design and development of such chips is hindered by several challenges including loss of sample to chamber walls due to dramatic increases in surface-to-volume ratio, evaporation in small volumes, and effective heat dissipation in order to achieve rapid thermal cycling. Micro-PCR devices have been fabricated in glass (Kopp *et al.*, 1998; Waters *et al.*, 1998; Lagally *et al.*, 2000; 2001), silicon (Wilding *et al.*, 1994; Belgrader *et al.*, 1998; 1999), and plastic (Boone *et al.*, 1998; Yu *et al.*, 2000; Kricka, *et al.*, 2002). Silicon, due to its superior thermal conductivity (approximately 10× that of glass and 700× that of polymers), allows for very fast temperature ramp times and results in short on-chip protocols (Belgrader *et al.*, 1999). Recently, with increasing emphasis on disposable devices, use of plastic and plastic fabrication methods have become popular in microreactor development. Such chips are inexpensive, optically transparent, and biocompatible (Alonso-Amigo, 2000; Grodzinski *et al.*, 2001). Despite all the advantages, plastic possesses a major challenge to a designer of PCR microreactors due to its poor thermal conductivity and resulting difficulty in achieving rapid thermal cycling. Recently, successful DNA amplification in polycarbonate chips was reported (Yang *et al.*, 2002). Thirty thermal cycles took 30 min, which, considering the poor thermal conductivity characteristics of polymers, is a significant achievement.

Low volume operation and rapid thermal cycling have been early motivators for on-chip PCR development. An infrared heating scheme with glass PCR devices containing a 1.7 µl microchamber was described, in which amplification was achieved by 15 thermal cycles in 4 min (Giordano *et al.*, 2001). In an integrated monolithic silicon-glass device, a submicroliter (280 nl) volume was thermally cycled as fast as 30 sec/cycle (Lagally *et al.*, 2001). The low volume limit for micro-PCR was further reduced by using a picoliter microchamber array (Nagai *et al.*, 2001). Using a real-time PCR device, PCR detection in 7 min of *Erwinia*, a vegetative bacterium was reported (Belgrader *et al.*, 1999). An integrated rapid PCR-detection system with amplification times in the 20 min range coupled with capillary electrophoresis analysis also was presented (Khandurina *et al.*, 2000; Lagally *et al.*, 2000).

Despite all these examples, sensitivity of micro-PCR chips has not been well studied. Most reports focused on achieving amplification *per se*, but the systematic evaluation of amplification yield and reaction sensitivity usually was not explored. The initial template concentration used to achieve fast and small volume amplification was usually high, ranging from 0.1 ng of phage DNA (Giordano *et al.*, 2001) to 100 ng of human genomic DNA (Cheng *et al.*, 1998c). The most sensitive micro-PCR assay demonstrated so far involved use of a single molecule DNA template, which was amplified in a glass-integrated microfluidic device (Lagally *et al.*, 2001; Fig. 28.6). For "real" sample (containing target cells, rather than purified DNA) analysis, the most sensitive silicon microstructure, which could perform rapid real-time PCR analysis from a sample con-

taining a low target concentration (*Erwinia*), was reported (Belgrader *et al.*, 1999). A positive amplicon signal was detected in less than 35 cycles (17 sec/cycle) with the starting template concentration as low as 5 cells. A systematic study on sensitivity of micro-PCR in plastic devices was reported for the first time (Yang *et al.*, 2002), demonstrating feasibility of amplifying template concentrations as low as 10 *E. coli* cells (50 fg of DNA) in the presence of blood. Similarly, multiplexing PCR was demonstrated within the same reactor chamber for four different bacteria species.

Investigators have worked on sample preparation using real-time PCR or hybridization in an array format as an assessment tool. An integrated, monolithic genetic assay device performing serial and parallel multistep molecular operations, including nucleic acid hybridization, was presented (Anderson *et al.*, 2000). More recently, a system with online PCR detection, on-chip spore preconcentration, lysing, and DNA purification was demonstrated (Taylor *et al.*, 2001). In addition, a prototype integrated semidisposable microchip analyzer for cell separation and isolation, PCR amplification, and amplicon detection was developed (Yuen *et al.*, 2001). The analyzer includes three glass-silicon microchips (an integrated sample preparation and PCR microchip, a denaturation microchip, and a reaction microchip) and a detection microarray made from a standard glass slide. The microchips and microarray are connected through a microfluidic network. The individual chips are assembled into a working system using three platforms: an upper plexiglas, a disposable lower plexiglas, and an aluminum platform.

Plastic, disposable chips for pathogen detection, PCR amplification, DNA hybridization, and hybridization in a single device were demonstrated (Liu *et al.*, 2002). DNA probes were immobilized in plastic channels to reduce target volume and to enable target motion to improve detection sensitivity. On-chip valving using phase-change Pluronics material was also implemented to facilitate separation of different stages of the assay. The chip was sandwiched between external Peltier heaters during PCR amplification, and sample transport was accomplished using external syringe pumps. The level of integration was expanded further to a complete self-contained biochip capable of magnetic, bead-based cell capture, cell preconcentration, purification, lysis, PCR amplification, DNA hybridization, and electrochemical detection of hybridization events (Liu *et al.*, 2003). The device is completely self-contained and does not require external pressure sources, fluid storage, mechanical pumps, or valves (see Fig. 28.7). It uses phase-transition paraffin valves and electrochemical pumps for efficient sample transport and isolation between adjacent assay stages.

It should be noted that a wide range of flow rates needs to be used, since large sample volumes (approximately 2 ml) have to be processed in short periods of time (<10 min). Pathogenic bacteria preconcentration and subsequent amplification and detection of genomic DNA as well as single

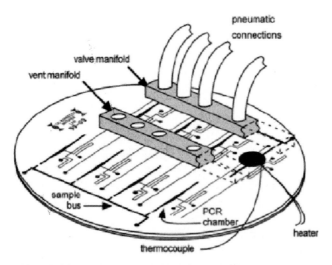

FIGURE 28.6 Microfluidic PCR-CE device. The PCR chambers are connected to a common sample bus through a set of valves. Hydrophobic vents at the other end of the PCR chambers are used to locate the sample and to eliminate gas. The PCR chambers are connected directly to the cross channel of the CE system for product injection and analysis. Two aluminum manifolds, one each for the vents and valves, are placed onto the respective ports and clamped in place using vacuum. The manifolds connect to external solenoid valves for pressure and vacuum actuation. Thermal cycling is accomplished using a resistive heater and a miniature thermocouple below the 280-nL chamber (adapted from Lagally *et al.*, 2001).

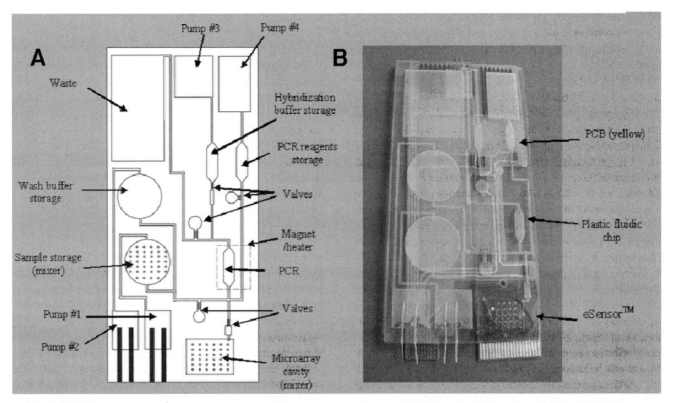

FIGURE 28.7 A. Schematic of the plastic fluidic chip. Pumps 1–3 are electrochemical pumps; pump 4 is a thermo-pneumatic pump. B. Photograph of the integrated device that consists of a plastic fluidic chip, a printed circuit board (PCB), and a Motorola eSensor™ microarray chip. The piezoelectric disks glued on the top of the sample storage chamber and microarray chamber that provide acoustic micromixing are not shown here (adapted from Liu *et al.*, 2003).

nucleotide polymorphism (SNP) analysis directly from blood was reported using this system.

An additional challenge remains quantitation of mRNA abundance in a micro-PCR format. Single-step RT-PCR using silicon dioxide-coated chips has been reported with an efficiency of 70% compared to that from a tube control. Analyses were conducted using an entangled-solution capillary electrophoresis (ESCE) system and laser-induced fluorescence (LIF) detection (Cheng and Mitchelson, 2001). Similarly, Anderson and colleagues (2000) developed chips, which, among others, performed reverse transcription reactions. Recently, PCR and RT-PCR in one chip using a continuous flow scheme was demonstrated (Obeid, 2003).

28.5 COMMERCIAL IMPLEMENTATION OF MOLECULAR ASSAYS WITH THE USE OF MICROELECTRONICS

Many of the research approaches just discussed have matured and grown to be commercialized, taking advantage of microfabrication and microelectronic technologies. In the following sections, a few primary examples will be discussed, including:

- Hitachi high-technologies microarray chips with thermal gradients
- Nanogen approach to electronic focusing of DNA in hybridization microarrays
- Motorola–Clinical Micro Sensors hybridization chips using electrochemical detection for SNP analysis
- CombiMatrix microarrays with *in situ* probe synthesis and electronic detection for immunoassay and DNA applications
- GeneOhm detection and direct DNA conductivity measurements
- Hybridization arrays with magnetic sensing
- Cepheid technology for real-time micro-PCR amplification and sample preparation
- Caliper capillary electrophoresis technology for DNA and protein separation

28.5.1 Thermal Gradient Chips

With the increasing complexity and size of the contemporary microarrays, the determination of optimum conditions for multiple sequences to be analyzed on the same chip is becoming increasingly difficult. Template/primer and temperature differences do not allow all sequences to have similar performances when being hybridized simultaneously. Several groups have tried to optimize primer length, melting temperature, GC content, buffer concentrations, wash conditions, and hybridization temperature in attempting to find one set of optimal hybridization conditions (Khrapko *et al.*, 1991; Sosnowski *et al.*, 1997; Gerry *et al.*, 1999; Gilles *et al.*, 1999; Lipshutz *et al.*, 1999; Fan *et al.*, 2000; Hirschhorn *et al.*, 2000; Mei *et al.*, 2000; Pastinen *et al.*, 2000; Dong *et al.*, 2001; Jobs *et al.*, 2002). These multiple attempts, which did not lead to an ultimate success, triggered a new research direction where chip platforms allowing for the independent control of the temperature at each individual array site were constructed (Kajiyama *et al.*, 2003). This independent temperature control of hybridization events allowed for multiple genes and sequences to be assayed together. Genotyping SNPs in two genes was demonstrated with this newly developed chip, and four clinically relevant loci were selected for testing: two in the Factor VII gene and two in the hemochromatosis gene (Kajiyama *et al.*, 2003).

A standard lithography process has been employed to design and fabricate the chip containing 100 individually controlled sites. Each site was 500×500 microns in size and contained a heater and temperature sensor. A diagram of the gradient chip, an individual reaction site diagram and images of the individual sites for reaction are shown in Fig. 28.8. Heat dissipation fins were added to the chip after initial testing to minimize any thermal effects one site may have on neighboring sites.

The thermal gradient chip has proven to be very effective in genotyping, as it allowed for multiple loci with different probe melting temperatures (T_m) to be typed on the same chip. The melting/hybridization temperature of a sequence is determined by several factors, including the GC content and size of the PCR product. A few degrees difference in T_m can determine success or failure of a successful analysis of the sequence of interest, providing that hybridization or wash conditions are not optimized for the individual locus. The GC content of the four loci studied varied from 43% to as high as 71%, and PCR product size ranged from 150 bp to 390 bp, and T_m from 51°C to 65°C. A Cy5 label was added to each PCR product to detect hybridization of the chip-bound normal or variant oligonucleotides. As expected, heterozygous samples hybridized to both chip-bound normal and mutant probes. The ability to control the temperature for hybridization for each locus facilitated genotyping of multiple loci simultaneously on the same chip. Such thermal gradient chips help minimize the number of ambiguous calls, making them potentially attractive for clinical laboratory testing.

28.5.2 Use of Electric Field to Accelerate Hybridization

Although microarrays are making a large impact on genomics, disease diagnosis, and biology in general, a significant problem still exists—hybridization is slow. Just as with Southern blotting, the solution containing the molecules to be hybridized with those attached to the surface typically takes 16 to 24 hours. Furthermore, the sample must be sufficiently concentrated in these molecules to permit hybridization even at these extended times. For example, concentration of about 4 nanomolar DNA usually is employed to achieve reasonably complete hybridization of 150 bp probes in 10 hours. Recommended protocols for mRNA use 2 μg or (assuming size of 1,000 to 10,000 bp) roughly 19 to 190 nM mRNA. Lower concentrations simply do not produce a measurable signal in any reasonable period of time. The reason for the relatively large difference is that larger molecules diffuse much more slowly than smaller ones.

Due to these deficiencies, numerous attempts to accelerate the hybridization process have been undertaken. They include target stirring or mixing with the hybridization cavity, dynamic DNA hybridization using paramagnetic beads (Fan *et al.*, 1999), rotation of the whole device (Chee *et al.*, 1996), and the use of a micro porous three-dimensional biochip with the hybridization solution being pumped continuously through it (Cheek *et al.*, 2001).

One of the most effective hybridization kinetics acceleration methods developed to date is to use an electric field to attract and concentrate DNA molecules at the chip surface.

NanoChip™ microelectronic arrays use electronic field control to drive the transport and concentration of nucleic acids to specific test sites within the array. The currently available array comprises 100 microelectrode test sites, each of which can be controlled individually and electronically to carry a positive, negative, or neutral charge. This flexibility differentiates NanoChip™ microelectronic arrays from other array platforms and enables analysis of multiple patient samples and/or multiple SNPs on a single array (Foglieni *et al.*, 2004).

Nanogen's (http://www.nanogen.com) electronic arrays are composed of silicon microchips coated with a synthetic hydrogel permeation layer containing streptavidin. By applying a positive charge to one or more test sites on the array, DNA, which is negatively charged, is attracted to those positions. Incorporation of a biotin moiety onto the DNA allows the sample to remain anchored at the test site after the electronic charge has been discontinued.

To perform a standard SNP genotyping assay, amplified, biotinylated genomic DNA is electronically addressed to one or more test sites. After all samples have been applied, fluorescently labeled reporter oligonucleotides that discriminate between the wild type and polymorphic alleles are

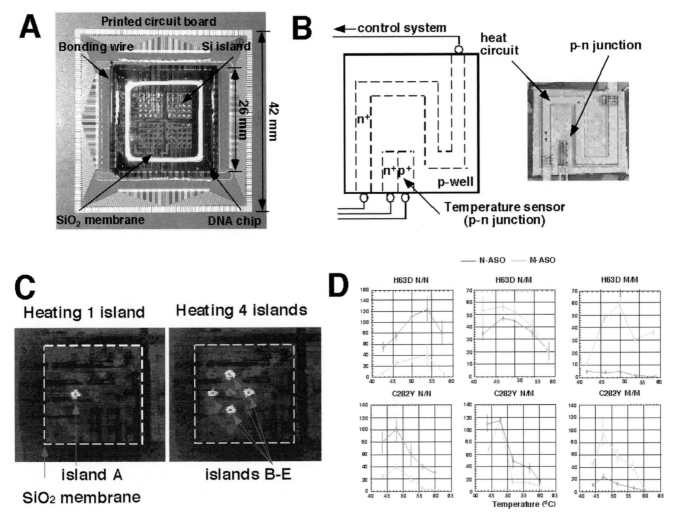

FIGURE 28.8 Thermal Gradient Chip Images. A. Completed thermal gradient chip wire-bonded to a small printed circuit board for testing. The silicon islands and supporting silicon dioxide support structure are visible in the center. B. Location and size of the heating and thermal sensing regions within a single island with photographic details. C. When four islands surrounding an unheated island are heated (right), their combined effect on the temperature of the central island is much greater than when a single island is heated (left). D. Genotype analyses for two SNPs in the gene for hemochromatosis: fluorescence units (y axis) as a function of chip island temperature (x axis) are shown for hybridization of Cy 5-labeled, denatured PCR, solution-phase targets to chip-bound normal (N-ASO, red) or mutant (M-ASO, green) oligonucleotide probes. Genomic DNA from either heterozygous (N/M, middle panels), wild type (N/N, left panels), or homozygous mutant individuals (M/M, right panels), encompassing either the H63D (upper three panels) or C282Y (lower three panels) SNPs in the gene for hemochromatosis, were amplified by PCR. Samples then were heat denatured and annealed to chips containing different thermally isolated islands. Numbers in brackets refer to the ratio of intensities of annealing of labeled targets to normal (red)/mutant (green) probe spots on the array (adapted from Kajiyama *et al.*, 2003). (See color plate)

hybridized to the DNA, indicating the genotype of each sample. This entire process is highly automated; subsequent to sample preparation, the electronic addressing, hybridization, fluorescence scanning, and data analysis steps are performed on the NanoChip™ Molecular Biology Workstation.

The Nanogen approach has been effective to reduce the hybridization duration events to several minutes; however, the design of these chips remains complex and thus chip cost is still high.

Investigators at the University of Pennsylvania (Su *et al.*, 2002) employed the electric field to attract DNA target to the chip surface, while drastically changing the chip design. They used standard glass slides coated commercially with a layer of indium and tin oxide (ITO) as the conductive material, which acts as an electrode. ITO is transparent and is used as the active material in liquid crystal displays (LCDs). It was found that standard methods for attaching oligonucleotides to a glass surface (silanization with an amino silane, reaction with a difunctional isothiocyanate, and reaction with an amino oligonucleotide) worked well with ITO surfaces. After oligonucleotide spots are attached to the conductive surface of one slide, a small amount of solution containing the complementary oligonucleotide is placed on the surface inside a thin gasket (typically around 25 µl) and the second glass slide with electrode facing the solution is clamped on. A low voltage (<1V) is applied between the electrodes such that the array side is at positive potential as compared to the second electrode. After a fixed time, the cell is disassembled and the hybridization is quantified with Cy5 dye labeling and laser scanning. Results are expressed as relative fluorescence intensity, defined as the ratio of (spot signal—background signal) to (reference spot signal—background signal).

A series of studies with PCR products of different sizes were performed to evaluate the improvement in hybridization times using this technique (also known as electric field-enhanced hybridization, or EFEH). The enhancements from about 40 to 200 times for DNA in the size ranges considered have been measured. In addition to improving speed of hybridization, EFEH showed an improved discrimination between perfectly matched and single-base mismatched hybrids. In this case, the field is applied in the reverse direction for a short time (sec) following the application in forward direction to drive target to the surface. The bond strengths for imperfect matches is lower than for perfect ones and therefore may be removed more easily by the reverse field.

The main value of the proposed EFEH technique is its simplicity. It uses glass slides (versus the more expensive silicon ones) and it moves away from individual addressing of the hybridization sites. Thus, fabrication of these devices is much less complex and less expensive. By eliminating the individual addressing scheme, this method loses some flexibility.

28.5.3 Electronic Detection of DNA Hybridization: Clinical Microsensors Technology

Electronic detection of DNA hybridization has been developing rapidly, since it offers potential toward further miniaturization of DNA diagnostics (Wang, 1999; 2000). Use of electrochemical detection schemes obviates the need for large and expensive fluorescent scanners. The detection chips can be produced inexpensively using Si- or printed circuit board-based fabrication technologies. Most importantly, detection schemes not requiring external labels can be devised, further simplifying signal collection and enabling studies of reaction kinetics.

Clinical Micro Sensors (http://www.cms.com) has developed electrochemical detection (Farkas, 1999; Umek *et al.*, 2001), low-density hybridization arrays for diagnostic applications, in particular, detection of SNPs. This so-called eSensor™ device is composed of a printed circuit board (PCB) chip and a plastic cartridge that is attached on the PCB chip using a double-side adhesive tape to form an 80 µl reaction chamber. The PCB chip consists of an array of gold electrodes modified with a multicomponent, self-assembled monolayer (SAM) that includes presynthesized oligonucleotide (DNA) capture probes that are covalently attached to the electrode through an alkyl thiol linker. When a sample solution containing target DNA is introduced into the detection cartridge, specific capture probes on an electrode surface encounter complementary DNA from the sample and hybridization occurs. Two ferrocene, electronic labels (capture probe and signaling probe) bind the target in a sandwich configuration.

Binding of the target sequence to both the capture probe and the signaling probe connects the electronic labels to the surface through the chains of molecular wires built into SAMs (Creager *et al.*, 1999). This adds a circuit element to the bioelectronic circuit on that electrode, and presence of hybridized (double-stranded) DNA can be detected using alternative current voltammetry (ACV). Since incoming target itself is not labeled, the washing step (to remove excessive, nonbound target) prior to the signal collection is not required. A continuous monitoring of the binding process with a quantitative measurement of the target accumulation is possible.

28.5.4 Hybridization Microarrays with *in situ* Probe Synthesis: CombiMatrix Technology

The use of on-chip *in situ* probe synthesis provides an attractive alternative to hybridization microarrays fabricated using presynthesized oligonucleotides and their physical deposition on the substrate. Chips with *in situ* probe synthesis are much more flexible to build, and valuable in the experiments where new sequences need to be introduced frequently. A method relying on photolithography and combinatorial chemistry allowing for parallel synthesis of probes nucleotide-by-nucleotide has been developed (Fodor *et al.*, 1991). The use of photolabile chemistry allows for deprotection of the linker under the UV light exposure and attachment of the nucleotide. The fabrication costs of photolithographic mask sets is, however, fairly high. Therefore, this technology has been modified by using Digital Light Processor (DLP) to develop the Maskless Array Synthesizer (MAS), which delivers focused light beams to selected portions of the substrate, thus eliminating need for the photomask (Nuwaysir *et al.*, 2002).

Another step toward miniaturization of diagnostic assays can be undertaken when electrochemical synthesis of probes is implemented. CombiMatrix (http://www.combimatrix.com) introduced electrochemical *in situ* synthesis using electronic CMOS devices (Dill *et al.*, 1999;

Oleinikov *et al.*, 2003). These chips contain high-density arrays of individually addressable microelectrodes. A typical chip has 1,024 microelectrodes that are each 100 μm in diameter (see Fig. 28.9). Each microelectrode in the array can be addressed independently. An activated electrode generates protons, resulting in a local change of pH. In turn, a pH-controlled chemical reaction can occur in the vicinity of a given electrode. The CombiMatrix biochips are coated with proprietary approximately 1 μm thick porous reaction layer material, which is used for immobilization and synthesis of biomolecules used for subsequent binding of target. Covalent linkage of the molecules within the porous layer is accomplished using reagents that are generated *in situ* by the microelectrodes.

In immunoassay format, biotin is immobilized in the porous layer. Subsequently, biotin-labeled sites then can conjugate with streptavidin, which in turn can capture biotin-labeled proteins or antibodies. Sandwich immunoassays can be used for larger entities and competitive immunoassays for smaller molecules. The technique was used for detection of saxitoxin, BG spores, murine IgG, and salmonella, among others (Dill *et al.*, 2001).

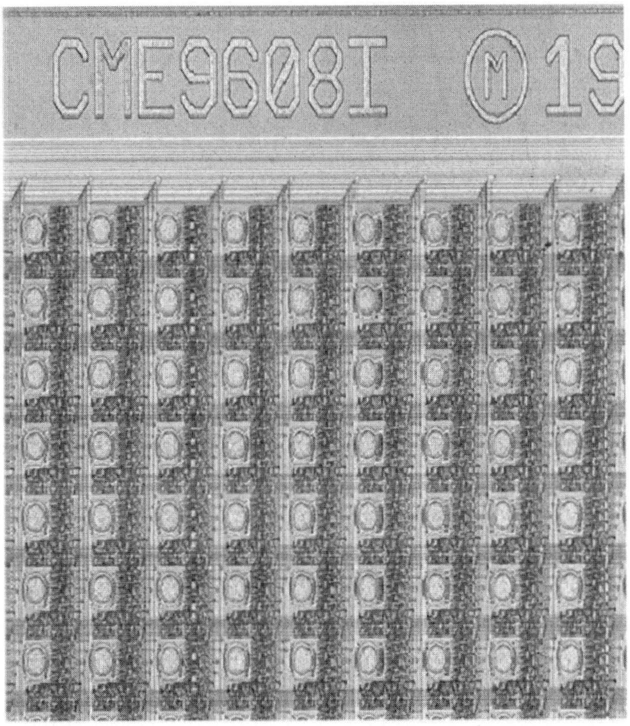

The use of phosphoramidite chemistry at the electrode sites, combined with sequential addressing of these electrodes in order to produce protons, results in the synthesis of oligonucleotides (Oleinikov *et al.*, 2003). CombiMatrix demonstrated an *in situ* synthesis of 50-mers oligonucleotides. A complete chip with such probes can be prepared within 24 hours. Using this technique, CombiMatrix demonstrated SNP and gene expression assays. Furthermore, they also showed protein recognition using oligonucleotide arrays. Proteins under study were tagged with matching (to the one synthesized at the electrode site) oligonucleotide through streptavidin-biotin linker. Hybridization of oligonucleotides allowed immobilization of protein on the chip surface (Oleinikov *et al.*, 2003).

At the initial stage of CombiMatrix development, detection was accomplished using fluorophore-tagged antibodies and epifluorescent microscopy. However, since the electrode sites on the chip are individually addressable, they also can be used to detect events occurring on the chip surface using electrochemistry (Oleinikov *et al.*, 2003).

28.5.5 E-Chem Detection and Direct DNA Conductivity Measurement: GeneOhm

GeneOhm (http://www.geneohm.com) uses electron flow through a double helical DNA molecule to generate a current, which is dependent upon complementary base pairing within the helix. Briefly, an oligonucleotide spanning the region to be interrogated is deposited on a gold surface array. PCR target DNA from patient is then heat-denatured and annealed to the array. Perfect matches allow electrons to flow through the helix generating a measurable current, wheras single-base mismatches limit current flow. Multiple loci can be interrogated since each position on the array has an independent current sensor at each register, thereby enabling parallel analysis (Boon *et al.*, 2000; 2002; Drummond *et al.*, 2003).

28.5.6 Hybridization Arrays with Magnetic Sensing

The use of magnetic devices in biological assays is not limited to cell separation. They can be also used for detection of hybridization events, when the target DNA is labeled with a magnetic bead. Pioneering work by the Naval Research Labs on their Bead Array Counter (BARC) series demonstrated feasibility of this approach (Edelstein *et al.*, 2000; Miller *et al.*, 2001) using giant magneto-resistance (GMR) multilayer magnetic sensors. Departure from optical or electrochemical detection methods allowed detector size to be considerably reduced while maintaining high sensitiv-

FIGURE 28.9 A section of the CombiMatrix CME9608 chip. Each of the small round disks is an electrode. Each electrode site is associated with CMOS circuitry for individually addressing and operating the electrode. CombiMatrix first generation CME9608 series chips have densities of over 1,000 sites per square centimeter. These chips are fabricated using a conventional three-micron CMOS process. The next generation of chips will have over 6,000 sites per square centimeter. CombiMatrix also has special-purpose designs that achieve densities well over one million sites per square centimeter (adapted from Dill *et al.*, 2001).

ity. This size reduction can occur due to feasibility of integrating the magnetic sensor with electronics for signal acquisition on a single chip. Furthermore, this approach allows for improvement of reaction kinetics and control of hybridization stringency. The former is achieved through utilization of magnetic field for target preconcentration at the probe location, and the latter through use of an AC field to repel mismatched targets.

28.5.7 On-Chip Amplification and Sample Preparation: Cepheid Technology

The importance of microchip PCR developments was previously described, the major advantages being the speed of operation and reduced reagent volumes. Cepheid (http://www.cepheid.com) has built its technology around rapid recognition of bacterial and viral pathogens using DNA amplification in microchip format (Belgrader *et al.*, 1998; 1999). Due to its market target applications within biodefense, food, and environmental testing, the company embarked on development of microdevices, which operate at rapid thermal cycling rates in order to reduce the PCR assay duration time to a minimum (Belgrader *et al.*, 1999). The microchip format technology, originally developed at Lawrence Livermore National Laboratory, uses polypropylene, single-use reaction vials (volume of 25–100 μl) equipped with a silicon ring heater for rapid thermal cycling. Silicon has a very high thermal conductivity and is a prime choice for a heater in such applications. The ramp rates achieved in solution are 10°C/sec for heating between 50° and 95°C and 2.5°C/sec for cooling at the same temperature range. In addition, to further reduce the duration of the assay, the chip is equipped with an optical window and online fluorescent detection to terminate thermal cycling as soon as the level of amplified signal reaches a detection threshold. This feature in combination with a modified two-temperature PCR protocol allowed for detection of *Erwinia* in 7 min (Belgrader *et al.*, 1999).

A similar assay for *B. anthracis* starting with 100 pg of DNA can be completed in 13 min, and using 0.1 pg DNA required 19 min. The original chip solution just described has evolved into the Smart Cycler II and Smart Cycler II TD instruments (Belgrader *et al.*, 2001), capable of amplification of 16 different samples at a time, with four-plex/well multiplexing capability and four-color fluorescence detection.

In order to facilitate sample preparation prior to DNA amplification, Cepheid developed a cell lysing unit. *B. anthracis* and *B. subtilis* spores are difficult to lyse using traditional thermal or chemical methods. They utilized sonication (Belgrader *et al.*, 2000), which relies on the use of ultrasonic horn operating at approximately 40 kHz to deflect a flexible wall of the sample container at high impact. In order to improve lysis efficiency, purified spores are mixed with glass beads prior to sonication. There was also a need for the use of large volumes of starting sample (100 μl to 1 ml), since initial concentrations of pathogens under study may be very low. This need for analysis of large sample volumes led to development of sample preconcentration units.

The individual units for sample preparation and PCR-based DNA detection are designed in a modular format using an I-CORE platform. The unit subsequently is assembled into the analytical system under the GeneXpert product name. GeneXpert is designed to purify, concentrate, detect, and identify targeted DNA sequences, taking unprocessed sample to result in less than 30 min (Taylor *et al.*, 2001).

28.5.8 LabChip: Caliper Technologies

Caliper Technologies Corporation (http://www.calipertech.com) designs and manufactures microfluidic chips for separating molecules. The chip incorporates electrokinetics and pressure to move molecules through microfluidic channels. For the analysis of cellular components, pressure is used for controlling movement of the cells through the chip. For DNA, RNA, and protein assays the charged molecules are moved through the channels using electrokinetics (Panaro *et al.*, 2000).

In collaboration with Agilent Technologies, a platform was developed for the analysis of DNA, RNA, proteins, or cells. The Agilent 2100 BioAnalyzer allows researchers to use small amounts of their sample for analysis and to see data within 30 minutes of sample loading. The software provided with the instrument analyzes the results for each sample. Samples are added to wells and are fluorescently labeled with an intercalating dye. They move through the microfluidic channels and are injected into a separation chamber for analysis. The labeled fragments of DNA, RNA, or protein are separated by molecular sieving and detected by fluorescence. Samples are injected and analyzed individually. The entire time from sample loading to completion of the assay is approximately 30 minutes for 12 samples. The RNA 6000 NanoChip has become the standard method in laboratories for quantitating RNA used in gene expression analysis.

Cells are analyzed on the instrument using two-color fluorescence in combination with flow cytometry. The preparation of cells for flow cytometry analysis normally involves a lengthy staining and labeling step; however, cells are stained on the chip before analysis to allow for a faster workflow. The BioAnalyzer allows researchers to use less cells than in traditional flow cytometry assays. Developed assays and protocols for cell analysis are provided with the Cell Fluorescence LabChip Kit, but can be adjusted for researchers wishing to make changes to the protocol. Six samples will take approximately 25 minutes to be analyzed on the system (http://www.chem.agilent.com).

28.6 CONCLUSIONS

The convergence of molecular biology and high microelectronic technology has been unprecedented in the last decade. This bidirectional utilization and the cross-enrichment of these techniques are leading to the development of integrated multifunctional genetic analytical systems, which will have high sensitivity and specificity and will be cheaper and faster to operate as compared to their bench-top predecessors.

The ability to produce large arrays of identical and miniature elements using lithography led to the birth of high density hybridization arrays and multichannel separation systems, and revolutionized high throughput analysis in gene discovery, mutation studies, and SNP analysis. Furthermore, the microfabrication techniques allow for delivering and harvesting of electrical and magnetic signals at the distinct locations of the analytical chip with an ability of choosing individual addresses in large sensor arrays. These developments allow for a design of localized sensors analyzing single cells or molecules immobilized in distinct locations on the chip. It is expected that further progression in capability of these techniques will be achieved through the emerging field of nanotechnology.

References

Ahn, C. H., and Allen, M. G. (1993). A planar micromachined spiral inductor for integrated magnetic microactuator applications. *J. Micromech. Microeng.* 3, 37–44.

Alonso-Amigo, G. (2000). Polymer microfabrication for microarrays, microreactors and microfluidics. *J. Assoc. Lab. Aut.* 5, 96–101.

Anderson, R. C., Su, X., Bogdan, G. J., and Fenton, J. (2000). A miniature integrated device for automated multistep genetic assays. *Nucleic Acids Res.* 28, E60.

Belgrader, P., Benett, W., Hadley, D., Long, G., Mariella, R., Jr., Milanovich, F., Nasarabadi, S., Nelson, W., Richards, J., and Stratton, P. (1998). Rapid pathogen detection using a microchip PCR array instrument. *Clin. Chem.* 44, 2191–2194.

Belgrader, P., Benett, W., Hadley, D., Richards, J., Stratton, P., Mariella, R., Jr., and Milanovich, F. (1999). PCR detection of bacteria in seven minutes. *Science* 284, 449–450.

Belgrader, P., Okuzumi, M., Pourahmadi, F., Borkholder, D. A., and Northrup, M. A. (2000). A microfluidic cartridge to prepare spores for PCR analysis. *Biosens. Bioelectron.* 14, 849–852.

Belgrader, P., Young, S., Yuan, B., Primeau, M., Christel, L. A., Pourahmadi, F., and Northrup, M. A. (2001). A battery-powered notebook thermal cycler for rapid multiplex real-time PCR analysis. *Anal. Chem.* 73, 286–289.

Berger, M., Castelino, J., Huang, R., Shah, M., and Austin, R. H. (2001). Design of a microfabricated magnetic cell separator. *Electrophoresis* 22, 3883–3892.

Boon, E. M., Ceres, D. M., Drummond, T. G., Hill, M. G., and Barton, J. K. (2000). Mutation detection by electrocatalysis at DNA-modified electrodes. *Nat. Biotechnol.* 18, 1096–1100.

Boon, E. M., Salas, J. E., and Barton, J. K. (2002). An electrical probe of protein-DNA interactions on DNA-modified surfaces. *Nat. Biotechnol.* 20, 282–286.

Boone, T., Hooper, H., and Soane, D. (1998). Integrated chemical analysis on plastic microfluidic devices. Hilton Head Island, South Carolina, June 8–11th.

Burns, M. A., Johnson, B. N., Brahmasandra, S. N., Handique, K., Webster, J. R., Krishnan, M., Sammarco, T. S., Man, P. M., Jones, D., Heldsinger, D., Mastrangelo, C. H., and Burke, D. T. (1998). An integrated nanoliter DNA analysis device. *Science* 282, 484–487.

Chalmers, J., Zborowski, M., Sun, L., and Moore, L. (1998a). Flow through immunomagnetic cell separation. *Biotechnol. Prog.* 14, 141–148.

Chalmers, J. J., Mandal, S., Fang, B., Sun, L., Zborowski, M. (1998b). Theoretical analysis of cell separation based on cell surface marker density. *Biotechnol. Bioeng.* 59, 10–20.

Chee, M., Yang, R., Hubbell, E., Berno, A., Huang, X. C., Stern, D., Winkler, J., Lockhart, D. J., Morris, M. S., and Fodor, S. P. (1996). Accessing genetic information with high-density DNA arrays. *Science* 274, 610–614.

Cheek, B., Matthew, A., Torres, P., Yu, Y., and Yang, H. (2001). Chemiluminescence detection for hybridization assays on the flow-thru chip, a three-dimensional microchannel biochip. *Anal. Chem.* 73, 5777–5783.

Cheng, J., Shoffner, M. A., Hvichia, G. E., Kricka, L. J., and Wilding, P. (1996). Chip PCR. II. Investigation of different PCR amplification systems in microfabricated silicon-glass chips. *Nucleic Acids Res.* 24, 380–385.

Cheng, J., Sheldon, E. L., Wu, L., Heller, M. J., and O'Connell, J. P. (1998a). Isolation of cultured cervical carcinoma cells mixed with peripheral blood cells on a bioelectronic chip. *Anal. Chem.* 70, 2321–2326.

Cheng, J., Sheldon, E., Wu, L., Uribe, A., Gerrue, L., Carrino, J., Heller, M., and O'Connell, J. (1998b). Preparation and hybridization analysis of DNA/RNA from E. coli on microfabricated bioelectronic chips. *Nat. Biotechnol.* 16, 541–546.

Cheng, J., Waters, L. C., Fortina, P., Hvichia, G., Jacobson, S. C., Ramsey, J. M., Kricka, L. J., and Wilding, P. (1998c). Degenerate oligonucleotide primed-polymerase chain reaction and capillary electrophoretic analysis of human DNA on microchip-based devices. *Anal. Biochem.* 257, 101–106.

Cheng, J., Mitchelson, M. (2001). In *Practical Applications of Capillary Electrophoresis*, Capillary Electrophoresis of Nucleic Acids, vol. 2, pp. 211–219. Humana Press, Totowa, NJ.

Chou, C.-F., Tegenfeldt, J. O., Bakajin, O., Chan, S. S., Cox, E. C., Darnton, N., Duke, T., and Austin, R. H. (2002). Electrodeless dielectrophoresis of single- and double-stranded DNA. *Biophysical J.* 83, 2170–2175.

Christenson, T. R., Klein, J., and Guckel, H. (1995). An electromagnetic micro dynamometer. IEEE Conf., pp. 386–391. Amsterdam, The Netherlands.

Creager, S., Bamdad, C., MacLean, T., Lam, E., Chong, Y., Olsen, G. T., Luo, J., Gozin, M., and Kayyem, J. F. (1999). Electron transfer at electrodes through conjugated "molecular wires" bridges. *J. Am. Chem. Soc.* 121, 1059–1064.

Dill, K., Stanker, L. H., and Young, C. R. (1999). Detection of salmonella in poultry using a silicon chip-based biosensor. *J. Biochem. Biophys. Methods* 41, 61–67.

Dill, K., Montgomery, D. M., Wang, W., and Tsai, J. C. (2001). Antigen detection using microelectrode array microchips. *Anal. Chim. Acta.* 444, 69–78.

Dong, S., Wang, E., Hsie, L., Cao, Y., Chen, X., and Gingeras, T. R. (2001). Flexible use of high-density oligonucleotide arrays for single-nucleotide polymorphism discovery and validation. *Genome Res.* 1, 1418–1424.

Drummond, T. G., Hill, M. G., and Barton, J. K. (2003). Electrochemical DNA sensors. *Nat. Biotechnol.* 21, 1192–1199.

Edelstein, R. L., Tamanaha, C. R., Sheehan, P. E., Miller, M. M., Baselt, D. R., Whitman, L. J., and Colton, R. J. (2000). The BARC biosensor applied to the detection of biological warfare agents. *Biosensors Bioelectr.* 14, 805–813.

Edman, C., Raymond, D., Wu, D., Tu, E., Sosnowski, R., Butler, W., Nerenberg, M., and Heller, M. (1998). Electric field directed nucleic acid hybridization on microchips. *Nucleic Acids Res.* 25, 4907–4914.

Fan, Z. H., Mangru, S., Granzow, R., Heaney, P., Ho, W., Dong, Q., and Kumar, R. (1999). Dynamic DNA hybridization on a chip using paramagnetic beads. *Anal. Chem.* 71, 4851–4859.

Fan, J. B., Chen X., Halushka, M. K., Berno, A., Huang, X., Ryder, T., Lipshutz, R. J., Lockhart, D. J., and Chakravarti, A. (2000). Parallel genotyping of human SNPs using generic high-density oligonucleotide tag arrays. *Genome Res.* 10, 853–860.

Farkas, D. H. (1999). Bioelectronic detection of DNA and the automation of molecular diagnostics. *J. Assoc. Lab. Autom.* 4, 125–129.

Fodor, S. P., Read, J. L., Pirrung, M. C., Stryer, L., Lu, A. T., and Solas, D. (1991). Light-directed, spatially addressable parallel chemical synthesis. *Science* 251, 767–773.

Foglieni, B., Cremonesi, L., Travi, M., Ravani, A., Giambona, A., Rosatelli, M. C., Perra, C., Fortina, P., and Ferrari, M. (2004). Beta-thalassemia microelectronic chip: a fast and accurate method for mutation detection. *Clin. Chem.* 50, 73–79.

Fortina, P., Delgrosso, K., Sakazume, T., Santacroce, R., Moutereau, S., Su, H., Graves, D., Mckenzie, S., and Surrey, S. (2000). Simple two-color array-based approach for mutation detection. *Eur. J. Hum. Genet.* 8, 884–894.

Freemantle, M. (1999). Downsizing chemistry. *Chem. Eng. News.* 77, 27–36.

Gascoyne, P. R. C., and Vykoukal, J. (2002). Particle separation by dielectrophoresis. *Electrophoresis* 23, 1973–1976.

Gerry, N. P., Witowski, N. E., Day, J., Hammer, R. P., Barany, G., and Barany, F. (1999). Universal DNA microarray method for multiplex detection of low abundance point mutations. *J. Mol. Biol.* 292, 251–262.

Gilles, P. N., Wu, D. J., Foster, C. B., Dillon, P. J., and Chanock, S. J. (1999). Single nucleotide polymorphic discrimination by an electronic dot blot assay on semiconductor microchips. *Nat. Biotechnol.* 17, 365–370.

Giordano, B., Ferrance, J., Swedberg, S., Huhmer, A., and Landers, J. (2001). Polymerase chain reaction in polymeric microchips: DNA amplification in less than 240 seconds. *Anal. Biochem.* 291, 124–132.

Grodzinski, P., Liu, R. H., Chen, H., Blackwell, J., Liu, Y., Rhine, R., Smekal, T., Ganser, D., Romero, C., Yu, H., Chan, T., and Kroutchinina, N. (2001). Development of Plastic Microfluidic Devices for Sample Preparation. *Biomed Microdevices* 3, 275.

Harrison, D. J., Fluri, K., Seiler, K., Fan, Z., Effenhauser, C. S., and Manz, A. (1993). Micromachining a miniaturized capillary electrophoresis-based chemical analysis system on a chip. *Science* 261, 895–897.

Hirschhorn, J. N., Sklar, P., Lindblad-Toh, K., Lim, Y. M., Ruiz-Gutierrez, M., Bolk, S., Langhorst, B., Schaffner, S., Winchester, E., and Lander, E. S. (2000). SBE-TAGS: an array-based method for efficient single-nucleotide polymorphism genotyping. *Proc. Natl. Acad. Sci. USA* 97, 12164–12169.

Jobs, M., Fredriksson, S., Brookes, A. J., and Landegren, U. (2002). Effect of oligonucleotide truncation on single-nucleotide distinction by solid-phase hybridization. *Anal. Chem.* 74, 199–202.

Joung, J., Shen, J., and Grodzinski, P. (2000). Micro pumps based on alternating high gradient magnetic fields. *IEEE Trans. Magnetics* 36, 36, 2012.

Kajiyama, T., Miyahara, Y., Kricka, L. J., Wilding, P., Graves, D. J., Surrey, S., and Fortina P. (2003). Genotyping on a thermal gradient DNA chip. *Genome Res.* 13, 467–475.

Khandurina, J., Meknight, T. E., Jacobson, S. C., Waters, L. C., Foote, R. S., and Ramsey, J. M. (2000). Integrated System for Rapid PCR-Based DNA Analysis in Microfluidic Devices. *Anal. Chem.* 72, 2995–3000.

Khrapko, K. R., Lysov, Y.-P., Khorlin, A. A., Ivanov, I. B., Yershov, G. M., Vasilenko, S. K., Florentiev, V. L., and Mirzabekov, A. D. A. (1991). Method for DNA sequencing by hybridization with oligonucleotide matrix. *DNA Sequence* 1, 375–388.

Kopp, M., De Mello, A., and Manz, A. (1998). Chemical amplification: continuous-flow PCR on a chip. *Science* 280, 1046–1048.

Kricka, L. (1998). Miniaturization of analytical systems. *Clin. Chem.* 44, 2008–2014.

Kricka, L., Fortina, P., Panaro, N., Wilding, P., Alonso-Amigo, G., and Becker, H., (2002). Fabrication of plastic microchips by hot embossing. *Lab on a Chip* 2, 1–4.

Lagally, E. T., Simpson, P. C., and Mathies, R. A. (2000). Monolithic integrated microfluidic DNA amplification and capillary electrophoresis analysis system. *Sens. Actuators* B 63, 138–146.

Lagally, E. T., Medintz, I., and Mathies, R. A. (2001). Single-molecule DNA amplification and analysis in an integrated microfluidic device. *Anal. Chem.* 73, 565–570.

Lipshutz, R. J., Fodor, S. P., Gingeras, T. R., and Lockhart, D. J. (1999). High density synthetic oligonucleotide arrays. *Nat. Genet.* 21 (Suppl), 20–24.

Liu, Y., Rauch, C., Stevens, R., Lenigk, R., Yang, J., Rhine, D., and Grodzinski, P. (2002). DNA Amplification and Hybridization Assays in Integrated Plastic Monolithic Devices. *Anal. Chem.* 74, 3063–3070.

Liu, R. H., Yang, J., Lenigk, R., Bonanno, J., Grodzinski, P., and Zenhausern, F. (2003). Self-contained, integrated biochip system for sample-to-answer genetic assays. *Proc. µTAS 2003 Symp.*, p. 1319. Kluwer Academic Publishers.

Manz, A., Harrison, D. J., Verpoorte, E., Fettinger, J. C., Paulus, A., Ludi, H., and Widmer, H. M. (1992). Planar chips technology for miniaturization and integration of separation techniques into monitoring systems: capillary electrophoresis on a chip. *J. Chromatogr.* 593, 253–258.

Mastrangelo, C., Burns, M., and Burke, D. (1998). Microfabricated devices for genetic diagnostics. *Proc. IEEE* 86, 1769–1787.

Mei, R., Galipeau, P. C., Prass, C., Berno, A., Ghandour, G., Patil, N., Wolff, R. K., Chee, M. S., Reid, B. J., and Lockhart, D. J. (2000). Genome-wide detection of allelic imbalance using human SNPs and high-density DNA arrays. *Genome Res.* 10, 1126–1137.

Miller, M. M., Sheehan, P. E., Edelstein, R. L., Tamanaha, C. R., Zhong, L., Bounnak, S., Whitman, L. J., and Colton, R. J. (2001). A DNA array sensor utilizing magnetic microbeads and magnetoelectronic detection. *J. Magnetism Magnetic Mat.* 225, 276–281.

Nagai, H., Murakami, Y., Morita, Y., Yokoyama, K., and Tamiya, E. (2001). Development of a microchamber array for picoliter PCR. *Anal. Chem.* 73, 1043–1047.

Nuwaysir, E. F., Huang, W., and Albert, T. J. (2002). Gene expression analysis using oligonucleotide arrays produced by maskless photolithography. *Genome Res.* 12, 1749–1755.

Obeid, P., Christopoulos, T., Crabtree, H., and Backhouse, C. (2003). Microfabricated device for DNA and RNA amplification by continuous-flow polymerase chain reaction and reverse transcription-polymerase chain reaction with cycle number selection. *Anal. Chem.* 75, 288–295.

Oleinikov, A. V., Gray, M. D., Zhao, J., Montgomery, D. D., Ghindilis, L. L., and Dill, K. (2003). Self-assembling protein arrays using electronic semiconductor microchips and in vitro translation. *J. Proteome Res.* 2, 313–319.

Panaro, N. J., Yuen, P. K., Sakazume, T., Fortina, P., Kricka, L. J., and Wilding, P. (2000). Evaluation of DNA fragment sizing and quantification by the Agilent 2100 Bioanalyzer. *Clin. Chem.* 46, 1851–1852.

Pastinen, T., Raitio, M., Lindroos, K., Tainola, P., Peltonen, L., and Syvanen, A. C. (2000). A system for specific, high-throughput genotyping by allele-specific primer extension on microarrays. *Genome Res.* 10, 1031–1042.

Pethig, R., Huang, Y., Wang, X.-B., and Burt, J. P. H. (1992). Positive and negative dielectrophoretic collection of colloidal particles using interdigitated castellated microelectrodes. *J. Appl. Phys.* 24, 881–888.

Simpson, P. C., Roach, D., Wooley, A. T., Thorsen, T., Johnston, R., Sensabaugh, G. F., and Matthies, R. A. (1998). High-throughput genetic analysis using microfabricated 96-sample capillary array electrophoresis microplates. *Proc. Natl. Acad. Sci. USA* 95, 2256–2261.

Sosnowski, R. G., Tu, E., Butler, W. F., O'Connell, J. P., and Heller, M. J. (1997). Rapid determination of single base mismatch mutations in DNA hybrids by direct electric field control. *Proc. Natl. Acad. Sci. USA* 94, 1119–1123.

Su, H.-J., Surrey, S., McKenzie, S. E., Fortina, P., and Graves, D. J. (2002). Kinetics of heterogeneous hybridization on indium tin oxide surfaces with and without an applied potential. *Electrophoresis* 23, 1551–1557.

Taylor, M. T., Belgrader, P., Joshi, R., Kintz, G. A., and Northrup, M. A. (2001). In *Micro Total Analysis Systems 2001*, J. M. Ramsey, A. Van den Berg, eds. Fully Automated Sample Preparation for Pathogen Detection Performed in a Microfluidic Cas-

sette, pp. 670–672. Kluwer Academic Publishers, Monterey, CA.

Tibbe, A. G., de Grooth, B. G., Greve, J., Liberti, P. A., Dolan, G. J., and Terstappen, L. W. (1999). Optical tracking and detection of immunomagnetically selected and aligned cells. *Nat. Biotechnol.* 17, 1210–1213.

Tibbe, A. G., de Grooth, B. G., Greve, J., Dolan, G. J., Rao, C., and Terstappen, L. W. (2002). Magnetic field design for selecting and aligning immunomagnetic labeled cells. *Cytometry* 47, 163–172.

Umek, R. M., Vielmetter, J., Terbrueggen, R. H., Irvine, B., Yu, C. J., Kayyem, J. F., Yowanto, H., Blackburn, G. F., Farkas, D. H., and Chen, Y. P. (2001). Electronic detection of nucleic acids: a versatile platform for molecular diagnostics. *J. Mol. Diagn.* 3, 74–84.

Wang, J. (1999). Electroanalysis and biosensors. *Anal. Chem.* 71, 328R–332R.

Wang, J. (2000). From DNA biosensors to gene chips. *Nucleic Acids Res.* 28, 3011–3016.

Wang, X.-B., and Cheng, J. (2001). In *Biochip Technology*, J. Cheng, and L. J. Kricka, eds. Electronic Manipulation of Cells on Microchip-based Devices, pp. 135–159. Harwood Academic Publishers.

Ward, M. D., Quan, J., and Grodzinski, P. (2002). Metal-polymer hybrid microchannels for microfluidic high gradient separations. *Eur. Cells Materials J.* 3, 123–126.

Waters, L. C., Jacobson, S. C., Kroutchinina, N., Khandurina, J., Foote, R. S., and Ramsey, J. M. (1998). Microchip device for cell lysis, multiplex PCR amplification, and electrophoretic sizing. *Anal. Chem.* 70, 158–162.

Wilding, P., Shoffner, M. A., and Kricka, L. J. (1994). PCR in a silicon microstructure. *Clin. Chem.* 40, 1815–1818.

Wilding, P., Kricka, L. J., Cheng, J., Hvichia, G., Shoffner, M. A., and Fortina, P. (1998). Integrated cell isolation and polymerase chain reaction analysis using silicon microfilter chambers. *Anal. Biochem.* 257, 95–100.

Woolley, A. T., and Mathies, R. A. (1994). Ultra-high speed DNA fragment separations using microfabricated capillary array electrophoresis. *Proc. Natl. Acad. Sci. USA* 91, 11348–11351.

Yang, J., Liu, Y., Rauch, C. B., Stevens, R. L., Liu, R. H., Lenigk, R., and Grodzinski, P. (2002). High sensitivity PCR assay in plastic micro reactors. *Lab on a chip* 2, 179–187.

Yu, H., Sethu, P., Chan, T., Kroutchinina, N., Blackwell, J., Mastrangelo, C., and Grodzinski, P. A miniaturized and integrated plastic thermal chemical reactor for genetic analysis. The Netherlands 2000; Kluwer Academic Publishers; 545–548.

Yuen, P., Kricka, L., Fortina, P., Panaro, N., Sakazume, T., and Wilding, P. (2001). Microchip module for blood sample preparation and nucleic acid amplification reactions. *Genome Res.* 11, 405–412.

CHAPTER 29

Miniaturization Technologies for Molecular Diagnostics

RONALD C. McGLENNEN

Department of Laboratory Medicine and Pathology, University of Minnesota, Access Genetics, Minneapolis, Minnesota, USA

29.1 MINIATURIZING MOLECULAR DIAGNOSTICS FOR THE CLINICAL LABORATORY

The growth of molecular diagnostic testing in the clinical laboratory has been steady, but at rates less than expected by the optimistic analysts of the biotechnology industry. The growth in test volumes for molecular genetic assays has been in excess of 25% per year, yet the impact of this segment is no more than 3% to 4% of the total laboratory market. One observation is that there are hardly any examples of gene-based assays useful in the everyday practice of medicine. Often these assays are complex and hence require more time and expertise to produce a result. The tests are also costly, and without exception, more expensive than comparable tests based on methods not involving nucleic acids. Lastly, the knowledge of genetics, and correspondingly, the interpretations of molecular assays by practicing physicians and laboratorians are rudimentary. These reasons, as well as those related to the state of art for the technologies used in molecular genetics, explain why this area of the clinical laboratory remains esoteric and a minor part of the practice of laboratory medicine. Despite these perceptions, the field of molecular diagnostics continues to promise a new plateau of what the laboratory can do for the understanding of a patient's disease and health.

This chapter will address several of the emerging innovations in molecular diagnostics that are proving to make these tests better, less costly, and more clinically relevant. A major focus in molecular genetics performed on devices the size of chips (i.e., microarrays); a concept that seemingly works in a world now comfortable with microelectronic chips in most aspects of the daily practice. The innovations in the creation, and industrial production of chips that do work, such as the pumping of fluids, the heating and cooling of chemical reactions, and the sensing of chemical and physical phenomenon, under the control of a computer processor, should go hand-in-hand with the fact that genetic testing is measurement at the molecular level.

Despite the logic to make genetic tests smaller and cheaper, however, there are practical considerations as to why miniaturization technologies also create new problems while solving others. Currently, the routine application of chip-based DNA testing remains only a promise. Today, as in the last 10 years, molecular diagnostics remains an area of the clinical laboratory with piecemeal innovation and integration. What may indeed be the basis to molecularize clinical medicine will not only be the innovation of miniaturization of testing technology, but equally important, the proven reduction of the cost of such tests as well as corresponding gains in test performance. This chapter first will focus on selected applications of gene-based testing as an introduction to the need for refinement of these technologies.

29.2 DRIVERS FOR TECHNICAL INNOVATIONS IN MOLECULAR DIAGNOSTICS

Notwithstanding the opportunities created when new and miniaturized technologies are made available to the molecular laboratory, the big opportunity lies in the vigilant search for the new applications of these tests. Innovations leading to improvements in cost, ease of use, and performance for gene testing technologies will be the principle drivers leading to increased use of these tests for patient care. However, the origins of the ideas for many of these innovations are appearing from unexpected directions. Bioterrorism and genetic modified organisms (GMOs; see also Chapter 19) and food safety are words hardly known less than a decade ago. Yet the pursuit of solutions to these new problems have led to new sources of research money for

such things as biochips and miniaturized mass spectrometers, and are the reasons why society is contemplating the use of gene chips for human diagnostics.

Perhaps the single greatest drive for innovation in molecular diagnostics has been the Human Genome Project (HGP). Outwardly the HGP was an effort principally focused on the cloning and sequencing of the human genome and that of other species. But an equally important driver of that public and private effort was the incentive to invent new technologies to improve the pace of genetic discovery. Major support for the HGP came from the United States Department of Energy, who provided grants to encourage entrepreneurship leading to tangible products now in use in many labs around the world. The emergence of new strategies of DNA sequencing, detections schemes for single nucleotide polymorphisms (SNPs), and advancements in more generalized bioanalysis, such as mass spectrometry (see also Chapters 17 and 18) are linked to these grants to industry. Figure 29.1 illustrates the growth of some key technologies consequent to the HGP.

Surveillance against bioterrorism and the monitoring of food safety are two key drivers of innovation in molecular diagnostics. The potential of a microbial threat from the air, water, or land were the impetus to create methods for DNA analysis from highly disparate samples and yet with high

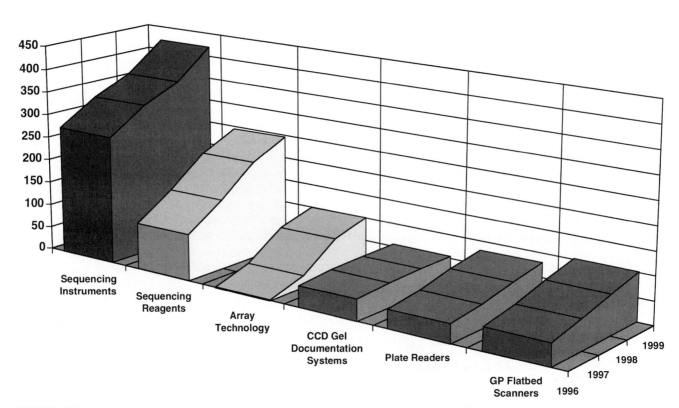

FIGURE 29.1 Estimated and realized output of technologies developed in the wake of the HGP. The variety of gene discovery and diagnostic devices together represents over 1 billion USD in revenue to market worldwide.

sensitivity. Cepheid Corp. was among the first companies to develop a portable platform for DNA testing using the polymerase chain reaction (PCR). Figure 29.2 illustrates one of the early versions of DNA in a suitcase that was field tested by the army and is the basis for the Smartcycler™, the commercial product derived from that early experience in the miniaturization of DNA testing.

In the case of food safety the focus has been on technologies with high throughput and very low cost. The prospect of testing thousands of samples each day for evidence of a wide variety of microbial contaminants as well as for other disease of the species creates a monumental task. In addition, systems that are portable and use instrumentation that are of low complexity are paramount. To this end, meat processing plants have developed their own in-house molecular testing capabilities, some that include technologies as sophisticated as mass spectrometry (Guo, 1999; Demirev et al., 1999).

Current laboratory procedures that rely on manual and highly purified DNA extraction methods are not practical to use in high throughput applications. One technology for sample collection involves a specially treated paper designed to capture cellular material and subsequently covalently affix the extractable DNA onto the interstices of the paper fibers. This product, known as FTA™, has been used principally in forensic investigations, and recently in the detection of pathogenic bacteria, has greatly reduced the complexity of sample procurement. At the same time it holds promise for miniaturization of the PCR reaction and other subsequent assay operations (Rogers and Burgoyne, 1997). The promise of a miniaturized testing platform that would permit the simultaneous analysis of key pathogens and contaminants is needed to create the throughput requirements.

29.3 STRATEGIES FOR THE MOLECULARIZING OF THE CLINICAL LABORATORY

Whereas research, the military, and agro business may have been the major drivers of innovation in molecular testing, a larger opportunity lies in applications for diagnostic medicine for people. Molecular diagnostics continues to be the fastest growing and most profitable for the large commercial references laboratories (Amos and Patnaik, 2002). Figure 29.3 highlights the expectation of the molecular diagnostics market and its distribution for various types of testing.

Despite this, the number of laboratories that have expertise in, or offer molecular genetics assays for patients are less than 500 across North America. The reasons for the slow rate of adoption may include the lack of genetic expertise, concerns over the clinical utility of such tests, and problems with reimbursement. But chief among the concerns is the state of the technology for performing these tests. In particular, there is a noted lack of automation and integration of the assays and the analytic technologies involved in the production of a genetic test result (see also Chapter 27). With the continued interest in genetic testing for patient care the question again can be raised—does molecular diagnostics represent the promise of the laboratory aspects of the science of genetics, or is it a compilation of technologies in need of a suitable place within the clinical laboratory?

A.

B.

FIGURE 29.2 The Smart Cycler™ XC (Xtreme Conditions) from Cepheid is a very rapid, highly efficient battery-operated thermal cycler with real-time optical detection. A. The system is composed of 16 independently programmable reaction sites, each with four channel multiplexed fluorometric detection. B. Sixteen different protocols can be processed and monitored simultaneously and each reaction can be terminated as soon as a positive signal threshold is reached. The Smart Cycler XC is capable of rapid, real-time, field-based pathogen detection or other nucleic acid probe analysis.

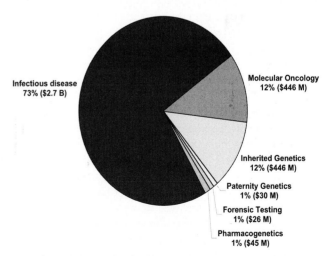

Molecular Diagnostics by Market Segment
$3.71 B in 2006

Infectious disease
73% ($2.7 B)

Molecular Oncology
12% ($446 M)

Inherited Genetics
12% ($446 M)

Paternity Genetics
1% ($30 M)

Forensic Testing
1% ($26 M)

Pharmacogenetics
1% ($45 M)

FIGURE 29.3 The size of the clinical molecular diagnostics market is estimated to be 3.6 billion USD by 2006. The distribution of the types of testing are largely unchanged from earlier years, and include infectious disease testing, especially for HIV, HCV, and more recently, HPV representing more than 70% of the market size.

With the emergence of miniaturization technologies for DNA testing, the concomitant focus on automation and integration will also occur. Correspondingly, with automation and integration the operations for DNA testing will become simpler, thereby reducing the technical barriers for implementation throughout many sectors of the clinical lab. This process of distributing the technology of the molecular diagnostics laboratory and the corresponding molecularization of other sectors of the clinical laboratory will greatly accelerate the adoption of DNA-based testing for patient care. Strategies for achieving this, however, will require the acknowledgement of those basic operations that are part of every genetic test result. These strategies must be understood by the engineers who will create, with their new technologies, the necessary innovations.

Molecular diagnostic laboratories are organized around operational procedures. Unlike other areas of the clinical laboratory with automated instrumentation that complete testing within a single piece of equipment, the molecular laboratory has at best, piece-meal automation and integration of those processes. In the case of any nucleic acid-based test, there are five distinct operational steps:

1. Specimen procurement
2. Nucleic acid extraction and characterization
3. Gene chemistry
4. Product detection
5. Interpretation

For each of these operations there exists the conventional methods and a series of emerging miniature technologies that eventually may replace the former. In the meantime, most clinical molecular genetics laboratories strive to perform each of these operational steps in the most efficient and cost-conserving manner using a combination of technologies, some of which are commercial and some of which are "home brew."

29.3.1 Specimen Procurement

The wide variety of molecular genetic tests also necessitates the use of numerous sample types. In the case of most forms of genetic testing, blood is the most commonly sought tissue type. However, there is no particular reason other sources of cells or tissue, including simple cellular scrapes, a sample of hair follicle, and a wide host of other body fluids cannot be used. Additionally, archived tissue fixed in formalin or alcohol and embedded in paraffin are the specimens used in up to 25% of molecular diagnostics tests. If the procurement of sample includes all costs that start from the labor to obtain the sample, the materials to procure them, and in some cases, the extra work to prepare them specially for the steps of nucleic acid extraction, this facet of genetic testing is the most costly, estimated to be as much as 70% of the total cost of the assay (Eng and Vijg, 1997). New technologies under

design and development to miniaturize the processes of DNA and RNA extraction should focus on the heterogeneity of specimens tested in the molecular laboratory and make this area of operations a primary target of innovation.

Two concerns in specimen procurement are paramount. First is the determination of whether DNA, RNA, or both types of nucleic acid are required for specific types of testing. Although most preservatives used in blood collection are satisfactory for the differential extraction of both RNA or DNA, most other types of tissue preservation including flash freezing, formalin fixation, and other techniques commonly used in anatomic pathology are usually unsatisfactory for routine retrieval of RNA. Recent improvements in RNA preservation are centered on the use of special collection vessels that contain liquids that inhibit RNA degradation (Medeiros *et al.*, 2003; Bhagwat *et al.*, 2003). The PAXgene™ System contains a proprietary blend of reagents that bring about immediate stabilization of RNA. This blend prevents the drastic changes in the cellular RNA expression profiles that normally take place *in vitro*, after blood collection. This product requires collection in a specialized tube and processing once the sample reaches the laboratory.

In addition to the technical aspects of specimen procurement is the consideration that not all tissue samples are representative of the disease process. One example is in the testing of genes involved in cancer, such as the mutational analysis of the tumor suppressor gene p53. In the process of malignant transformation, the p53 gene commonly accumulates point mutations in a series of exons that encode for important protein function. Malignant tissues are typically a mixture of tumor, stromal, and inflammatory cells. The latter do not harbor mutations in p53. The net effect is a relative dilution of the mutant p53 signal. Strategies for evaluating cancer genes must include methods for either improving the sensitivity of current detection methods or ways to enrich for the target genes in samples that are genetically heterogeneous. These practical aspects of specimen collection, in consideration of new technologies described later to make specimen procurement simpler and cheaper, need to be engineered and optimized.

Alternative sample collection strategies, such as specially treated filter paper, use of body fluids such as urine, or even swabs taken from specific body sites, are not only adequate but also preferred. In the future, specimen procurement will be designed as a part of an integrated genetic testing format, which may include a simple collection device, which in its packaging is associated with the other elements necessary to perform a complete DNA test. A prototype of such a device that incorporates FTA for collection of a blood droplet is linked to a prepackaged gene chemistry module containing lyophilized reagents for a selected PCR reaction (see Fig. 29.4).

One advantage of sampling for gene-based testing is that DNA and RNA are very durable and can be retrieved from

of DNA or RNA have been demonstrated at chip-sized dimensions. Here the principal innovations are the combination of fluidic channels and pumps. Figure 29.7 illustrates a set of microfabricated piezoelectric pumps that have been used in the delivery of aqueous lysis solutions to small reservoirs of cells. The displacement of a single stroke of each diaphragm is approximately 2–4 µL, cycling at 60–100 Hz. Pumps and other types of fluidic conduits described in other chapters have also been used to sieve whole blood to effect the sequestration of nucleated cells and the subsequent lysis and delivery of the whole solution to a chip-based thermocycler for PCR (Woolley *et al.*, 1996; Paegel *et al.*, 2003; Chung *et al.*, 2004).

29.3.3 Gene Chemistry

This section refers to the use of a variety of novel techniques for manipulating nucleic acid for the purpose of analyzing specific regions within those samples, whether they reside on a gene or a noncoding sequence that specifically relates to the diagnosis of genetic diseases or human identity. What are considered now traditional methods including Southern transfer, northern blotting, and other simple hybridization techniques like slot blot, dot blot, and reverse dot blots are still very much in use in clinical molecular genetics laboratory (see also Chapter 27). Depending on the particular medical condition or type of genetic testing, these techniques will continue to be used since they have the advantage of being robust and unambiguous. By contrast, the utility of various enzymatic manipulations of DNA and RNA are the dominant mode of genetic testing and are therefore the target of devices designed to make microscale such conventional instruments as the thermocycler.

Several gene chemistry platforms are now in common use in the clinical laboratory, and research directed at their application to chip-sized assays is ongoing. The stalwart of the molecular diagnostics laboratory has been the PCR, which has demonstrated to work well for a number of assays when performed in miniaturized thermocycling devices (Wilding *et al.*, 1994a; Kricka and Wilding, 1996; Woolley *et al.*, 1996; Kalinina *et al.*, 1997; Ramsay, 1998; Lagally *et al.*, 2001; Hawkins *et al.*, 2002; Trau *et al.*, 2002). One advance has been the technique of primer extension, which has been used in mass spectroscopy, based on MALDI-TOF (Guo, 1999; Meldrum, 2000; Wise *et al.*, 2003; see also Chapter 17).

In addition, however, gene chemistries based on ligase chain reaction (LCR), cleavase, or Invader®, and other variations on this theme including Q-beta replicase and self-sustained sequence replication also work when applied to chip-based devices (Cheng *et al.*, 1996; Wilkins Stevens *et al.*, 2001). Because these gene chemistries are not necessarily designed for the miniature environment, consideration of special adaptation of these technologies should be evaluated. Important to remember, however, is that PCR, and similarly other enzymatic amplification methods are used only to generate sufficient amounts of a DNA amplicon, so that they can secondarily be analyzed by any of a number of analytic technologies. The prospect of integrating the process of thermocycling reaction and direct product detection on miniaturized platforms is only now beginning to be explored.

Characteristics of a miniaturized thermocycling device are not well understood. The first demonstration of such a concept was achieved by Wilding and colleagues (1994a), who achieved the amplification of moderate-sized DNA fragments, easily visualized by polyacrylamide gel electrophoresis. Such a device was constructed from silicon, although subsequent devices have been fabricated from glass plastic and other composites. Multichannel and multiplexed devices have been created that are capable of up to 24 independent 1–2 µL PCR reactions (see Fig. 29.8). Early experience with such microinstruments has taught the importance of making the surface passivation, control of evaporation, and the role of surface tension to the fluid volume that makes up the PCR reaction (Wilding *et al.*, 1994a; Shoffner *et al.*, 1996; Kalinina *et al.*, 1997; Burke *et al.*, 1997; Ibrahim *et al.*, 1998; Hawkins *et al.*, 2002). Advantages of the chip-based thermocyclers include excellent temperature control, very rapid cycling times, and only nominal energy requirements. Figure 29.8c illustrates the performance of one such device. Ramping temperatures in excess of 25°C/sec and excellent thermal stability are the hallmark of these silicon-based devices.

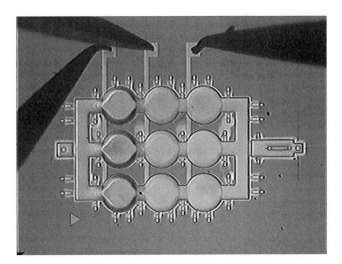

FIGURE 29.7 MEMS-fabricated micropumps are constructed from a freestanding diaphragm overlying silicon nitride base. The diaphragm is actuated either by electrostatic attraction and repulsion driven by an oscillatory circuit, or may be driven by the contraction and expansion of piezoelectric thin films such as PZT. The pump system shown can be differentially actuated to displace fluids in parallel, closing one pathway, while opening the other two, creating a valving of the fluid flow.

A

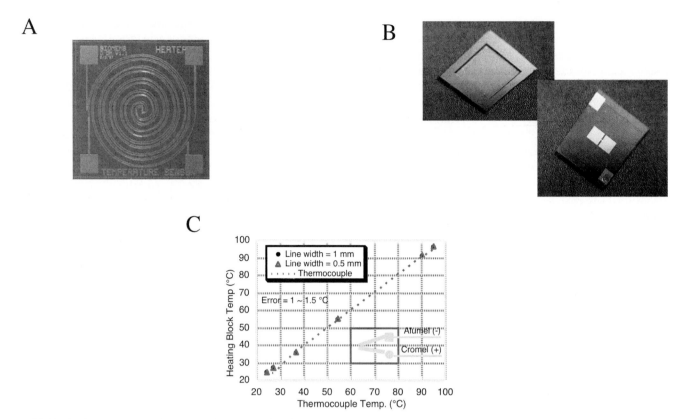

B

C

FIGURE 29.8 MEMS-based thermocycling chips. The devices shown vary in size and surface preparation as well as the means by which the chips are heated and the temperature measured. A. Thermocouple sensor chip. B. Chips with integrated thermodiode sensors. C. Sensor response for the chip in panel A, showing excellent linearity of heat temperature response. (See color plate)

Another version of a silicone chip-based thermocycler has been demonstrated, wherein secondary treatment of the surface with compounds such as bovine serum albumin as well as a variety of polymeric materials that increase surface hydrophobicity, work very well with PCR chemistry. These attempts have also focused on the integration of various thermal-sensing elements, including microfabricated thermodiodes, thermal couples, and placement of Peltier type coolers. Recently, exploitation of other silicon processing materials, such as various types of photoresist, materials used to protect and deprotect silicon structures during various chemical etching and/or metallization steps, can also be used to construct precisely defined reservoirs, channels, valves, and seals (Despont *et al.*, 1997). Use of these materials offer a convenient and cost effective means to structurally link multiple chip components with interconnecting microfluidic elements on a single piece of silicon (Burns *et al.*, 1996).

Corresponding to the construction of miniature PCR instruments are a number of test applications that perform well in most environments, including those for patient care (Kricka and Wilding, 2003; Kajiyama *et al.*, 2003). Specific assays for DNA analysis from a chip-based thermocycler include detection and genotyping of bacteria, viruses and fungi, diagnosis of human malignancy such as leukemia, and

the genotyping of common and complex inherited disorders (Crisan and Farkas, 1993; Hashimoto *et al.*, 1998; Hessner *et al.*, 1999; Hamels *et al.*, 2001; Huber *et al.*, 2001; Cooksey *et al.*, 2003). Although the limitations of the chip-based PCR reactions have been evaluated, the practical designs of a thermocycling module have been commercialized and are considered reliable technologies for routine clinical use (Wilding and Kricka, 1999; Nachamkin *et al.*, 2001).

29.3.4 Product Detection

The most commonly used analytic technique in the genetics laboratory is gel electrophoresis. For more than two decades, various preparations of polymer-based gel systems, such as agarose and polyacrylamide, have been used to isolate and separate restriction cut and PCR amplified DNA. Several variables can be controlled in routine gel electrophoresis to meet the requirements of the test including the running temperature, ionic strength of the buffer, and other chemical compounds that either denature or leave double-stranded the DNA as it migrates through the gel. Recently, the use of capillary electrophoresis has been examined as a means to achieve high throughput analysis for PCR-derived DNA. Capillary electrophoresis is similar to gel electrophoresis, except that the separation medium is confined to a small glass capillary (see also Chapter 7).

Recent innovations in automation technology to detect amplified DNA are most often variations on the theme of gel electrophoresis or capillary electrophoresis. Once fragments are separated they can be visualized. Conventional protocols employ chemicals like ethidium bromide to stain DNA, but these compounds gradually are being replaced by fluorescent dyes. Fluorescence detection involves scanning the gel or the capillary continuously with a low intensity laser as the DNA products migrate through the gel matrix. The detection is achieved by capturing the emitted fluorescence on a photodiode or a charge-coupled device (CCD). Incorporating several fluorescent dyes that emit at various wavelengths makes possible the analysis of multiple DNA products simultaneously. Hence, the schematic of electrophoretic separation and visualization whether manually or in an automated format is the way in which most genetic testing is performed.

Automation of this system introduces a means of interpreting the fluorescent data through a software algorithm that can catalog the signal strength and position to derive the fragment size and quantity. The addition of an auto sampler, which draws a small volume of the PCR product and loads it onto the medium, is most useful in the analysis of low complexity genetic tests, such as SNPs, where there is a predictable result.

The other facet of product detection lies in the innovation methods to detect directly DNA being created as the PCR goes forward, in real-time. For years most molecular genetic assays were based on end point determination; that is, the detection of the product at the termination of the gene chemistry reaction. Real-time assays have been developed for PCR, LCR, and the Invader® reactions. In general, real-time assays are based on the repeated measurement of fluorescence or an electrochemical signal during the process of amplifying or cutting the template DNA (see also Chapter 10).

The Invader™ assay takes advantage of a three-dimensional structure involving single-stranded template DNA hybridized to a primary oligonucleotide probe and a so-called Invader® oligonucleotide (Kwiatkowski *et al.*, 1999). The combination of the oligonucleotide probe and primer results in a structure of two overlapping DNA sequences and the creation of a flap, which is the structure recognized by the Cleavase® VII enzyme. Endonucleolytic digestion of the probe liberates the flap at or near the site of overlap liberating this flap into the media (Lyamichev *et al.*, 1999). In the specific case of the Invader® assay, the technology involves the use of a fluorochrome-labeled oligonucleotide probe that lies near a fluorescent quencher molecule. This double fluorochrome label strategy is called Fluorescent Resonant Energy Transfer or FRET, and is an increasingly common theme in bioanalytic assays. Upon cleavage of the flap, the fluorescent label is liberated from the quencher molecule,

and the result is fluorescence in the reaction solution. A single denaturation step and repeated annealing of new oligonucleotide probe results in a linear accumulation of a liberated fluorochrome-labeled DNA tail, which is measured spectrophotometrically at the completion of the reaction.

Another such technique, referred to as Taqman™ technology, takes advantage of properties of the exonuclease activity inherent to *Taq* polymerase (Desjardin *et al.*, 1998; see also Section 10.2.1). With Taqman, *Taq* polymerase binds to the forward and reverse DNA primer and begins to replicate the template DNA in a 5′ to 3′ direction. A third FRET oligonucleotide primer/probe is positioned to overlay a region of the amplified sequence or a particular point mutation. As *Taq* creates the elongating DNA strand, it will encounter the FRET probe, wherein the 5′ exonuclease activity of *Taq* will act to cleave the terminal nucleotides from one end of the FRET. This cleavage liberates each of the fluorochromes into solution, along with the de-inhibition of their respective fluorescent emissions. Fluorescence detectors in contact with the reaction tube or optical fibers placed into the reaction fluid monitor the fluorescence of the sample in real-time.

Variations on the theme of combining the amplification and detection phases into a single operational step are achieved with assays that use recombinant enzymes, such as thermal stable ligase and isothermic polymerases. Strand displacement amplification, developed by BD Biosciences, is one such system that combines the action of a unique isothermal polymerase and a restriction enzyme to create single-stranded DNA products. The fact that once genetic testing was imagined to be at least a four-step operational schema where assay setup and subsequent analysis were separate steps may soon give way to protocols that combine operations into a one- or two-step system. When such systems are available, the prospect of simple and miniaturized testing systems may be realized.

29.3.5 Interpretation and Reporting

The use of genetic information for patient care is still sufficiently novel; having expert interpretation of the data and construction of a thoughtful report is paramount. In most cases, clinical molecular genetics laboratories are dedicated to the performance of complex genetic tests that involve interpretation of a variety of genetic abnormalities including point mutations, deletions, rearrangements, and/or gene loss by scientists or physicians specifically trained in molecular genetics. Whereas the interpretation of a well-designed genetic test is easy, the complexity of each of the analytic operations and the chance of error in the many manual steps make the trust of an automated interpretation unappealing. However, a variety of simple genetic analyses with high clinical utility do serve as a model where automated interpretation and potentially clinical reporting can be explored.

Factor V *Leiden*, the common mutation in the gene for the Factor V coagulation protein, has a very high incidence in the Caucasian population. When present, patients with this mutation are at a markedly increased risk for deep vein thrombosis and other vascular complications. Several investigators have attempted to automate molecular genetic testing for Factor V *Leiden* in the hope that this simple model system will lead to strategies for other genetic testing applications (Lay and Wittwer, 1997; Enayat *et al.*, 1997; Gomez *et al.*, 1998).

In its simplest form, Factor V *Leiden* testing involves interpretation of a PCR-based assay that falls into one of three categories:

- Normal (i.e., no mutation)
- Heterozygote mutant
- Homozygote mutant

Detection of this two-allele system on a DNA microarray, where one spot, a second spot, or both spots are fluorescent is clearly a reduction of the complexity of performing serial techniques of PCR with restriction digestion followed by gel electrophoresis (see Fig. 29.9). Moreover, the multiplexing of a series of related thrombophilia gene markers onto the same DNA array, each with a simple yes/no interpretation, make likely the adoption of these types of tests in places other than in a specialized molecular diagnostics laboratory (Witowski *et al.*, 2000). At present, such systems are not commonly available, but as will be discussed in the subsequent section, the promise of several new and promising technologies centered around microminiaturization may lead to the eventual test bed of a fully automated genetic analysis system.

29.4 EMERGENCE OF TECHNOLOGIES FOR MINIATURIZATION OF MOLECULAR DIAGNOSTICS

Reduction of cost is a major driving force to improve the way molecular diagnostics are currently performed. As discussed earlier, to appreciate where costs accumulate in the operational schema of genetic testing requires a whole system analysis inclusive of sample collection. Based on current technology, molecular diagnostic laboratories have worked to optimize each of the operational steps, and through a combination of efforts involving computerization and/or piecemeal integration the overall testing schema can be semi-automated. Commensurate with the current practice of molecular diagnostics, however, is the emergence of engineering research directed at creating a fully integrated testing system (Beugelsdijk, 1991). This research has led principally to the conclusion that miniaturization comparable to the size of today's computer chips will afford the greatest likelihood achieving this integration and at the same

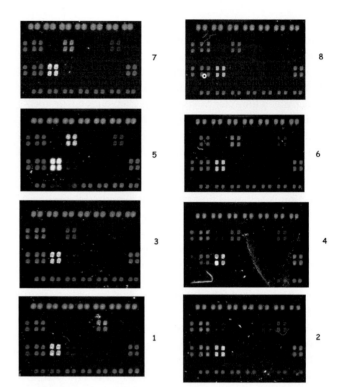

FIGURE 29.9 Home brew DNA microarray configured for a multiplex analysis of a series of genetic markers for inherited thrombophilia. The gene chemistry for this assay was based on the multiplex amplification by PCR of five independent gene loci, followed by ligase detection reaction genotyping of the corresponding SNPs at those loci. Each PCR product is tagged with a zip code; a heterologous length of DNA at the 3' end of one primer serves to anchor the amplicon to the microarray via its hybridization to its zip code complement. The opposite ligase primer has the 5' end labeled with the fluorochrome Cy-5, which provides the spot signal on the microarray. Panels 1–8 represent individual microarrays, each tested with DNA products from a single patient. The line of fluorescent spots along the top and bottom, as well as the single line of spots on the left side are used to orient the arrays. Each genetic marker is connoted as a quartet of fluorescent spots, the top row for the wild type sequence, and the second row for the mutant allele. Across the arrays the markers are Factor V Leiden, Prothrombin G20210A, Methylene tetrahydrofolate reductase C677T, Factor V HR2, Plasminogen activator inhibitor type 1 (PAI-I) 4G/5G. (See color plate)

time significantly decrease costs (Marshall and Hodgson, 1998).

The effort to create a highly portable and inexpensive gene amplification and detection system has been a convergence of expertise from at least two scientific areas for which there previously has been no practical connection. The first is in the area of biotechnology, with its focus in the refinement of the molecular biology and the *in vitro* application of key biochemical processes. The second area is in electrical engineering, material science, and in computers, with a specific focus on the design of microelectronic components and microelectromechanical systems (MEMS; Tang, 1997). Collectively, these two groups are working in the relatively new discipline of biomedical engineering. The interface between these two disciplines has led to some exciting new ideas that are only in the last few years being realized as devices with potential commercial viability.

In practice, PCR-based assays have become very small in dimension. Using finely calibrated and handheld pipettors, PCR reactions can easily be scaled to less than $10\,\mu L$. This volume, retrieved from a $100\,\mu L$ reaction tube, can then be analyzed on a gel or by capillary electrophoresis systems with high reliability and precision. Assays of this dimension are challenged by some of the physical limitations of handling specimens and reagents and are subject to the problem of evaporation, surface tension, and poor mixing. These problems are magnified further when attempting to use macroscale instrumentation to microscale devices. A related issue pertains to the fact that samples tested for certain rare or low concentration analytes risk being missed when tested on systems that use only microliter or nanoliter quantities of starting material. These problems, sometimes referred to as the "big hands, tiny device" dilemma are central to the challenge to miniaturize molecular genetic testing for routine or typical clinical assays. Hence, the design of such devices has recently steered toward the mesoscale. This strategy is proving to have abundant application for diagnostics and research, and the performance of such technologies is also superior.

The ability to exploit the production of silicon-based microchips as a platform for building an integrated genetic testing system is also a large part of the miniaturization of molecular diagnostics. Amalgamation with a personal computer controller leads this work to point-of-care testing for molecular diagnostics. Much of the development has evolved out of the field of MEMS (Joseph *et al.*, 1997). Microelectromechanics is a means to make microminiaturized actuators and sensors by processes identical to those used to fabricate microelectronic chips. Physical sensors are the primary commercial application of MEMS, although new approaches toward combining the mechanical features of MEMS with biologic materials are laying the foundation for a whole new class of biosensors. MEMS, combined with microelectric circuitry, is referred to as a smart sensor system, which is robust and can easily be fitted into highly portable and inexpensive handheld and bench top instrumentation.

Much of the progress in MEMS technology is borrowed from the optimization of techniques used in the integrated circuit technology arena, with the promise of miniaturization and batch fabrication. The fabrication of MEMS sensors is based on the processes of photolithography, surface micromachining, and deposition of novel materials on silicon. An analogy to the creation of a MEMS device can be found in the construction of a house. Similar to a draftsperson, who will outline a blueprint of each feature of

a house, designating the position of the walls and other key elements in that structure, the process of photolithography leads to a series of masks, each outlining one layer within the MEMS structure. Each photo mask is projected onto the surface of a silicon wafer resulting in a pattern of shadows and lighted areas.

Photoresists are epoxy-type materials used in photolithography to differentially protect and deprotect areas of the wafer. Exposure to specific wavelengths of light causes these epoxys to cure by cross-linking, or in the shadowed areas, become vulnerable to dissolution by organic solvents. The result of photolithography is to leave a footprint for each sequential step and the precise location for the placement of materials on the wafer, much like a bricklayer might first create the foundation with bricks and cement in preparation for the footings of the vertical walls.

Another process in MEMS fabrication is the deposition of thin films. Typically, this involves the use of high temperatures and mixtures of gases bathing the surface of the silicon wafer for precisely defined time and pressures. The result is a deposition of materials, which have either electrical or structural properties useful to the finished device. Micromachining refers to the use of a variety of techniques including dry and/or wet etching procedures to differentially remove materials from the silicon wafer and/or previously laid down thin films. Micromachining is performed in concert with serial photolithography steps to create the structures with intricate details including vertical wall, freestanding beams, or diaphragms, as well as to prepare planar surfaces for subsequent deposition of metals and/or biochemical layers. Figure 29.10 shows a summary of the various techniques used in MEMS fabrication, each of which requires highly specialized instrumentation and the performance of these techniques in specialized environments, such as high containment clean rooms.

The utility of MEMS in creating an integrated genetic device is highly focused on two broad areas. The first is in the creation of miniaturized chambers in which to carry out biochemical reactions such as PCR. In the next section, a summary of the various technologies involving MEMS and/or MEMS-like devices for gene amplification will be presented. The second area in which MEMS is employed includes the creation of unique sensors. Collectively referred to as biosensors, the advent of microaddressable arrays or DNA chips has been among the most remarkable and exiting areas of opportunity in the whole of the biotechnology industry.

In addition to these DNA chips, a variety of more classic sensor technologies are being evaluated as potential biosensors. One particularly exciting opportunity in the use of these other MEMS sensors is their compatibility with integrated microelectronic circuitry. With the goal of creating an integrated genetic testing system, which can be scaled to a device no bigger than a postage stamp, the need to provide control of these sensing and/or actuating elements mandates

1. Etch silicon wafer by RIE and fill it with PSG by LPCVD.

2. Deposit low stress Si_3N_4 and polysilicon by LPCVD.

3. Deposit PZT thin film by MOD and top electrode by sputtering.

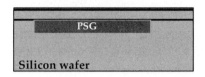
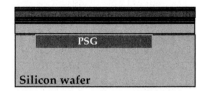

4. Patterning top electrode & PZT and bottom electrode by ion milling and deposit encapsulation layer.

5. Etching holes through PSG by RIE to etch PSG and patterning the beam structure.

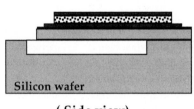
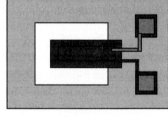

(Side view)   (Top view)

FIGURE 29.10   Overview of the various processes involved in the microfabrication of a silicon chip. MEMS use silicon wafers, and the various methods of micromachining finely detailed structures in silicon, similar to those used to create an integrated circuit, such as a computer chip. The combination of surface etching, reactive ion etching (RIE) with high energy plasma, and the subsequent deposition of materials, like metals via metal oxide deposition (MOD) and thin films of silicon oxide and silicon nitride via the technique of low pressure chemical vapor deposition (LPCVD) and polysilicon glass (PSG), result in tiny microminiaturized devices with both mechanical and electrical functionality. Piezoelectric thin films including ceramics such as lead zirconium titanate (PZT) expand and contract with application of a current creating mechanical motion.

comparably sized electronic controller units. These highbred sensor systems, or smart sensors, are also compatible with commonly available controlling software elements, such as that which might be found in any personal computer. In the next section, an overview of the industry's current sensor technologies will be presented.

29.4.1 DNA Chip Technologies

As is often the case with breakthrough technologies, there is a great deal of hyperbole that surrounds the development of chip-scaled technologies, popularly known as DNA or genetic chips or microarrays. In this case, however, much of this optimism is justified considering that in just a few short years basic research has led to a number of products that are at or near the commercial phase of their development. In this section, the focus is on technologies to miniaturize DNA testing onto platforms that involve microchips. Developers of DNA chips believe that in the near future these technologies will enable clinicians, and in some cases even patients themselves, to detect the presence of a wide variety of genetic-based diseases and conditions, including AIDS,

Alzheimer's disease, cystic fibrosis, and several forms of cancer, quickly and inexpensively.

In other arenas, this technology will also make it possible to develop inexpensive strategies for screening of new genes and proteins leading to new pharmaceutical discoveries. The principles underlying the design and operations of these chips are comparable to the macroscale versions in common use in laboratories today. But in their concurrent state, however, there is the need for considerable improvement to the routine performance of these technologies before there is acceptance in the clinical arena. To date, several companies have commercialized DNA chips and adapted these products for use in diagnostic applications. To a great extent, though, the clinical laboratory has not embraced these technologies due to their cost and because their performance is not as good as conventional protocols. Additionally, whereas the promise of chip technology lay in wholly integrated assay systems, the present state of the field is about a compilation of a device to analyze DNA subsequently to its processing off-chip; that is, using standard means of nucleic acid extraction, PCR, or other gene chemistries.

29.4.2 Basic Principles of Chip-Based DNA Analysis

As discussed in the earlier section, much of the DNA chip technology is based on advances in MEMS and microelectronics. The extrapolation of macroscale analytic technologies to that where the same is placed onto a silicon or glass chip is achievable, but with the consideration of how the physics and chemistry adjusts when assays are adjusted to such small sizes. Indeed, the concern of addressing issues such as surface preparation, surface tension of fluids, and electrical and chemical isolation and insulation are paramount. The following section summarizes the current state of several microchip technologies that in their own way demonstrate select strategies to the challenges of chip-based DNA analysis.

29.4.3 Microcapillary Electrophoresis

Electrophoresis refers to the migration of charged electrical species when dissolved, or suspended, in an electrolyte through which an electric current is passed. Cations migrate toward the negatively charged electrode (cathode) and anions are attracted toward the positively charged electrode (anode). In the example of the separation of DNA, migration is toward the anode. Conventionally electrophoresis has been performed on layers of gel or paper. Capillary electrophoresis refers to the process occurring in a thin channel, either enclosed, as in the case of a glass tube capillary, or on the surface of the chip, where the channel is constructed from a planar surface made of silicon or glass.

Among the first demonstrations of micro-fabrication technology for bioanalysis was microcapillary electrophoresis. Capillary electrophoresis has been used throughout the clinical laboratory, but in particular for the separation of serum proteins and more recently for DNA (Bosserhoff *et al.*, 2000; see also Chapter 7). The microfabricated devices are used in a manner identical to the mesoscale version of multichannel capillary electrophoresis, where the capillaries measure between 1–3 mm in diameter. These chips constructed from glass, plastic, and silicon have microchannels that are typically 50–100 micrometers in diameter (Paulus, 1998). Using conventional photolithography and surface micromachining, several designs demonstrate up to two meters of micromachined capillary into a space no larger than that of a postage stamp. At the proximal end of these microcapillaries is a fluid reservoir in continuity with micro-electrodes that provide the electromotive force (Wilding *et al.*, 1994b). Through the application of a voltage across these two electrodes, fluid will flow across the length of the microchannel and dissolved analytes will separate according to their electrophoretic mobility and the counterbalanced electro osmotic flow, as known as electroendoosmosis (see Fig. 29.11; Bruin, 2000).

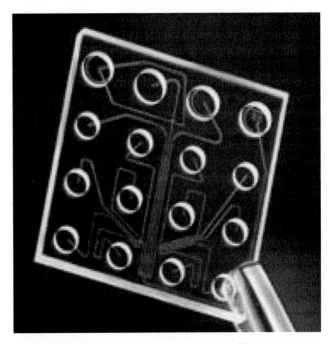

FIGURE 29.11 Photograph of the Labchip™, from Caliper Technologies (Palo Alto, CA). Caliper has borrowed manufacturing methods from the electronics industry that are conventionally used to produce microchips in the development of miniature, integrated biochemical processing systems. The design features interconnected channels etched into glass, silicon, quartz, or plastic. Serial flow of a liquid sample is routed through various positions on the chip, where different chemical reactions take place.

Technologies developed by Caliper Technologies incorporate microcapillary electrophoresis and an analytic strategy called electrokinetic flow. Electrokinetic flow involves the coordinated movement of fluid columns electrophoretically across intersecting microcapillaries by way of differential application of voltage across electrodes specific for each (Chiari *et al.*, 1998). Specifically, flow occurs when solvated cations, bound to a negatively charged microchannel wall, move toward the cathode under the influence of an applied voltage. The movement of the ions and the associated water molecules effectively pump fluid along the channel in the direction of the cathode. Further, by combining several fluid reservoirs, each containing separate buffer solutions and/or sample reservoirs, and through the coordinated application of current to the corresponding electrodes for each, there is the mixing of buffers and samples and electrophoretic separation across the longest length of capillary (Jacobson and Ramsey, 1996). This approach is the so-called lab-on-a-chip, and had the potential to involve several analytic assays to be combined in a very small and low cost platform. The Caliper lab chip has one current configuration as the disposable chip element in the Agilent Bioanalyzer System. A microcapillary electrophoresis chip, constructed wholly from MEMS processes is illustrated in Fig. 29.12.

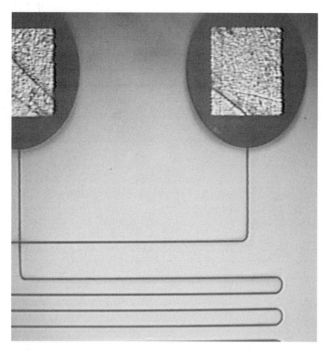

FIGURE 29.12 Creation of a micro fabricated capillary electrophoresis device fabricated from silicon for the transport of analytes in solution. Devices made from glass, plastic, and now silicon have their respective advantages and limitations. In the case of the chip shown, the typical system employs a laser-induced fluorescence via a confocal microscope. Additionally, integration of on-chip electronics highlights the potential of silicon devices to integrate with other on-chip functions.

This device is not only inexpensive to produce, but has the added advantage of integrated CMOS type microelectronic control elements to complete the miniaturized instrument package.

Coincident with the development of capillary technologies has emerged the better appreciation of the performance characteristics of bioanalytic processes at the micrometer to millimeter dimension. At the size of these chips, new concerns over the issues, such as the dynamics of fluid flow and the matter of electrokinetic mobility, become important. For example, the microfabrication of capillary structures may leave residue that requires removal to prevent unwanted adsorption of analyte as well as the potential of nonuniform performance due to the interaction of buffer solutions and carbon-silicon structures (Kaupp *et al.*, 2000). Several studies have investigated the advantages constructing the microchannel from plastic or other inert polymers instead of glass (Chiari and Cretich, 2001). Surface passivation with zwitterionic solutions was proven useful to prevent binding of protein to glass capillaries (Castelletti *et al.*, 2000). The addition of polyethylene oxide and the optimization of capillary fluid flow has been described, and are reported to enhance the performance of acrylamide and other thin-film sieving media including hydrocellulose and various types of

agar (Guttman and Ronai, 2000; Konig and Welsch, 2000; Heller, 2001; Kim and Yeung, 2001).

Based on these enhancements of capillary electrophoresis a number of studies have resulted that demonstrate the potential of clinical application. A simple version of a microcapillary electrophoresis device has been tested for detection of the bcr-*abl* gene transcripts associated with chronic myelogenous leukemia, separation of PCR fragments generated in the Factor V *Leiden* mutation assay, as well as a multiplexed reaction involving mutational analysis of the cystic fibrosis gene (Audrezet *et al.*, 1993; Kearney and Aumatell, 1997; Righetti and Gelfi, 1997). In each case, the clear advantage of microcapillary electrophoresis is a high throughput capability that markedly reduces the time of the electrophoretic analysis step.

Microcapillary detection of DNA without the need for PCR amplification is behind the technology being developed by US Genomics. Advances in the method to label genomic DNA with fluorochromes, as well as advances in the sensitivity of optical detection systems combine to potentially create assays that flow a solution of sample DNA through a microcapillary and directly detect the DNA sequence of interest. Accordingly, US Genomics claims to achieve single molecule (DNA) detection by a strategy of direct labeling. Correspondingly, the specificity of the technology is increased through the added confidence to interrogate thousands or potentially millions of fluorescent events through the microfluidic device. This approach is similar to that of a flow cytometer, where the analysis of many individual tagged cells permits the simultaneous and sensitive detection of many protein markers. The application of this strategy to molecular assays is made possible with bead technology, described next.

29.4.4 Microbead Technology

Microbead technology has emerged as a compelling strategy for the miniaturization of the DNA assays. The basis of the technology is the production of polystyrene or polypropylene beads onto which a wide variety of biologic materials can be attached. The porous surface of the bead permits a means for either covalent or noncovalent attachment of capture molecules, which serve as receptors for cognate analytes. In the case of DNA testing, oligonucleotides specific for a certain DNA sequence are linked to the bead by means of alkylthiol linkage, a noncovalent attachment strategy that is durable to both heat and buffers and washes of high ionic strength (Steinberg *et al.*, 2004). The beads can be fluorescent or the attached probes labeled with a variety of reporter molecules. When incubated in a solution containing the analyte of interest, the surface of the bead becomes coated with analyte hybridized to the surface attached probes. Such assays are read by measuring fluorescence from the collection of microbeads from individual wells in a microplate, from the stream passing through the detector of a miniature flow cytometer or sorting individual beads onto the surface of a microarray chip or slide.

Commercial instruments that use microbeads are described next. Although these technologies do not necessarily exemplify the dimensions of a miniaturized assay platform, the experience gained from the use of microbeads and the performance of these devices in clinical assays reveal much about the scalability of DNA chemistry when applied to other chip-based platforms.

29.4.4.1 ILLUMINA

An array of probe-coated microbeads at the tip of a fiber-optic bundle is the basis of the Illumina analyzer. The array is formed from a bundle of optical fibers cut to form a microwell of the dimension to cradle a microbead of 3 μm in diameter. Fluorescent beads, prepared in batches off-chip then are configured with a combination of oligonucleotides designed for identify differing DNA sequences. A bead array can then be constructed by linking a fiber optic bundle, containing upward of 50,000 individual fibers to a single bead, and then an array of beads assembled around a packet of fiber bundles. Bead arrays can be produced in a variety of common configurations, including those with 96 or 386 wells. A separate DNA reaction can be performed in each well, but multiplexed reactions are also possible.

Each assay is based on the fact that filtered light can be piped down the optical fiber bundles, where emitted photons evoke a fluorescent emission from the surface of the linked microbead. Fluorescence emitted from the surface of the beads is then transmitted back toward the incident light source along the fiber bundle, by means of the total internal reflectance, and redirected to a detector, typically a CCD or a photomultiplier tube (PMT; Epstein and Walt, 2003). Thus, the same group of fibers that provide the efferent light to stimulate the surface fluorescence serve as the conduit for the returning signal that represents the result of each DNA test. Figure 29.13 illustrates the principle of total internal reflectance, a property of a wave-guide material such as glass, plastic, and silicon derivatives.

Diagnostic applications of the Illumina technology have been reported, but most are demonstrations of schemes to detect SNPs (Oliphant *et al.*, 2002). In particular, the bead array technology has been demonstrated as a high throughput genotyping platform, which in its current design, has a capacity of one million SNP assays per day and is easily expandable. Kit-based diagnostic assays are under development and include ones for mutation detection for moderate to complex genetic diseases, such as cystic fibrosis, p53, and pharmacogenetic applications.

29.4.4.2 LUMINEX

Luminex Analyzers use color-coded microspheres, into 100 distinct sets. Each bead set can be coated with a reagent specific to a DNA marker assay, and within a single bead set, can produce upward of 100 distinct DNA marker results.

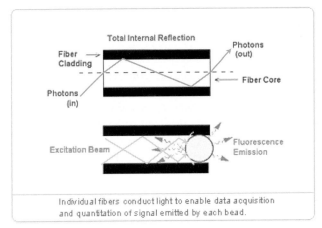

FIGURE 29.13 Total internal reflectance. The property of a waveform emitted from a monochromatic light source and transmitted through a medium, such as an optical fiber. Optical fibers can pipe photons to a specific location. In the case of the Illumina technology, the cut end of optical fibers are shaped into a microwell that cradles a microbead covered with fluorochrome-labeled DNA probes. Photons emitted from the excitation of the fluorochrome are captured by the optical fibers and that light is transmitted in the opposite direction to the efferent light source. In the figure, the dark gray line depicts photons transmitted through an optical fiber, where the photons striking the internal surface of the cladding for that fiber are reflected internally and further through the fiber. The lower panel depicts the same phenomenon, where the end of the fiber contacts a fluorescent microsphere. Photons emitted from the affixed fluorochromes on the surface of the microsphere are transmitted backward through the fiber, again by the principle of internal reflectance. The exiting fluorescent photons are ultimately detected by an electronic sensor, such as a CCD or a PMT.

The Luminex analyzer is based on measurement of a stream of individual beads past a laser that excites the internal dyes that identify each microsphere particle, and additionally any reporter dye captured during the assay. Thousands of beads are analyzed in each experiment, which provides the statistical power to establish the sensitivity and specificity of each assay. The combination of many fluorochromes within each bead set provides a broad spectrum of possible output signals that make possible the validation of multiplexed assays not possible using conventional readout technologies.

At this time a large number of assays are available to the clinical market. High throughput SNP analysis was demonstrated using a unique combination of PCR and the oligonucleotide ligation assay (PCR-OLA; see also Chapter 4). Bacterial and viral genotyping were shown to have high sensitivity and specificity as compared to gel electrophoresis-based assays (Ye *et al.*, 2001; Iannone *et al.*, 2003). In one application, six different viral or control PCR products were coupled to six microsphere sets. The surface oligonucleotide probes were biotinylated and the viral amplicons were detected with fluorescent streptavidin. The method is very sensitive, with a lower limit of detection of 50 virions/mL and with three logs of dynamic range. A recent comparison of a multiplex PCR assay for mutation screening of the cystic fibrosis gene (CFTR) on the Luminex system demonstrated near perfect concordance for a large cohort of samples from patients and carriers. The attributes of the system included the ability to analyze each sample for one or two mutant alleles; some gel-based assays have the limitation of detecting only the heterozygotes for many of the common mutations.

29.4.4.3 GOLD-COLLOID MICROSPHERES

The efficiency of probe capture technologies is improved when the support for those probes is of a similar dimension to the molecular size of the analyte being tested (Demers *et al.*, 2000). In the case of DNA, a highly sensitive and efficient capture strategy is based on gold nanoparticles. The proprietary product of the Nanosphere Corp., their assay platform called Verigene™ is based on the novel means of attaching oligonucleotide probes to the surface of gold particles of about 13 nanometers in diameter (Niemeyer *et al.*, 2003). These probes are composed of assay-specific complementary DNA sequences that hybridize to the cognate sequence from samples of genomic DNA, which in turn bind to a probe complementary to the DNA duplex on the surface of a simple microarray. Signal amplification is achieved when a secondary reaction occurs with the gold particle and a fluorescently labeled surface probe. Whether on a glass or silicon microarray or on a simple paper filter, a highly selective DNA assay can be built without the need to amplify the template before it is hybridized to the gold particles (Glynou *et al.*, 2003). Additionally, the unique properties of gold nanoparticle probes adapt to detection systems using optical, electrical, or magnetic technologies. Nanosphere's first system, Verigene ID, employs the optical method of molecular detection and produces unambiguous and precision results.

A further enhancement of the microbead technology is the ability to specifically tattoo each bead with one- or two-dimensional barcodes (Xu *et al.*, 2003). This feature expands the level of sensitivity of the corresponding DNA hybridization assay since admixtures of DNA sequence can be captured and counted by means of flow sorting or microarray-based readout systems. The potential of assays that find infectious isolates based on genotyping assays that also quantify those observations would have far-reaching impact on the selection of therapeutics and the monitoring of treatment effects.

29.4.5 Microaddressable Arrays

When the popular press refers to DNA chips, in most cases they are referring to variations on the theme of microaddressable arrays. Microaddressable arrays or DNA arrays represent a unique combination of technologies wherein microfabrication of silicon is combined with unique ways of affixing gene probes to a solid support system. The vanguard of this technology is based on research performed by Steven Fodor and colleagues, dating to the middle 1990s wherein conventional photolithography was used to construct oligonucleotide DNA probes directly from a silicon microchip surface (Fodor *et al.*, 1993; Pease *et al.*, 1994).

This so-called light-directed chemical synthesis encompasses two mature technologies: semiconductor-based photolithography and solid phase chemical synthesis. Synthesis linkers modified with photochemically removable protection groups are attached to the silicon substrate. Light is directed through a photolithographic mask to specific areas of the synthetic surface affecting localized photodeprotection. The first of a series of chemical building blocks, for example, a photo protected amino acid or hydroxyl is incubated with the surface and chemical coupling occurs at those sites, which were illuminated in the preceding step. Next, light is directed to a different region of the substrate through a new photo mask and the chemical cycle is repeated.

Complicated strategies can be employed that will generate a large number of oligonucleotide probes in a minimum number of chemical steps. For example, only $4 \times N$ chemical steps can produce a complete set of 4N oligonucleotides of length N, or any subset of the array. The pattern of illumination and the order of chemical reactants prescribe the synthesis of these products and their precise location on the chip. An alternative to photolithography is laser directed from the surface of a digitally directed micromirror array to photoactivate the surface of the DNA chip with pinpoint accuracy. This method is referred to as maskless microarray fabrication. The promise of this technique is rapidly changeable chip configurations and marked reduction in the cost of chip manufacture (Singh-Gasson *et al.*, 1999).

Affymetrix Corporation is commercializing chemical synthesis on a solid support system. GeneChip™ is a commercial name for the Affymetrix products, and has developed a variety of chips configured for assay for gene discovery as well as for diagnostic interrogation of selected genes for mutations (see also Chapter 15). The chip platform in the Affymetrix system serves as a convenient sample module that is read in the benchtop fluidics and optical station. Affymetrix's HIV GeneChip™ is designed to detect mutations in the HIV protease as well as in the virus' reverse transcriptase gene (Lipshutz *et al.*, 1995). Additional products include gene chip arrays for p53 and SNPs for genes encoding proteins involved in drug metabolism (Irizarry *et al.*, 2003; Cooper *et al.*, 2004). More recently, Affymetrix has also been involved in other inborn genetic disorders including detection of specific mutations in the human BRCA1 gene, cytochrome P450 gene polymorphisms, as well as in a host of other bacterial pathogens (de Longueville *et al.*, 2002; Wen *et al.*, 2003).

Alternative strategies for affixing oligonucleotide gene probes to a silicon surface have been commercialized by Nanogen and Genometrix. Nanogen Corporation (La Jolla,

California) involves the off-chip synthesis of oligonucleotide probes that are secondarily directed to microelectrodes through electrostatic attraction (Edman *et al.*, 1997; Sosnowski *et al.*, 1997). By differentially turning on each of these electrodes, different oligonucleotide probes can be directed to a precise location on the chip. Genometrix's approach involves the deposition of off-chip synthesized oligonucleotide probes through a microjet dispensing system to predetermined locations on the chip (Lamture *et al.*, 1994; Eggers and Ehrlich, 1995). Deposition of 5′-thiolated oligonucleotides onto glass slides using a bubble jet printer has been shown to be a very low cost, yet robust technique amenable for use with home-brew assays.

A third approach taken by Gerry and colleagues (1999) involves the use of hybrid polymeric materials to embed generic oligonucleotides called zip codes to a precise location on the surface of a glass or silicon chip by means of a microdispenser spotting tool (see Fig. 29.14). This research group describes the approach of using zip codes and combinations of PCR and the ligase detection reaction as creating universal microarrays because of their adaptability to various genetic mutations and the excellent signal-to-noise results produced even when performed on home-brew produced microarrays. To date, numerous clinical applications of simple and complex genetic assays have been clinically validated using this strategy (Gerry *et al.*, 1999; Favis and Barany, 2000; Favis *et al.*, 2000a; 2000b; Witowski *et al.*, 2000; DelRio-LaFreniere *et al.*, 2004).

Although several scientific studies examining the utility of these various DNA addressable arrays have been published, it is clear that a great deal of work remains to be done

before these technologies are ready for use by inexperienced individuals working either in a physician's office, or in the near future with a consumer-based product. One of the key problems associated with microaddressable arrays is the optimization of the hybridization chemistry. Those with experience in using conventional hybridization techniques such as Southern transfer, slot blot and/or dot blot, realize the delicacy these techniques require with regard to optimizing temperature, ionic strength, as well as an appreciation of the precise DNA sequence and the assessment of the relative binding efficiencies. Other detection modalities as outlined attempt to avoid some of the complications of these multiplexed array-based technologies by exploiting other physical properties of the DNA hybrid molecule.

29.4.6 Planar Wave Guide

A less costly design for a fluorescent array readout system is the intended strategy for the planar wave guide (PWG; Abel et al., 1996; Wulfman et al., 2003; Duveneck et al., 2003). Most array readers use a system where the spots on the array are illuminated via a laser and the resulting efferent fluorescence is detected through a series of lens and optical filters to capture an image that is reconstructed with a software interface (Cheung et al., 1999). The PWG requires no image collection or software analysis. Instead, the PWG works by illuminating specific regions on the surface of the glass microarray with a broad spectrum halogen light source exciting the fluorescent DNA spots to emit photons outward from the surface of the array, as well as the transmission of signal through the glass, to points at

A

B

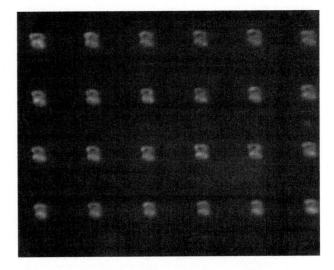

FIGURE 29.14 Preparation of DNA microarrays. A. A microspotter instrument is used to deliver minute amounts (approximately 2–10 nl) of solution containing DNA probes to precise locations on the surface of the glass slide-based microarrays. The spots (shown in panel B) are uniform in size, measuring approximately 50 μm in diameter. (See color plate)

the distant edge of glass and outward where the emitted signal is collected by a PMT (Hanning et al., 2000). As shown in Fig. 29.15, the glass that makes up the microarray serves as the wave guide for a fraction of the fluorescent signal through the property of total internal reflection (TIR). Simultaneous recording of position of the excitation light source, and the corresponding output of the PMT, provide two independent variables by which one can interpret each of the individual spots precisely and quantitatively. Figure

29.16 compares the data from a PWG reader output taken from the scan of a multispot array with the image taken from a laser scanner reader. Each of the peaks shown in the PWG output corresponds to a different fluorescent spot in the array. The magnitude of each peak is a measure of relative emission intensity.

Interpretation of the PWG signal from each fluorescent point source requires a mathematical algorithm to isolate each spot from the mixture of signals transmitting through the glass wave guide. The signal deconvolution algorithm is based on a logic triangulation; that is, the distribution of the signal source intensity across the side of the cut edge of the glass array. This algorithm provides the means to determine the fluorescent landscape of an array without a reconstruction of the image from the raw data. This technology has been shown to work even with high density arrays, where the output signals overlap; the algorithm can extract peak intensity data where image capture strategies would make spot resolution difficult or impossible. The deconvolution method also affords a means to analyze two-dimensional layouts.

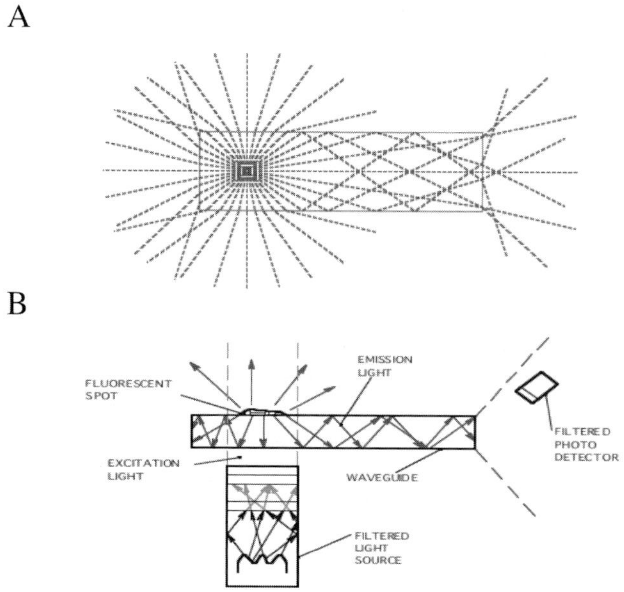

29.4.7 Chip-Based Biosensors Using Physical Properties of Nucleic Acids

29.4.7.1 ELECTRONIC DETECTION OF DNA PRODUCTS

The duplex DNA molecule normally is negatively charged. This is the basis for the migration of various molecular species in an electric field or in an electro-osmotic gradient. DNA also possesses the property to conduct electricity. Motorola Life Sciences has acquired a technology that aims to provide direct electronic detection of DNA/DNA hybridization on a chip. The essence of this company's approach is that the DNA probe is attached to electrode pads on the chip through molecular wires (made of a phenyl-acetylene polymer; Yu *et al.*, 2001). After hybridization,

A

B

FIGURE 29.15 Schematic of the function of a PWG. Fluorescence emitted from labeled DNA hybridized to its cognate probe on the surface of a glass microarray transmits a fraction of the efferent photons through the glass slide and secondarily emitted from the cut edge of the slide where the resulting signal is detected by a PMT. A. Top view of microarray. B. Side view with profile of the light source.

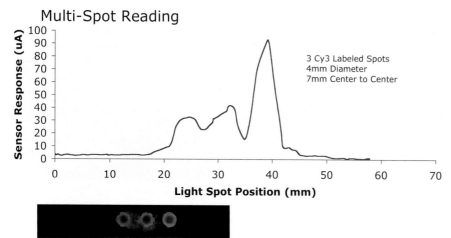

FIGURE 29.16 Signal output from a PWG (Photoimage inset). Image of fluorescent spots from the glass slide-based microarray analyzed in a laser-based microarray reader. Corresponding signals detected read on the PWG are shown in the graph on the upper panel. The signals are derived directly from the current measured from the photomultiplier tube. Individual spots from the microarray are resolved by the one-dimensional movement of the halogen light source with a fixed position PMT. (See color plate)

DNA linked to a ferrocene redox label is added. When the voltage is raised, a detectable current flows through the system. The aim of this approach is to be able to undergo direct DNA detection in an amplification-free system (Henke, 1999). For a typical clinical sample, that would mean detecting around 10,000 copies of a particular target DNA sequence. Current prototypes of this device are not that sensitive with detection limits of around 10^7 copies (Umek *et al.*, 2001).

Practical applications of electronic detection have been demonstrated. An evaluation of the so-called e-sensor technology involved the parallel evaluation of two cell lines carrying mutations in the gene for cystic fibrosis, as well as the four cell lines derived from patients with hereditary hemochromatosis. The study revealed perfect concordance with other methods, thereby illustrating the potential of this mode of electronic DNA sensing in clinical laboratory settings (Bernacki *et al.*, 2003).

Whereas the e-sensor platform detects the binding of a DNA molecule as a function of the change in electrical impendence, an alternative approach is to look for a change in the resistance to the flow of electrons, a reflection of the decrease in electrical potential across two electrodes. GeneOhm Sciences, an early stage company from San Diego, CA, has developed a simple, low-cost, and easy-to-use instrument for DNA detection, based on the unique electrical properties of nucleic acids to identify genetic mutations. Using DNA-modified microelectrode arrays, the platform interrogates individual reaction wells within the reaction plate and compares the change in electrical resistance to that of a reference cell. DNA products, produced in a reaction that combines PCR and secondary enzymatic manipulations will bind to a cognate probe immobilized to the surface of the electrodes within the reaction/readout plate. Specific binding of the desired analyte is associated with a decrease in electrical resistance, or increased current flow, across two electrodes proportional to the quantity of DNA product bound (Napier *et al.*, 1997). The signal output, measured in ohms, is referenced to a positive and negative control. Interpretation of these data is aided by the normalization of each sample result to the series of controls, which are expected to perform in a range of outputs.

Of potentially great advantage, however, is the fact that the assembly of the enzymatic reaction in a 96-well microassay plate is performed in a thermocycler, and the analysis of the resulting product can be performed without the need of sample transfer. This approach greatly reduces the risk of cross contamination, and is simply easier to use. Overall, the GeneOhm Sciences platform also promises to be a lower cost platform, a major feature for the greater adoption of molecular diagnostic testing.

29.4.7.2 SURFACE PLASMON RESONANCE

Surface plasmon resonance (SPR) is a technology initially developed for the measurement of rare protein and chemical analytes, but now adapted to detect nonamplified DNA samples through direct means. SPR is based on intermolecular changes in the refractive index over time taken at the surface of the sensor (Millot *et al.*, 1995). In this technique, a DNA template, amplified or not, is hybridized in a small reservoir containing an optical interface bound with specific oligonucleotide DNA probes. Through the process of complimentary-based hybridization, light illuminated through the sensor base will be refracted at an angle greater than in those sensors not demonstrated as specific hybridization (Kai *et al.*, 1999). This process can be calibrated by varying the temperature, and is dependent somewhat on time. The specific conditions and kinetics of this interaction, in turn, are characteristic of the degree of sequence complementarity between the DNA template and the cognate probe. Work by Nilsson and colleagues (1997) and others (Bianchi *et al.*, 1997; Feriotto *et al.*, 1999; Brooks *et al.*, 2000) has demonstrated the utility of SPR in the detection of clinical samples amplified by PCR for a series of clinically relevant gene markers, including p53.

A variation on this theme, and using a silicon microchip, is the technique to measure a change in the optical path length between bound and unbound analyte. An interferometer is a device consisting of an array of microsized pillars or finger-like projections created by etching deep into a silicone wafer. Incident light shown onto this area of micropillars will be reflected backward at a predictable angle of efferent light. When the effective thickness of these individual pillars is increased due to a coating of DNA hybridized to a series of cognate probes there is a change in the effective path length of this light creating what are called Perot-Fabry fringes (Lin *et al.*, 1997). The shift in the angle of the efferent light is proportional to the quantity of DNA hybridized to the affixed probes and thus is an effective way to quantitate DNA. This is achieved due to the specificity of DNA to hybridize only when complete complementarity is maintained.

29.4.7.3 MICRO-CANTILEVER DNA SENSORS

Exploitation of true MEMS devices only recently has been demonstrated to function also as a biosensor. In its truer sense, MEMS devices take advantage of certain mechanical properties of structures that have homologues in the macroscale world. One such device is a so-called quartz crystal microbalance (QCM), where bulk monocrystalline quartz or the MEMS version of a microfabricated cantilever-like structure is combined with certain probes to detect changes in mass when analyte is bound to its surface (Joseph *et al.*, 1997; Tuller and Mlcak, 1998). Quartz, having piezoelectric properties, can be put into a high frequency oscillatory mode where the frequency of this oscillation is dependent upon the mass of the structure. In this case, DNA probes serve as a receptor which then hybridizes its cognate DNA template, resulting in a structure of increased mass over that of devices not containing bound DNA (Fauver et al., 1998). The resulting device has a resonant frequency that is shifted downward or lower than those not displaying specific strand hybridization.

A QCM array has been constructed and has been demonstrated to show high sensitivity of DNA hybridization to as little as 10^{-18} molar or approximately 10^{-12} grams of target DNA (Caruso and Rodda, 1997). Other researchers, who have constructed a surface acoustic wave (SAW) sensor, employ thin film piezoelectric materials (Hussain *et al.*, 1997; Zhai *et al.*, 1997). In this case, DNA probes or other biomolecules are immobilized onto the surface and set into oscillatory mode with integrated oscillator circuit. Opposing the transmitter is a microfabricated receiver. Transmission of a wave function across the piezoelectric plane is met at the receiver at a prescribed time. Delays in that time are due to the impedance created by bound DNA. The utility of the so-called flexural plate wave sensor for the detection of

serum-based proteins through their affinity to a monoclonal antibody affixed to the piezoelectric sensor surface has been reported (Wang *et al.*, 1997).

McGlennen and colleagues have demonstrated how other types of piezoelectric thin films can be applied to structures such as microfabricated diaphragms and/or microcantilevers that detect mass changes with DNA hybridization (see Fig. 29.17). These sensors function similar to the SAW device, where a change in the resonant mode of the vibrating cantilever is a function of the quantity of DNA bound to its gold-covered surface. MEMS-based microcantilever sensors have been demonstrated to be sensitive to picograms of mass loading, but because of their construction from silicon have the added advantage of integration with CMOS control electronics onto a single device. Taken in total, these sensor systems can achieve high sensitivities not previously observed with addressable array-based approaches, which in

A

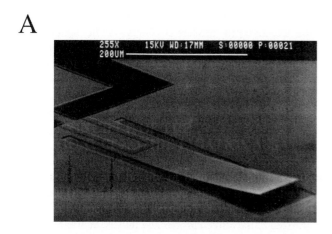

B

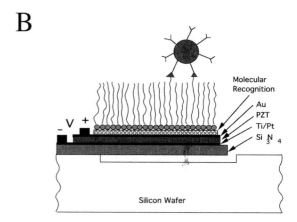

C

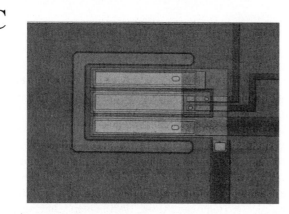

D

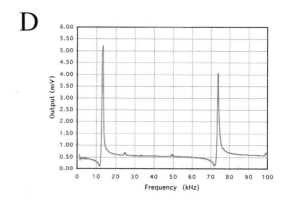

FIGURE 29.17 Overview of the microcantilever biosensor. A. A piezoresistive device that detects minute amounts of material bound to its surface through a change in the electrical resistance of a piezoresistive element. B. Schematic of the various layers of material involved in a MEMS biosensor. The surface, prepared with a molecular recognition material, can be adapted for detection of nucleic acid, peptides, proteins, or microparticulate. C. Top view of a piezoelectric microcantilever biosensor using PZT. D. One mode of operation is to apply a voltage to the cantilever, which sets up an oscillation that is adjusted to its fundamental resonant frequency. Material bound to the surface increases the mass of the oscillating body, and shifts the resonant frequency downward in a predictable way. This device is capable of detecting subpicogram quantities of bound material (DNA, peptide, antibody, and other molecules).

turn can be linked directly to microelectronic circuitry, resulting in a device that is highly compact and very inexpensive to manufacture.

29.4.7.4 HYBRID MECHANICAL AND BIOCHEMICAL SENSORS

Each of the preceding approaches involves in some way exploitation of either physical properties inherent to the DNA or the capability of light-based detection systems to resolve isolated areas of DNA hybridization. The last sensor system to be discussed involves the combination of complex nonliving biological systems with a silicone microchip interface, creating what is described as a truly universal platform for biosensors.

Nature provides perhaps the highest degree of analyte discrimination through the family of cell surface and nuclear receptors housed within living cells. A sensor system that takes advantage of naturally occurring receptor systems and the packaging of biomembranes has been reported (Cornell *et al.*, 1997). Briefly, a lipid bilayer is assembled onto a gold thin-film electrode, into which a known quantity of the artificial antibiotic gramicidin is embedded. Gramicidin functions when configured in a homodimer, where each monomer is essentially a linear peptide that transverses the lipid bilayer membrane. Dimeric gramicidin functions as a pore for the flow of cations across a membrane. To function as a biosensor on a subunit of gramicidin is covalently linked to specific cellular receptors, such as a thyroid releasing hormone receptor or a hapten directed against some analyte. Another hapten is tethered directed to the lipid membrane. When a ligand, such as thyroid releasing hormone, binds to its cognate receptor, there is competition between the tethered hapten and the one linked to gramicidin. The result is a disruption in the formation and continuity of the gramicidin transmembrane pores with a net loss of positively charged ions flowing across the membrane. Detection of this change in current can occur through a simple connection of a galvanometer, which measures the absolute loss or decrease in current flow and additionally can record the phenomenon as a function of time. The sensitivity of such a biosensor and the application to a broader base of sensing capabilities remains to be demonstrated. But the promise of such a hybrid design may be one of the simplest and yet robust currently shown.

29.5 WORKING TOWARD TOTAL SYSTEM INTEGRATION

The earlier sections focused on research and development of novel biosensor platforms. The effort demonstrates integration of several of the operations under the aegis of a single technology platform, which has been described for a system on a single chip and commercialized in a package that involves a combination of chips (Furcht *et al.*, 2001).

Perhaps most advanced among these is the SmartCycler System produced by Cepheid Corporation from Sunnyvale, CA. Cepheid's product, called the MicroBE Analyzer, is the product of nearly 15 years of development. Derived from research at national laboratories as well as academic institutions, it focuses on two key steps in the genetic testing format: a miniaturized thermal cycling device and an integrated detection system (Petersen *et al.*, 1998; Henke, 1999).

Regardless of the performance of individual components in the integrated genetic testing system, the key to success in the commercialization, and hence the success in their application to point-of-care gene-based testing, is the issue related to sample processing and microfluidics (Schomburg *et al.*, 1993; Zengerle and Sandmaier, 1996; Mastrangelo *et al.*, 1998). DNA assays for the detection of microorganisms or DNA mutations within cancers often exist at very low concentrations, often less than 100 copies per milliliter of raw biologic sample, such as blood or urine. Such sensitivities set fundamental physical limitations on the minimum quantities of starting material that will prove to be clinically useful. For example, with PCR, a 50 to 100 μl reaction containing 20–100 ng of sample DNA typically can detect as few as 20 copies of a target sequence. But in cases where the requirement is to achieve much higher levels of sensitivity—often the case in assays for infectious agents—either new detection modalities are required or a greater amount of sample needs to be processed. In the latter case, the advantages of using microchip-based instrumentation, with its inherent limitations to handle minute quantities of sample, are at odds with these clinical demands of high sensitivity in PCR-based assays. Hence, the consideration of sample processing, and microfluidic control in microchip-based integrated systems requires further discussion.

Under ideal conditions, microchip-based thermocyclers and detection systems are capable of processing sample sizes as small as 0.5 μl. In the clinical laboratory, and more importantly in other environments, such as point-of-care testing at the bedside or possibly even in at-home applications, the interface between the human hand and the struggle to work with such minute volumes will result in the avoidance of these new chip-based technologies.

Again, Cepheid Corporation is committed to developing and implementing fluidic systems that attempt to bridge the gap between the easy-to-use large sample volumes to the world of microstructures and microarrays. Their work is focused on attacking this problem on two fronts. First, their chip-based thermal cycling technology is based on performing high throughput PCR on relatively large samples. Their system can work with volumes as large as 100 μl. This is accomplished first by developing sample processing modules, which can handle large volumes of liquid sample in a flow-through format. Specifically, they have developed a DNA capture chip consisting of a dense array of microfabricated silicone pillars. Early work by Carlson and colleagues (1997) demonstrated that anticoagulated whole blood flowing across such a structure would tend to passively permit the flow of red blood cells through the microstructure where they could be captured and removed as waste. White blood cells are inherently sticky; finding the silicone surface to be abnormal would cause them to be retained and actually stick to the surface. This process involves, then, in a simple flow-through step, the marked enrichment of whole blood such that only the nucleated cellular components are retained. Secondary steps including flushing the chip with lysis buffers, electrical biasing to retain negatively charged DNA, and rinsing steps, all of which can be introduced to the device via prepackaged and/or stored extraction reagents. The net result is retention of high molecular weight and high integrity genomic DNA from samples ranging in volumes from 1 to 10 ml in size.

Transport of the enriched DNA fraction and other fluid elements following extraction has been the focus of several groups including Caliper Technologies and Aclara Biosciences. As previously discussed, Caliper's trademarked lab-on-a-chip approach is based on conventional microfabrication techniques of etching in both glass as well as in silicon to create a network of well-defined microcapillary conduits. Devices developed by Aclara Biosciences have focused on single-use flexible 96-well microplates, where each reservoir is connected to a series of microcapillaries. The company's proprietary microfluidics technology enables the accurate measurement, dispensing, and mixing of minute quantities of liquid with volumes as small as 10 to 0.1 µl. Plastic microchips are employed that also involve controlling elements and are trademarked under the name Lab Card™.

The challenge to understand the dynamic properties of fluids at this microscale is one potential advantage of microscale analyses. Microcapillary channels with a cross-sectional dimension less than 100 µm create a setting where fluid flows in a laminar (nonturbulent) manner (Gravesen *et al.*, 1993; Mastrangelo *et al.*, 1998). This phenomenon has a profound impact on the performance of these integrated devices, not only because of the speed by which the assay is completed, but also that certain benefits of the pattern by which certain molecules are diffused in a laminar flow environment can be exploited. On the negative side, microfluidic channels and reservoirs are inherently fraught with problems related to the formation of bubbles (Wilding *et al.*, 1994b). Whether in the case of PCR reactants within a thermal cycling chamber or the conduction of fluids between chip components in the integrated device, bubbles can prevent as well as completely occlude the system. Hence, the central challenge in the development of these systems is to create the optimized interface between the operator and fluidic components of the device in the effort to prevent or remove any types of bubbles consequent to pipetting and/or the motion of the fluid.

29.6 CONCLUSIONS

Although the prospects for an integrated gene testing device appear to be close at hand, a great deal of testing needs to be done before these platforms are ready for use in patient care. A great deal of this work has been supported to date by various agencies within the United States Department of Defense, the National Institute of Standards and Technology, as well as from venture capital sources, where the objective is to develop a microscale gene testing device that can be deployed not only into a battlefield environment or for the purposes of rapid detection of such agents such as those used in biologic warfare, but in point-of-care settings for human molecular diagnostics. In many respects, exhaustive testing of these platforms in the hands of military applications will be an appropriate test for the utility of these devices, since in general they will be evaluated in the hands of those who are generally inexperienced in molecular biology laboratory techniques. If such systems prove to be useful, simple, and robust in the hands of field soldiers, the pathway to application in the clinical arena will be made much more rapidly.

References

Abel, A. P., Weller, M. G., Duveneck, G. L., Ehrat, M., and Widmer, H. M. (1996). Fiber-optic evanescent wave biosensor for the detection of oligonucleotides. *Anal. Chem.* 68, 2905–2912.

Ahlquist, D. A. (2002). Stool-based DNA tests for colorectal cancer: clinical potential and early results. *Rev. Gastroenterol. Disord.* 2 Suppl. 1, S20–S26.

Amos, J., and Patnaik, M. (2002). Commercial molecular diagnostics in the U.S.: The Human Genome Project to the clinical laboratory. *Hum. Mutat.* 19, 324–333.

Audrezet, M. P., Costes, B., Ghanem, N., Fanen, P., Verlingue, C., Morin, J. F., Mercier, B., Goossens, M., and Ferec, C. (1993). Screening for cystic fibrosis in dried blood spots of newborns. *Mol. Cell. Probes* 7, 497–502.

Beaulieux, F., See, D. M., Leparc-Goffart, I., Aymard, M., and Lina, B. (1997). Use of magnetic beads versus guanidium thiocyanate-phenol-chloroform RNA extraction followed by polymerase chain reaction for the rapid, sensitive detection of enterovirus RNA. *Res. Virol.* 148, 11–15.

Bernacki, S. H., Farkas, D. H., Shi, W., Chan, V., Liu, Y., Beck, J. C., Bailey, K. S., Pratt, V. M., Monaghan, K. G., Matteson, K. J., Schaefer, F. V., Friez, M., Shrimpton, A. E., and Stenzel, T. T. (2003). Bioelectronic sensor technology for detection of cystic fibrosis and hereditary hemochromatosis mutations. *Arch. Pathol. Lab. Med.* 127, 1565–1572.

Beugelsdijk, T. J. (1991). The future of laboratory automation. *Genet. Anal. Tech. Appl.* 8, 217–220.

Bhagwat, A. A., Phadke, R. P., Wheeler, D., Kalantre, S., Gudipati, M., and Bhagwat, M. (2003). Computational methods and evaluation of RNA stabilization reagents for genome-wide expression studies. *J. Microbiol. Methods* 55, 399–409.

Bianchi, N., Rutigliano, C., Tomassetti, M., Feriotto, G., Zorzato, F., and Gambari, R. (1997). Biosensor technology and surface plasmon resonance for real-time detection of HIV-1 genomic sequences amplified by polymerase chain reaction. *Clin. Diagn. Virol.* 8, 199–208.

Bosserhoff, A. K., Buettner, R., and Hellerbrand, C. (2000). Use of capillary electrophoresis for high throughput screening in biomedical applications. A minireview. *Comb. Chem. High Throughput Screen* 3, 455–466.

Brooks, P. (2000). MutS-DNA interactions and DNase protection analysis with surface plasmon resonance. *Methods Mol. Biol.* 152, 119–132.

Bruin, G. J. (2000). Recent developments in electrokinetically driven analysis on microfabricated devices. *Electrophoresis* 21, 3931–3951.

Burke, D. T., Burns, M. A., and Mastrangelo, C. (1997). Microfabrication technologies for integrated nucleic acid analysis. *PCR Methods Appl.* 7, 189–197.

Burns, M. A., Mastrangelo, C. H., Sammarco, T. S., Man, F. P., Webster, J. R., Johnsons, B. N., Foerster, B., Jones, D., Fields, Y., Kaiser, A. R., and Burke, D. T. (1996). Microfabricated structures for integrated DNA analysis. *Proc. Natl. Acad. Sci. USA* 93, 5556–5561.

Carlson, R. H., Gabel, C. V., Chan, S., and Austin, R. H. (1997). Self-sorting of white blood cells in a lattice. *Phys. Review Letters* 15, 2149–2152.

Caruso, F., and Rodda, E. (1997). Quartz crystal microbalance study of DNA immobilization and hybridization for nucleic acid sensor development. *Anal. Chem.* 69, 2043–2049.

Castelletti, L., Verzola, B., Gelfi, C., Stoyanov, A., and Righetti, P. G. (2000). Quantitative studies on the adsorption of proteins to the bare silica wall in capillary electrophoresis. III: Effects of adsorbed surfactants on quenching the interaction. *J. Chromatogr. A.* 894, 281–289.

Cheng, J., Shoffner, M. A., Mitchelson, K. R., Kricka, L. J., and Wilding, P. (1996). Analysis of ligase chain reaction products amplified in a silicon-glass chip using capillary electrophoresis. *J. Chromatogr. A.* 732, 151–158.

Cheung, V. G., Morley, M., Aguilar, F., Massimi, A., Kucherlapati, R., and Childs, G. (1999). Making and reading microarrays. *Nat. Genet.* 21(1Suppl), 15–19.

Chiari, M., Damin, F., Melis, A., and Consonni, R. (1998). Separation of oligonucleotides and DNA fragments by capillary electrophoresis in dynamically and permanently coated capillaries, using a copolymer of acrylamide and beta-D-glucopyranoside as a new low viscosity matrix with high sieving capacity. *Electrophoresis* 19, 3154–3159.

Chiari, M., and Cretich, M. (2001). Capillary coatings. Choices for capillary electrophoresis of DNA. *Methods Mol. Biol.* 162–163, 125–138.

Chung, Y. C., Jan, M. S., Lin, Y. C., Lin, J. H., Cheng, W. C., and Fan, C. Y. (2004). microfluidic chip for high efficiency DNA extraction. *Lab. Chip* 4, 141–147.

Cooksey, R. C., Limor, J., Morlock, G. P., and Crawford, J. T. (2003). Identifying mycobacterium species and strain typing using a microfluidic labchip instrument. *Biotechniques* 35, 786–794.

Cooper, M., Li, S. Q., Bhardwaj, T., Rohan, T., and Kandel, R. A. (2004). Evaluation of oligonucleotide arrays for sequencing of the p53 gene in DNA from formalin-fixed, paraffin-embedded breast cancer specimens. *Clin. Chem.* 50, 500–508.

Cornell, B. A., Braach-Maksvytis, V. L., King, L. G., Osman, P. D., Raguse, B., Wieczorek, L., and Pace, R. J. (1997). Biosensor that uses ion-channel switches. *Nature* 387, 580–583.

Crisan, D., and Farkas, D. H. (1993). Bone marrow biopsy imprint preparations: use for molecular diagnostics in leukemias. *Annals Clin. Lab. Sci.* 23, 407–422.

de Longueville, F., Surry, D., Meneses-Lorente, G., Bertholet, V., Talbot, V., Evrard, S., Chandelier, N., Pike, A., Worboys, P., Rasson, J. P., Le Bourdelles, B., and Remacle, J. (2002). Gene expression profiling of drug metabolism and toxicology markers using a low-density DNA microarray. *Biochem. Pharmacol.* 64, 137–149.

DelRio-LaFreniere, S. A., Browning, M. K., and McGlennen, R. C. (2004). Low–density addressable array for the detection and typing of the human papillomavirus. *Diagn. Microbiol. Infect. Dis.* 48, 23–31.

Demers, L. M., Mirkin, C. A., Mucic, R. C., Reynolds, R. A. 3rd, Letsinger, R. L., Elghanian, R., and Viswanadham, G. (2000). A fluorescence-based method for determining the surface coverage and hybridization efficiency of thiol-capped oligonucleotides bound to gold thin films and nanoparticles. *Anal. Chem.* 72, 5535–5541.

Demirev, P. A., Ho, Y. P., Ryzhov, V., and Fenselau, C. (1999). Microorganism identification by mass spectrometry and protein database searches. *Anal. Chem.* 71, 2732–2738.

Desjardin, L. E., Chen, Y., Perkins, M. D., Teixeira, L., Cave, M. D., and Eisenach, K. D. (1998). Comparison of the ABI 7700 system (TaqMan) and competitive PCR for quantification of IS6110 DNA in sputum during treatment of tuberculosis. *J. Clin. Microbiol.* 36, 1964–1968.

Despont, M., Lorenz, H., Fahrni, N., Brugger, J., Renaud, P., and Vettiger, P. (1997). High-aspect-ratio, ultrathick, negative-tone near-UV photoresist for MEMS applications. *Proc. IEEE Micro Electro Mechan. Syst.* 518–522.

Duveneck, G. L., Bopp, M. A., Ehrat, M., Balet, L. P., Haiml, M., Keller, U., Marowsky, G., and Soria, S. (2003). Two-photon fluorescence excitation of macroscopic areas on planar waveguides. *Biosens. Bioelectron.* 18, 503–510.

Edman, C. F., Raymond, D. E., Wu, D. J., Tu, E., Sosnowski, R. G., Butler, W. F., Nerenberg, M., and Heller, M. J. (1997). Electric field directed nucleic acid hybridization on microchips. *Nucleic Acids Res.* 25, 4907–4914.

Eggers, M., and Ehrlich, D. (1995). A review of microfabricated devices for gene-based diagnostics. *Hematol. Pathol.* 9, 1–15.

Enayat, M. S., Williams, M. D., and Hill, F. G. (1997). Further simplifications of the factor V: Q506 mutation detection test [letter; comment]. *Blood Coagul. Fibrinolysis* 8, 205.

Eng, C., and Vijg, J. (1997). Genetic testing: the problems and the promise. *Nat. Biotechnol.* 15, 422–426.

Epstein, J. R., and Walt, D. R. (2003). Fluorescence-based fibre optic arrays: a universal platform for sensing. *Chem. Soc. Rev.* 32, 203–214.

Fauver, M. E., Dunaway, D. L., Lilienfeld, D. H., Craighead, H. G., and Pollack, G. H. (1998). Microfabricated cantilevers for measurement of subcellular and molecular forces. *IEEE Trans. Biomed. Eng.* 45, 891–898.

Favis, R., and Barany, F. (2000). Mutation detection in K-ras, BRCA1, BRCA2, and p53 using PCR/LDR and a universal DNA microarray. *Ann. N. Y. Acad. Sci.* 906, 39–43.

Favis, R., Day, J. P., Gerry, N. P., Phelan, C., Narod, S., and Barany, F. (2000). Universal DNA array detection of small insertions and deletions in BRCA1 and BRCA2. *Nat. Biotechnol.* 18, 561–564.

Feriotto, G., Lucci, M., Bianchi, N., Mischiati, C., and Gambari, R. (1999). Detection of the *deltaF508 (F508del)* mutation of the cystic fibrosis gene by surface plasmon resonance and biosensor technology. *Hum. Mutat.* 13, 390–400.

Fodor, S. P., Rava, R. P., Huang, X. C., Pease, A. C., Holmes, C. P., and Adams, C. L. (1993). Multiplexed biochemical assays with biological chips. *Nature* 364, 555–556.

Furcht, L. T., McGlennen, R. C., Polla, D. L., inventors; Reagent of the University of Minnesota, assignee. Integrated Microchip Genetic Testing Device. United States of America. 2001.

Gerry, N. P., Witowski, N. E., Day, J., Hammer, R. P., Barany, G., and Barany, F. (1999). Universal DNA array with polymerase chain reaction/ligase detection reaction (PCR/LDR) for multiplex detection of low abundance mutations. *J. Mol. Biol.* 292, 251–262.

Glynou, K., Ioannou, P. C., Christopoulos, T. K., and Syriopoulou, V. (2003). Oligonucleotide-functionalized gold nanoparticles as probes in a dry-reagent strip biosensor for DNA analysis by hybridization. *Anal. Chem.* 75, 4155–4160.

Gomez, E., van der Poel, S. C., Jansen, J. H., van der Reijden, B. A., and Lowenberg, B. (1998). Rapid simultaneous screening of factor V leiden and G20210A prothrombin variant by multiplex polymerase chain reaction on whole blood. *Blood* 91, 2208–2209.

Gravesen, P., Branebjerg, J., and Jensen, O. S. (1993). Microfluidics, a review. *J. Micromech. Microeng.* 3, 168–182.

Guo, B. (1999). Mass Spectrometry in DNA analysis. *Anal. Chem.* 71, 333R–337R.

Guttman, A., and Ronai, Z. (2000). Ultrathin-layer gel electrophoresis of biopolymers. *Electrophoresis* 21, 3952–3964.

Hamels, S., Gala, J. L., Dufour, S., Vannuffel, P., Zammatteo, N., and Remacle, J. (2001). Consensus PCR and microarray for diagnosis of the genus Staphylococcus, species, and methicillin resistance. *Biotechniques* 31, 1364–1372.

Hanning, A., Westberg, J., and Roeraade, J. (2000). A liquid core waveguide fluorescence detector for multicapillary electrophoresis applied to DNA sequencing in a 91-capillary array. *Electrophoresis* 21, 3290–3304.

Harding, J. D., Gebeyehu, G., Bebee, R., Simms, D., and Klevan, L. (1989). Rapid isolation of DNA from complex biological samples using a novel capture reagent-methidium-spermine-sepharose. *Nucleic Acids Res.* 17, 6947–6958.

Hashimoto, K., Ito, K., and Ishimori, Y. (1998). Microfabricated disposable DNA sensor for detection of hepatitis B virus DNA. *Sens. Actuat. B. Chem.* 220–225.

Hawkins, J. R., Khripin, Y., Valdes, A. M., and Weaver, T. A. (2002). Miniaturized sealed-tube allele-specific PCR. *Hum. Mutat.* 19, 543–553.

Heath, E. M., Morken, N. W., Campbell, K. A., Tkach, D., Boyd, E. A., and Strom, D. A. (2001). Use of buccal cells collected in mouthwash as a source of DNA for clinical testing. *Arch. Pathol. Lab. Med.* 125, 127–133.

Heller, C. (2001). Influence of polymer concentration and polymer composition on capillary electrophoresis of DNA. *Methods Mol. Biol.* 162, 111–123.

Henke, C. (1999). DNA-chip technologies Part 3: What does the future hold? *IVD Technology* 7, 37–42.

Hessner, M. J., Luhm, R. A., Pearson, S. L., Endean, D. J., Friedman, K. D., and Montgomery, R. R. (1999). Prevalence of prothrombin G20210A, factor V G1691A (Leiden), and methylenetetrahydrofolate reductase (MTHFR) C677T in seven different populations determined by multiplex allele-specific PCR. *Thromb. Haemost.* 81, 733–738.

Huber, M., Losert, D., Hiller, R., Harwanegg, C., Mueller, M. W., and Schmidt, W. M. (2001). Detection of single base alterations in genomic DNA by solid phase polymerase chain reaction on oligonucleotide microarrays. *Anal. Biochem.* 299, 24–30.

Hussain, I., Kumar, A., and Mangiaracina, A. (1997). Fabrication of piezoelectric sensors for biomedical applications. Materials for Smart Systems II. *Proc. Materials Research Society Symposium* 459, 501–506.

Iannone, M. A., Taylor, J. D., Chen, J., Li, M. S., Ye, F., and Weiner, M. P. (2003). Microsphere-based single nucleotide polymorphism genotyping. *Methods Mol. Biol.* 226, 123–134.

Ibrahim, M. S., Lofts, R. S., Jahrling, P. B., Henchal, E. A., Weedn, V. W., Northrup, M. A., and Belgrader, P. (1998). Real-time microchip PCR for detecting single-base differences in viral and human DNA. *Anal. Chem.* 70, 2013–2017.

Irizarry, R. A., Bolstad, B. M., Collin, F., Cope, L. M., Hobbs, B., and Speed, T. P. (2003). Summaries of Affymetrix GeneChip probe level data. *Nucleic Acids Res.* 31, E15.

Jacobson, S. C., and Ramsey, J. M. (1996). Integrated microdevice for DNA restriction fragment analysis. *Anal. Chem.* 68, 720–723.

Joseph, H., Swafford, B., and Terry, S. (1997). Medical applications for MEMS devices. *Proc. International Symposium on Test and Measurement* 399–404.

Kai, E., Sawata, S., Ikebukuro, K., Iida, T., Honda, T., and Karube, I. (1999). Detection of PCR products in solution using surface plasmon resonance. *Anal. Chem.* 71, 796–800.

Kajiyama, T., Miyahara, Y., Kricka, L. J., Wilding, P., Graves, D. J., Surrey, S., and Fortina, P. (2003). Genotyping on a thermal gradient DNA chip. *Genome Res.* 13, 467–475.

Kalinina, O., Lebedeva, I., Brown, J., and Silver, J. (1997). Nanoliter scale PCR with TaqMan detection. *Nucleic Acids Res.* 25, 1999–2004.

Kaupp, S., Bubert, H., Baur, L., Nelson, G., and Watzig, H. (2000). Unexpected surface chemistry in capillaries for electrophoresis. *J. Chromatogr. A.* 894, 73–77.

Kearney, P. P., and Aumatell, A. (1997). Rapid diagnosis of chronic myeloid leukaemia by linking PCR to capillary gel electrophoresis. *Clin. Lab. Haematol.* 19, 261–266.

Kim, Y., and Yeung, E. S. (2001). Capillary electrophoresis of DNA fragments using poly(ethylene oxide) as a sieving material. *Methods Mol. Biol.* 162, 215–223.

Konig, S., and Welsch, T. (2000). Moderation of the electroosmotic flow in capillary electrophoresis by chemical modification of the capillary surface with tentacle-like oligourethanes. *J. Chromatogr. A.* 894, 79–88.

Kricka, L. J., and Wilding, P. (1996). Micromachining—a new direction for clinical analyzers. *Pure Appl. Chem.* 68, 1831–1836.

Kricka, L. J., and Wilding, P. Microchip PCR (2003). *Anal. Bioanal. Chem.* 377, 820–825.

Kwiatkowski, R. W., Lyamichev, V., de Arruda, M., and Neri, B. (1999). Clinical, genetic, and pharmacogenetic applications of the Invader assay. *Mol. Diagn.* 4, 353–364.

Lagally, E. T., Medintz, I., and Mathies, R. A. (2001). Single-molecule DNA amplification and analysis in an integrated microfluidic device. *Anal. Chem.* 73, 565–570.

Lamture, J. B., Beattie, K. L., Burke, B. E., Eggers, M. D., Ehrlich, D. J., Fowler, R., Hollis, M. A., Kosicki, B. B., Reich, R. K., Smith, S. R., Varma, R. S., and Hogan, M. E. (1994). Direct detection of nucleic acid hybridization on the surface of a charge coupled device. *Nucleic Acids Res.* 22, 2121–2125.

Lay, M. J., and Wittwer, C. T. (1997). Real-time fluorescence genotyping of factor V Leiden during rapid-cycle PCR. *Clin. Chem.* 43, 2262–2267.

Lin, V. S. Y., Motesharei, K., Dancil, K. P., Sailor, M. J., and Ghadiri, M. R. (1997). A porous silicon-based optical interferometric biosensor. *Science* 278, 840–843.

Lipshutz, R. J., Morris, D., Chee, M., Hubbell, E., Kozal, M. J., Shah, N., Shen, N., Yang, R., and Fodor, S. P. (1995). Using oligonucleotide probe arrays to access genetic diversity. *Biotechniques* 19, 442–447.

Lyamichev, V., Mast, A. L., Hall, J. G., Prudent, J. R., Kaiser, M. W., Takova, T., Kwiatkowski, R. W., Sander, T. J., de Arruda, M., Arco, D. A., Neri, B. P., and Brow, M. A. (1999). Polymorphism identification and quantitative detection of genomic DNA by invasive cleavage of oligonucleotide probes. *Nat. Biotechnol.* 17, 292–296.

Marshall, A., and Hodgson, J. (1998). DNA chips: an array of possibilities. *Nat. Biotechnol.* 16, 27–31.

Mastrangelo, C. H., Burns, M. A., and Burke, D. T. (1998). Microfabricated devices for genetic diagnostics. *Proc. IEEE*, 86, 1769–1787.

Medeiros, M., Sharma, V. K., Ding, R., Yamaji, K., Li, B., Muthukumar, T., Valderde-Rosas, S., Hernandez, A. M., Munoz, R., and Suthanthiran, M. (2003). Optimization of RNA yield, purity and mRNA copy number by treatment of urine cell pellets with RNAlater. *J. Immunol. Methods* 279, 135–142.

Meldrum, D. (2000). Automation for genomics, part one: preparation for sequencing. *Genome Res.* 10, 1081–1092.

Miller Coyle, H., Shutler, G., Abrams, S., Hanniman, J., Neylon, S., Ladd, C., Palmbach, T., and Lee, H. C. (2003). A simple DNA extraction method for marijuana samples used in amplified fragment length polymorphism (AFLP) analysis. *J. Forensic Sci.* 48, 343–347.

Millot, M-C., Vals, T., Martin, F., Sebille, B., and Levy, Y. (1995). Surface plasmon resonance response of a polymer-coated biochemical sensor. *Proc. SPIE, The International Society for Optical Engineering* 2331:Photo-Optical.

Nachamkin, I., Panaro, N. J., Li, M., Ung, H., Yuen, P. K., Kricka, L. J., and Wilding, P. (2001). Agilent 2100 bioanalyzer for restriction fragment length polymorphism analysis of the Campylobacter jejuni flagellin gene. *J. Clin. Microbiol.* 39, 754–757.

Napier, M. E., Loomis, C. R., Sistare, M. F., Kim, J., Eckhardt, A. E., and Thorp, H. H. (1997). Probing biomolecule recognition with electron transfer—Electrochemical sensors for DNA hybridization. *Bioconjug. Chem.* 8, 906–913.

Niemeyer, C. M., Ceyhan, B., and Hazarika, P. (2003). Oligofunctional DNA-gold nanoparticle conjugates. Angew. *Chem. Int. Ed. Engl.* 42, 5766–5770.

Nilsson, P., Persson, B., Larsson, A., Uhlen, M., and Nygren, P. A. (1997). Detection of mutations in PCR products from clinical samples by surface plasmon resonance. *J. Mol. Recogn.* 10, 7–17.

Oliphant, A., Barker, D. L., Stuelpnagel, J. R., and Chee, M. S. (2002). BeadArray technology: enabling an accurate, cost-effective approach to high-throughput genotyping. *Biotechniques* Suppl. 56–61.

Paegel, B. M., Blazej, R. G., Mathies, R. A. (2003). Microfluidic devices for DNA sequencing: sample preparation and electrophoretic analysis. *Curr. Opin. Biotechnol.* 14, 42–50.

Paulus, A. (1998). Capillary electrophoresis of DNA chips using capillaries and micromachined chips. *Am. Lab.* 30, 59ff.

Pease, A. C., Solas, D., Sullivan, E. J., Cronin, M. T., Holmes, C. P., and Fodor, S. P. (1994). Light-generated oligonucleotide arrays for rapid DNA sequence analysis. *Proc. Natl. Acad. Sci. USA* 91, 5022–5026.

Petersen, K., McMillan, W., Kovacs, G., Northrup, A., Christel, L., and Purahmadi, F. (1998). The promise of miniaturized clinical diagnostic systems. *IVD Technol.* 6, 43–49.

Ramsay, G. (1998). DNA chips: state-of-the art. *Nat. Biotechnol.* 16, 40–44.

Raskin, S., Phillips, J. A., Kaplan, G., McClure, M., and Vnencak-Jones, C. (1992). Cystic fibrosis genotyping by direct PCR analysis of Guthrie blood spots. *PCR Methods Appl.* 2, 154–156.

Righetti, P. G., and Gelfi, C. (1997). Capillary electrophoresis of DNA for molecular diagnostics. *Electrophoresis* 18, 1709–1714.

Rogers, C., and Burgoyne, L. (1997). Bacterial typing: storing and processing of stabilized reference bacteria for polymerase chain reaction without preparing DNA—an example of an automatable procedure. *Anal. Biochem.* 247, 223–227.

Scherczinger, C. A., Bourke, M. T., Ladd, C., and Lee, H. C. (1997). DNA extraction from liquid blood using QIAamp. *J. Forensic Sci.* 42, 893–896.

Schomburg, W. K., Fahrenberg, J., Maas, D., and Rapp, R. (1993). Active valves and pumps for microfluidics. *J. Micromech. Microeng.* 3, 216–218.

Shoffner, M. A., Cheng, J., Hvichia, G. E., Kricka, L. J., and Wilding, P. (1996). Chip PCR. 1. Surface Passivation of Microfabricated Silicon-Glass Chips For PCR. *Nucleic Acids Res.* 24, 375–379.

Singh-Gasson, S., Green, R. D., Yue, Y., Nelson, C., Blattner, F., Sussman, M. R., and Cerrina, F. (1999). Maskless fabrication of light-directed oligonucleotide microarrays using a digital micromirror array. *Nat. Biotechnol.* 17, 974–978.

Song, K., Fendrick, A. M., and Ladabaum, U. (2004). Fecal DNA testing compared with conventional colorectal cancer screening methods: a decision analysis. *Gastroenterology* 126, 1270–1279.

Sosnowski, R. G., Tu, E., Butler, W. F., O'Connell, J. P., and Heller, M. J. (1997). Rapid determination of single base mismatch mutations in DNA hybrids by direct electric field control. *Proc. Natl. Acad. Sci. USA* 94, 1119–1123.

Steinberg, G., Stromsborg, K., Thomas, L., Barker, D., and Zhao, C. (2004). Strategies for covalent attachment of DNA to beads. *Biopolymers*, 73, 597–605.

Tang, W. C. (1997). Overview of microelectromechanical systems and design processes. *Proc. Design Automation Conference* 670–673.

Trau, D., Lee, T. M., Lao, A. I., Lenigk, R., Hsing, I. M., Ip, N. Y., Carles, M. C., and Sucher, N. J. (2002). Genotyping on a complementary metal oxide semiconductor silicon polymerase chain reaction chip with integrated DNA microarray. *Anal. Chem.* 74, 3168–3173.

Tuller, H. L., and Mlcak, R. (1998). Inorganic sensors utilizing MEMS and microelectronic technologies. *Curr. Opin. Sol. State Mat. Sc.* 3, 501–504.

Umek, R. M., Lin, S. W., Vielmetter, J., Terbrueggen, R. H., Irvine, B., Yu, C. J., Kayyem, J. F., Yowanto, H., Blackburn, G. F., Farkas, D. H., and Chen, Y. P. (2001). Electronic detection of nucleic acids: a versatile platform for molecular diagnostics. *J. Mol. Diagn.* 3, 74–84.

Wang, A. W., Kiwan, R., White, R. M., and Ceriani, R. L. (1997). A silicon-based ultrasonic immunoassay for detection of breast cancer antigens. *Proc. International Conference on Solid-State Sensors and Actuators* 191–194.

Wen, S. Y., Wang, H., Sun, O. J., and Wang, S. Q. (2003). Rapid detection of the known SNPs of CYP2C9 using oligonucleotide microarray. *World J. Gastroenterol.* 9, 1342–1346.

Wilding, P., Shoffner, M. A., and Kricka, L. J. (1994a). PCR in a silicon microstructure. *Clin. Chem.* 40, 1815–1858.

Wilding, P., Pfahler, J., Bau, H. H., Zemel, J. N., and Kricka, L. J. (1994b). Manipulation and flow of biological fluids in straight channels micromachined in silicon. *Clin. Chem.* 40, 43–47.

Wilding, P., and Kricka, L. J. (1999). Micro-microchips: just how small can we go? *Trends Biotechnol.* 17, 465–468.

Wilkins Stevens, P., Hall, J. G., Lyamichev, V., Neri, B. P., Lu, M., Wang, L., Smith, L. M., Kelso, D. M. (2001). Analysis of single nucleotide polymorphisms with solid phase invasive cleavage reactions. *Nucleic Acids Res.* 29, E77.

Wise, C. A., Paris, M., Morar, B., Wang, W., Kalaydjieva, L., and Bittles, A. H. (2003). A standard protocol for single nucleotide primer extension in the human genome using matrix-assisted laser desorption/ionization time-of-flight mass spectrometry. *Rapid Commun. Mass Spectrom.* 17, 1195–1202.

Witowski, N. E., Leiendecker-Foster, C., Gerry, N. P., McGlennen, R. C., and Barany, G. (2000). Microarray-based detection of select cardiovascular disease markers. *Biotechniques* 29, 936–942.

Woolley, A. T., Hadley, D., Landre, P., deMello, A. J., Mathies, R. A., and Northrup, M. A. (1996). Functional integration of PCR amplification and capillary electrophoresis in a microfabricated DNA analysis device. *Anal. Chem.* 68, 4081–4086.

Wulfman, D. R., Baas, T., and McGlennen, R. C. (2003). Planar Waveguide Genetic Assay readout device. *Adv. Bioeng.* I00679, 161–162.

Xu, H., Sha, M. Y., Wong, E. Y., Uphoff, J., Xu, Y., Treadway, J. A., Truong, A., O'Brien, E., Asquith, S., Stubbins, M., Spurr, N. K., Lai, E. H., and Mahoney, W. (2003). Multiplexed SNP genotyping using the Qbead system: a quantum dot-encoded microsphere-based assay. *Nucleic Acids Res.* 31, E43.

Ye, F., Li, M. S., Taylor, J. D., Nguyen, Q., Colton, H. M., Casey, W. M., Wagner, M., Weiner, M. P., and Chen, J. (2001). Fluorescent microsphere-based readout technology for multiplexed human single nucleotide polymorphism analysis and bacterial identification. *Hum. Mutat.* 17, 305–316.

Yu, C. J., Wan, Y., Yowanto, H., Li, J., Tao, C., James, M. D., Tan, C. L., Blackburn, G. F., and Meade, T. J. (2001). Electronic detection of single-base mismatches in DNA with ferrocene-modified probes. *J. Am. Chem. Soc.* 123, 11155–11161.

Zengerle, R., and Sandmaier, H. (1996). Microfluidics. *Proc. International Symposium on Micro Machine and Human Science* 13–20.

Zhai, J., Cui, H., and Yang, R. (1997). DNA based biosensors. *Biotechnol. Adv.* 15, 43–58.

CHAPTER 30

Human Gene Patents and Genetic Testing

TIMOTHY CAULFIELD

University of Alberta, Health Law Institute, Law Centre, Edmonton, Alberta, Canada

TABLE OF CONTENTS

30.1 INTRODUCTION

Since the start of the Human Genome Project, the patenting of human genetic material has been the focus of a great deal of social controversy. Concerns have ranged from claims that such patents infringe human dignity, to a possible adverse impact on the research environment. Despite the ongoing controversies, gene patents continue to be granted in jurisdictions throughout the world and those with an eye toward economic growth insist that they are an indispensable element in the innovation and technology transfer process.

This chapter provides an overview of the gene patenting controversy, particularly emphasizing how gene patents may impact, for better or worse, the development and provision of genetic testing technologies. The chapter begins with a discussion of benefits and social concerns often associated with human gene patents. This is followed by an analysis of gene patents in the health care context and the related patent reform recommendations that have been proposed by various policy-making entities.

30.2 BENEFITS FROM PATENTS

Patents are meant to benefit society by encouraging innovation and by granting the inventor an exclusive monopoly over the invention. This monopoly, which generally lasts 20 years, gives the inventor the ability to control the use of the invention, thus putting the inventor in a position to profit. Specifically, the patent holder may prevent anyone else from making, constructing, emulating, using, or selling the patented invention, or any other invention that achieves the same result in substantially the same manner. However, patents are also meant to ensure that the details of the invention are disclosed to the public. Indeed, one of the criteria for obtaining a patent is that the invention is described in sufficient a fashion that an individual with reasonable skill in the relevant art or technical field would have the ability to replicate the invention. This disclosure of information is meant to facilitate the advancement of further innovation and the avoidance of wasteful, duplicative research (ALRC, 2003).

Patents are viewed as an increasingly important part of the world economy. Many commentators have suggested that the world now lives in an era where ideas and innovation have come to dominate the new, knowledge-based economy. Just as the industrial revolution changed the way in which wealth was generated, so too are areas such as information technology, computers, biotechnology, and, perhaps one day, nanotechnology. In such fields, where ideas, as opposed to tangible goods, are often the primary commodity, intellectual property protection is seen to be fundamentally important to economic growth. Some commentators have gone so far as to suggest that the ability of a nation to generate patents is a strong indicator of the health and long-term prospects of the region's economy. For example, Juan Enriquez (2001) has suggested that "patents are a good window (although not the only window) on who might triumph and who might lose over the course of the next two decades."

Few other industries rely on patent protection as much as the biotechnology sector (Straus, 1998; Cook-Deegan and McCormack, 2001). Often, patents are the only assets that biotechnology companies have, and because potential products can take years to bring to market, investors believe they need strong intellectual property protection in order to provide a small sense of long-term security. The perceived importance of patents to the biotechnology sector is well illustrated by the way in which investment seems to follow strong intellectual property protection frameworks. When a rough map of the human genome was completed in 2000, United States President Clinton and British Prime Minister

Blair made a joint announcement suggesting that the "gene map belongs to all" (Evenson, 2000). The statement was not meant to implicate patent policy. Nevertheless, the fear that the two leaders would seek to diminish intellectual property protection sent stocks in biotechnology companies tumbling (Schehr and Fox, 2000). Similarly, when the Supreme Court of Canada held that a genetically modified mouse, known as the oncomouse, was not patentable (Harvard College vs. Canada (Comm'r of Patents), 2002), a case that makes Canada the only developed nation that has explicit jurisprudence not allowing the patenting of "higher life forms," there were many predictions that the decision would have dire consequences on the Canadian biotechnology sector (Abraham, 2002). For example, one Canadian commentator noted that: "While the technical impact of the decision will likely be minor, the negative effect on future biotechnology could be huge." (Gervais, 2002)

Although most in the biotechnology industry have strong views about the importance of patenting to the strength of the biotechnology sector, some commentators have noted that there is, in fact, little systematic research to support the claim. For example, Richard Gold and colleagues have noted that: "Despite the assumption within intellectual property systems that they are necessary to encourage research and development, there is only a modest body of empirical evidence to support this in the biotechnology industry." (Gold *et al.*, 2000; 2002a)

30.3 PATENTABILITY

Though the patenting of human genetic material has been the source of social controversy, there have never been any significant legal barriers to the patenting of human genetic material. Ever since the famous United States Supreme Court case of *Diamond Chakrabarty* [447 U.S. 303 (1980)], there have been few legal obstacles to the patenting of biologically based "inventions." In the *Chakrabarty* case, the plaintiff challenged the United States Patent Office after a patent on an oil-eating bacterium had been refused. The Patent Office took the position that the bacterium was not an invention but a product of nature. The Supreme Court sided with *Chakrabarty*, noting that so long as an element of human inventiveness was involved, the innovation could be patented.

Since the *Chakrabarty* case, the boundaries of patent law have been pushed even further. In the United States, for example, numerous patents have been awarded on plants and animals (Kevles and Berkowitz, 2001). The patenting of human genes is consistent with this trend in the jurisprudence. From the perspective of patent offices in Europe, Japan, the US, and Canada, human gene inventions are patentable so long as they satisfy the basic criteria of national patent regimes: the invention must be new, nonobvious, and useful. Though many may view human gene

inventions as the mere discovery of something that already exists in nature, if a degree of human inventiveness has been applied, such as isolating the gene to make it useful, the gene is potentially patentable. Just as a naturally occurring chemical can be patented if an inventor has devised a method of purifying and making the chemical useful, so too are human gene sequences. As summarized by Demaine and Fellmeth (2003), "the [United States] Patent and Trademark Office and federal courts now routinely hold discovered natural substances patentable if they are "isolated and purified" or otherwise insubstantially modified. Naturally occurring DNA and protein biomolecules have, consequently, become the subject of patent applications."

Similarly, the European Patent Office Guidelines state that to "find a substance freely occurring in nature is also a mere discovery and therefore unpatentable. However, if a substance found in nature has first to be isolated from its surroundings and a process for obtaining it is developed, that process is patentable. Moreover, if the substance can be properly characterized either by its structure, by the process by which it is obtained or by other parameters and is "new" in the absolute sense of having no previously recognized existence, then the substance per se may be patentable." (Bostyn, 2003)

Currently, there are tens of thousands of gene patent applications pending. As of the year 2000, over 25,000 DNA-based patents had been issued in the United States alone (Cook-Deegan and McCormack, 2001). Recently, Thomas and colleagues (2002) found that between 1996 and 1999, 6786 patents were filed. Most of these patents were filed in the United States (62%), Europe (20%), and Japan (10%). Given that much of the biotechnology industry remains concentrated in the United States, these figures are hardly surprising (OECD, 2001).

30.4 GENERAL CONCERNS

Human gene patents have become the source of a great deal of social controversy. Indeed, the appropriateness and potential impact of patents on human genetic material has emerged as one of the dominant social and ethical concerns in the area of human genetics. Jeremy Rifkin (1998), one of the most vocal opponents of human gene patents has stated, "the debate over life patents is one of the most important issues ever to face the human family. Life patents strike at the core of our beliefs about the very nature of life and whether it is to be conceived of as having intrinsic or mere utility value."

Many commentators have suggested that patenting human genes has the potential to facilitate the commodification of humans generally. The Australian Law Reform Commission recently summarized this concern stating, "Another objection to patents on genetic materials is that they may engender a lack of respect for human life and dignity (ALRC, 2003). On this view, to grant a patent—a proprietary right—on something suggests that it is a fit subject for such rights. Consequently, patents on genetic materials are thought to commodify parts of human beings by treating them as objects, or as something to be placed in the stream of commerce for financial gain. Commercialisation of parts of human beings is ethically problematic because it might affect how we value people (ALRC, 2003)."

Likewise, it has been suggested that the human genome should be considered a natural resource, part of the common heritage of humankind and, as such, should not belong to any one individual or corporation (Rifkin, 1998). This idea has gained a degree of traction with some policy-making entities. For example, the Human Genome Organization's (HUGO) Ethics Committee suggested that the outcomes of human genome research should be available to all (HUGO, 2000). UNESCO's 1997 Universal Declaration on the Human Genome and Human Rights states that "the human genome in its natural state shall not give rise to financial gains" (UNESCO, 1997). That said, it must be remembered that, to date, such policy declarations have had very little practical impact on domestic patent law. In other words, they have not affected how individual patent offices view the patentability of human genetic material.

The patenting of biological substances has also been associated with the notion of "biopiracy" (Rifkin, 1998). In general, this concern relates to the exploitation of one region's genetic heritage or genetic resources by researchers or industry from another region. This may include, for example, researchers from a developed country using genetic samples from a developing nation without appropriate consent or authorization. Concern about this kind of phenomenon led the HUGO Ethics Committee to recommend that organizations or individuals who profit from genetic research should have an obligation to return a percentage (1–3%) of the gross to the region and/or community involved in the research.

Though the stimulation of innovation is one of the stated goals of the patent system, there has also been a long-standing concern that the patenting process may, paradoxically, hurt research by promoting a more secretive and less collaborative research environment. Research conducted by Blumenthal and colleagues (1997) suggests academic-industry collaborations are "significantly associated with the tendency to withhold the results of research." They also found that the most commonly cited reason for the delay in publication was the need to allow time for filing a patent application. Similarly, many commentators have suggested that "up stream" patents, for example, on gene sequences of unclear function, may hinder the development of research on "downstream," clinically useful inventions, such as genetic tests or therapies. This concern is well illustrated by the controversy that surrounded the attempt by United States National Institutes of Health (NIH), in the early 1990s, to patent over 7,000 expressed sequence tags (ESTs). These were relatively random DNA fragments that had no identified function. Though the patents were never granted, the controversy raised questions about the degree to which a gene discovery must have a "known function" before it is patentable (Kevles and Berkowitz, 2001).

This concern is closely related to what has been called the "tragedy of the anti-commons." First articulated by Heller and Eisenberg (1998), it is suggested that patents may hurt innovation in the context of gene research because they may create a disincentive to research. Researchers may avoid investigating a particular gene or gene region for fear of infringing an existing patent, thus stifling what may be useful research (Knoppers, 1999). In addition, patents may increase the cost of research by compelling researchers to obtain licensing fees to use the patented invention. This latter concern is particularly problematic given the uncertainty surrounding the application of the research exemption in most relevant jurisdictions. For example, the recent United States case of *Madey vs. Duke University* in 2002 has severely limited the research exemption in that country. The court held that the research exemption is applicable only if the research is done "solely for amusement, to satisfy idle curiosity, or for strictly philosophical inquiry." The exemption does not apply if the alleged offender has a legitimate business interest in the patented invention. The United States Supreme Court has recently refused to review this decision.

Though the articulation of these practical research concerns has led to minor adjustments in the patent system, such as the United States Patent Office tightening its utility criteria, it remains a live issue in many jurisdictions (Kowalski, 2000; ALRC, 2003; Habeck, 2003). For example, concerns about the impact of gene patents on research led the UK Nuffield Council on Bioethics (2002) to suggest that, in the context of human gene patents, the application of the existing patent criteria of novelty, inventiveness, and usefulness has not been "sufficiently stringent." As such, the Nuffield Council (2002) recommended: "In general, patents which assert rights over DNA sequences as research tools should be discouraged. We have taken this view that the best way to discourage the award of such is by stringent application of the criteria for patenting, particularly utility." Other policy documents, such as the Ontario Government's Report to the Premiers recommends, *inter alia*, a clarification of patent criteria in relation to human genes and a clarification of the experimental and noncommercial exceptions (Ontario, 2002).

Despite the frequent articulation of such concerns, very little in the way of substantial patent reform has occurred in any jurisdiction. Indeed, many of these issues were identified over a decade ago by commentators like Hubbard and Ward (1993) who stated, "It is important not to lose sight of the fact that the pieces of DNA being sequenced are part of our bodies; they are not being invented by these researchers.

If the base sequences of the DNA can be patented, rather than remaining in the public domain, the rights to the commercial use of these sequences will belong to the NIH or to companies that buy them from the MRC. In the end, consumers will be the losers. They will first pat the costs of the research and patenting with their taxes, then pay prices inflated by monopolies." Nevertheless, the rules applying to the patenting of human genetic material have remained largely unchanged since the start of the Human Genome Project. The growing concern about the potential impact of gene patents on the sustainability and costs of health care systems has created new controversy and renewed efforts to reform the existing system.

30.5 CONCERNS RELATED TO THE PROVISION OF HEALTHCARE

As more and more genetic technologies move from the laboratory toward clinical use, new patent controversies have started to emerge. Specifically, that patenting may drive up the cost of health care and impede availability of new medical innovations (Henry *et al.*, 2002). Of course, concern about the potential impact of patents on health care systems is not unique to genetic technologies and services. In a recent report on the Canadian Health Care system, the problem was discussed in the context of pharmaceuticals: "On the one hand, [patents] protect the intellectual property of pharmaceutical companies and helps offset the considerable investment they make in researching and developing new drugs. On the other hand, it delays the introduction of low cost generic drugs" (Romanow, 2002). However, several high profile controversies have caused policy makers throughout the world to focus on the potential adverse implications of gene patenting in the context of health systems (Andrews, 2002).

As previously noted, patents give the inventor a monopoly over the use and control of the invention. This allows the patent holder to charge a premium for the use of the invention. Indeed, the ability to reap substantial rewards for the use of the invention is the primary incentive that is meant to encourage innovation. However, this monopoly also means that there are few ways to control costs. Unless a country has some kind of formal price control schemes in place, patent holders are well within their rights to charge whatever they deem appropriate. The down side of this loss of control is most readily apparent in countries, such as Canada, that have a publicly funded health care system where global budgets may not be able to accommodate the demanded monopoly price. In such situations, the patent may result in a loss of public access to a necessary health care service. This can happen either because the administrators of the public system decide that they will not pay for the patented test, or if the patent holder simply refuses to allow access (Caulfield *et al.*, 2000; 2002).

The practical impact of this situation is well illustrated by the controversy surrounding Myriad Genetics in Canada, though there have been, of course, a number of similar patenting controversies, including the patenting and exclusive licensing of the Canavan gene in Florida (Hahn, 2003), and the patenting of the SARS genome (Gold, 2003). Indeed, Williams-Jones (2002) characterized the case as follows: "The Myriad case is a harbinger of an increasing number of instances where gene patents provide companies with monopolies on the development, marketing, and provision of genetic tests and therapeutics. Not surprisingly, this case has become a focal point in Canada and Europe for debates about the social and ethical implications of DNA patenting and the commercialization of genetic tests."

Myriad Genetics controls the patents for the BRCA1/2 genes and for the related testing processes. These genes are associated with a predisposition to breast and ovarian cancer. Though there is still an ongoing debate regarding the clinical utility of testing for BRCA1/2 mutations (Healy, 1997), the provision of the test has become the standard of care in many regions throughout the world (Gold *et al.*, 2002b). In Canada, various provinces provided the test as part of the public health care system for a relatively reasonable cost, though there was a degree of variation in testing and sequencing techniques.

In the summer of 2001, Myriad Genetics decided to take steps to enforce its patents over the BRCA1/2 genes. Provincial health care ministries in Canada received a cease and desist letter from Myriad. They were told that all future genetic testing that utilizes the BRCA1/2 genes must be done through Myriad's laboratories. The Myriad test is quite expensive as compared to the testing process already being done in Canada. The Myriad test would cost approximately 3,800 Canadian dollars per test. In some cases, this was more than four times the cost of the testing being done within the provincial system. As a result, a number of Canadian provincial health ministers stated that the public system could not afford the Myriad test. Some provinces decided to ignore the patent and continue testing, others, at least temporarily, simply stopped offering the test.

These controversies highlighted a number of interesting policy issues. First, the Myriad dilemma illustrated the potential conflict between two key governmental priorities, access to affordable health care and the promotion of innovation and the economy (Caulfield, 2003). Over the past few decades, containing the cost of publicly funded health care systems has been a key policy issue for most OECD countries (Flood, 2000). The Myriad story stood as an example of how patents, at least initially, might drive up costs. However, as noted previously, patents are also viewed as an essential component of the commercialization process. Thus, for those in industry and government interested in economic growth, diminishing the strength of patent protection is hardly an attractive option. On the contrary, within universities and in many sectors of the government, the obtainment of biotechnology patents are aggressively promoted. As such, a strange policy paradox emerged where explicitly pro-biotechnology politicians, such as Ontario's former Premier Mike Harris, argued against gene patents. For example, shortly after the Myriad story broke, Premier Harris suggested: "The benefits of a world-wide effort such as the human genome project should not be the property of a handful of people or companies. Our genetic heritage belongs to everyone. We must share the benefits fairly and do what we can to make genetic tests and therapies affordable and accessible" (Benzie, 2001). One wonders what his position would have been if Myriad Genetics was located in Ontario.

Second, there is also concern that gene patents might inhibit the development of cheaper or more efficient and effective ways of doing the same testing. For example, there are those who believe that the Myriad test may not be the most comprehensive procedure for the determination of BRCA1/2 mutations, though this remains controversial (Gold *et al.*, 2002a). Researchers at the Institute Curie in France, and the Erasmus Medical Center in the Netherlands, claim that the Myriad technique misses 10% to 20% of expected mutations, seriously jeopardizing the quality of test results and usefulness of this information for patient care (Williams-Jones, 2002; Matthijs and Halley, 2002). If true, and there are more efficient and effective procedures available, the current Myriad patents would block their use. Again, this could result in increased costs to the health care system.

Recently, Cho and colleagues (2003) analyzed many of these concerns. In that study, researchers interviewed directors of genetic laboratories throughout the United States. They asked the directors various questions about the impact of gene patenting and licensing on clinical practice. The results showed that, in general, laboratory directors had a negative impression of gene patents. Indeed, the researchers noted that it "was striking that virtually no respondents, including those from commercial laboratories, thought the effects of patents and licenses on the costs, access, and development of genetic tests have been positive." Moreover, their findings are also consistent with concerns that "patents are inhibiting commercialization of genetic tests." Given that one of the justifications for patents is to facilitate the commercialization process and the dissemination of useful technologies, these findings raise interesting questions about the utility and effectiveness of the current patent regime in the context of genetic innovations. Cho and colleagues (2003) summarize their findings thus: "We conclude that patents and licenses have a significant negative effect on the ability of clinical laboratories to continue to perform already developed genetic tests, and that these effects have not changed substantially throughout the past 3 years. Furthermore, the development of new genetic tests for clinical use, based on published data on disease-gene associations, and

information sharing between laboratories, seemed to be inhibited."

Interestingly, there is some evidence that the public has similar concerns about the impact of gene patents on access to health care services. The public has the same divided view of patents as that represented in much of the literature. A survey of the Canadian public found that few of those surveyed had moral or religious objections to the patenting of human genes and a majority (63%) saw more benefits than risks associated with the patenting process (Pollara, 2000). However, in focus groups, it was found that there were major concerns based on issues of access and equity. In the context of health care, at least in Canada, access seems to be the dominant public consideration.

Another concern associated with patents, and the commercialization process in general, is the possible impact on utilization and uptake. The individuals and organizations that hold patents have a natural and understandable desire to see their invention utilized by as broad a market as possible. However, some have argued that this market pressure may cause commercial labs to offer their services to an inappropriately broad sector of the population (the broader the definition of at-risk, the larger the market; Martone, 1998; Caulfield *et al.*, 2001). Companies that control a patent for a particular disease gene may inappropriately stress the seriousness of a disorder in order to encourage people to get tested (Loeben *et al.*, 1998; Biesecker and Marteau, 1999). Such an approach may conflict with a more dispassionate assessment of the clinical utility for a given genetic service. For example, if one compares Myriad Genetics' indications for BRCA tests with those recommended by an independent body, one finds that the latter excludes women without a family history of breast or ovarian cancer, whereas Myriad's guidelines include these lower risk women (Willison and MacLeod, 2002).

Naturally, this kind of market pressure could create inappropriate utilization and unnecessary costs for the health care system. Controlling the dissemination and uptake of new technologies has always been a challenge, particularly because few health care systems have the necessary framework to make systemwide implementation policies. Utilization and uptake are the result of a complex mix of professional decision-making, market pressure, and patient preference and, in most countries, coverage by the public system. However, as research and health care become increasingly privatized, the market seems likely to have an increasingly important impact on the introduction of genetic tests into clinical practice. Blancquaert (2000) summarizes the problem as such: "Not only will commercialization increase public demand and affect the availability of genetic tests and services, it will certainly have an impact on the financial and human resources of national health care systems as well."

It should not be forgotten that if, as it is often predicted, pharmacogenetics and multiplex testing become a reality, then many of the previously noted concerns could be amplified (Evans and Relling, 1999). This is because the use of both pharmacogenetics and multiplex testing—that is, testing for many mutations at the same time—could implicate numerous, perhaps dozens, of patented gene sequences. As such, obtaining the necessary licenses to provide the genetic service could be a tremendous challenge. For example, pharmacogenetics is premised on the idea that we will be able to individualize pharmaceutical treatments to maximize effectiveness and minimize adverse events (Shah, 2003). However, this will necessarily involve testing an individual for the presence or absence of a variety of gene sequences. If these sequences are owned by another company or, more likely, several companies, the development of pharmacogenetic clinical protocols could become a complicated and expensive venture.

30.6 SUGGESTED REFORMS

Over the past few years, partly as a result of the previously noted controversies, a variety of governmental and policy entities have been seen to take on the gene patenting issues. Even though the UK Nuffield Council supported the use of patents on diagnostic tests as a means of "rewarding the inventor," they concluded that "in the future, the granting of patents that assert rights over DNA sequences should become the exception rather than the norm," and go on to make a number of specific reform suggestions, including applying existing patenting criteria more stringently (Nuffield Council, 2002). In addition, and perhaps most controversial, the Council also suggest that "in the case of patents that have been granted for diagnostic tests based on genes, compulsory licensing may be required to ensure reasonable licensing terms are available to enable alternative tests to be developed" (Nuffield Council, 2002).

Another interesting UK report, which explicitly endorses the Nuffield Council's recommendations, was issued in July of 2003 (Cornish *et al.*, 2003). This report was commissioned by the UK Department of Health because of emerging "serious concern about the impact of intellectual property upon research and the use of novel developments in genetics affecting health care." The authors provide a series of recommendations including the suggestion that the Department of Health consider "a robust central policy for 'licensing in' designed to moderate excessive demands by licensors by considering, as possible options, the use of compulsory licensing, competition law and Crown use."

In what appeared to be a direct response to the Myriad controversy, the Ontario government struck a policy group to examine the potential adverse social implications of gene patenting. The government's 2002 Report to the Premiers made a number of the same recommendations as the Nuffield Council. The Ontario report—which, on January 25th 2002, was adopted by the Premiers from all the Canadian provinces at the Premiers Conference on Healthcare in Vancouver—recommends *inter alia*, a clarification of patent criteria in relation to human genes, the exclusion of broad-based genetic patents covering multiple uses, a clarification of the experimental and noncommercial exceptions, and an expansion of the methods of medical treatment exclusion. The Ontario government has also decided to become an intervener in a plant patent case in hopes of raising issues associated with the impact of gene patents on health care policy. It was reported that the Ontario affidavit to the Supreme Court says the case "has important implications for the development of public policy in Ontario including the delivery of health care to its residents" (Bueckert, 2003).

The ALRC is also in the process of examining the patent issue. It has already released a discussion paper (ALRC, 2003) and hopes to issue a final report in 2004. The ALRC is considering a variety of possible reforms, including limiting patent terms, changing the requirements for patentability, creating a mechanism that would allow ethical issues to be considered in the assessment of a patent, and provisions for the use of compulsory licensing.

There are a number of common themes that run through these and other policy documents. Tightening patent criteria, clarifying exemptions, and the use of compulsory licensing are all suggestions that consistently emerge. Another common suggestion is to introduce a mechanism that would allow social and ethical issues to be considered in the assessment of patents (Gold and Caulfield, 2002). This could, for example, be similar to Article 53(a) of the European Patent Convention, which allows the rejection of inventions, the publication or exploitation of which would be contrary to *ordre public* or morality. Some worry, however, that such an approach would introduce too much uncertainty into a system that was never meant to address ethical issues. Crespi (2003), for example, emphasizes: "Most professional patent practitioners are somewhat dubious as to the wisdom or appropriateness of introducing ethical and moral questions into a law which is primarily based on the assessment of originality, innovativeness, and practical advances in science and technology."

30.7 CONCLUSIONS

Despite the continued articulation of concerns regarding the impact of gene patents and the publication of numerous policy recommendations, there are many reasons why altering the existing system will remain a challenge. International trade agreements, for example, have the potential to create technical barriers for those seeking patent reform. Trade-related Intellectual Property Rights (TRIPS) agreement of the World Trade Organization, which came into effect in 1995 and currently involves 146 nations, was envisioned as a mechanism for creating and maintaining strong patent protection within all member countries. Patent reform, which greatly alters or appears to erode patent protection, may be subject to challenge. At a minimum, existing international treaties add an element of uncertainty to the policy-making process. Moreover, the momentum of the biotechnology sector, and the imbedded perception that patent protection is crucially important to economic development, rightly or not, will create political barriers to significant patent reform. Finally, only the best possible evidence should inform any reform process. Currently, there is very little empirical data on the actual benefits and risks of patents. Thus, further research is clearly warranted.

As intellectual property becomes a more significant part of our social landscape, new challenges will undoubtedly emerge. Policy makers throughout the world must strive to create an intellectual property regime that balances a variety of laudable social goals, including the stimulation of the economy, the encouragement of innovation, and the facilitation of access to reasonably priced health care technologies.

Acknowledgments

I would like to thank my colleagues Lori Sheremeta and Richard Gold for their insight into the issues relevant to this area. Indeed, much of this chapter draws on work we have done together. I would also like to thank Suzanne Debow for her editorial help and Genome Canada, the Stem Cell Network, and the Alberta Heritage Foundation for Medical Research for the continued funding support.

References

Abraham, C. (2002). Mouse ruling may stall research. *The Globe and Mail*, December 6, 2002. A1.

Andrews, L. B. (2002). The gene patent dilemma: Balancing Commercial Incentives with Health Needs. *Houston J. Health Law Policy* 65, 106.

Australian Law Reform Commission. (2003). *Issue Paper 27: Intellectual Property Rights over Genetic Materials and Genetic and Related Technologies*. ALRC, Sydney.

Benzie, R. (2001). Ontario to defy U.S. patents on cancer genes: province will pay for $800 test, not $3,850 version by Myriad Genetic Laboratories Inc." *The National Post*, September 20, 2001. A15.

Biesecker, B., and Marteau, T. (1999). The Future of Genetic Counseling: an International Perspective. *Nat. Genet.* 22, 133–137.

Blancquaert, I. (2000). Availability of Genetic Services: Implementation and Policy Issues. *Community Genet.* 3, 179–183.

Blumenthal, D., Campbell, E. G., Anderson, M. S., Causino, N., and Louis, K. S. (1997). Withholding research results in academic life science. Evidence from a national survey of faculty. *JAMA* 277, 1224–1228.

Bostyn, S. J. R. (2003). The Prodigal Son: the Relationship between Patent Law and Health Care. *Med. Law Rev.* 11, 67–120.

Bueckert, D. (2003). Ontario government intervenes in high-profile gene patenting case. The Canadian Press. Tuesday, October 7, 2003.

Caulfield T., Gold, E. R., and Cho, M. K. (2000). Patenting human genetic material: refocusing the debate. *Nat. Rev. Genet.* 1, 227–231 .

Caulfield, T., Burgess, M. M., Williams-Jones, B., Baily, M-A., Chadwick, R., Cho, M., Deber, R., Fleising, U., Flood, C., Friedman, J., Lank, R., Owen, T., and Sproule, J. (2001). Providing Genetic Testing Through the Private Sector: A View from Canada. ISUMA. *Canadian J. Policy Res.* 2, 72–81.

Caulfield, T. A., Knoppers, B. M., Gold, E. R., Sheremeta, L. E., and Bridge, R. J. (2002). Genetic technologies, health care policy and the patent bargain. *Clin. Genet.* 63, 15–18.

Caulfield, T. (2003). Sustainability and the Balancing of the Health Care and Innovation Agendas: The Commercialization of Genetic Research. *Saskatchewan Law Rev.* 66, 629–645.

Cho, M. K., Illangasekare, S., Weaver, M. A., Leonard, D. G. B., and Merz, J. F. (2003). Effects of patents and licenses on the provision of clinical genetic testing services. *J. Mol. Diagn.* 5, 3–8.

Cook-Deegan, R. M., and McCormack S. J. (2001). Intellectual property. Patents, secrecy and DNA. *Science* 293, 217.

Cornish, W. R., Llewelyn, M., and Adcock, M. (2003). Intellectual Property Rights and Genetics: A Study into the Impact and Management of Intellectual Property Rights Within the Healthcare Sector. *Cambridge Genetics Knowledge Park* 1–134.

Crespi, R. S. (2003). Patenting and Ethics: A Dubious Connection. *Bio-Science Law Review*. February 26, 2003.

Demaine L., and Fellmeth, A. (2003). Patent Law: Distinguishing Natural Substances from Patentable Inventions. *Science* 300, 1375–1376 .

Enriquez, J. (2001). *As the Future Catches You: How Genomics & Other Forces Are Changing Your Life, Work Health & Wealth*. Crown Business, New York.

Evans W, and Relling M. (1999). Pharmacogenomics: translating functional genomics into rational therapeutics. *Science* 286, 487–491.

Evenson, B. (2000). Gene map belongs to all. *The National Post*. March 15, 2000. A1.

Flood, C. (2000). *International Health Care Reform: A Legal, Economic and Political Analysis*. London: Routledge.

Gervais, D. (2003). Eeek! the mouse confounds the house! *The Ottawa Citizen*, 14 December, B7.

Gold, E. R. (2000). Finding common cause in the patent debate. *Nat. Biotechnol.* 18, 1217–1218.

Gold, E. R., and Caulfield, T. A. (2002). The "moral tollbooth": a method that makes use of the patent system to address ethical concerns in biotechnology. *Lancet* 359, 2268–2270.

Gold, E. R., Castle, D., Cloutier, L. M., Daar, A. S., and Smith, P. J. (2002a). Needed: models of biotechnology intellectual property. *Trends Biotechnol.* 20, 327–329.

Gold, E. R., Caulfield, T. A., and Ray, P. (2002b). Gene patents and the standard of care. *Canadian Med. Assoc. J.* 167, 256–257.

Gold, E. R. (2003). SARS genome patent: symptom or disease? *Lancet* 361, 2002–2003.

Hahn, L. (2003). Owning a piece of Jonathan. *Chicago*, May.

Harvard College v. Canada (Commissioner of Patents), [2002] SCC 76; 219 D.L.R. (4th) 577.

Healy, B. (1997). BRCA genes: bookmaking, fortune telling, and medical care. *N. Engl. J. Med.* 336, 1448–1449.

Habeck, M. (2003). Brussels takes EU states to court over biopatent law. *Nat. Biotechnol.* 21, 960.

Heller, M. A., and Eisenberg, R. S. (1998). Can patents deter innovation? The anticommons in biomedical research. *Science* 280, 698–701.

Henry, M. R., Cho, M. K., Weaver, M. A., and Merz, J. F. (2002). Genetics. DNA patenting and licensing. *Science* 297, 1279.

Hubbard, R., and Wald, E. (1993). *Exploding the Gene Myth.* Beacon Press, Boston.

HUGO Ethics Committee (2000). Statement on Benefit Sharing. Genome Digest 6, 7–9, online, Human Genome Organization, http://www.hugo-international.org.

Kevles, D., and Berkowitz, A. (2001). The Gene Patenting Controversy: A Convergence of Law, Economic Interests, and Ethics. *Brooklyn Law Rev.* 67, 233–248.

Knoppers, B. M. (1999). Status, sale and patenting of human genetic material: an international survey. *Nat. Genet.* 22, 23–26.

Kowalski, T. J. (2000). Analyzing the USPTO's revised utility guidelines. *Nat. Biotechnol.* 18, 349–350.

Loeben, G. L., Marteau, T. M., and Wilfond, B. S. (1998). Mixed messages: presentation of information in cystic fibrosis screening pamphlets. *Am. J. Hum. Genet.* 63, 1181–1189.

Madey v. Duke University, 307 F.3d 1351 (Fed. Cir. 2002).

Martone, M. (1998). The Ethics of the Economics of Patenting the Human Genome. *J. Bus. Ethics* 17, 1679–1684.

Matthijs, G., and Halley, D. (2002). European-wide opposition against the breast cancer gene patents. *Eur. J. Hum. Genet.* 10, 783–785.

Nuffield Council on Bioethics (2002). *The Ethics of Patenting DNA*, Nuffield Council on Bioethics, London.

Organization for Economic Co-operation and Development (OECD). *Statistics in OECD Member Countries*, September 13, 2001. Available online, http://www.olis.oecd.org/olis/2001doc.nsf/c5ce8ffa41835d64c125685d005300b0/c1256985004c66e3c1256ac600350f21/$FILE/JT00112476.pdf.

Ontario (2002). *Genetics, Testing and Gene Patenting: Charting New Territory in Healthcare.* Available online, Ontario Ministry of Health and Long Term Care, http://www.health.gov.on.ca:80/english/public/pub/ministry_reports/geneticsrep02/report_e.pdf.

Pollara Research and Earnscliffe Research and Communications. (2000). Public opinion research into biotechnology issues, third wave. Biotechnology Assistant Deputy Minister Coordinating Committee, Ottawa. Available online, Canadian Biotechnology Strategy, http://www.biotech.gc.ca/docs/engdoc/3Wavexec-e.pdf.

Rifkin, J. (1998). *The Biotech Century.* Penguin Putnam, New York.

Romanow, R. J. (2002). Building on Values: the Future of Health Care in Canada. Commission on the Future of Health Care. 1–339. Commissioner, Saskatoon. Available online, Health Canada, http://www.hc-sc.gc.ca/english/pdf/care/romanow_e.pdf.

Schehr, R., and Fox, J. (2000). Human genome bombshell. *Nat. Biotechnol.* 18, 365.

Shah, J. (2003). Economic and regulatory considerations in pharmacogenomics for drug licensing and healthcare. *Nat. Biotechnol.* 21, 747–753.

Straus, J. (1998). Bargaining around the TRIPS agreement: the case for ongoing public-private initiatives to facilitate worldwide intellectual property transactions. *Duke J. Comp. Intern. Law* 9, 91–107.

Thomas, S., Hopkins, M. M., and Brady, M. (2002). Shares in the human genome—the future of patenting DNA. *Nat. Biotechnol.* 20, 1185–1188.

United Nations Educational, Scientific and Cultural Organisation (UNESCO). (1997). Universal Declaration on the Human Genome and Human Rights, 29th Sess., 29 C/Resolution 19.

Williams-Jones, B. (2002). History of a gene patent: tracing the development and application of commercial BRCA testing. *Health Law J.* 10, 123–146.

Willison, D. J., and MacLeod, S. M. (2002). Patenting of genetic material: are the benefits to society being realized? *CMAJ* 167, 259–262.

CHAPTER **31**

Genetic Counselling and Ethics in Molecular Diagnostics

MARY PETROU

Regional Hemoglobinopathy Genetics Centre (Perinatal Centre), University College Hospital NHS Trust, and University College London Medical School, Department of Obstetrics and Gynaecology, London, UK

TABLE OF CONTENTS

31.1 INTRODUCTION

Genetic counselling is inseparable from genetic diagnosis. It aims to replace misunderstandings about the causes of genetic disease with correct information, and to increase people's control of their own and their family's health by informing them of the resources available for diagnosis, treatment, and prevention. Though counselling has a role in many medical consultations, it is particularly important in medical genetics because of the often predictive nature of genetic information, the implications for other family members, the difficult choices that sometimes have to be made, and the important ethical problems that can be involved.

Genetic counselling has been defined as the process by which patients or relatives at risk of a disorder that may be hereditary are advised of the consequences of the disorder, and the probability of developing and transmitting it and the ways in which this may be prevented or ameliorated (Harper, 1988).

This definition requires:

- A correct diagnosis in the presenting family member.
- Explanation of the nature and prognosis of the disorder, the treatment available, and where to find it.

- Estimation of genetic risk for parents and family members. This requires drawing a family tree, and it may also call for investigations on other family members.
- Communication of genetic risks, and the options for avoiding them, including the chances for parents and other family members of passing the disorder on to other children, and explanation of the risk. The options for avoiding further affected children, are techniques of prenatal diagnosis, problems, risk of error, and complications.
- Supporting the individual or couple in making the decision that is right for them.
- Accessibility for long-term contact: people at risk often need counselling and support at several points in their life.

Therefore specialist genetic knowledge, training in counselling skills, time, ability to communicate, and back-up by a medical geneticist or trained genetic counsellor all are required. Because the author's expertise is in hemoglobin disorders, the remainder of this chapter will draw on experiences focused on these disorders.

Hemoglobinopathies constitute the most common monogenic disorders worldwide (reviewed in Weatherall and Clegg, 2001). They are caused by mutations, which affect the genes that direct synthesis of the globin chains of hemoglobin, and may result in reduced synthesis (thalassemia syndromes) or structural changes (hemolytic anemia, polycythemia, or more rarely, cyanosis).

Thalassemia mutations and various abnormal hemoglobins interact to produce a wide range of disorders of varying degrees of severity. There are four main categories of interactions associated with severe disease states, for which genetic counselling and prenatal diagnosis is indicated (Old, 1996): thalassemia major (co-inheritance of β- and/or $\delta\beta$-thalassemia mutations), sickle cell disease (and analogous interactions, e.g., Hb S/C, Hb S/β-thalassemia, Hb S/D Punjab, Hb S/O Arab, Hb S/Lepore), Hb E disease (co-inheritance of β-thalassemia mutations with Hb E or Hb O Arab), Hb Bart's Hydrops Fetalis syndrome (homozygous α^0-thalassemia), and (rarely) Hb H Hydrops Fetalis syndrome (α^0/α^T). These disorders are common in the United Kingdom, because of migration of ethnic minority populations.

31.2 PROBLEMS IN GENETIC COUNSELLING

Medical training rarely equips doctors to provide adequate genetic counselling and discuss complex issues with their patients, in order to help patients reach their own decisions. The responsibility involved in genetic counselling should not be underestimated. Genetic diagnosis is often difficult in view of the enormous diversity of the conditions involved, and misdiagnosis and misinformation can have disastrous consequences for individuals and their families.

For instance, when counselling at-risk couples for hemoglobin disorders the genetic counsellor should have a good knowledge of the molecular genetics of thalassemia and therefore be able to comprehend the molecular mechanisms, and to communicate this information to the families. This is very important in view of the phenotypic heterogeneity of thalassemia. Generally, the inheritance of two β-thalassemia mutations results in a blood transfusion-dependent thalassemia. However, there are mild β-thalassemia mutations resulting in thalassemia intermedia. For instance, when both partners carry the mild β^+ -88 (C→T) thalassemia mutation, the homozygous state generally results in a very mild clinical phenotype. In this case, should the parents consider prenatal diagnosis?

This type of mild thalassemia intermedia poses an ethical dilemma as far as termination of pregnancy is concerned. Health professionals also are faced with this ethical dilemma when couples request termination of pregnancy for a condition such as this. Although they have to accept the couple's decision when termination of pregnancy is requested for a possible mild condition, they themselves are often uneasy about such decisions. The situation is made even more complex, as there is a wide clinical phenotypic heterogeneity in thalassemia intermedia syndromes and it is often not possible to predict how mild the condition will be and therefore impossible to reassure at-risk couples that their affected child will be mildly affected.

However, there are conditions when the couple can be reassured, such as when there are silent β-thalassemia mutations; that is, the β^+ -101 (C→T) that produces mild or very mild clinical phenotypes in the homozygous state and when interacting with severe β-thalassemia mutations, or when one partner carries β-thalassemia trait and the other carries triplicated α-globin genes ($\alpha\alpha\alpha$) or hereditary persistence of fetal hemoglobin (HPFH). Prenatal diagnosis is not offered for these cases. However, there are other ameliorating factors when the clinical phenotype is not so consistent.

The counsellor should have the experience and expertise to communicate this information and the likely outcome to

the patients. Indeed as the knowledge of the molecular mechanisms that cause thalassemia continue to accumulate this scientific area will become even more complex.

In the United Kingdom, sickle cell disease is also common because of the ethnic minority groups present. Counselling couples who are at risk for sickle cell disorders is often perceived as relatively simple, but in fact it is quite complex because of the wide range of severity of sickle cell disorders, ranging from the very mild to the very severe. As a result, parents face considerable difficulty deciding whether or not to request prenatal diagnosis. A study performed in London showed decision-making was greatly influenced by the experiences of sickle cell disease within the family: a couple with an affected child was more likely to undergo prenatal diagnosis and selective termination of an affected fetus than a couple with no such experience. The study also found that uptake of prenatal diagnosis was influenced by the gestation at referral: a woman referred in the second trimester was less likely to proceed with prenatal diagnosis (Petrou *et al.*, 1992a).

In contrast, counselling couples at risk for α-thalassemia hydrops fetalis is more straightforward because of the usually hopeless prognosis for an affected fetus and the possibility of life-threatening obstetric risks for the mother (Petrou *et al.*, 1992b). It is rare for such couples to decline prenatal diagnosis. However even here new complexity is introduced by the possibility of initiating regular intra-uterine transfusion for an affected fetus, providing the diagnosis is made early enough (Ng *et al.*, 1998). More information is needed about outcomes and the best way to provide this service if it is requested. Perhaps an international collaborative study, where data on all cases undergoing intra-uterine transfusion including the fetal treatment provided will be collected, will issue guidelines on the most successful approaches. This information is essential before this treatment can be considered as an approach that could be offered widely. Until then the approach should be applied on a research basis. The main problem with surviving babies of intra-uterine transfusion therapy has been severe mental and physical handicap, though some appear to be doing well. However, is it justified to provide this treatment when the life of an α-thalassemia major fetus is saved, but then all the problems associated with the treatment of β-thalassemia major, such as regular blood transfusions and iron chelation therapy are created?

Therefore new medical possibilities created by the rapid advance of genetic technology need to be included in genetic counselling, often before the ethical and moral dilemmas they involve have been adequately considered in some societies. The most obvious example concerns the acceptability of prenatal diagnosis and selective abortion of an affected fetus.

In practice, peoples' options are greatly influenced by the stage in life at which they learn of their risk, and by whether prenatal diagnosis is available or not. If the risk is found before marriage then the options are to remain single, not to marry another carrier, or marry as usual. If, however, the risk is found after marriage, then the options are to separate and find a noncarrier partner, have few or no children, take the chance and have children as usual, or use prenatal diagnosis and selective termination of pregnancy if it is acceptable and available, in case the fetus is affected (see also Chapter 32).

Unless there is an active carrier screening program, it is unusual for a couple to learn they are at risk of having children with a hemoglobin disorder before marriage or before starting a family. At present, in countries where premarital or antenatal screening is not offered, most couples learn of their risk only after the diagnosis of an affected child, and this limits their choices.

Therefore there are challenges involved in genetic counselling, because all the available choices involve difficult moral and social problems. In most cases there appears to be no right answer, but on the other hand, once people understand their risk, they cannot escape from choosing an option. Even the decision not to choose constitutes a choice.

In most Western countries, prenatal diagnosis for couples at risk for hemoglobin disorders is available, with the option of selective abortion. It is recognized that selective abortion is not an optimal or easy solution, that early prenatal diagnosis is preferable to later prenatal diagnosis, and that not all couples at risk for having children with thalassemia feel that prenatal diagnosis is the right choice for them. However, these attitudes have evolved only gradually. Prenatal diagnosis became possible in the early 1970s, and at that time its acceptance in different Western countries depended largely on the status of family planning, and the existing abortion law in each country.

It is difficult for people to deal with the idea of prenatal diagnosis unless they are already used to the concept of controlling their own reproduction, a social change that takes time. When molecular diagnosis began to be developed in the 1950s, the options available to couples at reproductive risk were to ignore the information and hope for the best, to remain unmarried or separate, or to limit their reproduction using family planning (i.e., the same choices as those now available in countries where prenatal diagnosis is not available). Family planning was already becoming widely accepted in the West at that time, and it soon became clear that when options were so limited, most people chose either to avoid or ignore information on risk, or to limit the size of their family (Carter, 1974; Modell *et al.*, 1980).

31.3 OPTIONS AVAILABLE TO PEOPLE WITH A REPRODUCTIVE RISK

It is useful to examine the options available for people with a genetic reproductive risk in more detail. Because carriers of thalassemia can be detected and advised of reproductive risk either before or after marriage, this opens a wide range of choices for couples. There is extensive experience of the choices that people make in practice.

How can information be obtained on which approaches are acceptable, and which are unacceptable to a particular population? One possibility is through public meetings, discussions, and questionnaires addressed to interested professionals and the general public. However, this presupposes an informed public, which can be found in very few countries worldwide at present, so this approach also requires a substantial component of public education. The most extensive of such consultation exercises was carried out in the early 1990s by the Canadian Royal Commission on New Reproductive Technologies (Report of the Canadian Royal Commission, 1993). The results, like many other surveys, showed that the majority of people, including many who might not use the service themselves for religious or other reasons, approved of the availability of prenatal diagnosis and termination of pregnancy for serious genetic disorders.

Another possibility is to seek the views of informed people who are themselves at genetic risk. However, there can be a considerable difference between what people think they would do, and what they actually do when they have to make a choice. Therefore a more objective approach is to observe and report the choices that people at risk actually make. Data obtained in all these ways is referred to in the following discussion.

Though community-based programs for the prevention of hemoglobin disorders in Europe now include the option of prenatal diagnosis and selective abortion, carrier screening and genetic counselling was introduced in some countries either before prenatal diagnosis was possible (Stamatoyannopoulos, 1974; Angastioniotis and Hadjiminas, 1981; Gamberini *et al.*, 1991) or before it was legal (Gamberini *et al.*, 1991). This early experience produced useful information on the acceptability of some alternative approaches in Europe.

31.4 PREMARITAL SCREENING

It is often thought that affected births can be prevented if at-risk couples are identified prior to marriage, on the assumption that they will then decide to separate and each find another, noncarrier partner. However, it is not always easy to explain how most marriages come about to be able to make any valid assumptions about how choice of partner might be affected by genetic information. In many societies, marriage is a complex social phenomenon that involves many other family members besides the prospective couple, and marriage partners usually are selected either because of a strong personal preference, or for valid family or traditional reasons, or a mixture of all three.

If a planned marriage needs to be rearranged because both partners carry a genetic disorder, this causes social embarrassment or stigma to the young couple and their families, and there is a risk that the problem will recur if the new partners found are also carriers for the same disorder. For example, if population carrier frequency is 6%, the chance that one or both new partners will be a carrier is 12%. Therefore the recurrence risk for the couples is 12% (or even higher if the new potential partner is a relative).

A second possibility is to marry as planned, but avoid having children altogether. This is always a difficult choice, and is unrealistic in a strongly family-centered society. A more realistic option for married at-risk couples is to limit their family size. If they limit themselves to two healthy children, 56% of them will never have an affected child.

Other options for having children while avoiding a known genetic risk, such as artificial insemination by donor, egg donation, or adoption, so far have not proved popular in any society, and are unacceptable in many communities.

Another possibility is to marry and wait to have a family until appropriate methods for prevention become available. For example, when Cypriot couples at risk were informed that prenatal diagnosis would become available in the foreseeable future, many postponed conceiving until they could use the service (Angastiniotis and Hadjiminas, 1981).

A final possibility is to marry and have a family as usual, trusting in fate. This seems to be a common choice in the absence of alternatives that are acceptable to families.

Before prenatal diagnosis was feasible for hemoglobin disorders, Stamatoyannopoulos (1974) conducted a research study in the area of Arta (Greece), where 20% of the population are carriers of either thalassemia or sickle cell anemia. All young people in that area were screened and counselled, and counselling contact was maintained for a two-year period. When the pattern of marriages was assessed at the end of this period, there was no measurable effect on choice of partner (Stamatoyanopoulos, 1974).

In the same period, a similar approach was tried in Cyprus, with the added parameter that marriages between carriers were actively discouraged. This approach proved to be unacceptable to the population and was soon abandoned "because of evasions" (Angastiniotis *et al.*, 1986). Once prenatal diagnosis became possible for thalassemia, it was made available within the Cypriot health service. Soon after, confidential premarital screening was made mandatory among Greek Cypriots by the Greek Orthodox Church, and among Turkish Cypriots by the civil authorities. It was then found that 98% of at-risk couples detected just prior to marriage proceed to marry, even though Cypriot parents often have considerable influence on their children's choice of partner. Nevertheless, the annual number of new births of children with thalassemia major has decreased almost to zero in Cyprus, because couples use the information on genetic risk in a variety of ways to obtain a healthy family. It has been reported that less than 5% of the decrease in

thalassemia major births is due to separation of engaged couples, about 80% is due to prenatal diagnosis and selective abortion, and 20% is because at-risk couples have fewer children, on average, than couples not at risk (Angastiniotis *et al.*, 1986). In both Cyprus and Sardinia, the population is now very well informed, and has gained confidence that the thalassemia control programs involve little if any coercion. There is now increasing popular demand for carrier testing in high schools, so that young people and their families can take information on carrier status into account at an earlier stage in the choice of marriage partner (WHO, 1993). It will be of great interest to continue to follow the results of these Mediterranean social experiments.

In Canada, a program of information and screening for carriers of Tay Sachs disease or thalassemia in high schools in Montreal has proved highly acceptable, and also highly informative about pitfalls in foreseeing how people will use genetic information (Zeesman *et al.*, 1984). A sample of the young people screened were followed up with a questionnaire that included the questions, "Do you think that a couple planning to marry who found they were both carriers would change their marriage plans? Would you change your own marriage plans?" Interestingly, almost 80% thought that other couples would change their marriage plans, but only 10% thought they would change their own plans. Clearly, it is all too easy to underestimate the importance of other peoples' inner lives. Practical experience is the only reliable guide to how people are prepared to use information on genetic risk.

Therefore, there seems to be a strong case for early carrier diagnosis and genetic counselling, so that couples at genetic risk and their families can make an informed choice whether to separate or stay together. However, knowledge of a genetic risk may alone be insufficient to change a lifetime partner.

Prenatal diagnosis is now available in all the countries whose experience has been drawn on, earlier. In addition, assessment of its desirability from the medical, religious, and social points of view has now begun in many other countries, including a large number of Muslim countries, and indeed prenatal diagnosis is available in some of them, such as Pakistan and Iran (Alwan and Modell, 2003).

31.5 CAN PREVENTION PROGRAMS BE CONSIDERED EUGENICS?

The word *eugenics*, coming from the Greek word "ευγονική", meaning good birth, was coined by the British scientist Sir Francis Galton. Although it still carries its original meaning in some countries, usually it is associated with Nazi programs to eradicate so-called inferior groups.

The WHO has suggested another definition of eugenics: "A coercive policy intended to further a reproductive goal against the rights, freedoms and choices of the individual"

(WHO, 2001). For this definition coercion includes laws, regulations, positive or negative incentives, including the lack of accessibility to affordable medical services, put forward by states or other social institutions.

Globally there is little evidence for eugenics practice according to the new definition. It has been suggested that programs, available in countries such as Sardinia and Cyprus, of carrier screening and prenatal diagnosis are eugenic programs, because they have limited the birth of affected individuals. However, in Cyprus couples make their own decisions once mandatory screening takes place. Couples can still decide to marry and have children with thalassemia. Although most marry, very few choose to carry on a pregnancy diagnosed with thalassemia major. In Cyprus there is widespread approval of the program, which is also supported by the majority of the thalassemia major patients. This program gives individuals a choice, whereas if this service did not exist then that choice would not have existed either. Therefore using the suggested definition of the WHO, the Cyprus carrier screening program for thalassemia cannot be considered as eugenics.

The fate of many patients in some developing countries where carrier screening programs are not developed is dismal. In many developing countries the state does not have the resources to treat the affected patients. In Pakistan, there are 4,500 affected births every year and most of these patients die because their parents cannot afford treatment. Most of these families were not given a choice, as national screening programs for thalassemia were not available. However in Pakistan there is now religious approval for termination of pregnancy for a genetic disorder (Ahmed *et al.*, 2000) and prenatal diagnosis is available in private clinics (Ahmed *et al.*, 2000), although a national carrier screening program is not yet available.

Many countries now have introduced or are developing screening and counselling programs. In Iran, there are 1,000 births annually with over 20,000 children attending treatment centers. Iran has taken on the vast task of providing national premarital screening and genetic counselling. By the end of 1999, over 888,000 couples had been tested and over 2,600 at-risk couples identified and counselled. 50% of these couples proceeded with their marriage plans, 37% of at-risk couples separated, and the remainder were still struggling with their decision, when the data was collected (A. Samavat, personal communication). Therefore the majority of couples found it unacceptable to select a partner on the basis of genetic screening information and there was a high demand for prenatal diagnosis. It is noteworthy that at that time, prenatal diagnosis was available only at private clinics in Tehran, but was too expensive for most families. These results were considered at the highest religious and political levels and a fatwa was issued permitting first trimester abortion when a fetus is found to have a serious genetic disorder. A national network of molecular diagnostic laboratories has now been set up to make prenatal diagnosis available

within the health system. It remains to be seen how choices will differ, now that prenatal diagnosis is more easily and widely available.

Therefore prevention services are steadily spreading globally and services are now available in several underdeveloped countries (Petrou and Modell, 1995). The ethos in present day medical genetics is to help people make whatever voluntary decisions are best for them in the light of their own reproductive and other goals. There is, therefore, a decisive difference between present day medical genetics and yesterday's eugenics.

31.6 ETHICS AND RELIGION IN GENETIC COUNSELLING

Because the choices facing people at genetic risk can be so difficult, and can have life-long consequences, experienced genetic counsellors generally consider that informed individuals or couples are themselves the best judges of what to do. Genetic counselling should therefore be nondirective, and the genetic counsellor's main role is to provide people at risk with full information, give them time for consideration, and support them in making the decisions they feel to be morally right for themselves.

Some reasons why genetic counselling should be nondirective are:

- People at risk often have first-hand experience of the condition in question, unlike most of their advisers.
- They have to learn all the facts, think the issues through, and actually reach a decision that they must live with for the rest of their lives.
- The right choice for a given individual among the options actually available is likely to be determined by many factors, including their social and religious attitudes, personal experiences, economic and educational level, and family and reproductive history.
- Doctors and other professionals are no more, or less, qualified than their patients to make moral choices on issues associated with a person being made aware of their genetic risk.

Therefore, ethical practice in genetic counselling, as it has gradually evolved in the West during the past 20 years, requires the autonomy of the individual or couple, their right to complete information, and the highest standard of confidentiality (Zeesman *et al.*, 1984; Fletcher *et al.*, 1985).

Medical ethics are based on the moral, religious, and philosophical ideals, and the principles of the society in which they are practiced. It is therefore not surprising to find that what is considered ethical in one society might not be considered ethical in another. The wide range of family and social structures, religious and legal conventions, and economic resources within the Middle East may also lead to conclusions that differ between countries. Sensitive services like genetic counselling cannot be transported from one social context into another.

Truly ethical conduct consists of personal searching for relevant values that lead to an ethically inspired decision. Surveillance or audit shows how information on parental informed choice is obtained and this reflects what is ethically acceptable in a particular society (Modell *et al.*, 1997; 2000). It is necessary for practitioners and critics of conduct to be sensitive to such information before they make their judgments on what is acceptable. The ethical attitude of the individual, whether a patient or a genetic counsellor, is often colored by the attitude of their society. Social attitudes are influenced by theologians, demographers, family planning administrators, doctors, policymakers, sociologists, economists, and legislators. All these groups also should consider the fundamental ethical principles of genetic counselling, particularly the autonomy of the individual, and place emphasis primarily on informed parental choice by those at risk before making judgements for society.

All medical programs, including genetic prevention programs, must operate within existing legal and social frameworks. However, technology can develop rapidly, whereas legal, social, and religious attitudes evolve more slowly.

In most religions there is a range of opinions and dilemmas. The high uptake of prenatal diagnosis in Italy, Cyprus, and Greece clearly shows that people make their own choices for what is appropriate to them. There is no right answer. The right answer is the one that is acceptable to the individual for his/her own particular reasons.

As several Muslim countries already are offering prenatal diagnosis and selective abortion to at-risk couples or are at various stages of developing these services, it is important to comment on the information that is already available on their acceptability in these countries.

For many, Muslim religion is central to daily life and influences a lot of behaviors, attitudes, practices, and policymaking (Serour, 2000). Instructions regulating everyday activities of life to be followed by Muslims are called *Sharia*. The Sharia is not rigid or fixed, except for a few rules such as those concerning worship, rituals, and codes of morality. Islamic Sharia accommodates different honest opinions as long as they do not conflict with the spirit of its primary sources and are directed toward the benefit of humanity. In Muslim societies, the provisions and spirit of Islamic Sharia and the local and social conditions of the society should be taken into account when formulating rules and guidelines.

Fetal development has been viewed by Muslim theologians as occurring in three stages, each lasting 40 days: the sperm cell and ovum, the clump resembling a blood clot, and the lump of flesh (fetus). At the end of these stages, the fetus is ensouled. The belief that ensoulment starts only at 120 days does not change the fact that life starts at a much earlier stage of embryo life.

Therefore from a Muslim perspective, it is considered ethical to perform an abortion to protect the mother's life or health, or because of fetal anomaly incompatible with life (Glover, 1989; Serour *et al.*, 1995). However, the stage at which termination is deemed permissible seems to vary. Some Muslim jurists do not allow abortion in any circumstances, whereas others would permit abortion in the first 120 days of fetal life if there is some reason such as danger to the mother or the fetus. The time when abortion is allowed also varies, and some jurists would allow it only at 40 days and others at 90 days.

Therefore most Muslims accept that life does not begin at conception, but believe that human life requiring protection commences some weeks, perhaps two weeks or so, after development of the primitive streak (Serour, 2001; Serour and Dickens, 2001). Therefore preimplantation genetic diagnosis is encouraged, where feasible, as an option to avoid termination of pregnancy (Serour, 2001; Serour and Dickens, 2001). The importance of preimplantation genetic diagnosis was recognized at the international workshop at The International Islamic Centre for Population Studies and Research, Al-Azhar University, Cairo, on Ethical Issues in Assisted Reproductive Technology.

It should be recognized that ethical and religious reasoning on the same issue can justify different conclusions. Therefore, it is acceptable that adherents of one preferred outcome may well acknowledge that adherents of an alternative preferred outcome are applying approaches that result in different but equally ethical conclusions.

31.7 CONSANGUINEOUS MARRIAGE

Consanguineous marriage is usually defined as marriages between people who are second cousins or closer (Bittles, 1994). The chances of inheriting two identical genes, including neutral as well as pathological genes, at a particular locus are increased if parents are close relatives (Bodmer and Cavalli-Sforza, 1978).

The question of how to provide genetic counselling in the context of societies that favor consanguineous marriage can also create dilemmas for families and health workers. Any ideal genetics program must include a sensitive and realistic approach.

Over 20% of the world population lives in communities that favor cousin marriage, and worldwide at least 1 in 12 children are born to parents who are related (Modell and Darr, 2002). Cousin marriage has been customary in many parts of the world for thousands of years, and is not always associated with Muslim religion. A custom that has been so common for so long obviously has important social functions. Currently in Northern European populations about 0.5% of marriages are between first cousins.

The great majority of families where the parents are related suffer no adverse effect. The reported increase in average childhood mortality and morbidity in such populations is due largely to relatively severe effects in a limited number of families, which shifts the average figures for the group as a whole. Therefore medical attempts to help families reduce genetic problems should focus on identifying families at particularly high risk, and providing them with genetic counselling and access to appropriate services.

Nevertheless, efforts have been made in some countries to discourage consanguineous marriage through public information programs that emphasize the associated genetic risks. These programs arise from the perception that recessively inherited rare diseases are unusually common in these populations, and the need for prevention seems urgent, but appropriate genetic counselling services are rudimentary or nonexistent. It may then seem that altering the behavior of the population is the only possible way to reduce genetic disease incidence.

In view of the ignorance of the social causes and consequences of customary consanguineous marriage, attempts to reduce its frequency in the population as a whole on genetic grounds run the risk of doing more harm than good, by disturbing customary marriage arrangements when the majority of families would come to no harm in any case. Furthermore, the concept of nondirective genetic counselling is incompatible with a campaign against consanguineous marriage. Policies relating to consanguineous marriage should be firmly grounded in an understanding of its social role, and the possible consequences of attempts to disturb it.

Discussions in the UK with families who have had pressure against consanguineous marriage put on them led to the following conclusions:

- Pressure against cousin marriage rarely alters what people actually do, though it can make them feel uncomfortable about it.
- If people are told their children are sick because they are related, it causes great unnecessary distress, may alienate them, and makes it more difficult for them to understand the real explanation.
- Where cousin marriage is common, people are aware that most couples of cousins have perfectly healthy children. If they are told not to marry a cousin because their children may be sick, they may become confused and could lose confidence in medical advice.
- Avoiding cousin marriage does not guarantee that children will not have a congenital disorder. Unrelated couples who have affected children may lose confidence in medical advice.
- People do not attend genetic counselling if they think they will be criticized and their cultural conventions attacked (Modell and Kuliev, 1992; WHO EMRO, 1997).

31.8 WHY COUSIN MARRIAGE IS FAVORED BY SOME COMMUNITIES

Many non-European communities have a patrilineal kinship pattern (family name and property are inherited in the male line). The men and their descendants tend to stay together, especially when the family owns land and members of the extended family share responsibilities for each other. At marriage, women leave their family to enter their husband's family. Cousin marriage can soften the implications of this transition and contribute to family well-being by strengthening the woman's position within the family and promoting female networks. Darr and Modell (2002) made the following conclusions in this issue:

- Parents are able to remain closer to their children, particularly their daughters, when they marry a relative.
- A woman is likely to be comfortable with a mother-in-law who is also an aunt she has known since childhood.
- There is increased financial security, because if one partner dies the remaining partner is still a member of the family in their own right.
- Equal numbers of sons and daughters are needed within the extended family, so a daughter is not seen as a burden.
- In societies that practice segregation between the sexes, young people can get to know each other before marriage more easily if they are related.
- Expenses and exchange of property associated with marriage may be reduced.

A convention of cousin marriage could make family-oriented genetic counselling particularly effective, for two main reasons. First, unusually large numbers of carriers of the presenting disorder may be detected within the family. Thus carrier testing, for example, may permit early detection of many individuals at risk for some types of thalassemia and sickle cell anemia and identify newly affected individuals where surveillance and early treatment could be beneficial. Second, when cousin marriage is common within the family, carriers are at particularly high risk of making an at-risk marriage. Family studies following the diagnosis of a child with a disorder will identify many carriers, who should then be offered counselling about their reproductive risk. Many will already be married, and some of these may be at-risk couples, identified in time for prospective reproductive counselling. Many will be children or not yet married, and early information might permit carrier status to be taken into account to avoid further at-risk marriages within the family. This approach has been tested in Pakistan (Ahmed *et al.*, 2002).

Consanguineous marriage increases the couple's chance of both carrying the same mutation. Therefore from the genetic testing point of view a family orientated approach to genetic screening becomes more efficient, because it simplifies the challenges of DNA testing. In a recent study in

Jordan, where homozygous thalassemia births were studied, it was found that the homozygous genotype was approximately 5.5 times higher in consanguineous marriage than random mating, giving 62% homozygous births versus 11.5% by random mating (Qubbaj and Petrou, in preparation).

31.9 CONCLUSIONS

The medical profession's responsibility starts from the time prenatal diagnosis becomes available and therefore a new social responsibility is generated for every subsequent birth. This responsibility should be passed onto the parents through adequate information and genetic counselling. The availability of genetic diagnosis increases the responsibility of the medical service when providing this service to the population. It is then the parents' responsibility to make the decision. The most common single saying of couples at risk for thalassemia is: "We have no choice: we cannot knowingly bring a child into the world to suffer." The other face of responsibility is guilt. From counselling at-risk couples in the United Kingdom it is known that parents want to take this responsibility of life or death. Couples with a thalassemia major child frequently want to sue the obstetrician for negligence, because they have not been identified and counselled during pregnancy. Their complaint is that they have been deprived of information and choice, and this increases their difficulties because their child's suffering is now perceived as an injustice rather than as a simple misfortune. They are expressing a strong conviction that since knowledge is available, someone is responsible for the birth of their affected child and that the responsibility that should have been their own has been taken from them (Modell, 1988).

The ethical principles governing genetic counselling need to be assessed for each country, after taking into account the social and religious structures of that country. This is particularly important for issues surrounding prenatal diagnosis and counselling in relation to customary consanguineous marriage.

References

Ahmed, S., Saleem, M., Sultana, N., Raashid, Y., Waqar, A., Anwar, M., Modell, B., Karamat, K. A., and Petrou, M. (2000). Prenatal diagnosis of beta thalassaemia in Pakistan: experience in a Muslim country. *Prenat. Diagn.* 20, 378–383.

Ahmed, S., Saleem, M., Modell, B., and Petrou, M. (2002). Screening extended families for genetic haemoglobin disorders in Pakistan. *N. Engl. J. Med.* 347, 1162–1168.

Alwan, A., and Modell, B. (2003). Recommendations introducing genetic services in developing countries. *Nat. Rev. Genet.* 4, 61–68.

Angastiniotis, M. A., Kyriakidou, S., and Hadjiminas, M. (1986). How thalassaemia was controlled in Cyprus. *World Health Forum* 7, 291–297.

Angastiniotis, M. A., and Hadjiminas, M. G. (1991). Prevention of thalassaemia in Cyprus. *Lancet* 1, 369–370.

Bittles, A. H. (1994). The role and significance of consanguinity as a demographic variable. *Population and Development Review* 20, 561–584.

Bodmer, W. F., and Cavalli-Sforza, L. L. (1978). *Genetics Evolution and Man*. W.H. Freeman and Company, San Francisco.

Carter, C. O. Current status of genetic counselling and its assessment. (1974). In *Birth Defects*, A. Motulsky and W. Lenz, eds. Excerpta Medica. Amsterdam.

Final report of the (Canadian) Royal Commission on New Reproductive Technologies. Canada Communications Group. (1993). Proceed with Care—Publishing. Ottawa, Canada, K1A OS9.

Fletcher, J. C., Berg, K., and Tranoy, K. E. (1985). Ethical aspects of medical genetics: a proposal for guidelines in genetic counselling, prenatal diagnosis and screening. *Clin. Genet.* 27, 199–205.

Gamberini, M. R., Lucci, M., Vullo, C., Anderson, B., Canella, R., and Barrai, I. (1991). Reproductive behaviour of families segregating for Cooley's anaemia before and after the availability of prenatal diagnosis. *J. Med. Genet.* 28, 523–529.

Glover, J. (1989). *Ethics of New Reproductive Technologies: The Glover Report to the European Commission, Studies in Biomedical Policy*. Dekalb, IL: Northern Illinois University.

Harper, P. (1988). *Practical Genetic Counselling*, 3e. Bristol, Wright.

Modell, B., Ward, R. H. T., and Fairweather, D. V. I. (1980). Effect of introducing antenatal diagnosis on the reproductive behaviour of families at risk for thalassaemia major. *BMJ* 280, 1347–1350.

Modell, B. (1988). Ethical and Social aspects of Fetal Diagnosis for the haemoglobinopathies: a practical view. In *Prenatal diagnosis thalassaemia and the haemoglobinopathies*, D. Loukopoulos, ed. CRC Press Inc. Boca Raton, Florida.

Modell, B., and Kuliev, A. M. (1992). Social and genetic implications of customary consanguineous marriage among British Pakistanis. *Galton Institute Occasional Papers*, Second Series, No 4. The Galton Institute, 19 Northfields Prospect, Northfields, London.

Modell, B., Petrou, M., Layton, M., Varnavides, L., Slater, C., Ward, R. H., Rodeck, C., Nicolaides, K., Gibbons, S., Fitches, A., and Old, J. (1997). Audit of prenatal diagnosis for haemoglobin disorders in the United Kingdom: the first 20 years. *BMJ* 315, 779–784.

Modell, B., Harris, R., Lane, B., Khan, M., Darlinson, M., Petrou, M., Old, M., Layton, M., and Varnavides, L. (2000). Informed choice in genetic screening for thalassaemia during pregnancy: audit from a national confidential enquiry. *BMJ* 320, 337–341.

Modell, B. and Darr, M. (2002). Genetic counselling and customary consanguineous marriage. *Nat. Rev. Genet.* 3, 223–229.

Ng, P. C., Fok, T. F., Lee, C. H., Cheung, K. L., Li, C. K., So, K. W., Wong, W., and Yuen, P. M. (1998). Is homozygous alpha thalassaemia a lethal condition in the 1990s? *Acta Paediatr.* 87, 1197–1199.

Old, J. (1996). Haemoglobinopathies. *Prenat. Diagn.* 16, 1181–1186.

Petrou, M., Brugiatelli, M., Ward, R. H. T., and Modell, B. (1992a). Factors affecting the uptake of prenatal diagnosis for sickle cell disease. *J. Med. Genet.* 2, 820–823.

Petrou, M., Brugiatelli, M., Old, J., Ward, R. H. T., and Modell, B. (1992b). Alpha thalassaemia hydrops fetalis in the UK: The importance of screening pregnant women of Chinese, other South East Asian and Mediterranean extraction for alpha thalassaemia. *Br. J. Obstet. Gynaecol.* 99, 985–989.

Petrou, M., and Modell, B. (1995). Prenatal screening for haemoglobin disorders. *Prenat. Diagn.* 15, 1275–1295.

Serour, G. I., Aboulghar, M. A., and Mansour, R. T. (1995). Bioethics in medically assisted conception in the Muslim World. *J. Assist. Reprod. Genet.* 12, 559–565.

Serour, G. I. (2000). Ethical Considerations of Assisted Reproductive Technologies: a Middle Eastern perspective. *Middle East Fertil. Steril. J.* 5, 13–18.

Serour, G. I. (2001). Attitudes and Cultural perspectives on infertility and its alleviation in the Middle East Area. Paper presented at the WHO Expert Group Meeting on ART, Geneva, Sept. 2001.

Serour, G. I., and Dickens, B. M. (2001). Assisted reproductive developments in the Islamic World. *Int. J. Gynecol.* 74, 187–193.

Stamatoyannopoulos, G. (1974). Problems of screening and counselling in the hemoglobinopathies. In *Birth Defects*, A. G. Motulsky and W. Lenz, eds. pp. 268–276, Excerpta Medica, Amsterdam.

Weatherall, D. J., and Clegg, J. B. (2001). *The Thalassaemia Syndromes*, 4e. Blackwell Scientific Publications, Oxford.

WHO (1993). Unpublished report WHO/HDP/TIF/HA/93.1 Joint WHO/TIF meeting on the prevention and control of haemoglobinopathies (7th meeting of the WHO Working Group on the Control of Hereditary Anaemias). Nicosia, Cyprus 3–4 April 1993 (unpublished report WHO/HDP/TIF/HA/93.1).

WHO/EMRO Technical Publications Series 24 (1997). Community Control of Genetic and Congenital Disorders, Available from WHO Regional Office for the Eastern Mediterranean, Alexandria, Egypt.

WHO, Report of Consultants to WHO (2001). Review of Ethical Issues in Medical Genetics, WHO/HGN/ETH/00.4.

Zeesman, S., Clow, C. L., Cartier, L., and Scriver, C. R. (1984). A private view of heterozygosity: eight-year follow-up study on carriers of the Tay-Sachs gene detected by high-school screening in Montreal. *Am. J. Med. Genet.* 18, 769–778.

CHAPTER **32**

Genetic Testing and Psychology

ANDREA FARKAS PATENAUDE
Division of Pediatric Oncology, Dana-Farber Cancer Institute; Department of Psychiatry, Harvard Medical School, Boston MA, USA

32.1 INTRODUCTION

The completion of the comprehensive sequence of the human genome in the spring of 2003 was hailed as a "landmark event" and the beginning of a "revolution in biological research" (Collins *et al.*, 2003). The huge international cooperative effort that comprises the Human Genome Project (HGP) has produced a tool of enormous importance to basic scientists, clinical researchers, clinicians, and, ultimately, patients. Although a great deal already has been learned about basic biology from the HGP, including the fact that humans have only about 30,000 protein-coding genes, many fewer than originally estimated, the most interesting challenges lie ahead (International Human Genome Consortium, 2001). With the human genome sequence readily available to all, the challenge will be to "capitalize on the immense potential of the HGP to improve human health and well-being" (Collins *et al.*, 2003).

Genetic advances are going to change the practice of medicine in many significant ways. Though some may argue over the extent and the pace of changes to come as a result of the HGP (Holtzman and Marteau, 2000), it is clear that the ability to define, diagnose, detect, and prevent, treat, or cure many illness in the future will involve techniques derived from technologies that have their roots in recent genetic advances (Wolf *et al.*, 2000). On the occasion of completion of the human genome sequence, the United States National Human Genome Research Institute put forth a blueprint for future genomic research. The second of three major challenges laid out was labeled "Genomics to Health" and involves the translation of "genome-based knowledge into health benefits."

The goals of this part of the work of genomic scientists and clinical researchers are to:

- "Identify genes and pathways with a role in health and disease, and determine how they interact with environmental factors,
- Develop, evaluate, and apply genome-based diagnostic methods for the prediction of susceptibility to disease, the prediction of drug response, the early detection of illness, and the accurate molecular classification of disease, and
- Develop and deploy methods that catalyze the translation of genomic information into therapeutic advances" (Collins *et al.*, 2003).

It is argued that the achievement of these lofty aims is dependent on psychological factors in a variety of ways. Whether patients are willing to participate in genetic research, whether they opt to undergo genetic counseling (see also Chapter 31) or testing to determine if they have increased disease risk, whether they are willing to adopt screening or surveillance recommendations targeted to those known or suspected of being at high risk, or whether they opt for prophylactic surgery, depend in large measure on emotional, attitudinal, and behavioral factors. The recruitment of appropriate, diverse patients for genetic research studies will necessitate improved understanding of what patients bring to genetic counseling or testing from their own family experience with illness, their perception of disease risk, and their attitudes toward genetics research and the medical community. Individuals from different ethnic and cultural communities vary widely in their attitudes toward genetics, genetic testing, and genetic medicine.

Subjective feelings often play a large role in determining how people respond to situations they consider threatening. Research suggests that individuals often act on the basis of subjective feelings about their disease risk, even after receiving counseling that informs them that their risk perceptions are overestimates of actual risk (Iglehart *et al.*, 1998). How can such responses be modified to be in tune with actual risk? What kinds of counseling techniques need to be developed to guide patients toward selecting one of the appropriate choices for mutation carriers? How can people who are found to be at increased hereditary risk but who do not heed surveillance or screening recommendations be educated? How can people be helped to make optimal personal decisions about prophylactic surgery, which may be the risk-reduction option which conveys the greatest protection against developing disease but at significant emotional cost? What is known about how individuals in high-risk families communicate about hereditary disease predisposition with their relatives?

Much of the ultimate success of the HGP in improving human health rests on whether individuals, suspected or known to be at risk for hereditary illness, feel sufficiently comfortable with the use of genetic technology and with the emotional impact of genetic information to make use of the revolutionary advances which the HGP has made possible. This chapter attempts to summarize what is currently known about psychological factors as they affect individuals and families with concern about hereditary predisposition to illness.

With the advent of genetic testing, first for Huntington's disease (HD) and over the past decade for some types of cancer, we have been able to observe reactions to the increased availability of genetic information in individuals from hereditary disease families. In the United States and in several other countries, simultaneous with the molecular studies of the human genome, studies of the ethical, legal, social, and psychological impact of genetic testing have been undertaken. It is critical to learn under what circumstances genetic advances improve people's lives and health, as well as understanding under what circumstances adverse outcomes result from participation in genetic research, genetic counseling, and/or genetic testing.

32.2 GETTING TO THE TEST: AWARENESS, ACCESS, AND ADVERTISING

32.2.1 What Tests Are There?

Linkage analysis for HD has been possible since 1983 (Gusella *et al.*, 1983) and direct sequence analysis, a much more accurate form of genetic testing, became possible after the cloning of the *HD* gene in 1993 (Huntington's Disease Collaborative Research Group, 1993). The first cancer genetic testing was testing for *Rb1*, the gene responsible for hereditary forms of retinoblastoma, a relatively rare cancer of the eye, which was cloned in 1986 (Friend *et al.*, 1986). Testing for *p53* mutations in members of families with Li-Fraumeni Syndrome (LFS) began in the mid-1990s. LFS is a rare condition in which mutation carriers have a 90% lifetime risk of developing a number of common cancers including breast cancer, leukemia, sarcomas, and brain tumors, including a 40% chance of developing cancer in childhood (Williams and Strong, 1985). The first major cancer susceptibility genes that affected a large number of people and that achieved significant notice in the popular press were *BRCA1*, which was cloned in 1994 (Miki *et al.*, 1994), and *BRCA2*, cloned in 1995 (Wooster *et al.*, 1995). These two genes predispose female mutation carriers to greatly increased risks of developing both breast and ovarian cancer. Breast cancer risk for *BRCA1/2* carriers has been estimated as between 56% (Struewing *et al.*, 1997) and 85% (Ford *et al.*, 1994), with ovarian cancer risk estimates running from 27% (Ford *et al.*, 1998) to 60% (Easton *et al.*, 1995). Men who are mutation carriers also have an increased risk of breast cancer, although the risk for men is much smaller than for women. Hereditary cancers typically occur at younger-than-usual ages and it is not unusual for multiple cancers to occur.

Genetic testing is also now possible for a number of colon cancer genes, such as *APC*, which predisposes to Familial Adenomatous Polyposis (FAP), and *MLH1* and, *MSH2*, which predispose to Hereditary Non-Polyposis Colorectal Cancer (HNPCC, Giardiello *et al.*, 2001). Genetic testing is also possible for a number of quite rare cancer genetic syndromes, such as Von Hippel-Lindau Syndrome and Cowden's syndrome and for some members of families with significant history of melanoma. Genetic testing may not be far off also for hereditary prostate cancer. It is estimated that between 5% and 10% of all cancers have a hereditary etiology (Offit, 1998). Because most of the psychological studies of genetic testing have been done for *HD* and more recently for *BRCA1/2* and colon cancer genes, these studies will be the major focus of the discussion.

32.2.2 Genetic Testing

Genetic testing first was done in research settings and offered to members of families known to have had unusual numbers of affected individuals (that is, individuals diagnosed with the disease). Members of these families tended to have been aware for a long time, sometimes several generations, about the unusual concentration of disease in their families. Many were from families with extremely high concentrations of hereditary cancer, since it was the most affected families who first came to the notice of clinicians and disease registries. Family members had often observed several early deaths from the disease in their family and had grown up fearing that this would be their fate as well.

Prior to the availability of genetic testing, early research studies asked members of these hereditary disease families if they thought they would want to be tested when the relevant gene was found. Interest was high (typically over 70%) among members of HD and breast/ovarian syndrome families prior to cloning of the respective genes (Mastromauro *et al.*, 1987; Jacopini *et al.*, 1992; Lerman *et al.*, 1995).

Of great interest is the fact that with HD, the uptake of genetic testing since testing became available among members of affected families has been only about 5% to 20% (Harper *et al.*, 2000; Meiser and Dunn, 2001). HD is a devastating, neurodegenerative disease. The *HD* gene is 100% penetrant, meaning that knowledge of being a mutation carrier is equivalent to knowledge that one certainly will develop the disease. There are no preventive options or cures for HD. The earlier arguments put forth by at-risk individuals that testing would enable them to plan better for the future or to achieve certainty about whether or not they were at risk apparently paled in comparison with the possible pain of discovering, before symptoms occurred, that one would, with certainty, be subject to the fate of others in the family already affected with HD.

It had been conjectured that for *BRCA1/2*, uptake would be much more similar to projected interest, since there are some measures that are recommended for mutation carriers to try to prevent or detect cancer early. However, although the uptake for *BRCA1/2* testing—for example, 27% to 43% (Lerman *et al.*, 1996; Julian-Reynier *et al.*, 2000)—has been higher than for HD, it has not approached the levels reported for hypothetical interest before testing was actually possible.

Researchers are still in the process of understanding the decision-making of individuals who opt not to undergo genetic testing or, often, even genetic counseling for hereditary cancer. There is some research suggesting detrimental emotional effects, such as continuing depression, among those who refuse testing (Lerman *et al.*, 1998; Kash *et al.*, 2000), but further research is needed among larger and more diverse samples to fully understand the psychology of "not wanting to know." Some investigations have found it is patients with lower chances of being mutation carriers who are most likely to want to be tested, since testing, they hope, then can serve as reassurance that they are not at increased hereditary risk of disease. It is possible to say with some certainty, however, that hypothetical estimates of interest in genetic testing are not accurate predictors of eventual uptake.

Cancer genetic testing has moved beyond the research setting into clinical testing. Clinical testing makes it likely that individuals with less extensive family histories of disease will be tested; the impact of this change has not yet been fully evaluated. Some patients newly diagnosed with breast cancer are being offered up-front genetic testing, even

before treatment options are in place. It is unclear how these patients, many of whom had not previously considered they might carry an increased hereditary cancer risk, will react to testing. Myriad Genetics, Inc. has test kits for breast, colon, and melanoma genetic testing that can be accessed by any physician who wishes to test his/her patient, and has recently begun direct-to-consumer advertising. Some companies offer patients direct access to test kits via the Internet (Williams-Jones, 2003), which raises significant issues about assuring that individuals undergoing testing are sufficiently educated about the implications of learning their test result.

There also remain difficult issues about whether patients or even many medical professionals are able to interpret test results correctly (Giardiello *et al.*, 1997). Even though genetic counseling is recommended by most genetics professionals, there is no requirement that patients be counseled prior to receiving results. Some studies suggest frequent misinterpretation of results by patients, including counseled patients, particularly when test results are more ambiguous (Calzone, 2002; Hallowell *et al.*, 2002). An example of ambiguous findings is indeterminate results, which can occur when an individual is tested without prior identification of a familial mutation in an affected family member. This might be the case, for example, where an Ashkenazi Jewish individual is tested for the three mutations of *BRCA1/2* genes, which are found with high frequency in that ethnic group. It seems likely that many patients misinterpret the absence of these mutations to mean that they are not at increased risk. This cannot be assumed since there might be another mutation or gene that is causing hereditary cancers in that family. In responsible hands, patients at risk for cancer genes who receive indeterminate results are cautioned to continue to undergo screening or surveillance, which is appropriate for a mutation carrier, since they cannot be excluded as having a mutation.

Most of the patients coming forward for testing either in research or clinical settings have been well-educated, Caucasian patients from higher socio-economic strata. It is clear that specialized approaches will be necessary to provide genetic counseling to patients with less background knowledge of genetics and possibly greater skepticism about genetic privacy and the motivation of research scientists. It does seem, though, that patients with more genetics knowledge show some reservations about being tested after learning about the limitations of genetic testing, whereas less educated patients seem to overlook these limitations in embracing testing (Hughes *et al.*, 1997).

Genetic testing does not supply all the answers that patients wish to have. There are very real limitations to what genetic testing can tell patients about the likelihood they will develop the disease in question. A positive test result for a gene that is less than 100% penetrant does not indicate for certain that the mutation carrier will ever develop the disease, or when it will occur. In some cases, a single genetic mutation predisposes the individual to a number of related diseases; testing usually will not indicate which disease the person will develop or whether they will develop more than one disease within the syndrome of diseases to which they are predisposed. Cancer genetic testing is, after all, a trade of the uncertainty about whether one is a mutation carrier, for the uncertainty, in the event that the test is positive, about which cancers might appear and when they will emerge.

32.2.3 Access to Testing

Access to genetic testing is not equally available to all. Full-sequence testing for cancer genes costs several thousand dollars and patients may opt not to bill the costs to their health insurance, either because it is not a covered service in all plans or because patients are reluctant to share test results with their insurer for fear of later discrimination. In countries with a national health insurance system, discrimination is not an issue for health insurance. However, access may be limited by rationing of the number of genetic tests offered in a geographic region during a particular time period and/or very strict criteria for test eligibility. These practices may result in long waits for testing or may exclude patients with less extensive family histories.

Testing was free in most of the early research programs evaluating the uptake and reaction to genetic testing. Without free testing, it is not clear how much the cost of testing affects access or uptake. Minority populations are likely to be particularly hindered in access to genetic testing for a variety of reasons, including cost, lack of minority genetics professionals, and lack of non-English language educational materials.

32.3 INDIVIDUAL FACTORS INFLUENCING UTILIZATION OF GENETIC TESTING

Complex psychological factors govern an individual's uptake of genetic testing and the use they make of the information received. Because early research has focused on heavily affected families, much of what is known in this area pertains to members of what are often called research families or cancer families. It is not difficult to understand that personal experience with a family member who has had the disease in question and possibly died from it is likely to impact one's perception of personal risk, disease-related anxiety, and the desire to avoid a similar outcome. There are, however, other factors, including general personality characteristics, that can affect how actively a person seeks to confront knowledge of personal, hereditary disease risk and to undertake actions designed to prevent or detect illness.

Increasing knowledge of these factors will enable researchers and counselors to develop more effective genetic counseling and follow-up techniques and to better advise both patients and providers.

32.3.1 Risk Perception

Members of high-risk cancer families typically greatly overestimate their risk for developing cancer (Kash *et al.*, 2000). Those who have lost a parent to the disease are especially likely to overestimate the risk that they, themselves, will develop the cancer their parent died from. In an English study of women referred for genetic counseling to a Cancer Family History Clinic, 70% could not cite the population risk for breast cancer, but 76% perceived themselves to be at higher than population levels of risk for breast cancer (Hallowell *et al.*, 1998). A study by Zakowski and colleagues (1997) found that women undergoing mammography screening (but with no abnormal findings) whose mother had died of breast cancer estimated their lifetime risk of developing breast cancer as 70% versus the 53% risk estimate of women undergoing mammography who did not have a mother die of breast cancer and the 32% risk estimate of similarly-aged women in the general population not undergoing mammography studies. All these figures significantly overestimate breast cancer risk; lifetime risk for breast cancer for a woman in the United States general population is about 10%. Having a first-degree relative with breast cancer raises that risk somewhat, to 1.7 to 5 times (Offit, 1998), but not to 70%.

Overestimation of risk can provide motivation for seeking genetic counseling or testing. Perceived risk for colon cancer has been shown to be higher among first-degree relatives (FDRs) of colon cancer patients who accepted (versus declined) colon cancer genetic testing (Codori *et al.*, 1999). In that study, a third of those who declined testing thought their risk of colon cancer was below population levels and many of those individuals tended to avoid colonoscopy recommendations. Even among cancer patients with family histories suggestive of hereditary risk, risk estimates are often not accurate. In a study of 200 women who had had breast or ovarian cancer or both breast and ovarian cancer, 55% differed in their risk estimate by more than two quartiles from estimates made by an expert panel. Even after these women were given genetic screening tests and informed of their result, most continued to vastly overestimate their risk of carrying a *BRCA1/2* mutation (Iglehart *et al.*, 1998).

Interestingly, in HD families, prior to the advent of genetic testing, particular family members were often preselected as being likely to develop HD (Kessler, 1988). These individuals were thought of and treated as though their hereditary risk status was known, as though they were certain inheritors of mutated *HD* genes, often based on physical or behavioral resemblance to a parent with HD. In some cases, it was difficult for family members to alter this expectation when later testing found the individual not to carry the mutation, which predisposes to HD.

32.3.2 Distress

The perception that one has a high likelihood of developing a serious disease is often associated with elevated levels of emotional distress. Among both research and clinical samples of women at high risk for breast cancer, distress scores at or near clinical cut-offs have been found in a substantial proportion of subjects (Kash *et al.*, 1992; Audrain *et al.*, 1997). Distress is associated with higher breast cancer risk perception and lower perceptions of control over developing breast cancer. In an English study of 312 individuals reporting for counseling for hereditary breast/ovarian cancer, 22% of the women and 14% of the men scored at or above clinical cut-offs for a psychiatric disorder (Foster *et al.*, 2002). In a Dutch study, individuals at risk for HD scored higher on distress measures than those at risk for hereditary cancer syndromes (Dudokdewit *et al.*, 1997). Those whose scores were highest indicated that their distress was closely related to painful memories of family members who had had the familial disease.

Cognitive emotional factors have also been found to be related to distress around hereditary disease risk. Miller and colleagues (1995) have distinguished two styles of coping with threatening information. *Monitors* are those who tend to focus on the threat, to actively seek information related to the threat, and to have difficulty perceiving that all has been done which could be done to cope with the threat. *Blunters*, in contrast, are individuals who naturally try to avoid or shield themselves from threatening information. It has been shown that anxiety during the wait for genetic testing results is higher among women who are high monitors (Tercyak *et al.*, 2002), suggesting that these women may need extra support while awaiting their result. Cognitive style also appears to influence participation in screening, so that targeting of screening and surveillance recommendations according to the patient's cognitive style may be important to maximize participation (Miller *et al.*, 1996).

32.3.3 Health Beliefs

Misconceptions can affect one's ability to realistically evaluate the increase in risk imposed by being a mutation carrier and can also affect one's thinking about the value of screening, surveillance, or prophylactic surgery options. There are many misconceptions about inherited cancer risk even among those at increased risk. In a study of 200 well-educated women with family histories of breast cancer who had had breast cancer themselves, 86% erroneously believed that 1 in 10 women have a *BRCA1/2* mutation, 62% thought that half of all breast cancer is due to *BRCA1/2* mutations (correct estimate: 5–10%), and that 56% did not think fathers could transmit *BRCA1/2* mutations to their daughters as easily as mothers could (Iglehart, 1998).

Different ethnic or cultural groups tend to have differing values and may hold different beliefs related to medicine generally, and genetics, more specifically. African-Americans have a negative history of experimentation for sickle cell disease, another genetic condition, which may reduce motivation to participate in genetic studies (see also previous chapter). In fact, African-Americans have participated in genetic studies at a lower rate than their participation in medical research generally (Royal *et al.*, 2000). The fact that there are very few African-Americans genetics professionals may contribute further to the hesitancy of African-Americans to undergo genetic testing. On the other hand, a recent study of 407 African-American and Caucasian women who had at least one FDR with breast or ovarian cancer suggested that the African-American women had less knowledge about cancer genetics and genetic testing for *BRCA1/2*, but were more positive about the potential benefits of testing. A very small percentage of women tested to date, however, have been from minority populations. Research is underway on how to approach African-American and other minority populations about genetic testing (Hughes *et al.*, 1997; Meiser *et al.*, 2001; Thompson *et al.*, 2002). Much remains to be learned about the factors responsible for this lack of involvement of minorities in genetic medicine to date.

Cultural beliefs may also affect decision-making about uptake of risk-reduction options. Julian-Reynier and colleagues (2001) contrasted views on prophylactic surgery for *BRCA1/2* carriers in England, France, and French Canada. The French were the least accepting of prophylactic mastectomy of the three groups and the French and French-Canadians were less accepting of prophylactic oophorectomy (prophylactic removal of the ovaries) than the English.

32.3.4 Health Behaviors

It can be assumed that anyone known or suspected of being at high risk for disease would wish to undergo any recommended screening mechanism to detect disease early or to try to prevent the occurrence of the disease. This turns out to not always be the case. Anxiety is a powerful inhibitor and, as has been previously discussed, many individuals at high risk experience high levels of anxiety about developing cancer. As a result, compliance with recommended screening or surveillance practices is far from perfect in many groups of at-risk individuals. In a Canadian study of high-risk individuals, those who perceived themselves to be at high risk were five times less likely to adhere to screening recommendations (Ritvo *et al.*, 2002).

Although most women at high risk for breast or ovarian cancer report that they get regular mammograms, many do not do breast self-examination, as recommended (Kash *et al.*, 1992). A recent Australian study suggested that after women are informed via genetic testing that they are *BRCA1/2* mutation carriers, they were less likely to adhere to mammogram recommendations (Meiser *et al.*, 2002).

Colon cancer screening compliance among the general population is lower than screening for breast cancer, and high-risk populations are also below ideal levels for colon cancer screening. For individuals from HNPCC families, colonoscopy is recommended every three years beginning around age 25 (Vasen *et al.*, 1998). Among a group of adult patients studied after genetic testing for colon cancer genes, 45% to 48% had never had a colonoscopy and 28% had never had any colon cancer screening test (Johnson *et al.*, 2002). Genetic testing appeared to be a catalyst for those found to be positive to undergo regular colon screening, but appeared to deter those who were negative from getting regular colonoscopic screening, as recommended in population guidelines for those at normal risk. In another study of 42 individuals with a personal or family history fulfilling clinical criteria for hereditary colorectal cancer (HNPCC or FAP) who presented for genetic counseling, only 64% had undergone appropriate surveillance for colorectal cancer (Stoeffel *et al.*, 2003). Among the women in that study at risk for endometrial manifestations of hereditary colorectal cancer syndromes, only 25% had had appropriate screening. The authors felt for a variety of reasons that these were likely overestimates of compliance with screening and surveillance recommendations for those at risk for hereditary colorectal cancers.

It will be critical to address the cancer worry and general anxiety of those who are found to be mutation carriers. As evidence grows about the efficacy of various screening, chemopreventive, and surgical options for those at increased risk, it will be important to help high-risk individuals cope with the anxiety generated by their awareness of their genetic status in order to understand and, if desired, to take advantage of the available options. It is relatively recently that data has emerged showing that one of the more controversial preventive surgery options, prophylactic mastectomy, conveys 90% risk reduction for breast cancer (Hartmann *et al.*, 1999). Prophylactic oophorectomy has also been found not only to reduce the risk of ovarian or endometrial cancer, but has also been shown to have a risk-reducing effect on breast cancer (Rebbeck, 2001).

Increased data will help patients decide about which options they are willing to undertake, but it is clear that psychological factors will continue to be important in all such decisions. As a result, it will be critical that providers recognize the need for patients to discuss the risks and benefits of preventive options, but that they also have access to psychological counseling as needed to help evaluate the personal acceptability of various courses of surgical risk-reduction.

32.4 GETTING THE GENETIC TEST RESULT: PERSONAL IMPACT AND PROFESSIONAL COMMUNICATION

The disclosure of a genetic test result is a moment of high drama. It is a moment of importance not only for the individual patient present and their (often) accompanying spouse, but it has implications for children, born and unborn, and grandchildren. There may also be implications, though less directly, for siblings, cousins, and other relatives. It may be the end of many years of wondering if one has a hereditary predisposition or it may come as the response to a relatively recent education about hereditary illness, in some cases occasioned by a recent cancer diagnosis. Some results are more definitive than others. Learning one is truly freed from the family curse by virtue of a true negative result can be a moment of blissful relief. Learning, alternatively, that one is a mutation carrier, can answer for some people who have had cancer the question of why they became ill, but it opens many other questions about the subsequent risks for the patient him/herself and questions about risk to offspring. For unaffected patients, learning that they are mutation carriers can be a fearsome moment, conjuring images of ill or deceased family members. These individuals often express disappointment that they could not be released from their cancer worry as they had imagined they would be if they had tested negative. The intensity of their immediate, emotional response often makes it difficult for them to pay attention to the recommendations for mutation carriers offered by the genetic counselor. For others, the result they receive is indeterminate. There can be both disappointment that the answer they received is unclear and relief that at least, for the time being, a mutation was not found. The latter response is somewhat irrational, since an indeterminate finding does not assume that a mutation is not present, only that one could not be found with present testing methods. In each case, emotions are aroused and often patients recount that little is heard beyond the result itself.

32.4.1 Testing Positive

The model for much of how genetic testing has been conducted to date has been the HD research programs, which included a large number of interviews prior to taking blood for the genetic test and great care around the delivery of the result. There had been considerable worry when genetic testing for HD was initiated about whether patients learning that they were HD mutation carriers would experience extreme, adverse emotional reactions, including suicide. Concern was so high because of the unusually high suicide rate (4–8 times population rates) among HD patients

(Schoenfeld *et al.*, 1984; DiMaio *et al.*, 1993). Outcome studies from the HD programs have shown that among more than 4,000 patients tested worldwide for HD, less than 1% either have attempted or committed suicide following disclosure of their test result (Almqvist *et al.*, 1999); those who committed suicide already has symptoms of HD. These figures are relatively reassuring, but support efforts to define which patients are most likely to respond adversely and to develop psychological counseling models of use to such patients (Robins-Wahlin *et al.*, 2000). Factors predisposing to adverse outcomes following HD genetic test disclosure were having a psychiatric history of five years duration or less and being unemployed (Almqvist *et al.*, 1999).

The follow-up of patients undergoing cancer genetic testing has shown, too, that extreme adverse outcomes are rare (Butow *et al*, 2003). Most patients who are found to be carriers of deleterious mutations respond with sadness, but do not develop clinical symptoms of depression or anxiety. Mean scores of most samples of tested individuals have been within the normal range (Lerman *et al.*, 1996; Croyle *et al.*, 1997). Little difference has been noted in psychological outcomes between research samples and self-referred clinic samples (Schwartz *et al.*, 2002).

On the other hand, subclinical effects of increased intrusive thinking about cancer and cancer worry have been found in many individuals who have tested positive for a cancer predisposition gene (Meiser *et al.*, 2002). Also, about a quarter to a third of many of the samples of individuals undergoing cancer genetic testing have shown elevated, clinical levels of depression or anxiety following disclosure (Grosfeld *et al.*, 1996; Lodder *et al.*, 2001; Bonadona *et al.*, 2002). Some predictors of higher distress have been the test status of other family members. Lodder and colleagues (2001) found that noncarriers who had sisters who had been found to be mutation carriers were more depressed than other noncarriers. Similarly, Smith and colleagues (1999) found that men who were the first to be genetically tested in their families and had other siblings awaiting results were more distressed than carriers whose siblings were all test-negative. Female carriers who were the only positive members of their sibship (i.e., their siblings were all negative) scored as high as recently diagnosed cancer patients on the Impact of Events Scale (Horowitz *et al.*, 1979), a measure assessing cancer-related worry. Women carriers whose sisters had not yet received results also were quite distressed.

Disease status also affects psychological outcomes. The ability to correctly anticipate the level of one's emotional reaction to a forthcoming event is thought of as a predictor of good adjustment to that event. In a study of women tested for *BRCA1/2*, some of whom had had cancer and some who were unaffected (i.e., had not had cancer), the cancer patients found it significantly more difficult to anticipate their emotional reaction following disclosure. Cancer patients tended to feel that learning that they were mutation

carriers would not be as eventful for them as it turned out to be. Underestimation of the emotional impact of disclosure was associated with significantly increased distress levels six months later (Dorval *et al.*, 2000).

32.4.2 Testing Negative

Of interest is that in both the HD and cancer genetic testing outcome studies, a subset of individuals testing negative have been found to experience moderate levels of distress (Myers *et al.*, 1997). These paradoxical reactions were largely unanticipated by researchers who believed that one of the major benefits of genetic testing was that, for dominant genes, it could free 50% of the patients from fear of hereditary disease predisposition. Relief is prominent in the reactions of most individuals testing negative, although many (but not all) say they also feel guilty toward family members who have had the disease or who are mutation carriers. More extreme negative reactions have occasionally been seen in individuals who are significantly distressed after learning they are not mutation carriers. In some cases, particularly within HD samples, this distress has been associated with regret for having lived so many years with the expectation of getting HD and of having foregone opportunities for further education or other experiences because of that, now proved faulty, expectation (Huggins *et al.*, 1992).

There are also very rare cases reported in the literature of Munchausen Sydrome whereby individuals falsely claim to have family histories suggestive of hereditary syndromes for whatever secondary emotional gains they believe might accrue to at-risk individuals (Kerr *et al.*, 1998).

Another unexpected reaction of individuals testing negative for mutations in cancer genes or being told that their cancer risk is not sufficient to recommend genetic testing has been an unwillingness to accept the recommendations. Some of the people insist on getting tested or find it difficult to give up the increased screening they had received before testing when it was thought they could be mutation carriers. Individuals testing negative for FAP have, in some cases, stated their reluctance to discontinue the early and more frequent colonoscopic screening recommended for those at high risk, stating insufficient trust in the genetic test result (Rhodes *et al.*, 1991; Bleiker *et al.*, 2003). In another study, many of the women who were told on the basis of their family history that they were not eligible for *BRCA1/2* genetic testing because their risk levels were too low, reacted with anger rather than relief (Bottorff *et al.*, 2000).

32.4.3 Consequences of Testing

The majority of genetically tested individuals express no regret at being tested, even if they are found to be carriers. Carriers express sadness and disappointment, but also express the belief that knowing they are at increased hereditary risk will help motivate them to take good care of themselves physically and to be regularly screened by medical professionals. Concern for their offspring is a major factor in the guilt and sadness that they feel. Those who are negative feel particular relief on behalf of their children and grandchildren who are spared continuing worry about hereditary predisposition.

An important question, of course, is whether knowledge of one's genetic test status correlates with appropriate levels of screening behavior or with uptake of preventive options. Because this is a relatively new area of medicine, decisions about screening, surveillance, and preventive options are complicated by imperfect data about their efficacy in preventing cancer in mutation carriers. There are no preventive options for individuals at risk for HD. Nonetheless, it is of interest to learn how testing changes health behaviors. A recent Canadian 18-month follow-up study of 79 women who had tested positive for a *BRCA1/2* mutation found that 57% said that they had altered their screening practices following disclosure (Ritvo *et al.*, 2002). Especially younger (<age 50) women were likely to report change in health behaviors. Prophylactic mastectomy had been elected by 21% of the women and prophylactic oophorectomy by 40%. Lerman and colleagues (2000) showed how mutation carriers' initial intentions regarding prophylactic surgery did not translate, at least over the year following disclosure, into actions. At disclosure, 31% intended to consider prophylactic mastectomy and 40% intended to undergo prophylactic oophorectomy. Six months later, only 1% had undergone prophylactic mastectomy and 2% had had prophylactic oophorectomy. At one year, 3% of the women had had prophylactic mastectomy and 13% had undergone prophylactic oophorectomy. In that same group, only 21% had had a CA-125 test, which is recommended at least annually to *BRCA1/2* mutation carriers to try to detect ovarian cancer in its earliest stages. Two-thirds of the women had had a mammogram in the past year, a percentage unchanged from before testing.

Although many good intentions to optimize and personalize health screening behaviors in the face of disclosure of a positive genetic test were expressed just after disclosure, many of these intentions were not acted upon in the year following disclosure. This raises questions about the time frame in which we should expect patients to make such health behavior changes, but also about the need for follow-up counseling to encourage active screening and surveillance and to help individuals with consideration of more extreme preventive options. There is much that needs to be learned about how patients perceive recommendations for screening and surveillance, especially once it has been established that they are mutation carriers. In addition, researchers and genetic professionals also need to understand more about pretesting levels of distress (depression, anxiety, and disease-related worry) and their relation to utilization of health behaviors following disclosure.

32.5 FAMILY COMMUNICATION

Genetic medicine is family medicine. Genetic professionals advise patients to consider carefully the ramifications of genetic information on other family members. Both before and after genetic testing, there are reasons why open communication with family members about family history and hereditary risk may be advisable. In order to provide the medical information genetics professionals require to calculate the risk that an individual carries a mutation, it is often necessary to talk to relatives about which family members were affected by disease at what ages and with what outcomes. Women are typically the keepers of such family information. Because of the many premature deaths in families with hereditary illness, there are often gaps in knowledge about family history. An individual seeking genetic counseling may have to talk to relatives with whom they ordinarily have little contact in order to complete the family history to a point where the counselor will have enough information to calculate risk.

Gathering family history information in order to create a family illness pedigree may require talking to relatives on both sides of the family, which sometimes includes uncomfortable contacts with long-lost relatives. Family traditions may include not talking very directly about illness and doing so may open old emotional wounds. Some family members are completely estranged, in some cases, making completion of a full pedigree impossible. Because the hallmark of hereditary illness is the involvement of multiple generations of family members, family issues tend to surface more readily with regard to genetic testing than in other kinds of medical endeavors.

The relevance of one person's genetic information to other family members can set up difficult alliances or the tendency to avoid relatives with whom one is not comfortable discussing highly personal information, such as a genetic test result.

The accuracy of family information can affect genetic professionals' ability to correctly analyze the likelihood that there is a hereditary predisposition to disease in a particular family. Documentation of age of onset of disease is particularly important in such determinations, since, in many cases, early age of onset is a hallmark of hereditary illness. Also, knowledge of exactly which cancers a family member had may be critical as cancer syndromes link certain cancers to hereditary mutations. There is more confusion about some cancers than others. Knowledge about breast cancer within a family, for example, tends to be more accurate than colon cancer, which may be confused with other gastrointestinal cancers. In a study by Glanz and colleagues (1999), which identified colon cancer patients from a medical registry, 25% of the first-degree relatives of these patients denied that they had a relative who had had colon cancer. Schneider and colleagues (2004) found that over half of the members of families presented their family history with insufficient accuracy to have led providers to suspect that they were a LFS family. These reports suggest the need to verify diagnoses (other than breast cancer) from medical records of the family members, if possible, as part of the genetic counseling process.

Although most inaccuracies are due to misconceptions or unintentional error, there are occasions when family embarrassment about illness or other factors leads to deliberate falsification of the family history. One of the risks of gathering family history data is the risk of uncovering misassigned paternity; that is, that the putative father of a child is not, in fact, the actual father. Patients should be helped to understand that the experience of searching for family history data may be more emotional and/or upsetting than anticipated.

Once genetic test results are available and disclosed to the patient, considerations about the implications of the result for other family members come into play. Genetics professionals recommend quite strongly that results be shared with family members, especially when screening recommendations likely would be influenced by the genetic information. For example, if a man in his 50s is found to be a mutation carrier for HNPCC, it would be important for his children to know that they should initiate colonoscopic screenings at age 25, not at age 50, as recommended for the general population. Similarly, if the man is negative, his children could use that information to decide that they did not require colonoscopy until they are 50 years of age.

In most cases, information about genetic testing is readily shared, but this is not universal. There are rifts in many families that prevent parents and children from communicating about difficult subjects such as health risks. Some do not talk at all. Divorce, disgruntlement, and embarrassment can all inhibit or prevent open discussion of such mutually relevant medical information. Other family crises including disease-related deaths may impede immediate disclosure and discussion of genetic test results. It can be difficult to decide how and when to inform relatives. Since the testing of children for adult-onset disorders generally is discouraged, except when results would influence uptake of preventive options, which would begin in childhood, parents often postpone telling children about the parent's test result. Early research in this area, however, has shown that it is the family's style of communicating generally that typically governs whether and when children are told (Tercyak *et al.*, 2002).

Some family members do not wish to know about their own or others' genetic test status. This can raise problems when that family member is a twin or an obligate carrier of a mutation identified in another relative. Some family members are fearful that telling a relative could be quite upsetting and bring back unpleasant memories or arouse feelings of guilt for having passed on a deleterious mutation.

Providers need help in advising patients about the communication of results to family members. As the genetic component of many diseases is recognized, the complexity of such communication will increase. Both patients and providers will find it valuable to utilize CD-ROMs and other aids to communicate complicated genetic information to relatives, but this will not replace the need for sensitive patient-provider communication and awareness of the complex psychological responses, which the dissemination of genetic information within a family can evoke.

32.6 FUTURE CHALLENGES: MORE GENES, LESS CLARITY

The translation of twenty-first century genetic advances into clinical practice will necessitate ever-closer collaboration between molecular scientists, clinicians, and social scientists. Most of the genes for which genetic testing is currently available are inherited in Mendelian fashion, that is, in relatively simple ways. It is very likely that genes discovered in the future that convey susceptibility to a host of common conditions, such as heart disease, diabetes, and psychiatric disorders, will have their effects by virtue of complex combinations with other genes or in interaction with environment factors.

For patients to understand such complex patterns of inheritance will require carefully constructed teaching tools. Targeting of the information offered to the cognitive style of the individual may help as might the availability of adjunct materials, like CD-ROMs. But there will always be a need for a sympathetic and respectful clinician who can help the patient make links between their family history, their genetic risk, and the possibilities of reducing adverse outcomes of hereditary predisposition.

Genetic medicine holds great promise for identifying individuals at increased disease risk and for treating patients with greater efficacy and fewer side effects through pharmacogenetics (see also Chapter 20). Patient education about genetics will necessitate having well-informed, well-trained, culturally competent (Meiser *et al.*, 2001) professionals to address the different genetic counseling, medical, and psychological questions that arise. It is a challenge for professionals to educate themselves about this rapidly changing field (Emery and Hayflick, 2001). Professional organizations, such as the National Coalition for Health Professional Education in Genetics (NCHPEG) and Genetic Resources on the Web (GROW), are developing and disseminating resources to foster increased genetic literacy among health professionals in preparation for an expected exponential growth in genetic medicine in the decades to come.

In addition to the facts of hereditary risk and delineation of the risk and benefits of genetic testing, it will be important also to discuss the psychological factors governing patient and family response to genetic information. It is only when patients feel empowered and not threatened by such information that they can make optimal use of the advances in genetic medicine.

References

Almqvist, E. W., Bloch, M., Brinkman, R., Craufurd, D., and Hayden, M. R. (1999). A worldwide assessment of the frequency of suicide, suicide attempts, or psychiatric hospitalization after predictive testing for Huntington disease. *Am. J. Hum. Genet.* 64, 1293–1304.

Audrain, J., Schwartz, M. D., Lerman, C., Hughes, C., and Peshkin, B. N. (1997). Psychological distress in women seeking genetic counseling for breast-ovarian cancer risk: the contributions of personality and appraisal. *Ann. Behav. Med.* 19, 370–377.

Bleiker, E. M. A., Menko, F. H., Taal, B. G., Kluijt, I., Wever, L. D. V., Gerritsma, M. A., Vasen, H. F. A., and Aaronson, N. K. (2003). Experience of discharge from colonoscopy of mutation negative HNPCC family members. *J. Med. Genet.* 40, e55.

Bonadona, V., Saltel, P., Desseigne, F., Mignotte, H., Saurin, J-C., Wang, Q., Sinilnikova, O., Giraud, S., Freyer, G., Plauchu, H., Puisieux, A., and Lasset, C. (2002). Cancer patients who experienced diagnostic genetic testing for cancer susceptibility: Reactions and behavior after the disclosure of a positive test result. *Cancer Epidemiol. Biomarkers Prevent.* 11, 97–104.

Bottorff, J. L., Balneaves, L. G., Buxton, J., Ratner, P. A., McCullum, M., Chalmers, K., and Hack, T. (2000). Falling through the cracks: Women's experience of ineligibility for risk of breast cancer. *Can. Fam. Physician* 46, 1449–1456.

Butow, P. N., Lobb, E. A., Meiser, B., Barratt, A., and Tucker, K. M. (2003). Psychological outcomes and risk perception after genetic testing and counselling in breast cancer: a systematic review. *Med. J. Aust.* 178, 77–81.

Calzone, K. A., Biesecker, and B. B. (2002). Genetic testing for cancer predisposition. *Cancer Nurs.* 25, 15–25.

Codori, A-M., Petersen, G. M., Miglioretti, D. L., Larkin, E. K., Bushey, M. T., Young, C., Brensinger, J. D., Johnson, K., Bacon, J. A., and Booker, S. V. (1999). Attitudes toward colon cancer gene testing: factors predicting test uptake. *Cancer Epidemiol. Biomarkers Prevent.* 8, 345–351.

Collins, F. S., Green, E. D., Guttmacher, A. E., and Guyer, M. S. (2003). A vision for the future of genomics research. *Nature* 422, 835–847.

Croyle, R. T., Smith, K. R., Botkin, J. R., Baty, B., and Nash, J. (1997). Psychological responses to BRCA1 mutation testing: Preliminary findings. *Health Psychol.* 16, 63–72.

DiMaio, L., Squitieri, F., Napolitano, G., Campanella, G., Trofatter, J. A., and Conneally, P. M. (1993). Suicide risk in Huntington's disease. *J. Med. Genet.* 30, 293–295.

Dorval, M., Patenaude, A. F., Schneider, K. A., Kieffer, S. A., DiGianni, L., Kalkbrenner, K., Bromberg, J. I., Basili, L. A., Calzone, K., Stopfer, J., Weber, B. L., and Garber, J. E. (2000). Anticipated versus actual emotional reactions to disclosure of results of genetic tests for cancer susceptibility: Findings from *p53* and *BRCA1* testing programs. *J. Clin. Oncol.* 18, 2135–2142.

Dudok deWit, A. C., Tibben, A., Duivenvoorden, H. J., Frets, P. G., Zoeteweij, M. W., Losekoot, M., van Haeringen, A., Niermeijer, M. F., and Passchier, J. (1997). Psychological distress in applicants for predictive DNA testing of autosomal dominant, heritable, late onset disorders. *J. Med. Genet.* 34, 382–390.

Easton, D. F., Ford, D., Bishop, D. T., and the Breast Cancer Linkage Consortium (1995). Breast and ovarian cancer incidence in *BRCA1*-mutation carriers. *Am. J. Hum. Genet.* 56, 265–271.

Emery, J., and Hayflick, S. (2001). The challenge of integrating genetic medicine into primary care. *BMJ* 322, 1027–1030.

Ford, D., Easton, D. F., Bishop, Narod, S., and Goldgar, D. E. (1994). Risks of cancer in BRCA1-mutation carriers. Breast Cancer Linkage Consortium. *Lancet* 343, 692–695.

Ford, D., Easton, D. F., Stratton, M., Narod, S., Goldgar, D., Devilee, P., Bishop, D. T., Weber, B., Lenoir, G., Chang-Claude, J., Sobol, H., Teare, M. D., Struewing, J., Arason, A., Scherneck, S., Peto, J., Rebbeck, T. R., Tonin, P., Neuhausen S., Barkardottir, R., Eyfjord, J., Lynch, H., Ponder, B. A., Gayther, S. A., and Zelada-Hedman, M. for the Breast Cancer Linkage Consortium (1998). Genetic heterogeneity and penetrance analysis of the *BRCA1* and *BRCA2* genes in breast cancer families. *Am. J. Hum. Genet.* 62, 676–689.

Foster, C., Evans, D. G. R., Eeles, R., Eccles, D., Ashley, S., Brooks, L., Davidson, R., Mackay, J., Morrison, P. J., and Watson, M. (2002). Predictive testing for BRCA1/2: attributes, risk perception, and management in a multi-centre clinical cohort. *Br. J. Cancer* 86, 1209–1216.

Friend, S. H., Bernards, R., Rogelj, S., Weinberg, R. A., Rapaport, J. M., Albert, D. M., and Dryja, T. P. (1986). A human DNA segment with properties of the gene that predisposes to retinoblastoma and osteosarcoma. *Nature* 323, 643–646.

Giardiello, F. M., Brensinger, J. D., Petersen, G. M., Luce, M. C., Hylind, L. M., Bacon, J. A., Booker, S. V., Parker, R. D., and Hamilton, S. R. (1997). The use and interpretation of commercial APC gene testing for familial adenomatous polyposis. *N. Engl. J. Med.* 336, 823–827.

Giardiello, F. M., Brensinger, J. D., and Petersen, G. (2001). American Gastroenterology Association technical review on hereditary colorectal cancer and genetic testing. *Gastroenterology* 121, 199–213.

Glanz, K., Grove, J., Le Marchand, L., and Gotay, C. (1999). Underreporting of family history of colon cancer: correlates and implications. *Cancer Epidemiol. Biomarkers Prevent.* 8, 635–639.

Grosfeld, F. J. M, Lips, C. J. M., Ten Kroode, H. F. J., Beemer, F., Van Spijker, H. G., and Brouwers-Smalbraak, G. J. (1996). Psychosocial consequences of DNA analysis for MEN type 2. *Oncology* 10, 141–157.

Gusella, J. F., Wexler, N. S., Conneally, P. M., Naylor, S. L., Anderson, M. A., Tanzi, R. E., Watkins, P. C., Ottina, K., Wallace, M. R., and Sakaguchi, A. Y. (1983). A polymorphic DNA marker genetically linked to Huntington's disease. *Nature* 306, 234–238.

Hallowell, N., Statham, H., and Murton, F. (1998). Women's understanding of their risk of developing breast/ovarian cancer before and after genetic counseling. *J. Genet. Counsel.* 7, 345–364.

Hallowell, N., Foster, C., Ardern-Jones, A., Eeles, R., Murday, V., and Watson, M. (2002). Genetic testing for women previously diagnosed with breast/ovarian cancer: examining the impact of BRCA1 and BRCA2 mutation searching. *Genet. Test.* 6, 79–87.

Harper, P. S., Lim, C., and Craufurd, D. on behalf of the UK Huntington's Disease Prediction Consortium (2000). Ten years of presymptomatic testing for Huntington's disease: the experience of the UK Huntington's Disease Prediction Consortium. *J. Med. Genet.* 37, 567–571.

Hartmann, L. C., Schaid, D. J., Woods, J. E., Crotty, T. P., Myers, J. L., Arnold, P. G., Petty, P. M., Sellers, T. A., Johnson, J. L., McDonnell, S. K., Frost, M. H., and Jenkins, R. B. (1999). Efficacy of bilateral prophylactic mastectomy in women with a family history of breast cancer. *N. Engl. J. Med.* 340, 77–84.

Holtzman, N. A., and Marteau, T. M. (2000). Will genetics revolutionize medicine? *N. Engl. J. Med.* 343, 141–144.

Horowitz, M., Wilner, N., and Alvarez, W. (1979). Impact of Events Scale: A measure of subjective stress. *Psychosom. Med.* 41, 209–218.

Huggins, M., Bloch, M., Wiggins, S., Adam, S., Suchowersky, O., Trew, M., Klimek, M., Greenberg, C. R., Eleff, M., Thompson, L. P., Knight, J., MacLeod, P., Girard, K., Theilmann, T., Hedrick, A., and Hayden, M. R. (1992). Predictive testing for Huntington disease in Canada: adverse effects and unexpected results in those receiving a decreased risk. *Am. J. Med. Genet.* 42, 508–515.

Huntington's Disease Collaborative Research Group (1993). A novel gene containing a trinucleotide repeat that is expanded and unstable on Huntington's disease chromosomes. *Cell* 72, 971–983.

Hughes, C., Gomez-Caminero, A., Benkendorf, J., Kerner, J., Isaacs, C., Barter, J., and Lerman, C. (1997). Ethnic differences in knowledge and attitudes about BRCA1 testing in women at increased risk. *Patient Educ. Counsel.* 32, 51–62.

Iglhehart, J. D., Miron, A., Rimer, B. K., Winer, E. P., Berry, D., and Shildkraut, J. M. (1998). Overestimation of hereditary breast cancer risk. *Ann. Surg.* 228, 375–384.

International Human Genome Sequencing Consortium (2001). Initial sequencing and analysis of the human genome. *Nature* 409, 860–921.

Jacopini, G. A., D'Amico, R., Frontali, M., and Vivona G. (1992). Attitudes of persons at risk and their partners toward predictive testing. In *Psychosocial Aspects of Genetic Counseling*, G. Evers-Kiebooms, J.P. Fryns, J.J. Cassiman, eds. pp. 113–117, New York: Wiley-Liss.

Johnson, K. A., Trimbath, J. D., Petersen, G. M., Griffin, C. A., and Giardiello, F. M. (2002). Impact of genetic counseling and testing on colorectal screening behavior. *Genet. Test.* 6, 303–306.

Julian-Reynier, C., Sobol, H., Sevilla, C., Nogues, C., Bourret, P., and the French Cancer Genetic Network. (2000). Uptake of hereditary breast/ovarian cancer genetic testing in a French national sample of *BRCA1* families. *Psycho-Oncology* 9, 504–510.

Kash, K. M., Holland, J. C., Halper, M. S., and Miller, D. G. (1992). Psychological distress and surveillance behaviors of women with a family history of breast cancer. *J. Natl. Cancer Inst.* 84, 24–30.

Kash, K. M., Ortega-Verdejo, K., Dabney, M. K., Holland, J. C., Miller, D. G., and Osborne, M. P. (2000). Psychosocial aspects of cancer genetics: women at high risk for breast and ovarian cancer. *Sem. Surg. Oncol.* 18, 333–338.

Kerr, B., Foulkes, W. D., Cade, D., Hadfield, L., Hopwood, P., Serruya, C., Hoare, E., Narod, S. A., and Evans, D. G. (1998). False family history of breast cancer in the family cancer clinic. *Eur. J. Surg. Oncol.* 24, 275–279.

Kessler, S. (1988). Preselection: A family coping strategy in Huntington disease. *Am. J. Med. Genet.* 31, 617–621.

Lerman, C., Seay, J., Balshem, A., and Audrain, J. (1995). Interest in genetic testing among first-degree relatives of breast cancer patients. *Am. J. Med. Genet.* 57, 385–392.

Lerman, C., Narod, S., Schulman, K., Hughes, C., Gomez-Caminero, A., Bonney, G., Gold, K., Trock, B., Main, D., Lynch, J., Fulmore, C., Snyder, C., Lemon, S. J., Conway, T., Tonin, P., Lenoir, G., and Lynch, H. (1996). *BRCA1* testing in families with hereditary breast-ovarian cancer: A prospective study of patient decision making and outcomes. *JAMA* 275, 1885–1892.

Lerman, C., Hughes, C., Lemon, S. J., Main, D., Snyder, C., Durham, C., Narod, S., and Lynch, H. T. (1998). What you don't know can hurt you: adverse psychologic effects in members of BRCA1-linked and BRCA2-linked families who decline genetic testing. *J. Clin. Oncol.* 16, 1650–1654.

Lerman, C., Hughes, C., Croyle, R. T., Main, D., Durham, C., Snyder, C., Bonney, A., Lynch, J. F., Narod, S. A., and Lynch, H. T. (2000). Prophylactic surgery decisions and surveillance practices one year following *BRCA1/2* testing. *Prev. Med.* 31, 75–80.

Lodder, L., Frets, P. G., Trijsburg, R. W., Meijers-Heijboer, E. J., Klijn, J. G. M., Duivenvoorden, H. J., Tibben, A., Wagner, A., van der Meer, C. A., van den Ouweland, A. M., and Niermeijer, M. F. *et al.* (2001). Psychological impact of receiving a *BRCA1/BRCA2* test result. *Am. J. Med. Genet.* 98, 15–24.

Mastromauro, C., Myers, R. H., and Berkman, B. (1987). Attitudes toward presymptomatic testing in Huntington's disease. *Am. J. Med. Genet.* 26, 271–282.

Meiser, B., Butow, P., Friedlander, M., Barratt, A., Schnieden, V., Watson, M., Brown, J., and Tucker, K. (2002). Psychological impact of genetic testing in women from high-risk breast cancer families. *Eur. J. Cancer* 38, 2025–2031.

Meiser, B., and Dunn, S. (2001). Psychological effect of genetic testing for Huntington's disease: an update of the literature. *West. J. Med.* 174, 336–340.

Meiser, B., Eisbruch, M., Barlow-Stewart, K., Tucker, K., Steel, Z., and Goldstein, D. (2001). Cultural aspects of cancer genetics: setting a research agenda. *J. Med. Genet.* 38, 425–429.

Miki, Y., Swensen, J., Shattuck-Eidens, D., Futreal, P. A., Harshman, K., Tavtigian, S., Liu, Q., Cochran, C., Bennett, M. L., Ding, W., Bell, R., Rosenthal, J., Hussey, C., Tran, T., McClure, M., Frye, C., Hattier, T., Phelps, R., Haugen-Strano, A., Katcher, H., Yakumo, K., Gholami, Z., Shaffer, D., Stone, S., Bayer, S., Wray, C., Bogden, R., Dayananth, P., Ward, J., Tonin, P., Narod, S., Bristow, P. K., Norris, F. H., Helvering, L., Morrison, P., Rosteck, P., Lai, M., Barrett, J. C., Lewis, C., Neuhausen, S., Cannon-Albright, L., Goldgar, D., Wiseman, R., Kamb, A., and Skolnick, M. H. (1994). A strong candidate for the breast and ovarian cancer susceptibility gene *BRCA1*. *Science* 266, 66–71.

Miller, S. M., Roussi, P., Caputo, G. C., and Kruus, L. (1995). Patterns of children's coping with an aversive dental treatment. *Health Psychol.* 14, 236–246.

Miller, S. M., Shoda, Y., and Hurley, K. (1996). Applying cognitive-social theory to health-protective behavior: breast self-examination in cancer screening. *Psychol. Bull.* 119, 70–94.

Myers, R. H., Taylor, C. A., and Sinsheimer, J. A. (1997). Genetic testing for neuropsychiatric disease: Experiences from 8 years of genetic testing for Huntington's disease. In *Progress in Alzheimer's Disease and Similar Conditions*, L. L. Heston, ed. pp. 137–160. Washington DC: American Psychiatric Press.

Offit, K. (1998). *Clinical Cancer Genetics: Risk Counseling and Management.* New York: Wiley-Liss.

Rebbeck, T. R. (2001). Prophylactic oophorectomy in BRCA1 and BRCA2 mutation carriers. *J. Clin. Oncol.* 18, 100S–103S.

Rhodes, M., Chapman, P. D., Burn, J., and Gunn, A. (1991). Role of a regional register for familial adenomatous polyposis: experience in the northern region. *Br. J. Surg.* 78, 451–452.

Ritvo, P., Irvine, J., Robinson, G., Brown, G., Murphy, K. J., Matthew, A., and Rosen, B. (2002). Psychological adjustment to familial-genetic risk assessment for ovarian cancer: Predictors of nonadherence to surveillance recommendations. *Gynecol. Oncol.* 84, 72–80.

Robins-Wahlin, T. B., Backman, L., Lundin, A., Haegermark, A., Winblad, B., and Anvret, M. (2000). High suicidal ideation in persons testing for Huntington's disease. *Acta Neurol. Scand.* 102, 150–161.

Royal, C., Baffoe-Bonnie, A., Kittles, R., Powell, I., Bennett, J., Hoke, G., Pettaway, C., Weinrich, S., Vijayakumar, S., Ahaghotu, C., Mason, T., Johnson, E., Obeikwe, M., Simpson, C., Mejia, R., Boykin, W., Roberson, P., Frost, J., Faison-Smith, L., Meegan, C., Foster, N., Furbert-Harris, P., Carpten, J., Bailey-Wilson, J., Trent, J., Berg, K., Dunston, G., and Collins, F. (2000). Recruitment experience in the first phase of the African-American hereditary prostate cancer (AAHPC) study. *Ann. Epidemiol.* 10, S68–S77.

Schoenfeld, M., Myers, R. H., Cupples, L. A., Berkman, B., Sax, D. S., and Clark, E. (1984). Increased rate of suicide among patients with Huntington's disease. *J. Neurol. Neurosurg. Psych.* 47, 1283–1287.

Schwartz, M. D., Peshkin, B. N., Hughes, C., Main, D., Isaacs, C. and Lerman, C. (2002). Impact of *BRCA1/BRCA2* mutation testing on psychologic distress in a clinic-based sample. *J. Clin. Oncol.* 20, 514–520.

Schneider, K. A., DiGianni, L. M., Patenaude, A. F., Klar, N., Stopfer, J. E., Calzone, K. A., Li, F. P., Weber, B. L., and Garber, J. E. (2004). Accuracy of Cancer Family Histories: Comparison of Two Breast Cancer Syndromes. *Genetic Testing* 8, 222–228.

Smith, K. R., West, J. A., Croyle, R. T., and Botkin, J. R. (1999). Familial context of genetic testing for cancer susceptibility: Moderating effect of siblings' test results on psychological dis-

tress one to two weeks after *BRCA1* mutation testing. *Cancer, Epidemiol. Biomarkers Prev.* 8, 385–392.

Stoeffel, E. M., Garber, J. E., Grover, S., Russo, L., Johnson, J., and Syngal, S. (2003). Cancer surveillance is often inadequate in people at high risk for colorectal cancer. *J. Med. Genet.* 40, e54.

Struewing, J. P., Hartge, P., Wacholder, S., Baker, S. M., Berlin, M., McAdams, M., Timmerman, M. M., Brody, L. C., and Tucker, M. A. (1997). The risk of cancer associated with specific mutations of *BRCA1* and *BRCA2* among Ashkenazi Jews. *N. Engl. J. Med.* 336, 1401–1408.

Tercyak, K. P., Peshkin, B. N., DeMarco, T. A., Brogan, B. M., and Lerman, C. (2002). Parent-child factors and their effect on communicating BRCA1/2 results to children. *Patient Educ. Couns.* 47, 145–153.

Thompson, H. S., Valdimarsdottir, H. B., Duteau-Buck, C., Guevarra, J., Bovbjerg, D. H., Richmond-Avellanda, C., Amarel, D., Godfrey, D., Brown, K., and Offit, K. (2002). Psychosocial predictors of BRCA counseling and testing decisions among urban African-American women. *Cancer Epidemiol. Biomarkers Prev.* 11, 1579–1585.

Vasen, H. F., van Ballegooijen, M., Buskens, E., Kleibeuker, J. K., Taal, B. G., Griffioen, G., Nagengast, F. M., Menko, F. H., and Meera Khan, P. (1998). A cost-effectiveness analysis of colorectal screening of hereditary nonpolyposis colorectal carcinoma gene carriers. *Cancer* 82, 1632–1637.

Williams-Jones, B. (2003). Where there's a web, there's a way: commercial genetic testing and the Internet. *Community Genet.* 6, 46–57.

Williams, W. R. and Strong, L. C. (1985). Genetic epidemiology of soft tissue sarcomas in children. In *Familial Cancer*, H. Mueller and W. Webers, eds. Basel Switzerland: S. Karger.

Wolf, C. R., Smith, G., and Smith, R. L. (2000). Science, medicine, and the future: pharmacogenetics. *BMJ* 320, 987–990.

Wooster, R., Bignell, G., Lancaster, J., Swift, S., Seal, S., Mangion, J., Collins, N., Gregory, S., Gumbs, C., Micklem, G., Barfoot, R., Hamoudi, R., Patel, S., Rice, C., Biggs, P., Hashim, Y., Smith, A., Connor, F., Arason, A., Gudmundsson, J., Ficenec, D., Kelsell, D., Ford, D., Tonin, P., Bishop, D. T., Spurr, N., Ponder, B. A. J., Eeles, R., Peto, J., Devilee, P., Cornelisse, C., Lynch, H., Narod, S., Lenoir, G., Egilsson, V., Barkadottir, R. B., Easton, D. F., Bentley, D. R., Futreal, P. A., Ashworth, A., and Stratton, M. R. (1995). Identification of the breast cancer susceptibility gene BRCA2. *Nature* 378, 789–792.

Zakowski, S. G., Valdimarsdottir, H. B., Bovbjerg, D., Borgen, P., Holland, J., and Kash, K. (1997). Predictors of intrusive thoughts and avoidance in women with family histories of breast cancer. *Ann. Behav. Med.* 19, 362–369.

CHAPTER 33

Safety in Biomedical and Other Laboratories

WILLIAM E. GRIZZLE,[1] WALTER BELL,[1] JERRY FREDENBURGH[2]

[1] *Department of Pathology, University of Alabama at Birmingham, Birmingham, AL;*
[2] *Richard Allan Scientific, Kalamazoo, MI, USA*

TABLE OF CONTENTS

33.1 INTRODUCTION

The safe operation of a laboratory requires attention to many issues and details that are complex and depend on the activities of the specific laboratory. For example, if a laboratory works with human samples, then depending on the location of the laboratory, complex national regulations related to safety precautions may need to be followed. For example, in the United States, regulations specify approaches necessary to protect employees from laboratory chemicals, and blood borne pathogens. In contrast, if no radioactive material is stored or used in a laboratory, a safety plan dealing with radiological safety may be unnecessary for that laboratory. Clearly, issues related to fire, electrical, and physical safety affect all laboratories. However, complex, extensive, or detailed regulations related to these areas of safety primarily are promulgated by regional and local governments. Thus, each laboratory should determine which areas of safety affect it and develop a safety program to protect its employees from all laboratory hazards to which they are exposed. This includes determining the local, regional, national, and international regulations related to safety that apply to any given laboratory and/or organization.

The approaches to safety discussed in this chapter are based upon regulations governing safety in the United States. Nevertheless, the approaches to safety discussed can be generalized to biological laboratories located internationally.

Operating a biological laboratory safely is difficult, and no single source of information will provide all the information necessary to develop an adequate safety program/plan. Therefore, the limited approaches and information presented in this chapter are only a starting point; this starting point is not static in that new issues related to laboratory safety are developing almost daily. The authors do not maintain that these approaches and information presented here are adequate to ensure the safety of laboratory personnel or to ensure that a laboratory is able to meet regulatory or accreditation standards in safety.

33.2 INTERNATIONAL, NATIONAL, REGIONAL, AND LOCAL SAFETY REGULATIONS

In developing safety plans for a laboratory, there are more aids available than in other areas of laboratory operation because of the extensive national, regional, and local regulations that must be met in order to protect the health and safety of employees. Along with these regulations, many national, regional, and local governmental organizations provide guidance concerning how to meet specific regulations related to safety. For example, the web-based aids to understanding the regulations of the United States and some other countries concerning safety are listed in Table 33.1. Table 33.2 contains general informational resources that may aid in developing a safety program, and several reviews devoted to safety issues have been published in the scientific literature (Grizzle and Polt, 1988; Sewell, 1995; Richmond *et al.*, 1996; Beekmann and Doebbeling, 1997; Cardo and Bell, 1997; Padhye *et al.*, 1998; Grizzle and Fredenburgh, 2001). Laboratories outside the United States will also find many of these educational aids to be very useful in developing safety programs.

TABLE 33.1   Internet resources on safety.

Web site	Organization	Topics
General Safety		
http://www.osha.gov	Occupational Safety and Health Administration, Department of Labor, USA	Current regulations; regulations under development; technical information; prevention information, training information; links
http://www.rmlibrary.com/db/lawosha.htm		Occupational safety laws of all 50 states
http://www.med.virginia.edu	University of Virginia, International Health Care Worker Safety Center	Surveillance data
http://www.medcntr/centers/epinet	Exposure Prevention Information Network (EpiNet)	Surveillance information
http://www.cap.org	College of American Pathologists	General and technical information; lab management
http://www.lbl.gov/ehs/pub3000	Berkeley Lab Health and Safety	Health and safety manual
http://www.nccls.org	National Committee for Clinical Laboratory Standards	General and technical information; forums; links
Biosafety		
http://www.cdc.gov	Centers for Disease Control and Prevention, Atlanta, GA	Surveillance data; prevention information, technical information; links. Proposed guidelines for working safety with *M. tuberculosis*
http://www.cdc.gov/ncidod	National Center for Infectious Diseases	General and technical information; case information, teaching materials; research and resources
http://www.fda.gov/cber	Food and Drug Administration, Center for Biologics Evaluation and Research	Information on recalls, withdrawals; and safety issues concerning biologics
http://www.absa.org	American Biological Safety Association	Technical information
http://www.ace.osrt.edu	National Antimicrobial Information Network of Oregon State University and the EPA	Technical information on disinfectants; links
http://www.maaf.gov.uk	Ministry of Agriculture, Fisheries and Food, UK	Surveillance data on BSE in Europe; technical information; links
http://www.cjd.ed.ac.uk	UK Surveillance Unit for CJD	Surveillance data on vCJD; technical information; links
Chemical safety		
http://www.cdc.gov/niosh/database.html	National Institute for Occupational Safety and Health (NIOSH)	Databases and information resource links and publications
http://www.ilo.org/public/english/protection/safework/cis/products/icsc/dtasht/index.htm	International Occupational Safety and Health Information Center	Chemical database; International Chemical Safety Cards (ICSC)
http://response.restoration.noaa.gov/chemaids/react.html	Chemical Reactivity Worksheet	Chemical database of reactivity of substances or mixtures of substances
http://www.cdc.gov/niosh/chem-inx.html	Master Index of Occupational Health Guidelines for Chemical Hazards (NIOSH)	Guidelines for chemical hazards of specific chemicals

TABLE 33.1   Internet resources on safety.—Cont'd

Web site	Organization	Topics
Electrical safety		
http://ehssun.lbl.gov/ehsdiv/pub3000/CH08.html	Berkeley Laboratory Health and Safety	Electrical safety program
http://www.princeton.edu/~ehs/labmanual/sec7-7.html	Princeton University	Laboratory electrical safety program
http://www.ehs.uconn.edu/Word%20Docs/Electrical%20Safety%20in%20the%20Lab.pdf	University of Connecticut Environmental Health and Safety	Electrical safety in the laboratory
Fire safety		
http://ehssun.lbl.gov/ehsdiv/pub3000/CH12.html	Berkeley Laboratory Health and Safety	Fire prevention and protection program
http://www.ehs.stonybrook.edu/policy/LabFireSafetyHazardAssessment.pdf	Stony Brook Environmental Health and Safety	Laboratory fire safety hazard assessment and work practices
Radiological safety		
http://ehssun.lbl.gov/ehsdiv/pub3000/CH21.html	Berkeley Laboratory Health and Safety	Radiation safety program
http://www.jmu.edu/safetyplan/radiology/advisorycommittee.shtml	James Madison University	Radiation protection program

TABLE 33.2   Books and reference material on safety.

- Block, S. S., ed. (1991). *Disinfection, Sterilization, and Preservation*, 4e. Lea and Febiger, Philadelphia, PA
- Bloom, B. R., ed. (1994). *Tuberculosis, Pathogenesis, Protection and Control*. American Society for Microbiology Press, Washington, DC, pp. 85–110
- Centers for Disease Control/National Institutes of Health. (1999). *Biosafety in Microbiological and Biomedical Laboratories*, 4e. US Government Printing Office: US Department of Health and Human Services, Public Health Service, CDC and NIH, Washington, DC (available online from CDC)
- Fleming, D. O., Richardson, J. H., Tulis, J. J., and Vesley, D., eds. (1995). *Laboratory Safety: Principles and Practices*, 2e. American Society for Microbiology, Washington, DC
- Fredenburgh, J. L., and Grizzle, W. E. *Safety and Compliance in the Histology Laboratory: Biohazards to Toxic Chemicals* (available only at workshops)
- Furr, A. K., ed. (2000). *CRC Handbook of Laboratory Safety*. CRC Press, Boca Raton, FL
- Heinsohn, P. A., Jacobs, R. R., and Concoby, B. A., eds. (1996). *AHIA Biosafety Reference Manual*. American Industrial Hygiene Association, Fairfax.
- Kent, P. T., and Kubica, G. P. (1985). *Public Health Mycobacteriology. A Guide for the Level III Laboratory*. US Department of Health and Human Services, Public Health Service, CDC, Atlanta, GA
- Kubica, G. P., and Dye, W. E. (1967). *Laboratory Methods for Clinical and Public Health Mycobacteriology*. Public Health Service Publication, Number 1547. US Department of Health, Education, and Welfare. United States Government Printing Office, Washington, DC
- Lieberman, D. F., ed. (1995). *Biohazards Management Handbook*. Marcel Dekker, New York. *Biosafety in the Laboratory: Prudent Practices for the Handling and Disposal of Infectious Materials*. (1989) National Academy Press, Washington, DC
- Miller, B. M., ed. (1986). *Laboratory Safety Principles and Practices*. American Society for Microbiology, Washington, DC
- *Preventing Bloodborne Pathogen Infections: Improved Practice Means Protection*. NCCLS
- Richmond, J. Y., ed. (2000) *Anthology of Biosafety: I. Perspectives on Laboratory Design*. American Biological Safety Association
- Richmond, J. Y., ed. (1997) *Designing a Modern Microbiological/Biomedical Laboratory: Lab Design and Process and Technology*. American Biological Safety Association
- Richmond, J. Y., ed. (2000). *Anthology of Biosafety: II. Facility Design Considerations*. American Biological Safety Association
- Richmond, J. Y., ed. (2000). *Anthology of Biosafety: III. Application of Principles*. American Biological Safety Association
- Wald, P. H., and Stave, G., eds. (1994). *Physical and Biological Hazards of the Workplace*. Van Nostrand Reinhold, New York. *Preventing Occupational Disease and Injury*. (1991) American Public Health Association, Washington, DC

33.3 GENERAL CONSIDERATIONS IN LABORATORY SAFETY

Safety plans should be developed and used to prevent or to minimize injuries to employees in the work environment. In order to develop an effective safety plan, the likelihood and source of specific injuries for each employee must be identified. The sources and likelihood of specific injuries depend upon the procedures/activities that employees perform (i.e., their jobs) as well as the locations in which the employees are likely to spend time. For example, though secretarial or office personnel usually would not be expected to be exposed to chemical hazards, chemical exposure might occur if the secretary/office personnel pick up material to be typed, filed, or transferred in a room in which chemicals are utilized and fumes or surface contaminations with chemicals are present. Thus, safety plans need to be specific for the individual. Similarly, the environment around the work area from the standpoint of dangers it might impose on employees should be considered.

Each person and their supervisor should identify potential sources of injury and how the likelihood of injury from these sources can be minimized via changes in procedures or engineering changes, which would include the use of safety equipment or the improvement of ventilation within a specific area.

33.4 Training in Safety

Training in safety has the same areas of focus as concerns in safety. Of the general areas of safety training, requirements for training in biohazards, chemical hazards, and radiological hazards are the most demanding.

Training in each area of safety should be given to employees before they begin their work in the laboratory. The training should be updated periodically for all employees according to governing regulations. In the United States, training of employees in safety must be updated annually. Training should be led by knowledgeable trainers in a language that is appropriate for the employees being trained. The training should be at a level that is appropriate for the educational background of each employee and for the risks to which each employee may be exposed. Thus there may be a need for different levels of training in safety based upon the needs and requirements of specific employees. Records of employee training should be maintained according to regulations; in the United States they must be kept for at least three years.

33.5 Safety Infrastructure

The Chief Executive Officer (CEO) of each institution has total responsibility for the safe operation of all components of the institution; depending on the country, the CEO may be subject to civil and criminal penalties depending on safety violations and the extent of any injuries resulting form safety violations/problems.

The CEO usually appoints a safety committee (SC), which is responsible for the overall safety plan of the institution, for periodic monitoring of the success of the safety plan (SP), for changing the safety plan to correct problems with safety, and for training in safety. The SC appoints a safety officer to administer the safety program. While a working safety infrastructure is important, the responsibility for safe operation of an organization falls primarily on each and every employee. Very large organizations that have many separate large divisions may have separate divisional safety committees and safety officers. This is especially true when specific hazards are limited to only one division. For example, a large clinical laboratory that is part of a university may sometimes have a separate safety committee, safety officer, and safety plan. Usually the administrator responsible for areas with increased safety concerns appoint these separate safety officers.

The safety officer is responsible for day-to-day issues related to safety; the safety officer establishes a training program in safety, monitors and maintains compliance with the overall safety program, evaluates safety incidents and injuries, and recommends to the safety committee changes to the safety program to prevent recurrence of incidents and injuries. Evaluations by the safety officer of safety incidents and actions to prevent recurrence are submitted to the safety committee, which evaluates needed changes to the safety plan. In the day-to-day operations of the safety program, the safety officer works closely with area supervisors to ensure local safety.

Biomedical laboratories are potentially very dangerous working environments compared to other working environments because of the many existing potential dangers. Not only are there biohazards associated with specimens of human and/or animal tissues but also with cell lines or cellular components developed or produced from humans, animals, and animal tissues as well as micro-organisms. The type of laboratory may limit the type of biohazards to which employees are exposed. Similarly, it is the rare biomedical laboratory that does not contain a wide variety of chemicals—some may be potential carcinogens and allergens such as formaldehyde; others may be teratogens, toxins, and irritants; some may be flammable/explosive. Similarly, each biomedical laboratory will have physical hazards, electrical hazards, and fire hazards. Also, many laboratories will use some type of radioactivity, and all forms of radioactivity are dangerous.

If a laboratory operates within safety infrastructure as described, the safety officer together with the laboratory supervisor(s) will monitor safety in the laboratory. If such an infrastructure does not exist at a specific organization, it may be necessary to establish mini-infrastructure in safety that follows the safety outlines described previously.

Specifically, a senior technologist needs to be appointed as a safety officer and this person will be responsible for developing a safety plan and for training of laboratory personnel in safety.

The safety plan plus the general details of the safety requirements and national regulations related to safety should be available to all employees. Providing specific actions and details to ensure safety in laboratories, especially those handling human and/or animal tissues, chemicals, and radioactivity, are beyond the scope of this chapter. Multiple books and articles are devoted to specific safety information, which are appropriate for safety training and for establishing or improving a safety program.

33.5.1 Biological Safety

Animal and especially all human tissues are inherently dangerous and must be handled with universal precautions (Grizzle and Fredenburgh, 2001). All employees of organizations handling or processing tissues must be educated in the dangers of tissues as well as governmental regulations that apply to handling or being exposed to human and animal tissues. Similarly, those transferring tissues to other individuals or laboratories should require that anyone who receives the tissues as well as all those at the receiving site who may handle or contact the tissues are educated in the potential dangers of human tissue (e.g., cuts, sticks, splashes, oral, and respiratory transmissions). They must also be familiar with applicable governmental regulations related to exposure to human tissues.

All human tissues and to a lesser extent animal tissues, whether fixed, paraffin embedded, fresh-frozen, or freeze-dried should be considered as biohazardous. As the extent of alteration of tissue increases (e.g., fresh → frozen → fixed → paraffin embedded), the risks from various infective agents usually are reduced. However, certain agents such as prions (e.g., the infective agents for Creutzfeldt-Jacob Disease, Mad Cow Disease, Deer/Elk Wasting Disease, Scappie) may still be infective even when tissues are fixed and processed to paraffin blocks. Also, spores of certain bacteria such as anthrax in animal tissues (e.g., pelts) as well as specific soils may be infective for decades. Thus, all human and animal tissues independent of their physiochemical state should be treated with universal precautions; that is, they should be handled as if infected with agents that may be pathogenic to humans (Grizzle and Fredenburgh, 2001).

Other biological hazards in laboratories may include systems for transfecting cells with specific genetic products. Biological hazards may result when the transfection system includes self-replicating viral vectors. The safety of such systems should be always considered.

Similarly, with the advent of international travel, newly identified infectious agents frequently spread rapidly through populations. An example is Severe Acute Respiratory Syndrome (SARS), which developed in China and spread rapidly to multiple countries; and monkey pox and West Nile Virus, which have been acquired in or transferred to North America from exotic pets. Similarly, with the possibility of bioterrorism, agents that may be encountered very rarely such as anthrax and small pox could represent agents to which laboratory personnel are exposed. When there is notice of a "new" disease which may be studied and diagnosed or to which employees may be exposed, a plan to deal with this condition should be developed. These laboratory safety issues are in addition to those prescribed for the usual pathogens potentially encountered in a laboratory dealing with human tissues.

In the United States, the major federal regulation (29CFR Part 1910.1030 Occupational Exposure to Bloodborne Pathogens) addresses requirements for laboratories to protect against bloodborne pathogens. Any laboratory that deals with human tissues may need to meet the requirements of this regulation. The development of a safety program in biohazards is outlined in Table 33.3.

General approaches to biosafety include identifying potential hazards without regard to standard operational procedures (SOPs) or to the use of safety equipment. After potential hazards/dangers are identified, SOPs should be modified to reduce the likelihood of injuries from biohazards. SOPs should include requirements for frequent washing of hands, for using safety equipment, for providing employees with hepatitis B vaccinations, and for related preventive medical support. Similarly, all employees should be trained with respect to minimizing biohazards. Medical support should be provided when employees are injured and work practices should be changed based on the analysis of safety incidents to prevent future injuries. The safety plan

TABLE 33.3 The key steps in developing a biosafety program.

1. Identify requirements related to biohazard safety promulgated by governmental and laboratory accrediting organizations and likely sources of up-to-date information regarding biosafety. Use this information in developing or updating an overall safety program and in training programs related to biohazards.
2. Develop the organizational infrastructure necessary to develop and maintain a safety program.
3. Identify risks and general issues of biosafety in the laboratory; this includes identification of work activities and the safety issues of each activity as well as risks in various working spaces.
4. Develop written guidelines to ensure biosafety based on published information, federal, state, and local regulations as well as local and consultant experience. These guidelines should be reviewed and updated periodically and modified as soon as possible to correct any identified problems. Maintain records of personnel safety incidents.
5. Develop and implement a training program of which a major focus is biosafety and maintain records of employee training.

should be evaluated yearly and improved to prevent the recurrence of injuries. Clear, detailed records of these approaches to biosafety should be maintained.

Most laboratories that handle human tissues for research will at some time be queried as to whether or not an individual from whom tissues were obtained is infected with HIV, Hepatitis B, C, or D, West Nile Virus, Creutzfeldt-Jakob prions, and so on. This query may be prior to use of the tissues and/or after an employee is exposed to infection by the tissue. Organizations that provide tissues for research usually will not have permission to test patients and/or their tissue for such agents. In addition, a negative test does not insure safety of the tissues for that pathogen, and for some pathogens there are no tests available. Thus, recipients of tissue specimens must agree to educate employees as to the dangers of tissues with which they come into contact, and not to test tissues for human pathogens. When an employee is exposed (cut or stick) to human biohazards immediate medical care is necessary. If the hepatitis B titer is low, a revaccination should be considered. Also, treatment with hyperimmune gamma globulin might be warranted. After acute medical care and appropriate medical advice is provided, the employee is then monitored for evidence of those human pathogens for which testing is available.

In addition to considering biological dangers associated with human tissues and materials that may be used within the laboratory, some biohazardous conditions may develop within the laboratory independent of the use of human or animal tissues. For example, the ventilation system, cooling system, condensation drains, drains of sinks, animal debris, bedding, incubators, cell culture or other growth media, refrigerators/freezers, and/or unconsumed food may develop colonization with fungal species or other biohazardous agents that especially may affect laboratory personnel who may have compromised immune systems (e.g., employees with HIV infections or those taking immunosuppressive agents such as steroids for asthma or arthritis or chemotherapeutic drugs for cancer). Also, some common fungi such as Cryptococcus may affect apparently healthy individuals. These same agents via circulating spores also may compromise biological experiments. Therefore, all biological laboratories should maintain strict standards of cleanliness, for example, with periodic decontamination of the drains of sinks and prohibition of use of food and drinking material in the laboratory. Material and debris should be cleared from the laboratory.

33.5.2 Chemical Safety

There are several federal regulations related to chemical safety that may affect repositories. These include Occupational Exposure to Hazardous Chemicals in Laboratories (29CFR 1910.1450), the Hazard Communication Standard (29 CFR 1910.1200), and the Formaldehyde Standard (29

CFR 1910.1048). Most repositories must abide by the Occupational Exposure to Hazardous Chemicals in Laboratories law (29CFR 1910.1450). This law mandates that employers develop a written chemical hygiene plan; it is the core of the standard. The chemical hygiene plan must be capable of protecting employees from hazardous chemicals in the laboratory and capable of keeping chemical exposures below the action level or in its absence the permissible exposure limit (PEL). Table 33.4 outlines the designated elements that a chemical hygiene plan must include. Laboratories dealing with fixed tissues should also understand the Formaldehyde Standard.

Most laboratories deal with relatively small amounts of most chemicals; nevertheless, even small amounts of specific chemicals may be very dangerous. Tiny amounts of some chemicals may be toxic or may be carcinogens or teratogens. Similarly ether, especially old ether that has oxidized after opening, or picric acid that has dried out, may constitute explosive hazards. Combinations of chemicals may also cause spontaneous combustion as well as accelerated heating, which may cause boiling and splashing, or even explosions. Personnel should avoid direct contact with even small quantities of carcinogens, teratogens, and/or highly toxic agents such as cyanides.

All chemicals used in repositories should have material safety data sheets (MSDS) available for reference for

TABLE 33.4 Mandated elements of a chemical hygiene SOP plan relevant to safety and health must be developed, and the following procedures need to be addressed.

1. A written emergency plan to address chemical spills. The plan should include consideration of prevention, containment, clean-up, waste disposal, and disposal of chemically contaminated materials used during the clean-up.
2. A policy to monitor the effectiveness of ventilation and to minimize exposure to potentially dangerous vapors.
3. A policy in reference to ventilation failure, evacuation, medical care, reporting of chemical exposure incidents, and chemical safety drills.
4. Policies prohibiting eating, drinking, smoking, gum chewing, and application of cosmetics in the laboratory.
5. Policies to prohibit storing food and/or beverages in storage areas or laboratory refrigerators.
6. Mouth pipetting and mouth suctioning for starting a siphon must be prohibited.
7. Personal protection should be mandated; all persons including visitors must wear appropriate eye protection. Suitable gloves should be worn where there is a potential for contact with toxic chemicals. Gloves should be inspected before using and washed before removing. The use of contact lenses in the laboratory should be avoided.
8. After handling hazardous materials, hands and other possible areas of exposed skin should be washed.

employees who potentially will come into contact with these chemicals. MSDS sheets are prepared by the manufacturer and are available from the manufacturer of the chemical. The MSDS lists the various hazards of the specific chemical to which the MSDS apply. This includes toxicity, explosive potential, and categories of danger (e.g., strong oxidizers). The MSDS also specifies procedures to minimize toxic exposures as well as contact information so that additional information on the chemical can be obtained rapidly.

Even though the dangers of most chemicals are identified in MSDS, combinations of chemicals that may be serious hazards may not be specified clearly. It may be obvious that concentrated strong acids (HCl) should never be combined directly with concentrated bases, especially strong bases (e.g., NaOH) without dilution and extreme care; however, mixing strong oxidizes (e.g., potassium permanganate) with materials with high carbon content (e.g., ethylene glycol) may cause spontaneous fires and such mixes also should be approached with great care. These are just a few of the examples that should be included in the educational program in chemical safety.

When individuals work with chemicals, the chemical safety plan should ensure that potential injuries are avoided by proper work procedures, by proper clothing and safety equipment, and by extensive education. Such approaches to chemical safety are regulated in the United States by the Department of Labor and are addressed by laws such as the Occupational Exposure to Hazardous Chemicals in Laboratories (29CFR1910.1450) and the Hazard Communication Standard (29CFR1910.1200).

There also are special laws that address particularly dangerous chemicals such as formaldehyde (the Formaldehyde Standard 29CFR1910.1048). This law has specific safety requirements for the preparation and use of formaldehyde.

Although laboratories outside the United States may be governed by separate regulations, the previously mentioned laws may provide aids in developing a chemical safety program. Also, MSDS are international and understanding such sheets is a great step in laboratory safety.

33.5.3 Electrical Safety

Electrical injuries can be avoided by ensuring that all equipment is grounded. This is ensured by testing all equipment when first purchased, and yearly thereafter. Similarly, all electrical base plugs must be in good condition and grounded. Electrical work should be done with great care, ensuring that all areas are protected by removal of fuses and with written warnings at the fuse box. Frequently personal electrical appliances such as radios, hairdryers, and so on, may be ignored when testing for grounding, and represent significant dangers. Also, great care should be taken with electrical appliances/equipment around water sources, especially sinks and bathrooms/showers.

33.5.4 Fire Safety

Fire safety can be evaluated by arranging an inspection by the local fire department. Prior to such inspections and at least yearly, fire drills should be practiced and emergency exit pathways should be posted at all room exits. Obviously, emergency exits should never be blocked, obstructed, or locked, and hallways must not be obstructed or cluttered. Similarly, access to fire blankets, showers, and fire equipment must not be impeded.

Flammable agents should be stored appropriately including storage of large amounts of flammable agents only in fire cabinets if more than several quarts of flammable agents are stored in one area. Fire cabinets or areas where small amounts of flammable chemicals are stored should not be located near the exits of rooms/areas.

Smoking should be regulated carefully; similarly, furniture, rugs, and equipment should be constructed of nonflammable material. Regulations for types of doors to serve as fire barriers should be followed as should fire requirements for construction of buildings that house specific activities (e.g., laboratories).

33.5.5 Physical Safety

The physical safety of employees is a safety consideration that must be taken into account in all organizations and for all employees. Physical safety ranges from preventing falls to ensuring employees are not physically injured or intimidated by other individuals, either employees or nonemployees. Much of a plan for ensuring physical safety involves careful maintenance of the physical plant and facilities. Tears in rugs, broken steps, and water, soap, paraffin, and other slippery substances on floors, and inappropriate use of ladders or chairs as ladders, all may lead to unnecessary falls. Similarly, unrestrained gas cylinders, unbalanced file cabinets, and inadequately secured shelves all can lead to injuries via falling or moving agents or structures. Also, included in causes of physical injuries include repetitive action injuries and back injuries secondary to inappropriate lifting as well as temperature burns both cold (e.g., liquid nitrogen) and hot.

Great care should be taken with the overall security of the workplace; this includes limiting access to the workplace by unauthorized personnel. There should be no tolerance of threats to employees, especially by other employees. The protection of employee safety from others may extend outside the direct work environment to areas surrounding the workplace, including parking areas.

Physical injuries that are more difficult to avoid include minor cuts (e.g., paper), bumps, and strains due to inattentive actions. However, such minor injuries should not be compounded by exposure, for example of broken skin, to biohazards. The other hazards that can be prevented or ameliorated (use of gloves to avoid burns) should be addressed in the overall safety program.

33.5.6 Radiological Safety

Laboratories that purchase, store, and/or use radioactive material should have a radiological safety plan. For laboratories requiring a radiological safety plan, the personnel who utilize or come into contact with radioactive material require extensive training as well as the availability of specific equipment to monitor for the extent of radiation and radioactive contamination. As with other areas of safety, the training must be periodically updated and should be appropriate for language and education of personnel. Everyone who has access to the areas where radiation is used or stored should have appropriate training. Depending on the danger of the radioisotopes used, some employees, usually those with professional degrees, may hold a local license to use specific isotopes, and these licensed personnel are responsible for the direction and supervision of all employees who work under their supervision and all spaces to which they are assigned.

Although radiation safety is a very specialized area, it can be generalized (see Table 33.5). Surveys of all laboratory areas, which can detect types of radioactive material being used in the laboratory, should be performed periodically. Gamma survey meters are an aid indicating contamination by strong β and γ emitters but periodic surveys of work, floor, and storage areas using swipes are critical to laboratory safety. Contamination can be easily spread throughout a building and even to the homes of laboratory personnel.

When occupying unsurveyed laboratory space or using unsurveyed equipment (e.g., refrigerators and freezers), these should be certified as radiation free before occupancy or use.

Personnel should be shielded from exposure to radiation. This is especially important for strong β and γ emitters (e.g., ^{32}P, ^{60}Co). Not only does this require shielding but also appropriate laboratory clothing and safety equipment. Contamination is avoided by quickly and effectively containing all radioactive spills and cleaning up all spills or release of radioactivity.

33.6 CONCLUSIONS

The safe operation of a laboratory depends upon all employees working to maintain a safe working environment. This requires establishing a safety infrastructure including developing and monitoring of an effective safety plan and training as to the safety hazards associated with special types of laboratory work, such as biohazards, chemical hazards, and radiation hazards. All laboratories should consider issues in physical, electrical, and fire safety. Laboratory safety can be maintained with the careful evaluation of all safety incidents and modification of the safety plan to prevent similar incidents.

NOTE:

At the time of this publication, a U.S. regulation for protection from tuberculosis is no longer being developed.

TABLE 33.5   Essential elements of a radiological safety program.

1. Train in radiological hazards all personnel who have access to areas where radioactivity is used and/or stored and update training periodically.
2. Monitor usage of radioactivity and maintain accurate records as to the usage, disposal, and inventory of radioactive nuclides.
3. Develop a plan to prevent and to minimize radioactive contamination.
4. Develop a plan to survey periodically for radioactive contamination all areas of the laboratory.
5. Monitor radiation exposure of all employees exposed to radiation.
6. Develop a plan to minimize the exposure of all personnel to radiation using appropriate shielding, safety equipment, and optimal procedures.
7. Develop a plan to ensure safe and secure storage of radioactive material.
8. Develop a plan to contain and clean up major radioactive spills.
9. Record and evaluate all incidents involving radioactivity.
10. Maintain careful records of all aspects of radiological safety plan.

References

Beekmann, S. E., and Doebbeling, B. N. (1997). Frontiers of occupational health. New vaccines, new prophylactic regimens, and management of the HIV-infected worker. *Infect. Dis. Clin. North Am.* 11, 313–329.

Cardo, D. M., and Bell, D. M. (1997). Bloodborne pathogen transmission in health care workers. Risks and prevention strategies. *Infect. Dis. Clin. North Am.* 11, 331–346.

Grizzle, W. E., and Polt, S. S. (1988). Guidelines to avoid personnel contamination by infective agents in research laboratories that use human tissues. *J. Tissue Cult. Methods* 8, 191–200.

Grizzle, W. E., and Fredenburgh, J. (2001). Avoiding biohazards in medical, veterinary and research laboratories. *Biotech. Histochem.* 76, 183–206.

Padhye, A. A., Bennett, J. E., McGinnis, M. R., Sigler, L., Fliss, A., and Salkin, I. F. (1998). Biosafety considerations in handling medically important fungi. *Med. Mycol.* 36, 258–265.

Richmond, J. Y., Knudsen, R. C., Good, R. C. (1996). Biosafety in the clinical mycobacteriology laboratory. *Clin. Lab. Med.* 16, 527–550.

Sewell, D. L. (1995). Laboratory-associated infections and biosafety. *Clin. Microbiol. Rev.* 8, 389–405.

CHAPTER 34

Quality Management in the Laboratory

DAVID BURNETT

Lindens Lodge, Bradford Place, Penarth, United Kingdom

TABLE OF CONTENTS

34.1 INTRODUCTION

The introduction of international standards for quality and competence in laboratories and a growing interest in the accreditation of medical laboratories throughout the world has led to an increasing need in medical laboratories for an understanding of quality management and quality management systems.

The International Organization for Standards (ISO) has defined *quality* as "the degree to which a set of inherent characteristics fulfils requirements" (ISO 9000:2000). This sort of definition is almost impenetrable to those not familiar with the world of standards, but relating it to a clinical situation can help bring it to life. For example, a patient in a hypoglycemic coma has a requirement for a blood glucose measurement and the inherent characteristics of that measurement are that it be done on the correct specimen in an accurate and timely manner and be properly interpreted in order to provide a quality result or service.

Quality management in the laboratory involves the identification of the requirements posed by differing clinical situations and ensuring that the inherent characteristics required for a laboratory measurement or examination are fulfilled. A quality management system is the mechanism for quality management and aims to ensure quality. The preamble to Standard A4 in the Clinical Pathology Accreditation (UK) Ltd "Standards for the Medical Laboratory" (www.cpa-uk.co.uk; Burnett *et al.*, 2002) describes a quality management system as providing "... the integration of organisational structure, processes, procedures and resources needed to fulfil a quality policy and thus meet the needs and requirements of users." It is this all embracing concept of a quality management system that this chapter seeks to emphasize.

34.2 INTERNATIONAL STANDARDS AND THEIR ROLE IN ACCREDITATION

34.2.1 Elements of Accreditation

The author, in his book, *A Practical Guide to Accreditation in Laboratory Medicine*, recognizes four elements in accreditation systems (Burnett, 2002): the *accreditation body*, which oversees the assessments and grants accreditation; the *standards*, with which a laboratory has to comply in order to gain accreditation; the *assessors*, who establish compliance with the standards by conducting an assessment; and the *laboratory*, which is required, or voluntarily seeks, to comply with the standards. Central to any effective system of accreditation are clearly written and objectively verifiable standards.

34.2.2 International Standards for Quality and Competence

The purpose of international standards is to "reflect the quest for quality and promote harmonisation of practice from

TABLE 34.1   International standards for quality and competence.

Title of the standard	Application
ISO 9001:2000 Quality management systems—Requirements	For certification of quality management systems
ISO/IEC 17025:1999 General requirements for the competence of testing and calibration laboratories	For accreditation of the quality and competence of testing and calibration laboratories
ISO 15189:2003 Medical laboratories—Particular requirements for quality and competence	For accreditation of the quality and competence of medical laboratories

laboratory to laboratory, from country to country" (www.iso.org). There are three standards published by different groups within ISO that deal with quality and competence and are useful in different situations. These standards and their application are shown in Table 34.1.

ISO 9001:2000 is conceptually important in establishing a working quality management system but does not include competence requirements, whereas ISO/IEC 17025:1999 does include competence requirements but does not address issues specific to medical laboratories. The standard of choice for medical laboratories is ISO 15189:2003. It takes into account the special constraints imposed by the medical environment, the pre- and postexaminations aspects of the work, the role of diagnostic manufacturers, and emphasises the essential contribution that medical laboratory services make to patient care through advisory, interpretative, and educational services.

The relationship between the three standards can be seen as similar to a set of three "Russian dolls," the inner doll being ISO 9001:2000, the middle doll being ISO/IEC 17025:1999, and the outer, all-embracing doll being ISO 15189:2003. In the introduction to ISO/IEC 17025:1999, its relationship to ISO 9001:2000 is clearly stated, "testing and calibration laboratories that comply with this International Standard will therefore act in accordance with ISO 9001." The relationship between ISO 15189:2003 and ISO/IEC 17025:1999 and ISO 9001:2000 is indicated by their citation as normative references in ISO 15189:2003. Within ISO conventions this means that compliance with ISO 15189:2003 includes compliance with the requirements of ISO/IEC 17025 and ISO 9001:2000.

34.2.3   Accreditation Bodies and Standards

Until the official publication of ISO 15189:2003, accreditation bodies throughout the world were adopting ISO/IEC 17025:1999 as the basis for assessing performance of testing

and calibration laboratories. It was seen as applicable to all laboratories whether they tested for pesticide residues in fresh fruit or examined a renal biopsy for the evidence of disease. Some accreditation bodies have been reluctant to adopt a standard specific for medical laboratories, but at a recent General Assembly of the International Laboratory Accreditation Cooperation (www.ilac.org) an earlier resolution, which stated that when ISO 15189 is published, "medical laboratories may be accredited to that standard as an alternative to ISO/IEC 17025," was confirmed. The resolution is a compromise between those accrediting bodies that wanted a single generic standard for all laboratories and those who understand the desire of the medical laboratory community to be accredited to a standard specifically designed for medical laboratories.

34.2.4   The Principle of Subsidiarity and Guidelines to Standards

Although harmonization of practice in medical laboratories is important there is more than one way to achieve a particular goal. In the European community, the principle of subsidiarity requires the community to act "only if and so far as the objectives of the proposed action cannot be sufficiently achieved by Member States, and can therefore, by reason of the scale or the effects of the proposed action, be better achieved by the Community." Subsidiarity in its original philosophical meaning* as being "concerned with fostering social responsibility" can be translated as saying that architects of standards and of accreditation schemes must be wary of aggregating to themselves rights that may not be acceptable to the "fourth element of accreditation" (Burnett and Blair, 2001; Burnett, 2002), the laboratories to be accredited.

Different situations require different solutions. After national accreditation bodies have decided to adopt a particular standard as a basis for accreditation they often develop secondary documents that facilitate the assessment process and take into account local practice. A good example of these are the standards developed by CPA(UK)Ltd (www.cpa-uk.co.uk). They are cross-referenced to ISO 15189:2003 and other applicable standards. In countries with limited resources or where accreditation is voluntary, benefit can be gained from development of standards or guidelines for national use and this principle of subsidiarity can then be seen as promoting the long-term aim of harmonization.

---

* The principle of subsidiarity in its original philosophical meaning was expressed by Pope Pius XI in his Encyclical letter in 1931: "It is an injustice, a grave evil and disturbance of right order for a larger and higher association to arrogate to itself functions which can be performed efficiently by smaller and lower sections."

34.3 A PROCESS-BASED APPROACH TO QUALITY MANAGEMENT SYSTEMS

The starting point for developing a framework for process-based quality management of a medical laboratory lies in the introduction to ISO 9001:2000. It promotes the adoption of "a process approach when developing, implementing and improving the effectiveness of a quality management system" in order "to enhance customer satisfaction by meeting customer requirements." Process is described as "an activity using resources, managed in order to enable the transformation of inputs into outputs." In the context of a medical laboratory this translates into consultation with users, receiving a request for an examination, carrying out the work, and reporting the results, with interpretation where appropriate.

Within any organization (e.g., a medical laboratory) there are numerous interrelated or interacting processes, and it is "the identification and interactions of these processes and their management" that is referred to as a process approach. It is the adoption of this approach that creates a process-based quality management system. Figure 34.1, adapted from ISO 9001:2000, represents a model of such a system.

It is helpful to translate some of the terms used in this model into language more familiar to medical laboratory professionals. It can be viewed in two different ways. First, the user (*customer*) has requirements that are formulated in consultation with laboratory management (*5. Management responsibility*) and the laboratory responds by carrying out preexamination, examination, and postexamination processes (*7. Product realization*) to produce a report (*product*) for the user. Depending on whether their requirements have been met or not, users may be defined as satisfied or dissatisfied.

The second view is that of a process model in which laboratory management (*5. Management responsibility*) creates a quality system (*4. Quality management system*) and uses resources, staff, equipment, and so on (*6. Resource management*) to carry out preexamination, examination, and postexamination processes (*7. Product realization*) to fulfil the requirements of the user. The preexamination, examination, and postexamination processes are evaluated continually and improvements made as appropriate (*8. Measurement, analysis and improvement*). Evaluation and continual improvement activities include for example, assessment of users needs and requirements, internal audit of the examination processes, and review of participation in external quality assessment schemes.

One of the unsatisfactory aspects of ISO 15189:2003 is that the management (quality management system) and technical (competence) requirements of the standard are presented in two separate sections. This has the unfortunate effect that it is very difficult for laboratories to discover the dynamic relationships between quality and competence requirements. Figure 34.2 resolves this dilemma by placing the appropriate clauses of the standard into a simplified version of the process-based model presented in Fig. 34.1.

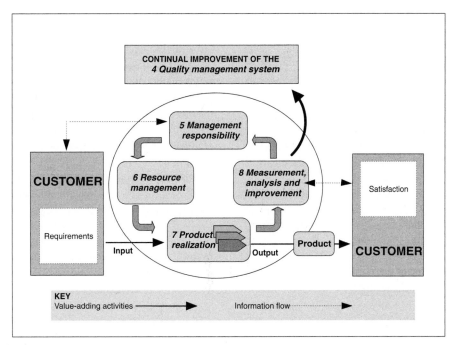

FIGURE 34.1 The ISO 9001:2000 model for a process-based quality management system. The numbers 4–8 in the figure correspond to the main clauses of ISO 9001:2000.

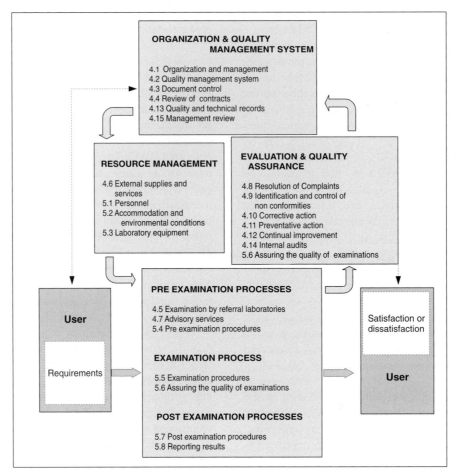

FIGURE 34.2 The requirements of ISO 15189:2003 placed in a process-based quality management system. The numbers 4.1–4.14 and 5.1–5.8 in the figure correspond to the main clauses of ISO 15189:2003.

34.4 BUILDING A QUALITY MANAGEMENT SYSTEM

In defining a *management system* as a "system to establish policies and objectives and to achieve those objectives," ISO 9000:2000 "Quality management system—Fundamentals and vocabulary" draws attention to the fact that the overall management system of an organization can include different management systems, for example, systems for quality, environmental, or financial management. Thus it defines a *quality management system* as a "management system to direct and control an organization with regard to quality." This section discusses the steps to be taken to establish, control, review, and improve a quality management system (QMS).

The National Pathology Accreditation Advisory Council (NPAAC) has published a valuable document, "Guidelines for quality systems in medical laboratories" (available at www.health.gov.au/npaac). Figure 34.3 is adapted from the quality system flowchart illustrated in the document and shows a structured approach to the establishment, control, review, and improvement of a QMS.

Although the NPAAC document was designed as a practical guide to ISO 17025:1999 it is equally applicable to ISO 15189:2003. The figure shows four elements of building a quality management system—establishment, control, review and improvement—as a cycle. In each element a distinction is drawn between management responsibility, which focuses on the establishment and review elements; and organizational responsibility, which focuses on the control and improvement elements. In a medical laboratory, management responsibility would equate with laboratory management, the equivalent term in ISO 9001:2000 being top management. This responsibility is executive in character, in contrast to organizational responsibility, which is corporate and is the responsibility of an organization such as the medical laboratory. In the next two sections the different elements of building a quality management system are establishment, control, review, and improvement.

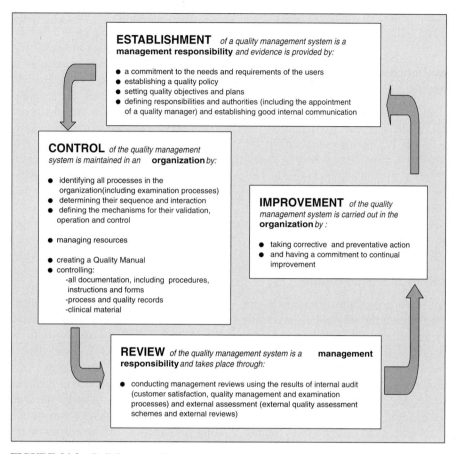

FIGURE 34.3   Building a quality management system.

34.5 ESTABLISHMENT AND CONTROL

34.5.1 Action in Quality Management

As has been seen within the establishment and control stages of creating a QMS, there is a sequence of action in quality management. This sequence is illustrated in a pyramidal form on the left-hand side of Fig. 34.4. The first step in the sequence is the creation of policies that can be defined as the overall intentions and direction of an organization. The second step, objectives and plans, involves making plans and setting objectives to enable the fulfilment of the intentions expressed in the policies. The third step, processes, involves the definition of the activities needed to carry out the intentions, and the fourth step, procedures, are the practical way in which intentions are translated into action. The fifth and final step, records, (made on forms) provide evidence, on a day-to-day basis, that procedures have been carried out correctly and that intentions have been fulfilled.

In terms of a medical laboratory this sequence would translate as follows. The quality policy of the laboratory includes a commitment to the reporting of results of exam-

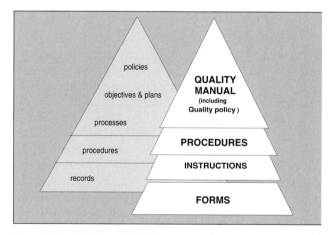

FIGURE 34.4   Hierarchy of documentation.

inations in a timely manner. The supplier of the laboratory computer system announces the release of a module for reporting of results. Laboratory management establishes the installation of this module as an objective for the next financial year. Planning for this development requires the inclusion of the resource implications in the business plan. Its impact on postexamination processes is defined, and procedures and forms reviewed and revised.

34.5.2 Documentation and the Need for Evidence

Evidence of action in quality management is adduced from the documentation that is produced and illustrated on the right-hand side of Fig. 34.4. The primary requirement for evidence is to enable the laboratory to reconstruct its examination and other processes as a result of questions asked by users of the laboratory concerning its performance. The other side of the evidence coin is the need of assessors to obtain evidence to enable them to assess a laboratory's compliance with standards.

34.5.3 The Hierarchy of Documentation

Figure 34.4 shows that at the top of this hierarchy is the *quality manual*, the road map to the whole documentation of the laboratory. In practical terms it should be no more than 25 pages in length. It should contain the *quality policy(s)*, and describe the *processes* that take place in the laboratory in order to fulfil the requirements of particular standards. Examples of such processes are the procurement of equipment, the examination of specimens, and the reporting of results. A policy can be defined as "setting out the commitment of an organization to follow a particular course of action." A pathology laboratory can have a single policy statement, which is inclusive of all aspects of its work, or there can be a number of separate policies relating to different aspects of the way in which a laboratory works.

The quality policy itself will be subject to periodic review but is unlikely to change significantly unless the primary purpose of the laboratory were to change. However, in order to pursue and maintain a particular course of action, *objectives* have to be set and *plans* executed. In contradistinction to the quality policy, objectives and plans are constantly changing in response to the changing needs and requirements of the users. The processes involved should be set out in the quality manual, particularly in relation to the annual management review. Throughout the quality manual there should be references to *procedures,* which form the second level in the hierarchy of documentation. Procedures are the practical way in which policies are translated into action and are often called SOPs, or standard operating procedures. The quality policy should refer to management, quality evaluation, health and safety, laboratory methods, and so on, and procedures are needed that relate to the same areas.

In the same way that the quality manual refers to procedures, so procedures can contain references to (*working*) *instructions*. This third level of documentation involves the practical day-to-day work instructions that are needed near the work situation for easy reference. For example, they might describe starting up or closing down a haematology analyser, a microbiology plate-pouring machine, or a recipe for staining a slide. Instructions can be part of a procedure or can be referred to in a procedure and published separately or both in the document and separately. The advantage of having them separate is that any changes to instructions do not require a change to the procedure.

The final level in the hierarchy of documentation are the *forms*. These forms (and the records created using them) are a crucial part of quality management; they are the evidence that a procedure and/or related instructions have been carried out.

If the procedure or instructions require something to be recorded on a form, the form should be referred to in the procedure. The forms or records do not necessarily have to be created as hard copy (a paper record). An electronic record can be created by completing a form on a computer screen in the laboratory or a consultant's office, by anybody who has the correct authorization identity. Records, whether hard copy or electronic, have to be readily accessible for inspection. In a medical laboratory, request forms and test reports are an example of such documentation. Records of any information or data, such as patients, results, minutes of meetings, quality control data, or the result of an audit should be made on forms of an approved format and not on the backs of envelopes or the cuffs of laboratory coats!

A practical example of this hierarchy of documentation would be a statement in a quality policy requiring "the use of examination procedures that will ensure the highest achievable quality of all tests performed." A procedure produced as a result of such a policy statement might be a procedure for measuring hemoglobin (Hb) A1c. In the quality manual, reference would be made to where a list of examination procedures can be found. The procedure might refer to working instructions for starting the HbA1c analyser and for closing it down and these could be published separately and displayed near the analyser for easy reference. If the analyser is interfaced to a laboratory computer, then an example of a form would be a computer-generated worksheet to assist with checking in samples. Additionally, a computer file that holds the patient details and results should be regarded as a record. Such computer-held data needs to be as easily accessible on demand as any paper record.

All the documents referred to in this hierarchy must be subject to control as described next. The concept of documents can be extended to include specifications, calibration tables, charts, text books, posters, notices, memoranda, software, drawings, and so on. They can be on various media, whether hard copy or electronic, and they may be digital, analogue, photographic, or written.

The preparation of required documentation might appear to be a daunting task for a medical laboratory, but if approached in a practical manner it provides the basis of effective quality management of the laboratory.

34.5.4 Document Control

Control of documents requires that they are approved for adequacy prior to issue, are reviewed and updated as required, are available at point of use, remain legible and uniquely identifiable, and that unintended use of obsolete documents is prevented. The purpose of regularly reviewing documents is to ensure that they remain fit for their intended purpose.

An inherent part of document control is a document register or master index of documentation. It is important to decide at an early stage whether the document register should be a manual paper record, a homemade spread sheet or database, or an off-the-shelf (albeit customisable) commercial product. This is perhaps the most important decision that any laboratory can make in building a QMS. The author has experience of all three approaches but would unequivocally opt for the off-the-shelf commercial product providing it was well tested and a reasonable price. One such product is Q-Pulse produced by Gael Quality that in the author's experience is robust, reasonably priced, user friendly, and is being progressively developed. It can be downloaded and used on a 30-day trial basis from the Gael Quality (www.gaelquality.com).

34.5.5 Control of Records and Clinical Material

A major feature of all quality management systems is the need to control process and quality records and, in the case of medical laboratories, clinical material. ISO 9001:2000 says that "records shall be established and maintained to provide evidence of conformity to requirements and of the effective running of the QMS." It further stipulates that they shall remain legible, readily identifiable and retrievable, and that a procedure be established to define the controls necessary for identification, storage, protection, retrieval, retention time, and disposition of records. ISO 9001:2000, being a quality standard, simply refers to retention and storage of quality records, whereas the laboratory focused standards ISO 17025:1999 and ISO 15189:2002 refer to the retention and control of quality and process records, and in the case of ISO 15189:2002, clinical material as well. Whether the requirement is for control of clinical material or records, there are three distinct issues to be considered:

1. Are the records being retained going to serve a useful purpose, for example to reconstruct an examination, or to audit corrective action?
2. What are the relevant retention times?
3. How should the material be kept?

34.6 REVIEW AND IMPROVEMENT

34.6.1 Evaluation and Continual Improvement

Inspection of the review and improvement sections of building a QMS (Fig. 34.3) indicate any laboratory should be evaluating its activities constantly and seeking to continually maintain and improve quality. Evaluation and continual improvement could be regarded as synonymous with quality assurance, but it seems increasingly uncertain what is meant by the term "quality assurance." The difficulty seems to arise from the meanings of the words assure and ensure. To try to ensure the quality of something is "to make sure or certain" of its quality, whereas to assure, "to give confidence to oneself or others," seems a relatively impotent activity if it is seen from the point of view of the user clinician.

34.6.2 Internal Audit and External Assessment

ISO defines audit as "systematic independent and documented process for obtaining evidence and evaluating it objectively to determine the extent to which audit criteria are fulfilled."

Three different types of audit can be distinguished. The first is an *internal audit* conducted by the laboratory itself (or occasionally on behalf of a laboratory by an outside auditor) on some aspect of laboratory activity such as the accuracy of transcription of data from a request form into the laboratory information system, or whether all members of staff have up-to-date job descriptions. The second is *external audit* (sometimes termed assessments), conducted by some person or bodies interested in the organization such as a purchasing authority, or by external independent organizations such CPA(UK)Ltd or a regulatory authority. A third type of audit, not shown in orthodox classifications, is *cooperative audit* conducted between the laboratory and another party for mutual benefit. Examples of cooperative audit are clinical audit or customer satisfaction surveys and benchmarking activities. Schemes for external quality assessment that are run on a primarily educational basis can in some senses be regarded as cooperative audit or equally well classified as external audit. Audits provide an important mechanism for the detection and investigation of nonconformity.

34.6.3 Nonconformities/Corrective and Preventative Action

A *nonconformity* can arise in two distinct ways: one, from an (reactive) *audit* resulting from a problem in the conduct of a process, leading to the need for *corrective* and/or *preventive action* and thus contributing to the maintenance of quality or to *continual improvement*; and two, a proactive *audit* produces a *nonconformity* that again requires *correc-*

tive and/or *preventive action*, thus contributing to the maintenance of quality or to *continual improvement*.

An example of the first situation might be as follows. An inspection of the first results from a new batch of quality control material being introduced on an analyser showed that results at all three levels for each analyte were approximately 20% lower than expected (*a nonconformity*). Investigation (*an audit*) revealed that although the freeze-dried material had been reconstituted with 5 ml of reconstituting fluid as per the documented procedure, the manufacturer had changed the reconstitution volume from 5 ml to 4 ml without sending out a notice to this effect. All vials wrongly reconstituted were immediately removed (*corrective action*). Following this incident all personnel involved had the matter drawn to their attention and the procedure was altered and an adverse incident report dispatched to the Medical Devices Agency UK, with a copy to the manufacturer (*preventive action*). These actions contribute to ensuring the quality of examinations (*continual improvement*). An example of a proactive audit would be a good housekeeping audit, and such audits are at the core of maintaining a program of continual improvement.

34.6.4 Continual Improvement

Examples of approaches to continual improvement are shown in Fig. 34.5 as what the author has termed cycles of continual improvement. The intention of the diagram is to represent at the center, the management review as the core focus of all continual improvement activity. The circles around the central circle represent individual circles of continual improvement focused on specific topics; for example, with *Personnel*, the activity is the annual joint review of staff, with Internal audit of examination processes, the vertical audit of examinations and with *Equipment and diag-*

nostic systems, the procurement of In Vitro Diagnostic Devices (IVDs).

An important question to answer at this point is when and how often should these activities take place. These circles of continual improvement should carry on throughout the year and most of the nonconformities discovered have to be resolved in a reasonably short time span for the process to be effective.

The nonconformities that are thrown up during the day-to-day activities of quality management are the "grist to the mill" (defined in common English usage as "anything that can be turned to profit or advantage") of continual improvement, or the cogs in the cycles of continual improvement.

However, during the course of a year, issues that require the formal setting of new objectives and detailed planning will be identified and these properly go forward as items for consideration at the (annual) *management review*. If the results from an External Quality Assessment Scheme (or Proficiency Testing Scheme) indicate a problem with an examination, it is no good waiting until the management review for its resolution, whereas the requirement for new service provision may have to wait for the capital purchase of the appropriate IVD or the recruitment of new staff.

34.6.5 Management Review

All the standards referred to in Table 34.1 have clauses entitled *management review(s)*. In ISO 9001:2000, the responsibility for its conduct is defined as being with top management, in ISO 17025:1999, with the laboratory's executive management, and in ISO 15189:2003 with laboratory management. The time interval for this activity is not defined in ISO 9001:2000, but notes in the other standards suggest that a typical period for conducting a management review is once every twelve months. It is important that the review should be seen as having inputs and outputs and the CPA(UK)Ltd, Standard A11 "Management review" has specific information regarding required inputs; Table 34.2 is based on these requirements.

The management review is a crucial part of a quality management system of a laboratory. It sets overall objectives for the following year and within the laboratory they are translated into objectives for the staff and thus into the

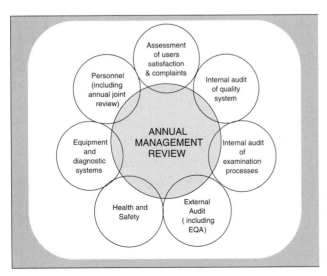

FIGURE 34.5   Cycles of continual improvement.

TABLE 34.2   Inputs to a management review.

- Assessment of user satisfaction and complaints
- Internal audit of quality management system
- Internal audit of examination processes
- External quality assessment reports
- Reports of assessments by outside bodies
- Status of preventative, corrective, and improvement actions
- Major changes in organization and management, resource (including staff), or process

staff joint reviews that identify the training needs of those staff. Continual improvement underpins the continuing provision of a quality service that aims to meet the needs and requirements of the user.

References

Burnett, D., and Blair, C. (2001). Standards for the medical laboratory-harmonization and subsidiarity. *Clin. Chim. Acta* 309, 137–145.

Burnett, D. (2002). A Practical Guide to Accreditation in Laboratory Medicine. ACB Venture Publications. London (www.acb.org.uk).

Burnett, D., Blair, C., Haeney, M. R., Jeffcoate, S. L., Scott, K. W., and Williams, D. L. (2002). Clinical Pathology accreditation: standards for the medical laboratory. *J. Clin. Pathol.* 55, 729–733.

Guidelines for quality systems in medical laboratories. (2001). National Pathology Accreditation Advisory Council, Canberra, Australia (www.health.gov.au/npaac).

ISO 9000:2000. Quality management systems—Fundamental and vocabulary.

Glossary

1. **5′ nucleotidase:** A structure-dependent single-stranded nuclease, used for genotyping and sequence copy number determination reactions, generally related to the flap endonucleases, such as FEN-1.
2. **Acrylamide:** A material that is polymerized to make electrophoretic gels for separation of mixtures of macromolecules.
3. **Allele drop-out:** Failure of amplification of one allele when performing PCR on a single cell, which can be detected only in heterozygous cells.
4. **Allele:** A version of a gene or any DNA sequence that differs from the corresponding normal gene or DNA sequence by one or more base change(s), brought about by a mutation.
5. **Allele-specific oligoprimer:** A pair of oligonucleotide primers that contain basically the same sequence, but one part of the pair differs from the other in that one bears the normal allele sequence and the other bears the DNA change that specifies a mutated allele.
6. **Amplicon:** Fragment of DNA flanked by PCR primers, thus amplified by the PCR.
7. **Amplification refractory mutation system (ARMS):** A PCR-based mutation detection method, which allows amplification of a single allele, depending on the primer used.
8. **Amplimer:** One of a pair of primers needed to amplify DNA by the PCR.
9. **Aneuploid cell:** A cell, whose number of chromosomes differs from the normal chromosome number for the species by one or more chromosomes.
10. **Aneuploidy:** An irregular number of chromosomes of chromosomal regions that is not an exact multiple of the haploid chromosome number for a cell or organism, which can be higher or lower compared to the wild-type situation.
11. **Annealing:** Spontaneous alignment of two single DNA strands, resulting in the formation of a double-stranded DNA molecule.
12. **Autoradiography:** A process in which a pattern is formed onto a film, which corresponds to the location of radioactive materials incorporated into cell structures or DNA molecules, when these cell structures or DNA molecules are placed next to a film.
13. **Autosome:** Any chromosome other than a sex chromosome.
14. **Base pair (bp):** A pair of complementary nucleotide bases in a duplex DNA or RNA molecule.
15. **Blunt ends:** Ends of a duplex DNA molecules with no overhangs due to restriction or removal by single-strand specific nucleases.
16. **Broad spectrum PCR:** Pivotal PCR tests that aim at amplification of DNA sequences that are universally conserved among a variety of bacterial species for the characterization of microbial communities or the identification of novel microbial species.
17. **Cap:** A 7-methylguanosine molecule, which is added 5′ to the pre-mRNA.
18. **CEL I nuclease:** A mismatch-specific nuclease isolated from celery, representative of a family of plant bifunctional DNase/RNase, induced during senescence.
19. **Cell separation and concentration methods:** Methods allowing to arrest selectively cell subpopulations from the complex sample matrix through the use of electric, magnetic, or acoustic fields.
20. **Centromere:** The constricted region of the chromosome where the sister chromatids are joined and where the spindle fibers are associated for chromosome movement during mitosis and meiosis.
21. **Charge-coupled device (CCD) Camera:** A camera whose silicon chip is divided into light sensitive pixels that can be counted upon photon exposure.
22. ***Chlamydia trachomatis*:** Intracellular bacterial species best known as a sexually transmitted agent, which can cause infertility in women due to pelvic inflammatory disease, one of the most prevalent sexually transmitted diseases.
23. **Chorionic villus sampling (CVS):** A placental sampling procedure for obtaining fetal tissue for chromosome and DNA analysis to assist in prenatal diagnosis of genetic disorders.
24. ***Cis*-acting element:** A DNA (or RNA) sequence, on which sequence-specific DNA (or RNA) binding proteins are bound.
25. **Cleavage stage embryo:** Stage in between the fertilized egg and the morula stage (day 1–day 3 after fertilization). The divisions of the cells are characterized by the lack of the G-phase, so that the cells become smaller and the total volume of the embryo does not increase.
26. **Cloning vector:** The DNA vehicle (plasmid, cosmid, or phage chromosome) used to carry the cloned DNA fragment.

27. **COBAS Amplicor:** Completely automated system suited for the detection of bacterial pathogens, such as *C. trachomatis*, by PCR.

28. **Codon:** A specific DNA or corresponding RNA sequence of three base pairs, which encode for a particular amino acid, start, or termination signal.

29. **Comparative genomic hybridization (CGH):** Molecular cytogenetic whole genome scanning technique based on the cohybridization of labeled test- and control-DNA to normal metaphase spreads, followed by the computer-assisted quantification of fluorescence intensities over the entire length of each chromosome, resulting in a "copy number karyotype" reflecting chromosomal imbalances.

30. **Competitive oligoprimers:** A pair of allele-specific oligonucleotides that are both used in a polymerase chain reaction so that, by the resulting competition between them for annealing to the substrate, amplification of only (or mainly) the "correct" (that is, the fully matching) sequence is promoted by each competitive oligoprimer.

31. **Composite exoninc regulatory elements of splicing (CERES):** Exonic sequences with overlapping splicing enhancer and silencer function.

32. **Core database:** A collection of published mutations from all genomic loci.

33. **Cystic fibrosis:** A potentially lethal autosomal recessive genetic disorder, affecting the secretory glands.

34. **Degenerate genetic code:** A genetic code in which some amino acids are encoded by more than one codon each.

35. **Deletion:** Removal of a DNA fragment, from a single nucleotide to a chromosomal segment or even an entire chromosome.

36. **Denaturation:** The separation of two DNA strands in a DNA double helix, using heat, chemicals, or extremes of pH.

37. **Deoxyribonucleotide (dNTP):** A, C, T, or G triphosphates, used as building blocks by DNA polymerase to make a new DNA strand copied from a template.

38. **Diploid:** A cell or organism with two complete sets of homologous chromosomes.

39. **DNA cloning:** Insertion of a DNA fragment into a vector molecule, such as a plasmid, cosmid, and so on, which is then replicated to generate many copies.

40. **DNA ligases:** A group of enzymes that catalyze the formation of a phosphodiester linkage between the 3′- and 5′-ends of two DNA fragments.

41. **DNA polymerases:** Several enzymes that can synthesize new DNA strands from a DNA template, by adding nucleotides to the growing DNA chain in a 5′ to 3′ direction.

42. **DNA sequencing:** A technique for determining the array of nucleotides in a fragment of DNA.

43. **Electrochemical detection:** Method to quantitate an analyte by measuring the electrical current or voltage generated as a result of a chemical reaction involving the analyte.

44. **Electrophoresis:** Migration of charged molecules in an electric field and the separation of the components of a mixture of proteins, DNA, or RNA within an agarose or polyacrylamide gel.

45. **Endonuclease:** An enzyme that catalyses the hydrolysis (cleavage) of phosphodiester bonds between nucleotides in a sequence of DNA or RNA.

46. **Enzyme-linked immunosorbent assay (ELISA):** A variety of assays in which an enzyme is attached to the antigen (substance of interest) or to the antibody, and the enzyme converts a substrate into a colored or fluorescent product to demonstrate that the antigen-antibody binding has occurred in the reaction.

47. **Ethidium bromide:** A molecule that intercalates into the DNA double helix, which mostly serves to visualize DNA fragments under ultraviolet (UV) light transillumination.

48. **Exon:** The coding sequence of a gene, which is transcribed into mRNA and translated into protein.

49. **Expressed sequence tag (EST):** A sequence-tagged site, derived from a cDNA clone, used to identify genes in genomic analysis.

50. **Expression profiling:** Identification of transcriptional differences between two different RNA samples, usually providing a unique pattern, also known as expression signature, for certain disease types, such as tumors.

51. **Familial adenomatous polyposis (FAP):** An autosomal dominant human disease, caused by germline mutations in the adenomatous polyposis coli (APC) gene.

52. **Fluorescence *in situ* hybridization (FISH):** Molecular cytogenetic approach based on the duplex formation of complementary, single-stranded, and fluorescence-labeled nucleic acid probes with target material (interphase cell; metaphase spread) allowing the exploration of the presence, number, and distinct location of genetic material *in situ*.

53. **Fluorescence resonance energy transfer (FRET):** A detection method, based on a distance-dependent interaction between the electronic excited states of two dye molecules, in which the energy of an excited fluorophore (the donor) is passed to a nearby fluorophore (the acceptor) through resonance, then released as a photon that is detected by a fluorimeter.

54. **Forensic science:** The application of science in the investigation of legal matters.

55. **Frameshift mutation:** The insertion or deletion of nucleotide(s), resulting in a disruption of the translational reading frame.

56. **Gene therapy:** The insertion of genetically corrected or wild-type genes into cells especially to replace

defective genes in the treatment of genetic disorders or to provide a specialized disease-fighting function.

57. **Gene:** The fundamental physical and functional unit of heredity, which carries information from one generation to the next, consisting of the coding region and a regulatory sequence that make its transcription possible.

58. **Genetic locus:** A gene or gene family, or a noncoding DNA sequence that resides at a specific position of a specific chromosome of an organism.

59. **Genetic test:** A test that is capable of revealing the zygosity of the specific allelic constitution at a genetic locus in the diploid cells of an organism.

60. **Genetically modified organism (GMO):** An organism containing an additional trait encoded by an introduced gene or genes, which generally produces a protein that confers the trait of interest.

61. **Genotype:** The genetic composition of the entire cell or, more commonly, of a set of genes.

62. **Gradient:** A gradual change in a quantitative property over a specific distance.

63. **Haploid:** A state in which only one chromosome or chromosome set resides in a cell, such as the Y chromosome in human males.

64. **Haplotype:** Array of alleles or loci on a single chromosome.

65. **Hemizygous gene:** A gene that is present in only one copy in a diploid organism, such as the X-linked genes in a male organism.

66. **Hemoglobin (Hb):** The oxygen-transporting protein in the blood of most animals.

67. **Hereditary hemochromatosis:** A human autosomal recessive disease, resulting in iron overload and irreversible tissue damage, due to increased iron absorption.

68. **Hereditary nonpolyposis colorectal cancer (HNPCC):** One of the most common predispositions to cancer.

69. **Heteroduplex:** A double-stranded DNA molecule that contains one or more nucleotides that are not in the correct base pairing conformation, formed by annealing of single strands from different sources, which can show abnormalities such as loops or buckles.

70. **Heterozygosity:** A measure of the genetic variation in a population, indicated as the frequency of heterozygotes for a specific gene.

71. **Heterozygote:** An individual with a heterozygous gene pair.

72. **High-throughput PCR:** A PCR system that can be applied simultaneously in a large number of samples (large-scale) for the amplification of a normal or mutant genetic locus.

73. **Homologous chromosomes:** Chromosomes that pair with each other at meiosis.

74. **Homozygote:** An individual with a homozygous gene pair.

75. **Huntington disease:** A lethal autosomal dominant human disease, characterized by nerve degeneration, with late-age onset.

76. **Hybrid minigene:** A simplified version of a gene that contains the sequences to be studied (i.e., exon and flanking intron sequences) placed in an heterologous context.

77. **Hybridization:** The annealing of complementary, single-stranded nucleic acids (DNA, RNA).

78. **Insertional translocation:** The insertion of a segment from one chromosome into another nonhomologous one.

79. **Interphase:** The cell cycle stage between nuclear divisions, consisting of the G1, S, and G2 phases, when chromosomes are extended and functionally active.

80. **Intracytoplasmic sperm injection:** A procedure that involves sperm injection into the oocyte's cytoplasm to obtain fertilization, developed to treat severe male infertility and used to prevent contamination during preimplantation genetic diagnosis and failure of fertilization.

81. **Intron or intervening sequence:** A DNA fragment within a gene, which is initially transcribed, but not included in the mRNA, due to its removal by splicing.

82. **Karyotype:** The entire chromosome complement of an individual or cell, as can be seen during mitotic metaphase.

83. **Kilobase (kb):** 1,000 nucleotides.

84. **Linkage:** The association of genes on the same chromosome, tending due to their close physical proximity to be inherited together.

85. **Locus-specific database:** A collection of all published and unpublished mutations from a specific gene.

86. **Loss of heterozygosity (LOH):** Loss of one allele at a chromosomal locus, which may imply the presence of a tumor suppressor gene at that site.

87. **Melting curve:** Originally used to denote the hypochromic shift between double-stranded and single-stranded nucleic acids, caused by greater electron delocalization in the double-strand state accounted for by stacking forces, a surrogate measurement now commonly obtained being a thermal profile of SybrGreen (A-T intercalation) fluorescence differences between the double- and single-stranded states.

88. **Melting temperature (T_m):** The temperature in which a double-stranded DNA fragment is denatured.

89. **Messenger RNA (mRNA):** A mature RNA molecule, resulting from gene transcription by RNA polymerase, which specifies the order of amino acids during its translation to protein.

90. **Metaphase:** The intermediate stage of nuclear division in the cell cycle in which the highly condensed chromosomes align along the equatorial plane between the two poles of the dividing cell, a feature that is

exploited for the identification of chromosome aberrations.

91. **Microarrays:** An orderly arrangement of usually thousands of defined DNA molecules, immobilized onto a solid surface, such as glass or membrane.

92. **Microfluidics:** Networks of interconnected channels and chambers equipped with isolation valves and fluid pumps and fabricated in polymer, glass, silicon, or ceramic substrates, developed for miniaturized chemical analysis and field-deployable sensor applications.

93. **Micro-PCR:** Performing DNA amplification in miniature microfluidic chambers (chips) using polymerase chain reaction leading to the reduction of the sample volume and acceleration of thermal cycling, and integration of the amplification stage with subsequent detection platforms.

94. **Microsatellite Instability (MSI):** A frequent, if not obligatory, surrogate marker of underlying functional inactivation of one of the human DNA mismatch repair genes, characterized by length alterations of oligonucleotide repeat sequences that occur somatically in human tumors.

95. **Minimal residual disease (MRD):** Remnant of a tumor or cancer after primary, potentially curative therapy, which can by detected by PCR with a sensitivity of one in 10^4 to 10^5 cells.

96. **Minisatellite DNA:** Repetitive DNA sequence, based on a repeated sequence core, used for DNA fingerprinting.

97. **Mismatch:** A position within a double-stranded DNA molecule, in which the two bases found opposite each other do not comply with the DNA pairing rule of A-G and C-T.

98. **Missense mutation:** A mutation that alters a codon, resulting in a different amino acid.

99. **Mitochondrial DNA:** A closed circular, extranuclear chromosome, approximately 16,569 base pairs in size in humans, residing in the mitochondrion, and containing genes for two ribosomal RNAs, 22 transfer RNAs, and 13 proteins, predominately for oxidative phosphorylation.

100. **Molecular cytogenetics:** Application of molecular biology techniques to cytogenetic preparations as metaphase spreads and interphase nuclei for the identification of chromosome abnormalities and RNA expression.

101. **Molecular diagnostics:** The identification of a genetic disorder at the molecular (DNA) level, using molecular biology techniques.

102. **Molecular genetics:** The study of the molecular processes, which govern gene structure and function.

103. **Mosaicism:** The presence of two or more genetically different cell populations in one organism.

104. **Multilocus sequence typing (MLST):** A method of bacterial genetic polymorphism identification by comparative sequence analysis of several housekeeping genes, in which point mutations can be used to calculate precise genetic distances between isolates of any given bacterial species.

105. **Multiplex amplification:** Coamplification of multiple target sequences in one single amplification reaction.

106. **Multiplex PCR:** A PCR containing sets of forward and reverse amplification primers for more than one allele of a genetic locus or for different genetic loci.

107. **Mutagenesis:** The process that produces a gene or a chromosome set differing from the wild type.

108. **Mutant allele:** An allele differing from the wild type one.

109. **Mutation:** A transmittable change in the DNA sequence of a gene or anywhere in the genome, with relation to a reference sequence, with or without consequences for the phenotype of the organism.

110. **MutHLS nuclease:** An evolutionally conserved enzyme complex of *Escherichia coli* used in DNA replication linked mismatch repair.

111. **Nitrocellulose filter:** A type of filter used to attach DNA fragments for hybridization.

112. **Nonsense mutation:** A mutation that produces a stop codon, resulting in the premature termination of the protein chain.

113. **Northern blot:** Transfer of electrophoretically separated RNA molecules from a gel onto a filter, which is then immersed in a solution containing a labeled probe that will bind to the RNA of interest.

114. **Nucleotide:** The basic building block of nucleic acids, composed of a nitrogen base, a sugar, and a phosphate group, joined in pairs by hydrogen bonds.

115. **Oligonucleotide ligation assay (OLA):** An assay in which genetic variants are analyzed by utilizing the covalent joining of oligonucleotide probes by DNA ligase and the detection of ligation by specific labels attached to the probes.

116. **Oligonucleotide:** Short linear sequence of nucleotides, often used as probes to hybridize to the target DNA in genetic tests or as primers to promote DNA synthesis in a polymerase chain reaction.

117. **On-chip sample preparation:** An integrated sample processing within microfluidic channels and chambers, consisting of cell concentration, lysing, nucleic acid purification, and amplification, leading to the process automation and assay miniaturization.

118. **Open reading frame (ORF):** A DNA segment, beginning with a start codon and ending with a stop codon, which is presumed to be the coding sequence of a gene.

119. **Perpendicular electrophoresis:** The first step of the optimization process of a TGGE or DGGE analysis, in order to verify the reversible melting behavior of the DNA fragment and to determine its T_m under the experimental conditions.

120. **Pharmacogenetics:** The assessment of clinical efficacy, safety, and tolerability profile of a drug in groups of individuals that differ with regard to certain DNA-encoded characteristics, which if indeed associated with a differential response or phenotype may allow prediction of individual drug response.

121. **Pharmacogenomics:** The evaluation of the differential effects of a number of drugs or chemical compounds in the process of drug discovery, with regard to inducing or suppressing the expression of transcription of genes in an experimental setting.

122. **Phenotype:** The observed outward manifestations of a specific genotype.

123. **Phenylketonuria (PKU):** An autosomal recessive human metabolic disease, caused by a mutation in a gene encoding a phenylalanine-processing enzyme, which leads to mental retardation if not treated.

124. **Philadelphia chromosome:** A translocation between the long arms of chromosomes 9 and 22, often found in the white blood cells of patients with chronic myeloid leukemia.

125. **Phosphodiester bond:** The bond between a sugar group and a phosphate group, resulting in the sugar-phosphate backbone of DNA.

126. **Photolithography:** A method for etching silicon wafers using a LASER (light amplification through stimulated emission of radiation) in conjunction with a mask to perform feature addressing.

127. **Physical mapping:** The identification of the positions of cloned genomic fragments.

128. **Plasmid:** Autonomously replicating extrachromosomal DNA molecule, which serves as DNA cloning vehicles.

129. **Polar bodies:** By-products of the female meiosis. The first is extruded at maturation of the oocyte, leaving the oocyte haploid, and the second is extruded after fertilization and contains chromatids.

130. **Polymerase chain reaction (PCR):** A method for the exponential amplification of a specific DNA fragment from a template by multiple rounds of DNA synthesis.

131. **Polymerase Chain Reaction-Restriction Fragments Length Polymorphism (PCR-RFLP):** A two-primer polymerase chain reaction followed by digestion with a class II restriction enzyme (one lacking restriction modification activity), allowing for variant visualization through sizing on an electrophoretic gel.

132. **Polymorphism:** A gene or any DNA sequence that appears with more than one allele in a population of an organism, namely the normal allele, or reference sequence, and one or more mutated alleles, provided that the least frequent allele is found with an incidence of at least 1% in this population.

133. **Primer extension:** Method of using a short deoxynucleotide primer precisely complementary to genomic or mRNA sequence, in the presence of an enzyme with $3' \rightarrow 5'$ polymerase activity, cofactors and deoxynucleoside triphosphates, to extend the oligonucleotide to a $5'$ template terminus, in which successful polymerization can be used to discern sequence variants from each other, since extension generally fails if the primer does not precisely match the $3'$ base in the initial template.

134. **Primer:** A short single-stranded RNA or DNA that is used as the starting point for chain elongation by the DNA polymerase, when bound to a single-stranded template.

135. **Probe:** Defined nucleic acid fragments that can be used to identify specific DNA molecules bearing the complementary sequence, usually through autoradiography.

136. **Promoter:** A regulator region in short distance from the $5'$ end of a gene that acts as the binding site for RNA polymerase and other transcription factors.

137. **Proteome:** The complete set of proteins in an organism.

138. **Pulsed-field gel electrophoresis (PFGE):** An electrophoretic technique in which the gel is subjected to electrical fields alternating between different angles, allowing efficient separation of very large DNA fragments through the gel.

139. **Pyrosequencing:** A real-time sequencing method that uses light release as the detection signal for nucleotide incorporation into a target DNA strand, allowing for the direct assessment of DNA sequence.

140. **Quality control management:** A process involving the identification of the requirements posed by differing clinical situations and ensuring that the inherent characteristics required for a laboratory measurement or examination are fulfilled.

141. **Real-time PCR:** A PCR assay in which the PCR product is measured continuously throughout the amplification process, for example the TaqMan Assay.

142. **Recessive allele:** An allele whose phenotypic effect is expressed only in the homozygous state.

143. **Reciprocal translocation:** The situation when parts of chromosomes are exchanged.

144. **Reference material:** A standardized material to compare the quality and performance of testing regimen.

145. **Rehybridization:** The anti-parallel realignment of two complementary DNA single strands to form the double helix.

146. **Resolvase:** An enzyme that hydrolyses phosphodiester bonds in a DNA strand crossover complex undergoing DNA recombination to separate the DNA strands.

147. **Restriction endonucleases:** A variety of enzymes, extensively used in genetic engineering, that recognize specific target DNA sequences and hydrolyses the phosphodiester bond at these points.

148. **Restriction fragment length polymorphism (RFLP) analysis:** Detection of different sizes or numbers of restriction fragments, often as a result of presence or absence of restriction sites, which can be used as markers in chromosome mapping.

149. **Retroviral integration site analysis:** Analysis of the junction between the retroviral vector sequence and the genomic sequence to identify the retrovirus integration site in retroviral-mediated gene transfer.

150. **Retroviral vector:** A artificial DNA construct derived from a retrovirus, used to insert sequences into an organism's chromosomes.

151. **Retrovirus:** An RNA virus that replicates by first being converted into double-stranded DNA.

152. **Reverse dot blots:** Allele-specific oligonucleotide hybridization matricies involving covalent or hydrogen bonding attachment of interrogator sequences to a solid support, such as a nylon or plastic-backed membrane. In general, muliplexed PCR products are incubated with the prepared matrix after denaturation, permitted to hybridize, and visualized using radioactive, chemiluminescent, or chromogenic techniques.

153. **Reverse transcription-PCR (RT-PCR):** RNA amplification by PCR, following copying of the RNA to cDNA by reverse transcription.

154. **RNA interference (RNAi):** A general mechanism for inhibiting gene expression, by small antisense RNA molecules, also known as small interfering RNAs (siRNA).

155. **Robertsonian translocation:** The situation when two acrocentric chromosomes translocate to become one large chromosome.

156. **Rolling circle amplification:** An amplification system in which DNA polymerase enzyme with strand displacement activity amplifies a probe, circularized by ligation, in a rolling cycle fashion by releasing a chain of amplicons.

157. **Sampling:** A statistical method to collect samples containing the desired organisms.

158. **Satellite DNA:** Any type of highly repetitive DNA.

159. **Short tandem repeat (STR) loci:** Also known as microsatellites, STR loci are highly repetitive DNA elements of individual repeat motifs, each motif ranging from two to seven base pairs in length, whose difference among individuals is the basis for the population variation of these loci.

160. **Sickle cell anemia:** A potentially lethal autosomal recessive human inherited disorder, caused by a mutation in the gene encoding for the oxygen transporting hemoglobin molecule, resulting in sickle-shaped red blood cells.

161. **Single nucleotide polymorphism (SNP):** Any polymorphic variation at a single base pair position between individuals of the same species, including restriction fragment length polymorphisms and transitions or transversions that reach a frequency of greater than 1% in any given population.

162. **Solid phase:** Attachment of DNA to solid support, such as silica beads, which remains intact during reactions and purification steps.

163. **Southern blot:** Transfer of electrophoretically separated DNA fragments from a gel to a filter, which is then immersed in a solution containing a labeled probe that will bind to a fragment of interest.

164. **Splicing enhancer and silencer:** Short RNA nucleotide sequences, generally six to eight base pairs in length, located in the exon or in the intron that enhance or inhibit splicing, respectively.

165. **Splicing:** The reaction that removes introns and joins exons together, resulting in the mature mRNA.

166. **SR proteins:** Regulatory splicing factors that contain a serine-arginine rich region.

167. ***Staphylococcus aureus*:** A bacterial species, known for its capacity to colonize the human nostrils, that in case of immune-compromisation, is the best known opportunistic pathogen, also well known for its capacity to collect a variety of antimicrobial resistance traits. Methicillin resistant *S. aureus* or MRSA is the best-known hospital pathogen worldwide.

168. **Stop codons:** The terminating signals for translation of an mRNA into protein.

169. **Telomere:** Specialized DNA structure, capping the end of a chromosome involved in replication and stability of linear DNA molecules.

170. **Tool Command Language (Tcl) and Toolkit (Tk):** An interpreted software script syntax commonly used to instruct automata movements; Tk for graphical user interfaces provides a complementary graphical user interface to widgets being instructed.

171. **Transcription:** A process by which a DNA template is copied to RNA by RNA polymesare.

172. **Transition:** A type of nucleotide substitution involving the replacement of a purine with another purine or of a pyrimidine with another pyrimidine, for example G→A or C→T.

173. **Translocation:** Relocation and interchange of a chromosomal segment after breakage and attachment to a different nonhomologous chromosome in the genome.

174. **Transversion:** A type of nucleotide substitution involving the replacement of a purine with a pyrimidine or vice versa, for example, G→T or C→A.

175. **Tumor suppressor gene:** A gene encoding for a protein that suppresses tumor formation, speculated to function as negative regulators of cell proliferation.

176. **Variable number tandem repeats (VNTR):** Variations in the number of tandem repeats of DNA sequences found at specific loci in different populations, often used as markers.

177. **Vector FISH:** Chromosomal mapping of proviral sequences by FISH, using a retroviral vector plasmid DNA as a probe.

178. **Western blot:** Transfer of protein molecules, separated by electrophoresis, from a gel to a filter, which can be probed with a labeled antibody to detect a specific protein.

179. **Wild type:** The genotype or phenotype that is found in nature or in the standard laboratory stock for a given organism.

180. **Y chromosome:** The smallest of the human chromosomes, approximately 60 million base pairs including the gene conveying the male gender, parentally transmitted from father to son.

181. **Zona pellucida:** Transparent protein structure surrounding the cleavage stage embryo.

182. **Zygosity:** The state of genetic identity, that is, normal versus mutant, of the two alleles of a gene or DNA sequence, in the somatic cells of a diploid organism, specified as either normal, heterozygous, or homozygous mutant.

183. **Zygote:** The unique diploid cell, formed by the fusion of an oocyte and a sperm cell, that will divide mitotically to create a differentiated diploid organism.

Index